ANTIMALARIAL DRUGS

AGE OF THE ARTEMISININS

PHARMACOLOGY - RESEARCH, SAFETY TESTING AND REGULATION

Additional books in this series can be found on Nova's website
under the Series tab.

Additional e-books in this series can be found on Nova's website
under the e-books tab.

TROPICAL DISEASES - ETIOLOGY, PATHOGENESIS AND TREATMENTS

Additional books in this series can be found on Nova's website
under the Series tab.

Additional e-books in this series can be found on Nova's website
under the e-books tab.

PHARMACOLOGY - RESEARCH, SAFETY TESTING AND REGULATION

ANTIMALARIAL DRUGS

AGE OF THE ARTEMISININS

QIGUI LI
AND
PETER J. WEINA

New York

NOTICE TO THE READER

Library of Congress Cataloging-in-Publication Data

Antimalarial drugs : age of the artemisinins / editors, Qigui Li and Peter J. Weina.
     p. ; cm.
  Includes bibliographical references and index.
  ISBN: 978-1-62948-013-8 (softcover)
  1. Artemisinin. I. Li, Qigui. II. Weina, Peter J.
  [DNLM: 1. Antimalarials--therapeutic use. 2. Artemisinins--therapeutic
use. QV 256]
  RC159.A7A58 2010
  616.9'362061--dc22
                         2010033094

Published by Nova Science Publishers, Inc. † New York

Contents

Preface — vii

Book Commentary — ix

Abstract — 1

Chapter I   Current Status and Policies of Antimalarial Drugs — 3

Chapter II   Antimalarial Drugs Cost and Cost-Effectiveness — 61

Chapter III   Efficacy and Chemotherapy of Antimalarial Drugs — 111

Chapter IV   Antimalarial Drugs in Chemoprophylaxis — 195

Chapter V   Influence Factorsof Antimalarial Drug Efficacy — 243

Chapter VI   Toxicity of Antimalarial Drugs — 329

Chapter VII   Antimalarial Drug Toxicity — 403

Chapter VIII   Antimalarial Drugs Withdrawn and in the Pipeline — 465

Appendix I — 555

References — 563

Index — 619

Preface

This phenomenal comprehensive review of antimalarial chemotherapy entitled "Antimalarial Drugs: Age of the Artemisinins," was meticulously compiled by two extraordinary clinician scientists, Qigui Li, MD, PhD, and Peter J. Weina, MD, PhD, from the Division of Experimental Therapeutics, Walter Reed Army Institute of Research (WRAIR), Silver Spring, Maryland, USA. This review differs from many other reviews in that these two scientists lived the experience of the artemisinin drugs with their combined service of over 67 years to the WRAIR. Moving from bench top to bedside these scientists navigated a Bermuda triangle (time, risk, and funding) of research woes on a day in and day out basis. Their writings reflect real time challenges associated with drug discovery and development. While Pete took a "short" break from research for medical school, internal medicine training and high risk military deployments, Qigui plowed tremulous pharmacokinetic ground for the World Health Organization (WHO) resulting in Arteether approval in the Netherlands. A potential class effect resulting in neurotoxicity and brainstem lesions in rodents and primates was finally sorted out when the properties of lipid versus water soluble drugs was solved by the team. Dr. Weina was challenged to be the chair of Pharmacology during his second year of his infectious disease fellowship. The hurdles were monumental both scientifically and politically considering the lack of a co-development partner for muddling through the Chemical and Manufacturing Control (CMC) issues of a highly unstable molecule. I have reserved a special place for this book next to the classics of Professor Wallace Peters on the front of my desk. The most important place, though, will be in my heart and soul for the tremendous dedication and loyalty of the Artesunate team. Every student or fellow I mentor will see this book and hopefully enjoy my reminiscing about the trials, tragedies, and triumphs of drug development with Pete and Qigui.

Wil Milhous, PhD F(AAM) D(ABMM)
Associate Dean and Professor of Global Health Infectious Disease
College of Public Health, The University of South Florida

Book Commentary

Malaria is a common and serious tropical disease. Antimalarial drugs are medicines that prevent or treat malaria, a disease which takes a great toll on human health and well-being, particularly in tropical regions including Africa south of the Sahara, South and Southeast Asia, Oceania, and parts of the Americas. In recent years, strains of *Plasmodium* have become increasingly resistant to more antimalarial drugs and researchers have stepped up efforts to revise antimalarial drug policies and develop new antimalarial strategies. Resistance has arisen to all classes of antimalarials (chloroquine, amodiaquine, mefloquine, and sulfadoxine-pyrimethamine) except, as yet, definitively to the artemisinin derivatives. In order to prevent widespread resistance, the concept of antimalarial combination therapy (CT) has been employed and a global resistance surveillance system (World Antimalarial Resistance Networks) has been established. Arguably, the recent action of selecting artemisinin-based combination therapies (ACTs) to replace other CTs was one of the most important steps. The artemisinins are antimalarials derived from the Chinese herb, *Artemisia annua*. These compounds clear the parasites from the blood more rapidly than other antimalarial agents and have recently been recommended by the World Health Organization as first line therapy in the fight against this age old killer. Intravenous formulations of artemisinins have been used in much of the world and represent an improvement in efficacy and safety for severe malaria.

ACTs, currently, successfully cure patients suffering from uncomplicated malaria with superior efficacy and lower toxicity. Although ACTs are 10-20 times more expensive than that of the traditional antimalarial drugs currently in use, the high cost-effectiveness and global subsidy plan could benefit every patient treated with ACT. In the treatment of severe malaria, there still remains a huge challenge (high mortality rate), but a new clinical trials have demonstrated that intravenous artesunate (AS) can reduce by one-third the mortality as compared to the current standard of care, quinine injection. AS used either in intravenous, intramuscular, or rectal formulations have some theoretical and some demonstrated superiority to artemether and quinine injections in the treatment of severe malaria. New Guidelines for the Treatment of Malaria published in 2006 and 2008 by the WHO with new policies (including the withdrawal of less effective drugs and monotherapies, the increase of ACTs in overall use, and the selection of AS-injection in first-line with sequential therapy for severe malaria) sign a beginning of artemisinin age and the ending of chloroquine time. These new strategies could make artemisinin derivatives continue to be the mainstay of treatments of falciparum malaria for some time to come. For the next transitional period of 8–10 years or more, no alternative medicines to the artemisinins are able to offer similar high levels of therapeutic efficacy nor are expected to enter the market. The Walter Reed Army Institute of Research has a long history with the artemisinins which place members of this group, who have been at the forefront of antimalarial drug development since World War II, to assist in summarizing the data of drug costs, efficacy, and safety for the scientific community.

This book reveals a new age of artemisinins as a mainstay in antimalarial drugs for current malaria treatment and prevention. Eight chapter topics are described as follows:

1.   Current Status and Policies of Antimalarial Drugs
2.   Antimalarial Drug Cost and Cost-effectiveness
3.   Efficacy and Treatment of Antimalarial Drugs
4.   Antimalarial Drugs in Chemoprophylaxis
5.   Influences on Antimalarial Drug Efficacy
6.   Toxicity of Antimalarial Drugs
7.   Antimalarial Drug Toxicity
8.   Antimalarial Drugs Withdrawn and in the Pipeline

Chapter 1 expresses that current situation of malaria and relevant policies.

Antimalarial drugs treat or prevent malaria, a disease that occurs in tropical, subtropical, and some temperate regions of the world. There is currently no vaccine that will prevent malaria, but this is an active field of research. Many researchers argue that prevention of malaria may be more cost-effective than treatment of the disease in the long run, but the capital costs required are out of reach of many of the world's poorest people. Currently, drug resistance has arisen to all classes of antimalarials except, as yet, to the artemisinin derivatives. This has increased the global malaria burden and is a major threat to malaria control. Widespread and indiscriminate use of antimalarials places a strong selective pressure on malaria parasites to develop high levels of resistance. Resistance to antimalarials has been documented for *P. falciparum, P. vivax,* and recently, *P. malariae.*

The only currently available drugs for which the problem of resistance of *P. falciparum* has not arisen are the artemisinin therapies. For other medications (amodiaquine, chloroquine, mefloquine, quinine and sulfadoxine– pyrimethamine), the degree of resistance varies from drug to drug. The geographical distributions and rates of spread have varied considerably. Resistance of *P. falciparum* to chloroquine, the cheapest and the most used drug has been reported in almost all the endemic countries, except in Central America and Hispaniola. Resistances to the combination of sulfadoxine-pyrimethamine and to mefloquine are found. Sporadic cases of prophylactic failure of mefloquine in travelers and therapeutic failure have been reported in Africa, South America, and in other Asian countries. *P. falciparum* has shown the development of resistance towards all antimalarial medicines when used as monotherapy: time periods vary from 12 years (chloroquine), to 5 years (mefloquine), to 1 year (proguanil) up to even less than 1 year (sulfadoxine/pyrimethamine, atovaquone). Experience in Thailand with successive drug regimens of monotherapy drugs has shown substantial decreases in cure rates.

In recent years, the treatment and prevention policies of antimalarial drugs have been revised based on new developments, such as drug resistance, artemisinin drugs, artemisinin based combination therapy (ACT), artemisinin drugs in severe malaria therapy and reducing cost on antimalarial drugs. As a response to the antimalarial drug resistance situation, WHO recommends that treatment policies for *falciparum* malaria in all countries experiencing resistance to monotherapies should use the combination therapies; preferably, treatments with artemisinin-based derivatives. The current WHO policy on antimalarial treatment is based on the recommendations and conclusions of two guidelines of "WHO Guidelines for the Treatment of Malaria 2006" and "WHO Policy Brief, March 2008, Global Malaria Programme."

Chapter 2 illustrates that the costs and cost-effectiveness of antimalarial drugs.

Economics is the study of the allocation of scarce resources among competing ends. It is not surprising, therefore, that economic considerations should loom large in health policy, including the provision of effective pharmaceuticals. For more than 40 years, the system had been largely based on a single agent – chloroquine - which was at one time very effective and remarkably cheap. Even in the poorest countries, at 10 cents per retail course, most people can still afford it. Moreover, the drug is familiar to the populace, and has been used - both within and outside of organized health care systems - well enough to prevent many malaria deaths and suppress (if not completely cure) acute attacks of the disease. Over time, however, resistance to chloroquine emerged worldwide, first leading to treatment failures in Southeast Asia, then to treatment failures in large parts of east Africa. It is now believed that chloroquine will be useless against most life-threatening *falciparum* malaria infections in fairly short order. In the meantime, replacement artemisinins have been introduced, but they too have quickly lost ground. Over the last 25 years, artemisinin derivatives have proved highly effective in Asian and African countries while no artemisinin resistance has surfaced. Partnering artemisinin-based combination therapy (ACT) confers even greater protection against the development of drug-resistant mutants, also offer therapeutic advantages over single antimalarial drugs.

However, ACT is relatively, though not absolutely, expensive. At present, ACTs cost about US$2 a treatment, roughly 20 times the price of chloroquine. The historical slowness in producing new antimalarials reflects the way in which new drugs are developed in a market system. Research, development, and testing of drugs engender a large upfront cost. Economic evaluation of a new antimalarial treatment requires an analysis of its respective costs and benefits, or at least a comparison of its reduction of malaria morbidity and mortality vs. other therapies. In the case of ACTs, the costs were high and the relative efficacy was so high that inquiring into the benefits in greater detail was reasonably cost-effectiveness. Many researchers argue that prevention of malaria may be more cost-effective than treatment of the disease in the long run, but the capital costs required are out of reach of many of the world's poorest people.

Chapters 3, 4, and 5 describe that the efficacy of antimalarial drugs.

First antimalarial drug, chloroquine, silently saved millions of lives and cured billions of debilitating episodes of malaria. *Falciparum* malaria has always been a major cause of death and disability, particularly in Africa, but chloroquine offered a measure of control even in the worst-affected regions. At roughly 10 cents a course and readily available from drug peddlers, shops, and clinics, it reached even those who had little contact with formal health care. Treatment of the disease with single drugs, such as chloroquine, sulfadoxine/pyrimethamine or mefloquine, has led to the emergence of resistant *P. falciparum* parasites that lead to the most severe form of the illness. Chief among these new treatments are the "artemisinins," paradoxically both an ancient and a modern class of drug that can still cure any form of human malaria. If *falciparum* malaria sufferers had the same broad access to artemisinins that currently exists for chloroquine, the rising burden of malaria in the world today would halt or reverse.

The challenge, therefore, is twofold: to facilitate widespread use of artemisinins while, at the same time, to preserve their effectiveness for as long as possible. Artemisinin-based combination therapies (ACTs) are currently recommended by WHO for the treatment of uncomplicated *P. falciparum* malaria. Artemisinin and semisynthetic derivatives, including artesunate, artemether, and dihydroartemisinin, are short-acting antimalarial agents that kill parasites more rapidly than conventional antimalarials, and are active against both the sexual and asexual stages of the parasite cycle. Artemisinin fever clearance time is shortened to 32 hours as compared with 2-3 days with older agents. To delay or prevent emergence of resistance, artemisinins are combined with one of several longer-acting drugs - amodiaquine, mefloquine, sulfadoxine/pyrimethamine or lumefantrine - which permit elimination of the residual malarial parasites. Therefore, the treatment of *falciparum* malaria is under a new age of artemisinin.

Malaria chemoprophylaxis is the prevention of malaria disease by giving healthy pregnancy, children and travelers medication prior to exposure to infective mosquitoes. Chemoprophylaxis is given to reduce the risk of malaria but is not perfectly effective, primarily because of non-compliance with medication regimens or parasite drug resistance. The principles of the use of antimalarial drugs for protection against malaria and the treatment of malaria were reviewed in International Travel and Health 2008 and Guidelines for the Treatment of Malaria 2006 by World Health Organization. These chapters consider the particular requirements for chemoprophylaxis in pregnant women and for chemoprophylaxis and stand-by treatment in travelers. It also comments on the management of severe malaria, the treatment of *vivax* malaria and the need for antimalarial formulations for pediatric use.

The aims of malaria chemoprophylaxis, in broad terms are to alleviate symptoms by the erythrocytic phase, to prevent relapses by the hypnozoites of *vivax* and *ovale*, and to prevent further transmission of the parasite. The principals of treatment depend on four factors, these being: 1) the type of infection, 2) the severity of infection, 3) the status of the host, and 4) any associated conditions or diseases present in the host. These chapters aim to present the current situation regarding malaria chemoprophylaxis, indicate what new drugs might be available in the future and discuss some of the common problems encountered when prescribing malaria chemoprophylaxis. Chemoprophylaxis is recommended for (a) pregnant women, (b) children in high-risk areas, and (c) travelers including service personnel who temporarily go on duty to high malarious areas. The efficacious new drugs and artemisinin-based combination therapy are available to prevent malaria, despite the rapid advance of drug resistance. The better tolerated regimens that require less compliance from the pregnancy, children, and travelers may be the key elements for improved malaria chemoprophylaxis in the future.

Chapters 6 and 7 show the toxicities of antimalarial drugs.

Various drugs are widely used in the prophylaxis and treatment of malaria. All antimalarial drug toxicity is viewed in the treatment and prophylaxis of *P. falciparum* malaria, which has a high mortality if untreated; a greater risk of adverse reactions to antimalarial drugs is inevitable. Antimalarial drug toxicity is one side of the risk-benefit equation and is viewed differently depending upon whether the clinical indication for drug administration is malaria treatment or prophylaxis. Research that leads to drug registration tends to omit two important groups who are particularly vulnerable to malaria – very young children and pregnant women. Prescribing in pregnancy is a particular problem for clinicians because the risk-benefit ratio is often very unclear.

All drugs cause toxicity that was assessed and evaluated. Type A adverse effects result from excessive responses to a drug; these adverse effects are predictable from the known effects of the drug and are dose or concentration related. In contrast, type B adverse effects are not predictable from the known effects of the drug; there may be an immunological basis to the adverse effect, and there is often no clear relationship with the dose or concentration of drug. In recent years, treatment of malaria using combination of available antimalarial drugs raises the hope that it will have enhanced efficacy along with reduction in rate of spread of resistance to antimalarial drugs. However, the risk of combination-drug-related toxicity must be assessed against the potential outcome and benefit of the treatment or prophylaxis. Antimalarial drug toxicity must be acceptable to patients and cause less harm than the disease itself. Cost and access to treatment often limit therapeutic choices in the tropics; these limitations do not generally apply in temperate zones. Drug-related toxicity and its risk must be balanced against the likely outcome of malaria treatment or prophylaxis and the circumstances of clinical practice. In this regard, there are no contra-indications to prescribing an effective drug that will save a life.

Antimalarial drug treatment policy in most malaria-endemic countries centers on the use of single agents used in sequence once the first-line drug line is failing. This policy will become unsustainable because of the limited number of affordable antimalarial drugs, and the small number of drugs in development. A new therapeutic strategy has been proposed recently, namely, combining standard or new antimalarial drugs with an artemisinin derivative with the aim of increasing drug efficacy, reducing transmission, and retarding the development of resistance. This approach has a similar rationale to the use of multiple drugs in the treatment of tuberculosis and HIV infection. It has two major implications. First, for drug-related toxicity, will two drugs be more toxic than one? Second, older drugs whose efficacy has been eroded by resistance may be given a new lease of life. These chapters have reviewed the adverse effects of the widely used traditional antimalarial drugs, the artemisinin derivatives, and drug combinations newly introduced.

Chapter 8 depicts a problem in the shortage of high efficacious antimalarial drugs.

There are a number of antimalarial drugs withdrawn from current use due to drug resistance, severe adverse effects, and inappropriate dose regimens. Because of widespread and unsupervised use of malaria drugs, chloroquine-resistant *P. falciparum* emerged in the early 1960s and rapidly spread around the world. Today there are reported cases of *Plasmodium* parasite resistance to most of the currently available antimalarial therapies, thus necessitating the development of new antimalarial treatments. Only class of antimalarials - artemisinin derivatives - has been shown to be highly efficacious against parasites resistant to other antimalarial drugs, and currently becomes the last line of defense against malaria.

Faced with malaria as the greatest scourges of humanity, there can be no doubt that the need for new antimalarial drugs is being taken very seriously both by the scientific community and the pharmaceutical industry. Even a cursory glance at the primary scientific literature reveals that a wide range of novel chemical structures are being assessed for such activity. When discussing issues relating to drug discovery and development, it is necessary to keep in mind the many disciplines and resources that have to come together to deliver success. Genomics can provide many potential molecular targets, but only those targets that can be readily manipulated and tested are likely to provide an opportunity to discover a lead molecule. For example, by high-throughput screening, that is worth taking on to a full medicinal chemistry project. Thus, many potential targets will never be advanced because no chemistry lead is identified.

Once a chemistry effort is started, many factors need to be considered over and above improving enzyme inhibition and efficacy against the parasite in culture. The molecules finally selected for development must also be easy to manufacture (low cost is crucial for antimalarials), stable, readily formulated, bioavailable (that is, extensively adsorbed from the gut and avoiding first pass metabolism in the liver to achieve effective concentrations in the systemic circulation), have an appropriate half-life, and not show any overt toxicity. It is essential that academic scientists better understand these issues as they increasingly have to take the lead in drug discovery efforts against malaria and other neglected diseases. An analysis of the reasons why candidate drugs fail to achieve registration and reach the market has been undertaken recently. In 39% of cases, drugs failed after entering development because of 'biopharmaceutical' issues such as oral bioavailability and formulation, and in 21% of cases because of toxicity. These issues are equally as important as drug efficacy, for which 29% of cases failed.

The antimalarial drugs are used in three distinct modes: as causal prophylactics to prevent the development of blood parasitemia; as blood schizonticides to kill blood parasites; and as anti-relapse agents to kill the hypnozoite stages in the liver. Also, there are four species of human malaria parasite, including *P. falciparum* which causes acute disease and is responsible for most malaria deaths and *P. vivax* which causes a much more chronic disease, typified by frequent relapses. Current drug discovery and development efforts focus on identifying molecules that will be active in different treatment due to resistance molecular mechanism in uncomplicated malaria. The ideal product profile comprises orally active compounds that can cure the disease with a 3-day regimen using once-a-day dosing. As a strong portfolio develops, however, it is anticipated that other, more specific malarial indications may be targeted, such as adjunct treatments to improve the outcome of severe malaria cases; intermittent treatment of malaria during pregnancy to protect both the mother from the disease and the unborn child from a higher risk of being born underweight; and long half-life drugs to treat malaria with a single dose in complex emergency situations. There is also the prospect of combining compounds from several projects into combination products. This could both enhance efficacy and reduce the likelihood of drug resistance development.

Abstract

Antimalarial drugs are medicines that prevent or treat malaria, a disease which takes a great toll on human health and well-being, particularly in tropical regions including Africa south of the Sahara, South and Southeast Asia, Oceania, and parts of the Americas. In recent years, strains of *Plasmodium* have become increasingly resistant to more antimalarial drugs and researchers have stepped up efforts to revise antimalarial drug policies and develop new antimalarial strategies. Resistance has arisen to all classes of antimalarials (chloroquine, amodiaquine, mefloquine and sulfadoxine-pyrimethamine) except, as yet, definitively to the artemisinin derivatives. In order to prevent widespread resistance, the concept of antimalarial combination therapy (CT) has been employed and a global resistance surveillance system (World Antimalarial Resistance Networks) has been established. Arguably, the recent action of selecting artemisinin-based combination therapies (ACTs) to replace other CTs was one of the most important steps. Currently, ACTs successfully cure patients suffering from uncomplicated malaria with superior efficacy and lower toxicity, but there still remains a huge challenge (high mortality rate) in the treatment of severe malaria. New clinical trials have demonstrated that intravenous artesunate (AS) can reduce by one-third the mortality as compared to the current standard of care, quinine injection. AS used either in intravenous, intramuscular, or rectal formulations have some theoretical and some demonstrated superiority to artemether and quinine injections in the treatment of severe malaria.

Although, artemisinin derivatives as a class of drugs are poorly efficacious at curing malaria when used as monotherapy, this shortfall has been overcome by using ACTs in uncomplicated therapy or AS injection in severe malaria therapy by sequential administration with slower acting antimalarial drugs. Artemisinins provide a clear-cut advantage over other antimalarials currently in malaria therapeutic use. Although, ACTs are 10-20 times more expensive than that of the traditional antimalarial drugs currently in use, the high cost-effectiveness and global subsidy plan could benefit every patient treated with ACT. New Guidelines for the Treatment of Malaria published in 2006 by the WHO with new policies (including the withdrawal of less effective drugs and monotherapies, the increase of ACTs in overall use and the selection of AS-injection in first-line with sequential therapy for severe malaria) sign a beginning of artemisinin age and the ending of chloroquine times. These new strategies could make artemisinin derivatives continue to be the mainstay of treatments of *falciparum* malaria for some time to come. For the next transitional period of 8–10 years or more, no alternative medicines to the artemisinins are able to offer similar high levels of therapeutic efficacy nor are expected to enter the market. Since current therapy mainly relies on ACT, the emergence and subsequent spread of resistance against artemisinins would be a disaster. Therefore, it would be necessary, to develop new antimalarials with novel mechanism of actions. New and great efforts have to be made to identify novel targets and to develop drug candidates as a pipeline against these targets to be prepared for the time when artemisinins begin to lose their effectiveness.

Chapter I

Current Status and Policies of Antimalarial Drugs

Malaria is a common and serious tropical disease. It is a protozoal infection transmitted to human beings by mosquitos biting mainly between sunset and sunrise. In total there are nearly 120 species of *Plasmodia*, including at least 22 found in primate hosts, and 19 in rodents, bats and other mammals. About 70 other *Plasmodia* species have been described in birds and reptiles. Human malaria is caused by four species of Plasmodium: *Plasmodium falciparum, P. vivax, P. ovale,* and *P. malariae.* There are of course multiple zoonoses that occur as well in humans and the recent recognition of *P. knowlesi* as the 5th human malaria parasite is noteworthy.

Malaria is a public health problem in over 100 countries worldwide, inhabited by some 40% of the world population, i.e., over 2 billion people. It has been estimated that the incidence of malaria in the world may be in the order of 300 million clinical cases each year. Countries in tropical Africa account for more than 90% of these. Malaria mortality is estimated at almost 1 million deaths worldwide per year. The vast numbers of malaria deaths occur among young children in Africa, especially in remote rural areas with poor access to health services. Other high risk groups include women during pregnancy, and non-immune travelers, refugees, displaced persons, or labor forces entering into endemic areas.

Antimalarial drugs are medicines that prevent or treat malaria. Treatment of malaria involves supportive measures, as well as specific antimalarial drugs. When properly treated, someone with malaria can expect a complete cure. Methods used to prevent the spread of disease, or to protect individuals in areas where malaria is endemic, include prophylactic drugs, mosquito eradication, and the prevention of mosquito bites. There is currently no vaccine that will prevent malaria, but this is an active field of research. Many researchers argue that prevention of malaria may be more cost-effective than treatment of the disease in the long run, but the capital costs required are out of reach of many of the world's poorest people.

In recent years, the treatment and prevention policies of antimalarial drugs have been revised based on new developments, such as drug resistance, artemisinin drugs, artemisinin based combination therapy (ACT), artemisinin drugs in severe malaria therapy and reducing cost on antimalarial drugs. As a response to the antimalarial drug resistance situation, WHO recommends that treatment policies for *falciparum* malaria in all countries experiencing resistance to monotherapies should be combination therapies, preferably those containing an artemisinin derivative as artemisinin-based combination therapy (ACT). The current WHO policy on antimalarial treatment is based on the recommendations and conclusions of two guidelines, "WHO Guidelines for the Treatment of Malaria 2006" and "WHO Policy Brief, March 2008, Global Malaria Programme" [1, 2].

1.1. Threat of Current Antimalarial Drug Resistance

1.1.1. The Widespread and Increasing Resistance

Resistance has arisen to all classes of antimalarials except, as yet, to the artemisinin derivatives. This has increased the global malaria burden and is a major threat to malaria control. Widespread and indiscriminate use of antimalarials places a strong selective pressure on malaria parasites to develop high levels of resistance. Resistance to antimalarials has been documented for *P. falciparum, P. vivax,* and recently, *P. malariae.*

The only currently available mainstream drugs for which the problem of resistance of *P. falciparum* has not arisen are the artemisinin therapies. For other medications (amodiaquine, chloroquine, mefloquine, quinine and sulfadoxine– pyrimethamine), the degree of resistance varies from drug to drug. The geographical distributions and rates of spread have varied considerably (Figure 1). Resistance of *P. falciparum* to chloroquine, the cheapest and the most used drug has been reported in almost all the endemic countries, except in Central America and Hispaniola. Resistance to the combination of sulfadoxine-pyrimethamine, which was already present in South America and in South-East Asia, is now becoming highly prevalent in Africa. Resistance to mefloquine is found mostly in Cambodia, Myanmar, Thailand, and Vietnam. Sporadic cases of prophylactic failure of mefloquine in travelers and therapeutic failure have been reported in Africa, South America, and in other Asian countries. *P. falciparum* has shown the development of resistance towards all antimalarial medicines when used as monotherapy: time periods vary from 12 years (chloroquine), to 5 years (mefloquine), to 1 year (proguanil), and even less than 1 year (sulfadoxine/pyrimethamine, atovaquone). Experience in Thailand with successive drug regimens of monotherapy drugs has shown substantial decreases in cure rates as shown in Figure 2 [1].

P. vivax has developed resistance rapidly to sulfadoxine–pyrimethamine in many areas. Cases of prophylactic and therapeutic failures have been reported in several places in the world. But, resistance to chloroquine in *P. vivax* is difficult to absolutely confirm and may need additional tests, in particular chloroquine blood concentrations. During the 2001 meeting, a protocol for assessing the efficacy of chloroquine for the treatment of *P. vivax* was established. The most convincing chloroquine resistance is confined largely to Indonesia, East Timor, Papua New Guinea, and other parts of Oceania (Figure 3). There are also documented reports from Peru. *P. vivax* remains largely sensitive to chloroquine in South-East Asia, the Indian subcontinent, the Korean peninsula, the Middle East, north-east Africa, and most of South and Central America [2].

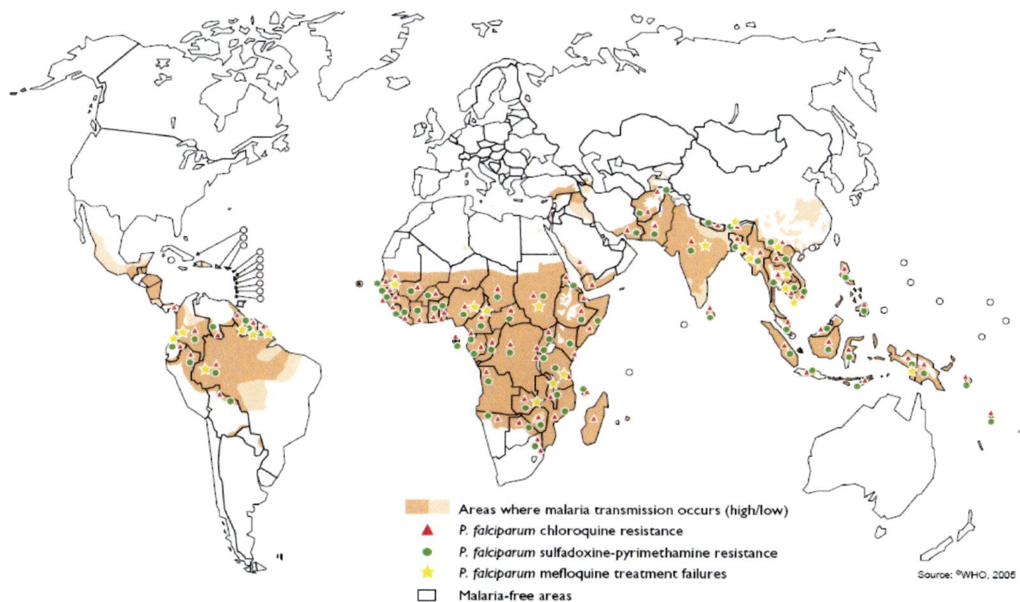

Figure 1. Malaria transmission areas and reported *P. falciparum* resistance in 2005 [2]. The designations employed and the presentation of material on this map do not imply the expression of any opinion whatsoever on the part of the World Health Organization concerning the legal status of any country, territory, city or area or of its authorities, or concerning the delimitation of its frontiers or boundaries. Dashed lines represent approximate border lines for which there may not yet be full agreement.

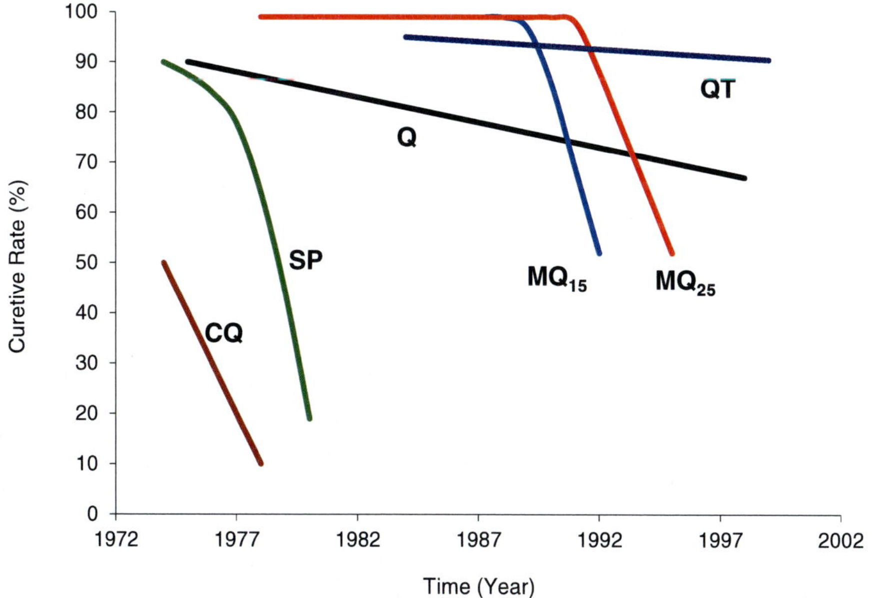

Figure 2. Drug efficacy reductions were observed in western Thailand [1]. CQ = chloroquine; SP = sulfadoxine-pyrimethamine; Q = quinine; QT = quinine and tetracycline; M_{15} = mefloquine at 15 mg/kg; and M_{25} = mefloquine at 25 mg/kg.

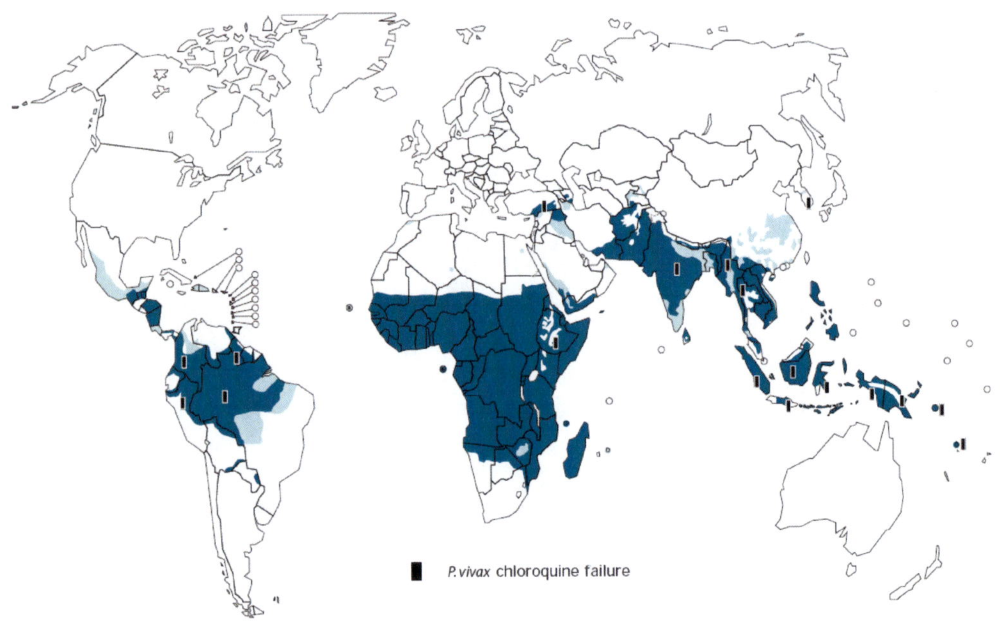

Figure 3. Malaria transmission areas and reported *P. vivax* resistance in 2004 [3].

The increasing resistance to most of the antimalarials can be reviewed from a historical perspective to appreciate potential future antimalarial drug resistance.

Resistance of *P. falciparum* to chloroquine appeared almost simultaneously in Colombia and on the frontier between Thailand and Cambodia in 1950s. In Asia, chloroquine resistance was initially confined to the Indochinese peninsula, until the 1970s, when it spread westwards and towards the neighboring islands in the south and east. The advent of chloroquine resistance in Africa occurred much later, and it took a decade to cross the continent. Today, only countries in Central America north of the Panama Canal and on the island of Hispaniola have not documented chloroquine-resistant *P. falciparum* malaria. Despite cross-resistance between chloroquine and amodiaquine, amodiaquine remains more effective than chloroquine in areas of chloroquine resistance. Nevertheless, amodiaquine could rapidly lose its efficacy if it is used intensively in areas in which chloroquine resistance is widespread or very high.

The sulfadoxine–pyrimethamine combination was used as a replacement for chloroquine in most countries. At the beginning of the 1980s, however, that treatment became almost totally ineffective in Thailand and neighboring countries, and resistance to the treatment spread rapidly in South America. In 1993, Malawi was the first country in East Africa to change from chloroquine to the sulfadoxine–pyrimethamine combination as the first-line drug, and other African countries followed this example in the late 1990s. Because of extensive use of the combination, however, resistance also spread in East Africa.

Mefloquine resistance was first observed near the Thai-Cambodian border in the late 1980s. The advent of mefloquine resistance in Thailand may have been influenced by the heavy use of the chemically related drug chloroquine and quinine just before the introduction of mefloquine. Mefloquine alone is no longer effective on the Thai-Myanmar and Thai-Cambodian borders, although it was operationally useful in most other endemic areas in and around Thailand during 1992-1997, showing field efficacy of more than 75%. Several studies have shown a diminution in sensitivity *in vitro*, and studies *in vitro* in West Africa showed the existence of strains with decreased sensitivity to mefloquine even before its introduction into the region for therapeutic use. Resistance to quinine is often overestimated, as the dosage of 24 mg base per kg for 7 days is rarely respected, and the threshold for resistance *in vitro* has not been clearly defined. So far, no resistance to artemisinin or artemisinin derivatives has been definitively reported, although some decrease in sensitivity *in vitro* has been reported in China and Vietnam. *In vivo* decreases in sensitivity to the artemisinins have also recently been reported and are being followed very closely for evolution into true classic resistance [4]. In WHO database [2], covering 1996–2004, of clinical outcomes on day 14 after treatment (which grossly underestimates failures), chloroquine failures >10% were reported in 93% of countries in Africa and 100% in Asia and the Americas. The corresponding values for pyrimethamine-sulfadoxine were 30% in Africa and 80% in Asia and the Americas (Table 1).

Resistance to antimalarial drugs has increased the global cost of controlling the disease. Therapeutic failure necessitates consultation at a health facility for further diagnosis and treatment, resulting in loss of work for adults and school absences for children. Studies in East Africa suggest that ineffective treatment causes anemia, which renders children's health more fragile. In Central Africa, the appearance of chloroquine resistance led to an increase in hospital admissions because of severe attacks of malaria. Similarly, increasing mortality trends were found at the community level in Senegal. The impact of drug resistance can also be illustrated by the changes in the proportion of *P. falciparum* relative to other species of malaria parasites. For example, in India, since the advent of drug resistance, *P. falciparum* accounts for more than 50% of all malaria attacks, instead of the previously reported 23%.

**Table 1. Clinical failure rates on day 14 after treatment with chloroquine
or pyrimethanmine-sulfadoxine in the world [2]**

| Drugs | Day 14 Clinical Failure (%)* | | | No. of |
Region	**< 10**	**10-25**	**> 25**	**Countries**
Chloroquine				
Africa	7	38	55	42
Asia	0	11	89	19
Americas	0	0	100	5
Sulfadoxine-pyrimethamine-				
Africa	71	24	6	34
Asia	20	53	27	15
Americas	20	80	0	5

* Data are % of countries. Source was from WHO/HTM/MAL 2005.1103.

1. Resistance to antimalarial drugs arises as a result of spontaneously-occurring mutations that affect the structure and activity at the molecular level of the drug target in the malaria parasite or affect the access of the drug to that target. Mutant parasites are selected if antimalarial drug concentrations are sufficient to inhibit multiplication of susceptible parasites but are inadequate to inhibit the mutants, a phenomenon known as drug selection. This selection is thought to be enhanced by

subtherapeutic plasma drug levels and by a flat dose-response curve to the drug. The important factors associated with resistance are: Longer half-life;

2. Single mutation for resistance;
3. Poor compliance;
4. Host immunity; and
5. Number of people using these drugs.

The factors influencing the rate of spread of antimalarial drug resistance, once it has emerged or been introduced in a given area, are still not defined. Theoretically, parasite resistance can emerge for any antimalarial drug, and the occurrence of de novo mutations and drug selection pressure (frequent drug use and long drug elimination half-life) are critical and essential prerequisites. Nevertheless, vector, parasite, and human host factors probably play an important role [4].

1. Antimalarial resistance results in enormous public health burden because of prolonged or recurrent illness and progression to severe malaria which is associated with increased hospitalization and death. Even lower levels of resistance which cause recrudescence of infection are associated with return of illness, prolonged or worsening anemia and increased gametocyte carriage which fuels transmission, particularly of the resistant parasite and a higher risk of treatment failure in subsequent infections. Impact of drug resistance is more in areas of low transmission where diagnostic facilities and infrastructure are inadequate and resources not allocated to respond to a sudden increase in treatment needs. The impacts of antimalarial drug resistance on health are listed as follows [3, 4]Prolonged or recurrent illness;
2. Increased outpatient cases;
3. More progression to severe malaria;
4. Increased cerebral malaria-neurological sequel;
5. Prolonged or worsening anemia and effect of anemia;
6. Increased hospital admissions and death;
7. Increased gametocyte carriage;
8. Increased demand on diagnosis; and
9. Higher cost and fewer cost-effectiveness of combination treatment.

Initially, at low levels of resistance and with a low prevalence of malaria, the impact of resistance to antimalarials is insidious. The initial symptoms of the infection resolve and the patient appears to be better for some weeks. When symptoms recur, usually more than two weeks later, anemia may have worsened and there is a greater probability of carrying gametocytes (which in turn carry the resistance genes) and transmitting malaria. However, the patient and the treatment provider may interpret this as a newly acquired infection. At this stage, unless clinical drug trials are conducted, resistance may go unrecognized. As resistance worsens the interval between primary infection and recrudescence shortens, until eventually symptoms fail to resolve following treatment. At this stage, malaria incidence may rise in low transmission settings and mortality is likely to rise in all settings.

To combat and control malaria, in addition to vector control measures, surveillance for drug resistance and a drug policy that is based on accurate surveillance data, development of simple, inexpensive diagnostic tests, new drugs, drug combinations and drug delivery systems and operational research are needed.

1.1.2. The Risk of Resistance of Artemisinin Derivatives

Although, the artemisinin resistance has not arisen, an emergence of artemisinin resistance could be devastating for individuals with malaria, unless timely detection by surveillance systems allows evaluation of new treatments and interventions to counter the spread of resistance. Developing an early warning system for resistance to artemisinin and other antimalarial drugs should be an urgent priority of malaria control programs and their sponsors.

Stable, therapeutically significant resistance to the artemisinin derivatives has not yet been identified and cannot be induced yet in the laboratory, which suggests that it may be a very rare event [5]. But it would be foolish to consider this will not happen, and should it arise, it would be a global disaster. For mutual protection against the emergence of drug resistance, these drugs should be used only in combination with other antimalarials. Artemisinin derivatives are particularly effective in combinations because of their very high killing rates (parasite reduction ratios 10,000-fold per cycle), lack of adverse effects, and absence of significant resistance [6].

Against these advantages, other factors could promote resistance. Artemisinin used alone or in subtherapeutic doses favors resistance by reducing, but not eliminating, parasitemia [7]. Taking artemisinin monotherapy for less than the 7 days required to eliminate parasitemia occurs because symptoms resolve rapidly and artemisinins remain expensive in many areas. WHO recently asked pharmaceutical companies to end the marketing and sale of artemisinin monotherapy; often common in some malaria-endemic areas. While 15 companies agreed to comply [8], many more have not, including at least 12 in Asia and several in Africa and Europe that continue to manufacture tablets with artemisinin derivatives alone [9]. Governments share the blame; according to WHO, 48 of 76 countries that need ACT permit marketing of artemisinin monotherapy [8, 9] (several companies promised to stop manufacturing monotherapy when countries withdraw marketing authorization).

Drug counterfeiting also may promote resistance. There are at least 12 different types of counterfeit artesunate, some produced by sophisticated, industrial-scale operations. Some counterfeits contain small amounts of artesunate, not enough to resolve infection, but possibly enough to select for artesunate-resistant strains and to register favorable results on simple, field-use drug quality tests [10]. Where ACT cost remains high, a thriving counterfeit antimalarial drug industry can be expected to sell substandard tablets inexpensively in the private sector. In general, national malaria control programs lack the sophisticated laboratory equipment and personnel to monitor ACT quality.

Another potential factor in development of artemisinin resistance is resistance to the partner drug in ACT, effectively resulting in monotherapy in the case of complete resistance. A large meta-analysis of randomized trials comparing artesunate plus a standard antimalarial drug to the standard drug alone showed, overall, that adding artesunate makes standard regimens more effective [11]. However, failure rates in the ACT group remained unacceptably high with less-effective partner drugs. For example, sulfadoxine-pyrimethamine plus artesunate was more effective than sulfadoxine-pyrimethamine alone in East Africa where sulfadoxine-pyrimethamine resistance is common, but the ACT groups still experienced high 28-day failure rates (approximately 30%–45%) [11].A humble perspective on the history of antimalarial drug resistance also calls for caution. Some antimalarial drugs encountered resistance shortly after introduction (e.g., sulfadoxine-pyrimethamine, atovaquone, and proguanil almost immediately because of preexisting reduced susceptibility), while others retained effectiveness longer (e.g., mefloquine, approximately 5 years; chloroquine, approximately 12 years) (Figure 2). Experimental studies showed that genetically stable reduced artemisinin sensitivity can be induced and transmitted by mosquitoes in a rodent model [5], and that a single amino acid substitution in *P falciparum* confers reduced artemisinin sensitivity [12], the relevant question appears to be when, not whether, artemisinin will become less effective.

While there is currently little compelling evidence that artemisinin resistance has already occurred [13], WHO has called for enhanced monitoring based on data, including possible treatment failures in Thailand and India [13, 14]. In another recent study, Jambou *et al.* [15] obtained *in vitro* drug sensitivity profiles for *P. falciparum* isolates from French Guiana, Senegal, and Cambodia. The study did not assess clinical outcomes. Although the correspondence between *in vitro* artemisinin sensitivity as measured by 50% inhibitory concentration and clinical response is not established, maximum 50% inhibitory concentration values from French Guiana and Senegal were higher than any reported previously. A single-nucleotide polymorphism in a possible artemisinin target gene was strongly associated with reduced sensitivity. The least-sensitive infections came from areas where artemisinin sale and use are poorly controlled [16].

1.1.3. Improving Surveillance for Antimalarial Drug Resistance

World malaria monitor efforts toward controlling malaria are greatly challenged by the increasing spread of antimalarial drug resistance. Use of ineffective antimalarials is thus considered partly responsible for the difficulties in reducing malaria morbidity and mortality. This is a problem particularly in sub-Saharan Africa, where susceptibility of *Plasmodium falciparum* to previously used cheap and commonplace antimalarials, such as chloroquine and sulfadoxine–pyrimethamine (SP), has been declining rapidly in many high-transmission areas. To ensure that malaria control strategies and malaria treatment policies rely on the deployment of effective antimalarials, there is a need for systematic monitoring of antimalarial drug efficacy and drug resistance. Artemisinin-based combination treatments (ACTs) are today the only consistently effective antimalarials available, and as ACTs now become more and more widely used, careful monitoring of their therapeutic efficacy and of any emerging resistance is needed. Preventing development of artemisinin resistance requires that the short-acting artemisinin component is "protected" by a longer-acting partner drug of known high efficacy. Therefore, it is important also to keep an eye on the efficacy of the different partner drugs currently recommended for use in artemisinin-based combinations. The available three tools were developed recently for current monitoring of drug efficacy and early detection of resistance for standardized and coordinated use [17]. The available and improving surveillances include first of all the therapeutic efficacy test *in vivo*, which involves the repeated assessment of clinical and parasitological outcomes of treatment during a fixed period of follow-up to detect any reappearance of symptoms and signs of clinical malaria and/or parasites in the blood. Other available methods include *in vitro* tests of parasite susceptibility to drugs in culture and studies with molecular marker methods of gene mutations or gene amplifications associated with parasite resistance (Table 2).

Table 2. Current tools for monitoring antimalarial drug efficacy and drug resistance [17]

Groups	Therapeutic efficacy test	*In vitro* sensitivity assay	Molecular markers
Definition	• Treatment of symptomatic *P. falciparum*-infected patients with a standard dose of an antimalarial drug and subsequent follow-up of parasitemia and clinical signs and symptoms over a defined period (response of the host–parasite system to the drug)	• Cultivation of *P. falciparum* parasites *in vitro* with a range of antimalarial drug concentrations (response of the parasites to the drug)	• Detection of gene mutation(s) or amplification that modify drug-target (enzymes) or drug-transporter functions or affinities (genetic characterization of drug targets or transport)
Indications	• Gold standard for monitoring antimalarial drug efficacy and for guiding drug policy	• Detect reduced parasite response to antimalarial drug • Early warning system (adjunct to therapeutic efficacy test)	• Detect resistance-related mutations or amplification • Early warning system (adjunct to therapeutic efficacy test)
Advantages	• Easily interpretable results • Simple method with minimal training required (except microscopy) • Minimal equipment and supplies required • Relatively inexpensive to conduct (depending on local conditions) if integrated into the national malaria control programs	• Avoids host confounding factors • Accurate for detecting true drug resistance • Provides quantitative results • Multiple tests can be performed with a single isolate, and several drugs can be assessed simultaneously • Experimental drugs can be tested (except prodrugs) • *In vitro* resistance precedes *in vivo* resistance	• Avoids host confounding factors • Accurate for detecting true drug resistance • Samples on filter paper easily obtained, transported, and stored • Multiple tests can be performed with a single filter paper, and molecular targets of several drugs can be characterized • If known, targets of new and experimental drugs can be tested (e.g., atovaquone) • Mutations precedes *in vivo* resistance

Table 2. (Continued)

Drawbacks	• Interference of immunity, previous drug intake, variation of drug absorption or metabolism • Misclassification of re-infection and recrudescence • Treatment failures do not reflect the level of true drug resistance • Difficult to conduct in areas of low transmission given the limited numbers of eligible patients • Overestimation of early treatment failures for slowly acting drugs • Numerous local adaptations and modifications result in poor ability to compare between sites • Long duration of patient monitoring may result in high patient loss to follow-up	• Correlation with therapeutic efficacy test not fully established • Presence of mixed population with different drug sensitivity phenotypes • Expensive equipment and supplies required • Training required • Numerous available methods but not always comparable • Lack of standardized *in vitro* protocol • Threshold of resistance not validated	• Correlation with therapeutic efficacy test not fully established • Presence of mixed population with mixed alleles • Expensive equipment and supplies required • Training required Identified for a limited number of antimalarial drugs • Lack of standardized PCR protocol, including sample collection and DNA extraction

1.1.3.1. In Vivo *Assessment of Therapeutic Efficacy*

The first test systems for evaluating the response of *P. falciparum* to drugs *in vivo* were developed in 1965, shortly after the first occurrences of chloroquine resistance were observed in this species. These test systems were revised in 1967 and remained largely unchanged until the WHO Scientific Group on the Chemotherapy of Malaria and Resistance to Antimalarials modified them in 1972 [18]. The standardized tests were originally developed for chloroquine. Performance of these tests relied on adherence to set criteria for administration of a standard treatment regimen of the appropriate drug and regular examination of blood for the stipulated period, i.e., 7 or 28 days for chloroquine. Their use in the field was constrained by the need for daily blood sampling during the first week, followed by weekly tests if follow-up was extended beyond 7 days. In addition, these tests were conceived primarily for assessing the parasitological response of *P. falciparum* in areas with low-to-moderate malaria transmission (ideally in a setting free of malaria transmission, in particular for the extended 28-day field test). Therefore, little attention was paid to the clinical response to the drugs or to the immunity of patients. In view of the lack of clinical information, which is generally required by policy-makers, it was decided to introduce a simplified test system, in which the number of parasitological observations was reduced and the test was complemented by standardized clinical observations [14].

1) *Therapeutic Efficacy Testing in* P. Falciparum

The therapeutic efficacy test remains the gold standard for determining antimalarial drug efficacy for the management of *P. falciparum* infections. It provides policy makers and national malaria control programs a straightforward indicator of the efficacy of an antimalarial drug or a combination treatment in a given population at risk, to suggest whether a drug is still appropriate as first- or second-line treatment. Studies of therapeutic efficacy are relatively straightforward to set up, although they tend to be lengthy and may be costly, unless integrated into a national malaria control program; in particular, medical and technical personnel (especially microscopists) must be trained. To interpret the results uniformly to follow trends over time and to compare levels of resistance between different regions, studies must be carried out with the same standardized protocol, at the same sentinel sites, in the same age groups, and, if possible, at the same time of year in a given site.

The WHO standard protocol is meant for the evaluation of antimalarial drugs or drug combinations (chloroquine, amodiaquine, quinine, SP, ACTs, etc.) for treatment of uncomplicated *P. falciparum* malaria [14]. The design is simple: a one-armed prospective study of clinical and parasitological responses after administration of antimalarial treatment to children of age 6–59 months with a degree of immunity that is unlikely to have much impact on the outcome of the test. In areas of low or moderate transmission in which it is difficult or time-consuming to enroll enough children in this age group, children > 5 years old and adults can be included, although it should be borne in mind that the results will be biased, as adults always respond better than children. To avoid the inclusion of asymptomatic carriers, only patients with 2000 parasites per µL or more (1000 parasites per µL in areas of low or moderate transmission) are included. Calculation of the sample size required by estimating prevalence is now preferred to the lot quality assurance sampling method. If the selected confidence interval is 95% with a precision of 10%, the sample size ranges between 50 and 100 patients. A minimum of 50 patients must be enrolled in order for the sample to be representative. This sample size was appropriate to detect failure cases during routine monitoring of the efficacy of currently recommended ACTs after their implementation. The recommended duration of follow-up is 28 days in areas of intense as well as low-to-moderate malaria transmission. For treatment with drugs such as amodiaquine, chloroquine, and SP, a 28-day follow-up is considered appropriate; for slowly eliminated antimalarials (lumefantrine, mefloquine, piperaquine, pyronaridine), recrudescences may occur after 28 days, and so 42- or even 63-day follow-up periods are recommended.

Several changes to the WHO protocol have been introduced based on the feedback from the field since 1996. The main difference between the protocols for areas of high transmission and for areas of low-to-moderate transmission, apart from the inclusion criteria, was in the management of parasitological failures without clinical signs. The differences reflect regional program priorities (treatment of clinical cure or for radical cure). In 2006, the malaria treatment guidelines stressed that the objective of antimalarial treatment, even in areas of high transmission, should be radical cure of the disease. From a clinical point of view, the persistence of parasites resulting from treatment failure can lead to anemia, gametocyte carriage, and risk of recurrent clinical signs and symptoms. These 3 harmful consequences of treatment failure are taken into consideration in the modified protocol of WHO in 2008 [7].

Table 3. Classification of treatment outcome according to the WHO Protocol 2005 for all endemic areas [17]

Early treatment failure
- Danger signs or severe malaria on day 1, 2, or 3, in the presence of parasitemia;
- Parasitemia on day 2 higher than on day 0, irrespective of axillary temperature;
- Parasitemia on day 3 with axillary temperature $\geq 37.5°$ C;
- Parasitemia on day 3 $\geq 25\%$ of count on day 0.

Late clinical failure
- Danger signs or severe malaria in the presence of parasitemia on any day between days 4 and 28 (42 or 63), without the patient previously meeting any of the criteria of early treatment failure;
- Axillary temperature $\geq 37.5°$ C in the presence of parasitemia on any day between days 4 and 28 (42 or 63), without the patient previously meeting any of the criteria of early treatment failure.

Late parasitological failure
- Presence of parasitemia between days 7 and 28 (42 or 63) with temperature $< 37.5°$ C, without the patient previously meeting any of the criteria of early treatment failure or late clinical failure.

Adequate clinical and parasitological response
- Absence of parasitemia on day 28 (42 or 63), irrespective of axillary temperature, without the patient meeting any of the criteria of early treatment failure, late clinical failure, or late parasitological failure.

Patients included in the efficacy study and not lost to follow-up or excluded (e.g., because of self-medication, development of concomitant febrile infections, or refusal to continue participation) are classified in one of the following categories: early treatment failure, late clinical failure, late parasitological failure, or adequate clinical and parasitological response. These classifications rely on the presence or absence of fever or other signs of clinical malaria and/or presence of parasitemia during the course of follow-up, as listed in Table 3 [17]. The rates of total failure are used to define cut-off points for drug-policy change. Monitoring antimalarial drug efficacy is based on presence or absence of asexual parasites detected by microscopy at admission and during the follow-up, and the use of rapid diagnostic tests has been suggested for the screening of febrile patients. So far, however, rapid tests cannot be used during the follow-up because quantitative parasitemia is needed to classify patient outcome as early treatment failure, and some tests show false-positive results related to the persistence of circulating antigens or presence of gametocytes, despite complete parasite clearance [19].

2) Therapeutic Efficacy Testing in P. Vivax, P. Ovale, and P. Malariae

Relapse, re-infection, and recrudescence cannot be distinguished reliably in the infections with *P. vivax* and *P. ovale*. Nevertheless, *in vivo* assessments of chloroquine susceptibility can be performed using the same protocol format as for *P. falciparum*, with a follow-up period of 28 days and preferably accompanied by measurement of whole-blood chloroquine and desethychloroquine levels at the day of failure. Recurrent infections within this period presenting with whole-blood chloroquine + desethychloroquine concentrations exceeding 100 ng/ml are currently considered as resistant whether they are a relapse, a recrudescence, or even a new infection, as this concentration should be suppressive. This definition will most probably need to be reconsidered according to the on-going trials [17].

3) WHO Database on Therapeutic Efficacy Studies

To facilitate the use of data from such standardized assessments, a comprehensive database has been established by WHO, where reported standardized drug-efficacy results are compiled and summarized to better inform endemic countries and their national malaria control programs about the current situation and of changing patterns of resistance. The data in the database originate from 3 sources: 1) published studies; 2) unpublished studies (available in reports by ministries of health, national control programs, or nongovernmental organizations, consultant reports, theses, and papers, or posters presented at national and international conferences); and 3) regular data from surveillance studies conducted according to the WHO standard protocol and sent from countries to WHO for validation. On the basis of this database, detailed estimates of global levels of drug efficacy in the period from 1996 to 2004 were recently reported as mentioned earlier [14].

1.1.3.2. In vitro *Tests of Parasite Susceptibility to Drugs in Culture*

To support the evidence of a failing antimalarial, an *in vitro* test can be used to provide a more accurate measure of drug sensitivity under controlled experimental conditions. Parasites are exposed to precisely known concentrations of an antimalarial drug and are observed for inhibition of maturation into schizonts. Several *in vitro* tests exist, and they differ primarily in how their results are interpreted. These include microscopic examination of blood films for the WHO Mark III test, the radioisotopic test, and the enzyme-linked immunosorbent assay with antibodies directed against Plasmodium lactate dehydrogenase or histidine-rich protein II and more recently the Sybr Green test [14, 20].

In vitro tests overcomes some of the many confounding factors influencing the results of *in vivo* tests, such as sub-therapeutic drug concentrations and the influence of host factors on parasite growth (e.g., factors related to acquired immunity), and therefore provide a more accurate picture of the " true" level of resistance to the drug. Multiple tests can be performed on parasite isolates, using several drugs and drug combinations simultaneously. However, in part because *in vitro* tests do not include host factors, the correlation between results of *in vitro* and *in vivo* tests is not consistent and is not well understood. *In vitro* tests have proven useful as part of epidemiological monitoring, including monitoring of cross-resistance patterns between different antimalarials in a region, monitoring of baseline drug sensitivity before it is introduced as part of national policy, in studies of temporal and spatial changes for early warning and guidance on need for therapeutic efficacy studies, and for the validation of molecular markers of resistance. It is often difficult to compare results, even from laboratories where the same type of test is used, because the results— which are usually expressed as the 50% inhibitory concentration (IC_{50}), the 90% inhibitory concentration (IC_{90}), or the MIC— are the expression of more than 10 factors that are rarely identical in different laboratories. To date, few thresholds of resistance have been correctly validated. Moreover, a cut-off point validated with a given test is valid only for that assay system, with its specific *in vitro* factors and cannot be extrapolated to another test. To validate the *in vitro* cut-off point for resistance, the results of *in vitro* tests and of therapeutic efficacy tests conducted in a nonimmune population (children or travelers) must be compared in a sufficiently large sample.

Furthermore, patient follow-up must be sufficiently long, and every effort must be made to confirm that failures are not due to insufficient drug absorption, re-infection, or other causes unrelated to drug resistance. Therefore, results should not be expressed as percentage resistance, especially when the thresholds of resistance are not validated. Instead, they should be expressed as a geometric mean of the IC_{50} or MIC, which allows a more precise quantitative comparison of sites in a given country and over time [17].

1.1.3.3. Molecular Marker Method

1) Genetic Markers of Resistance

As an additional means of detecting changing patterns of resistance, molecular tests have been developed in recent years for the detection of parasite gene mutations or amplifications associated with resistance to a number of antimalarials. These tests are based on PCR analyses of only small amounts of parasite DNA material in finger-prick blood dried on filter paper, one sample allowing for multiple tests to be performed and molecular targets of several drugs to be characterized. The tests are easier to run than *in vitro* tests and are thus more readily deployed routinely by malaria control programs, although technical capacity is still required. Information on the prevalence of gene mutations may give an indication of the level of drug resistance in an area as a mean of early warning, and relatively well-defined molecular markers of resistance have been established for pyrimethamine (dihydrofolate reductase, *dhfr*), sulfadoxine (dihydropteroate synthase, *dhps*), and chloroquine (*P. falciparum* chloroquine-resistance transporter, *Pfcrt*) [21]. An increased number of *P. falciparum* multidrug-resistance gene 1 (*Pfmdr*1) has been identified as a marker of mefloquine resistance, and mutations in Pfmdr1 have also been implicated in modification of the sensitivity to amino alcohols and artemisinins, although field studies have provided contradictory results [22, 23]. Treatment failures observed with atovaquone– proguanil have been associated with a specific point mutation in the cytochrome b gene at codon 268, but other reported treatment failures with this drug were not associated with this mutation [24, 25].

Although monitoring of molecular markers of resistance from a programmatic point is simpler than *in vitro* sensitivity testing, the molecular methods also have their drawbacks. An important problem is that the mutations and therapeutic efficacy do not always correlate well, as many factors determine the therapeutic response in addition to parasite sensitivity to the antimalarial treatment. To be useful as a public-health measure, the molecular markers should reliably predict the clinical and parasitological outcome of treatment, but at present the markers do not give an accurate prediction in all epidemiologic settings. However, serial assessment of molecular markers can be a useful guide to the emergence of resistance, especially if used consistently over time in comparable study populations to detect trends. In using newly developed high-throughput methods [26, 27], more comprehensive population-based analyses will be possible, which may lead to a better understanding of the different genetic markers of resistance and their use in prediction of drug efficacy [28].

2) Genetic Markers of Recrudescence and Re-infection

Detection of genetic parasite markers also provides a valuable tool in support of monitoring programs that rely on therapeutic efficacy testing. When monitoring of therapeutic efficacy is extended beyond 14 days in an intense transmission area or with 28-day follow-up in a low-to-moderate transmission area, and as most antimalarial drugs have no effect on the liver stages of *P. falciparum*, methods are needed to distinguish cases of new infection from recrudescence. PCR methods are relatively simple means of analyzing genetic diversity by analyzing 3 highly polymorphic genes: merozoite surface proteins 1 and 2 (msp-1 and msp-2) and glutamate-rich protein (*glurp*). However, PCR analyses are still limited primarily to research-oriented monitoring projects and have not yet become part of routine efficacy monitoring in many places. Thus, there is a need to build technical and human capacity at national levels to provide this testing service for sentinel sites [17].

The need for high-quality monitoring of antimalarial drug resistance has increased in recent years. It is increasingly important to know if the required high levels of efficacy of the new and costly combination treatments are maintained, and a more proactive approach to malaria treatment policy change is considered a key to effective malaria control. Technical advances in recent years have made it possible to provide more accurate efficacy results than previously, but the technical requirements in routine monitoring place an additional burden on malaria control efforts, such as the increase in the recommended duration of follow-up and the need of PCR analysis, which will inevitably increase the cost of routine monitoring. Sustained funding for routine monitoring activities is becoming a major issue. In the last decade, countries had to carry out a limited numbers of studies to demonstrate that the monotherapies used as first- and second-line drugs were failing and to compile the baseline data on ACT efficacy. Monitoring ACTs is becoming crucial because the long-acting partner drug of some of the recommended ACTs are marketed as monotherapies, and their efficacy is compromised by high resistance rates, exposing the artemisinin derivatives to the development of resistance.

Incorporation of routine therapeutic efficacy testing into all national malaria control programs is an essential first step. But integrating efficacy monitoring with laboratory-based approaches provides the best chance of detecting artemisinin resistance early [29]. Several antimalarial resistance monitoring networks involve partnerships between ministries of health, which conduct efficacy studies, and malaria research laboratories, which perform *in vitro*, pharmacokinetic, or molecular tests.

1.1.4. World Antimalarial Resistance Networks (WARN)

The request for better artemisinin resistance surveillance stems from long-standing attention to antimalarial drug resistance. To allow for the collection of accurate and comparable data on drug resistance, WHO recommends a systematic and uniform organization of drug-efficacy monitoring, coordinated by ministries of health and national malaria control programs. Many countries are already collecting drug-efficacy data in a systematic and timely manner, and monitoring systems in other countries can be improved, even if only limited resources are available. National monitoring of drug efficacy mainly requires a well-planned and coordinated system for resistance-data collection, including established sentinel sites with the necessary trained staff, some minimum laboratory facilities, and agreed standards for data collection, analysis, and reporting of results according to the WHO standard efficacy protocol.

The study by Jambou *et al.*, [15] drew on ministry of health collaborations with Pasteur Institutes, to comprise a global laboratory network. The East African Network for Monitoring Antimalarial Treatment (EANMAT) serves as a good example of such coordinated efforts [30]. The US Department of Defense Global Emerging Infections Surveillance and Response System and other international networks are also including partnerships with malaria research laboratories to form great collaborations. By firm commitment to improve drug resistance surveillance by major sponsors of malaria control programs, partnerships between public health agencies and malaria research laboratories could lay the groundwork for a global antimalarial drug resistance surveillance system. A possible model for this system is the WHO Global Influenza Surveillance Network, which includes national surveillance programs and national and global reference laboratories. Drawing on this example, a global antimalarial drug resistance surveillance program could include a national laboratory in each affected country with advanced surveillance tools, such as *in vitro* testing capabilities. The national laboratories would work closely with ministries of health conducting efficacy tests and connect with regional or global referral laboratories that establish standards and provide specialized capabilities.

An essential initial step in developing a global antimalarial drug resistance surveillance system is establishing a case definition for artemisinin resistance. Without a standard case definition for national and regional surveillance programs, data comparisons and consensus on when global public health responses are required will be problematic. Achieving agreement across surveillance systems that now function, for the most part, independently, will be challenging, especially where therapeutic efficacy tests alone are used to identify resistance to artemisinins used in combination therapy. The proposed development of a global antimalarial drug resistance database [31] is a promising attempt to share information. If implemented, the database could provide a rich empirical foundation for these discussions.

The proliferation of antimalarial drug trials in the last ten years provides the opportunity to launch a concerted global surveillance effort to monitor antimalarial drug efficacy. The diversity of clinical study designs and analytical methods undermines the current ability to achieve this. Recently, the global community is going to establish a comprehensive clinical database of World Antimalarial Resistance Network (WARN), that aims to standardize estimates of antimalarial efficacy and resistance which can be derived and monitored over time from diverse geographical and endemic regions. WARN will consist of four linked open-access global databases containing clinical, *in vitro*, molecular and pharmacological data, and networks of reference laboratories that will support these databases and related surveillance activities. WARN will serve as a public resource to guide antimalarial drug treatment and prevention policies and to help confirm and characterize the new emergence of new resistance to antimalarial drugs and to contain its spread [32].

Two important challenges now emerge. First, how can the accumulated information on the history of resistance to the old drugs to be used to understand how resistance was selected and spread? Second, how can the tools developed and lessons learned to be leveraged to mobilize systems for effective, worldwide surveillance of resistance to the newly introduced drugs?

Here, an approach is proposed to respond to these challenges. The broad outlines were agreed by an ad hoc group (37 institutions of the scientists involved) in October 2006. These plans are described here and the details for a Worldwide Antimalarial Resistance Network (WARN) are presented separately in a series of papers in the *Malaria Journal* [32–36]. For the plan to work, it will need to be a collective effort that involves the whole malaria community. In cooperation with the WHO, a central repository of information on resistance to antimalarial drugs could serve to integrate and make accessible the important information that is being generated by the many current and proposed programs that study and monitor drug efficacy and resistance. The Malaria in Pregnancy Consortium, Artemisinin Combination Therapy Consortium, INDEPTH network, the Multilateral Initiative against Malaria network and national and regional networks, such as the EANMAT, are all examples of projects than can both contribute to and benefit from WARN. Comment sections linked to detailed descriptions of each component of WARN in the *Malaria Journal* can be used to lodge comments, additions and criticisms (*www.malariadrugresistance.net/warn*). These comments can then be incorporated so that the plan evolves in a dynamic way, involving and serving the whole malaria community [37].

The WARN database will be modular, with data from four linked components: clinical efficacy trials, *in vitro* measures of parasite chemosensitivity, clinical pharmacology and molecular-resistance markers. Building the inter-relationships among these different types of data will be challenging but will add significant value to the database. An organized plan for surveillance of ACTs (four components) system is needed, to incorporate clinical, *in vitro*, pharmacological, and molecular measures of drug resistance and efficacy [37].

1.1.4.1. Clinical Data

Clinical-efficacy trials will provide the core data on clinical and parasitological-drug efficacy. The precise definitions of therapeutic responses and analytical approaches profoundly affect the estimates of efficacy that serve as the basis for policy decisions. Neither comparison of aggregate data nor a meta-analysis approach is adequate to make reliable estimates of antimalarial-drug efficacy from studies done at different times and locations. To avoid these problems; first, WARN will use individual patient data, not summary statistics, to enable direct comparisons between studies and to provide more comprehensive and accurate estimates of regional efficacy. Second, the primary outcome will also be the first reappearance of parasites, unless an early treatment failure has occurred. Third, survival analysis – the time between treatment and reappearance of parasites – will be used as the most flexible and useful metric for comparison of drug efficacy, particularly when pooling and comparing studies with different duration of follow up [38]. Fourth, an agreement on how best to distinguish and compare initial and recurring infections after treatment needs to be reached. The details of how the clinical module of the WARN database could be developed and used are described in Price *et al.* [33]. A simple prototype has been designed, which can be accessed at *www.malariadrugresistance.net/warn/clinical*. Members of the malaria community can comment on the proposal and try the database with their clinical data without downloading the data into the central database.

1.1.4.2. In vitro *Surveillance of Drug Sensitivity of Clinical Isolates*

A surveillance network of comparable *in vitro* chemosensitivity data from current patient isolates is needed to document parasite responses to the individual components of the various ACTs in use. The WARN *in vitro* group proposes that disparities among different test sites can be reduced greatly by agreement on four parameters of *in vitro* culture and testing: culture conditions used, time of incubation, starting parasite density and hematocrit. If these parameters are standardized, then a variety of growth measurements can yield comparable data. Quality control is a key: assays should include common standard strains whose drug responses are known, so that the *in vitro* test data from any laboratory would be assured by the results achieved with the standard strains. The *in vitro* laboratories will also need to adapt to culture and cryopreserve putatively resistant isolates, so that results can be validated in other laboratories. Finally, common sources of both the drugs to be tested and, for some drugs, their active metabolites will need to be available for use in the reference laboratories. Using these common elements, the *in vitro* network will be able to provide data that can be compared over time and across countries.

The advantage of the *in vitro* analysis is that small decreases in drug susceptibility of the parasites from a region will be visible. In addition, both partners in an ACT can be tested individually and any hints of resistance can be attributed to one or both of the partner drugs. By identifying decreased susceptibility to the longer-acting partner in the ACTs at the earliest possible time, a change can be made to protect the artemisinin component. This network should be established now so that reliable baseline data will be available before changes in sensitivity occur in areas where partner drugs remain effective, although all of the current WHO-recommended ACT partner drugs are already compromised by a degree of resistance in some areas. If that is done, early indications of resistance can be detected and prompt responses will be possible. The details of the *in vitro* component of the WARN database are described in Bacon *et al.* [34].

1.1.4.3. Pharmacological Surveillance

Pharmacokinetic surveillance must be added to the clinical-efficacy studies. When a patient fails drug treatment, the drug concentration in the blood might not have reached or sustained adequate levels. This therapeutic failure is not a result of true parasite resistance but is a consequence of poor drug quality, inadequate dosing or absorption or too rapid clearance of that drug. This is especially true in the ACT era [39]. The plethora of new combinations and the rapidity with which they have been introduced mean that information on the proper dose of a drug is frequently not available. This lack is most acute in high-risk populations: infants, pregnant women, the malnourished and people living with HIV. These are the groups most likely to differ in the pharmacological big four – absorption, distribution, metabolism, and excretion– and studies of pharmacokinetics and pharmacodynamics in these high-risk populations are required urgently. Moreover, the drug level in individual patients should be determined preferably in the context of clinical efficacy studies, so that we learn how the pharmacokinetic properties of each drug influence overall therapeutic response. Obtaining the appropriate foundation studies, a blood sample taken at day 7 could provide sufficient information to assess adequacy of drug exposure for the longer acting components in the majority of treatments.

Accurate quantification of drug levels is challenging, however, there are laboratories in each of the endemic regions that are equipped for these assays. A practical system of standardized protocols and quality assurance is needed so that results from different laboratories can be compared. Therefore, a system of regional pharmacology reference laboratories must be established to assure quality control. Without this quality-control program, there would be rather limited high-quality data available and pooling and comparing results would be near impossible. The details of this pharmacokinetic component of the WARN database can be found in Barnes *et al.* [36].

1.1.4.4. Identification and Monitoring of Genetic Correlates of Drug Resistance

Molecular markers that predict resistance to most of the drugs used in ACTs have not been defined, although experience with the old drugs provides a clear path for identifying and validating resistance markers. If patient-derived isolates show decreased susceptibility to drugs *in vitro*, then genome-wide comparison of these isolates with susceptible isolates from the same region could identify candidate loci that carry single-nucleotide polymorphisms (SNPs) or copy-number changes associated with the resistance [40]. This approach has clearly identified loci already known to be responsible for resistance to the old drugs and could also be effective in identifying candidate loci for resistance to the new drugs. At least three centers are involved currently in such comparative genomic analyses and they are pooling data on both reference and field strains of *P. falciparum* [41]. This focus on comparative genomics will establish the baseline population structure and diversity needed to interpret genomic comparisons between putative resistant strains and sensitive strains. The molecular component of the WARN database is detailed in Plowe *et al.* [35].

1.1.5. Prevention of Resistance by Antimalarial Combination Therapy

Resistance to these old drugs precipitated a transformation in antimalarial treatment approaches, through the recognition that malaria, similar to tuberculosis (TB) and HIV/AIDS, should be treated with combinations of drugs that act on different targets. By 2003, the WHO, the Roll Back Malaria Partnership and the Global Fund to Fight AIDS, TB, and Malaria recommended combination therapy for malaria as standard [42]. Most recommended combinations include derivatives of artemisinins: highly potent, rapid-acting drugs based on ancient Chinese herbal treatments for fever. Artemisinin combination therapies (ACTs) have since become the recommended first-line treatment, essentially worldwide [7], and both the public and private sectors have invested huge sums in the purchase and distribution of ACTs [37].

Combinations of chemotherapeutic agents can accelerate therapeutic response, improve cure rates and protect the component drugs against resistance. The theory underlying if two drugs are used with different modes of action, and therefore different resistance mechanisms, then the per-parasite probability of developing resistance to both drugs is the product of their individual per-parasite probabilities. This is particularly powerful in malaria, because there are only about 10^{17} malaria parasites in the entire world. For example, if the per-parasite probabilities of developing resistance to drug A and drug B are both 1 in 10^{12}, then a simultaneously resistant mutant will arise spontaneously every 1 in 10^{24} parasites. As there is a cumulative total of less than 10^{20} malaria parasites in existence in one year, such a simultaneously resistant parasite would arise spontaneously roughly once every 10,000 years — provided the drugs always confronted the parasites in combination. Thus, the lower *de novo* per-parasite probability of developing resistance could result in greater delay in the emergence of resistance [4].

Stable, therapeutically significant resistance to the artemisinin derivatives has not yet been identified and cannot be induced yet in the laboratory, which suggests that it may be a very rare event. But it would be foolish to bank on it not happening, and should it arise, it would be a global disaster. For mutual protection against the emergence of drug resistance, these drugs should be used only in combination with other antimalarials.

Artemisinin derivatives are particularly effective in combinations because of their very high killing rates, lack of adverse effects, and absence of significant resistance [6]. The ideal pharmacokinetic properties for an antimalarial drug have been greatly debated. From a resistance prevention perspective, the combination partners should have similar pharmacokinetic properties. Rapid elimination ensures that the residual concentrations do not provide a selective filter for resistant parasites, but these drugs (if used alone) must be given for 7 days, and adherence to 7-day regimens is poor. Even 7-day regimens of artemisinin derivatives are associated with approximately 10% failure rates. In order to be highly effective in a 3-day regimen, terminal elimination half-lives of at least one drug component need to exceed 24 hours.

Combinations of artemisinin derivatives (which are eliminated very rapidly) given for 3 days, with a slowly eliminated drug such as mefloquine (artemisinin combination treatment) provide complete protection for the artemisinin derivatives from selection of a *de novo* resistant mutant if adherence is good (i.e., no parasite is exposed to artemisinin during one asexual cycle without mefloquine being present). But this does leave the slowly eliminated "tail" of mefloquine unprotected by the artemisinin derivative. The residual number of parasites exposed to mefloquine alone, following two asexual cycles, is a tiny fraction (less than 0.00001%) of those present at the peak of the acute symptomatic infection. Furthermore, these residual parasites are exposed to relatively high levels of mefloquine, and, even if susceptibility was reduced, these levels are usually sufficient to eradicate infection. The long "tail" of the mefloquine elimination phase does, however, provide a selective filter for resistant parasites acquired from elsewhere and, as described earlier, contributes to the spread of resistance once it has developed. Yet on the northwestern border of Thailand, an area of low transmission where mefloquine resistance had developed already, systematic deployment of artesunate-mefloquine combination therapy was dramatically effective, both in stopping resistance and also in reducing the incidence of malaria. This strategy would be expected to be effective at preventing the *de novo* emergence of resistance at higher levels of transmission, where high-biomass infections still constitute the major source of *de novo* resistance [4].

Given the practical problems of quinine, particularly low patient compliance from frequent adverse events such as tinnitus, artemisinin compounds are increasingly used for *falciparum* treatment. Derived from the Chinese pharmacopoeia and a type of wormwood plant, artemisinin compounds exist in a variety of derivatives most of which appear to have as their active metabolite dihydroartemisinin. The major advantages of artemisinin compounds are their fast action in clearing the blood of parasites and preventing the emergence of the gametocyte transmission stages. Both of these characteristics minimize the development of drug-resistance and encourage further use of artemisinin compounds in combination with other drugs as ACT. There are several drugs used in combination with artemisinins to include mefloquine, pyronaridine, piperaquine, amodiaquine, chlorproguanil-dapsone, and lumefantrine, which have all undergone clinical field trials, in some cases with many thousands of patients. Artesunate is available in parental and well as oral form and has been shown superior to quinine for severe malaria. It is realistic to hope that these drug combinations will be available in the prevention of resistance.

1.2. Artemisinin Evaluations and Artemisinin-based Combination Therapy (ACT)

1.2.1. Superior and Inferior Antimalarial Profiles of Artemisinins

1.2.1.1. Superior Antimalarial Profiles of Artemisinins

The artemisinin and its derivatives were discovered in China. A crude extract of the wormwood plant *Artemisia annua* (qinghao) was first used as an antipyretic 2000 years ago, and its specific effect on the fever of malaria was reported in the 16th century [43]. The active constituent of the extract was identified and purified in the 1970s, and named qinghaosu, or artemisinin. Although artemisinin proved effective in clinical trials in the 1980s, a number of semi-synthetic derivatives were developed to improve the drug's pharmacological properties and antimalarial potency [44]. The pharmacological and clinical evaluations of artemisinin group of drugs have been taken place for 30 years and four advantages have been evaluated.

1) Rapid Action and High Efficacy with Multi-drug Resistant P. Falciparum

Each artemisinin derivative is highly active against asexual forms of the four species of *Plasmodium* that infect humans. The initial reduction in parasitemia is the most rapid of all available antimalarial drugs. Their half-lives are short relative to the duration of their effect on parasite clearance. This suggests the equivalent of a "post-antibiotic effect," where there is persistent suppression of bacterial growth following limited exposure to an antimicrobial agent.

Of the available derivatives, artesunate has the most favorable pharmacological profile for use in ACT treatment of uncomplicated malaria. The presence of a hemisuccinate group in the molecule confers water solubility and relatively high oral bioavailability. It is rapidly and quantitatively converted *in vivo* to the potent active metabolite dihydroartemisinin (DHA). Artemisinin, dihydroartemisinin, and artemether are all poorly water-soluble, which means that absorption is slower and less complete. Nevertheless, consistent with the wide therapeutic index, the plasma concentrations achieved after oral administration of conventional doses of all of the artemisinin derivatives are substantially greater than those currently required to kill the parasite [44].

2) No Artemisinin Drug Resistance yet

The cellular target for artemisinins has remained controversial. A commonly proposed theory states that iron contained in parasite heme reacts with artemisinins' peroxide moiety, leading to release of free radicals and damage to parasite structures. Further, it has been suggested that target resistance against a drug with such a non-specific mechanism of action is unlikely to emerge; resistance could only emerge via mutations in drug transporters such as *pfmdr1* and *pfcrt*. Increased *pfmdr1* copy number is associated with elevated IC_{50} for artemisinins in Southeast Asia, but this almost certainly relates to mefloquine, not artemisinin pressure, and the small associated increase in artemisinin IC_{50} has probably occurred as a 'bystander' effect. As for arylaminoalcohols, *pfmdr1* SNPs are generally associated with artemisinin hypersensitivity in laboratory studies, transfection and field isolates from Africa and Southeast Asia, although these are not uniform findings. Similarly, *pfcrt* appears to have a minor modulating effect on artemisinin IC_{50}; this effect is either slight or non-existent depending on parasite background. A study of several transporters revealed an association between a polymorphism in another predicted transporter (gene G7, a member of the ABC transporter family) and artesunate IC_{50}, but this area has not been explored in further detail. Recently a hypothesis has been advanced describing the sarco/endoplasmic reticulum calcium-dependent ATPase (SERCA)-type PfATP6 protein as the target of artemisinins. No association was found between SNPs in *PfATP6* and artesunate IC_{50} in Southeast Asia, although these field studies have been carried out in areas where there was no evidence of significant *in vitro* or *in vivo* artemisinin resistance. The ability of malaria parasites to develop stable resistance to artemisinins via other means is of interest, although the system used (rodent malaria parasite *P. chabaudi*) may limit the applicability of these findings to human infection [45].

3) Low Toxicity (Safety Profile)

The therapeutic index of the artemisinin derivatives is wide. The most common reported adverse effects include nausea, vomiting, bowel disturbance, abdominal pain, headache, and dizziness — symptoms that can result from malaria infection itself. Mild and reversible hematological and electrocardiographic abnormalities, such as neutropenia and first-degree heart block, are observed infrequently.

Neurotoxicity, principally in the form of brainstem lesions, was first identified in animals receiving high doses over long periods, especially with oil-soluble artemether (AM) and arteether (AE). In contrast, few neurotoxic effects are induced by water-soluble artesunate (AS) and artelinate (AL), which have never produced any CNS side-effects in animal species after intravenous injection. Pharmacokinetic profiles demonstrate that drug exposure time plays a more important role in producing neurotoxicity than drug exposure level, production of the more active metabolite (DHA), or drug location in the CNS system. Neurological side-effects such as ataxia, slurred speech, and hearing loss have also been reported in a small numbers of humans, but are unlikely to be of clinical importance, especially as there is limited penetration of artemisinin derivatives into the cerebrospinal fluid [44, 45]. In addition, it has been demonstrated that at a below human therapeutic dose, injectable AS can cause fetal death due to a buildup of high peak concentration of AS during the sensitive period of the early gestation days (GD) in animals. To date, however, the neurotoxicity and reproductive toxicity have not been convincingly established in humans treated with any of the artemisinins. Fatal neurotoxicity can be avoided by reducing the drug exposure time that varies greatly between the different artemisinin compounds. Also reproductive toxicity can be avoided by limiting the exposure of artemisinins during this sensitive period in fetal development [44].

4) Gametocidal Effect (Prevent the Interacted Transmission of Malaria)

The artemisinin derivatives are also active against the sexual form of the parasites (gametocytes) taken up by the mosquito and can therefore reduce transmission rates. Their exact mechanism of action is unknown. An endoperoxide moiety, which is essential for antimalarial activity, may cause destructive free-radical generation within the parasite and, through the formation of covalent bonds, alter the function of key parasite proteins, including membrane transporters.

**Table 4. Summary of artemisinin derivatives by monotherapy
in patients from uncomplicated, severe and complicated malarias
in 54 clinical trials [44]**

Drug and Regimen	PCT (hr)	FCT (hr)	% of Recrude-scence	Clinical Trials No.
AE in sesame oil (Artemotil)				
(IM 3.2 mg/kg-1.6 mg/kg x 4)	70.2	81.0	40.0	1
(IM 4.8 mg/kg-1.6 mg/kg x 5)	59.2	61.5	22.2	1
AM				
(IM 3.2 mg/kg-1.6 mg/kg x 4)	28.3 ± 13.0	38.8 ± 4.4	8.6 ± 6.2	8
(oral 200 mg-100 mg/kg x 5)	39.7 ± 2.2	34.8 ± 6.9	5.2 ± 2.2	9
AS				
(IV & oral in various regimens)	30.3 ± 10.7	35.9 ± 11.1	7.5 ± 3.1	23
DHA				
(oral in various regimens)	49.1 ± 7.5	33.8 ± 7.7	6.0 ± 1.8	7
AE in peanut oil				
(IM 150 mg daily for 3 days)	37.8 ± 12.2	49.9 ± 16.2	9.4 ± 4.1	5

Values are given mean ± SD.
DHA = Dihydroartemisinin, AM = Artemether, AE = Arteether, AS = Artesunate, IM = intramuscular, IV = intravenous, PCT = parasite clearance time, FCT = Fever clearance time.

The artemisinin derivatives do not kill mature gametocytes of *P. falciparum*. However, they prevent gametocyte development by their activity against both the young and the more mature asexual-stage parasites (the precursors of the sexual stages) and also early (stage I-III) gametocytes. The antimalarial drugs that act later in the asexual life cycle, such as the quinoline antimalarials, may not interrupt this process so effectively. The benefits of the artemisinin derivatives are not restricted to areas where multidrug-resistant parasites are already prevalent. Even when patients were treated successfully and parasite clearance was rapid, gametocyte carriage was reduced at least five-fold by artemisinin treatments compared with halofantrine, mefloquine, or quinine [46].

1.2.1.2. Inferior Antimalarial Profiles of Artemisinins

Although there has been no evidence for clinically relevant *in vivo* artemisinin resistance when the class is used as a partner in combination therapy, artemisinin monotherapy is well described as having poor efficacy. In any case, WHO has recommended cessation of monotherapy to try to limit emergence of artemisinin resistance [47]. Recrudescence with monotherapy tends to occur rather late, and studies following patients up to day 42 detect a higher proportion of recrudescences than those stopping at 28 days. Recent estimates of recrudescence rates after artemisinin monotherapy at 28 day follow-up are 20-40% in Africa and about 20% in Southeast Asia. Duration of therapy has been reported to be of critical importance in efficacy of artemisinin-based monotherapies, with extension to 7 days improving cure rates significantly. However, these earlier suggestions have not been confirmed in a separate study, where PCR correction was applied to differentiate reinfection from recrudescence [45].

Table 5. Countries (68) that have adopted ACTs and the ACT choices made by them in 2006 [55, 63]

Continent	Countries	ACT choice	Indica-tion
Africa (41 countries)	*Burundi, Cameroon, Congo, Côte d'Ivoire, Democratic Republic of Congo, Eq. Guinea, Gabon, Ghana, Guinea, Liberia, Madagascar, Malawi, Mauritania, Senegal, Sao Tomé & Principe, Sierra Leone, Sudan (S), Tanzania (Zanzibar)*	AS + AQ	First-line
		AL	First-line
	Angola, Benin, Burkina Faso, Central African Republic, Chad, Comoros, Ethiopia, Gambia, Guinea Bissau, Kenya, Mali, Namibia, Niger, Nigeria, Rwanda, Uganda, South Africa (Kwa Zulu Natal), Tanzania (Mainland), Togo, Zambia, Zimbabwe	AL	Second-line
	Côte d'Ivoire, Djibouti, Gabon, Malawi, Mozambique, Sudan (N), Sao Tomé & Principe, Tanzania (Zanzibar)	AS + SP	First-line
	Mozambique, Djibouti, Somalia, South Africa (Mpumalanga), Sudan (N)		
Asia (19 countries)	*Cambodia, Malaysia, Myanmar, Thailand*	AS+MQ	First-line
	Bangladesh, Bhutan, Laos, Saudi Arabia	AL	First-line
	Indonesia	AS+AQ	First-line
	Afghanistan, India (5 Provinces), Iran, Tajikistan, Yemen	AS + SP	First-line
	Vietnam	DP	First-line
	Papua New Guinea	AS + SP	Second-line
	Iran, Philippines, Solomon Islands	AL	line
S. America (8 countries	*Ecuador, Peru*	AS + SP	First-line
	Bolivia, Peru, Venezuela	AS+MQ	First-line
	Brazil, Guyana, Suriname	AL	First-line

Countries that have started deployment of these medicines in the general health services are shown in italics (situation as of July 31, 2006).

AQ, amodiaquine; AL, artemether/lumefantrine; AS, artesunate; DP, dihydroartemisinin/piperaquine; MQ, mefloquine; SP, sulfadoxine/pyrimethamine.

The therapeutic potentials of all artemisinin drugs in monotherapy with 54 clinical trials have been summarized in the treatment of malaria with various dose regiments (Table 4). Successful (or 100% cure) treatment is found neither in the 3-5 day treatments nor the high dose (4.8 mg/kg) regimens with monotherapy [44]. The curative rate, fever (FCT), and parasite (PCT) clearances are traditional evaluations for the efficacy of the antimalarial drugs in humans. These three parameters exhibit the therapeutic potential of intravenous, intramuscular and oral agents in various regimens. Results have shown that AE in sesame oil vehicle is not superior to that of the oral DHA, oral and intravenous AS, intramuscular AM, or even intramuscular α/β-arteether formulated with peanut oil.

Revealed by this data, and their known mode of rapid action, the artemisinin group of antimalarials is considered to be highly and independently effective compared with other traditional malaria drugs. Although there is no evidence that drug resistance has occurred *in vivo*, short treatment courses (3-5 days) are associated with high recrudescence rates. The data that has emerged from all available artemisinin drugs is that given a dose sufficient to induce initial parasite clearance and clinical recovery, the duration of therapy is the ultimate determinant for recrudescence. The rates of recrudescence (30-50% for 3 day regimen, 6-15% for 5 day regimen, and 2-6% for 7 day regimen) continue to be a big problem for all these compounds described previously [48]. It has been suggested that preventing recrudescence does not have the highest priority, but that a recrudescence should rather be treated with the same regimen again. Eventually, all patients could be radically cured with this strategy. The potential risk of development of resistance and continued recrudescence urges us to look for combination regimens that offer first time radical cure instead.

1.2.2. ACT is a "Policy Standard" and First Line in Malaria Treatments

Antimalarial combination therapies can improve treatment efficacies of failing individual components and provide some protection for individual components against the development of higher levels of resistance. ACTs have been advocated as the best available option, and are the most commonly adopted regimen in countries changing antimalarial policy in the last decade. ACTs are most preferred for their enhancement of efficacy [49-51], lower malaria incidence and their potential to lower the rate at which resistance emerges and spreads [52, 53]. Four ACTs recommended by a WHO Expert Consultative Group in 2001 are AM-lumefantrine (Coartem), AS-mefloquine (Artequin), AS-amodiaquine, and AS-sulfadoxine/pyrimethamine [47]. Recently, WHO has endorsed ACTs as the "policy standard" for all malaria infections in areas where *P. falciparum* is the predominant infecting species [54].

Artemisinins rapidly reduce parasitemia, but have poor efficacy as short course monotherapy. When used in combination with another agent, the rapid reduction in parasite numbers results in relatively few parasites being exposed to the second drug (to which significant resistance may already exist), theoretically preventing emergence of additional resistance mutations [4]. Furthermore, since artemisinins themselves are not required to mediate final cure, there should also be little opportunity for artemisinin resistance to develop. In addition, reduction of the drug combinations in gametocyte carriage may also reduce transmission of resistant parasites by increasing cure rates and inhibiting formation of gametocytes [46]. Most currently recommended drug combinations for *falciparum* malaria are variants of ACT where a rapidly acting artemisinin compound is combined with a longer half-life drug of a different class. Artemisinins used include DHA, AS, AM and companion drugs include mefloquine, amodiaquine, sulfadoxine/pyrimethamine, lumefantrine, piperaquine, pyronaridine, and chlorproguanil/dapsone. The standard of care must be to cure malaria by killing the last parasite. Combination antimalarial treatment is vital not only to the successful treatment of individual patients but also for public health control of malaria.

ACTs continue to be the mainstay of treatment of uncomplicated *falciparum* malaria. For the next 8–10 years, no alternative medicines to the artemisinin derivatives able to offer similar high levels of therapeutic efficacy are expected to enter the market. For this reason, WHO has focused its efforts not only to increase access to quality ACTs, but also to contain the risk of development of *falciparum* resistance, mostly created by the large-scale use of oral monotherapies for treatment of uncomplicated malaria [1, 2].

In January 2006, WHO appealed to manufacturers to stop marketing oral artemisinin monotherapies and instead to promote quality ACTs in line with WHO policy (*www.who.int/mediacentre/news/releases/2006/pr02/en/index.html*). The position has been widely disseminated via WHO Offices, WHO briefings to staff and to representatives of national health authorities at regional and inter-country meetings. Major procurement and funding agencies and international suppliers have accepted WHO recommendation and agreed not to fund or procure oral artemisinin monotherapies. In April 2006, the Global Malaria Programme of WHO provided a technical briefing to 25 pharmaceutical companies involved in the production and marketing of artemisinin monotherapies. Out of these, 15 declared their willingness to stop marketing artemisinin monotherapies over a short period of time, but 10 companies did not disclose their marketing plans for the future (meeting report available at: *www.who.int/malaria/docs/ Meeting_briefing19April.pdf*). In addition, some countries, like China and Pakistan, have been visited by WHO delegations to address multiple domestic manufacturers involved in this sector. The evolving position of manufactures and of National Drug Regulatory Authorities (NDRA) of malaria endemic countries is monitored and displayed on the WHO Global Malaria Programme website front-page: *http://malaria.who.int/*.

In May 2007, the 60th World Health Assembly resolved to take strong action against oral monotherapies and approved the resolution WHA60.18:

1. urges Member States to cease progressively the provision, in both the public and private sectors, of oral artemisinin monotherapies, to promote the use of artemisinin-combination therapies, and to implement policies that prohibit the production, marketing, distribution and the use of counterfeit antimalarial medicines;
2. requests international organizations and financing bodies to adjust their policies so as progressively to cease to fund the provision and distribution of oral artemisinin monotherapies, and to join in campaigns to prohibit the production, marketing, distribution, and use of counterfeit antimalarial medicines;

These mentioned benefits of ACTs make them an important tool for malaria treatment and control that has recently led to a big push for their deployment. The ACT choices made by countries for the public sector are presented in Table 5. By 2006, most countries (65 of 67 countries), including 41 in Africa, adopted ACTs as their first-line treatment of uncomplicated *falciparum* malaria. Only two countries adopted ACTs exclusively as second-line treatment [1, 55].

1.2.3. Adoption and Deployment of ACTs in Asia since 1994

Morbidity and mortality from *P. falciparum* malaria are increasing in many tropical areas. This increase results partly from the spread of drug resistance in *P. falciparum* and the absence of an effective strategy to combat it. In Southeast Asia, increasing resistance to chloroquine and sulphadoxine-pyrimethamine led to the introduction of mefloquine in 1984. The drug was used initially in combination with sulphadoxine-pyrimethamine, later on its own at a higher dose, and finally, in certain areas, in combination with AS taken place in 1994 [56]. Since the general deployment of the artesunate-mefloquine combination in 1994, the cure rate increased again to almost 100% from 1998 onwards, and there has been a sustained decline in the incidence of *P. falciparum* malaria in the study area. *In vitro* susceptibility of *P. falciparum* to mefloquine has improved significantly (p = 0.003). In this area of low malaria transmission, early diagnosis and treatment with combined AS and mefloquine has reduced the incidence of *P. falciparum* malaria and halted the progression of mefloquine resistance.

Despite the high prevalence of mefloquine-resistant *P. falciparum* before use of this regimen, combination of artesunate and mefloquine has achieved sustained high cure rates (>95%), reduced *P. falciparum* transmission and the incidence of *falciparum* malaria, and halted the progression of resistance to mefloquine. In 1998, experts agreed that WHO and the Special Programme in Research and Training in Tropical Diseases (WHO/TDR) should co-ordinate trials to assess the safety and effects of artemisinin combination treatment, concentrating on antimalarial drugs used such as chloroquine, amodiaquine, and sulfadoxine-pyrimethamine. Evidence of the benefits of ACTs has been obtained in Southeast Asia. Mefloquine resistance was reported to be reversed by the addition of AS with the combination retaining a cure rate of over 95% in patients without hyperparasitemia [57].

Soon afterwards, four ACTs recommended by a WHO Expert Consultative Group in 2001 are artemether-lumefantrine (Coartem), artesunate-mefloquine, artesunate-amodiaquine, and artesunate-sulfadoxine/pyrimethamine [47]. However, despite being recommended by WHO since 2001, overall deployment of ACT has been slow. Limiting factors are high cost, limited knowledge, and public awareness on the concept of combination therapy (CT) and ACT; in particular, limited knowledge on safety of ACTs in pregnancy, operational issue such as inappropriate drug use, lack of suitable drug formulations, lack of post-marketing surveillance systems, and the imbalance between demand and supply. Through concerted efforts of multilateral organizations, the local scientific community with involvement of policy-makers progress has been on several fronts leading to improved ACT uptake rates recently.

In northwest Thailand, the use of a combination of mefloquine and artesunate since 1994 as a standard treatment of uncomplicated *falciparum* malaria has stopped the decline in the efficacy of mefloquine and reduced the incidence of malaria [58]. Similar decreases in malaria transmission following widespread use ACTs have been reported from Vietnam and South Africa [59]. These important public health benefits are due to the direct action of artemisinins on gametocytes and the reduction of parasite biomass. While the impact on transmission has been reported from areas of low to moderate malaria transmission, it is unclear if similar effects would occur under intense malaria transmission. Considering the fact that artemisinin derivatives reduce post-treatment infectivity dramatically but do not abolish it completely, there is urgent need to determine conditions and modalities under which an impact can be achieved.

WHO and the Roll Back Malaria (RBM) Partnership in collaboration with National Malaria Control Programmes (NMCPs) and the local scientific community have always facilitated selection processes by providing technical inputs and guidance to ensure an informed policy decision. Despite its urgency, the uptake and implementation of ACT policies was so slow between 2001 and 2003 that a group of scientists and public health experts expressed concern over the gloomy picture [60]. The rest of this paper focuses on the challenges and opportunities towards quick adoption and implementation of ACT policies in Southeast Asia and Africa.

Countries in Southeast Asia where multi-drug resistance was highly prevalent had begun using artemisinins in the treatment of malaria in the early 1990s [58]. In China and Vietnam, artemisinins were used mainly as monotherapy, except for a triple combination of DHA-piperaquine-trimethoprim combined with primaquine (known as CV8), which was deployed on a limited scale in Vietnam. In Thailand and Cambodia, AS and mefloquine were co-administered in some geographical regions of the country. However, by 2000, no country had adopted ACTs as their national malaria treatment policy.

Since 2000, an increasing number of endemic countries, numbering 67 at present [55], have adopted ACTs in the treatment of uncomplicated *falciparum* malaria (Figure 4). Having a time-lag of 12–18 months between policy adoption and implementation, the proportion of countries actually deploying ACTs in the general health services has also increased. Initially, most of these countries began by limiting the deployment of ACTs to either selected geographical areas (e.g., districts or provinces supported by grants or donors), to the most vulnerable population groups (e.g., children younger than 5 years of age), and/or to certain levels of the health care system (e.g., where microscopy is available). Cambodia as the exception, ACTs has not yet been introduced beyond health facilities into home-based management programs.

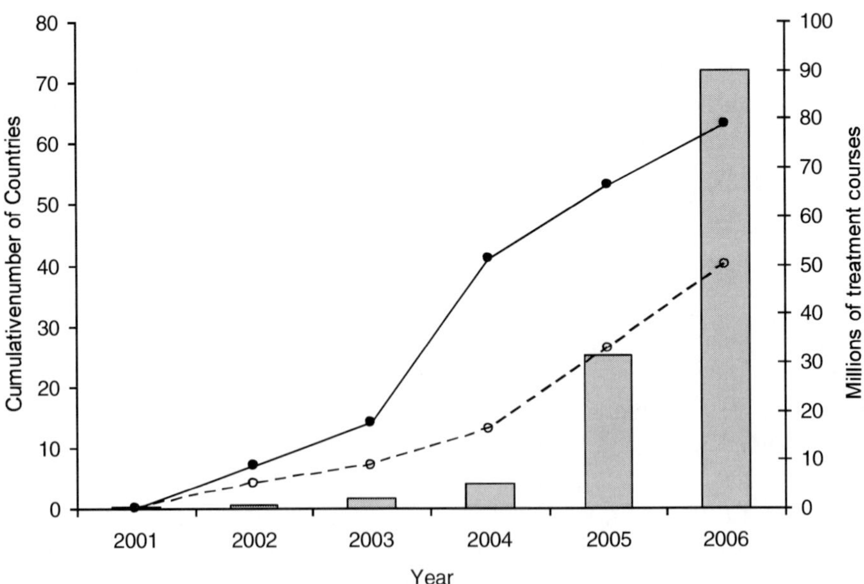

Figure 4. Cumulative number of countries adopting (continuous line) and deploying (interrupted line) ACTs as first-line treatment of malaria from January 2001 to July 2006. The annual ACT orders for the public sector in period 2001–2005 (solid bars) and the demand forecast (hatched bar) for 2006 [55].

The ACT choices made by countries for the public sector are presented in Table 5. Most countries (65 of 67), including 41 in Africa, adopted ACTs as their first-line treatment of uncomplicated falciparum malaria. Only two countries adopted ACTs exclusively as second-line treatment. There are nine more endemic countries with recorded treatment failure rates of their first-line medicines exceeding 10%, which, according to WHO guidelines [2], should adopt ACTs (Botswana, China, Colombia, Eritrea, Nepal, Pakistan, Sri Lanka, Swaziland, Vanuatu).

Thus far (in 2007), 65 endemic countries (24 in Asia and 41 in Africa) are deploying ACTs. Among the ACTs, artemether/lumefantrine has been adopted as first-line treatment by 28 countries, artesunate plus amodiaquine by 19 countries; artesunate plus sulfadoxine/pyrimethamine by 11 countries; artesunate plus mefloquine by 7 countries; and dihydroartemisinin/piperaquine by 1 country [55].

1.2.4. Adoption and Deployment of ACTs in Africa since 2000

Antimalarial drug resistance is a growing and important public health challenge, especially in African countries, where over 90% of the global mortality of malaria occurs. Since 1979 when chloroquine resistant *P. falciparum* was first documented in East Africa, chloroquine-resistant strains have spread across the continent. Sulfadoxine-pyrimethamine (SP) and mefloquine were until recently the obvious successor to chloroquine. However, resistance to SP and mefloquine are also developing rapidly in those parts of East Africa where it has been introduced as a first-line drug in malaria treatment. Strategies are being developed to halt these trends [61].

One strategy to improve cure rates and limit the development of drug resistance is the use of artemisinin-based combination therapy (ACT) [47]. The African leaders at the Abuja Summit on Roll Back Malaria held in April 2000 endorsed the laudable goal of having, by 2005, at least 60% of African malaria patients on prompt access to affordable and appropriate treatment within 24 h of the onset of symptoms. There is a growing international consensus that wide scale, systematic implementation of ACT is one of the few effective measures that will enable malaria-endemic countries to achieve the ambitious goals set in Abuja in 2000 to 'roll back malaria,' particularly that of halving malaria morbidity and mortality by 2010. By June 2006, 37 countries in Africa had adopted ACT as the first- or second-line treatment policy [62]. Currently there are a number of registered co-formulated ACTs that are produced to internationally recognize good manufacturing practice standards that are used in sub-Saharan Africa. These include; artemether-lumefantrine (AL), artesunate-MQ (AS + MQ), artesunate-amodiaquine (AS + AQ), and artesunate-SP (AS + SP). Despite the importance of ACT in the treatment of uncomplicated malaria, a comprehensive review of the impact of its usage and the public health implications in sub-Saharan Africa is essentially lacking.

The International Artemisinin Study Group [11] has indicated higher cure rates, faster parasite clearance and decreased gametocyte carriage with ACT compared with monotherapy. WHO has stated that AL demonstrates cure rates above 95%, even in areas of multiple drug resistance [47]. Preliminary results from Africa indicate that combinations of AS + AQ or AS + SP are highly effective, although efficacy may be compromised in areas with moderate to high levels of resistance to SP. Hence, the combination AS + AQ is not only more effective than AS + SP at present but also may remain effective for an extended period of time.

Combinations with AS have proved very effective and could, in the right circumstances, protect drugs from the progressive development of resistance, as has been shown with the implementation of ACT elsewhere. AS + AQ has now been adopted as first-line antimalarial treatment in 15 countries in Africa [63]. A number of studies in sub-Saharan Africa have shown the combination of AS + AQ to have a 28-day corrected efficacy of over 90%. However, in another study the efficacy of this combination was below 90%. Results indicated that the combinations were safe, had rapid fever reduction time, had 100% clinical efficacy and reduced gametocytaemia incidence during the 14-day follow-up time [64].

In addition, the primary aim of malaria treatment is saving lives. Prompt and effective treatment in the early stages of *falciparum* malaria reduces the risk of death as much as 50-fold, whereas effective treatment after progression to severe illness produces only a five-fold reduction in the risk of dying [52]. However, malaria treatment also can reduce malaria transmission in endemic areas. This section reviews the role of treatment as a control measure capable of reducing malaria transmission, as well as past and present chemoprevention strategies in residents of malaria-endemic areas.

Ingwavuma district in northern KwaZulu Natal carried the highest malaria burden in South Africa until 2000. This changed after improvements in vector control in KwaZulu Natal and neighboring southern Mozambique during 2000, and the implementation of an artemisinin combination therapy (ACT), artemether-lumefantrine (AL), in January 2001. These interventions were followed by a 78% decrease in malaria case notifications in the province within 1 year. There was a further 75% decrease in notifications in 2002 [59], and this decrease was sustained throughout 2003 (Figure 5).

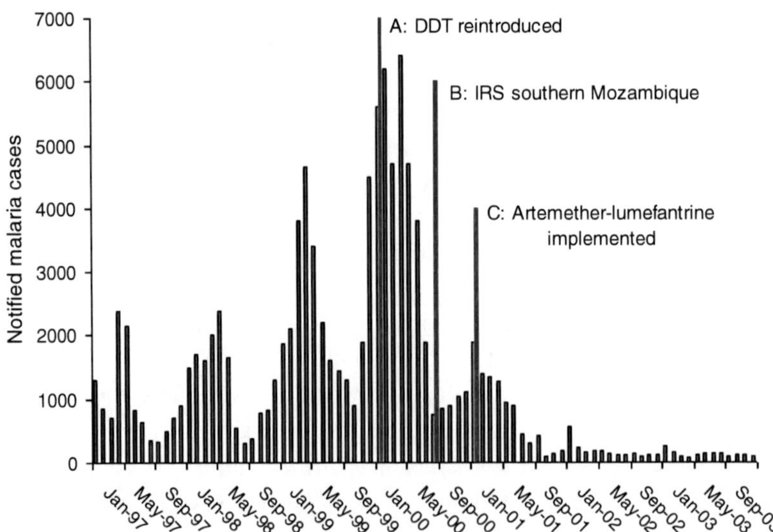

Figure 5. The synergistic effect of improved vector control and artemisinin combination therapy, artemether-lumefantrine, on malaria transmission in KwaZulu Natal, South Africa [59].A. Reintroduction of DDT for indoor residual spraying (IRS) of traditional structures in KZN in March 2000.

B. Introduction of IRS in southern Mozambique in October 2000.

C. Implementation of artemether-lumefantrine (AL) for the treatment of uncomplicated *falciparum* malaria in KZN in January 2001.

The success of this dual intervention is further reflected by the 89% decrease in both malaria-related hospital admissions (range 81-92%), and deaths (range 78-93%) across the three rural district hospitals that serve the Ingwavuma community. Despite KwaZulu Natal also carrying South Africa's highest burden of the HIV/AIDS epidemic, the total number of "all cause" hospital admissions also was reduced following improved malaria control. Dr. Hervey Vaughan Williams, medical superintendent of Mosvold hospital, a rural district hospital in Ingwavuma, reports that "during the peak of the 2000 malaria epidemic, hospital admissions were averaging three patients per bed." Following the reduction in the overall number of hospital admissions, it is expected that the standard of care for all patients could be improved.

Thus, the simultaneous use of two or more drugs is a chemo-therapeutic strategy for improving treatment outcome and retarding the development of resistance to the individual drugs in the combination. This concept is already standard practice in multiple-drug therapy for leprosy, tuberculosis, cancer therapy, and more recently in antiretroviral treatments. Antimalarial drug combination therapy (CT) is defined as the simultaneous administration of two or more blood schizonticidal antimalarial drugs. The drugs, which may either be co-formulated or co-administered, should have independent modes of action and different biochemical targets in the parasite. It should be noted that multiple-drug therapies in which neither of the components has significant schizonticidal effect when used alone (e.g., SP) is not considered as CT. Neither is the use of a blood schizonticidal drug with a tissue schizonticidal drug, or with a non-antimalarial drug to enhance its action considered antimalarial combination therapy. The potential value of artemisinin-based combination therapy (ACT) to improve therapeutic efficacy and delay the development and selection of drug resistant parasites has been demonstrated in areas of low to moderate transmission in Southeast Asia. However, experience with antimalarial combination therapy in areas of intense malaria transmission of Africa is very limited, although it is now the recommended approach to malaria case management in the region. Table 6 compares factors in favor of and current limiting factors to the introduction of artemisinin-based combination therapy in the African region.

The number of ACT treatment courses procured by national governments of endemic countries for use in their public sector increased from around half a million in 2001 to 31.3 million in 2005 (Figure 4)—25.5 million of these being for countries in Africa [55]. More than 80% of these orders were placed through WHO. From January 2005 to July 2006, the total ACT procurements for the public sector made by WHO, UNICEF, and to a lesser extent, by Crown Agents has been 62.6 million treatment courses: 80.8% of these orders were for artemether/lumefantrine, 12.5% were for artesunate plus amodiaquine, and 6.7% were for artesunate plus sulfadoxine/pyrimethamine [55].

A majority of countries have adopted a single ACT, but a few adopted different ACTs for different geographical areas of the country (e.g., Peru, South Africa, Sudan, and Tanzania). Some countries (Nigeria and Uganda) have formally adopted two ACTs as first-line treatment, although only one is procured by the government.

Table 6. ACT use in Africa

Factors in favor of
- Current inadequacy of first-line treatments in many countries, including possible hidden burdens of disease and death from malaria and chronic anemia.
- Potential for protecting effective and affordable antimalarial drugs currently available for future use.
- High clinical and parasitological cure rates.
- Absence of documented parasite resistance.
- Potential for reducing in transmission in areas of low to moderate transmission due to a reduction in gametocyte carriage.

Limiting factors
- Potential misuse of artemisinin derivatives in view of their value in the treatment of severe malaria.
- Problems of compliance with non co-formulated combinations, particularly at the household level.
- Higher cost of artemisinin derivatives.
- Lack of evidence for ACT delaying resistance in areas of high transmission.
- Effort and cost of policy change.

1.2.5. Advantages and Limitations of ACTs

The particular features of ACT relate to the unique mode of action of the artemisinin component, that includes the following:

- rapid and substantial reduction of the parasite biomass,
- rapid parasite clearance,
- rapid resolution of clinical symptoms,
- effective action against multidrug-resistant P. falciparum, and
- reduction of gametocyte carriage, which potentially reduces transmission of resistant alleles.

There are also few reported adverse clinical effects [65], although preclinical toxicology data on many artemisinin derivatives are limited. Because of the short half-life of artemisinin derivatives, their use as monotherapy requires daily doses over a period of 7 days. Combination of one of these drugs with a longer half-life partner antimalarial drug allows a reduction in the duration of antimalarial treatment while at the same time enhancing efficacy and reducing the likelihood of resistance development. The major immediate effect of the artemisinin component is to reduce the parasite biomass. The residual biomass is exposed to maximum concentrations of the partner drug, well above its minimum inhibitory concentration, resulting in a lesser likelihood of resistant mutations breaking through.

Antifolates, quinolines, and artemisinins have been deployed in various combination therapies with each other to both to increase the efficacy of treatment, as well as delaying the insurgence of drug resistance. In most of the artemisinin-based combinations currently in use or being evaluated, e.g., artesunate-mefloquine, the partner drug is eliminated slowly. The partner drug is therefore unprotected once the artemisinin has been eliminated from the body and operates a selective pressure on new infections. The implications of this "pharmacokinetic mismatch" [66] are not fully understood at present, particularly in areas of high transmission in Africa. The safest approach is to use a drug partner that has a residual half-life as short as possible, while still enabling parasite clearance with a 3-day treatment. However, this is difficult to achieve given the limited range of antimalarial drugs available.

Two aspects are important when choosing drug combinations with the aim of containing the development and spread of drug resistance: different yet synergistic modes of actions and a good pharmacokinetic match. As recently discussed by Hastings and Watkins [66], mismatched drug combinations do not significantly impact the spread of resistance and may even risk jeopardizing the efficacy of components against, which resistance has not been recorded (such as artemisinins). Accordingly, drug combination need to have similar half-lives in order to keep a constant dual pressure on current and reinvading parasites. Ideally, drugs with short half-lives should be preferred, in order to reduce the exposure of re-invading parasites to suboptimal drug levels that may induce the selection for tolerance and eventually the development of resistance [67].

ACT has become the first treatment of choice in many countries. The principal combinations being adopted consist of a pairing of an artemisinin with a quinoline or an antifolate drug. Combinations with other drugs have also been tested, but not yet adopted on a wide scale. The combination of artemisinins with quinolines has given mixed results. While their actions differ, the two components should guarantee a good synergy, the pharmacokinetic half-life differences between the short-lived artemisinins and the long-lived quinoline is a cause for concern regarding the long-term development and spread of drug resistance.

Artemisinins have been most prominently combined with the quinoline amodiaquine, mefloquine and lumefantrine. Artesunate-mefloquine has proved particularly effective in field trials [56, 58]. However, the combination is expensive and presents a pharmacokinetic mismatch (mefloquine having a half-life between 14 - 21 days), which could drive the spread of mefloquine resistance and jeopardize the efficacy of artemisinins in the long-term. Furthermore it has already shown significant failure rates in areas of high mefloquine resistance [68]. Pharmacokinetically, the combination of artemether and lumefantrine (half-life 3-6 days) is better matched [69]. However, it is also rather expensive and appears to be slightly less efficacious than AS-mefloquine in areas with a high prevalence of multidrug resistant parasites [70]. Artesunate-amodiaquine (ASAQ) has been recently introduced as an inexpensive combination therapy directed especially at Sub-Saharan Africa. ASAQ also suffers from a pharmacokinetic mismatch due to the long half-life (9-18 days) of its active metabolite (desethylamodiaquine) [71]. Further, some African studies have already indicated a significant failure rate due to the presence of amodiaquine resistance. Resistance is likely to spread more rapidly as the combination is being adopted as the main treatment in several African countries [67]. The main currently used antimalarial combination therapies with advantages and disadvantages are summarized in Table 7.

**Table 7. Main currently used antimalarial combination therapies
with advantages and disadvantages [67]**

Drug combinations	Advantages	Disadvantages
Sulfadoxine + Pyrimethamine (SP)	Synergistic activity, inexpensive	Widespread resistance, medium-long half-lives, moderate PK mismatch
Chloproguanil + Dapsone	Synergistic activity, good PK match, short half-lives	Cross-resistance with SP
SP + Amodiaquine	Similar pharmacokinetic profiles	Resistance to both components spreading, medium-long half-lives
SP + Artesunate	Efficacious	Significant PK mismatch, efficacy dependent on level of SP resistance
Proguanil/Dapsone + Artesunate	Efficacious, similar PK profiles, short half-lives	Cross-resistance with SP may limit its application
Artesunate + Mefloquine	Efficacious	PK mismatch, resistance to mefloquine spreading
Artesunate + Amodiaquine	Efficacious	PK mismatch, resistance to amodiaquine spreading
Artesunate + Fosmidomycin	Efficacious, excellent PK match, Short-half-lives	Not suitable for prophylaxis
Artesunate + Clindamycin	Efficacious, similar PK match, Short-half-lives	Not suitable for prophylaxis, treatment failure observed
Proguanil + Atovaquone	Synergistic activity, efficacious	Moderate PK mismatch, resistance easy to acquire

PK = pharmacokinetic.

There is a growing interest in using antimalarial combinations containing an artemisinin derivative as first-line treatment. The aim is to provide efficacious and safe antimalarial drug treatment while probably delaying the onset and spread of resistance to both drugs in the combination. This interest results from experience with the combination of artesunate and mefloquine on the Thai-Myanmar border [58]. Following the introduction of the combination there have been four principal clinical and epidemiological effects:

- the efficacy of the combination has exceeded 95% at a time when high-dose mefloquine was showing a failure rate of approximately 25%,
- this high efficacy has been sustained over the past 7 years,
- the transmission of *P. falciparum* has been reduced (with reduced gametocyte carriage from the artesunate),
- the *in vitro* sensitivity of mefloquine has increased, suggesting that the combination has reversed the previous decline in mefloquine sensitivity.

Table 8. Factor for and against the introduction of artemisinin-based combination therapy [72]

Factors for	Against
▪ The need to replace inadequate drug regimens that are leading to increased malaria-related mortality and morbidity. ▪ Potential avoidance of the loss of available effective and affordable antimalarial drugs, especially in Africa. ▪ Excellent efficacy (both clinical and parasitological clearance) of artemisinin derivatives, with no resistance reported from Southeast Asia despite extensive use. ▪ Potential reduction in transmission (especially of resistant mutants) due to the gametocytocidal effect of artemisinin derivatives.	▪ Highest cost ▪ Problems of adherence to non-fixed combinations and their rational use, particularly in the home. ▪ Lack of extensive clinical experience with most of the combinations currently under investigation. ▪ Lack of evidence so far in Africa of its effectiveness in delaying the development of resistance. ▪ Importance of not missing artemisinin derivatives in view of their role in the treatment of severe malaria

These studies have been conducted in areas where there is a high level of medical service provision and malaria transmission is low. It is not yet known whether similar results can be achieved in Africa and other high-transmission regions. Moreover, evidence of the effectiveness of ACT in delaying the development of resistance is not yet available in Africa. Clinical trials using combinations of artesunate with amodiaquine, chloroquine, sulfadoxine-pyrimethamine or mefloquine are currently in progress to assess the efficacy and safety of ACT for treating uncomplicated *falciparum* malaria in Africa, South America, and Asia.

The factors for and against the introduction of advantages and disadvantages of ACT are summarized in Table 8. ACT should be considered in two different settings. In places where the combination is presently more efficacious than available monotherapies, such as parts of South-East Asia and Africa, it has a clear role. In areas where monotherapy is still efficacious but where this may change if the drug concerned is not protected from resistance, the justification for introducing ACT is less clear at the operational level. Although the general theory behind the promotion of ACT is widely accepted, doubts remain about the quantitative impact it will have in real-life situations. Ministries of health are therefore reluctant to commit to a high-cost ACT strategy where the merits of the different options available remain unclear and significant operational barriers still need to be overcome.

1.3. Artesunate in Treatments of Complicated and Severe Malaria

Severe malaria kills over a million people every year with annual death rates of up to one in a 100 children under the age of five [73]. A small proportion of children may also suffer from long-term neurological disability as a consequence of severe malaria. Severe malaria occurs when infection with the *P. falciparum* parasite is complicated by serious organ failure or metabolic abnormalities; cerebral malaria, unrousable coma not attributable to any other cause, is a specific type of severe malaria that even with correct treatment can have a mortality rate approaching 20% [73, 74]. A small proportion of survivors of cerebral malaria are left with persistent neurological sequel. Severe malaria occurs most commonly in those with limited immunity to malaria. In highly endemic areas, young children are therefore at most risk of severe disease and death, whereas in areas of lower endemicity and travelers, both adults and children get severe disease [75].

There are three recommended treatments for severe malaria: artesunate, artemether, and quinine, although in many countries only quinine is available.

1.3.1. Limitation of Quinine in Treatments of Severe Malaria

Intravenous quinine is currently the most widely used agent in the treatment of severe *falciparum* malaria, usually formulated as a dihydrochloride salt. In the USA, quinidine gluconate (the dextrorotatory optical diastereoisomer of quinine) is the only available intravenous antimalarial agent, and it may be used instead of quinine. The standard treatment for severe malaria is an intravenous infusion of quinine or quinidine [73]. Quinine may also be administered as an intramuscular injection. A loading dose of 20 mg/kg is recommended to reduce the time needed to reach effective concentrations in the blood [75]. A Cochrane Review found a significant reduction in fever clearance time and parasite clearance time with a loading dose compared with no loading dose but concluded that data were insufficient to demonstrate an impact on risk of death [76].

For nearly 400 years quinine has been the principal drug used to treat severe malaria. Despite its long history of efficacy, quinine has significant limitations. Even with prompt administration, case-fatality rates in severe malaria often exceed 20%, especially, in areas of South East Asia [77]. Currently, drug resistance is probably the major problem for malaria control countries where malaria is endemic. This extract from Clinical Evidence defined chloroquine and sulfadoxine-pyrimethamine as drugs of "unknown effectiveness"; in the light of the widespread resistance to chloroquine and the emerging resistance to sulfadoxine-pyrimethamine, these two drugs should not be considered in severe cases.

Slow, constant intravenous infusion is the preferred route for giving quinine. This is not always possible and quinine can also be given by deep intramuscular injection into the anterior thigh. Intragluteal injection should be avoided because of the risk of sciatic nerve damage, and the absorption is slow and uncertain. A few studies have shown good efficacy and tolerability for rectal administration, without the problems of the intramuscular route or the complexity of intravenous administration [78].

Adverse effects resulting from quinine therapy are common. Cinchonism (symptoms of quinine overdose) often occurs at conventional dose regimens. This usually mild and reversible symptom complex consists of tinnitus, deafness, dizziness, and vomiting, and may affect adherence. Hypoglycaemia is a less common but more serious adverse effect. Some people are allergic to quinine and develop skin rashes and edema even with small doses. Toxic levels of quinine can occur following rapid intravenous administration and can result in heart rhythm disturbances, blindness, coma, and even death; hence the recommendation for routine cardiac monitoring during parenteral treatment [77]. In accordance with WHO policy of antimalarial drugs in severe malaria treatment [73], quinine remains the drug of choice, although possesses certain drawbacks. Quinine has narrow therapeutic ratio, causing hyperinsulinaemic hypoglycaemia (more frequent and severe in pregnancy) and prolongs the QTc interval when given parenterally, particularly if infused too rapidly. Intramuscular quinine is effective, but can cause local toxicity, as well as hypoglycaemia in patients who may not have intravenous access [79]. Especially, in Southeast Asia there is evidence of increasing quinine resistance [80] so that artemisinins (mainly artesunate) are now increased to use as first line treatment for severe malaria in most areas [74].

Quinidine is the only US FDA approved drug for treatment of severe malaria. While it is commercially available and effective against malaria, it is not an ideal drug. Similar to quinine, quinidine is associated with sudden cardiac death, principally via cardiac arrhythmias, and, because of its short half-life, must be administered 2-3 times a day. Quinidine is more cardiotoxic than quinine and should be administered in an intensive care unit with continuous electrocardiographic and frequent blood pressure monitoring [81]. Quinidine-related cardiovascular adverse effects are potentially serious and may be more frequent if the drug is administered rapidly. The risk of cardiotoxicity is increased with bradycardia, hypokalemia, and hypomagnesemia; and if the patient has received other drugs that may prolong the QTc interval (e.g., quinine, mefloquine, or macrolide antibiotics) [82].

1.3.2. Artemether Failed to Show Superiority over Quinine

Although still uncommon, resistance to quinine is being reported more and more often. Development of artemisinins for treatment of severe malaria has been conducted since 1987 [83]. The most important question in antimalarial drug treatment of severe malaria is whether the artemisinin derivatives can reduce mortality better than quinine. Recently, intramuscular artemether has been the focus of many comparative studies versus parenteral quinine with over 3,000 cases having been reported in Asia and Africa. For clearance of parasitemia current results document the greater efficacy of a five-day treatment using intramuscular artemether as compared to the standard seven-day treatment using intravenous quinine bichlorhydrate [84]. Approved and available in most countries with endemic malaria, intramuscular artemether is included in the WHO List of Essential Drugs. The indication for use only in cases of severe malaria must be strictly respected to avoid development of resistance.

Large randomized comparisons of intramuscular artemether and quinine in Gambian children [85], and Vietnamese adults [86], and a meta-analysis of individual data from 1919 patients in 11 trials of parenteral therapy [77], identified no significant difference in efficacy between these agents. However, in the meta-analysis, the subgroup of adults had lower mortality when treated with artemether. Initially, a number of trials of parenteral artemether against quinine were undertaken in both children and adults with severe malaria [85–87]. In these trials, quinine was administered either intravenously or intramuscularly and artemether was administered intramuscularly. Artemether is a lipid-soluble artemisinin derivative that can only be administered intramuscularly and is now known to be subject to more erratic uptake and a slower time to peak plasma concentrations than artesunate [88]. Although the results of these trials showed that artemether was safe and easier to use than quinine, none of these trials showed a survival benefit individually. Yet, a survival benefit was demonstrated in subgroup analysis for a meta-analysis of adults in Southeast Asia [77]. This evidence was not sufficiently convincing to prompt a change of policy from quinine to artemether.

Most significantly, artemether may not have been the best choice of artemisinin to study in the first place, as the results of this study show that in terms of preventing deaths, artemether is as effective as quinine, which is comparable to findings of the 11 trials [77]. Compared with artesunate, artemether is less completely biotransformed to the more potent dihydroartemisinin and has slow, erratic absorption after intramuscular administration; in fact the ability of artemether to provide equivalent benefit to quinine is probably testament to the antimalarial potency of the artemisinin derivatives as a group. Attention has therefore switched to artesunate.

1.3.3. Artesunate in Treatments of Severe Malaria and SEAQUAMAT Trials

Artesunate is a more attractive formulation than artemether for pharmacokinetic reasons, as it can be administered either intravenously or intramuscularly, and it achieves therapeutic plasma concentrations rapidly when administered by either route [44]. Theoretically, therefore, artesunate may be more effective than artemether, because artemether is as effective as quinine. Initial trial results in adults have been encouraging, with some evidence of survival benefits compared with quinine, albeit with small numbers [89]. For this reason the SEAQAMAT trial was undertaken between 2003 and 2005.

SEAQUAMAT Trials

This was an open label, randomized, multicenter trial with centers in India (Rourkela, Orissa), Bangladesh (Chittagong), Myanmar (Myitkyina, Lashio, Taungoo, Tathon, and Rangoon) and Indonesia (Timika). Patients were randomized into either an artesunate or a quinine arm, as described by:

- Artesunate arm: intravenous artesunate 2.4 mg/kg on admission then at 12 h, 24 h and daily until able to take medication orally, thereafter oral artesunate salt 2 mg/kg/day to 7 days total;
- Quinine arm: intravenous quinine 20 mg/kg followed by 10 mg/kg every 8 h until able to take orally, then 10 mg/kg every 8 h orally for 7 days in total.

Both regimens were followed by doxycycline 100 mg daily for 7 days in Myanmar and Indonesia but not in Bangladesh or India. Inclusion criteria included a positive histidine-rich protein II (HRP2) malaria antigen test and 'clinical severe disease' according to a checklist. Patients who had received pretreatment with quinine or artesunate for 24 h or longer or who were known to be allergic to either drug were excluded. The primary end point was death from severe malaria and the analysis was intention to treat.

A large, randomized comparison of intravenous artesunate and quinine in 1461 patients (including 202 children) in Asia showed a significant survival benefit with artesunate, 730 to the artesunate arm and 731 to the quinine arm. A total of 41 patients were subsequently excluded from the artesunate arm and 38 from the quinine arm owing to negative blood films for malaria parasites. Baseline characteristics and severity of disease were similar between the two groups. Mortality was 22% with quinine, as compared with 15% with artesunate, a risk reduction of 34.7% [90]. There was a large heterogeneity in overall mortality rates between countries with 28% in the Bangladeshi site compared with 9% in Indonesia. The mortality advantage of artesunate over quinine was preserved in all the study sites. The number needed to treat with artesunate to avoid one death ranged from 11.1 to 20.2 depending on the treatment site. Most of the mortality benefit occurred in the first few days of admission, with comparable mortality in both arms thereafter. The treatment effect was similar in children compared with adults but they constituted a much smaller group and results did not reach statistical significance [91]. Treatment with artesunate had a relatively mild side-effect profile; hypoglycemia was significantly more common with the use of quinine. A systematic review of five randomized trials comparing the efficacy of intravenous quinine with that of artesunate and one additional trial of intramuscular artesunate demonstrated the superiority of artesunate, with significant reductions in the risk of death (relative risk, 0.62), incidence of hypoglycemia, and parasite clearance time, as compared with quinine.

Neurological sequel were extremely rare, providing further reassurance of its safety. As a result of this study, parenteral artesunate is now recommended by WHO as the drug of choice for the treatment of severe malaria in low-transmission areas and in pregnancy in the second and third trimesters [2]. The current recommendation for treating severe malaria patients in high-transmission areas is either quinine or an artemisinin derivative. In these areas, severe malaria is mainly a pediatric disease, and in children mortality often occurs within the first 24 hours after admission, leaving a smaller time window for artesunate to deliver a clinical advantage. Additionally, African malaria parasites are more quinine-sensitive than those in Asia. Earlier studies comparing parenteral quinine with the oil-soluble derivative artemether showed that artemether was associated with a modest but significantly lower mortality than quinine in Asian adult patients, although this effect was not seen in African children [77]. Intramuscular artemether has since been shown to be erratically absorbed in Vietnamese adults with severe malaria; it is unclear if this can explain the lack of effect on mortality in African children. Hence there is a need for an additional trial of the water-soluble artesunate in African children, which is currently underway.

The effective currently available artesunate formulation produced by Guilin Pharmaceutical factory in China is manufactured to Chinese GMP standards, but is not 'international GMP certified'. Further certification is under review. A new GMP formulation of artesunate is currently being developed by Walter Reed Army Institute of Research (WRAIR), USA and should obtain FDA approval in the near future. Even if parenteral artesunate is widely available in hospitals, it will not be available to the majority of patients developing severe malaria living in rural tropics far from the nearest health clinic or hospital. For these patients, artesunate suppositories offer the prospect of effective early treatment, preventing clinical deterioration and buying time to reach hospital. The WHO has sponsored the development of GMP-manufactured rectal artesunate capsules, which are well absorbed and effective for the treatment of moderately severe malaria [92]. A large phase III study of early deployment has been completed and will be reported soon. Home- or village-based deployment of rectal artesunate in rural areas of Africa and Asia may play an important role in reducing malaria-associated morbidity and mortality.

Severe malaria remains a major global health problem and, despite advances in prevention, it is likely to remain so for the foreseeable future. The SEAQUAMAT study demonstrates convincingly that treatment of severe adult malaria acquired in Asia with artesunate is associated with a reduction in mortality compared with quinine. No ethical committee is likely to approve another similar study in adults with severe malaria in Asia and the SEAQUAMAT trial should be taken as definitive in this context. Can the results be assumed to also apply to children in Asia, and also to adults and children in Africa, where the bulk of severe *falciparum* malaria occurs? There are various reasons to be cautious about extrapolating the SEAQUAMAT data to all these groups. For adults with severe malaria, artesunate should be the treatment of choice. The drug is more effective than quinine, is simple to administer, and is safe. The only serious adverse effect that has been clearly linked with artesunate is a rare type-1 hypersensitivity reaction, which arises in about one in 3000 treated patients [38]. By contrast, quinine is locally toxic (particularly as the acidic dihydrochloride salt) after intramuscular injection (often causing sciatic nerve damage), cannot be given by bolus intravenous injection, requires three-times daily administration, is associated with potentially serious hypoglycaemia, and is less effective than artesunate. Unfortunately, despite 20 years of use in East Asia, there is still no generally available formulation of parenteral artesunate made to Good Manufacturing Practices standards. Quality assured and affordable parenteral artesunate should be made widely available in malaria endemic areas as a matter of urgency [92].

1.3.4. Advanced Effect of Artesunate Compared to Artemether and Quinine

Artesunate is a semisynthetic derivative of artemisinin whose water solubility facilitates absorption and provides an advantage over artemisinin because it can be formulated for oral, rectal, intramuscular, and intravenous preparations. Artesunate is rapidly hydrolyzed to dihydroartemisinin, which is the most active schizonticidal metabolite. Extravascular administration of artesunate results in a more rapid systemic availability of artesunate compared with intramuscular artemether. This pharmacokinetic advantage may provide a clinical advantage in the treatment of severe malaria [44]. Rectal artesunate has been shown to be absorbed rapidly, with a considerable inter-individual variability. Artesunate is highly effective against multidrug-resistant *falciparum* malaria and severe malaria in Vietnam, Thailand, China, and Myanmar [93]; however, limited studies have been carried out in Africa.

1.3.4.1. Intravenous Artesunate in Replacement of Quinine for Adults and Pregnant Woman

As reported earlier, the South East Asian Quinine Artesunate Malaria Trial (SEAQUAMAT) Group published results of almost 1500 patients with severe malaria showing that artesunate had superiority over quinine using mortality as an endpoint for the trial. This remarkable study added fuel to the fire that quinine, and in the U.S., quinidine, should be replaced by artesunate in the treatment of severe malaria. This study lead to the WHO's guidelines at the end of 2005 and confirmed to the WRAIR development team that they had indeed made the right choice of drug candidate for development three years earlier. These results also provided encouragement for the team to redouble their efforts in getting this product to market as quickly as possible [44, 94]. The problem to date had been that of getting a commercial partner. This drug, as noted earlier, had little commercial interest because it was not patentable. Few pharmaceutical companies would be knocking at the door to share co-development costs with the Army in getting this product to market, much less take the chance of providing this drug over a sustained period after licensure. Luckily, there are new pharmaceutical companies emerging that specialize in niche markets such as orphan drugs. It was while the Army's program was within a year or two of approaching its goal of filing a New Drug Application with the FDA that WRAIR researchers met the Italian subsidiary in the U.S., Sigma-Tau Pharmaceuticals, an Italian subsidiary in the U.S. that specializes in rare diseases and orphan drugs. This partnership will be sealed in early 2009 and should bring this critically needed product not only to the U.S. market, but also hopefully to the European and broader world market in 2008 [95].

In pregnant patients, first trimester pregnancy remains a concern with artesunate. Animal data have suggested an association of artemisinin drugs with fetal resorption and fetal abnormalities when administered early in pregnancy, although no increase in abortion or stillbirth rates were demonstrated in an observational study of 461 pregnant women on the Thailand–Myanmar border. The effects of artesunate in early pregnancy remain a concern in non-severe malaria but, since severe malaria is frequently associated with major adverse outcomes in pregnancy, the priority in treating a pregnant patient with malaria must be to administer an effective antimalarial quickly and, where quinine resistance is a possibility, this is likely to be artesunate. Rates of fetal death were similar in both treatment groups in the SEAQUAMAT trial, although only small numbers of patients were included and the outcomes of the surviving pregnancies were not reported. Balancing the potential risks of artemisinins to the fetus against the real risks of severe malaria to both of mother and fetus, artesunate should be recommended in all pregnant women from Asia with severe malaria. In pregnant women in their first trimester in Africa, the risks are more balanced; the potential teratogenicity remains the same while there is no proven advantage [91].

1.3.4.2. Intramuscular Artesunate in Replacement of Artemether or Quinine for Children

Intravenous artesunate and intramuscular artemether have been highly efficacious for the treatment of severe malaria. Since artemether failed to show superiority over quinine intravenously or intramuscularly [85-88], intramuscular artesunate is also effective and may be of value in settings with limited resources [87].

The intramuscular route for quinine administration has recently regained favor. It reduces the risk of dosing errors that can cause cardiotoxicity, although the risk of hypoglycemia is unaltered. For these reasons, there is an urgent need to develop alternative safe antimalarials that are efficacious in patients with severe malaria. Artemether is the most widely assessed of the parenteral artemisinin derivatives. Although it is an oil based formulation that can be given only by the intramuscular route, it has proved at least as effective as intramuscular quinine in the treatment of severe malaria in large randomized trials that have enrolled collectively nearly 2,000 patients both in Asia and in Africa [85-87]. Despite the generally good clinical results with artemether, intramuscular artemether only provides equivalent benefit to quinine [77]. Compared with artesunate, artemether is less completely converted to the more potent dihydroartemisinin and has slow, erratic absorption after intramuscular administration; in fact the ability of artemether is showing much less antimalarial potency when compared to artesunate [44].

Artesunate is a water-soluble hemisuccinate semisynthetic derivative of artemisinin that is one of the most rapidly acting antimalarials available. It can be administered by the various routes, providing an opportunity to carry out bioavailability studies with patients with malaria of various degrees of severity. Intramuscular artesunate were performed many trials in uncomplicated and severe malaria patients [96-98]. Results showed when an Africian child with severe malaria is hospitalized, many factors influence the eventual outcome. These include both the rapidity of diagnosis and the speed with which effective treatment is given, especially as most deaths occur within the first 12 h after admission. At present there is no alternative therapy to quinine given in a loading dose by either the intramuscular route or the intravenous route. However, our study shows that intramuscular artesunate is an important alternative to quinine in Africa [98]. Also, intramuscular artesunate proved safe and effective in the treatment of severe and complicated childhood malaria in Asia [96].

The public health implications of suppository treatment of a life-threatening disease that kills millions of children in rural Africa each year are of great importance and warrant further clinical assessment of acceptability and implementation, although the possible consequences of widespread and relatively uncontrolled use of artemisinin and its derivatives also need to be assessed and taken into consideration.

1.3.4.3. Rectal Artesunate in Replacement of Quinine for Children

Many children in Africa die of malaria before reaching formal healthcare services. One attractive procedure for administration of artesunate in severe disease is the rectal method. Rectal artesunate is unlikely to challenge either parenteral artesunate or quinine in hospital practice, but it may have a significant role in pre-hospital settings, especially in rural parts of Africa where the bulk of malaria deaths occur. A trial in Uganda comparing rectal artemether with intravenous quinine showed a trend towards lower mortality, faster parasite clearance and faster time to effervescence in the artemether arm compared with the quinine arm; however, these results failed to reach statistical significance [99] and rectal artesunate may well fare better [92]. A recent multicenter trials of rectal artesunate used in a rural setting has recently been completed and will be reported shortly.

One systematic review (search date 2008) comparing rectal artemisinins with intravenous quinine in severe malaria [100]. This review addresses the lack of any data directly comparing the therapeutic efficacy and safety of the different rectal preparations of artemisinin derivatives. The pooled analysis of individual patient data suggests that artemisinin and artesunate suppositories rapidly eliminate parasites and are safe. There is far less evidence for artemether [99] and no studies of dihydroartemisinin suppositories were available to be included in this analysis. The results indicate that both artemisinin and artesunate, whether as single or multiple dose regimens, induce a superior parasitological response than parenteral quinine over the 24 hours following initiation of treatment. Regimens employing a higher single dose of rectal artesunate were five times as likely to result in >90% parasite reductions at 24 hours than were multiple lower doses of rectal artesunate or than a single administration of artemether. These results imply that dosage regimens that result in immediate high blood concentrations of drug [92, 101] are those best able to reduce parasitemia in patients with evolving severe malaria and that sustained drug exposure achieved by sequential treatment with moderate doses offers no therapeutic advantage [102].

A recent trial comparing intravenous artesunate with parenteral quinine demonstrated a 35% lower mortality with artesunate, confirming that more rapid initial parasite clearance may translate to reduced mortality in severe adult malaria [90]. Therefore, although the current pooled analysis was not powered to assess mortality as an endpoint, the differences in parasite clearance rates between rectal artemisinins and parenteral quinine, and between different rectal artemisinin dosing regimens, should be regarded as important indicators of possible real differences in therapeutic efficacy and clinical benefit. It should also be noted that any future study designed to use mortality as an endpoint to compare different rectal artemisinins would require such a large sample size that it is unlikely ever to be implemented. Therefore the surrogate marker of parasite clearance used in this analysis is likely to remain the best available evidence on which to base comparisons of treatment efficacy for the rectal artemisinins.

The evidence from this analysis shows rectal artesunate has superior effect in reducing parasite densities compared to quinine (intravenous or intramuscular) at 12 and 24 hours after administration. The result supports the WHO recommendation for the use of artesunate and artemisinin as initial pre-referral treatment [2]. Basis of systematic review, the WHO recommendation is to use artemisinins rectally for complete treatment only when parenteral antimalarial treatment is not possible [100].

1.3.4.4. Artesunate is the Best Choice for the Severe Malaria Therapy

These trials conducted in intravenous, intramuscular, and rectal methods using artesunate, artemether, and quinine demonstrated that artesunate should now become the therapy of choice for the severe malaria in adults at least, although more information was needed in children living in high transmission settings [90, 103-105]. Severe malaria is a medical emergency. After rapid clinical assessment and confirmation of the diagnosis, full doses of parenteral antimalarial treatment should be started without delay with whichever effective antimalarial is first available. Recently, WHO recommended that artesunate should be in the first selection for treatment of severe malaria.

Artesunate 2.4 mg/kg bodyweight intravenous (i.v.) or intramuscular (i.m.) given on admission (time = 0), then at 12 h and 24 h, then once a day is the recommended choice in low transmission areas or outside malaria endemic areas. For children in high transmission areas, the following antimalarial medicines are recommended as there is insufficient evidence to recommend any of these antimalarial medicines over another for severe malaria:

- artesunate 2.4 mg/kg bodyweight i.v. or i.m. given on admission (time = 0), then at 12 h and 24 h, then once a day;
- artemether 3.2 mg/kg bodyweight i.m. given on admission then 1.6 mg/kg bodyweight per day;
- quinine 20 mg salt/kg bodyweight on admission (i.v. infusion or divided i.m. injection), then 10 mg/kg bodyweight every 8 h; infusion rate should not exceed 5 mg salt/kg bodyweight per hour.

A major concern that has been expressed is that the current production of artesunate does not reach good manufacturing practice (GMP) standards, although there is no evidence that this translates into reduced efficacy. This is likely to be a temporary issue, although these concerns may give rise to delayed uptake in some settings, especially in treating patients from Africa where the evidence is less certain. GMP products are currently under development, both in China and in a collaboration involving the Walter Reed Army Institute of Research (WRAIR) [95].

1.4. Impact of Drug Cost in Current Malaria Treatments

ACTs are the best anti-malarial drugs available and recommended by WHO for first-line use now [47]. Artemisinin enhances efficacy and has the potential of lowering the rate at which resistance emerges and spreads. Under low transmission intensity, ACTs have an additional public health benefit of reducing the overall malaria transmission and studies are urgently needed to investigate modalities of attaining similar benefits under high transmission. Despite being recommended by WHO since 2001, overall deployment of ACT has been slow due to its high cost and other limiting factors [72].

1.4.1. Influence of Malaria Cost on the National Economy

The costs of malaria are also enormous when measured in economic terms. Highly malarious countries are among the very poorest in the world, and typically have very low rates of economic growth; many have experienced outright declines in living standards in the past thirty years. Malaria has played a significant role in the poor economic performance of these countries. The evidence strongly suggests that malaria obstructs overall economic development. Statistical analysis shows that during the period 1965-1990, highly malarious countries suffered a growth penalty of more than one percentage point per year (compared with countries without malaria), even after taking into account the effects of economic policy and other factors that also influence economic growth. The annual loss of growth from malaria is estimated to range as high as 1.3 percentage points per year. If this loss is compounded for fifteen years, the GNP level in the fifteenth year is reduced by nearly a fifth, and the toll continues to mount with time.

These considerations indicate that the cost of malaria is substantially greater than economists have previously estimated. Traditional estimates have looked at some of the short-run costs of malaria without taking into account the longer-term effects of malaria on economic growth and development. Short-run costs - including lost work time, economic losses associated with infant and child mortality and morbidity, and the costs of treatment and prevention - are typically estimated to be higher than one percent of a country's gross national product. These estimates, however, neglect many other short-run costs. For instance, very few studies include the economic costs of the pain and suffering associated with the disease. Yet researchers have found that households might be willing to pay several times the direct income loss caused by malaria in order to avoid it, suggesting that the pain, suffering, and uncertainty associated with the disease is very high and should certainly be included among its short-term costs.

Furthermore, these short-run costs are likely to have risen in recent years due to the increasing number and complexity of cases in many countries. Moreover, the spread of drug-resistant malaria is substantially raising the costs of treatment in many cases, as well as the burdens of morbidity and mortality. Children and adults needing blood transfusions due to malaria are too often inadvertently infected with HIV, hepatitis C virus, and other infectious agents who taint the blood supply.

Beyond these high and rising short-run costs, malaria impedes economic growth and long-term development in many ways. Malaria may impede the flows of trade, foreign investment, and commerce, thereby affecting a country's entire population. Multinational firms choosing the location of foreign investments shun regions with high malaria, as might many potential tourists. Also, the economic effect of malaria on infected individuals may greatly exceed the direct costs of any single episode of the disease. Repeated bouts of malaria tend to hinder a child's physical and cognitive development, and may reduce a child's attendance and performance at school. Furthermore, repeated bouts of malaria may expose individuals to chronic malnutrition, anemia, and increased vulnerability to other diseases [106].

1.4.2. ACTs Are Expensive than Old Routine Antimalarial Drugs in Use

ACTs are now the officially recommended treatment for *P. falciparum* malaria in most malaria-endemic countries [2]. However, there are significant challenges in the actual implementation of this change in antimalarial drug policy [107-109]. In the era of cheap monotherapies, namely chloroquine and sulphadoxine-pyrimethamine (SP), treatment for malaria was largely based on inaccurate clinical diagnosis and often obtained by patients from sources outside of the formal health sector. Currently, ACTs costs around five to 10-20 times more than the traditional monotherapies, there needs to be a radical change in the approach to how they are funded and supplied in order to ensure that they are accessible and affordable to those who most need them. At a global level, support is gathering for the setting up of the Affordable Medicines Facilities -malaria, a global subsidy for ACTs, as recommended by the Institute of Medicine report "Saving lives, buying time" [42]. It is hoped that such a subsidy would enable ACTs to flow through the private sector, as well as the public sector, to reach the poor and remote communities who are most vulnerable and, if subsidized heavily enough, displace monotherapies and sub-standard drugs. However, there is still significant uncertainty about how to best maximize access, while limit the wastage and risks associated with widespread inappropriate use of ACTs by those who do not have malaria [72]. Although the process and costs of policy change itself has been described [110], there is little experience of nationwide implementation and the cost of operationalizing such a policy change is largely unknown.

Indeed, affordability is the immediate problem of ACT. Chloroquine - the standard, effective drug for decades - cost is about 10 U.S. cents per course of treatment for an adult. The new effective drugs - artemisinin combination therapy (ACT) - today cost US$1.50-2.40 per course wholesale, and can be marked up to five times that amount in pharmacies in Africa. There are limited alternatives to ACTs for effectiveness and safety for widespread use, and their adoption is recommended by WHO and virtually all other expert bodies.

Unfortunately, even 10 cents is too much for the poorest of the poor in every endemic country.

The rationale for combining two or more drugs is that doing so dramatically reduces the odds that malaria strains resistant to any of the drugs in the combination would survive to be transmitted. In this report, ACT is used mainly to describe a drug "co-formulation" (i.e., two drugs in one pill), and not just two separate drugs taken at the same time. Multiply the per-dose difference in price by the millions or tens of millions courses used per year in each country in Africa, and the money adds up. Continent-wide, an estimated 200-400 million courses of treatment are used per year, and an additional 100 million courses in the rest of the world. Realistically, the endemic countries are able to contribute very little of the incremental cost, and that is not likely to change for many years, given the pace of economic development in Africa and in poor nations elsewhere. One positive change that is almost certain, however, is that several different ACTs will become available, and their price will come down by more than half, to about 10 times the current price of chloroquine, or about US$1 per adult course and US$0.60 for an average child's course. However, that price is still too high.

1.4.3. Cheap Antimalarial Drugs Can Save Million Lives

For over a decade, malaria treatment in sub-Saharan Africa and other endemic regions has been in crisis. Conventional antimalarial drugs such as chloroquine and sulphadoxine/pyrimethamine (SP) have become increasingly obsolete in the face of growing drug resistance. The region has the largest number of people exposed to stable malaria transmission and constitutes the greatest burden of malaria morbidity and mortality in the world with pregnant women and children less than five years being the most affected groups. Current debates favor using artemisinin-based combination therapy (ACT) regimens [47]. ACTs are highly efficacious and offer potential to check the progression of drug resistance. They are also expensive with prices as much as 10 to 20 times greater than conventional monotherapies.

Cost is inevitably a factor in determining whether and when a country should switch malarial drug policy, with most attention falling on the cost of the drugs [72, 110, 111]. Through concerted efforts of multilateral organizations, the local scientific community with involvement of policy-makers progress has been on several fronts leading to improved ACT uptake rates in the last 2 years. By early 2006, 34 (79%) countries in sub-Saharan Africa had adopted ACTs as first-line treatment for malaria [63], while 18 (42%) adopted the policy in 2004 [72]. Therefore, cheaper ACTs are alternative ways for increasing adoption and deployment of ACTs in more countries and saving more lives, otherwise most policies will remain adopted on paper.

In addition, each year between 1 and 2 million children die from acute *P. falciparum* malaria [2], and the majority is in sub-Saharan Africa. Some die rapidly at home before any effective treatment can be instituted; others arrive at their local health facility too late for lifesaving treatment. Children with severe malaria are often unconscious, convulsing or vomiting, so cannot take oral medication. In many cases the parenteral route of antimalarial administration is either impossible, due to lack of equipment or adequately trained staff, or potentially hazardous if dirty needles and syringes are used.

Rectal administration of drugs is both easy and safe. A generous blood supply from the middle and inferior hemorrhoidal veins ensures rapid absorption and much of the absorbed drug enters the systemic circulation directly, without passing through the portal circulation, thereby avoiding first pass clearance. Most of the estimated deaths from malaria occur in children, and most of these deaths occur outside hospitals. Rectal administration of antimalarial drugs may allow effective treatment to be instituted early in outlying rural areas, and gain valuable time in order to seek appropriate medical care.

These drugs are very well tolerated and no serious adverse effect has been reported yet in humans [99]. Suppository formulations have proved to be effective in acute *falciparum* malaria [42, 101] and cerebral malaria [42, 100, with clinical and parasitological responses similar to those following parenteral administration. If cheap suppositories are available for physician patient use, in rural areas where medical facilities are lacking these drugs will allow antimalarial therapy to be instituted earlier in the course of the disease and may therefore save lives [96].

1.4.4. Reduction and Subsidy Plans for Cutting Drug Prices

Why are ACTs so much more expensive than chloroquine? Even at prices anticipated after the massive scale-up that would be necessary to supply the African market - should funding become available for large-scale purchase - the price as sold by the producer will likely remain at about 10 times the price of chloroquine. These prices reflect the production costs of artemisinins without a premium for any exclusivity related to patents. They are high because the process involves growing the source plant, *Artemisia annua*, extracting the active moiety, and creating the desired artemisinin derivative (artesunate, artemether, etc.). Co-formulation with the companion drug follows. The price of the finished product is driven mainly by the cost of the artemisinin derivative, but also is affected by the companion drug (e.g., lumefantrine, the companion drug in Coartem, is a relatively expensive drug on its own, which contributes to the high price of the co-formulation).

There is something of a chicken-and-egg quality about the current price of artemisinins, and the prospect for significantly lower prices. Lower prices can be expected in response to large-scale demand, which, in turn, will induce competition among producers. ACTs, manufacturers will not have the incentive to scale up production, and prices will not drop. In addition to competition driving prices down (i.e., companies being willing to accept lower profits per dose with higher volumes), technological improvements in the process could bring down the actual production costs, which would be passed along to purchasers in a competitive environment. The real price breakthrough will likely occur only when a fully synthetic artemisinin is developed, eliminating the growing and extraction process. The Medicines for Malaria Venture (MMV) has such a compound under development, which they predict could be available in 5-6 years. The ultimate price is not known, but it should be significantly less expensive than current artemisinins (assuming no premium for exclusivity). If the synthetic product is better than, or at least as effective as the extracted ones, the market would change dramatically. A global subsidy might still be needed, but it could be less than what is needed now.

In this case, as malaria-endemic countries start to roll-out ACTs and the global community considers setting up a global subsidy, it is important to consider the non-drug costs of roll-out. Subsidies of public money increase access to goods or services that, from a societal perspective, are necessary or highly desirable but would not otherwise reach intended users because of cost. In cases where the item being subsidized also produces "positive externalities" that benefit everyone, the rationale for subsidizing individual access is strengthened. Artemisinin combination therapies (ACTs) have proven effective at the individual level in every setting where they have been tried. Used in combination with another effective antimalarial - as opposed to an artemisinin alone - a "global public good" results from the reduced risk of early development of resistance to this class of drug. A global subsidy, therefore, benefits not only those using the drugs directly but the global community.

Without subsidies, a large proportion (in some countries, a majority) of residents of sub-Saharan Africa and the poorer malaria-endemic countries of Asia will not be able to afford appropriate courses of ACTs. Moreover, evolving patterns of use are likely to create a negative externality—an increased chance that artemisinin-resistant strains will develop and spread, to the detriment of all. To round out the arguments for a global subsidy, ACTs are cost effective by global standards, and the full subsidy has a reasonable global price tag when held against the human and economic burden of malaria.

A global high-level subsidy was the best way to make effective antimalarial drugs - currently; artemisinin-combination therapies (ACTs) - widely available at affordable prices and at the same time substantially delay the emergence and spread of artemisinin-resistant strains of *falciparum* malaria. The subsidy would be available to manufacturers of all ACTs meeting pre-specified efficacy, safety, and quality criteria. Buyers would pay very low prices, allowing drugs to flow through existing channels, with the aim of reaching consumers at a similar price to chloroquine, the most frequently used (although no longer effective) malaria drug. Unsubsidized artemisinin monotherapies would be more expensive than subsidized ACTs (co-formulations), thereby largely eliminating their use through market forces. Conditions favoring the emergence of artemisinin-resistant malaria would be greatly reduced. The global high-level subsidy is a powerful idea that is moving from economic concept to pragmatic reality [112].

1.5. WHO Recommended New Antimalarial Drug Policies (AMDPs)

The rapid spread of antimalarial drug resistance over the past few decades has necessitated increased monitoring for further resistance, to ensure proper management of clinical cases and early detection of changing patterns of resistance for revision of national malaria treatment policies. A national antimalarial treatment policy is a set of recommendations and regulations concerning the availability and rational use of antimalarial drugs in a country [47]. In 2001, WHO issued "Antimalarial Drug Combination Therapy: Report of a Technical Consultation" to recommend 4 ACTs in uncomplicated malaria. There are artemether-lumefantrine, artesunate plus amodiaquine, artesunate plus SP in areas where SP efficacy remains high, and SP plus amodiaquine in areas where efficacy of both amodiaquine and SP remain high [47].

The objective of a national antimalarial treatment policy is to enable the population at risk of malaria infection to have access to safe, good quality, effective, affordable, and acceptable antimalarial drugs, in order to:

- ensure a rapid and long lasting clinical cure for individual malaria patients
- prevent progression of uncomplicated malaria to severe disease and death
- shorten clinical episodes of malaria and reduce the occurrence of malaria-associated anemia in populations residing in areas of high malaria transmission
- reduce consequences of placental malaria infection and maternal malaria-associated anemia through chemoprophylaxis or preventive intermittent treatment during pregnancy
- delay the development and spread of resistance to antimalarial drugs.

More recently (November 8, 2006 and March 2008), a new Guideline for the Treatment of Malaria [2, 113] was issued by WHO to revise all old antimalarial drug policies (AMDPs) due to increasing resistance and discovery of the novel drug (ACTs). The national antimalarial treatment policies aim to offer antimalarials that are highly effective. The main determinant of policy change is the therapeutic efficacy and the consequent effectiveness of the antimalarial in use. Other important determinants include: changing patterns of malaria-associated morbidity and mortality; consumer and provider dissatisfaction with the current policy; and the availability of new products, strategies and approaches.

1.5.1. Current Brief AMDPs in Malaria Chemotherapy

Global malaria control is being threatened on an unprecedented scale by rapidly growing resistance of *P. falciparum* to conventional monotherapies such as chloroquine, sulfadoxine-pyrimethamine (SP), and amodiaquine. Multi-drug resistant *falciparum* malaria is widely prevalent in South-East Asia and South America. Now Africa, the continent with highest burden of malaria is also being seriously affected by drug resistance. As a response to the antimalarial drug resistance situation, WHO recommends that treatment policies for *falciparum malaria* in all countries experiencing resistance to monotherapies should be combination therapies, preferably those containing an artemisinin derivative (ACT - artemisinin-based combination therapy).

1.5.1.1. Treatment Policy for Uncomplicated Malaria

The objective of treating uncomplicated malaria is to cure the infection. This is important as it will help prevent progression to severe disease and prevent additional morbidity associated with treatment failure. Cure of the infection means eradication from the body of the infection that caused the illness. In treatment evaluations in all settings, emerging evidence indicates that it is necessary to follow patients for long enough to document cure. In assessing drug efficacy in high-transmission settings, temporary suppression of infection for 14 days is not considered sufficient by the group. The public health goal of treatment is to reduce transmission of the infection to others, i.e., to reduce the infectious reservoir.

A secondary but equally important objective of treatment is to prevent the emergence and spread of resistance to antimalarials. Tolerability, the adverse effect profile and the speed of therapeutic response are also important considerations. The brief policies [113] for treatment of uncomplicated *falciparum* malaria are listed:

- Artemisinin-based combination therapies (ACTs) are the treatment recommended for all cases of uncomplicated *falciparum* malaria including:
  - in infants,
  - in people living with HIV/AIDS
  - for home-based management of malaria
  - pregnant women in the 2nd and 3rd trimesters
      Exception: 1st trimester of pregnancy*
      *only use when there are no alternative effective antimalarials
- The following ACTs are presently recommended:
  - artemether-lumefantrine
  - artesunate + amodiaquine
  - artesunate + mefloquine
  - artesunate + sulfadoxine-pyrimethamine
- Second-line treatment:
  - an effective alternative ACT (efficacy of ACTs depend on efficacy of the partner medicine, therefore it is possible to use 2 different ACTs as 1st and 2nd line options)
  - quinine + tetracycline or doxycycline or clindamycin
      Note: The artemisinin derivatives (oral, rectal, or parenteral formulations) and partner medicines of ACTs are not recommended as monotherapy for uncomplicated malaria.

1.5.1.2. The Options Recommended for Treatment of Uncomplicated Falciparum *Malaria in Pregnancy*

These options are recommended for treatment of uncomplicated *falciparum* malaria in pregnancy [113]

- 1st Trimester:
  - Quinine + clindamycin
- 2nd and 3rd trimesters:
  - any of the recommended ACTs listed
  - artesunate + clindamycin
  - quinine + clindamycin

1.5.1.3. Current AMDPs in Treatment of Severe Malaria

The primary objective of antimalarial treatment in severe malaria is to prevent death. Prevention of recrudescence and avoidance of minor adverse effects are secondary. In treating cerebral malaria, prevention of neurological deficit is also an important objective. In the treatment of severe malaria in pregnancy, saving the life of the mother is the primary objective. These policies are recommended for treatment of severe and complicated *falciparum* malaria [113, 114]:

- Any of the following antimalarial medicines are recommended for initial treatment.
    - artesunate (i.v. or i.m.)
    - artemether (i.m.)
    - quinine (i.v. infusion or i.m. injection).
- Follow-on treatment:
    once the patient recovers enough and can tolerate oral treatment, the following options can be used to complete treatment:
    - full course of an ACT or
    - quinine + clindamycin or doxycycline

Over the last three years (2004-2007) around 67 countries (41 in Africa) have updated their treatment policies to include ACTs as 1st-line or 2nd-line treatment of malaria (Table 5). This was based on WHO advice, and was made possible with the participation of RBM partners and increased mobilization of international funding [55]. Consistent with WHO recommendations, malaria endemic countries which are experiencing resistances that currently use antimalarial drug monotherapies (chloroquine, sulphadoxine/pyrimethamine or amodiaquine) should change treatment policies to the highly effective artemisinin-based combination treatments (ACTs).

1.5.2. Current AMDPs in Malaria Chemoprophylaxis

The principles of the use of antimalarial drugs for protection against malaria of uncomplicated and severe malaria were reviewed in 2004 [115]. Malaria control aims to reduce illness and death from malaria. The WHO's global malaria control strategy recommends a multi-pronged control approach that combines multiple preventive interventions with prompt diagnosis and treatment of symptomatic persons with efficacious antimalarial drugs [116]. Antimalarial drugs have been used in various ways to prevent malaria in the resident populations of endemic areas for nearly 100 years. The primary aim of most early studies was to interrupt transmission. This was rarely achieved, but administration of antimalarial drugs either through medication of salt or by mass administration frequently led to a marked reduction in the prevalence of malaria infection and in the incidence of clinical attacks. Chemoprophylaxis is highly effective in reducing mortality and morbidity from malaria in young children and pregnant women living in endemic areas, but is difficult to sustain and, in some studies, has impaired the development of naturally acquired immunity. Intermittent preventive treatment, in which full therapeutic doses of a drug are given at defined intervals, has the potential to provide some of the benefits of sustained chemoprophylaxis in pregnant women and young children without some of its drawbacks and is a promising new approach to malaria control (Table 9).

Table 9. Approaches to the administration of antimalarial drugs as a means of preventing malaria in population of malaria-endemic communities [115]

Approach	Comments
Treatment of clinical cases	Requires that a high proportion of infections are symptomatic and that patients have easy access to treatment
	Likely to be more effective if treatment includes a gametocidal drug
Medication of salt	Produces sub-therapeutic but cumulative blood levels
	Highly likely to induce resistance
Mass drug administration	Involves administration of drug, usually in therapeutic doses, to infected and non-infected subjects
	Usually has only a transitory effect on levels of parasitemia and clinical malaria
Chemoprophylaxis	Involves the repeated administration of drug, usually at sub-therapeutic doses, over a sustained period so as to obtain persistent protective blood levels
Intermittent preventive treatment	Involves the administration of a therapeutic dose of drug over an intermittent and defined period

Cochrane Reviews have confirmed the effectiveness of insecticide-treated nets in reducing malaria morbidity and mortality in preschool children and pregnant women [116], but coverage of this intervention in most sub-Saharan African countries lags far behind global targets. By 2005, less than a third of the endemic countries in this region had attained 30% coverage for children under five years; far below the Roll Back Malaria targets of 60% and 80% for 2005 and 2010 respectively [116]. Indoor residual spraying is another vector control measure recommended by the WHO for community protection, but it is expensive and requires high coverage to be effective [117]. Such high levels of coverage would be difficult to attain in many endemic areas, especially those with high perennial transmission.

For chemoprophylaxis, the guiding principles of malaria prevention are based on WHO's recommendations and policies [116, 118].

1) Chemoprophylaxis to areas with Chloroquine-resistant *P. falciparum*
   - Atovaquone/proguanil (Malarone™)
   - Doxycycline (brand names and generics)
   - Mefloquine (Lariam™ and generic)
   - Primaquine
2) Chemoprophylaxis to areas with mefloquine-resistant *P. falciparum*
   - Mefloquine-resistant *P. falciparum* is present in eastern Burma (states of Shah, Kayin, and Kayah), the western provinces of Cambodia that border Thailand, and all malaria-risk areas in Thailand.
   - Either atovaquone/proguanil or doxycycline can be used by travelers to these areas.
3) Chemoprophylaxis to Areas with Chloroquine-sensitive *P. falciparum*

- In areas where chloroquine-resistant *P. falciparum* has not been reported, either chloroquine phosphate (Aralen and generic) or hydroxychloroquine sulfate (Plaquenil) may be used. Less evidence exists on hydroxychloroquine sulfate's effectiveness as an antimalarial drug.
- Chloroquine prophylaxis should begin 1-2 weeks before travel to malarious areas. It should be continued once a week, on the same day of the week, during travel to malarious areas and for 4 weeks after a traveler leaves such areas.
- Travelers unable to take chloroquine should take atovaquone/proguanil, doxycycline, mefloquine, or primaquine; these drugs are also effective against chloroquine-sensitive *P. falciparum.*

For a traveler visiting highly endemic areas for malaria (mainly sub-Saharan Africa), the use of a chemoprophylaxis has to be considered as mandatory in addition to exposure prophylaxis measures (including in migrants largely over-represented among returning travelers with imported malaria). The choice of the appropriate drug depends mainly on the visited area with regard to the level of resistance to chloroquine.

1.5.3. Current AMDPs in Special Populations

1.5.3.1. Guiding Principles of Chemoprophylaxis in Travelers

The spread and intensification of drug resistance worldwide has greatly complicated recommendations for the prevention of malaria in travelers. Travel to malarious areas is on the increase, while many countries are experiencing a resurgence of malaria. As a short-term measure, chemoprophylaxis is recommended for international and national travelers from non-endemic areas, and for soldiers, police and labor forces serving or working in highly endemic areas. Detailed recommendations for the protection of travelers against malaria are updated and published annually by WHO in International travel and health: vaccination requirements and health advice [119].

Malaria chemoprophylaxis should be selected on the basis of an individual risk assessment of the traveler, an assessment of the safety and efficacy of potential chemoprophylactic regimens, and drug resistance and the extent of malaria transmission in the region to be visited. Weekly prophylactic antimalarial regimens should normally be started one week before travel. Daily drugs such as proguanil and doxycycline should be started the day before travel. Drugs should then be taken with unfailing regularity for the duration stay in the area of malaria risk, and continued for 4 weeks after leaving the endemic area. The exception is atovaquone/proguanil, which can be stopped one week after leaving the area with malaria risk. Mefloquine prophylaxis should preferably be started 2-3 weeks before departure, so that adequate blood levels are attained, and adverse reactions can be detected before travel, allowing consideration of alternative drug regimens. Antimalarial drugs should be taken with food and swallowed with plenty of water.

1.5.3.2. Guiding Principles of Chemoprophylaxis in Pregnancy

Malaria infection in pregnancy poses a substantial risk to the mother, the fetus and the newborn infant. Pregnant women are less capable of coping with and clearing malaria infections. In areas of low transmission of *P. falciparum*, where levels of acquired immunity are low, women are susceptible to attacks of severe malaria, which may result in stillbirths or spontaneous abortions, or the death of the mother. In areas of high *P. falciparum* transmission, levels of acquired immunity tend to be high and women may have asymptomatic infections, which may result in maternal anemia and placental parasitemia. Both of these conditions can lead to low birth weight, an important contributor to neonatal mortality.

In programs for the prevention or treatment of malaria in pregnant women, two major issues are the safety and effectiveness of the antimalarial drug regimen. The programmatic effectiveness of a given drug is determined by the efficacy of that drug against the parasite and by the drug's characteristics, including affordability, availability, acceptability to the target population, and deliverability in terms of dosing requirements and incorporation into existing antenatal care delivery systems.

Chloroquine phosphate or hydroxychloroquine sulfate are recommended for pregnant women traveling to areas with chloroquine-sensitive *P. falciparum*. Chloroquine has not been found to have any harmful effects on the fetus when used in the recommended doses for weekly prophylaxis. Mefloquine is currently the only medication recommended for prophylaxis during pregnancy. Studies indicate that prophylactic use during second and third trimesters is not associated with adverse fetal or pregnancy outcomes. More limited data suggest it is also safe during the first trimester. Because of insufficient data regarding the use during pregnancy, atovaquone/proguanil is not currently recommended for the prevention of malaria in pregnant women. Doxycycline is contra-indicated during pregnancy because of the known risks of tetracycline on fetal development, including discoloration and dysplasia of the teeth and inhibition of bone growth. Primaquine should not be used during pregnancy because the drug may pass to a G6PD-deficient fetus and cause hemolytic anemia in utero.

1.5.3.3. Guiding Principles of Chemoprophylaxis in Children

Children are at special risk of malaria since they may rapidly become seriously ill. Persuading young children to take antimalarial medications may be difficult because of the lack of pediatric formulations and the bitter taste of many drugs. Furthermore, some chemoprophylactic drugs are contra-indicated in children. Chloroquine remains the drug of choice in areas where malaria remains sensitive to this drug, while mefloquine is the preferred agent in chloroquine-resistant areas. Although the manufacturer recommends that mefloquine should not be given to children who weigh less than 5 kg, it should be considered for chemoprophylaxis of all children at high risk of acquiring chloroquine-resistant *P. falciparum* malaria. Atovaquone-proguanil may be a safe and effective chemoprophylactic alternative to doxycyline for children under 8 years of age who weigh more than 11 kg and are travelling to mefloquine-resistant areas.

Children of any age can contract malaria; all children traveling to a malaria-risk area should take an antimalarial drug. The antimalarial drugs are available only in tablet form and may taste quite bitter. Pediatric dosages should be carefully calculated based on the child's current weight; children's dosages should never exceed adult dosage.

Full-service (compounding) pharmacists can pulverize tablets, weigh out the precise dose, and place the dose in a gelatin capsule. Advise parents to open the gelatin capsule and mix the drug with something sweet such as applesauce, chocolate syrup, or jelly and to give the drug on a full stomach to minimize stomach upset and vomiting. Parents should allow sufficient time before travel to allow preparation of these dosages.

In Africa, particularly in areas of high transmission, children under 5 years of age are the most affected by malaria, leading to a high case fatality rate in this age group. In spite of the importance of the disease in children and the fact that they are the major targets for antimalarial drugs, there are problems with existing pediatric formulations and regimens.

1.5.3.4. Guiding Principles of Chemoprophylaxis in Area with Severe Malaria

Treatment with antimalarial drugs has a major role to play in preventing severe malaria and death. It reduces fever promptly and effectively to interrupt the progression of infection or mild illness to severe disease, and reduces fatality rates in severe malaria. Since the great majority of patients with fever or other symptoms suggestive of malaria receive their initial treatment at home, improving home management of fever is a critical component of this process. Patients and their families need up-to-date and practical guidance on when and how to use antimalarial drugs at home, how to recognize when a patient is not responding to therapy in order to seek medical attention; and on the importance of correctly following, while at home, the recommendations and treatments that are given in health centers and hospitals. This guidance should be complemented with more user-friendly treatment regimens, improved formulations, especially for the treatment of children, and pre-packaging of antimalarial tablets.

At the health post or health center level, availability of effective drugs is crucial. Health workers need clear guidelines on how antimalarial drugs should be used and how to deal with potential adverse reactions. In addition, there should be facilities to administer fluids, glucose, antibiotics, and anticonvulsants to severely ill patients. In suspected cases of severe malaria, rectal formulations of the artemisinin drugs and other preparations can be used as an emergency pre-referral treatment when parenteral antimalarial therapy is not available, and have the potential to reduce early mortality. In hospitals, prompt confirmation of diagnosis, rapid assessment of the severity of disease and the administration of prompt specific and supportive treatment, including safe blood for transfusion, are all critical.

1.5.3.5. Guiding Principles of Chemoprophylaxis in Area with Vivax Malaria

P. vivax is the predominant malaria species in most of Asia (including the Indian subcontinent), Oceania, North Africa, and Central and South America and is estimated to account for about 55% of the total malaria incidence outside subtropical Africa. In recent years, there has been a major resurgence of *vivax* malaria in Eastern Europe and central Asia, areas which had been free of malaria for several decades. The major threat to the control of *P. vivax* today is the emergence and spread of chloroquine-resistant strains in Guyana, India, Indonesia (Irian Jaya), Myanmar, and Papua New Guinea [117]. The existence of strains of *P. vivax* that differ in their relapse patterns and their innate sensitivity to primaquine influences the choice of regimen for radical cure. For strains of *P. vivax* from Papua New Guinea, Solomon Islands and Vanuatu, and parts of Indonesia, a total dose of primaquine base of 7 mg/kg (equivalent to 420 mg in an adult) given as 30 mg of primaquine base daily for 14 days, is required to achieve 100% cure rates. Strains from China, South-East Asia, central

Asia, the Middle East, northern Africa, and Central and South America can be cured with half this dose. There is limited evidence that strains from the Indian subcontinent may respond to a 5-day course of 15 mg of primaquine base [120].

1.5.4. Revision of AMDPs for Chemotherapy and Chemoprophylaxis

The new global and national antimalarial drug policies (AMDPs) for chemotherapy and chemoprophylaxis updated in 2008 by World Health Organization are listed in Appendix 1. These malaria treatment guidelines recommend that antimalarial treatment policy should be changed at treatment failure rates considerably lower than those recommended previously. This major change reflects the availability of highly effective drugs, and the recognition both of the consequences of drug resistance, in terms of morbidity and mortality, and the importance of high cure rates in malaria control.

It is now recommended that a change of first-line treatment should be initiated if the total failure proportion exceeds 10%. However, it is acknowledged that a decision to change may be influenced by a number of additional factors, including the prevalence and geographical distribution of reported treatment failures, health service provider and/or patient dissatisfaction with the treatment, the political and economical context, and the availability of affordable alternatives to the commonly used treatment.

- In therapeutic efficacy assessments, the cure rate should be defined parasitologically, based on a minimum of 28 days of follow-up. Molecular genotyping using PCR technology should be used to distinguish recrudescent parasites from newly acquired infections.
- Review and change of the antimalarial treatment policy should be initiated when the cure rate with the current recommended medicine falls below 90% (as assessed through monitoring of therapeutic efficacy).
- A new recommended antimalarial medicine adopted as policy should have an average cure rate $\geq 95\%$ as assessed in clinical trials.

Changing antimalarial treatment policy in countries requires concerted action among all stakeholders and continuous stewardship by the Ministry of Health. The key evidence of the need for treatment policy change is the failing therapeutic efficacy of the antimalarial drugs in use, assessed according to standard WHO protocols. WHO's current recommendation is to change a treatment policy when the:

- Treatment failure of >10% (as assessed through monitoring of therapeutic efficacy at 28 days).
- Similarly, an antimalarial medicine should only be selected as a new policy treatment option only when the medicine has an average cure rate of > 95% as assessed in clinical trials.

Guiding principles for changing a treatment or prevention policy, which are sometimes based on parasitological or economic parameters, are often determined arbitrarily. Such suggested cut-off points are indicative, and decision makers at the national level should feel free to initiate change at any time. In the past, many countries were too slow or too cautious in deciding on policy change, despite a high failure rate with chloroquine reported in the field. The WHO malaria treatment guidelines from 2005 suggest that, with the introduction of more effective combination therapies, efficacy of an antimalarial treatment should reach 95% and that a policy change should be seriously considered if the efficacy is < 90% on day 28 [2]. Between 2001 and 2006, 65 countries (41 in Africa) have changed their policies. Although it might be clear from surveillance at sentinel sites that a first-line treatment is ineffective, the choice of a replacement drug or drug combination at consensus meetings has sometimes proved difficult. According to the 2001 WHO recommendations, an endemic country in which drug resistance to monotherapy is observed should change to a combination therapy based on artemisinin. The currently recommended options are artemether–lumefantrine, artesunate + amodiaquine, artesunate + SP, amodiaquine + SP, and artesunate + mefloquine [47].

Results of tests for therapeutic efficacy are the most important information for determining whether an antimalarial drug is still effective. Surveillance of therapeutic efficacy over time is an essential component of malaria control. All policy changes have been based on the results of *in vivo* tests. In South Africa, data on *in vitro* drug sensitivity were also taken into consideration when chloroquine was abandoned as a first-line drug. In Mali, the results of use of molecular markers were determinant in deciding to use sulfadoxine–pyrimethamine instead of chloroquine during an epidemic outbreak. *In vitro* tests and detection of genetic markers of resistance conducted over time can provide early warning of impending resistance before it becomes clinically apparent and can help guide therapeutic efficacy studies. These tests are also useful for monitoring changes over time in susceptibility to a drug that has been withdrawn. The usefulness of these tests has become evident with the ever increasing use of combination therapy. It is often impossible to conduct therapeutic efficacy tests for each component, owing to ethical problems, non-availability of the drug as a single therapy, and the need to study a large number of patients. *In vitro* tests can be used to monitor susceptibility to each drug in a combination [121].

Chapter II

Antimalarial Drugs Cost and Cost-Effectiveness

Economics is the study of the allocation of scarce resources among competing ends. It is not surprising therefore, that economic considerations should loom large in health policy, including the provision of effective pharmaceuticals. It was clear before the most recent IOM Committee met that the existing antimalarial drug supply was starting to fail. For more than 40 years, the system had been largely based on a single agent – chloroquine - which was at one time very effective and remarkably cheap. Even in the poorest countries, at 10 cents per course (retail), most people can still afford it. Moreover, the drug is familiar to the populace, and has been used - both within and outside of organized health care systems - well enough to prevent many malaria deaths and suppress (if not completely cure) acute attacks of the disease. Over time, however, resistance to chloroquine emerged worldwide, first leading to treatment failures in Southeast Asia, then to treatment failures in large parts of east Africa. It is now believed that chloroquine will be useless against most life-threatening *falciparum* malaria infections in fairly short order. In the meantime, replacement artemisinins have been introduced, but there is the fear that they too will quickly lose ground. Over the last 25 years, artemisinin derivatives have proved highly effective in Asian and African countries while no definitive artemisinin resistance has surfaced. Partnering artemisinin-based combination therapy (ACT) confers even greater protection against the development of drug-resistant mutants, as well as offering therapeutic advantages over single antimalarial drugs [2, 44, 47].

Unfortunately, ACT is relatively, though not absolutely, expensive. At present, ACTs cost about US$2 a treatment, roughly 20 times the price of chloroquine. The historical slowness in producing new antimalarials reflects the way in which new drugs are developed in a market driven system. Research, development, and testing of drugs engender a large upfront cost. Economic evaluation of a new antimalarial treatment requires an analysis of its respective costs and benefits, or at least a comparison of its reduction of malaria morbidity and mortality vs. other therapies. In the case of ACTs, the costs were high and the relative efficacy was so high that inquiring into the benefits in greater detail was reasonably cost-effectiveness.

2.1. The Cost of Antimalarial Drugs in Use

2.1.1. Current Malaria Situation and Antimalarial Drug Choice

Recent estimates of the global malaria burden have shown increasing levels of malaria morbidity and mortality, reflecting the deterioration of the malaria situation in Africa during the 1990s. About 80% of all malaria deaths occur in Africa south of the Sahara, and the great majority of them in children under five. Key among the factors contributing to the increasing malaria mortality and morbidity is the widespread resistance of *P. falciparum* to conventional antimalarial drugs, such as chloroquine, sulfadoxine–pyrimethamine (SP), and amodiaquine. Multidrug-resistant *falciparum* malaria is widely prevalent in south-east Asia and South America. Now Africa, the continent with highest burden of malaria, is also affected.

For more than 50 years, chloroquine silently saved millions of lives and cured billions of debilitating episodes of malaria. *Falciparum* malaria has always been a major cause of death and disability—particularly in Africa—but chloroquine offered a measure of control even in the worst-affected regions. At roughly 10 cents a course and readily available from drug peddlers, shops, and clinics, it reached even those who had little contact with formal health care.

Resistance to inexpensive monotherapies such as chloroquine and SP has developed or is developing rapidly, with increased mortality as a result. The inappropriate use of antimalarial drugs during the past century has contributed to the current situation: antimalarial drugs were deployed on a large scale, always as monotherapies, introduced in sequence, and were generally poorly managed in that their use was continued despite unacceptably high levels of resistance. Chief among these new treatments are the "artemisinins," paradoxically both an ancient and a modern class of drug that can still cure any form of human malaria. If *falciparum* malaria sufferers had the same broad access to artemisinins that currently exists for chloroquine, the rising burden of malaria in the world today would halt or reverse.

The crisis is both economic and biomedical. On the economic side, to state the obvious, the era of cheap and effective antimalarial treatment may have ended, but poverty in sub-Saharan Africa and malarious countries elsewhere, has not. Thus, although artemisinins at a dollar or two per course are both inexpensive by U.S. or European standards and highly cost-effective by any norm, neither national governments nor consumers in most malaria-endemic countries can afford them in quantities that remotely approach the world's current consumption of chloroquine—roughly 300-500 million courses of treatment per year. Biomedically, today's malaria situation also is precarious. The artemisinins are the only first-line antimalarial drugs appropriate for widespread use that still work against all chloroquine-resistant malaria parasites. If resistance to artemisinins is allowed to develop and spread before replacement drugs are at hand, malaria's toll could rise even higher.

Over the past decade, a new group of antimalarials – the artemisinin compounds, especially artesunate, artemether, and dihydroartemisinin – have been deployed on an increasingly large scale. These compounds produce a very rapid therapeutic response (reduction of the parasite biomass and resolution of symptoms), are active against multidrug resistant *P. falciparum*, are well tolerated by the patients and reduce gametocyte carriage (and thus have the potential to reduce transmission of malaria). To date, no resistance to artemisinin or artemisinin derivatives has been reported, although some decrease in sensitivity *in vitro* has been detected in China and Vietnam [47]. If used alone, the artemisinins will cure *falciparum* malaria in 7 days, but studies have shown that in combination with certain synthetic drugs they produce high cure rates in 3 days with higher adherence to treatment. Furthermore, there is some evidence that use of such combinations in areas with low to moderate transmission can retard the development of resistance to the partner drug [63].

In the case of any antimalarial drug, the new development of drug resistance is a rare event: a chance genetic change in a single parasite in a single patient. But once that single malaria parasite generates to multiple descendants, the math will be changed. Now, mosquitoes can acquire resistant parasites from a single individual and transmit them to other people. The subsequent spread of a robust, resistant clone would be similar to the spread of any malaria strain. The way around this dismal scenario derives from the fact that a nascent resistant parasite will not survive and proliferate in an infected person's bloodstream if a second effective antimalarial drug with a different mechanism of action is simultaneously present. The key, therefore, to preserving the artemisinins is to eliminate their routine use as "monotherapies," and to treat patients with uncomplicated malaria (the vast majority of cases) with ACTs instead. ACTs equal an artemisinin derivative plus another unrelated antimalarial drug, ideally co-formulated in a single pill so that individual drugs cannot be knowingly or inadvertently used as monotherapies. Combining drugs in this way - already an accepted practice in the treatment of tuberculosis and HIV/AIDS infection - minimizes the likelihood that a single parasite with drug resistance will propagate and spread.

Although still relatively new to malaria, combination therapy is no longer controversial: within just the past few years, it has been endorsed by the World Health Organization, the Global Fund to fight AIDS, Tuberculosis and Malaria, and by malaria experts worldwide. In addition, combinations are not only good for preventing resistance; they also confer advantages to patients. In the case of the artemisinins, the treatment regimen is shortened from 7 days using monotherapy to 3 days using a combination [2].

In order for antimalarial combination treatment to keep drug resistance at bay over the long term, however, it must be used as first-line treatment for uncomplicated *falciparum* malaria as widely as possible. Should monotherapies persist in some locales and drug resistance result, there would be no way to contain the resistant parasites to one country or continent. Artemisinin monotherapy is, in fact, widely used in Asia today in the same areas where resistance to chloroquine and SP originally surfaced not that long ago. There is an urgent need for Asian-based production of artemisinin monotherapy to convert to ACT production, aided by a global subsidy. The window of opportunity to create a global public good - years of extended effective antimalarial drug life - is open now, but it may not remain open very long.

The equal urgency for change in Asia and Africa may seem counter-intuitive but antimalarial drug resistance historically has emerged in areas of low transmission—mainly Southeast Asia and South America—and then spread to high-transmission areas, mainly Africa. This pattern fits with current understanding of the biology of drug resistance in low- and high-transmission areas. As a practical matter, it is therefore essential that ACTs quickly dominate the market in low-transmission areas in Asia and Africa, as well as penetrate Africa's most endemic locales [42].

As a response to increasing levels of resistance to antimalarial medicines, WHO recommends that all countries experiencing resistance to conventional monotherapies, such as chloroquine, amodiaquine, or sulfadoxine–pyrimethamine, should use combination therapies;

preferably, those that contain artemisinin derivatives (ACTs – artemisinin-based combination therapies) for *falciparum* malaria.

As yet another step towards combating drug resistance in Africa, WHO has lowered the resistance-threshold recommended for treatment policy change from 25% to 10% as assessed by standard WHO protocols in children less than 5 years of age , meaning that a more effective treatment should be adopted when the proportion of treatment failures to the old treatment reaches 10%.

WHO currently recommends the following combination therapies (in alphabetical order):

1) artemether/lumefantrine
2) artesunate plus amodiaquine
3) artesunate plus mefloquine
4) artesunate plus sulfadoxine/pyrimethamine.

Note: Amodiaquine plus sulphadoxine-pyrimethamine may be considered as an interim option where ACTs cannot be made available, provided that efficacy of both is high [63].

2.1.2. The Economics and Affordability of Artemisinins and ACTs

WHO provides technical cooperation to ministries of health on all aspects of national treatment policy change – monitoring the therapeutic efficacy of medicines, updating, implementing and monitoring ACT-based treatment policies. The countries that are currently deploying ACT in the general health services to some extent are listed in Table 5. Since 2001, a total of 67 countries have adopted one of the WHO recommended artemisinin-based combination therapies [47], several as first-line treatment and a few as second-line in 2006 [55].

By Western standards, artemisinin derivatives are already inexpensive. They also are well within the bounds of what is internationally acknowledged as cost effective for health care interventions in low-income countries. Incentives of a secure and large market, producers already are promising that wholesale prices for a course of ACTs will fall to US$ 0.50-1.00 (or possibly lower) within 2 years. However, that projected price is still 5 to 10 times higher than the price of chloroquine or SP in Africa, making artemisinins essentially unaffordable for most sub-Saharan governments and individuals, especially the poor, rural families whose children are most likely to die from malaria.

Without substantial and sustained subsidies, in some countries well over half of all malaria patients needing treatment - including most who currently are able to acquire chloroquine and SP - will not have access to artemisinin-based drugs. Children will die and many more will suffer. Once scaled up, the additional cost of ACTs versus currently failing drugs for the world is expected be US$ 300-500 million per year (a closer approximation is difficult because estimates of malaria cases - or fevers assumed to be malaria - are imprecise). This cost estimate assumes that ACTs will be used to treat up to half a billion episodes per year, which roughly equals the number currently treated by chloroquine or SP. The estimate also assumes a competitive market in which ACT prices fall over time.

In the meantime, there is a chicken-and-egg dilemma surrounding ACT supply and demand. Without an assured market, potential manufacturers will not commit to adequate

ACT production, nor will farmers expand the cultivation of *Artemisia annua*, the source plant. There is a critical need to jump-start ACT production. To do this, the global community must provide sufficient funds to encourage investments by manufacturers guarantee purchases of ACTs and generally stimulate a robust world market. This must be done through a visible, centralized mechanism, ideally using existing national and international organizations (e.g., UNICEF, WHO), which can quickly take on the task.

In one sense, the economics of antimalarial drugs are simple: currently effective drugs - those recommended by the global community for most of Africa and Asia - are not affordable by the endemic countries, so few people are getting them. This is in contrast to the previous generation of effective drugs that were exceedingly cheap by any standards, but which no longer work because drug resistance is so widespread. As a result, malaria mortality rates are rising in Africa, and more cases are inadequately treated. Again, chloroquine - the standard, effective drug for decades - costs about 10 U.S. cents per course of treatment for an adult. The new effective drugs—artemisinin combination therapy (ACT) - today cost US$2.40 per course wholesale, and can be marked up to five times that amount in pharmacies in Africa. There are limited alternatives to ACTs for effectiveness and safety for widespread use, and their adoption is recommended by WHO and virtually all other expert bodies. Even 10 cents is too much for the poorest of the poor in every endemic country.

The rationale for combining two or more drugs is that doing so dramatically reduces the odds that malaria strains resistant to any of the drugs in the combination would survive to be transmitted. In this report, ACT is used mainly to describe a drug "co-formulation" (i.e., two drugs in one pill), and not just two separate drugs taken at the same time. Multiply the per-dose difference in price by the millions or tens of millions courses used per year in each country in Africa, and the money adds up. Continent-wide, an estimated 200-400 million courses of treatment are used per year, and an additional 100 million courses in the rest of the world. Realistically, the endemic countries are able to contribute very little of the incremental cost, and that is not likely to change for many years, given the pace of economic development in Africa and in poor nations elsewhere. One positive change that is almost certain, however, is that several different ACTs will become available, and their price will come down by more than half, to about 10 times the current price of chloroquine, or about US$1 per adult course and US$0.60 for an average child's course as current affordability of artemisinins and ACTs [55].

2.1.3. The Cost of Antimalarial Drugs in Use

ACTs cost higher than most currently used monotherapies and alternative non-artemisinin-based combinations. Most cost up to 20 times that of conventional monotherapies and in view of the current high demand than supply, the costs are likely to remain high for several years to come. Approaches to facilitate program implementation include preferential reduction of market price for ACTs to developing countries and subsidizing implementation costs through bilateral financial assistance. The current prices of antimalarial drugs in use are listed in Table 10.

Table 10. Cost, convenience, cost-effectiveness, and primary clinical application of oral antimalarial therapies

Therapy	Cost ($)*	Doses	Duration of Therapy	Dosage mg	Cost-effectiveness	Application	WHO Status
Chloroquine	0.12	Daily	3 days	250 x 6	No in most regions	BS schizonticide	P
Sulfadoxine-pyrimethamine	0.12-0.24	Daily	3 days	500+25 x 6	No in most regions	BS schizonticide	C E E
Quinine	0.21-0.49	Daily	7 days	300 x 7	No with shortcoming	BS schizonticide	N C
Mefloquine	1.60-2.55	Daily	1 day	250 x 4	No in Asian region	BS schizonticide	P E
Atovaquone-proguanil	58.27	Daily	3 days	)+100 x4	Yes	BS schizonticide	P E
Artesunate	1.02-2.00	Daily	2 days	50 x 12 50 x 12	Yes	BS schizonticide	E E E
Artemether	3.96	Daily	2 days	20+120 x	Yes	BS schizonticide	
Artemether-lumefantrine	3.87-8.04	Daily	3 days	24 50x6, 250x2100	Yes	BS schizonticide, gametocytocide	
Artesunate-mefloquineArtesu nate-SP	2.10	Daily	1 day	x3, 500/25x3	Yes	BS schizonticide, gametocytocideBS schizonticide,	
	1.45-2.50	Daily	3 days	50x6, 153x640x 6, 320x6	Yes	gametocytocideBS schizonticide,	
Artesunate-amodiaquine	1.27-1.90	Daily	3 days	15 x 7-14	Yes	gametocytocide BS schizonticide, gametocytocide	
DHA-piperaquine							
Primaquine	1.00*	Daily	3 days		Yes	LS schizonticide, gametocytocide	
	0.14-1.20	Daily	1-2 weeks		Yes	gametocytocide	

Unless otherwise indicated, the cost shown is the cost, in 2007 U.S. dollars, of medication for one adult treatment regimen, purchased in bulk, according to the International Drug Price Indicator Guide (IDPIG) in 2007 (http://erc.msh.org/mainpage.cfm?file=1.0.htm&module=dmp&language=English). SP = sulfadoxine-Pyrimethamine; DHA = dihydroartemisinin; BS = blood stage; LS = liver stage. *The price is according to Myint et al. 2007 [432].

Most developing nations license and make available only the antimalarial drugs that are provided through national health programs. This approach often excludes relatively expensive or risky therapies, even for patients who may be able to afford a given drug and have access to medical supervision. The main factor affecting availability is economic — the ability or inability to purchase a drug for broad distribution and the potential reluctance to use an agent because of an inability either to screen users or to monitor its quality.

Chloroquine and sulfadoxine–pyrimethamine cost less than 15 percent of the cost of the least expensive alternative agents and approximately 1 to 2 percent of the cost of many of the agents marketed in the developed world (Table 10). The developing world requires distribution strategies for effective therapies that overcome the availability of cheap but ineffective drugs. The U.S. National Academies made a core recommendation [42] that governments and international financial institutions should commit $300 to $500 million per year within five years to subsidize artemisinin-combined therapies (an emerging family of antimalarial drugs, including artemether, artesunate, and dihydroartemisinin) to achieve prices for the end user in the range of 10 to 20 cents. Unless this recommendation is followed, market forces will continue to drive the use of ineffective therapies [121].

In this consideration, negotiations between WHO and Norvatis Pharma has reduced the cost of artemether–lumefantrine (Coartem) to around US$ 0.9–1.4 for a treatment course of a child up to 7 years old and around US$ 2.4 per adult treatment dose [106]. Despite this reduction, prices are still high for most households. For example, in Tanzania where the mean monthly income of rural households is only US$ 13.4 and US$ 29.0 for urban households outside Dar-es-Salaam, one full treatment course for a child takes 1.4/13.4 (10%) of the monthly income. On average, one child may have four malaria episodes in a year and another four episodes that may be wrongly treated as malaria leading to a significant drain on the family income. Money is the main barrier in this malaria fight, as correctly pointed out in the Malaria World Report released by WHO and UNICEF on May 3, 2005. Unless cost is sufficiently addressed both locally and internationally, deploying ACTs could lead to adverse results ranging from an increase in potentially fatal delays in infected people presenting to medical services to systematically excluding poorest malaria sufferers from receiving treatment [122]. This unfortunate scenario should be prevented by the developed world taking responsibility for the humanitarian disaster. Proven interventions like ACTs and insecticide-treated nets (ITNs) should reach many more people in order to register a real impact and roll back malaria.

The Global Fund to fight AIDS, Tuberculosis and Malaria (GFATM) has taken lead in granting funds to countries to implement ACT policies. To function well, it needs constant injection of funds from funding agencies, independence from politics of big donors and sound scientific advice on disease control. These have contributed to a slow start of the Fund. To date, early deficiencies within WHO and other RBM partners in giving appropriate technical advice to CCMs, and also deficiencies within the Fund's Technical Review Panel that led to funding some ineffective chloroquine AMDPs (either singly or in combination with SP) have been addressed and the Fund has gained momentum. However, delays in cash flow to countries after approval of grants remains a major issue that require immediate solution from all concerned (The Fund, WHO and RBM Partners, CCMs and National Malaria Control Programmes) to facilitate quick ACT policy implementation [72].

In addition, direct drug costs and operational costs for policy implementations are high and probably the biggest obstacle in convincing policy-makers in poor developing countries to adopt ACTs without assured sources of funding. In the human capital approach, the immediate costs of treating or preventing an episode of illness consist of the "direct" costs—money spent for malaria prevention and treatment, including, for example, the cost of bed nets and mosquito repellents in the first instance, and of consulting a health provider, buying drugs, and paying for transportation, in the second. Indirect costs represent loss of income (or productive labor, even if the benefits are not monetized, e.g., lost agricultural production because of an inability to plant or harvest crops) due to illness. Indirect costs also include periods of being too sick to work, time spent caring for others who are sick, and time spent seeking care. Some drawbacks of the method, as used, are discussed after a review of the findings of recent studies.

2.1.3.1. Drawback in the Human Capital Method for Estimating Costs of Malaria

The studies reported here vary widely in the details of the methodology, their sources of data, and in their perspectives. Comprehensive studies of this type are difficult to conceive and to carry out, given inherent limits on data available in the places most affected, general difficulties in carrying out research in such places, and a lack of funding for such studies. Even when done well, however, the methods often require assumptions about such things as employment levels (which affect whether wages are actually lost), the value of leisure time, and substitutability of work by other families. Some of the aspects not generally captured by the human capital method relate to coping strategies adopted by families, which affect their economic well-being in ways not usually measured through direct and indirect costs usually measured.

2.1.4. The Cost of Research and Development of Antimalarial drugs

Managed well, artemisinins could remain the first-line antimalarial for many decades. Resistance has not yet developed in Asia, even following extensive monotherapy use. Eventually, however, the artemisinins will begin to lose effectiveness and new drugs will be needed. Even before that happens, drugs that can be manufactured more cheaply, or are superior to artemisinins (e.g., effective in a single dose) could supplant or join artemisinins on the front line.

Table 11. ACT procurement from 2001 to 2007 [1]

Therapies	2001	2002	2003	2004	2005	2006	2007
AS	155,800	39,120	1,682,600	1,540,200	5,617,800		
A/L	227,503	82,873	1,339,346	3,626,670	10,468,712	61,406,282	94,000,000
AS + AQ					10,048,719	14,446,246	20,000,000
AS + SP					2,298,000	6,671,212	10,500,000
AS + MQ						251,000	300,000
Total ACTs	227,503	82,873	1,339,346	3,626,670	22,815,431	82,774,740	124,000,000

AS = artesunate; A/L = artemether-lumefantrine; AS + AQ = artesunate-amodiaquine; AS + SP = artesunate-sulfadoxine-pyrimethamine; AS + MQ = artesunate-mefloquine.

For the first time in decades, the R&D pipeline for antimalarials has been invigorated through the Medicines for Malaria Venture (MMV), a public-private partnership begun in 1998. MMV, the WHO Special Programme on Research and Training in Tropical Diseases (TDR) and the Walter Reed Army Institute of Research (WRAIR) are the main innovators in antimalarials. Current annual funding needs for these organizations total an estimated US$60 million dollars, rising another US$20 million per year as more drugs reach the stage of clinical and field trials. These projected investments represent only a small addition to the cost of supplying artemisinins globally, but are several times the current collective budgets. Agencies supporting malaria control should allocate sufficient funds to these organizations to meet the MMV goal of one new antimalarial drug every 5 years, and WRAIR's similar target [42].

2.1.5. The Cost of Producing ACTs and Future Prices

Currently, 70 out of 78 countries in need have adopted ACTs in their treatment policies, among them, 45 are deploying ACTs mainly due to a general lag time of 12 to 24 months between the country's ACT adoption and the implementation. All new antimalarial products expected until the year 2011 are ACTs. Procurement of ACTs for the public sector on a global level was relatively low in 2001 and 2002 (Table 11). It started to increase slightly in 2003 (1.3 million treatments) and 2004 (3.6 million treatments). However, the total need of artemether-lumefantrine in 2004 could not be satisfied, mainly due to inadequate supply of artemisinin in relation to increasing demand. In 2005, a strong increase in the procurement of ACTs was noticeable, that nearly quadrupled in 2006. For 2007 are expected to sum up to a total of approx. 124 million treatment courses [1].

ACTs remain a small share of the total antimalarial market. In 2006, the total volume of demand was estimated at 546 million treatments, of which 140 million treatment courses went to the public (ACTs: 82–90 million) and 406 million to the private sector (ACTs: 8–10 million; artemisinin-based monotherapies: 25 million).

Following the global shortage of Coartem® in 2004 and the consequent sharp increase in the price of artemisinin (from US$ 230/kg in 2003 to US$ 1,100/kg in 2004–2005) there has been a major expansion of plantations (Figure 6) and of artemisinin producers in China (10 producers in 2004, more than 80 in 2006), Vietnam (3 producers in 2004, 20 in 2006) and East Africa in 2006 [1]. The mismatch between increased offer and tempered demand for ACTs in the public sector (to which also the low approval rates in GFATM rounds 5 and 6 have contributed), created a major reduction in the price of artemisinin (currently at US$ 180/kg, below the production costs). The consequent withdrawal of farmers and artemisinin producers from this market will create a reduction of the artemisinin inventory for 2007, with high risk of ACT shortages in 2008–2009.

The price of artemisinin is influenced by many different factors. To make cultivation of the *Artemisia annua* plant still profitable for growers, prices for dry leaves should not go beyond certain levels. Likewise, the content of artemisinin substance in the leaves may vary due to agronomic factors, and it is also affected by storage. Artemisinin costs for extraction from the dry leaves, as well as its purification require investments from the manufacturers in facilities and equipment. Processes should be efficient to guarantee yield and a high level of quality.

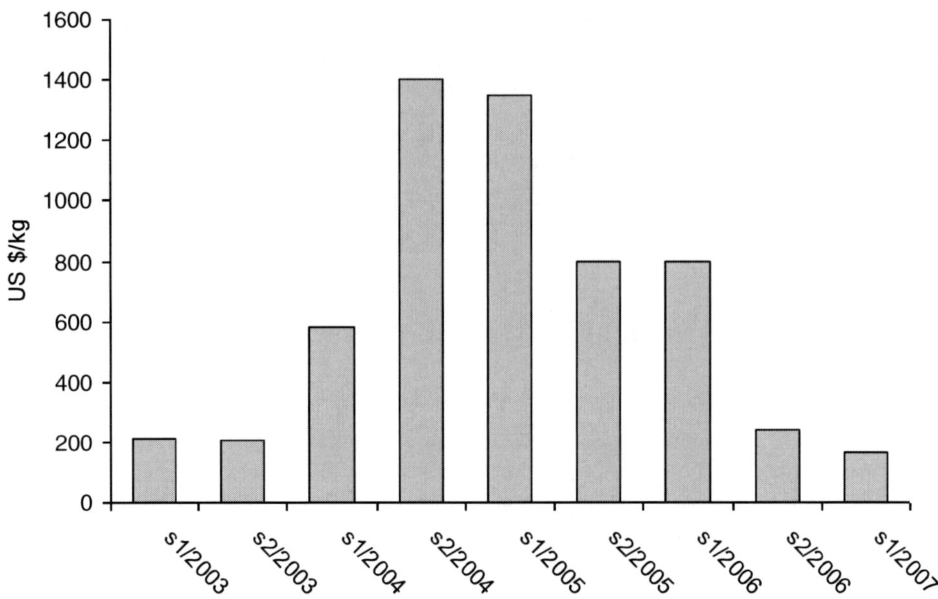

Figure 6. Artemisinin prices (US $/kg) over time [1].

The extraction from *Artemisia annua* plants will remain during the next years; before 2011, at the earliest, no new sources of artemisinin for ACTs are expected. Results of the studies on high yield seeds are not expected before 2010. Investment in new extraction processes should be required, as the efficiency and safety of current extraction is not satisfactory with the present conditions. Overstock and low prices in 2007 may lead to further withdrawals from producers and thus the risk of a new shortage in the next coming years. The implementation of high quality artemisinin safety stocks is needed. Prequalification, not only on the production but also on the API producers' level, should be strengthened to guarantee good quality and fair competition between the producers. To minimize financial risks, reliable forecasts before the planting season, covering two years, are needed, to allow ACTs producers to contract with upstream.

In 2008, several new fixed-dose combinations and other co-blistered ACTs will appear in the market: The Drugs for Neglected Disease Initiative (DNDi) is developing artesunate-amodiaquine and artesunate-mefloquine combinations and the Medicines for Malaria Venture (MMV) is developing pyronaridine-artesunate, chloroproguanil-dapsone-artesunate (CDA), and dihydroartemisinin-piperaquine (results likely in the end of 2007 till the beginning of 2009). By 2008–2009, fixed-dose combinations (artesunate-naphthoquine, artesunate-piperaquine) are also expected to be available in the market at internationally recognized regulatory standards [1].

Why are ACTs so much more expensive than chloroquine? Even at prices anticipated after the massive scale-up that would be necessary to supply the African market—should funding become available for large-scale purchase—the price as sold by the producer will likely remain at about 10 times the price of chloroquine (Table 12). These prices reflect the production costs of artemisinins without a premium for any exclusivity related to patents. They are high because the process involves growing the source plant, *Artemisia annua*, extracting the active moiety, and creating the desired artemisinin derivative. Co-formulation with the companion drug follows. The price of the finished product is driven mainly by the cost of the artemisinin derivative, but also is affected by the companion drug (e.g., lumefantrine, the companion drug in Coartem, is a relatively expensive drug on its own, which contributes to the high price of the co-formulation) [42].

There is something of a chicken-and-egg quality about the current price of artemisinins, and the prospect for significantly lower prices. Lower prices can be expected in response to large-scale demand, which, in turn, will induce competition among producers. Coartem has some patent protection, but Novartis is selling it in developing countries for less than production costs, i.e., at a loss. The real price breakthrough will likely occur only when a fully synthetic artemisinin is developed, eliminating the growing and extraction process. The Medicines for Malaria Venture (MMV) has such a compound under development, which they predict could be available in 5-6 years. The ultimate price is not known, but it should be significantly less expensive than current artemisinins (assuming no premium for exclusivity). If the synthetic product is better than, or at least as effective as the extracted ones, the market would change dramatically. A global subsidy might still be needed, but it could be less than what is needed now.

2.2. Cost Effectiveness of Artemisinins and ACTs

2.2.1. Conception and Determination of Cost-effectiveness

Antimalarial drugs are not the only pressing health care need in sub-Saharan Africa, where AIDS, tuberculosis, and a host of other infectious and other diseases also claim lives prematurely. And resources allocated to malaria can be spent on bed nets, indoor spraying, or other measures, and not necessarily on antimalarial drugs. Thus drugs must compete for both public and private funds on grounds of their value for the dollars spent: their cost-effectiveness. All link cost-effectiveness judgments to some measure of national wealth. Cost-effectiveness is a relative concept: what is considered a bargain in a high-income country may be completely out of range for a low-income country.

The question most relevant to this study is the cost-effectiveness of treating versus not treating acute episodes of malaria (the latter includes using an ineffective drug, as well as no treatment). In the short term (possibly the next 5 years), it also is relevant to ask whether it would be beneficial to switch from failing but affordable chloroquine, to an intermediate drug that is as inexpensive as chloroquine, but which can only last a few years because of rapidly evolving resistance. In this scenario, a switch to more expensive ACTs would be expected within a few years, but in the meantime, considerable funds might be saved. The only drug that currently meets the conditions for such an intermediate is SP, but the analysis might apply in the future to other drugs. We look at both the cost-effectiveness of treatment versus no treatment, and of switching to a stopgap drug and then to ACTs.

Table 12. Wholesale prices for artesunate and ACTs [42]

Drug	Range of prices for standard adult dose
Artesunate	Asian suppliers:
	US$0.50 (not yet in production); less for greater quantities
	US$0.63; 15% discount for greater than 1 million treatments
	US$1.25 (not yet in production)
	US$1.26; decreasing to US$1.01 at 3 million treatments
	US$1.35
	European suppliers:
	US$2.68 (2.10 Euros)
	US$2.42 (1.90 Euros)
	African suppliers:
	US$5.36
Coartem	US$2.40 for public services of developing countries (tiered pricing; higher for all other buyers)
CV8 (8 tablets)	US$1.21 for <1 million treatments; US$0.97 for > 1 million treatments
Artekin II	US$1.00 at retail
AS + AQ blister	Asian suppliers:
	US$1.50 (not yet in production)
	<US$1.91 (1.50 Euros) (not yet in production)
	European suppliers:
	US$1.53 (1.2 Euros)
	US$1.91 (1.5 Euros)
	US$2.68 (2.10 Euros) for<500,000 treatments; US$1.39 (1.09 Euros) for >500,000 treatments
AS + SP blister (6 tablets)	European suppliers:
	US$2.42 (1.90 Euros) for<500,000 treatments; US$1.24 (0.97 Euros) for >500,000 treatments

AQ = Amodiaquine; AS = Artesunate; SP = Sulfadoxine-Pyrimethamine; CV8 = a new combination of dihydroartemisinin, piperaquine, trimethoprim and primaquine.
Data From: J.M. Kindermans, Médecins Sans Frontières, 2003.

ACTs, manufacturers will not have the incentive to scale up production, and prices will not drop. In addition to competition driving prices down (i.e., companies being willing to accept lower profits per dose with higher volumes), technological improvements in the process could bring down the actual production costs, which would be passed along to purchasers in a competitive environment

2.2.1.1. Are Benchmarks for Cost-effectiveness?

There are no firm benchmarks or cutoff points to separate interventions that are deemed worth the cost ("cost-effective") and those that are not, but there are some guidelines. At one end, a 1996 WHO Ad Hoc Committee recommended that, for low-income countries with a GDP per capita of less than $765, interventions that cost $25 or less per disability-adjusted life-year (DALY, *which is a summary measure of disease burden that incorporates years lost by premature death with years of life lived with disability. It allows comparison of the influence of different diseases whose effects on mortality, morbidity, and disability differ. One DALY can be considered as one year of "healthy" life*) were "highly attractive," and those costing $150 or less per DALY were "attractive" (all figures in 1996 dollars) [123]. In 2001, the Commission on Macroeconomics and Health suggested that interventions costing less than three times GDP per capita for each DALY averted represent good value for money [124]. WHO's World Health Report for 2002 built on this principle, defining "very cost-effective" interventions as those that cost less than GDP per capita to avert each additional DALY, and "cost-effective" interventions as those where each DALY averted costs between one and three times GDP per capita [125].

2.2.2. Measuring and Comparing Effectiveness

It bears noting that cost-effectiveness analysis is an inexact science; more so in cases, such as this one, where many assumptions must be made on scant data. Uncertainty bounds are very wide. At its best, cost-effectiveness can help to define the relative positions of interventions by grouping them according to reasonably similar cost-effectiveness. Then, broad policy questions related to health care spending can be addressed, e.g., why are some interventions supported that cost more than three times GDP per capita, while others that appear to be better buys are not? One important aspect, relevant to ACTs, is that cost-effectiveness analysis does not address the total cost of an intervention for a country. Something that is very cost-effective − because it costs little and works well − could be very expensive per capita if a large proportion of the population needs it, which is the case for malaria treatment. It may not be affordable without additional resources, even though it is desirable and relatively cost-effective.

There are two simplified models to examine the consequences of treating versus not treating (i.e., treating with a wholly ineffective antimalarial): switching drugs once - directly to ACTs - versus twice - to a less expensive but less durable alternative first, then to ACTs. In the analysis of treating versus not treating, the evolution of resistance is not introduced (although the model from which this was adapted does include resistance), the aim being to define a basic level of benefit from effective treatment. The second analysis (one change versus two) is based on a different model, which does, by necessity, take into account the evolution of resistance. It also incorporates the effects of partial coverage within a population. Neither analysis attempts to model large-scale spatial effects if certain countries adopted effective antimalarials but others did not.

The model used to generate cost-effectiveness estimates for malaria treatment with ACTs was adapted from an earlier model looking at other aspects of malaria treatment (Table 13). Here, study compare ACT treatment with no treatment under two sets of treatment conditions. Treatment is either presumptive (i.e., everyone with fever suspected to be malaria is treated) or only after diagnosis of symptomatic individuals with a "hypothetical rapid diagnostic test" (RDT, 100% accurate). Under the assumptions of this model, treatment with ACTs saves lives, and at a cost that is well under even the low cutoff for a "highly attractive" or "very cost-effective" intervention for children under 5, at less than US$8 per DALY with presumptive treatment, and US$6.23 with an RDT. A life saved for US$209 or US$171 (without and with RDT) is also considered a bargain. For the over-5s, the answer from this model is not as clear-cut, though the point estimate of the cost-effectiveness ratio still falls within the "attractive" or "very cost-effective" range, at US$112 per DALY with presumptive treatment, and US$81.61 with RDT. Saving the life of someone over 5 (the model uses "adult" assumptions for this whole group, so this really applies to the adult population) costs about US$2,800 with presumptive treatment, and US$2,027 with RDT. It is important to note that these benefits apply only to reduced mortality, except for a very small portion for permanent disability caused by severe malaria. The model does not give any value to reduced morbidity; while the benefit per case may be small, for adults, the loss of a few days to 2 weeks of work are not trivial financially, and in the aggregate, can be very large [42].

**Table 13. ACTs versus no treatment: Key points in simplified model
of the cost-effectiveness analysis [42]**

Assumptions	
Patients	45% of people coming to the clinic with fever have malaria. The average age of patients under 5 years is 2 years. The average age of patients over 5 years is 27 years
Natural History of Malaria	Without treatment, 5% (3-7%) of malaria cases in under-5s, and 1% (0.5-1.5%) in over-5s progress to severe malaria. 50% of under-5s (40-60%), and 25% (20-30%) of over-5s with severe malaria die from the episode. 1.3% of under-5s (0.4-2.2%), and 0.5% (0.25-0.75%) of over-5s with severe malaria develop neurological squeal.
Effectiveness of RDTs and ACTs	RDT has perfect accuracy (no cases missed or misdiagnosed). ACT is 100% effective if a full course is taken—no resistance has yet developed. 70% (the midpoint of the estimated range of 60-80%) of patients take the full course. If less than the full course is taken, the patient will be "cured" (an "adequate clinical response") 20% (10-30%) of the time; 80% (70-90%) of the time the patient will not be cured and will either have fever >37.5°C or detectable malaria in the blood 14 days after treatment.
Costs	International wholesale cost of a course of ACTs: US$0.50 for a child, US$1 for an adult. Cost of transporting the drug from producer to clinic: 20% of drug cost (US$0.10 for a child course; US$0.20 for an adult course). Clinic cost (staff, equipment, and capital): US$1.14 per patient. Cost of one RDT: US$0.50
Effects of ACTs	Disability-adjusted life-years (DALYs) averted for each death avoided by treatment. This is roughly the difference between dying from the episode and living out an average lifespan, plus a small number of years lived with a disability for those who would have had long-term neurological effects from severe malaria. Separate calculations were done for under-5s and over-5s. A discount rate of 3% was applied (so future "years" are considered less important and add progressively smaller increments to the total). DALYs averted for each under-5 death avoided: 27.5; DALYs averted for each over-5 death avoided: 24.9

Key Results		Presumptive Treatment	With Diagnosis by RDT
Under 5 years of age	Effect per person (DALYs)	0.262	0.262
	Costs per person	US$ 2	US$ 1.64
	Cost-effectiveness ratio (CER) per DALY	US$ 7.62	US$ 6.23
	CER per death averted	US$ 209.23	US$ 171.21
Over 5 years of age	Effect per person (DALYs)	0.023	0.023
	Costs per person	US$ 2.60	US$ 1.90
	CER per DALY	US$ 112.11	US$ 81.61
	CER per death averted	US$ 2,784.16	US$ 2,026.72
Notes on Results	• Level of effect for adults is much lower than for children because adults are much less likely to die. • Costs for adults are higher because drug costs are higher. • Comparing presumptive diagnosis with use of the RDT, the effects are the same because, in both cases, everyone with malaria is treated. Average costs are lower because RDT costs less than the drug and 55% of "fever" patients do not have malaria in this model.		

Note: The analysis was carried out two ways: 1) giving ACTs to all patients who come to the clinic with fever (presumptive treatment), and 2) using a rapid diagnostic test (RDT; a "dipstick") for all fever patients, but treating only those with a positive test.

* This adaptation differs from the original model in having a no-treatment comparison.

** The costs of ACTs and RDTs were estimated from data available to the committee.

The analyses based on presumptive treatment should be closer to the current and near-term reality in Africa, where RDTs are seldom used. RDTs are used routinely in at least some parts of Asia, but the variables in this model (e.g., the proportion with fever that has malaria and the probabilities of death at different ages) are not applicable to most of Asia without significant modification. Another way the model has been simplified is to assume that ACT resistance does not exist, thus the costs and benefits would remain perpetually the same. The model from which this was adapted does build in the spread of drug resistant malaria. As time goes on, costs increase (because of the cost of re-treating initial failures) and benefits decrease. It is reasonable to assume that, even if ACTs are used well, malaria strains resistant to ACTs will eventually develop and spread (as discussed elsewhere in this report), but it also is reasonable to anticipate that ACTs could remain.

ACTs are recommended by WHO as the drug of choice for policy change in countries with significant chloroquine resistance. Within a few years, this will include every country in sub-Saharan Africa. But if these countries were to switch to an inexpensive intermediate drug or an inexpensive combination that does not include an artemisinin (e.g., chloroquine + SP) as an interim measure, this would delay the higher treatment cost of ACTs. As recently as a year ago, changing to SP was a potentially viable option, with the understanding that resistance was likely to develop within a few years. A model was developed by a member of the IOM committee (Laxminarayan, forthcoming) to compare the costs and effects of a direct change to ACTs versus changing first to SP and then to ACTs. The model incorporates aspects of malaria transmission, levels of immunity, and drug resistance, and the costs of a malaria episode (including the cost of the drug, but no dispensing costs; and the costs of morbidity and mortality, set at US$0.50 per patient per day with malaria). Although based on SP as the intermediate, the model is indicative of a change to any relatively cheap alternative antimalarial with the potential for rapid development of resistance.

Characteristics that affect both the costs and effects of the two strategies are the level of coverage (i.e., the number and proportion of the population who get treatment for malaria when needed); the level of immunity to malaria in the population, which is related directly to the intensity of malaria transmission; and how quickly patients are treated after becoming symptomatic (this is dependent on access). Where transmission is most intense, the level of immunity is highest and becomes evident at the earliest ages. Where infective bites are rare (e.g., in much of Asia), very little immunity develops. Immunity to malaria is a temporary phenomenon, waxing and waning with exposure.

For the purposes of this model, an initial frequency of drug-resistant parasites of 10–12 (a number very close to 0) was arbitrarily chosen for the artemisinin-based combinations, and for SP, 10–3 (one in one thousand) (Laxminarayan, forthcoming). Over time, treatment selection pressure leads to a greater prevalence of infected individuals who carry the drug-resistant strain relative to those carrying a drug-sensitive strain. The model was used to compare the economic consequences of two strategies:

1) Replacing chloroquine (CQ) with ACTs, and
2) Replacing CQ with SP and waiting for resistance (to a level of 20%) to develop before introducing ACTs. A 3% discount rate is used for all relevant aspects of the model.

Using either strategy, the total costs of infection decrease with increasing levels of coverage with either strategy. This is attributable to faster cure rates, lower morbidity, and less need for re-treatment because of initial failures. At very low levels of treatment coverage (and low drug pressure), resistance to the intermediate drug is not a problem, so the least expensive drug gives good results for less money. At high levels of treatment coverage (and high drug pressure), resistance evolves so rapidly regardless of which strategy is followed that the faster acquisition of immunity with a less effective drug plays a critical role in determining the superior strategy.

The bottom line is that if one were interested in only the short term, using the less expensive drug makes better economic sense since the costs of resistance-related morbidity do not enter into the considerations. However, for longer planning horizons, a direct switch to ACTs is advantageous, given the costs of higher morbidity associated with increasing resistance to the intermediate drug. In theory, with higher intensity of disease transmission, the benefit of switching to ACTs directly may be diminished because of greater immunity associated with higher transmission, and hence a lower risk of resistance developing to the intermediate drug, whatever it is. Resistance to the intermediate would be expected to take longer to develop and, therefore, the benefits of switching to using it first, then switching to ACTs, increase. Other models to evaluate the cost-effectiveness of antimalarial drugs including ACTs are introduced.

2.2.3. Cost-Effectiveness of Non-artemisinins in Current Use

Because chloroquine (CQ)-resistant *P. falciparum* has now spread throughout most of Africa, the efficacy and practicability of other drugs such as amodiaquine (AQ), and sulfadoxine-pyrimethamine (SP), for the treatment of fever needs to be assessed. A decision-analysis model was used to compare the cost and effectiveness of CQ, AQ, and SP in various studies. The variables considered were the probability of *P. falciparum* infection, drug compliance, minor and lethal side-effects of the drug, the level of drug resistance in the community, and case-fatality rates associated with treatment. The measures of effectiveness were the number of malaria-related fever episodes cured parasitologically with each treatment and the number of malaria deaths prevented in children and adult patients. Cost-effectiveness comparisons were made for cases cured and deaths prevented.

Wilkins *et al.* established a model that was developed to access the cost-effectiveness of replacing CQ with SP as first-line treatment in Mpumalanga, South Africa, where malaria is seasonal and the population is non-immune. The relative cost-effectiveness of chloroquine (CQ) and sulfadoxine-pyrimethamine (SP) as first-line antimalarial therapy in southern Africa is of great interest to policymakers, clinicians, and researchers in the sub-region. *In vivo* drug resistance levels were used to derive a resistance variable for each drug, which was used to compare the costs to the public healthcare provider associated with either therapeutic option. Using chloroquine and SP as first-line drugs, and the probability of drug resistance in new malaria cases for both modeled drugs, the resistance variable R was derived (Table 14).

Costs including drugs, staff time, transport, maintenance, utilities, training, and consumables were determined and subjected to Monte Carlo simulation and subsequent analysis to generate an average cost-effectiveness ratio (ACER) with confidence intervals for each drug. SP was found to be 4.8 (95% CI 3.3-6.7) times more cost-effective than CQ in Mpumalanga at 1997 resistance levels and costs, despite the far greater cost per treatment course of SP (US$ 4.02 as opposed to US$ 0.22 for CQ) in South Africa. The result indicated that SP use as a first-line agent at public healthcare facilities in the Tonga district was almost 5-fold more cost-effective than the use of chloroquine, despite the approximately 20-fold lower cost of chloroquine (US$ 0.22 for an adult chloroquine treatment course vs. US$ 4.02 for SP). At the price of SP in Kenya and Uganda (US$ 0.47-4.80 per treatment course), the ACER for SP does not change materially, increasing to between 5.1 and 5.6. Resistance emerged as the factor that most influenced the ACER of a specific drug. Indirect costs, compliance, changes in effectiveness and costs over time and costs of adverse events were not included in the model owing to paucity of data and logistical difficulties. Since most of these are likely to be similar in both drug models, the relative ACER is unlikely to be significantly altered by their inclusion. Therefore, SP was 5-fold more cost-effective than the use of chloroquine in southern Africa in 2002 [126].

In year 2001, the Tanzanian Government officially changed its malaria treatment policy guidelines whereby CQ – the first-line drug for a long time was replaced with SP. This policy decision was supported by research evidence indicating parasite resistance to CQ and clinical CQ treatment failure rates to have reached intolerable levels as compared to SP and amodiaquine (AQ). Mubyazi and Gonzalez-Block reviewed the cost-effectiveness studies for CQ, SP, and AQ in Tanzania in 2005 [127]. Although SP was also facing rising resistance trend in 2004, the need for a more effective drug was indispensable, for an interim 5–10 year period it was justifiable to recommend SP because SP was relatively more cost-effective than

CQ and AQ in this review. The government launched the policy change considering that studies (ethically approved by the Ministry of Health) on therapeutic efficacy and cost-effectiveness of artemisinin drug combination therapies were underway. Nevertheless, the process of communicating research results and recommendations to policy-making authorities involved critical debates between policy makers and researchers, among the researchers themselves and between the researchers and general practitioners, the speculative media reports on SP side-effects and reservations by the general public concerning the rationale for policy change, when to change, and to which drug of choice.

Table 14. Data used for the calculation of the cost-effectiveness of chloroquine or sulfadoxine-pyrimethamine as first-line treatment for malaria in South Africa [126]

Variable	Chloroquine	Sulfadoxine-pyrimethamine	Source
Drugs probability of drug resistance	47.92%	5.50%	*In vivo* studies conducted by South Africa Medical Research Council, 1996-97 (Freese *et al.*, unpublished report; Govere *et al.*, 1999).
Drug resistance variable (R)	0.335 ± 0.07	0.039 ± 0.01	Calculation, assuming patient return rate of 0.7 ± 0.15.
Clinical staff: Treatment failures referred to hospital (of 507 patients treated)	119.04 ± 41.91	13.66 ± 4.84	Estimated using R.
Clinical staff: Hospital admissions for i.v. quinine treatment.	51.02 ± 41.91	5.86 ± 4.84	Proportion of referrals admitted estimated by hospital clinicians based upon their experience.

Similar data was found out by Sudre *et al.* as early as 1992. The cost-effectiveness comparisons were made for cases cured and deaths prevented. For treating 100,000 febrile episodes, CQ, SP, and AQ cost US$1812, US$2622, and US$3044, respectively. Cost of the drug, compliance, and the level of CRPF had the greatest effect on the cost-effectiveness ratio. The prevalence of high-level drug resistance (RIII) was the most important determinant of the cost-effectiveness. In a scenario with high-level CRPF, treatment with CQ costs US$0.47 to cure one patient and US$2.29 to prevent one death compared with US$0.05 and US$1.52 for treatment with SP. When the prevalence of RIII-level CRPF is greater than 14-31% (depending on the level of compliance), the most cost-effective treatment is SP, despite it was 45% greater cost. Decision analysis models will be useful for malaria control planners as strategies are reconsidered in the 1990s [128].

In response to recommendation by WHO to countries facing classical antimalarials to implement artemisinin-based derivatives combination therapy [47], the African governments through the policy has already explicitly announced its strategy for replacing SP monotherapy with ACTs in effect from 2003-2007 [55]. Such a policy move faces similar challenges such as reservations/inadequate faith by the research and drug prescribing community who anticipate immediate parasite resistance to ACT mainly due to the majority poor residents potential failure to comply with the treatment schedules because of the drug cost, individual tastes and preferences driven by their perceptions of the drug with the sulpha component and availability of alternative drugs in the liberal retail market (as some people may prefer monotherapy such as AQ or other drugs). The concern about cost related barriers both to the provision and utilization of ACTs related services have been documented by other authors [129-134].

2.2.4. Cost-effectiveness of Current Artemisinins and ACTs

Over the recent past, health information and various surveys have revealed that widespread treatment failure precipitated a rise in malaria mortality and morbidity.National data collected by the National Malaria Control Centre confirmed that a considerable decline in the therapeutic efficacy of SP and CQ was responsible for the high and widespread treatment failure rates [47]. This situation had significant implications since treatment with antimalarial drugs has been the only tool used in fighting malaria from the late 1970s. In terms of prospects for young children, most of who are treated at home, failure of the only possible defense against malaria meant rising mortality. Widespread treatment failure was an Africa-wide phenomenon which was observed from the late 1980s and spread rapidly from then on. In response to growing criticism against the use of failing monotherapies, the World Health Organization and Roll-Back Malaria led a global campaign to replace SP and chloroquine with artemisinin combination therapies (ACT) as first-line treatment [47]. To date, not all countries in Africa have implemented ACTs as first-line treatment for malaria. The two most widely considered ACT brands in Africa presently are the fixed-dose combination artemether-lumefantrine (AL) and the co-packaged combination of amodiaquine and artesunate (AQ+AS). These combinations have proved highly efficacious in carefully controlled phase III clinical trials in areas with moderate to high levels of SP or CQ resistance in Africa [129].

One result suggested that AL produces successful treatment at less cost than SP, implying that AL is more cost-effective. While it is acknowledged that implementing national ACT program will require considerable resources, the study demonstrates that the health gains (treatment success) from every dollar spent are significantly greater if AL is used rather than SP. The incremental cost-effectiveness ratio is estimated to be US$4.10. When the costs of second line treatment are considered the ICER of AL becomes negative, indicating that there is greater resource savings associated with AL in terms of reduction of costs of uncomplicated and complicated malaria treatments [129, 130].

Muheki *et al.* did same cost-effective study to replace AL from SP, and demonstrated that the findings also highlight the greater effectiveness and significant cost savings associated with the implementation of AL. These cost savings result from the improved clinical cure rates and decrease in malaria transmission achieved with AL. There is growing international evidence that ACT is one of the few effective measures available to 'Roll Back Malaria.' However, concerns about the costs and affordability of ACT are obstacles to its widespread implementation. This study explores some economic aspects of the implementation of AL to replace SP in the KwaZulu Natal (KZN) province, South Africa. Recurrent and capital costs for malaria treatment were compared at baseline and post-intervention for nine clinics and a sentinel rural district hospital. Changes in the unit costs of, and total expenditure on, malaria services were calculated and the cost effectiveness of AL relative to SP was assessed. The number of outpatient malaria cases and inpatient admissions both declined by 94% between 2000 and 2002. After accounting for the role of concurrent improvements in vector control, it was conservatively estimated that 36% of the decline in outpatient cases and 46% for inpatient admissions was attributable to changing the first-line drug to AL. Although AL is considerably more expensive than SP, its improved cure rate and reduced malaria transmission resulted in an estimated US$ 201,065 cost saving in 2002 alone for the sub-district studied [59].

More economic analyses show that ACTs are more than 95% likely to be cost-effective under most conditions, other than very low levels of initial resistance to SP and a five-year time frame. In an area of high drug resistance, there is evidence that AL and AQ+AS are the most cost-effective drugs despite being the most expensive, because they are significantly more effective than other options and therefore reduce the need for further treatment. This is not necessarily the case in parts of Africa where recrudescence following SP and AQ treatment (and their combination) is lower so that the relative advantage of ACTs is smaller, or where diagnostic services are not accurate and as a result much of the drug goes to those who do not have malaria.

The studies showed that the effectiveness was measured in terms of resource savings and cases of malaria averted (based on parasitological failure rates at days 14 and 28). All costs to providers and to patients and their families were estimated and uncertain variables were subjected to univariate sensitivity analysis. Incremental analysis comparing each combination to monotherapy (AQ) revealed that from a societal perspective AL was most cost-effective at day 14. At day 28 the difference between AL and AQ+AS was negligible; both resulted in a gross savings of approximately US$1.70 or a net saving of US$22.40 per case averted. Varying the accuracy of diagnosis and the subsistence wage rate used to value unpaid work had a significant effect on the number of cases averted and on program costs, respectively, but this did not change the finding that AL and AQ+AS dominate monotherapy [131, 132].

A new ACT, dihydroartemisinin-piperaquine (DP), has been tested the cost-effectiveness with artesunate-mefloquine (AS+MQ), and the purpose is to alternative to AS+MQ. AS+MQ is widely recommended in Southeast Asia, but its high cost and tolerability profile remain obstacles to widespread deployment. To assess whether DP is a suitable alternative to AS+MQ, the study compared the safety, tolerability, efficacy, and effectiveness of the two regimens for the treatment of uncomplicated *falciparum* in western Myanmar (Burma). The study did an open randomized comparison of 3-day regimens of AS+MQ (12/25 mg/kg) versus DP (6·3/50 mg/kg) for the treatment of children aged 1 year or older and in adults with uncomplicated *falciparum* malaria in Rakhine State, western Myanmar. Within each group, patients were randomly assigned supervised or non-supervised treatment. The primary endpoint was the PCR-confirmed parasitological failure rate by day 42. Failure rates at day 42 were estimated by Kaplan-Meier survival analysis.

Findings of 652 patients enrolled, 327 were assigned DP (156 supervised and 171 not supervised), and 325 AS+MQ (162 and 163, respectively). 16 patients were lost to follow-up, and one patient died 22 days after receiving DP. Recrudescent parasitemias were confirmed in only two patients; the day 42 failure rate was 0·6% (95% CI 0.2–2.5) for DP and 0 (0–1.2) for AS+MQ. Whole-blood piperaquine concentrations at day 7 were similar for patients with observed and non-observed DP treatment. Gametocytaemia developed more frequently in patients who had received DP than in those on AS+MQ: day 7, 18 (10%) of 188 versus five (2%) of 218; relative risk 4.2 (1.6–11.0, p=0.011). In conclusion, DP is a highly efficacious and inexpensive treatment of multidrug resistant *falciparum* malaria and is well tolerated by all age groups. The effectiveness of the unsupervised treatment, as in the usual context of use, equaled its supervised efficacy, indicating good adherence without supervision. DP is a good alternative to AS+MQ [133]. The summary of the cost-effectiveness evaluations for antimalarial drugs including ACTs in current use are listed in Table 10.

2.3. Drug Cost and Cost-Effectiveness for Special Groups

2.3.1. Burden of Antimalarial Drug Resistance

Non-artemisinin antimalarial drugs including chloroquine (CQ), amodiaquine (AQ), sulfadoxine-pyrimethamine (SP), mefloquine (MQ), and quinine all have some degree of resistance. The drugs had cost-ineffectiveness in most of Asian and African regions that varied from drug to drug (Figure 2). Even today, the geographical distributions and rates of spread have varied considerably. Resistance of *P. falciparum* to CQ, the cheapest and the most used drug has been reported in almost all the endemic countries, except in Central America and Hispaniola. Resistance to the combination of SP, which was already present in South America and in South-East Asia, is now becoming highly prevalent in Africa. Resistance to mefloquine is found mostly in Cambodia, Myanmar, Thailand, and Vietnam. Sporadic cases of prophylactic failure of mefloquine in travelers and therapeutic failure have been reported in Africa, South America, and in other Asian countries. *P. falciparum* has shown the development of resistance towards all antimalarial medicines when used as monotherapy: time periods vary from 12 years (chloroquine), to 5 years (mefloquine), to 1 year (proguanil) up to even less than 1 yea (SP, atovaquone).

Increasing resistance of *P. falciparum* malaria to antimalarial drugs is posing a major threat to the global effort to "Roll Back Malaria." Chloroquine and SP are being rendered increasingly ineffective, resulting in increasing morbidity, mortality, and economic and social costs. Resistance to affordable drugs in Africa, which carries an estimated 90% of the burden of malaria, has reached critical levels [134]. In order for national governments, donor countries, and international institutions to make rational decisions on drug policy, there is a need to clarify how much of a burden antimalarial resistance causes currently, how much it is likely to cause in the future under different control strategies, and how much these strategies will cost and save. Models are increasingly being used to help explore policy options such as these, where outcomes are uncertain and decisions complex. They are ideally suited to exploring both biologic and economic influences on outcomes. Moreover, they can be used to produce the estimates of the cost-effectiveness of policy options which are now accepted to be a vital input to decision making in the health sector.

2.3.1.1. The Nature of the Burden of Drug Resistance

The burden of disease caused by malaria and its consequences has been documented in terms of childhood mortality, anemia, maternal and infant morbidity and mortality, neurologic disability, and economic and social costs. The burden caused specifically by antimalarial drug resistance is more difficult to quantify. Recent estimates based on the best available data from Africa suggest that the demise of CQ is the "most plausible single factor contributing to the change in malaria specific mortality," which has been estimated to have at least doubled over the last 15 years.

The definite relationships between therapeutic response to treatment and *ex vivo* measures such as resistance *in vitro* and the carriage of resistance genes is generally poorly defined, but of the drug, parasite, and host factors that contribute to therapeutic outcome, particularly important is the immunity of the host. Thus, in areas of high transmission, adults are relatively immune and tend to self cure irrespective of the effectiveness of the drug or indeed whether an antimalarial drug is taken at all. Everyone has malaria parasites in their blood all the time, but usually at densities below that causing illness. The sensitivity of the parasite to the drug affects the clinical outcome more noticeably where the hosts are non-immune and unable to control the infection themselves. In these circumstances, taking an ineffective drug can result in severe or protracted disease and even death. If an effective second-line drug is easily accessible, this impact may be limited although costs, particularly to the patient, may increase. Unfortunately, however, effective second line drugs are often not readily available and therefore first line drug failure can lead directly to an increase in morbidity and mortality.

The burden of antimalarial drug resistance in terms of increased costs includes the direct cost of more expensive second-or third-line treatments and hospital admissions, the costs of seeking treatment, and the indirect costs of lost productivity. In addition, there are broader costs at the household and macroeconomic levels as well as intangible costs such as psychological stress and loss of confidence in a health system that fails to deliver a cure. Much of this burden falls on the poor, exacerbating already existing inequities, since the more expensive, effective antimalarials are accessible only to patients affluent enough to obtain them through informal sources, and remain out of reach to the majority of the rural poor who carry the largest burden of disease [134].

2.3.1.2. Measuring the Burden of Drug Resistance

Evaluating the impact of antimalarial drug resistance is difficult. The impact may not be recognized until it is severe, especially in high transmission areas. This is partly because routine health information systems grossly underestimate the magnitude of the problem. Until the prevalent malaria parasites become completely resistant, patients will usually have an initial response to the drug, with a reduction in parasite numbers. However, as drug levels fall below the minimum inhibitory levels of the infecting resistant parasite in the individual patient, the parasite population once again begins to expand. By the time the parasite biomass increases back to a level causing symptoms, and is microscopically detectable, several days or weeks may have passed and the symptoms may not be associated with the initial illness, by either the patient or the health worker, and as a consequence may not be recorded as a treatment failure. This lack of reliable information is exacerbated by the fact that in many settings the majority of patients with malaria symptoms do not access the public sector where such information can be collected.

Since routine health information systems cannot often be relied upon for accurate data on drug efficacy, there is a need for systematic surveillance and monitoring. Much effort has recently been put into setting up national and regional networks that use standardized methodology for monitoring *in vivo* drug efficacy. In recent years, there has been a trend towards 14-day assessments in high transmission areas. Ideally, follow-up should be for at least 28 days, since failures do not become apparent within 14 days of taking treatment until drug resistance has reached very high levels and following patients up only until the 14th day results in a serious overestimation of the efficacy of failing drugs. One of the arguments for the 14-day test was the difficulty in differentiating between recrudescence and re-infection, especially in areas of high transmission. Now that the genotypes of parasites in initial infections can be compared with those in any subsequent infections and it is therefore possible to differentiate confidently between re-infections and recrudescent infections, longer follow-up studies can and should be undertaken [134].

2.3.1.3. Reducing the Burden of Drug Resistance

In considering possible strategies for the reduction of the burden of antimalarial drug resistance, it is useful to differentiate between the current burden of drug resistance and the potential burden in the future resulting from the continued emergence and spread of drug resistance. Central to achieving a reduction in both current and future burdens is an improvement in drug usage by patients and providers so that good quality drugs are available and taken at the correct dose and for a sufficient length of time to affect a radical cure and reduce the likelihood that partially resistant parasites will survive.

Improving drug use is most effective where the parasite is still sensitive to the drug. Where resistance has rendered the drug ineffective, the current burden of resistance can only be reduced by replacing the failing drug regimen with one that is effective. The difficulty lies in deciding which drug regimen to switch to, since the choice of drug or drug combination will determine the subsequent development of drug resistance. Reducing the future burden of resistance requires that effective antimalarial drugs continue to be available in the future and requires the continuous search for and development of potential new antimalarial drugs (WHO/TDR, MMV, and WRAIR). However, the complete drug development process can take 10−15 years, making it imperative that the currently available drugs are deployed in a way most likely to maximize their lifespan by decreasing the likelihood that resistance will develop. The key strategy put forward to do this is to use available drugs in combination to prevent the emergence and spread of resistance [134].

2.3.1.4. Fitness Burden of Drug Resistance

Mutations that are associated with resistance to antimalarial drugs and deviate from the evolutionarily optimal biochemical structure are likely to be disadvantageous in the absence of drug pressure, and natural selection will act against the changes induced by mutation. Thus, the fitness of a mutant parasite could be reduced in relation to a non-mutated parasite in a drug-selection-free environment. This reduction is termed 'fitness costs of drug resistance' [135, 136]. Quantifying the costs of antimalarial drug resistance will provide an estimate that is needed urgently for modeling the evolution and spread of drug resistance. More importantly, such models might be applied to public health measures predicting whether (and how quickly) drug resistance might vanish once drug pressure has been eliminated by a change of drug policy.

Molecular monitoring of single nucleotide polymorphisms (SNPs) associated with CQ resistance in natural *P. falciparum* populations from Malawi has shown that the prevalence of the mutant allele pfcrt-76T dropped from 85% detected in 1992 before discontinuation of CQ treatment to 13% detected in the year 2000 after replacing CQ with antifolates. These data indicated that *in vivo* fitness of parasites carrying the *pfcrt* mutations is substantially compromised in the absence of drug pressure. Similarly, it has recently been shown [137] that artemether-lumefantrine seems to select for wild-type parasites in *pfmdr1*. Thus, elimination of CQ-resistant genotypes is enhanced under treatment with artemether-lumefantrine. Experimental support comes from findings of Hayward *et al.* [138], who conducted competition experiments between drug-sensitive and -resistant *P. falciparum in vitro* culture strains that had been modified by allelic exchange of wild-type and mutant *pfmdr1* alleles. A reduced fitness of resistant parasites harboring three mutations at codons 1034, 1042, and 1246 in the *pfmdr1* gene was found in the absence of drug pressure, and the relative fitness of the less fit strain in competition experiments was quantified by allele-specific quantitative PCR. Loss of fitness for these three mutations was calculated to be 25% per generation.

Suitability of *P. falciparum* parasites is intrinsically linked to transmission success of individual parasite clones in an environment where competitive interactions in multiclonal infections are the rule. The fitness costs incurred by the mutation probably lead to an increased rate of elimination for mutant parasites and, thus, to a decrease of infectiveness. Thus, the loss of the drug-resistant subset of the parasite population can be used as an indicator of the burden of the resistance trait on the fitness of the parasite.

It is clear that the resistance to antimalarial drugs is resulting in avoidable morbidity, mortality, and financial losses. Urgent measures are needed now to reduce the current and future burden of disease. There is little justification for the continued use of ineffective drugs because effective drugs are currently available. The decisions of which drug regimen to change to, and how to implement the change in a way that maximizes potential benefit, are more difficult, but delaying a decision to switch because of these difficulties can only result in increased morbidity and mortality. Furthermore, delaying a switch to ACTs potentially puts at risk one of the key advantages of this strategy, which is to delay the emergence of resistance. The longer the decision is delayed, the more entrenched will become the unregulated use of the artemisinins and partner drugs as monotherapies. Partly because of the uncertainties, there is still significant reluctance to take action amongst potential funders and some national governments, both of whose commitment is essential for the success of any change in policy.

2.3.2. Cost-effectiveness of Antimalarial Drugs during Pregnancy

The economic impact of malaria in pregnancy, a major public-health problem in the developing world, is not well documented. The symptoms and complications of malaria during pregnancy differ with the intensity of transmission and thus with the level of immunity the pregnant woman has acquired, with likely implications for the economic burden. People in epidemic or unstable transmission areas lack a substantial level of immunity and therefore usually become ill when infected with malaria. Pregnant women living in these areas are at a two to three-fold higher risk of developing severe malaria and face a higher risk of mortality than non-pregnant adults living in the same area. Severe malaria and malaria-related severe anemia may result in maternal death or adverse pregnancy outcomes such as spontaneous abortion, neonatal death, and low birth weight, which is a leading cause of poor infant survival and development [139]. Malaria-related low birth weight has a fatality rate of 37·5%

and is responsible for at least 13% of malaria mortality in African children under 5 years of age [140].

In stable malaria transmission areas, as a result of acquired immunity, *P. falciparum* infection in pregnant women does not usually result in fever or other clinical symptoms. The main effect of infection is malaria-related anemia in the mother and the presence of parasites in the placenta, which impairs fetal nutrition and thus contributes to low birth weight and its consequences.

2.3.2.1. The Economic Burden of Malaria in Pregnancy

Devoid of estimates of the human and economic costs of malaria in pregnancy, it is hard to prioritize the use of scarce resources and to quantify the benefit of potential interventions. The lack of reliable data on the global burden of malaria in pregnancy makes it difficult, if not impossible, to estimate its economic cost. In particular, there is a paucity of information from Asia and the Americas, and little data on the burden caused by *non-falciparum* malaria. There are two possible approaches to estimating the economic burden of malaria in pregnancy. Microeconomic approaches are used to measure the effect of a disease on an individual or household, while macroeconomic approaches measure the effect of disease on an entire society. Taking a traditional "micro" level approach, economic costs can be categorized as direct, indirect, and intangible, and can be measured from the perspectives of the government (mainly Ministry of Health) and households [140].

The direct costs of malaria in pregnancy can be divided into 1) the cost arising from interventions targeted at all pregnant women in malaria endemic settings, and 2) the additional costs arising as a consequence of malaria infection in pregnant women. Direct costs to the health service arising from specific interventions for preventing or treating malaria in pregnancy include the cost of IPTp. Direct costs associated with malaria infections in pregnant women include the immediate costs of maternal infection and also the immediate and long-term costs of treating the consequences of maternal infection on the infant, most of which relate to mitigating the consequences of low birth weight. Immediate costs are those of additional outpatient consultations, hospitalization, staff time, diagnostic tests, drugs, and other supportive treatment. The costs incurred by the mother (or her household) include those of obtaining additional health care such as transport, drug costs, and consultation fees. A large portion of these costs are likely to be incurred purchasing goods and services from the private sector. For example, one study found that in Sudan, Kenya, and Uganda, of the total spent by the household on maternal and child health and family planning services, 85%, 37%, and 53%, respectively; and were spent in the private sector.Pregnancy has been shown to be an expensive time for households, and the incidence of malaria when resources are already stretched could force poor and vulnerable households into even worse economic circumstances.

The immediate consequences of low birth weight are well described and include complications such as hypothermia, hypoglycemia, and increased risk of mortality. The associated costs therefore include those of prolonged hospital stay, incubation, and nutritional support. The long-term consequences include impaired neurological and metabolic development, poor growth, and chronic health problems in adult age (e.g., respiratory and renal disease). Each of these will have future cost implications, which have been investigated in detail elsewhere. One study found that children born with low birth weight were two to four times more likely to experience failure at school. The effect of low birth weight has even been shown to impact negatively upon the next generation since the children of low birth weight babies may be adversely affected. Additionally, the death of a mother increases the child's risk of death within the first year of life, especially for female children. Additional direct costs may be incurred at the household level in the case of maternal or child mortality, including the brought forward cost of funerals, and the cost of replacing the role of the mother [140].

The indirect costs of malaria in pregnancy incurred at the household level are those related to the consequences of time lost because of morbidity, treatment seeking, and premature mortality for either the mother or child. Indirect costs to the household and the economy relating to lost production because of morbidity or premature mortality can in theory be estimated using the human capital method (which estimates the value of forgone productive activities); however, this has not been done specifically for malaria in pregnancy. This method may fail to accurately capture the intangible elements of women's roles (e.g., as caregivers), and thus underestimate the true cost of malaria in pregnancy. Many of the weaknesses in estimates of the economic burden of malaria on households13 are pertinent when considering malaria in pregnancy household labor substitution or sale of (productive) assets to obtain cash. This could in part explain why micro studies of the cost of malaria generally reach lower estimates than macro studies.

The total cost of malaria in pregnancy would be the sum of all direct and indirect costs incurred by households, enterprises, and government, plus any additional costs affecting the economy as a whole. An estimate of costs of malaria in pregnancy to the economy would have to take into account the long-term and inter-generational effect of malaria in pregnancy, which could have profound effects on educational attainment, future productivity, and resulting economic growth. For example, the reproductive choices of whole communities might be affected by increased risk of mortality as a result of malaria in pregnancy, or school performance might be systematically depressed by the effects of malaria on child development. In theory, these effects can be estimated by a macroeconomic approach, examining the overall effect of malaria on gross domestic product using multi-country time-series data and regression analysis to control for confounding factors. Because of these described factors, it is possible that much of the long-term economic impact of malaria is a result of malaria in pregnancy; however, the available macroeconomic studies of malaria do not explicitly examine or quantify its impact. Indeed, it would be difficult using this methodology to separate the effect of malaria in pregnancy from the overall effect of malaria.

Thus, understanding the full economic burden of malaria in pregnancy helps to identify the potential benefits of effective prevention. For example, the economic benefit from reducing the incidence of low birth weight status in newborn babies is estimated at US$510 per newborn baby. These benefits result from lower mortality rates and medical costs, and increased learning and productivity [140].

2.3.2.2. Intermittent Preventive Treatment in Pregnancy (IPTp)

IPTp refers to giving treatment doses of antimalarials to pregnant women intermittently (at least twice), regardless of whether or not they are infected. Although this may appear inefficient, it has demonstrated benefits to mother and child and is necessary because febrile case management is inadequate for preventing malaria related low birth weight in areas of high transmission because of lack of symptoms. For example, a Kenyan study found 25-30% of malaria-infected women reported no fever at any time during pregnancy. Safety and efficacy data on IPTp are currently available only for chloroquine and sulfadoxine-pyrimethamine. Both drugs - sulfadoxine-pyrimethamine in particular - have long half-lives resulting in a prophylactic effect. However, widespread resistance has rendered CQ ineffective and sulfadoxine-pyrimethamine is now also increasingly threatened. Newer effective drugs, in particular the ACTs, are much more expensive and therefore likely to substantially alter the cost-effectiveness of IPTp.

An incremental cost of adding IPTp was estimated with sulfadoxine-pyrimethamine to an existing antenatal care service for primigravidae (in high transmission areas the effects of malaria in pregnancy are more marked for women in their first pregnancy) to be $1.13 per pregnancy in very low-income countries. When service overheads were added to intervention costs, the mean average cost of IPTp with sulfadoxine-pyrimethamine was $2.24. The authors also examined the cost of providing IPTp to all mothers regardless of parity. This would substantially increase total cost, but would represent only a relatively minor addition to existing government health expenditure and would be easier to implement than targeting. Where there is no pre-existing antenatal care infrastructure, the incremental cost would be much higher, though Worrall *et al.* [140] and Goodman [141] advised that community-based delivery could be an alternative if health workers are well organized, well supervised, and motivated.

Current approximations of sulfadoxine-pyrimethamine drug costs for all pregnant women attending antenatal clinics at least twice during pregnancy in malarious parts of sub-Saharan Africa (average of 2.5 doses per women to account for extra doses needed for HIV-positive women) is $3 million [141]. However, an expert panel recently concluded that sulfadoxine-pyrimethamine in pregnancy was beneficial only up to 50% parasitological failure (drug resistance to sulfadoxine-pyrimethamine measured in children). Cost implications are substantial if pregnant women require quinine (or artemisinin-based combination therapies in the future), both of which cost over ten times more than sulfadoxine-pyrimethamine [142].

Four studies have examined the cost-effectiveness of IPTp [140-144]. The most recent and comprehensive is the Disease Control Priorities Project, which provides evidence of the cost-effectiveness of a range of malaria control interventions including IPTp. The authors modeled cost-effectiveness, allowing for changes such as increasing drug or insecticide resistance, and used ranges and probabilistic sensitivity analysis to account for uncertainty. Two to three doses of sulfadoxine-pyrimethamine during pregnancy were assumed to be delivered at a prenatal clinic to primigravidae women, allowing for the probability of attending each visit, the level of drug resistance, and compliance. The model was based upon that of Goodman and colleagues but was extended to include benefits to the mother, as well as infants. Additional doses were assumed to be needed for HIV-positive women, and benefits to the mother resulting from changes in the incidence of severe anemia were included. The incremental cost-effectiveness ratio for IPTp in a low-income sub-Saharan African setting was reported as a (90% CI) range between $9–21 per disability-adjusted life year (DALY) averted (mean $13 per DALY averted). The average total cost-effectiveness ratio (which includes overhead costs) was reported to be between $16–35 (90% CI) (mean $24 per DALY averted). The community-based delivery increased access and adherence to IPTp and was cost-effectiveness [145].

2.3.2.3. Malaria Case Management during Pregnancy
1) Treatment Coverage and Access

Access to prompt and effective malaria treatment is limited for many at-risk populations by poverty, geographical inaccessibility, low quality of services, and knowledge of and attitudes to malaria. However, pregnant women face additional barriers such as the belief that fever is a normal sign of pregnancy or that bitter-tasting substances can provoke abortion. Such perceptions can deter access to treatment. Economic factors within the household such as male control of spending decisions and women's lack of cash might compound this problem even in households of relatively high socioeconomic status. Few studies examine the effect of these issues, although Shulman and colleagues found that low birth weight was more common in younger women and women of lower socioeconomic status [140, 144].

2) Cost and Cost-effectiveness of Improving Treatment

Although there is some data on the costs and treatment outcomes for pregnant women, further analysis is required to put them together to come up with estimates of cost-effectiveness of improving malaria treatment for pregnant women. For all malaria patients, Goodman and colleagues examined the cost-effectiveness and affordability of interventions to improve malaria case management using a model of treatment of outpatients with suspected uncomplicated malaria [110]. The interventions and results are reported here to show the approximate costs of improving treatment. No evidence was identified on interventions to increase the demand for malaria treatment by pregnant women (i.e., increase coverage), or the cost-effectiveness of interventions to improve treatment sought outside the formal health system, although a study found that training shopkeepers could improve malaria treatment at a cost of $4.00 per additional appropriately treated case of childhood fever [146].

Despite the lack of evidence on the economic burden of malaria in pregnancy, it is highly likely that a substantial cost is imposed on the health service, households, and the economy more broadly. The available evidence identifies a number of interventions to be cost effective, including improving treatment at less than $5 per DALY averted, and IPTp at less than $24 per DALY averted. These interventions are attractive even though the estimation of benefits is conservative. They are also generally low costs in terms of both per capita and total costs compared with other interventions, since the target population is relatively small. There is a reasonable amount of information on the cost-effectiveness of insecticide-treated nets and IPTp although analyses need updating as interventions are scaled-up or when drugs are altered. Evidence on interventions to improve treatment is especially scarce, and fails to capture adequately the benefits of improved treatment of malaria in pregnancy. There is also lack of data on cost-effectiveness of other interventions especially outside of Africa, in low transmission settings, and for *non-falciparum* malaria [140].

2.3.3. Cost-Effectiveness of Antimalarial Drugs in Children

Morbidity, mortality, and disability from malaria (and other diseases) are now being summarized in a new metric, disability-adjusted life years (DALYs), and an aggregate measure of premature mortality, morbidity, and disability, which can also be used for analyzing cost-effectiveness of the major interventions in children. The efficacy and relative cost-effectiveness of insecticide-treated nets (ITNs) for the control of malaria in children under 5 years of age have recently been demonstrated by several large-scale trials. However, it has been suggested that long-term use of ITNs in areas of high transmission could lead to mortality rebound in later childhood, which would reduce the cost-effectiveness of the intervention, and at the extreme could lead to negative overall effects.

A model is presented in which the cost and DALYs per child aged 1–119 months were estimated for a sub-Saharan African population with and without an ITN intervention. The rebound rate, defined as the percentage increase in age-specific all-cause mortality and malaria specific-morbidity, was varied to estimate the threshold at which the intervention was no longer cost-effective. Rebound was considered over two possible age ranges: 5–9 years and 3–6 years. For mortality and morbidity reductions due to ITNs in children aged 1–59 months and rebound in the 5–9 years age class, one could be reasonably certain that the cost per DALY averted is below $150 up to a rebound rate of 39%. Up to an 84% rebound rate it is highly likely that the intervention will be DALY-averting, that is the DALYs averted by the intervention outweigh DALYs incurred through rebound effects. These thresholds are sensitive to the age range over which reductions and rebound in morbidity and mortality occurs. In reductions confined to children aged 1–35 months and rebound in the 3–6 years age class, the cost per DALY is highly likely to fall below $150 only up to a 2.5% rebound rate, and with a rate in excess of 11% one can no longer be reasonably certain that the intervention is DALY-averting. These rates apply to the whole population. If there is no rebound among children who did not comply with the intervention, the actual increases in morbidity and mortality required to reach these thresholds amongst compliers would be much higher. The age range over which rebound occurs is a critical determinant of the thresholds at which one can no longer be reasonably certain that ITNs remain cost-effective in the long term.

The impact of a rebound in morbidity and mortality among older children sleeping under ITNs since early birth has been quantified from a cost-effectiveness perspective. The results indicate that the age range over which rebound occurs is a critical determinant of the threshold rebound rate where one can no longer be reasonably certain that ITNs remain cost-effective in the long term. This is the result of both a reduction in the age range over which positive effects are experienced, and higher underlying mortality rates in younger children, meaning that a given percentage increase in mortality results in a much larger absolute increase in the 3–6 years age group than in the 5–9 age group. Based on empirical estimates of age-specific malaria mortality in SSA, it appears unlikely that this threshold would be reached if rebound occurs over the 5–9 years age range. By contrast, if rebound occurs over the ages of 3–6 years, the increase in mortality rates required to reach this threshold falls within the observed range of malaria-specific mortality rates for this age group. Two implications of the findings for the implementation of ITN programs are:

- it is essential that long-term surveillance is included as a component of any ITN intervention, with particular attention to the age range over which any rebound effect occurs and the variation in this age range in different transmission zones;
- the possibility of increased incidence of severe malaria in older children highlights the importance of a coordinated approach to malaria control, which encompasses both prevention and strategies to improve the accessibility of effective treatment.

Based on empirical estimates of age-specific malaria mortality in sub-Saharan Africa, it appears unlikely that this threshold rate would be reached if rebound occurs over the 5–9 years age range. By contrast, if rebound occurs over the ages of 3–6 years, the increase in mortality rates required to reach this threshold falls within the observed range of malaria-specific mortality rates for this age group. It is essential that long-term surveillance is included as part of ITN interventions, with particular attention to the age range over which rebound may occur [147].

ACTs are a major breakthrough in the clinical management of malaria in children. By 2006, 41 countries in Africa had revised their national drug policies to incorporate ACTs into malaria case management guidelines [55]. Many countries in Africa are considering a change to combination treatment for *falciparum* malaria because of the increase in drug resistance. However, there are few effectiveness data for these combinations. Our aim was to study the effectiveness of three drug combinations that have proven efficacious in east Africa compared with amodiaquine monotherapy.

A randomized trial of antimalarial drug combinations was conducted in children (aged 4–59 months) with uncomplicated malaria in Muheza, Tanzania, an area with a high prevalence of resistance to sulfadoxine-pyrimethamine and chloroquine. Children were randomly allocated 3 days of amodiaquine (n = 270), amodiaquine, sulfadoxine-pyrimethamine (n = 507), or amodiaquine-artesunate (n = 515), or a 3-day six-dose regimen of artemether-lumefantrine (n = 519). Drugs were taken orally, at home, unobserved by medical staff. The primary endpoint was parasitological failure by day 14 assessed blind to treatment allocation. Secondary endpoints included day 28 follow-up and gametocyte carriage. Analysis was by intention to treat.

Results showed that the amodiaquine group was stopped early by the data and safety monitoring board in 3158 children and 1811 who were randomly assigned treatment and 1717 (95%) reached the 14-day follow-up. By day 14, the parasitological failure rates were 103 of 248 (42%) for amodiaquine, 97 of 476 (20%) for amodiaquine, sulfadoxine-pyrimethamine, 54 of 491 (11%) for amodiaquine-artesunate, and seven of 502 (1%) for artemether-lumefantrine. By day 28, the parasitological failure rates were 76% (182 of 239), 61% (282 of 476), 40% (193 of 472), and 21% (103 of 485), respectively. The difference between individual treatment groups and the next best treatment combination was significant (p < 0.001) in every case. Recrudescence rates by day 28, after correction by genotyping, were 48·4%, 34·5%, 11·2%, and 2·8% (Figure 7), respectively [148].

Cheap and effective treatment for malaria with one drug is no longer an option for most countries in Africa because of the rapid emergence of drug resistance. This trial provides evidence from a head-to-head effectiveness comparison of three drug combinations that are available in Africa, have reasonable efficacy and safety data to lend support to their use in children, and stand a realistic chance of being deployed. The results of the trial draw attention to the impending malaria treatment crisis in the sub-region of east Africa where resistance to chloroquine (CQ), amodiaquine (AQ), and sulfadoxine-pyrimethamine (SP) is established. Combinations of available drugs, such as CQ+SP and CQ+AS, have not proven effective, even in closely observed efficacy trials in areas where CQ resistance is common. Furthermore, the combination of SP+AS has been shown to be disappointing in areas where the level of SP resistance is high. Addition of an artemisinin to a failing drug has not proven effective in Africa before and did not seem effective in this trial. The very high rates of reinfection in this part of Tanzania led to day 14 being taken as the primary endpoint, but the day 28 results show a widening difference between the groups and this difference in recrudescence rates probably would have continued even beyond this time.

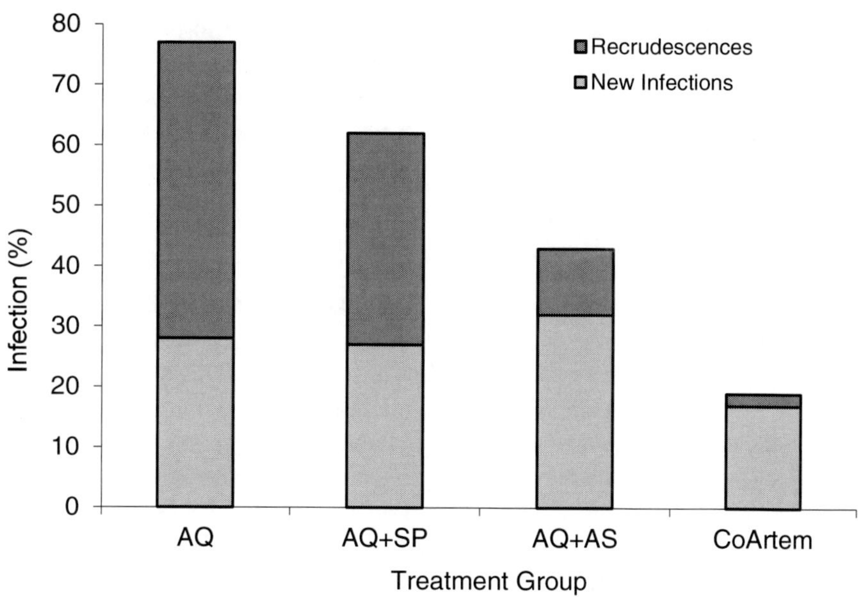

Figure 7. Parasitological failure rates of amodiaquine (AQ), AQ combined with sulfadoxine-pyrimethamine (SP), AQ combined with artesunate (AS), and artemether combined with lumetantrine (CoArtem) by day 28 [148].

The effectiveness of the combination of AQ+AS could have been affected by the fact that there is no packaging of this combination for young children and infants. A co-formulated version is being developed and might improve adherence, especially if it is well packaged. Appropriate packaging of antimalarials improves adherence and hence probably clinical outcome, although this hypothesis has not been investigated directly. The results obtained with amodiaquine/artesunate in this study are, however, in keeping with some efficacy data from areas of east Africa where amodiaquine resistance rates are high; thus this combination should not be used in areas where resistance to amodiaquine is already high. However, the combination has been shown to be efficacious in areas where resistance to amodiaquine is moderate or low, which is the case in much of West Africa at the moment [149, 150].

The study suggests the need to test any combination in effectiveness trials with the packaging that will be used in practice before any drug is adopted as national policy. The study has shown how limited the options are for antimalarial treatment in children in the parts of east Africa where levels of resistance are high. The artemether-lumefantrine combination works at present, although the long half-life of lumefantrine might make this combination vulnerable to selection pressure. The cost of the drug means that it is likely to reach only a fraction of those who need it, unless the price is substantially reduced either through market mechanisms or (more realistically) through subsidy. Supply is currently a major problem. Drug combinations that can help to fill the gap in the medium term are being developed, of which piperaquine-dihydroartemisinin [151] and chlorproguanil-dapsone-AS are among the most promising. Chlorproguanil-dapsone can be effective where sulfadoxine-pyrimethamine has failed. None of the artemisinin combinations is cheap, and this is a serious limitation; addressing this issue in a sustainable way will not be easy [42]. Current efforts by the Global Fund and others will provide cheap artemisinin containing combinations for several countries, but it is too early to assess the sustainability and scope of these efforts. Present study shows that in areas where CQ, AQ, and SP have failed badly, use of any of these drugs in combination is unlikely to work [148]. The study shows how few the options are for treating malaria where there is already a high level of resistance to SP and AQ. WHO-packaged six-dose regimen of artemether-lumefantrine is effective taken unsupervised, although cost is a major limitation [2].

2.3. 4. Cost-effectiveness of Antimalarial Drugs in Prophylaxis

Chloroquine remained effective in Africa and in major parts of the other continents for nearly 30 years. It became so easy to prevent malaria that drug prophylaxis was considered more or less mandatory once a traveler entered endemic areas. Indeed, chloroquine was often recommended despite the fact that a particular traveler would not be at risk as he or she only visited a non-malarious area in a country with malaria within its borders. The habit of recommending prophylactic drugs even when the risk for malaria was only theoretical prevailed, even when resistance to chloroquine became a problem. In addition, the traditional prophylaxis with chloroquine led to a dogma that drug prophylaxis should be given regardless of risk as soon as a traveler entered endemic areas. This prevailed also when resistance to chloroquine and adverse effects of alternatives became a problem. A cost/benefit analysis of the risk for malaria versus risk for adverse effects and cost of the recommended drug is not uniformly applied and drug prophylaxis is still advocated even when the risk for severe adverse effects greatly exceeds the risk for malaria, which is unethical.

The drug of choice was then switched to a combination of pyrimethamine and dapsone (Maloprim) or one containing pyrimethamine and a long-acting sulfonamide (Fansidar). Unfortunately, agranulocytosis was soon reported during use of pyrimethamine/dapsone, and there was a small but significant risk for extremely severe allergic reactions with sulphonamides-pyrimethamine. Risks associated with antimalarials for prophylaxis. Several studies have indicated that the risk for adverse events during the use of antimalarial drugs for prophylaxis is in the range of 30 to 40%. Most studies of travelers are retrospective and open. Only a few are prospective. As there is no comparison with placebo, a proportion of the reported adverse events are not related to the drug. Composed of 623 non-immune travelers to sub-Saharan Africa, study is blinded, prospective, and, in part, placebo-controlled because it includes a group of travelers who took a placebo before the trip. The outcome of this study is therefore especially interesting. Even at the baseline visit, which took place before the trip, 16% of travelers noted adverse events with the placebo compared with 15 and 16% for atovaquone/proguanil and doxycycline, and 22 and 24% with mefloquine and chloroquine plus proguanil, respectively. No statistical difference was noted between the groups. Overall, no less than 85% of the sample population reported at least a single adverse event during the study period. Similarly, there were no significant differences between groups concerning travelers who withdrew from the study owing to adverse events.

What does this study tell us? It indicates that the vast majority of adverse events from current antimalarials used for prophylaxis is minor and is not related to the drug intake. Adverse events prominent enough to make the traveler discontinue the drug were reported in 2 to 5%; however, a proportion of these might also have been due to placebo effects. The actual risk of severe adverse effects likely differs significantly between drugs, but it would be quite difficult to show in a prospective study since the number of participants in such a study would have to be extremely large. Chloroquine and mefloquine appear to have a similar risk for severe adverse effects such as psychosis, which probably affects 1 in 10,000 to 20,000 users. Chloroquine is extremely toxic when an overdose occurs. Atovaquone/proguanil (Malarone) and doxycycline are probably safer. The experience from long-term use with atovaquone/proguanil is limited, and the cost makes prolonged use prohibitive for many travelers. Doxycycline is associated with phototoxicity from sun exposure [149]. However, to select right prophylactic drug the best way should be also by drug effectiveness evaluation.

For prophylactic drugs the effectiveness is a must - a drug that does not protect should not be recommended. It is right in theory but may not be in reality. There are diverging opinions between the Centers for Disease Control and Prevention (CDC) in the United States, the World Health Organization (WHO), British experts, and those in Sweden. Scientists at the CDC argue that "a drug that does not protect should not be used." But protection is not always black or white. There are shades of gray also. Fatality rates are reduced when any prophylactic drug is taken, whether it is fully effective or not. This is explained by the fact that the drug might have a retarding effect even if it does not eradicate the parasites. Chloroquine often slows down the progress of disease in chloroquine-resistant areas and provides time for the traveler to seek medical attendance.

It has recently been suggested that drug prophylaxis should only be used in areas where the risk in the local population exceeds 10 cases of *P. falciparum* malaria per 1,000 inhabitants per year, approximately 1 in 100 people per year [152]. Rombo prefers to recommend that prophylaxis with drugs be given if the risk for malaria despite other precautions exceeds 1 in 10,000 travelers [149]. These suggestions differ in the presented numbers but are still similar. Assume that a visitor will spend 1 month in an area where the API is 1 in 100. As the time spent is only 1 month (and usually in the dry season), the personal risk is less than 1 in 1,200. Personal protection with repellents and bed nets or air conditioning should reduce the risk further. Assume that the malaria risk for a traveler is 1 in 10,000.The risk of dying from *P. falciparum* malaria is in the range of 1%.This implied that you would have to give drugs to 1 million (100 x 10,000 travelers to save 1 from dying from malaria (if your drug was 100% effective).By doing so, you would have at least 50 travelers experience psychosis owing to chloroquine or mefloquine, of which some might commit suicide. The disadvantages with mefloquine or chloroquine in this case are higher than the benefit, and these drugs should not be recommended. Such severe adverse effects are not described with doxycycline or atovaquone/proguanil. However, the cost to save 1 life when the risk for malaria is 1 in 10,000 is still staggering—at least US$15,000,000 for doxycycline and substantially more for atovaquone/proguanil [149].

Policy makers need information on cost-effectiveness and on the way cost-effectiveness changes with increasing drug resistance to decide on the priority to attach to antenatal prevention. Although a few studies are available [141], they relate to a limited number of settings and intervention specifications, and the results are not comparable with cost-effectiveness estimates for other malaria control interventions because of the partial cost of methodology and/or intermediate outcome measures used. A standardized modeling framework was used to provide estimates of the cost-effectiveness of the CQ and SP regimens in an operational setting with moderate to high malaria transmission. The model combined data from the Cochrane meta-analysis of strategies for preventing malaria in pregnancy with information on costs and compliance from a range of published and unpublished sources. To facilitate comparisons with other methods of malaria control and interventions for other health problems, the full incremental costs of the interventions were included, and the benefits were expressed in terms of a generic health outcome measure. The impact on cost-effectiveness of different levels of antimalarial drug resistance and changes in the intervention specification were explored.

Antimalarial chemoprophylaxis during pregnancy significantly increases the birth weight of babies born to primigravidae, but coverage in sub-Saharan Africa is very limited. This analysis assessed whether increasing coverage is justified on cost-effectiveness grounds. A standardized modeling framework was used to estimate ranges for the cost per discounted year of life lost averted by weekly CQ chemoprophylaxis and intermittent SP treatment for primigravidae in an operational setting with moderate to high malaria transmission. The SP regimen was found to be more cost-effective than the CQ regimen, because of both lower costs and higher compliance. Both regimens appear to be a good value for money in comparison with other methods of malaria control and based on rough cost-effectiveness guidelines for low-income countries, even with high levels of drug resistance. However, extending the SP regimen to all gravidae and increasing the number of doses per pregnancy could make the intervention significantly less cost-effective.

CQ prophylaxis and SP intermittent treatment for primigravidae are cost-effective interventions compared to other methods of malaria control and to interventions for other health problems. This remains the case up to high levels of parasitological resistance to either drug. In addition to being cost-effective, the interventions are also relatively affordable for sub-Saharan African governments. The analysis lends weight to the argument that there should be a drive to extend coverage of preventive strategies in pregnancy to all women in malarious areas of Africa. It also shows that modifying the intervention in response to increasing HIV prevalence could substantially reduce its cost-effectiveness, although the impact of HIV on the burden of disease, the effectiveness of the interventions, and the costs of any intervention modifications required needs further exploration. Finally, because resistance to SP is expected to grow rapidly as it is more widely used in sub-Saharan Africa, research is urgently needed to identify potential replacement drugs to enable the provision of effective antenatal malaria prevention in the future [141].

2.4. Effects of Lack of Global Purchasing Power and Cost Reduction

2.4.1. Global Financing for Malaria Control

Families bear much of the financial burden of malaria treatment in sub-Saharan African and elsewhere. Funds deployed by governments come from internal country resources, as well as from the major donors of development assistance, including support from individual countries (bilateral aid), and from international organizations, most importantly, the International Development Association of the World Bank, and other United Nations agencies. Nongovernmental organizations (particularly church-run medical services and groups like MSF) have for decades directly provided significant amounts of medical care, including drugs, in some cases (e.g., Zambia) acting as part of the public medical care system. Newer foundations—in particular, the Bill and Melinda Gates Foundation—and the beginning of operations by the Global Fund have provided significant increments of new funding, but also have highlighted the very low levels available in comparison to stated needs.

Although difficult to track accurately, the level of external support for malaria control from the key bilateral and multilateral agencies appears to have remained relatively static in recent years. The Commission on Macroeconomics and Health [153] reported on contributions to health development assistance by bilateral and multilateral donors in the late 1990s, which totaled about US$1.7 billion for all diseases and programs. The biggest share - about US$287 million per year - went to HIV/AIDS. Malaria-specific funding was US$87 million, and for tuberculosis, US$81 million. These are the funds earmarked for specific diseases, but it is likely that other general funds were also used for them. A more recent estimate for international funding of malaria control, including bilateral aid and low-cost loan assistance from multilateral organizations for the years 2000 and 2002, totals about US$100 million per year [154]. This figure is for all aspects of control (insecticide-treated nets, household spraying, and other preventive measures), and not just treatment. The amount that would have gone toward the cost of antimalarial drugs would be some fraction of the total.

Foundations have, for many years, contributed to global public health initiatives. In recent years, the Bill and Melinda Gates Foundation has entered at a level surpassing any other such efforts, spending close to US$1 billion per year on a variety of programs, some funded independently, and some through existing programs (e.g., the Gates Foundation is a major donor to the Global Fund, MMV, GAVI, and a number of other initiatives). This single entity is spending well over half of the amount spent by all governments and multilateral institutions together.

The Global Fund was created in 2002 to finance a dramatic turnaround in the world's response to this challenge, providing developing countries with the resources they need to turn the tide against the three diseases. In just four years (2003-2007), the Global Fund has become a leading force in the fight against the diseases. It has committed US$ 6.6 billion to over 460 programs, making it the largest international financer of efforts to control tuberculosis and malaria and among the three largest funders of HIV/AIDS programs. With investments in 136 countries, its reach is truly global.

Every year, between 350 and 500 million people get sick with malaria. More than one million of these die (most of them infants and children in Africa). For half the world's population, malaria poses an enormous health, social and economic burden - perpetuating poverty, hampering development, and reducing investment. After decades of hoping only to barely contain a growing epidemic, three new developments are now making it possible to rapidly drive down malaria mortality to levels below international targets: the development of new, long-lasting insecticide-treated bed nets, along with a renewed use of indoor residual spraying; development of new effective medicines called artemisinin-based combination therapies (ACTs) and large, additional sources of funding to finance comprehensive national malaria control campaigns.

The Global Fund to fight AIDS, tuberculosis, and malaria is the world's largest external source of finance for malaria control programs, providing two-thirds of all international financing. To date of 2008, the Global Fund has approved grants with a total value of US$ 2.6 billion over five years to 117 programs in 85 countries to support aggressive interventions against malaria. US$ 833 million has been disbursed so far (*http://www.theglobalfund.org/ documents/publications/factsheets/malaria_information_sheet_en.pdf.*).

2.4.2. WHO/UNICEF Procurement of ACTs with Price Cuts

In January 2006, WHO appealed to manufacturers to stop marketing oral artemisinin monotherapies and instead to promote quality ACTs in line with WHO policy [1]. The position has been widely disseminated via WHO Offices, WHO briefings to staff and to representatives of national health authorities at regional and inter-country meetings. Major procurement and funding agencies and international suppliers have accepted WHO recommendation and agreed not to fund or procure oral artemisinin monotherapies. In April 2006, the Global Malaria Programme of WHO provided a technical briefing to 25 pharmaceutical companies involved in the production and marketing of artemisinin monotherapies. Out of these, 15 declared their willingness to stop marketing artemisinin monotherapies over a short period of time, but 10 companies did not disclose their marketing plans for the future [155]. In addition, some countries, like China and Pakistan, have been visited by WHO delegations to address multiple domestic manufacturers involved in this sector. The evolving position of manufactures and of National Drug Regulatory Authorities (NDRA) of malaria endemic countries is monitored and displayed on the WHO Global Malaria Programme website front-page: *http://malaria.who.int/*.

In May 2007, the 60th World Health Assembly resolved to take strong action against oral monotherapies and approved the resolution WHA60.18 [1]:

- urges Member States to cease progressively the provision, in both the public and private sectors, of oral artemisinin monotherapies, to promote the use of artemisinin-combination therapies, and to implement policies that prohibit the production, marketing, distribution, and the use of counterfeit antimalarial medicines;
- requests international organizations and financing bodies to adjust their policies so as progressively to cease to fund the provision and distribution of oral artemisinin monotherapies, and to join in campaigns to prohibit the production, marketing, distribution, and use of counterfeit antimalarial medicines;

WHO/ United Nations Children's Fund (UNICEF) joint procurement of ACTs was initiated in 2003. At present, the list of available antimalarial products for WHO/UNICEF procurement comprises:

- Artemether-lumefantrine (Novartis);
- Artesunate + amodiaquine (Guilin, Ipca, Strides);
- Artesunate + sulfadoxine-pyrimethamine (Guilin, Ipca);
- Artesunate + mefloquine (Mepha);
- Artemether injectable forms (Dafra, Strides);

- Artesunate injectable forms (Guilin);
- Artesunate rectal preparations (Mepha).

Currently, WHO and UNICEF are preparing a joint Request for Proposal (RFPDAN-2007-500332) for the procurement of ACTs in 2008. Acceptable products have to meet a set of criteria:

- Compliance with WHO Guidelines for the Treatment of Malaria (i.e., product selection, dosage, treatment regimens, strength formulation, therapeutic indication);
- GMP compliance certified by WHO or stringent drug regulatory authority;
- Evaluation of the Interagency Pharmaceutical Product Questionnaire and support documentation submitted by the supplier on the following:

1) registration information,
2) regulatory (licensing) situation,
3) finished product specifications and compliance with international,
4) pharmacopoeia standards,
5) stability testing data (both accelerated and real time studies in Zone IV),
6) labeling information,
7) active pharmaceutical ingredient (API) characteristics and certification.

The new tender will also include as new element the submission of the Product Dossier to the WHO Prequalification Programme. Review of the documentation submitted by suppliers is affected by a WHO Team consisting of GMP, QSM, and Procurement Services. The recommendations of the team are presented to the respective Contract Review Committee for endorsement. On behalf of each Organization Letters of Long Term Agreement (LTA) will be issued to the selected suppliers and the products will then be indicated on the RBM website respectively in the Organization's WebBuy Catalogue at: *http://www.who.int/malaria/ pages/performance/antimalarialmedicines* (Table 15). This WHO/UNICEF-approved list of ACTs is used as reference list to guide procurement by other entities such as the World Bank, USPMI.

Table 15. Antimalarial Medicines available for procurement by WHO

International Non-Proprietary Name (INN)	Strength	Dosage form	Supplier	Packaging
Artemether /lumefantrine	20mg/120mg	Fixed dose tablets	Ajanta, IndiaCipla, IndiaNovartis, ChinaNovartis, USA, Cipla, India	30 blisters/box Single blisters
Artesunate + amodiaquine	50mg + 150mg (base) 50mg + 153mg (base)	Co-blistered tablets	Guilin, China; Cipla, India, Ipca, India, Strides, India	25 blisters/box Single blisters 10 treatments/box 100 treatments/box 10 treatments/box

Table 15. (Continued)

Artesunate + mefloquine	200mg+250mg	Co-blistered tablets	Mepha, Switzerland	Single blisters
Artesunate + sulfadoxine/pyrimeth amine	50mg+500/25mg	Co-blistered tablets	Cipla, India	Single blisters
Artemether	20mg/ml 80mg/ml	Injection	Dafra, Belgium	3 Ampoules/pack 10 Ampoules/pack 5 Ampoules/pack

WHO website: *http://www.who.int/malaria/pages/performance/antimalarialmedicines*.

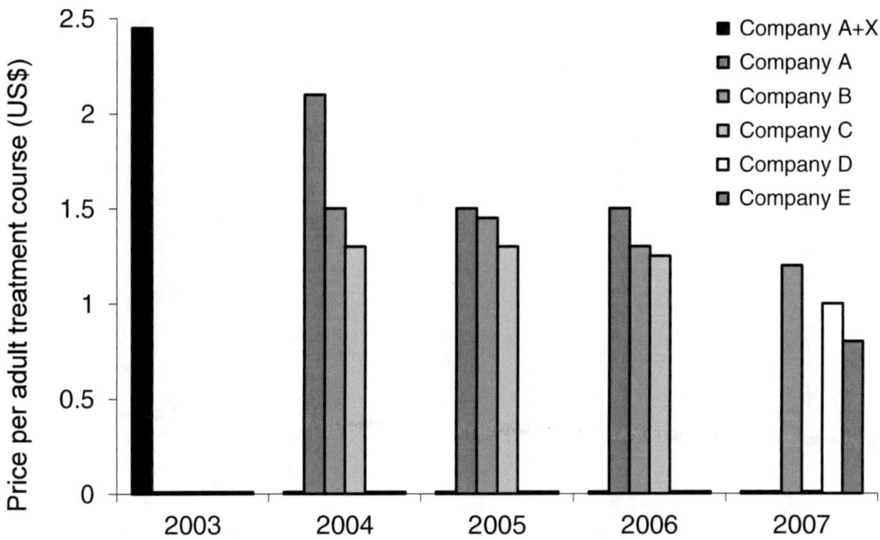

Figure 8. AS + AQ price reduction over five consecutive years by six WHO/UNICEF tenders (price per adult treatment course (in US$) from 2003 to 2007 [1].

The market effect of tenders can be illustrated by the example of artesunate + amodiaquine (Figure 8). Over the past five consecutive WHO/UNICEF tenders a substantial reduction in prices has been obtained for this combination. In 2003, when no co-blistered product was available on the market, the total price for the two components was at US$ 2.44/treatment course. The prices for the co-blistered products started at a range of US$ 2.10–1.35 in 2004 and went down to US$ 1.52–1.25 in 2006. In 2007, two companies did not respond to the WHO/UNICEF tender, but two new companies responded. Prices currently range between US$ 1.19–0.80 per adult treatment course; compared to the lowest price valid in 2005, this represents a price reduction of 20% [1].

2.4.3. Global Purchasing Power and Reduce Price

The overall market size for artemisinin-based treatments in 2007 is estimated at 150 million treatment courses, 20% of which is in the private sector. The market and supply of artemisinin raw material is still unstable. Following the global shortage in 2004 and the consequent sharp increase in the price of artemisinin (from US$ 230/kg in 2003 to US$ 1100/kg in 2004-2005) there has been a major expansion of plantations and of artemisinin producers in China, Vietnam, and East Africa in 2006. The mismatch between increased offer and tempered demand for ACTs in the public sector (due to low approval rates in GFATM Rounds 5 and 6, and delayed disbursement and utilization of funds from Round 4) created a major reduction in the price of artemisinin (currently at US$ 180/kg, below its production costs). There is growing concern that the withdrawal of farmers and artemisinin producers from this market will create a relative reduction of the artemisinin inventory and risk of shortages in 2008-2009. Only few manufacturers are establishing stockpiles of artemisinin to prevent the effects of a possible shortage of raw materials, and the production of ACT will continue to depend from plant extraction.

The pre-qualification of artemisinin pharmaceutical products by WHO is hindered by the lack of "originator products" for most ACTs, i.e., medicines approved by stringent drug regulatory authorities against which generic companies are compared through bioequivalence studies. Several ACT products are under evaluation, and the status of product assessment is available at *http://mednet3.who.int/prequal/*. Technical support to manufacturers is required to accelerate the pre-qualification of antimalarial medicines.

The number of ACT treatment courses procured by national governments of endemic countries for use in their public sector increased from around half a million in 2001 to 31.3 million in 2005 (Figure 4)—25.5 million of these being for countries in Africa. More than 80% of these orders were placed through WHO. From January 2005 to July 2006, the total ACT procurements for the public sector made by WHO, UNICEF, and to a lesser extent, by Crown Agents has been 62.6 million treatment courses: 80.8% of these orders were for artemether/lumefantrine, 12.5% were for artesunate plus amodiaquine, and 6.7% were for artesunate plus sulfadoxine/pyrimethamine. At the time when the ACT market was quite new and unstable, two key interventions helped to prevent ACT prices from escalating: a differential pricing agreement between WHO and Novartis Pharma AG, the manufacturer of artemether/lumefantrine (at that time a single-source product) to supply the medicine at cost price for public sector use in endemic developing countries and the joint call by UNICEF and WHO for annual tenders for multi-source generic ACTs (artesunate plus amodiaquine, artesunate plus sulfadoxine/pyrimethamine, and artesunate plus mefloquine) meeting minimum quality standards. The latter not only increased market competition among ACT manufacturers but also led to quality ACTs entering the market. It also contributed significantly to reducing the price of artesunate + amodiaquine from USD 2.60 in 2003 (ex-factory price of adult treatment course with the two compounds in separate blister packs) to USD 1.68 in 2004 (for the co-blistered product), and even further, to the current price range of 1.35–1.55 (Table 16).

Key among the interventions during the transition period of market instability was WHO and UNICEF maintaining a dialog with the pharmaceutical industry providing regular updates of global estimates of ACT requirements. After a period of great market instability and as a result of these multiple interventions, the ACT market is now stable to the extent that the current global production capacity for quality ACTs is adequate to meet the global demand. The transition phase, however, was replete with difficulty.In 2004, the rapid increase in demand led to a global shortage of the single-source ACT, artemether-lumefantrine, manufactured by Novartis Pharma AG. The public sector demand for artemether/lumefantrine increased rapidly from 2001/2002, when WHO ordered only 0.32 million treatment courses, to 2004/2005. During the period from October 2004 to March 2005, WHO and UNICEF ordered a total of 4.5 million treatment courses of treatment, but the company had artemisinin supplies to produce only 2.4 million treatment courses.

Table 16. Price per adult treatment course of ACTs paid by WHO and UNICEF as of July 2006 [55]

Medicine	Manufacturer/supplier	FCA price/adult Rx (US $)
Artemether/lumefantrine	Novartis	2.40
Artesunate + amodiaquine	CIPLA	1.49
	IPCA	1.35
	Sanofi/Aventis	1.55
Artesunate + mefloquine	Mepha	5.27
Artesunate + SP	Efficacious	1.41

* Only artemether/lumefantrine is currently available as a co-formulated product. The others are available as co-blistered products. FCA, free carrier (ex-factory); RX, treatment.

The reason for the shortage was an insufficient supply of artemisinin, the raw material extracted from the plant *Artemisia annua*, needed to manufacture artemether. Cultivation of *A. annua* requires a minimum period of 6 months, and extraction, processing, and manufacturing of the final product requires an additional 3–5 months. The surge in the global demand for artemisinin created a temporary market crisis, with a consequent global shortage of artemether-lumefantrine that lasted until the end of 2005. Since then, the manufacturing company has invested in sourcing stocks of artemisinin raw materials and has expanded its manufacturing facilities to increase production capacity to meet the global demand. The cultivation of *A. annua* and the extraction of artemisinin raw material, which, until recently were restricted to a few countries - notably Vietnam and China, has now been extended to several countries, including in Africa. The multi-source ACTs that are manufactured by several companies were never subject to shortages. The production capacity for ACTs was in 2006 by the six major pharmaceutical producers (Novartis, Guilin, Ipca, Mepha, Dafra, and Strides), and estimated at 130 million treatment courses, exceeds the public sector demand for ACTs, which are currently estimated at 90 million treatments for 2006. The global forecast for 2007 and 2008 for the public sector requirements of ACTs, taking into account existing financing mechanisms and deployment rates in countries, is in the order of 160 and 200 million treatment courses, respectively [55].

Antimalarial R&D affects the lock of global purchasing power.

Malaria has never been an attractive target for the research-based pharmaceutical industry for the very reason the current crisis exists: lack of funds to pay for even modestly priced drugs. Most of the antimalarial drugs developed in the past half century are the products of publicly financed R&D. Up to now; the largest number of new products has come from R&D conducted by the Chinese government and the U.S. military. Most recently, the early development of the artemisinins was launched by Chinese government scientists, in response to concerns of the North Vietnamese government over the toll of chloroquine-resistant malaria on its soldiers during the conflict with the United States. The U.S. military continues its research program (though at a considerably lower level than its peak, during the Vietnam War), but in the past 5 years, has been up-staged - in a positive way - by the Medicines for Malaria Venture (MMV), an international "public-private partnership" with the sole mission of developing new antimalarial drugs. MMV today has the most active pipeline of research and products in various stages of development that has ever existed (see chapter 10 for a more detailed discussion of antimalarial drug R&D).

The current inaction over funding ACTs could call into question the sense of continuing to allocate public funding and solicit private contributions toward R&D for new antimalarials. The unstated implication of the failure to purchase ACTs is that only drugs that are about as inexpensive to produce as chloroquine are worth developing for first-line treatment. There may well be new drugs, cheaper than ACTs, in another 10 years, but they may still be unaffordable to most users. Rationalizing current R&D expenditures is not a reason to establish a global subsidy for ACTs, but the failure to do so may well have consequences for the R&D process that are detrimental and difficult to reverse [42].

2.4.4. More Efforts on ACT Cost-effectiveness

The current product profile of ACTs requires improvements to simplify prescription and promote higher adherence to treatment: some ACTs have relatively complex regimens (twice daily for three days for artemether-lumefantrine) or are available mainly or exclusively as co-blistered products (artesunate in combination with amodiaquine, mefloquine or sulfadoxine-pyrimethamine). Current formulations have generally only 2 years shelf life. This, together with different age-specific course-of-therapy packaging, makes supply management of ACTs very difficult. Moreover, there are too few ACT pediatric formulations on the market or under development, and the co-administration of both ACT components in separate pediatric formulations (suspension, powder for dissolution, granulate in sachets) is suboptimal. There is a need for highly effective ACTs in fixed-dose combinations, with simplified dosage regimens, in stable formulations with appropriate packaging for large-scale use, as well as new pediatric formulations to facilitate administration to infants.

The market of artemisinin pharmaceutical products is very dynamic, with the number of generic companies active in this sector increasing every year. In particular the number of companies involved in the production and marketing of oral artemisinin monotherapies has increased from 17 companies identified in 2005, to 38 in 2006, up to the total of 67 companies known to WHO by August 2007. In addition, there are 16 companies involved in manufacturing exclusively ACTs and 25 companies of uncertain status are involved in rebranding, re-packaging, re-selling products manufactured by other companies, without disclosing product information to the public. Overall, WHO has recorded and classified 94 oral artemisinin products, currently being sold for the treatment of uncomplicated malaria, mainly in the private sector of malaria endemic countries. A major challenge for the implementation of the withdrawal of oral artemisinin monotherapies from the market has been the rapid occupation of the market by other manufacturing companies not following WHO recommendations. In unregulated settings, responsible manufacturers who are phasing out their monotherapies provide market opportunities to companies which may promote substandard products. Intensified support should be given to national drug regulatory authorities of malaria endemic countries to withdraw the marketing authorizations of all oral artemisinin monotherapies, withdraw/withhold licenses for manufacturing and exports, to establish national laws to prohibit imports, and to enforce strict control of borders for importation of medicines and promote active recall of the products already in the market. Measures should also be put in place at country level to mitigate the financial losses of manufacturers (domestic and international), importers and distributors, with agreed timeframes between private sector representatives and the national health authorities of the country concerned.

Increasingly high levels of funds are being mobilized for malaria control, and ACT procurement will be an important component of several new initiatives, including the GFATM Round 7, US PMI, World Bank and, more recently, UNITAID. A Global ACT Subsidy (recently renamed as "Affordable Medicine Facility for malaria") is under definition with increasing attention and support from international and bilateral development agencies as mechanism to ensure access to quality ACT at affordable prices in both the public and the private sectors. All these initiatives, however, and their potential high impact on the malaria burden is threatened by the risk of development of resistance to artemisinin and its derivatives, which is enhanced by the wide scale use of oral artemisinin monotherapies, and partially effective ACTs (e.g., ineffective companion medicine, suboptimal dosage or substandard quality). The recent findings at the Thai-Cambodia border, where several cases of ACT treatment failures have been documented, point to real risks of development of artemisinin resistance in areas where oral artemisinin monotherapies and/or partially effective ACTs, have been deployed on a large-scale. Efforts to replace oral artemisinin monotherapies with highly effective ACTs need to be accelerated.

The overall market size for artemisinin-based treatments has been progressively expanded and, by 2007, it has reached an estimated 150 million treatment courses, 20% of which in the private sector. With increased mobilization of international funds, the size of the ACT market is expected to progressively expand in both the public and private sectors, and it will continue to be dependent on extraction from *Artemisia annua*, since bio-synthetic artemisinin or chemical analogues will not become available before 2011. The market of artemisinin raw material has undergone major price fluctuations over the last two years and present serious risks of supply shortages in the short term. In order to minimize supply shortages and to contain the risk that sub-standard quality artemisinin and API will be sourced at the critical time of supply shortages, international support needs to be mobilized soon for the creation of artemisinin safety stocks of high quality. Artemisinin offers the possibility of longer shelf life up to 4–5 years, through a process of re-crystallization, and its early establishment could serve the ACT production needs for multiple years.

The process of pre-qualification of artemisinin pharmaceutical products by WHO is the international reference standard to advice UN and international procurement agencies on the procurement of quality medicines for HIV, tuberculosis and malaria. Differently from ARV medicines, the process of pre-qualification of ACTs is not satisfactory as only one fixed-dose combination ACT has met the required standards, i.e., artemether-lumefantrine supplied by Novartis Pharma. Acceleration of the review process, consistency in follow-up of technical questions raised by PQ program, more opportunities for face-to-face meetings between manufactures and evaluators, as well as intensified technical support to manufacturers are all required to increase the number of antimalarials which will be prequalified in the future.

In the absence of multiple pre-qualified products, two main systems are in use, i.e., the WHO/UNICEF joint Request for Proposals and the GFATM compliance list for single source or limited source medicines (classified as "Ci"), supplemented by pre-shipment QC of samples ex-manufacturer. Over the recent months, multiple funding institutions and bilateral development agencies have expressed concerns regarding significant important differences in quality assessment criteria for antimalarial medicines in case of non-prequalified medicines. A general trend is emerging between multiple Agencies to agree on common quality criteria for the procurement of antimalarial medicines in the absence of prequalified products. The more efforts on ACTs for price cutting and more cost-effectiveness are going on.

1) Priority efforts for WHO
- WHO will promote a universal campaign to ensure access to highly effective ACTs for the treatment of uncomplicated malaria in the public and private sectors as well as in the community, based on the most cost-effective treatment selected according to the resistance profile of *P. falciparum* in the country.
- WHO will develop a new system of ACT forecasting based on tracking of international funds allocated for procurement, actual disbursements to recipients, and expected orders according to funded procurement plans, taking into account the operational targets and the implementation capacity of the countries concerned.
- WHO will invest in ACT market analysis, to improve the understanding of production capacity of manufacturers of finished pharmaceutical products, API suppliers, including extractors and derivatives, and better understanding of dynamics affecting the agricultural sector and production of *Artemisia annua*.
- WHO will explore with international partners and funding agencies the feasibility of establishment of buffer stocks of high quality artemisinin, in order to reduce the risks of shortages due to recent high price fluctuations that have discouraged a large proportion of raw material suppliers from this market.
- WHO will expand the pre-qualification of manufacturers of active pharmaceutical ingredients (API) for ACTs, including extractors of artemisinin and suppliers of arteether, artemether, artesunate, and dihydroartemisinin, in order to ensure sourcing of quality API for ACT manufacturers.
- WHO will consider multiple approaches to increase the pre-qualification of antimalarial medicines, including the following: i) additional training workshops & materials for NDRAs and manufacturers, ii) expanded pool of consultants providing on-site technical advice to manufacturers on pre-qualification requirements, iii) increased opportunities for face-to-face meetings between manufacturers and inspectors, and iv) pre-qualification of individual medicinal components for copackaged ACT products, when only one meets the pre-qualification requirements.
- WHO will support National Drug Regulatory Authorities to take necessary measures to withdraw the authorization of oral artemisinin monotherapies and

will be fully supportive of even more stringent actions by the national health authorities, such as recall of these products from the market, once ACTs have become widely available and affordable to the people affected by malaria.

- WHO will seek information from collaborating companies willing to share confidential evidence of poor marketing practices (i.e., marketing oral artemisinin monotherapies, ineffective antimalarial medicines not in line with the WHO Guidelines for the Treatment of Malaria, or misusing WHO name/logo for commercial purposes) as part of the implementation of the new code for artemisinin marketing practices (CAMP).
- WHO will also collate, as part of the implementation of the new code for artemisinin marketing practices (CAMP), confidential information provided to manufacturers on results of analysis of substandard artemisinin products, including poor stability data, as well as alert reports on counterfeit products. A pilot reporting system already developed by WPRO (Rapid Alert System, see *http://218.111.249.28/ras/default.asp*) will be considered for possible extension to serve global monitoring of antimalarial counterfeits.
- WHO will accelerate the current system for reviewing new antimalarial medicines as part of the updating of the Guidelines for the Treatment of Malaria, and, in addition to the planned cycle of regular updates every two years, it will also consider convening ad hoc expert reviews during the interim periods, and making the results of such reviews publicly available.
- WHO will also promote high international quality standards for the procurement of artemisinin-derivatives by multiple international organizations involved in the supply of these medicines for use in disease endemic developing countries [1].

2) To improve agricultural practices, selection of high-yielding hybrids, microbial production, and the development of synthetic peroxides will lower the prices of ACTs [156].

3) The University of California, Berkeley is using full synthetic biology to help reduce the cost of artemisinin [157].

2.5. Low-cost ACTs Available Today and Future Opportunities

2.5.1. Cheaper ACT of Artesunate Plus Amodiaquine (ASAQ)

The first affordable combination anti-malarial drug tailored for children will soon be available across Africa, potentially saving millions of lives, the nonprofit organization and the pharmaceuticals giant who worked to develop it said recently. The new medication, known as ASAQ, combines two of the most effective drugs known to treat malaria, artesunate and amodiaquine. ASAQ is the result of a $21 million, two-year project by the Drugs for Neglected Diseases Initiative and pharmaceutical manufacturer Sanofi-Aventis. To make the drugs available where they are most needed, the new drug is not patented and will be available to anyone who wants to manufacture it. "This drug will save lives," said Dr. Bernard

Pecoul, executive director of the Drugs for Neglected Diseases Initiative. "The aim of this project was to develop a product as adapted as possible to the malaria situation in Africa."

The drug already has been registered in Morocco, and has been authorized in 10 other African countries. Every year there are as many as 500 million cases of malaria worldwide with more than 1 million deaths. The disease primarily affects children under 5 in sub-Saharan Africa. If ASAQ and other drugs like it are made widely available, along with other anti-malaria weapons such as bed nets impregnated with insecticide, the number of malaria cases in Africa could be dramatically cut. Instead of cutting two different pills intended for adults into child-appropriate sizes - a less-then precise method - health care workers will now be able to give children a single pill. The drug comes in four formulations, including versions for infants and children over 13, or adults. ASAQ merges two pills into one, making it easier both to prescribe and to take. Children will need to take one tablet per day for three days. "Having a fixed-dose combination is a significant advance," said Dr. Chris Hentschel, president and CEO of the Medicines for Malaria Venture. "The fewer pills people have to take, the more chance they will actually take them."

In addition, the problem of resistance developing should be lessened. Experts worry that when patients have multiple pills to take, they may not take all of both sets of pills, which can lead to the development of resistance. But because ASAQ combines two drugs in one, there are fewer concerns about patients taking their full doses. ASAQ is the first of numerous malaria treatments in the drug development pipeline, which marks a new era in antimalarials. "We need a greater diversity of drugs to drive the prices down to affordable levels," said Dr. Sylvia Meek, technical director at the Malaria Consortium.

As a result of new partnerships between public and private sectors, as well as an injection of funds from organizations such as the Bill & Melinda Gates Foundation, there is a newfound drive to develop new drugs for Africa, which has previously been seen as unprofitable. ASAQ is cheaper than other currently available fixed-dose combinations, and will be available at-cost price to countries battling malaria, international organizations, and non-governmental organizations. The pills cost less than US$ 0.50 cents for children and less than US$1.00 for older children and adults. Even at these relatively low costs, however, that may still be too expensive for Africa. "Only 5 percent of people who need anti-malaria drugs get them, "said Dr. Nick White, a malaria expert at Mahidol University, Bangkok." That's pathetic and we need to do better." Malaria is one of the top three killer diseases in Africa. It is transmitted to people from mosquitoes and usually causes symptoms including fever, vomiting, headaches, and other flu-like symptoms. If left untreated, the disease can be fatal, as announced by Marla Cheng in London on March 1, 2007.

2.5.2. Cheaper ACT of Artemether Plus Lumefantrine (AL)

Artemether plus lumefantrine (AL) has recently been introduced in South Africa as the first line of treatment for malaria, in areas where chloroquine and sulfadoxine/pyrimethamine (SP) have become resistant. In September 2006, Novartis reduced the average price (children and adults) of Coartem from US$ 1.57 to US$ 1.00 per treatment to help expand access to medicine; maybe this agreement means the price can possibly go down even further? However, a malaria treatment at US$ 1.00 is still quite expensive compared to the cents we paid for chloroquine and SP.

Recently, Novartis announces 20% average reduction in average price of malaria drug Coartem® from US$ 1.00 to US$ 0.80 to further accelerate access in malaria-endemic regions (Basel, April 23, 2008, at: *http://www.mmv.org/IMG/pdf/Coartem_press_ release_April_2008.pdf*). Malaria affects 300 to 500 million people every year and is the biggest killer of children in Africa – every 30 seconds a child dies of malaria. Coartem, a highly-effective artemisinin-based fixed-dose combination, produces cure rates up to 95%. Since 2001, Novartis has provided more than 160 million treatments without profit to those most in need. Price reduction made possible through efficiency gains in producing Coartem. Starting this Friday, which is World Malaria Day, this price reduction will increase access to Coartem for millions of malaria patients, especially children in low income regions of Africa.

To ensure a dependable supply of Coartem and to meet rising demand, Novartis has invested heavily to expand production capacity at state-of-the-art facilities in China and the United States. The recent efficiency increases in producing Coartem mean that the public sector price can now be reduced by an average of 20% compared to the 2007 price. The price reduction applies to all Coartem dosages including those for children and adolescents, who account for nearly 75% of patients taking Coartem. With the new lower price of US$ 0.37 for children's doses, countries will now be able to treat many more children than before.

Coartem, the only fixed-dose ACT that has been approved by a stringent, internationally-recognized health authority, is indicated for the treatment of acute uncomplicated *falciparum* malaria, the most dangerous form of malaria. Coartem is highly effective and well-tolerated, providing cure rates of up to 95% even in areas of multi-drug resistance. Combining two or more malaria drugs has the potential to prevent or delay the development of resistance.

Many clinical trials conducted in children and adults have been confirmed that the ASAQ (Coarsucam™) is as effective as artemether-lumefantrine, AL, (Coartem®) in the African countries where the 2 combinations are recommended by the WHO. Tolerance proved satisfactory in all treatment groups in these clinical trials.

The first clinical trial comparing the efficacy of ASAQ and AL that involved an extended, 42-day follow-up period, polymerase chain reaction-adjusted parasitological cure rates (PCR APCRs), and systematic analyses of genetic markers related to quinoline resistance was conducted by Martensson *et al.* in 2005 [158]. A total of 408 children with uncomplicated *P. falciparum* malaria in Zanzibar, Tanzania, were enrolled. Children who were 6-8 months of age and/or who weighed 6-8 kg were assigned to receive ASAQ for 3 days. Children who were 9-59 months of age and who weighted > or =9 kg were randomly assigned to receive either ASAQ or AL for 3 days in standard doses. Intention-to-treat analyses were performed. Results indicated that age- and weight-adjusted PCR-APCRs by follow-up day 42 were 91% (188 of 206 patients) in the ASAQ group and 94% (185 of 197 patients) in the AL group (odds ratio [OR] for the likelihood of cure, 2.07; 95% confidence interval [CI], 0.84-5.10; P=.115). A total of 5 and 7 recrudescences occurred after day 28 in the ASAQ and AL groups, respectively. On the assumption that 10 malaria episodes with uncertain PCR results were recrudescences, PCR-APCRs decreased to 88% in the ASAQ group and to 92% in the AL group. Unadjusted cure rates by day 42 were 56% (116 of 206 patients) in the ASAQ group versus 77% (151 of 197 patients) in the AL group (OR, 2.55; 95% CI, 1.66-3.91; P<.001). Rates of re-infection by day 42 were 36% (65 of 181 patients) in the ASAQ arm versus 17% (31 of 182 patients) in the AL arm (OR, 0.37; 95% CI, 0.22-0.60; P<.001). A significant selection of *P. falciparum* multidrug resistance gene 1 allele 86N was found in isolates associated with re-infection after AL treatment, compared with isolates at baseline (2.2-fold increase; P<.001).

Both treatments were highly efficacious, the cure rate is 91% for ASAQ and 94% for AL, but AL provided stronger prevention against re-infection. The high proportion of recrudescences found after day 28 and the genetic selection by the long-acting partner drug underlines the importance of long follow-up periods in clinical trials. A long follow-up duration and performance of PCR genotyping should be implemented in programmatic surveillance of antimalarial drugs [158].

Another clinical trial was reported by Guthmann *et al.* in 2006. Children 6-59 months of age with uncomplicated *P. falciparum* (137 patients) malaria were randomized to receive either AL (Coartem) or ASAQ. After 28 days of follow-up, there were 2/61 (3.2%) recurrent parasitemias in the AL group and 4/64 (6.2%) in the ASAQ group (P = 0.72), all classified as re-infections after PCR genotyping (cure rate = 100% [95%CI: 94-100] in both groups). Only one patient (ASAQ group) had gametocytes on day 28 versus five (AL) and three (ASAQ) at baseline. Compared with baseline, anemia was significantly improved after 28 days of follow-up in both groups (AL: from 54.1% to 13.4%; ASAQ: from 53.1% to 15.9%). The findings are in favor of a high efficacy of both combinations in Caala. Now that AL has been chosen as the new first-line anti-malarial, the challenge is to insure that this drug is available and adequately used [159].

Recently, this study compared the new fixed-dose combination of ASAQ jointly developed with the fixed-dose combination of AL. It was conducted in 5 centers in Africa, two of which are in Senegal, 1 in Mali, 1 in Cameroon and 1 in Madagascar, from March 2006 to January 2007. Three groups of patients were enrolled – two of these groups were treated using ASAQ taken once or twice a day and one was treated with AL administered twice daily in accordance with recommended dosage. The primary objective was to demonstrate, 28 days after treatment initiation (D28), the non-inferiority of ASAQ in terms of clinical and parasitological efficacy compared with AL.

The two secondary objectives were to compare the clinical and parasitological efficacy of the two treatments on D28 in the sub-group made up of the children aged less than 5 years, which is the WHO's main target population, as well as to evaluate the treatment regimens of artesunate + amodiaquine administered once or twice daily. Clinical and biological tolerability was also evaluated for all 3 groups. Overall, total 3153 patients were included in the study, with the following results:

- On D28, clinical and parasitological cure was achieved in 95.2-97.1% of the patients treated with ASAQ once daily, and in 95.5-98.2% of the patients in the AL group.
- Non-inferiority was established for all treatment groups, including children aged less than 5 years.
- Clinical and biological tolerability was good with a comparable profile between the treatment groups; no unexpected side effects were observed.

AS+AQ and AL were efficacious for treatment of children with uncomplicated malaria in Africa and drug-related adverse events were rare in treated subjects during one year of follow-up. The high prevalence of potentially AQ resistant parasites raises questions about the utility of AQ as a partner drug for ACT in Africa. In view of the high prevalence of potentially AQ-resistant mutant parasites in the study area, the efficacy of the AS+AQ combination, the first-line treatment for uncomplicated malaria in Africa, requires continuous monitoring and evaluation. These results support the recommendation for a minimum follow-up period of 28 days in the assessment of efficacy of ACTs, with PCR-genotyping of all recurrences and the use of the Kaplan-Meier product-limit formula to estimate efficacy of treatment in all correctly randomized patients [160-162].

2.5.3. Opportunities of ACTs

2.5.3.1. More Funds for Malaria Research and Policy Implementation

In the last decade, more funding has been availed by several International agencies for malaria research. These include the Global Fund, Bill & Melinda Gates Foundation, USAID, DFID, WHO/TDR, MMV, etc. Within the same period, insecticides, ITNs and ACTs have been identified as best malaria control tools. Further funding should now be used in finding the best possible ways of deploy these tools for effective malaria control. Additionally, funds should be invested in building and strengthening local capacities (personnel, resources, and infrastructure) so that developing countries have enabling environments and a critical mass of experts to fully implement policies. Appropriate use of granted funds, as well as clear policies within recipient countries and of funding agencies is strongholds for success. Any fund misappropriation and lack of clear policies together with political influences of powerful donors require quick resolution.

The Global Fund is the financial engine behind global scaling up of ACTs. Prior to granting funds WHO and the other RBM partners give technical advice on each application submitted to the Fund through the Country Coordinating Mechanism (CCM). Unclear guidance for submission and approval of applications is believed to have led to the Fund funding some ineffective antimalarial drug policies (AMDPs) prior to 2003 [163]. Currently, the mode operand for CCMs is clearer, WHO/RBM and WHO Country Offices are providing adequate technical advice to CCMs, and the Fund's Technical Review Panel has been improved in composition and in its decision making process. Consequently, the number of countries funded appropriately to deploy ACTs has increased significantly in 2004. However, quick flow of cash from the Fund to countries has remained slow. This is one reason why despite high adoption of ACTs in 2004, only nine countries were implementing the policies [164]. Reasons for the delays should be identified and sorted out quickly otherwise grants will remain on paper.

2.5.3.2. Interest in Antimalarial Drug Development

Interest in anti-malarial drug development among some drug companies has increased in the last 5 years. Parallel to this, collaboration between public and private sectors has resulted in less expensive drugs affordable to developing countries. For example, Chlorproguanil/dapsone (LapDap) was developed through such collaboration among WHO, GlaxoSmithKline (GSK), the UK Department for International Development (DFID), and UK researchers. LapDap was registered in UK in August 2003 and since then several African countries have also registered it. GSK has agreed to make the drug available at preferential prices [165] across sub-Saharan Africa as soon as local approval is granted. The cost to governments and non-governmental organizations is projected at US\$ 0.29 for a 3-days course for adults and US\$ 0.15 for children (http://bmj. com/cgi/content/full/327/741/360-b). In meeting CT requirements, developmental plans are underway to produce LapDap–artesunate as another ACT. It is assumed that addition of artesunate will not make significant increase in the cost.

1) Artesunate-amodiaquine and Artesunate-sulfadoxine/pyrimethamine

Age-based blister packs of artesunate-sulfadoxine/pyrimethamine (AS-SP) and artesunate-amodiaquine (ASAQ) have been developed and used in clinical trials in Africa. However, rapidly increasing parasite resistance to SP and amodiaquine in Africa is a major limitation on their future usefulness. In a recent study in Tanzania, amodiaquine- SP and ASAQ were significantly less effective than artemether–lumefantrine in treating uncomplicated malaria in an area with moderate resistance to SP and amodiaquine [148]. Similarly, drug efficacy studies for surveillance of resistance conducted regularly by EANMAT in East Africa since 1998 indicate gradual decline in efficacy for these combinations (*http://www. eanmat.org*). Despite, ASAQ and artesunate-mefloquine are being developed as fixed dose combinations by WHO/TDR (Drugs for Neglected Diseases/Populations), which will minimize noncompliance as a possible cause of ineffectiveness. Their development and deployment should be fast before they become useless in most of Africa. Equally important, studies are required to identify threshold resistance levels of component drugs beyond which it would not be advisable to develop and deploy respective combinations.

2) Dihydroartemisinin-Piperaquine (Artekin)

Artekin was developed in China and is registered in China and Cambodia. It has a cost advantage over other ACTs (US\$ 1 for an adult treatment) that brings it within reach of many, making it a potential best candidate for deployment in Africa and other poor countries. Non-GMP Artekin has been evaluated in clinical trials in China, Cambodia [166, 167], Vietnam [151], and Thailand [168]; where it has proved tolerable and efficacious, including in treating multi-drug resistant strains [169]. GMP–Artekin is now produced by Sigma-Tau Farmaceutiche Riunite ApA in Italy and shortly Phase III clinical trials to assess its safety and efficacy prior its registration will start in Africa. A developmental program has been agreed between Holleyin Pharmaceuticals and Guangzhou University (China), The University of Oxford, Malaria Medicine Venture (MMV) and Sigma-Tau Industrie Farmaceutiche Riunite ApA to support international registration of the drug. To avoid shortage like the case with Coartem, there is need to pre-qualify several other pharmaceutical companies for accelerated production.

3) Other Drugs in Development

Bayer is developing Artemisone, a metabolically stable semi-synthetic derivative of artemisinin. Several trioxanes obtained by total synthesis are now available and are being assessed for further development by MMV [72].

Chapter III

Efficacy and Chemotherapy
of Antimalarial Drugs

First antimalarial drug, chloroquine, silently saved millions of lives and cured billions of debilitating episodes of malaria. *Falciparum* malaria has always been a major cause of death and disability, particularly in Africa, but chloroquine offered a measure of control even in the worst-affected regions. At roughly 10 cents a course and readily available from drug peddlers, shops, and clinics, it reached even those who had little contact with formal health care. Now chloroquine is increasingly impotent against *falciparum* malaria, and malaria's death toll is rising. Treatment of the disease with single drugs, such as chloroquine, sulfadoxine/pyrimethamine or mefloquine, has led to the emergence of resistant *P. falciparum* parasites that lead to the most severe form of the illness. Chief among these new treatments are the "artemisinins," paradoxically both an ancient and a modern class of drug that can still cure any form of human malaria. If *falciparum* malaria sufferers had the same broad access to artemisinins that currently exists for chloroquine, the rising burden of malaria in the world today would halt or reverse.

The challenge, therefore, is twofold: to facilitate widespread use of artemisinins while, at the same time, to preserve their effectiveness for as long as possible. Artemisinin-based combination therapies (ACTs) are currently recommended by WHO for the treatment of uncomplicated *P. falciparum* malaria [47]. Artemisinin and semisynthetic derivatives, including artesunate, artemether and dihydroartemisinin, are short-acting antimalarial agents that kill parasites more rapidly than conventional antimalarials, and are active against both the sexual and asexual stages of the parasite cycle. Artemisinin fever clearance time is shortened to 32 hours as compared with 2-3 days with older agents. To delay or prevent emergence of resistance, artemisinins are combined with one of several longer-acting drugs - amodiaquine, mefloquine, sulfadoxine/pyrimethamine or lumefantrine - which permit elimination of the residual malarial parasites. Therefore, the treatment of *falciparum* malaria is under a new age of artemisinin.

3.1. Antimalarial Activity of Antimalarial Drugs

3.1.1. Malaria Disease and Ancient Medicine

Malaria, a disease transmitted by the female Anopheles mosquito, has had devastating effects on human populations for more than 4000 years. The WHO definition of malaria is an infectious disease caused by protozoa of the genus Plasmodium; episodes of illness are "attacks of chills, fever, and sweating caused by Plasmodium infection." Malaria is a major global health problem and affects more than 2400 million people, over 40% of the world's population, in more than 100 countries in the tropics from Asia to Africa. Every year 300 million to 500 million people suffer from this disease (90% of them in sub-Saharan Africa, two-thirds of the remaining cases occur in six countries- India, Brazil, Sri Lanka, Vietnam, Colombia, and Solomon Islands). WHO forecasts a 16% growth in malaria cases annually. About 1.5 million to 3 million people die of malaria every year (85% of these occur in Africa), accounting for about 4-5% of all fatalities in the world. One child dies of malaria somewhere in Africa every 20 sec., and there is one malarial death every 12 sec somewhere in the world. Malaria kills in 1 year what AIDS killed in 15 years. In 15 years, if 5 million have died of AIDS, 50 million have died of malaria. Malaria ranks third among the major infectious diseases in causing deaths - after pneumococcal acute respiratory infections and tuberculosis. It is expected that by the turn of the century malaria would be the number one infectious killer disease in the world.

It accounts for 2.6 percent of the total disease burden of the world. It is responsible for the loss of more than 35 million disability-adjusted life-years each year. Every year ~ 30000 visitors to endemic areas develop malaria and 1% of them may die. Estimated global annual cost (in 1995) for malaria: US$ 2 billion (direct and indirect costs, including loss of labor). Estimated worldwide expenditure on malaria research: US$ 58 million, one thousandth of the US$ 56 billion spent globally on health research annually. Estimated annual expenditure on malaria research, prevention and treatment: $ 84 million. Estimated worldwide expenditure per malaria fatality: $ 65; as compared to $ 3274 for HIV/AIDS and $ 789 for asthma. That is to say, one HIV/AIDS death is equal to about 50 malaria deaths [170].

Indeed, close relatives of the human malaria parasites remain common in chimpanzees, our closest relatives. References to the unique periodic fevers of malaria are found throughout recorded history, beginning in 2700 BC in China. The term malaria originates from Medieval Italian: *mala aria* — "bad air"; and the disease was formerly called ague or marsh fever due to its association with swamps. To fight against the fever, a plant known as qinghao had been used by Chinese in the early second century BCE (before Common Era). The earliest description of qinghao herb for treatment of malaria-related symptoms is found in the writings of Ge Hong (281-340 AD) who was a fourth century Chinese physician. In spite of extraction problems, Chinese physicians have used *A. annua* and *A. apiacea* for producing drugs with antimalarial and other apparently therapeutic properties since at least the second century BCE. However, while the literature on the chemistry of artemisinin and clinical trials reporting its effects is vast, research into the history of the Chinese medical drug qinghao is scarce, or even non-existent.

The name qinghao sometimes refers to the 'drug,' and sometimes to the 'kind' of plant from which the drug is derived. In the history of qinghao, *A. annua* and *A. apiacea* were not always kept apart. This distinction became definitive in the work by a famous physician and natural historian Li Shizhen (1518—1593), whose Encyclopedic Classified Materia Medica was published posthumously in 1596 and represents a landmark in the history of the Chinese materia medica. Li Shizhen did away with the main term caohao, probably because he considered it too general and vague, and instead differentiated between two kinds, qinghao (blue-green hao) and huang huahao (yellow blossom hao). Modern botanists have identified the former as *A. apiacea*, the latter as *A. annua*.

This leads to the theme, which concerns the preparation of the Chinese medical drug qinghao. Ge Hong described an ingenious method of preparing qinghao, although it was not until 1596 that the method was recorded within the genre of the Chinese materia medica by Li Shizhen. Unfortunately, no one has published this method in a Western language, although it may have important practical implications for future antimalarial applications of the fresh plant. Ge Hong's method consisted of soaking the fresh plant in water, and then wringing out the whole plant and ingesting its juice in its entirety. The soaking of the entire fresh plant in water and its subsequent wringing out must have resulted in an emulsion of water, flavonoids and other etherical oils contained in stem and leaves (as mentioned earlier, one of the names of the plant was 'fragrant hao'), which may have facilitated the extraction of the Artemisia sesquiterpenes.

It was this emulsion of the plant's expressed juice that should be ingested. Ge Hong's text reads as follows: Another recipe: take a bunch of qinghao and two sheng (2×0.2 l, one sheng amounted to 0.2 l at the time) of water for soaking it, wring it out to obtain the juice and ingest it in its entirety (Emergency Prescriptions kept in one's Sleeve, chapter 3.16). Ge Hong was the first in medical history to recommend the drug qinghao for the treatment of 'intermittent fevers.' The term 'intermittent fevers' designated many different morbid conditions, and in all likelihood, also referred to intermittent fevers due to malaria. It is possible that his method of extraction, which is likely to have yielded artemisinin in larger quantities than other methods recorded in the Chinese materia medica, was directly linked to his recommendation of treating acute episodes of intermittent fever [43, 171].

The discovery of the Chinese medical drug qinghao was part of a series of complex historical processes that began with the systematic screening of the Chinese materia medica in the late 1960s and led to the first demonstration of the antimalarial effects of an extract of qinghao in 1971. The chemical structure of the active substance in this extract was identified in the late 1970s and was found, totally unexpectedly, to be a sesquiterpene lactone peroxide. The active ingredient of the herb, now known as artemisinin, was found to have this remarkable antimalarial action because of its unique chemical structure, quite unlike previously known antimalarials. In chemical parlance, it is the presence of peroxide forming a kind of bridge in the molecular structure of artemisinin that seems to give the drug its antimalarial activity. Pharmaceutical production of marketable drugs began in 1986. The World Health Organization (WHO) started to investigate artemisinin and its derivatives in the early 1990s and has promoted them on a large scale since 2004.

These artemisinin derivatives include dihydroartemisinin (DHA), artemether and artesunate, originally developed by Chinese scientists, and arteether and artelinic acid. Artemisinin derivatives have been used by over several million patients and are well tolerated [2].

Beside the artemisinin, scientific studies on malaria made their first significant advance in 1880, when a French army doctor working in Algeria named Charles Louis Alphonse

Laveran observed parasites inside the red blood cells of people suffering from malaria. He therefore proposed that malaria was caused by this protozoan, the first time protozoa were identified as causing disease [172]. For this and later discoveries, he was awarded the 1907 Nobel Prize for Physiology or Medicine. The protozoan was called *Plasmodium* by the Italian scientists Ettore Marchiafava and Angelo Celli [173]. A year later, Carlos Finlay, a Cuban doctor treating patients with yellow fever in Havana, first suggested that mosquitoes were transmitting disease to and from humans. However, it was Britain's Sir Ronald Ross working in India who finally proved in 1898 that malaria is transmitted by mosquitoes. He did this by showing that certain mosquito species transmit malaria to birds and isolating malaria parasites from the salivary glands of mosquitoes that had fed on infected birds [174]. For this work Ross received the 1902 Nobel Prize in Medicine. After resigning from the Indian Medical Service, Ross worked at the newly-established Liverpool School of Tropical Medicine and directed malaria-control efforts in Egypt, Panama, Greece, and Mauritius [175]. The findings of Finlay and Ross were later confirmed by a medical board headed by Walter Reed in 1900, and its recommendations implemented by William C. Gorgas in the health measures undertaken during construction of the Panama Canal. This public-health work saved the lives of thousands of workers and helped develop the methods used in future public-health campaigns against this disease.

**Table 17. Drug classes of current antimalarial drugs in according
to chemical structure [170]**

	Generic drug classes	Drug example
1	Arylaminoalcohols:	Quinine, quinidine (cinchona alkaloids), mefloquine, halofantrine, lumefantrine,
2	4-aminoquinolines	Chloroquine, amodiaquine (bisquinoline piperaquine)
3	8-aminoquinolines	Primaquine, tafenoquine (WR238605), 4-methyl primaquine diphosphate. (WR181023)
4	Folate synthesis inhibitors (Antifolates)	Type 1 - competitive inhibitors of dihydropteroate synthase – sulphones (dapsone), sulphonamides (sulphadoxine) Type 2 - inhibit dihydrofolate reductase - biguanides like proguanil and chloroproguanil; diaminopyrimidine like pyrimethamine; cycloguanil, trimethoprim
5	Peroxides (Sesquiterpene lactones)	Artemisinin (Qinghaosu) derivatives - dihydroartemisinin artemether, arteether, artesunate, artelinic acid
6	Antimicrobials:	Tetracycline, doxycycline, clindamycin, azithromycin, fluoroquinolones, chloramphenicol
7	Naphthoquinones (Respiratory chain inhibitors)	Atovaquone
8	Iron chelating agents	Desferrioxamine
9	Drug combinations	▪ Sulphadoxine-pyrimethamine ('Fansidar') ▪ pyrimethamine + sulphadoxine + mefloquine ('Fansimef') ▪ atovaquone + proguanil ('Malarone')

The first effective treatment for malaria was the bark of cinchona tree, which contains quinine. This tree grows on the slopes of the Andes, mainly in Peru. This natural product was used by the inhabitants of Peru to control malaria, and the Jesuits introduced this practice to Europe during the 1640s where it was rapidly accepted [176]. However, it was not until 1820 that the active ingredient quinine was extracted from the bark, isolated and named by the French chemists Pierre Joseph Pelletier and Jean Bienaime Caventou [177]. In the early twentieth century, before antibiotics, patients with syphilis were intentionally infected with malaria to create a fever, following the work of Julius Wagner-Jauregg. By accurately controlling the fever with quinine, the effects of both syphilis and malaria could be minimized. Although some patients died from malaria, this was preferable than the almost-certain death from syphilis [178]. Although the blood stage and mosquito stages of the malaria life cycle were established in the 19th and early 20th centuries, it was not until the 1980s that the latent liver form of the parasite was observed [179, 180]. The discovery of this latent form of the parasite finally explained why people could appear to be cured of malaria but still relapse years after the parasite had disappeared from their bloodstreams.

3.1.2. Classification of Current Antimalarial Drugs

The antimalarial drugs can be classified according to their chemical structures (Table 17). The traditional antimalarial drugs can also be classified according to their activity in the human body and their effective/ineffectiveness.

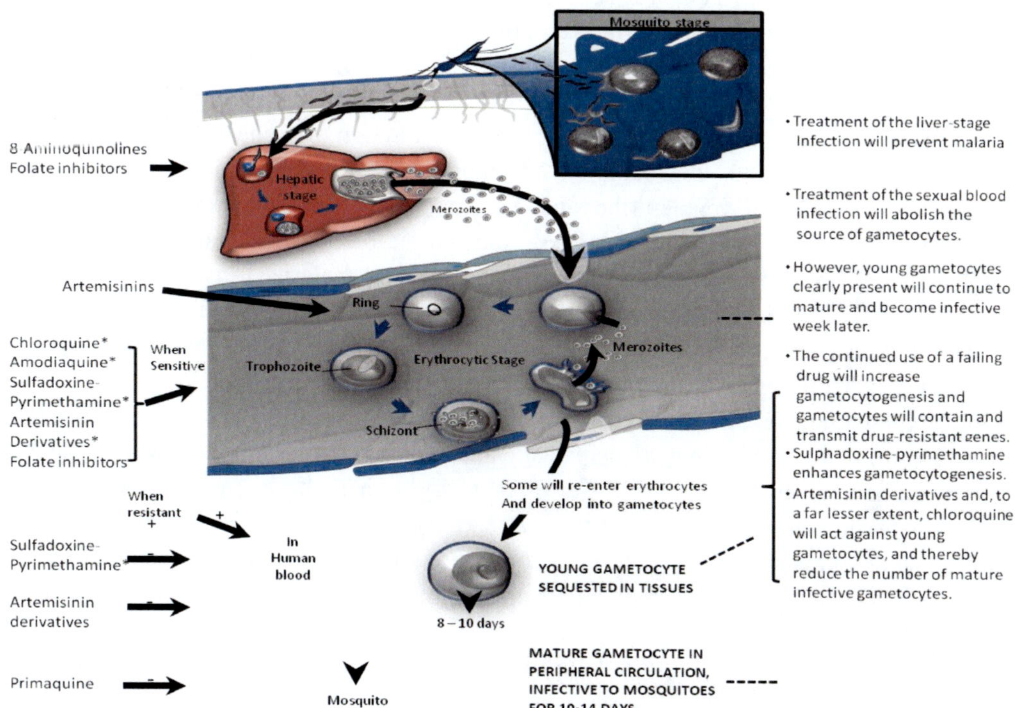

Figure 9. Antimalarial drugs and their effects on the various stages in life-cycle of *P. falciparum* parasite in human.

3.1.2.1. According to Antimalarial Activity on the Life Cycle of Malaria Parasites

The life cycle of malaria parasites in the human body and the various stages in this process are discussed in Figure 9. Malaria in humans develops via two phases: an exoerythrocytic (hepatic) and an erythrocytic phase. When an infected mosquito pierces a person's skin to take a blood meal, sporozoites in the mosquito's saliva enter the bloodstream and migrate to the liver. Within 30 minutes of being introduced into the human host, they infect hepatocytes, multiplying asexually and asymptomatically for a period of 6–15 days. Once in the liver these organisms differentiate to yield thousands of merozoites which, following rupture of their host cells, escape into the blood and infect red blood cells, thus beginning the erythrocytic stage of the life cycle [181]. The parasite escapes from the liver undetected by wrapping itself in the cell membrane of the infected host liver cell [182].

Within the red blood cells the parasites multiply further, again asexually, periodically breaking out of their hosts to invade fresh red blood cells. Several such amplification cycles occur. Thus, classical descriptions of waves of fever arise from simultaneous waves of merozoites escaping and infecting red blood cells. Some *P. vivax* and *P. ovale* sporozoites do not immediately develop into exoerythrocytic-phase merozoites, but instead produce hypnozoites that remain dormant for periods ranging from several months (6–12 months is typical) to as long as three years. After a period of dormancy, they reactivate and produce merozoites. Hypnozoites are responsible for long incubation and late relapses in these two species of malaria.

The parasite is relatively protected from attack by the body's immune system because for most of its human life cycle it resides within the liver and blood cells and is relatively invisible to immune surveillance. However, circulating infected blood cells are destroyed in the spleen. To avoid this fate, the *P. falciparum* parasite displays adhesive proteins on the surface of the infected blood cells, causing the blood cells to stick to the walls of small blood vessels, thereby sequestering the parasite from passage through the general circulation and the spleen. This "stickiness" is the main factor giving rise to hemorrhagic complications of malaria. High endothelial venules (the smallest branches of the circulatory system) can be blocked by the attachment of masses of these infected red blood cells. The blockage of these vessels causes symptoms such as in placental and cerebral malaria. In cerebral malaria the sequestrated red blood cells can breach the blood brain barrier possibly leading to coma.

Although the red blood cell surface adhesive proteins (called *PfEMP1*, for *Plasmodium falciparum* erythrocyte membrane protein 1) are exposed to the immune system they do not serve as good immune targets because of their extreme diversity; there are at least 60 variations of the protein within a single parasite and perhaps limitless versions within parasite populations. Like a thief changing disguises or a spy with multiple passports, the parasite switches between a broad repertoire of *PfEMP1* surface proteins, thus staying one step ahead of the pursuing immune system. Some merozoites turn into male and female gametocytes. If a mosquito pierces the skin of an infected person, it potentially picks up gametocytes within the blood. Fertilization and sexual recombination of the parasite occurs in the mosquito's gut, thereby defining the mosquito as the definitive host of the disease. New sporozoites develop and travel to the mosquito's salivary gland, completing the cycle. Pregnant women are especially attractive to the mosquitoes, and malaria in pregnant women is an important cause of stillbirths, infant mortality and low birth weight.

Antimalarial drugs can treat malaria and help bring about a reduction in malaria transmission by their effect on parasite infectivity. This can be a direct effect on the gametocytes, the infective stages found in human infections (gametocytocidal effect) or, when the drug is taken up in the blood meal of the mosquito, an effect on the parasite's development in the insect (sporonticidal effect) (Figure 9). Chloroquine acts against young gametocytes but has no suppressive effects on mature infective forms. Chloroquine has even been shown to be capable of enhancing the infectivity of gametocytes to the mosquito. In contrast, sulfadoxine-pyrimethamine increases gametocyte carriage but, provided there is no resistance, reduces the infectivity of gametocytes to mosquitoes. Artemisinins are the most potent gametocytocidal drugs among those currently being used to treat an asexual blood infection. They destroy immature gametocytes, preventing new infective gametocytes from entering the circulation, but their effects on mature gametocytes are less and so they will not affect the infectivity of those already present in the circulation at the time a patient presents for treatment [170].

1) Tissue schizonticides for causal prophylaxis: These drugs act on the primary tissue forms of the plasmodia which after growth within the liver, initiate the erythrocytic stage. By blocking this stage, further development of the infection can be theoretically prevented. Pyrimethamine and Primaquine have this activity. However, since it is impossible to predict the infection before clinical symptoms begin, this mode of therapy is more theoretical than practical.

2) Tissue schizonticides for preventing relapse: These drugs act on the hypnozoites of *P. vivax* and *P. ovale* in the liver that cause relapse of symptoms on reactivation. Primaquine is the prototype drug; pyrimethamine also has such activity.

3) Blood schizonticides: These drugs act on the blood forms of the parasite and thereby terminate clinical attacks of malaria. These are the most important drugs in anti malarial chemotherapy. These include chloroquine, quinine, mefloquine, halofantrine, pyrimethamine, sulfadoxine, sulphones, tetracyclines, and artemisinins etc.

4) Gametocytocides: These drugs destroy the sexual forms of the parasite in the blood and thereby prevent transmission of the infection to the mosquito. Chloroquine and quinine have gametocytocidal activity against *P. vivax* and *P. malariae*, but not against *P. falciparum*. Primaquine has gametocytocidal activity against all plasmodia, including *P. falciparum*. Artemisinins have also the gametocytocidal activity.

5) Sporontocides: These drugs prevent the development of oocysts in the mosquito and thus ablate the transmission. Primaquine and chloroguanide have this action

Thus in effect, treatment of malaria would include a blood schizonticide, a gametocytocide and a tissue schizonticide (in case of *P. vivax* and *P. ovale*). A combination of artemisinin-based drug is thus needed in all cases of malaria.

3.1.2.2. According to the Chemical Structure of Drugs

In many instances, there have been reports that the malaria parasite is becoming resistant to drugs such as chloroquine, which has been the staple of malaria treatment since the 1960s. More recently, the most effectively used drugs have been based on artemisinins. This is a molecule which contains a pharmacophoric peroxide bond in a unique 1, 2, 4-trioxane heterocycle. Available evidences suggest that artemisinin and related peroxidic antimalarial drugs exert parasiticidal activity subsequent to reductive activation by heme released as a result of hemoglobin digestion by the malaria-causing parasites. Unfortunately, artemisinin family of drugs suffer from

- highest cost,
- problems of adherence to non-fixed combinations and their rational use, particularly in the home,
- lack of extensive clinical experience with most of the combinations currently under investigation,
- lack of evidence so far in Africa of its effectiveness in delaying the development of resistance, and
- importance of not missing artemisinin derivatives in view of their role in the treatment of severe malaria.

The recent discovery in the area of antimalarial drugs deals with the synthesis of RBX-11160 or OZ-277, which has been developed by the Indian pharmaceutical company Ranbaxy in a multinational project venture [183]. This synthetic drug is somewhat structurally related to the natural product artemisinin, as the derivatives of the latter are the most potent drugs that are available which kill the drug-resistant *P. falciparum* parasite rapidly as a class of drug listed in Table 17. Seven main classes and their activity of antimalarial drugs according to chemical structure are described [184].

1) Arylaminoalcohols

    The mechanism of action of arylaminoalcohols seems to be different from that of 4-aminoquinolines but is not known exactly. Arylaminoalcohols seem to interfere with the heme digestion [185]. Amplification of the *pfmdr1* gene appears to be the main factor in arylaminoalcohol resistance [45]. This gene codes for another transport protein (*P. falciparum* multi drug resistance 1 – PfMDR1) located in the membrane of the digestive vacuole which has been shown to transport arylaminoalcohols into the digestive vacuole [186].

2) 4-Aminoquinolines

    4-Aminoquinolines form complexes with ferriprotoporphyrine IX (FPPIX), thereby preventing its polymerization into non-toxic hemazoin. FPPIX is formed in large quantities during the parasite's digestion of hemoglobine and represents a toxic waste product to the parasite [187]. Resistance against 4-aminoquinolines results from a mutation K76T in the gene of a transport protein (chloroquine resistance transporter, CRT) located in the membrane of the digestive vacuole which facilitates the removal of 4-aminoquinolines from the digestive vacuole.

3) 8-Aminoquinolines

    The only 8-Aminoquinolines currently in use is primaquine, which is different from other antimalarial drugs as it is active against the liver and the sexual blood stages of different plasmodia. The 8-aminoquinolines are the only compounds that have been shown to be efficacious as anti-relapse therapy.

4) Antifolates

Antifolates target two enzymes of the biosynthesis of tetrahydrofolate, the dihydropteroate synthase (DHPS) and dihydrofolate reductase (DHFR) [188]. Sulfonamides like sulfadoxine or the sulphone act as competitive inhibitors of 4-aminobenzoic acid or false substrates against dihydropteroate synthase. DHPS inhibitors have only weak antimalarial activity but are synergistic with dihydrofolate reductase inhibitors. Stepwise accumulation of mutations has led to considerable resistance against dihydropteroate synthase inhibitors. Pyrimethamine and cycloguanil inhibit dihydrofolate reductase. In therapy, not cycloguanil but its open-chain biguanide prodrug proguanil is given which is transformed into cycloguanil via oxidative ring closure. Main determinant of the therapeutic success of antifolate combinations is the mutational status of DHFR [189]. With this enzyme too, stepwise accumulation of mutations has let to considerable resistance [190]. Most important is the so-called triple mutant from which is insensitive to pyrimethamine and proguanil but is still sensitive to chlorcycloguanil, which has a second chloro-substituent in the meta-position of DHFR, chlorcycloguanil's activity is too weak to be of therapeutic relevance [191].

5) Peroxides (Artemisinin)

The most widely used artemisinin derivatives artemether and artesunate are hemisynthetic derivatives of artemisinin. Both compounds are rapidly transformed into dihydroartemisinin, which has a same short elimination half-life of 40-60 minutes. Key structural feature of artemisinins is the endoperoxide which is believed to be cleaved by intraparasital iron-II sources to yield carbon-centered radicals. Whether these radicals un-specifically modify multiple targets like proteins and heme in the digestive vacuole [192] or whether they specifically inhibit an ER-located calcium pump (PfATP6) [193] is a matter of debate [194]. Artemisinins act predominantly against the late ring stages, but in contract to other antimalarials also against the small ring stages present in the erythrocytes a few hours after infection [195]. Artemisinins are highly active reducing the parasite biomass 10000-fold in a single asexual cycle [79]. This makes artemisinins the most active and rapid acting antimalarial drugs known today.

Up to now, there is no clinically relevant resistance to artemisinins. Reduced sensitivity (2- to 3-fold) has been associated with multiple copies of the *pfmdr1* gene, what is also responsible for resistance against arylaminoalcohols [45]. Furthermore, West-Africa isolates with significantly reduced sensitivity in the *pfatp6* gene. So, resistance against artemisinins appears as a real threat.

6) Antimicrobials

Several antibiotics display an antimalarial effect through their action on the prokaryote-like protein biosynthesis machinery of the mitochondrion and/or the apicoplast [196]. The apicoplast is the remnant of algae in which heme, fatty acid, and isopentenyl diphosphat biosyntheses are located [197]. Characteristically, many antibiotics do not exert any visible effect in the first intracellular cycle, but during the second cycle the parasites are killed after the invasion of the new host cell. This phenomenon is known as "delayed death phenotype" or "delayed kill effect" [198]. Due to this effect, fever and parasite-clearance times are significantly longer than with classical antimalarials when antibiotics are employed as single agents. Since this delay may be fatal in non-immune patients, antibiotics are used only in combination with a faster acting drug (quinine, artesunate, or fosmidomycin) for the therapy of acute malaria. In such combinations, faster acting "classical" antimalarial should rapidly reduce the parasites including those possible showing reduced sensitivity against the faster acting antimalarial. Up to now, no clinically relevant resistance of malaria parasites against antibiotics has been reported.

7)   Inhibitors of the respiratory chain

The hydroxynaphthoquinone atovaquone inhibits the mitochondrial electron transport chain and leads to a rapid breakdown of the mitochondrial membrane potential. It binds to the ubiquinone binding side of the cytochrome bc_1 complex and blocks the movement of an iro-sulfur cluster containing a protein domain which is normally involved in electron transport [199]. Application of atovaquone at single agent showed rapid selection of resistance strains, resulting in therapy failure rates of 30% [200]. In such resistant strains of the Q_Q site has been altered by the exchange of one amino acid reducing the sensitivity of the cytochrome bc_1 complex towards atovaquone more than 1000-fold [201]. There is a strong synergism between atovaquone and the non-metabolized biguanides proguanil [202] and chlorproguanil [203]. Recently, a theory regarding the mechanism of this synergy between atovaquone and proguanil has been offered: when electron transport, which is normally dominant in establishing the mitochondrial membrane potential, is inhibited by atovaquone, an alternative pathway involving ATP hydrolysis and exchange of generated ADP3- against ATP+ by the ATP/ADP transporter, becomes apparent. This pathway is inhibited by proguanil resulting in a rapid breakdown of membrane potential [204].

3.1.3. Therapeutic History and Development of Antimalarial Drugs

Antimalarial drugs are designed to prevent or cure malaria. There are 9 main classes of antimalarial drugs in use (Table 17) and they have been long discovery and development history for more than 2000 years. Quinine is the oldest and most famous antimalarial. Some antimalarial agents, particularly artemisinins, are also used in the treatments of tumors [205].

In the ancient times, antimalarial drugs were all discovered to treat fevers, which have always haunted mankind and several ingenious remedies were tried to combat the fevers. Limb blood-letting, emesis, amputation, and skull operations were tried in the treatment of malarial fever. In England, opium from locally grown poppies and opium-laced beer were tried. Even the help of astrology was sought as the periodicity of malarial fevers suggested a connection with astronomical phenomena. Claudius Galenus of Pergamum (131-201 AD), more popularly known as Galen, was an ancient Greek physician who worked in Rome from 162 AD. He suggested that the normal humoral balance should be restored by bleeding, purging, or both. Vomiting accompanying malaria was believed to be the patient's attempt to expel poisons. The bleeding supposedly rid the body of corrupt humors. These tenets were accepted without question for the next fifteen hundred years. Countless malaria patients were subjected to blood-letting and purgation with disastrous results: repeated bleedings only made the anemia of malaria much worse and the powerful purgatives on top of the debilitating effects of the disease itself often finished off most sufferers in a short time.

Table 18. Time line for malaria treatment

Time	Treatments for malaria
340AD	Anti-fever properties of qinghao first described by Ge Hong in China
1620-1630AD	Spanish Jesuit missionaries in Peru learn the healing power of a tree bark
1632	Jesuit Barnabé de Cobo takes Cinchona bark to Europe
1633	Properties of the bark in the treatment of malaria first written by Father Antanio de la Calancha
1640	Juan de Lugo first employed the tincture of the cinchona bark for treating malaria in Countess of Chinchon
1658	The first prescription of cinchona in England by Robert Brady
1670s	Robert Talbor develops an infusion of cinchona powder in white wine and uses it as a 'secret remedy'
1712	Fransesco Torti writes a book on the therapeutic properties of the bark
1742	Linnaeus, a Swedish botonist, classifies the Peruvian bark and names the tree cinchona
1820	French chemists Joseph Pelletier and Jean Biename Caventou isolate quinine
1844, 1910	Sporadic resistance to quinine reported
1934	Resochin (chloroquine) synthesized at Bayer, Germany by Hans Andersag
1944	Proguanil or Paludrine (chlorguanide hydrochloride) developed by Curd, Davey and Rose
1945, 1950	Camoquin (amodiaquin) and Primaquine (Elderfield) developed
1952	Pyrimethamine developed
1971	The active ingredient of qinghao (artemisinin) isolated by Chinese scientists
1974-1975	Mefloquine jointly developed by the U.S. Army Medical Research and Development Command, World Health Organization and Hoffman-La Roche
1989	Artemisinin and its directives were recommended to be used in antimalarial therapy by WHO
1992, 1998	Atovaquone becomes available in 1992; a combination of proguanil and atovaquone, called Malarone, becomes available in Australia in 1998
2001	Four artemisinin-based combination therapies (ACTs) recommended by WHO Expert Consultative Group in 2001 are AM-lumefantrine (Coartem), AS-mefloquine (Artequin), AS-amodiaquine, and AS-sulfadoxine/pyrimethamine.
2006	Artesunate IV injection has been recommended as first-line agent to treat severe malaria by WHO

To fight against the fever, the plant known as qinghao had been used by Chinese in the early second century BCE (before Common Era). The earliest description of qinghao herb for treatment of malaria-related symptoms is found in the writings of Ge Hong (281-340 AD) who was a fourth century Chinese physician. In spite of extraction problems, Chinese physicians have used *A. annua* and *A. apiacea* for producing drugs with antimalarial and other apparently therapeutic properties since at least the second century BCE. The discovery of the Chinese medical drug qinghao was part of a series of complex historical processes.

Then, the first effective treatment for malaria was the bark of cinchona tree, which contains quinine. This tree grows on the slopes of the Andes, mainly in Peru. This natural product was used by the inhabitants of Peru to control malaria, and the Jesuits introduced this practice to Europe during the 1640s where it was rapidly accepted. However, it was not until 1820 that the active ingredient quinine was extracted from the bark, isolated and named by the French chemists Pierre Joseph Pelletier and Jean Bienaime Caventou [177]. The time line for malaria treatment is shown in Table 18 and a more detailed therapeutic history and development for each antimalarial drug are summarized as follows.

Quinine

Quinine is the chief alkaloid of cinchona bark (known as 'Fever Bark'), a tree found in South America. It has a colorful history of more than 350 years. Calancha, an Augustinian monk of Lima, first wrote about the curative properties of cinchona powder in "fevers and tertians" as early as in 1633. By 1640, the bark had already found its way into Europe, thanks to the Jesuit fathers (hence the name 'Jesuit's bark'). Eminent philosopher Cardinal de Lugo popularized the bark in Rome (hence it is also called Cardinal's bark). In 1820, Pelletier and Caventou isolated quinine and cinchonine from cinchona. Even today, quinine is obtained entirely from the natural sources due the difficulties in synthesizing the complex molecule. Quinine acts rapidly, targeting the blood borne asexual stages of all malaria species.

Chloroquine

Many drugs were developed to protect the troops from malaria, particularly during World War II. Chloroquine, Primaquine, Proguanil, amodiaquine, and Sulfadoxine/Pyrimethamine were all developed during this time.

During World War I, Java and its valuable quinine stores fell into Japanese forces. As a result, the German troops in East Africa suffered heavy casualties from malaria. In a bid to have their own antimalarial drugs, the German government initiated research into quinine substitutes and entrusted it to Bayer Dye Works. Most of the work was done at Bayer Farbenindustrie A.G. laboratories in Eberfeld, Germany. Several thousands of compounds were tested and some were found to be useful. Plasmochin naphthoate (Pamaquine) in 1926 and quinacrine, mepacrine (Atabrine) in 1932 were the first to be found. Plasmochin, an 8 amino quinoline, was quickly abandoned due to toxicity, although its close structural analog primaquine is now used to treat latent liver parasites of *P. vivax* and *P. ovale*. Atabrine, although found superior and persisting in the blood for at least a week, had to be abandoned due to side-effects like yellowing of the skin and psychotic reactions. The breakthrough came in 1934 with the synthesis of Resochin (chloroquine) by Hans Andersag, followed by Sontochin or Sontoquine (3 methyl chloroquine). These compounds belonged to a new class of antimalarials known as 4 amino quinolines. But Farben scientists overestimated the compounds' toxicity and failed to explore them further. Moreover, they passed the formula for Resochin to Winthrop Stearns, Farben's U.S. sister company, in the late 1930s. Resochin was then forgotten until the outbreak of World War II.

During the German invasion of Holland and the Japanese occupation of Java, the Allied forces were cut off from quinine. This stimulated a renewed search for other antimalarials both in the United Kingdom and in the United States. After the Allied occupation of North Africa, the French soldiers raided a supply of German manufactured Sontochin in Tunis and handed it over to the Americans. Winthrop researchers made slight adjustments to the captured drug and this new formulation was called chloroquine. Later, it was found to be identical to the older and supposedly toxic Resochin. However, it was not available for the troops until the end of the War. But following World War II, chloroquine and DDT became the two principal weapons in the global malaria control campaign. However, after only about ten to twelve years of use, chloroquine resistance appeared in *P. falciparum*. Two initial foci of resistance developed simultaneously in Colombia and on the Cambodia-Thailand border. From these loci, resistance spread throughout South America and southern Asia. By the late 1970s chloroquine resistance had reached Africa and has since spread across sub-Saharan Africa [170].

Amodiaquine

Amodiaquine (trade names Camoquin, Flavoquine) is a 4-aminoquinoline compound related to chloroquine. Amodiaquine was first introduced as an alternative to chloroquine since it appeared to have activity against chloroquine-resistant *P. falciparum* parasites. Amodiaquine has been shown to be more effective than chloroquine in treating CRPF (chloroquine-resistant *Plasmodium falciparum*) malaria infections and may afford more protection than chloroquine when used as weekly prophylaxis. The drug was actively used for malaria chemoprophylaxis in travelers between 1948 and 1990. However, it fell out of favor because it caused adverse effects on bone marrow and liver when used for prophylaxis. Amodiaquine is currently being reevaluated as a co-formulation partner with artesunate. Concerns over toxicity remain. Although licensed, this drug is not marketed in the United States but is widely available in Africa. Its use, therefore, is probably more practicable in long-term visitors and persons who will reside in Africa. The drug is now being actively used in combination with artesunate for the treatment of uncomplicated malaria.

Artemisinin

The herb *Artemisia annua* (sweet wormwood) was known to the Chinese as *qing-hao* for more than 2000 years. The Mawanhgolui Han dynasty tombs, dating to 168 BC, mention it as a treatment for hemorrhoids. In 340 AD, the anti-fever properties of *qinghao* were first described by Ge Hong of the East Yin Dynasty. An appeal for help from Ho Chi Minh to Zhou En Lai during the Vietnam War triggered the work on this herb and in 1967, the Chinese scientists set up Project 523. The active ingredient of *qinghao* was isolated by Chinese scientists in 1971. An ethyl ether extract of *qinghao* fed to mice infected with the rodent malaria strain, *Plasmodium berghei*, was found to be as effective as chloroquine and quinine at clearing the parasite. The human trials were published in the *Chinese Medical Journal* in 1979. Many active derivatives of artemisinin have since been synthesized and it is today a very potent and effective antimalarial drug, particularly against drug resistant malaria in many areas of Southeast Asia. So far, clinically relevant genetic resistance to artemisinin has not been reported, although tolerance has been noted. Artemisinins include the compounds artesunate, artemether, arteether, and dihydroartemisinin derived from the sesquiterpene lactone principle (artemisinin) of the plant *Artemisia annua* recently during the last two decades.

Antifolate Combination Drugs

These combination drugs include various combinations of dihydrofolate reductase inhibitors (proguanil, chlorproguanil, pyrimethamine, and trimethoprim) and sulfa drugs (dapsone, sulfalene, sulfamethoxazole, sulfadoxine, and others). Current combinations include sulfadoxine-pyrimethamine (SP) introduced in Thailand in 1967, sulfalene/pyrimethamine, and sulfamethoxazole/trimethoprim in 1969. Proguanil has also been used in combination with chloroquine for prophylaxis in areas of moderate chloroquine resistance, although it confers only minimal added benefit, especially with prolonged exposure. The latest antifolate combination is chlorproguanil and dapsone, also known as Lapdap.

Mefloquine

Mefloquine was jointly developed by the U.S. Army Medical Research and Development Command, the World Health Organization (WHO/TDR), and Hoffman-La Roche, Inc. After World War II, about 120 compounds were produced at the Walter Reed Army Institute of Research (WRAIR) and WR142490 (mefloquine), a 4-quinoline methanol was developed. Mefloquine is structurally related to quinine, the drug effective against all forms of malaria and used for treatment of complicated malaria. Its efficacy in preventing and treating resistant *P. falciparum* was proved in 1974-75 and was useful for the US Army in Southeast Asia and South America. By the time the drug became widely available in 1985, evidence of resistance to mefloquine also began to appear in Asia.

Halofantrine

Halofantrine was also developed in the 1960s by the WRAIR. It is a phenanthrene methanol structurally related to quinine.

Lumefantrine

Lumefantrine is an arylamino alcohol that belongs to the same family as quinine, mefloquine and halofantrine. It has been synthesized by the Academy of Military Medical Sciences in Beijing, China. A combination of artemether and lumefantrine was registered for the treatment of uncomplicated *P. falciparum* malaria in 1992 and further developed by Novartis Pharmaceuticals. It was first approved for use (including standby emergency use for travelers in regions where malaria is prevalent) in Switzerland in 1999.

Malarone (Atovaquone/proguanil)

In 1998, a new drug combination was released in Australia called Malarone. This is a combination of proguanil and atovaquone. Atovaquone became available 1992 and was used with success for the treatment of *Pneumocystis carinii*. More popularly known as proguanil, this drug was developed by British antimalarial research in 1945. It is a biguanide derivative that is converted to an active metabolite called cycloguanil pamoate. The synergistic combination with proguanil is found to be an effective antimalarial treatment.

Primaquine

Primaquine, an 8-aminoquinoline, acts against malaria parasites in the liver, thereby reducing the likelihood of *P. vivax* or *P. ovale* relapse. Primaquine was synthesized by

Elderfield in 1950. Primaquine is not routinely used to prevent malaria in travelers, and is only used as such when no other alternatives are appropriate. It is not licensed for this use in the U.S. or UK.

Tafenoquine

Tafenoquine, a synthetic analog of primaquine by WRAIR (also called WR-238605 or SB-252263), is currently being tested in clinical trials. It is manufactured by GlaxoSmithKline. It is highly effective against both liver and blood stages of malaria. Because of its long half-life (14 days versus 6 hours for primaquine), tafenoquine may prove to be a valuable chemoprophylactic drug. As with primaquine, tafenoquine can produce acute hemolytic anemia in patients with G6PD deficiency.

Pyronaridine

Pyronaridine has been used in China for over 20 years. While it was reportedly 100 percent effective in a single trial in Cameroon, the drug was only 63 to 88% effective in Thailand. Further testing is required before pyronaridine can be recommended for widespread use.

Piperaquine

Piperaquine is an orally active bisquinoline discovered in the early 1960s and developed for clinical use in China in 1973. *In vitro* testing in several laboratories has shown that piperaquine approximates chloroquine's effects against sensitive parasites and is significantly more effective than chloroquine in treating resistant *P. falciparum*. In China, Vietnam, and Cambodia, piperaquine is now available as a fixed combination with dihydroartemisinin (it has also been combined with trimethoprim and primaquine). Prospective clinical trial data are pending.

Clindamycin

Clindamycin is an antibiotic with weak antimalarial activity. It should only be used in combination with a fast-acting schizonticide, such as quinine or artemisinins, especially when treating patients with little or no immunity to malaria. Clindamycin is marketed alone and in combination with other drugs under various trade names, including Dalacin and Cleocin (manufactured by Pfizer), in a foam as Evoclin and Duac (with benzoyl peroxide, made by Stiefel), and a gel form by the name of Benzaclin, manufactured by Sanofi Aventis. It is also available as a generic drug.

Tetracycline

Tetracycline, an antibiotic, and its derivatives such as doxycycline may be paired with other drugs for treatment or used as single agents for prophylaxis. In areas where quinine efficacy is diminished, tetracyclines are often added to quinine to improve cure rates. Tetracyclines are also used with shortened courses of quinine to decrease quinine-associated side-effects.

Piperaquine

Piperaquine (a bisquinoline analogue of chloroquine) was initially formulated with dihydroartemisinin (DHA), trimethoprim and primaquine phosphate and marketed under the trade name China-Vietnam (CV4®). CV4® was later reformulated as CV8®, which contained different quantities of the same four components. After evaluation of its safety and efficacy in Vietnam, CV8® was introduced into the Vietnamese National Malaria Control Programme in 2000. The combination was changed yet again because of concerns including the following: (a) the role for trimethoprim in CV8® was unclear; (b) the total dihydroartemisinin dose was lower than the recommended dihydroartemisinin dose when used alone for the initial treatment of acute malaria; and (c) there was a fear that high rates of glucose-6-phosphate dehydrogenase deficiency amongst Asian populations could potentially lead to primaquine-induced cell hemolysis. The combination was modified to exclude primaquine (Artecom®) and more recently changed to exclude both primaquine and trimethoprim and only DHA and piperaquine left as a new ACT [133].

Other Antimalarial Drugs

The formula of Atabrine (mepacrine, a 9-amino-acridine), was also soon solved by Allied chemists and it was produced in large scale in the U.S. It immediately gained widespread acceptance as an excellent therapeutic agent. After the experiments of Brigadier N. Hamilton Fairley in Australia in 1943, it was also found to be useful as a prophylactic agent, protecting the troops in malarious areas. It is no longer used in view of many undesirable side-effects. The success of chloroquine led to the exploration of many (nearly 15000) compounds in the United States and another 4-aminoquinoline Camoquin (amodiaquin) was discovered. Meanwhile, British investigators at ICI also carried out extensive studies on malaria drugs and Curd, Davey and Rose synthesized antifolate drugs proguanil or Paludrine (chlorguanide hydrochloride) in 1944 and Daraprim or Malocide (pyrimethamine) was developed in 1952. However, resistance to proguanil was observed within a year of introduction in Malaya in 1947. *P. falciparum* strains resistant to pyrimethamine and cross-resistant to proguanil emerged in 1953.

3.1.4. Action Mechanisms of Antimalarial Drugs

There are several families of drugs used to treat malaria. A traditional antimalarial drug, chloroquine, is very cheap and, until recently, was very effective, which made it the antimalarial drug of choice for many years in most parts of the world. However, resistance of *P. falciparum* to chloroquine has spread recently from Asia to Africa, making the drug ineffective against the most dangerous Plasmodium strain in many affected regions of the world. In those areas where chloroquine is still effective it remains the first choice. Unfortunately, chloroquine-resistance is associated with reduced sensitivity to other drugs such as quinine and amodiaquine. Recently, antimalarial drugs derived from natural *Artemisia annua* have many advantages: quick reduction of fevers, fast clearing parasites in blood (90% of malaria patients recovered within 48 hrs) and no significant side-effects. Experimental and clinical studies reveal that artemisinin, artemether, and artesunate are not only the potent antimalarial drugs but also the useful agents for combination, especially combined with long-acting agents. Combination chemotherapy, using two or more drugs together with different mechanisms and sites of action, is proposed as a mechanism for slowing the process of development of resistance. Therefore, individual mechanism of each drug is introduced as follows.

Chloroquine

Chloroquine is the prototype antimalarial drug, most widely used to treat all types of malarial infections. The mechanism of action of chloroquine is unclear. Being alkaline, the drug reaches high concentration within the food vacuoles of the parasite and raises its pH. It is found to induce rapid clumping of the pigment. Chloroquine inhibits the parasitic enzyme heme polymerase that converts the toxic heme into non-toxic hemazoin, thereby resulting in the accumulation of toxic heme within the parasite. It may also interfere with the biosynthesis of nucleic acids and with parasite heme detoxification. Other mechanisms suggested include formation of drug-heme complex, intercalation of the drug with the parasitic DNA etc. It is highly effective against erythrocytic forms of *P. vivax*, *P. ovale*, and *P. malariae*, sensitive strains of *P. falciparum* and gametocytes of *P. vivax*. It rapidly controls acute attack of malaria with most patients becoming a febrile within 24-48 hours. It is more effective and safer than quinine for sensitive cases. Resistance is related to genetic changes in transporters (*PfCRT, PfMDR*), which reduce the concentrations of chloroquine at its site of action, the parasite food vacuole.

Amodiaquine

Amodiaquine is a Mannich base 4-aminoquinoline with a mode of action similar to that of chloroquine. *In vitro*, amodiaquine has been shown to inhibit hemoglobin proteolysis and to interfere with the accumulation of an insoluble polymer (hemozoin) by forming a complex with ferriprotoporphyrin IX. The drug also inhibits the glutathione-dependent destruction of ferriprotoporphyrin IX, resulting in the accumulation of this toxic peptide. It is effective against some chloroquine-resistant strains of *P. falciparum*, although there is cross-resistance.

Artemisinins

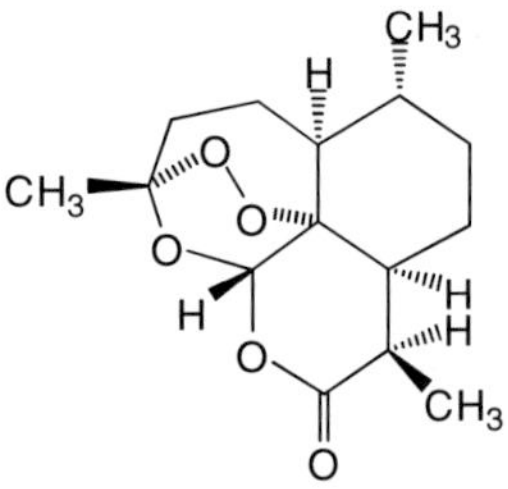

Artemisinin (QHS) 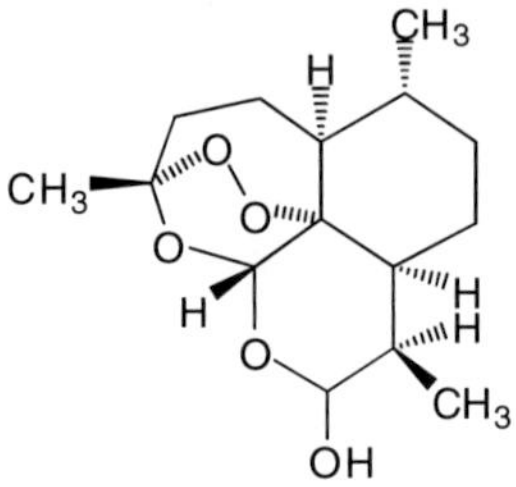   Dihydroartemisinin (DHA) 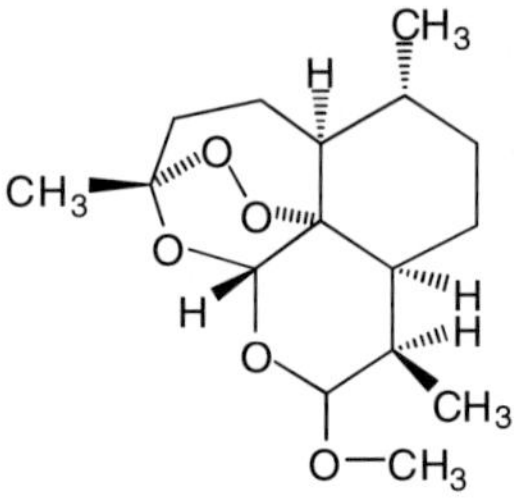   Artemether (AM)

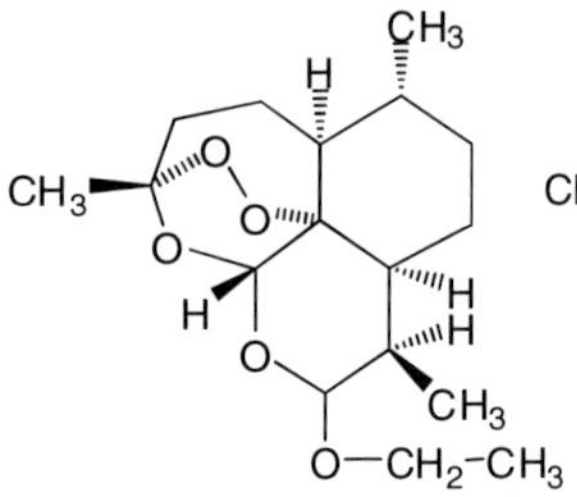

Arteether (AE) 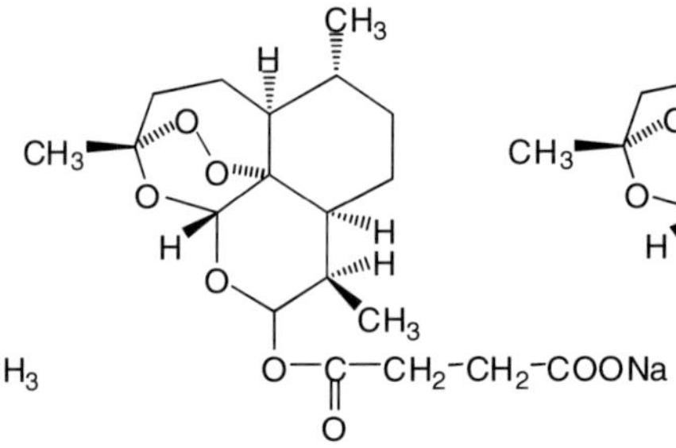   Artesunate (AS) 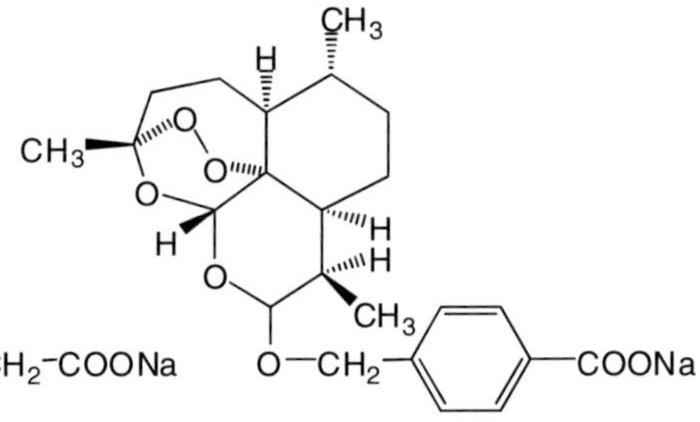   Artelinate (AL)

Artemisinin (qinghaosu, QHS) is the active principal of the Chinese medicinal herb *Artemisia annua*. The WHO accorded high priority to the development of fast acting artemisinin derivatives for the treatment of cerebral malaria, as well as for the control of multi-drug resistant *P. falciparum* malaria. A water soluble ester called artesunate, two oil soluble preparations called artemether and arteether, as well as their parent compound, dihydroartemisinin, have now been developed. These drugs should be given as combination therapy to protect them from resistance. They act by inhibiting a *P. falciparum*-encoded sarcoplasmic-endoplasmic reticulum calcium ATPase, and not by inhibiting the heme metabolic pathway as previously supposed. Most clinically important artemisinins are metabolized to dihydroartemisinin, in which form they have comparable antimalarial activity. However, their use in monotherapy is associated with high incidences of recrudescent infection, suggesting that combination with other antimalarials might be necessary for maximum efficacy.

It is the fastest acting anti malarial available. It inhibits the development of the trophozoites and thus prevents progression of the disease. Young circulating parasites are killed before they sequester in the deep microvasculature. These drugs start acting within 12 hours. These properties of the drug are very useful in managing complicated *P. falciparum* malaria. These drugs are also effective against the chloroquine resistant strains of *P. falciparum*. It has also been reported that artemisinin drugs cleared parasites faster than quinine in patients with severe malaria but fever clearance was similar. Artemisinin compounds have been reported to reduce gametocytogenesis, thus reducing transmission of malaria, this fact being especially significant in preventing the spread of resistant strains.

These drugs prevent the gametocyte development by their action on the ring stages and on the early (stage I-III) gametocytes.

Quinine

Quinine acts as a blood schizonticide although it also has gametocytocidal activity against *P. vivax* and *P. malariae*. Because it is a weak base, it is concentrated in the food vacuoles of *P. falciparum*. It is said to act by inhibiting heme polymerase, thereby allowing accumulation of its cytotoxic substrate, heme. As a schizonticidal drug, it is less effective and more toxic than chloroquine. However, it has a special place in the management of severe *falciparum* malaria in areas with known resistance to chloroquine. Quinine is an alkaloid derived from the bark of the Cinchona tree. Four antimalarial alkaloids can be derived from the bark: quinine (the main alkaloid), quinidine, cinchonine, and cinchonidine. Quinine is the L-stereoisomer of quinidine. The mechanisms of its antimalarial actions are thought to involve inhibition of parasite heme detoxification in the food vacuole, but are not well clear.

Mefloquine

Mefloquine was born during the Vietnam War, as a result of research into newer antimalarials, to protect the American soldiers from the multi drug resistant *falciparum* malaria. Nothing much has happened after that and hence this 'new' drug should be restricted for use against multi drug resistant *falciparum* only. Mefloquine has been found to produce swelling of the *P. falciparum* food vacuoles. It may act by forming toxic complexes with free heme that damage membranes and interact with other plasmodial components. It is effective against the blood forms of *falciparum* malaria, including the chloroquine resistant types. It is a quinoline-methanol derivative of quinine that can be used for treatment or prevention in most areas with multidrug resistant malaria. In addition, mefloquine has shown efficacy in an *in vitro* assay against Progressive Multifocal Encephalopathy (PML). Biogen Idec has recently announced that a trial of Mefloquine in HIV related PML is beginning.

Halofantrine

Halofantrine was developed in the 1960s by the Walter Reed Army Institute of Research. It is a phenanthrene methanol structurally related to quinine. Its mechanism of action may be similar to that of chloroquine, quinine, and mefloquine; by forming toxic complexes with ferritoporphyrin IX that damage the membrane of the parasite. This synthetic anti malarial is effective against multi drug resistant (including mefloquine resistant) *P. falciparum* malaria. It is used in the treatment of chloroquine resistant and multi-drug resistant, uncomplicated *P. falciparum* malaria. The clinical response to treatment may be unpredictable due to the variable drug absorption. Halofantrine is only used to treat malaria. It is not used to prevent malaria (prophylaxis) because of the risk of toxicity and unreliable absorption.

Sulphadoxine

Sulfadoxine inhibits the utilization of para-aminobenzoic acid in the synthesis of dihydropteroic acid. The combination of pyrimethamine and sulfa thus offers two step synergistic blockade of plasmodial division. Sulfonamides are structural analogues and competitive antagonists of p-aminobenzoic acid. They are competitive inhibitors of dihydropteroate synthase, the bacterial enzyme responsible for the incorporation of paminobenzoic acid in the synthesis of folic acid.

Pyrimethamine

Pyrimethamine inhibits the dihydrofolate reductase of plasmodia and thereby blocks the biosynthesis of purines and pyrimidines, which are so essential for DNA synthesis and cell multiplication. This leads to failure of nuclear division at the time of schizont formation in erythrocytes and liver. It is a slow-acting blood schizonticide and is also possibly active against pre-erythrocytic forms of the malaria parasite and inhibits sporozoite development in the mosquito vector. It is effective against all four human malarias, although resistance has emerged. Pyrimethamine is a diaminopyrimidine used in combination with a sulfonamide usually sulfadoxine or dapsone. Sulphadoxine/pyrimethamine (SP) is very useful adjuncts in the treatment of uncomplicated, chloroquine resistant, *P. falciparum* malaria.

Chloroguanide (Proguanil)

Proguanil exerts its antimalarial action by inhibiting parasitic dihydrofolate reductase enzyme. Proguanil is a biguanide compound that is metabolized in the body via the polymorphic cytochrome P450 enzyme CYP2C19 to the active metabolite, cycloguanil. Cycloguanil inhibits plasmodial dihydrofolate reductase. The parent compound has weak intrinsic antimalarial activity through an unknown mechanism. It is possibly active against pre-erythrocytic forms of the parasite and is a slow blood schizontocide. Proguanil also has sporontocidal activity, rendering the gametocytes non-infective to the mosquito vector. It has causal prophylactic and suppressive activity against *P. falciparum* and cures the acute infection. It is also effective in suppressing the clinical attacks of *vivax* malaria. However it is slower compared to 4-aminoquinolines.

Atovaquone

A synthetic hydroxynaphthoquinone developed in the early 1980s, atovaquone has been found to be useful against the *Plasmodia* (as well as *Toxoplasma* and *Pneumocystis carinii*). It has a highly lipophilic molecule that supposedly interferes with the mitochondrial electron transport and thereby ATP and pyrimidine biosynthesis and in Plasmodia, it is found to target cytochrome bc1 complex and disrupt the membrane potential. Its bio-availability after oral administration is poor and may be increased by a fatty meal. Atovaquone plus proguanil: A fixed dose combination of atovaquone and proguanil hydrochloride (Malarone™) is now approved for both treatment and prophylaxis of malaria. It has been shown to be highly efficacious in the treatment of uncomplicated malaria caused by *P. falciparum*, including malaria that has been acquired in areas with chloroquine-resistant or multidrug-resistant strains.

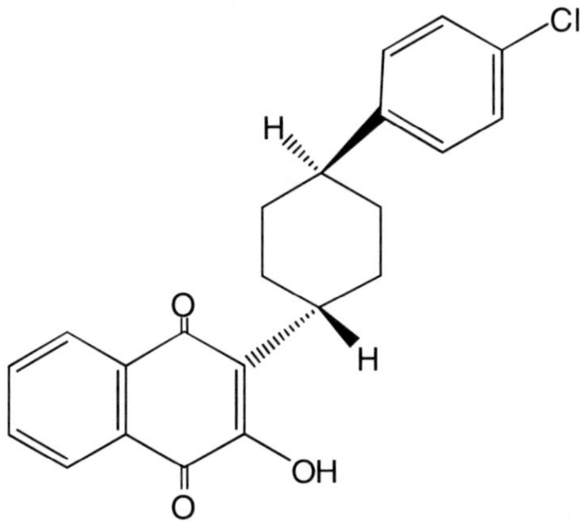

Chlorproguanil

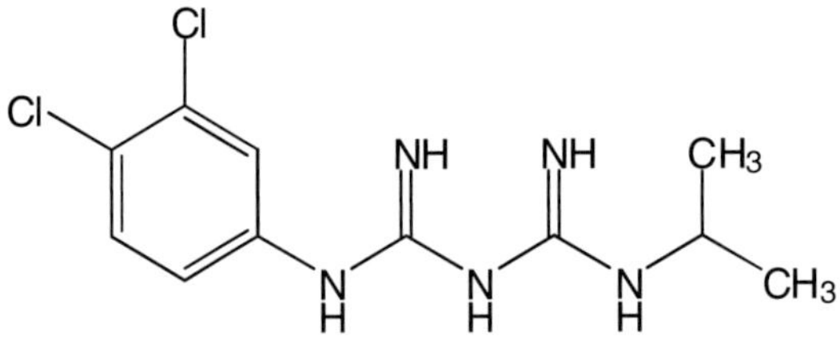

Chlorproguanil is a biguanide and is given as the hydrochloride salt. Its actions and properties are very similar to those of proguanil. It is available only in combination with a sulfone such as dapsone (co-formulated as Lapdap).

Dapsone

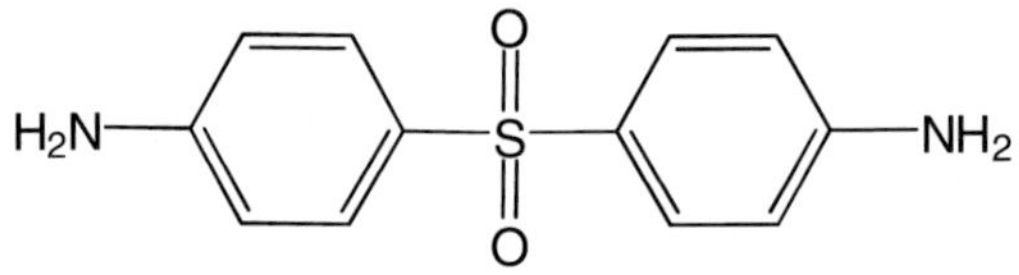

Dapsone is a sulfone widely used for the treatment of leprosy, and sometimes also for treatment or prophylaxis of Pneumocystis carinii pneumonia, and treatment of toxoplasmosis, cutaneous leishmaniasis, and actinomycetoma. For malaria, dapsone is given in combination with another antimalarial. It is co-formulated with chlorproguanil (as Lapdap). Dapsone inhibits plasmodial dihydropteroate synthase.

Primaquine

Primaquine is an 8-aminoquinoline and is effective against intrahepatic forms of all types of malaria parasite. It is highly effective against the gametocytes of all plasmodia and thereby prevents spread of the disease to the mosquito from the patient. It is also effective against the dormant tissue forms of *P. vivax* and *P. ovale* malaria, and thereby offers radical cure and prevents relapses. It has insignificant activity against the asexual blood forms of the parasite and therefore it is always used in conjunction with a blood schizonticide and never as a single agent. Mechanism of action is not well understood. It may be acting by generating reactive oxygen species or by interfering with the electron transport in the parasite.

Tafenoquine

Tafenoquine is an 8-aminoquinoline drug, such like primaquine, manufactured by GlaxoSmithKline that is currently being investigated as a potential treatment for malaria, as well as for malaria prevention. The main advantage of tafenoquine is that it has a long half-life and therefore does not need to be taken as frequently as primaquine. Like primaquine, tafenoquine causes hemolysis in people with G-6-PD deficiency.

Other Drugs for Chemotherapy of Malaria

Many drugs have been tested for their potential antimalarial effects. Research into newer antimalarials being scanty, such attempts might throw up one or two candidates for use in malaria; however, these drugs are yet to find a place in standard antimalarial regimen. Clindamycin, fluoroquinolones like ciprofloxacin and Norfloxacin, azithromycin, etc., have been found to be effective against malarial parasites. Desferrioxamine; Pyronaridine; Piperaquine; WR-288605; and 566C80 are drugs undergoing trials.

Clindamycin: Clindamycin is a lincosamide antibiotic and it acts by inhibiting the protein synthesis by binding to the 50s subunit of ribosomes. It can be used for drug resistant malaria along with quinine. Clindamycin is considered safe for use in pregnant women and very young children.

Azithromycin: Azithromycin is found to have anti malarial activity and has been found to be useful as a causal prophylactic agent. It was found to be effective at the dose of 300 mg stat, followed by 250 mg daily for 7 days as a prophylactic agent against chloroquine resistant *P. falciparum* infection.

Tetracyclines: One among the first antibiotics to come into use in human beings, these drugs have stood the test of time and are continuing to be useful in treating a broad range of infections, including malaria. Tetracyclines are bacteriostatic agents, supposedly acting by inhibiting protein synthesis by binding to the 30s ribosome subunit. These anti microbials are useful in the treatment of drug resistant *P. falciparum* malaria. They act relatively slowly and hence should always be combined with a faster acting drug like quinine. It works by inhibiting action of the prokaryotic 30S ribosome, by binding the 16S rRNA thereby blocking the aminoacyl-tRNA. However, bacteria strains can acquire resistance against tetracycline and its derivates by encoding a resistance operon.

Doxycycline: Doxycycline is a tetracycline derivative with uses similar to those of tetracycline. It may be preferred to tetracycline because of its longer half-life, more reliable absorption and better safety profile in patients with renal insufficiency, where it may be used with caution.

Lumefantrine is an aryl alcohol related to quinine, mefloquine, and halofantrine that is devoid of cardiac toxicity of halofantrine. It has a similar mechanism of action. Lumefantrine is a racemic fluorine derivative developed in China. It is only available in an oral preparation co-formulated with artemether. This ACT is highly effective against multidrug-resistant *P. falciparum.*

Pyronaridine: Structurally, it resembles amodiaquine and has been found to be highly effective against chloroquine resistant strains in China.

Piperaquine: Its activity is similar to that of chloroquine. A combination with artemisinin is undergoing studies.

3.2. Chemotherapy for Uncomplicated Malaria with ACTs

To counter the threat of resistance of *P. falciparum* to monotherapies, and to improve treatment outcome, combinations of antimalarials are now recommended by WHO for the treatment of *falciparum* malaria.

3.2.1. Uncomplicated Malaria and ACTs

3.2.1.1. Uncomplicated Malaria

Uncomplicated malaria is defined as symptomatic malaria without signs of severity or evidence of vital organ dysfunction. In acute *falciparum* malaria there is a continuum from mild to severe malaria. Young children and non-immune adults with malaria may deteriorate rapidly. Detailed definitions of severe malaria are available to guide practitioners and for epidemiological and research purposes but, in practice, any patient whom the attending physician or health care worker suspects of having severe malaria should be treated as such initially. The risks of under-treating severe malaria considerably exceed those of giving parenteral or rectal treatment to a patient who does not need it.

In clinical symptoms, the classical (but rarely observed) malaria attack lasts 6-10 hours. It consists of: a cold stage (sensation of cold, shivering), a hot stage (fever, headaches, vomiting; seizures in young children), and finally a sweating stage (sweats, return to normal temperature, tiredness). Classically (but infrequently observed) the attacks occur every second day with the "tertian" parasites (*P. falciparum*, *P. vivax*, and *P. ovale*) and every third day with the "quartan" parasite (*P. malariae*).

More commonly, the patient presents with a combination of the following symptoms: fever, chills, sweats, headaches, nausea and vomiting, body aches, and general malaise. In countries where cases of malaria are infrequent, these symptoms may be attributed to influenza, a cold, or other common infections, especially if malaria is not suspected. Conversely, in countries where malaria is frequent, residents often recognize the symptoms as malaria and treat themselves without seeking diagnostic confirmation ("presumptive treatment").

Physical findings may include: elevated temperature, perspiration, weakness, and enlarged spleen. In *P. falciparum* malaria, additional findings may include: mild jaundice, enlargement of the liver and increased respiratory rate. Diagnosis of malaria depends on the demonstration of parasites on a blood smear examined under a microscope or rapid diagnosis tests (RDTs). In *P. falciparum* malaria, additional laboratory findings may include mild anemia, mild decrease in blood platelets (thrombocytopenia), elevation of bilirubin, and elevation of aminotransferases, albuminuria, and the presence of abnormal bodies in the urine (urinary casts).

3.2.1.2. Artemisinin-based Combination Therapy (ACT)

Combination chemotherapy, using two or more drugs together with different mechanisms and sites of action, is proposed as a mechanism for slowing the process of development of resistance. This is standard practice in the treatment of HIV and tuberculosis. The concept is based on the potential of two or more simultaneously administered schizontocidal drugs with independent modes of action to improve therapeutic efficacy and also to delay the development of resistance to the individual components of the combination.

1. Rationale for Antimalarial Combination Therapy

Some of the studies reviewed clearly indicated that artemisinins are a particularly effective partner drug because they are more active than any other antimalarial, reducing the number of parasites by approximately 10,000 per asexual cycle and therefore reducing the number of parasites that are exposed to the partner drug alone. In addition, artemisinins have broad stage-specificity and can be used to treat severe as well as uncomplicated malaria, as they inhibit the production of gametocytes and therefore have the potential to reduce transmission [64]. To date, there has been no evidence of stable resistance either in therapeutic use or in experimental systems. The rationale for combining antimalarials with different modes of action is two-fold: (1) the combination is often more effective; and (2) in the rare event that a mutant parasite that is resistant to one of the drugs arises *de novo* during the course of the infection, the parasite will be killed by the other drug. This mutual protection is thought to prevent or delay the emergence of resistance. To realize the two advantages, the partner drugs in a combination must be independently effective. The possible disadvantages of combination treatments are the potential for increased risk of adverse effects and the increased cost.

The probability of mutations arising spontaneously to both drug A and drug B is equal to the probability of a mutation arising to drug A multiplied by the probability of a mutation arising to drug B. Artemisinins have been suggested as ideal drugs for use in combination therapies; this treatment is known as artemisinin combination therapy (ACT). Artemisinins have the broadest antimalarial activity against parasites, from the ring stage to early schizonts, and cause the fastest decline in parasite numbers of all the antimalarial drugs. Artemisinins as monotherapy must be taken for 7 days for radical cure. Studies have demonstrated that adherence to this regimen is extremely low in malaria endemic regions of sub-Saharan Africa, as most people discontinue treatment when they feel better, usually after a couple of days, and this can result in late recrudescence. 3-day course of artemisinins in combination with another effective drug is sufficient, encourages compliance and protects both drugs against resistance [64].

2. Artemisinin-based Combination Therapy (ACT)

Artemisinin and its derivatives (artesunate, artemether, artemotil, and dihydroartemisinin) produce rapid clearance of parasitemia and rapid resolution of symptoms. They reduce parasite numbers by a factor of approximately 10,000 in each asexual cycle, which is more than other current antimalarials (reduces parasite numbers 100- to 1000-fold per cycle). Artemisinin and its derivatives are eliminated rapidly. When given in combination with rapidly eliminated compounds (tetracyclines, clindamycin), a 7-day course of treatment with an artemisinin compound is required; but when given in combination with slowly eliminated antimalarials, shorter courses of treatment (3 days) are effective. The evidence of their superiority in comparison to monotherapies has been clearly documented.

In 3-day ACT regimens, the artemisinin component is present in the body during only two asexual parasite life-cycles (each lasting 2 days, except for *P. malariae* infections). This exposure to 3 days of artemisinin treatment reduces the number of parasites in the body by a factor of approximately one hundred million ($10^4 \times 10^4 = 10^8$). However, complete clearance of parasites is dependent on the partner medicine being effective and persisting at parasiticidal concentrations until all the infecting parasites have been killed. Thus the partner compounds need to be relatively slowly eliminated. As a result of this the artemisinin component is "protected" from resistance by the partner medicine provided it is efficacious and the partner medicine is partly protected by the artemisinin derivative. Courses of ACTs of 1–2 days are not recommended; they are less efficacious, and provide less protection of the slowly eliminated partner antimalarial.

The artemisinin compounds are active against all four species of malaria parasites that infect humans and are generally well tolerated. The only significant adverse effect to emerge from extensive clinical trials has been rare (circa 1:3000) type 1 hypersensitivity reactions (manifested initially by urticaria). These drugs also have the advantage from a public health perspective of reducing gametocyte carriage and thus the transmissibility of malaria. This contributes to malaria control in areas of low endemicity [2].

3. Non-artemisinin Based Combination Therapy

Drug combinations such as sulfadoxine–pyrimethamine, sulfalene–pyrimethamine, proguanil-dapsone, chlorproguanil-dapsone and atovaquone-proguanil rely on synergy between the two components. The drug targets in the malaria parasite are linked. These combinations are operationally considered as single products and treatment with them is not considered to be antimalarial combination therapy. Multiple-drug therapies that include a non-antimalarial medicine to enhance the antimalarial effect of a blood schizontocidal drug (e.g., chloroquine and chlorpheniramine) are also not antimalarial combination therapy.

Non-artemisinin based combinations (non-ACTs) include sulfadoxine–pyrimethamine with chloroquine (SP + CQ) or amodiaquine (SP + AQ). However, the prevailing high levels of resistance have compromised the efficacy of these combinations. There is no convincing evidence that SP + CQ provides any additional benefit over SP, so this combination is not recommended; SP + AQ can be more effective than either drug alone, but needs to be considered in the light of comparison with ACTs. The evidence is summarized here.

4. Current Choice of ACTs for Uncomplicated Malaria Therapy by WHO

Although there are some minor differences in oral absorption and bioavailability between the different artemisinin derivatives, there is no evidence that these differences are clinically significant in current formulations. It is the properties of the partner medicine that determine the effectiveness and choice of combination. ACTs with amodiaquine, atovaquone-proguanil, chloroquine, clindamycin, doxycycline, lumefantrine, mefloquine, piperaquine, pyronaridine, proguanil-dapsone, sulfadoxine–pyrimethamine, and tetracycline; have all been evaluated in trials carried out across the malaria-affected regions of the world. Some of these are studies for product development. Though there are still gaps in our knowledge, there is reasonable evidence on safety and efficacy on which to base recommendations.

The following ACTs are currently recommended (alphabetical order):

- artemether-lumefantrine (dose fixed),
- artesunate + amodiaquine,
- artesunate + mefloquine,

- artesunate + sulfadoxine–pyrimethamine.

Note: amodiaquine + sulfadoxine–pyrimethamine may be considered as an interim option where ACTs cannot be made available, provided that efficacy of both is high.

5. Rationale for the Exclusion of Certain Antimalarials

Several available drugs that were considered by the Technical Guidelines Development Group are currently not recommended.

- Chlorproguanil-dapsone has not yet been evaluated as an ACT partner drug, so there is insufficient evidence of both efficacy and safety to recommend it as a combination partner.
- Atovaquone-proguanil has been shown to be safe and effective as a combination partner in one large study, but is not included in these recommendations for deployment in endemic areas because of its very high cost.
- Halofantrine has not yet been evaluated as an ACT partner medicine and is not included in these recommendations because of safety concerns.
- Dihydroartemisinin (artenimol)-piperaquine has been shown to be safe and effective in large trials in Asia, but is not included in these recommendations as it is not yet available as a formulation manufactured under good manufacturing practices, and has not yet been evaluated sufficiently in Africa and South America.

Several other new antimalarial compounds are in development but do not yet have a sufficient clinical evidence to support recommendation here by WHO [2].

3.2.2. Therapeutic Principles of Uncomplicated Malaria

3.2.2.1. WHO Policy of Antimalarial Drugs for Uncomplicated Malaria

The objective of treating uncomplicated malaria is to cure the infection. This is important as it will help prevent progression to severe disease and prevent additional morbidity associated with treatment failure. Cure of the infection means eradication from the body of the infection that caused the illness. In treatment evaluations in all settings, emerging evidence indicates that it is necessary to follow patients for long enough to document cure. In assessing drug efficacy in high-transmission settings, temporary suppression of infection for 14 days is not considered sufficient by the group. The public health goal of treatment is to reduce transmission of the infection to others, i.e., to reduce the infectious reservoir.

A secondary but equally important objective of treatment is to prevent the emergence and spread of resistance to antimalarials. Tolerability, the adverse effect profile and the speed of therapeutic response are also important considerations. The brief policies [113] for treatment of uncomplicated *falciparum* malaria are listed here:

- Artemisinin-based combination therapies (ACTs) are the treatment recommended for all cases of uncomplicated *falciparum* malaria including:
  - in infants,

- in people living with HIV/AIDS
- for home-based management of malaria
- pregnant women in the 2nd and 3rd trimesters
    Exception: 1st trimester of pregnancy*
    *only use when there are no alternative effective antimalarials

- The following ACTs are presently recommended:
  - artemether-lumefantrine
  - artesunate + amodiaquine
  - artesunate + mefloquine3
  - artesunate + sulfadoxine-pyrimethamine
- Second-line treatment:
  - an effective alternative ACT (efficacy of ACTs depend on efficacy of the partner medicine, therefore it is possible to use 2 different ACTs as 1st and 2nd line options)
  - quinine + tetracycline or doxycycline or clindamycin
    Note: The artemisinin derivatives (oral, rectal, or parenteral formulations) and partner medicines of ACTs are not recommended as monotherapy for uncomplicated malaria.

The following options are recommended for treatment of uncomplicated *falciparum* malaria in pregnancy [113]:

- 1st Trimester:
  - Quinine + clindamycin

- 2nd and 3rd trimesters:
  - any of the recommended ACTs as listed earlier
  - artesunate + clindamycin
  - quinine + clindamycin

3.2.2.2. Treatment of Malaria Depends on the following Factors

1) Type of infection

Treatment obviously depends on the type of infection. Patients with *P. falciparum* malaria should be evaluated thoroughly in view of potential seriousness of the disease and possibility of resistance to anti malarial drugs.

P. vivax: Only Chloroquine 25 mg/kg + Primaquine for 14 days.

P. falciparum: Treat depending on severity & sensitivity. ACTs as blood schizonticides and gametocytocidal is a must to prevent spread.

Mixed infections: ACTs as blood schizonticides for *P. falciparum* and Primaquine as blood schizonticides for *P. vivax.*

2) Severity of infection

All patients with malaria should be carefully and thoroughly assessed for complications of malaria. Acute, life-threatening complications occur only in *P. falciparum* malaria. Malaria is probably the only disease of its kind that can be easily treated in just 3 days, yet if the diagnosis and proper treatment are delayed, it can kill the patient very quickly and easily.

- All cases of severe malaria should be presumed to have *P. falciparum* malaria.
- If there is any uncertainty about the drug sensitivity of the parasite, it is safer to treat these cases as chloroquine resistant malaria with drugs like quinine or artemisinin.
- All cases of severe malaria should be admitted to the hospital for proper evaluation, treatment, and monitoring.
- All cases of severe malaria should be treated with injectable antimalarials (artesunate, artemether, and quinine) so as to ensure adequate absorption and plasma drug levels. It is better to use two blood schizonticidal drugs, one for fast acting and another for slow acting, to ensure complete treatment. Newer drugs available for only oral administration (e.g., Mefloquine, Halofantrine) should be avoided.
- All associated conditions should be carefully assessed and treated.

3) Status of the host
Treatment of malaria is also dependent on host factors.

- Patient's age and weight should be recorded so as to administer adequate doses of anti malarial drugs.
- Functional capacity- independent, dependent, bed ridden etc. All patients with severe prostration and who are looking ill should be admitted to a hospital and monitored.
- Patients with nausea and vomiting should be given anti emetic drugs to ensure adequate treatment. While high-grade fever frequently stimulates vomiting, this may be further aggravated by anti malarial drugs. Therefore it is better to avoid administration of oral antimalarials at the height of fever. One can wait for the fever to subside before taking the drugs. If the patient vomits within one hour of taking the anti malarial drugs, the same should be re-administered. In case of persistent vomiting, patient should be admitted and vomiting should be controlled with parenteral anti emetics. Parenteral antimalarials are needed only in cases of severe malaria or uncontrolled vomiting.
- Adequate hydration should be ensured.

4) Associated conditions/diseases
Treatment of malaria may have to be modified due to certain associated conditions/diseases. Therefore, all such should be carefully assessed before starting the patient on antimalarial treatment.

- *Pregnancy*: Treatment of malaria in pregnancy may prove to be difficult due to contra indication for use of certain antimalarials. Chloroquine can be used safely in all trimesters of pregnancy. Artemisinin is not shown to have any deleterious effects on the fetus in animal studies and therefore can be considered if the situation demands. Quinine can be used in pregnancy, but one should be watchful about hypoglycemia. Whereas mefloquine is contra-indicated in the first trimester of pregnancy, pyrimethamine/ sulphadoxine is contra-indicated in the first and last trimesters. Halofantrine, tetracycline, and doxycycline are absolutely contra-indicated in pregnancy. Primaquine is also contra-indicated in pregnancy, and therefore pregnant women with *P. vivax* malaria should be started on 500 mg of chloroquine weekly as suppressive chemoprophylaxis against relapse of malaria. (see more detail in section 3.4.2.)

- *Children*: There are important differences between infants and older children in the pharmacokinetics of many medicines for choosing antimalarial drugs. Accurate dosing is particularly important in infants. Despite this, few clinical studies focus specifically on this age range, partly because of ethical considerations relating to the recruitment of very young children to clinical trials, and also because of the difficulty of repeated blood sampling. In the majority of clinical studies, subgroup analysis is not used to distinguish between infants and older children. Infants are more likely to vomit or regurgitate antimalarial treatment than older children or adults. Taste, volume, consistency, and gastrointestinal tolerability are important determinants of whether the child retains the treatment (see more detail in section 3.4.3.).

- *Travelers*: Travelers who acquire malaria are often non-immune adults either from cities with little or no transmission within endemic countries, or visitors from non-endemic countries. Both are likely to be at a higher risk of malaria and its consequences because they have no immunity to malaria. Within the malaria endemic country they should in principle is treated according to national policy. Travelers who return to a non-endemic country and then develop malaria present particular problems and, have a high case fatality rate. Doctors may be unfamiliar with malaria and the diagnosis may be delayed, relevant antimalarials may not be registered and/or therefore available. If the patient falls ill far from a major health facility, availability of antimalarials can be a life threatening issue despite registration.

- *Epilepsy*: Malaria as well as antimalarials can trigger convulsions. Mefloquine is better avoided in these patients. See C.N.S. Disease and malaria

- *Cardiac disease*: High-grade fever of malaria can exacerbate left ventricular failure and therefore, in all such patients' energetic management of malaria is called for. Fever should be controlled with antipyretics and tepid sponging. Chloroquine, artemisinin, pyrimethamine/ sulphadoxine, tetracyclines and primaquine can be safely used in these patients. Quinine can also be used carefully. Mefloquine and halofantrine are better avoided in patients with known cardiac illness. See C.V.S. Disease and malaria

- *Hepatic insufficiency*: None of the antimalarial drugs have any direct hepatotoxic effect. However, chloroquine is not advisable in patients with severe hepatic insufficiency. See liver disease and malaria

- *Renal failure*: The initial dose of antimalarial drugs need not be reduced in patients with renal failure. However, if the patient requires parenteral antimalarials even after three days and continues to be sick, then the dose can be reduced by one third to half of usual dose. See renal disease and malaria

- *Dermatitis*: Concomitant use of chloroquine with gold salts and phenyl butazone should be avoided because all the three can cause dermatitis.

5) Recommendations of ACTs

The choice of ACT in a country or region will be based on the level of resistance of the partner medicine in the combination:

- in areas of multidrug resistance (South-East Asia), artesunate + mefloquine or artemether-lumefantrine
- in Africa, artemether-lumefantrine, artesunate + amodiaquine; artesunate + sulfadoxine-pyrimethamine.

Note: The artemisinin derivative components of the combination must be given for at least 3 days for an optimum effect. Artemether-lumefantrine should be used with a 6-dose regimen. Amodiaquine + sulfadoxine-pyrimethamine may be considered as an interim option in situations where ACTs cannot be made available.

3.2.2.3. Modes of Treatment

Two important concepts in the treatment of malaria are suppressive and radical treatments.

1. *Suppressive treatment:* The symptoms of malaria can be alleviated by suppressing the erythrocytic stage of the parasitic development. Suppressive therapy involves administration of appropriate blood schizonticidal drugs. In all cases of *P. vivax* malaria and in most cases of *P. falciparum* malaria, it involves administration of chloroquine. In areas with high transmission of malaria, it is advisable to practice presumptive treatment for malaria.

   What is presumptive treatment?

   Presumption - In an area with high transmission of malaria, it should be presumed that all cases of fever are due to malaria.

   Treatment - First loading dose of ACT should be administered immediately after collecting the blood specimen, even without waiting for its report. If the fever is indeed malaria, this treatment alleviates symptoms early, may be well before the test result is available. If it is malaria, ACT also prevents the spread of malaria by destroying the gametocytes of *P. vivax* (the more common malaria). If it is not malaria, nothing is lost, for ACT at this dose is safe and has no adverse effects!

   Presumptive treatment in areas with high transmission of *P. falciparum* malaria: In areas with high transmission of *P. falciparum* malaria, ACT as presumptive treatment may not be enough. ACT does not sterilize the gametocytes of *P. falciparum* and thus will not prevent its spread. Therefore, in areas with high transmission of *P. falciparum* malaria, it is recommended to administer the single dose of Primaquine also, along with ACT. Therefore, presumptive treatment must be practiced by all - it cures early and more important, it prevents spread of *P. vivax* malaria.

2. *Radical treatment:* Radical treatment is administration of ACTs to all confirmed cases of malaria. In *P. vivax* malaria, 2 weeks' therapy with primaquine completely cures the infection in the host by its tissue schizonticidal activity and thereby prevents relapses. In *P. falciparum* malaria, a single dose of primaquine destroys the gametocytes, thereby prevents the spread of the infection into the mosquito.

Therefore, administration of primaquine is a must in all proven cases of malaria, (a two weeks' course in *P. vivax* malaria and a single dose in *P. falciparum* malaria).

3.2.2.4. Additional Features of Clinical Management [2]

1) *Patient without oral activity*: Some patients cannot tolerate oral treatment, and will require parenteral or rectal administration for 1–2 days until they can swallow and retain oral medication reliably. Although such patients may not show signs of severity, they should receive the same antimalarial dose regimens as for severe malaria (see section 4.2.).

2) *Hyperparasitemia*: Some patients may have no signs of severity but on examination of the blood film are found to have very high parasitemia. The risks associated with high parasitemia vary depending on the age of the patient and on transmission intensity. Thus cut-off values and definitions of hyperparasitemia also vary. Patients with high parasitemias are at an increased risk of treatment failure and of developing severe malaria, and therefore have an increased risk of dying. These patients can be treated with the oral ACTs recommended for uncomplicated malaria. However, they require close monitoring to ensure that the drugs are retained and that signs of severity do not develop, and they may require a longer course of treatment to ensure cure. Details of definitions and management are provided in sections 8.1 and 8.15.

3) *Use of antipyretics*: Fever is a cardinal feature of malaria, and is associated with constitutional symptoms of lassitude, weakness, headache, anorexia, and often nausea. In young children, high fevers are associated with vomiting, including medication, and seizures. Treatment is with antipyretics and, if necessary, tepid sponging. Care should be taken to ensure that the water is not too cool as, paradoxically, this may raise the core temperature by inducing cutaneous vasoconstriction. Paracetamol (acetaminophen) 15 mg/kg bodyweight every 4 h is widely used; it is safe and well tolerated given orally or as a suppository. Ibuprofen (5 mg/kg bodyweight) has been used successfully as an alternative in malaria and other childhood fevers, although there is less experience with this compound. Acetylsalicylic acid (aspirin) should not be used in children because of the risks of Reye's syndrome. There has been some concern that antipyretics might attenuate the host defense against malaria, as their use is associated with delayed parasite clearance. However, this appears to result from delaying cytoadherence, which is likely to be beneficial. There is no reason to withhold antipyretics in malaria.

4) *Use of antiemetics:* Vomiting is common in acute malaria and may be severe. Antiemetics are widely used. There have been no studies of their efficacy in malaria, and no comparisons between different antiemetic compounds, although there is no evidence that they are harmful.

5) *Management of seizures*: Generalized seizures are more common in children with *falciparum* malaria than in those with the other malarias. This suggests an overlap between the cerebral pathology resulting from malaria and febrile convulsions. Sometimes these seizures are the prodrome of cerebral malaria. If the seizure is ongoing, the airway should be maintained and anticonvulsants given (parenteral or rectal benzodiazepines or intramuscular paraldehyde). If it

has stopped, the child should be treated, as indicated in section 7.6.3, if core temperature is above 38.5°C. There is no evidence that prophylactic anticonvulsants are beneficial in otherwise uncomplicated malaria.

3.2.2.5. Operational Issues in Treatment Management

To optimize the benefit of deploying ACTs, and to have an impact on malaria, it will be necessary to deploy them as widely as possible – this means at most peripheral health clinics and health centers, and in the community. Deployment through the formal public health delivery system alone will not reach many of those who need treatment. In several countries, they must also be available through the private sector. Ultimately, effective treatment needs to be available at community or household level in such a way that there is no financial or physical barrier to access. The strategy to secure full access must be based on an analysis of the national and local health systems, and will often require adjustment based on program monitoring and operational research. The dissemination of clear national treatment guidelines, use of appropriate information, education and communication materials, monitoring both of the deployment process, access and coverage and provision of adequately packaged and presented antimalarials are needed to optimize the benefits of providing these new effective treatments widely [2].

1) *Health education:* At all levels, from the hospital to the community, education is vital to optimizing antimalarial treatment. Clear guidelines in the language understood by the local users, posters, wall charts, educational videos and other teaching materials, public awareness campaigns, education of and provision of information materials to shopkeepers and other dispensers can all improve the understanding of malaria and the likelihood of improved prescribing and adherence, appropriate referral, and minimizing the unnecessary use of antimalarials.

2) *Adherence to treatment*: To achieve the desired therapeutic effectiveness, a drug must be intrinsically efficacious and must be taken in the correct doses at the proper intervals. Patient adherence is a major determinant of the response to antimalarials, as most treatments are taken at home without medical supervision. There have been few studies of adherence. These suggest that 3-day regimens of medicines such as ACTs are adhered to reasonably well, provided that patients or careers are given an adequate explanation at the time of prescribing. Prescribers, shopkeepers or vendors should therefore give a clear and comprehensible explanation of how to use the medicines. Co-formulation is probably a very important contributor to adherence. User-friendly packaging, such as blister packs, also encourages completion of the treatment course and correct dosing.

3) *Quality assurance of antimalarial medicines*: Many of the antimalarials available in malaria endemic areas are substandard in that the manufacturing processes have been unsatisfactory, or their pharmaceutical properties do not meet the required pharmacopoeial specifications. Counterfeit tablets and ampoules containing no antimalarials are a major problem in some areas. These may result in fatal delays in appropriate treatment, and may also give rise to a mistaken impression of resistance. WHO, in collaboration with other United Nations agencies, has established an international mechanism to pre-qualify manufacturers of artemisinin compounds and ACTs on the basis of compliance with internationally-recommended standards of

manufacturing and quality. It is the responsibility of national ministries of Health and regulatory authorities to ensure the quality of antimalarials provided through both the public and private sectors, through regulation, inspection and law enforcement.

4) *Pharmacovigilance*: Rare but serious adverse effects are often not detected in clinical trials and can only be detected through pharmacovigilance systems operating in situations of wide population use. Since there is little data from prospective Phase IV post-marketing studies but potentially serious adverse effects of antimalarials. Chloroquine has the best-documented adverse effect profile. The safety profiles of the artemisinin derivatives, mefloquine and sulfadoxine–pyrimethamine are supported by a reasonable evidence base, but mainly from large clinical trials.

The neurotoxicity observed in animals treated with artemisinin derivatives has prompted large prospective assessments in humans, but no evidence of neurotoxicity has been found. Concerns over the risk of severe liver or skin reactions to sulfadoxine-pyrimethamine treatment have receded with increasing numbers of negative reports. More data is needed on the newer drugs and on amodiaquine as well. There is also an urgent need to obtain more information on the safety of antimalarials, in particular the ACTs, in pregnancy. It is recommended that countries or regions should consider establishing pharmacovigilance systems if they have not already done so (see section 5.2).

3.2.2.6. Management of Treatment Failures

1) *Failure within 14 days*: Treatment failure within 14 days of receiving an ACT is very unusual. Of 39 trials of artemisinin or its derivatives, which together enrolled 6124 patients, 32 trials (4917 patients) had no failures at all by day 14. In the remaining 7 trials, failure rates at day 14 ranged from 1% to 7%. The majority of treatment failures occur after 2 weeks of initial treatment. In many cases failures are missed because patients presenting malaria symptoms are not asked whether they have received antimalarial treatment within the preceding 1–2 months. Recurrence of *falciparum* malaria can be the result of a re-infection, or a recrudescence (i.e., failure). In an individual patient it may not be possible to distinguish recrudescence from re-infection, although if fever and parasitemia fail to resolve or recur within 2 weeks of treatment then this is considered a failure of treatment. Wherever possible treatment failure must be confirmed parasitologically – preferably by blood slide examination (as *HRP2*-based tests may remain positive for weeks after the initial infection even without recrudescence). This may require transferring the patient to a facility with microscopy; transfer may be necessary anyway to obtain second-line treatment. Treatment failures may result from drug resistance, poor adherence or unusual pharmacokinetic properties in that individual. It is important to determine from the patient's history whether he or she vomited previous treatment or did not complete a full course. Treatment failures within 14 days should be treated with a second-line antimalarial (see section 3.2.2.1.).

2) *Failure after 14 days*: Recurrence of fever and parasitemia more than 2 weeks after treatment, which could result either from recrudescence or new infection, can be retreated with the first-line ACT. Parasitological confirmation is desirable but not a precondition. If it is a recrudescence, then the first-line treatment should still be effective in most cases. This simplifies operational management and drug

deployment. However, reuse of mefloquine within 28 days of first treatment is associated with an increased risk of neuropsychiatric sequel and, in this particular case; second-line treatment should be given. If there is a further recurrence, then malaria should be confirmed parasitologically and second line treatment given.

3) *Recommended second-line antimalarial treatments*: On the basis of the evidence from current practice and the consensus opinion of the Guidelines Development Group, the following second-line treatments are recommended, in order of preference:
   - alternative ACT known to be effective in the region,
   - artesunate + tetracycline or doxycycline or clindamycin,
   - quinine + tetracycline or doxycycline or clindamycin.

   The alternative ACT has the advantages of simplicity, and where available co-formulation to improve adherence. The 7-day quinine regimes are not well tolerated and adherence is likely to be poor if treatment is not observed. Summary of recommendations on second-line antimalarial treatment for uncomplicated *falciparum* malaria are follows:
   - Alternative ACT known to be effective in the region.
   - Artesunate (2 mg/kg bodyweight once a day) + tetracycline (4 mg/kg bodyweight four times a day) or doxycycline (3.5 mg/kg bodyweight once a day) or clindamycin (10 mg/kg bodyweight twice a day). Any of these combinations to be given for 7 days.
   - Quinine (10 mg salt/kg bodyweight three times a day) + tetracycline or doxycycline or clindamycin. Any of these combinations to be given for 7 days [2].

3.2.3. Efficacy of Current ACTs in Uncomplicated Malaria Therapy

Artemisinin-based combination therapy (ACT) is defined as the simultaneous administration of two or more blood schizonticidal antimalarial drugs. The drugs, which may either be co-formulated or co-administered, should have independent modes of action and different biochemical targets in the parasites. It should be noted that multiple-drug therapies in which neither of the components has significant schizonticidal effect when used alone (e.g., sulfadoxine–pyrimethamine) is not considered as combination therapy. Neither is the use of a blood schizonticidal drug with a tissue schizonticidal drug, or with a non-antimalarial drug to enhance its action considered antimalarial combination therapy. The potential value of ACTs to improve therapeutic efficacy and delay the development and selection of drug resistant parasites has been recommended treatments for uncomplicated *falciparum* in Asia and Africa by WHO [2, 47]. The following ACTs are currently recommended: artemether-lumefantrine, artesunate + amodiaquine, artesunate + mefloquine, artesunate + sulfadoxine-pyrimethamine.

The choice of ACT in a country or region will be based on the level of resistance of the partner medicine in the combination: 1) in areas of multidrug resistance (South-East Asia), artesunate + mefloquine or artemether-lumefantrine; 2) in Africa, artemether-lumefantrine, artesunate + amodiaquine; artesunate + sulfadoxine-pyrimethamine. The artemisinin derivative components of the combination must be given for at least 3 days for an optimum effect. Artemether-lumefantrine should be used with a 6-dose regimen. Amodiaquine + sulfadoxine-pyrimethamine may be considered as an interim option in situations where ACTs cannot be made available.

Table 19. Dosing schedule for artemether-lumefantrine (AL)

Body weight in kg (age in years)		Number of tablets at approximate timing of dosing*					
		0 h	8 h	24 h	36 h	48 h	60 h
5 - 14	(< 3)	1	1	1	1	1	1
15 – 24	(≥ 3 – 8)	2	2	2	2	2	2
25 – 34	(≥ 9 – 14)	3	3	3	3	3	3
> 34	(> 14)	4	4	4	4	4	4

* The regimen can be expressed more simply for ease of use at the program level as follows: the second dose on the first day should be given any time between 8 h and 12 h after the first dose. Dosage on the second and third days is twice a day (morning and evening) [2].

Table 20. Dosing schedule for artesunate + amodiaquine (ASAQ)

Age	Dose in mg (Number of tablets)					
	Artesunate (50 mg)			Amodiaquine (153 mg)		
	Day 1	Day 2	Day 3	Day 1	Day 2	Day 3
5 – 11 months	25 (1/2)	25	25	76 (1/2)	76	76
≥ 1 – 6 years	50 (1)	50	50	153 (1)	153	153
≥ 7 – 13 years	100 (2)	100	100	306 (2)	306	306
> 13 years	200 (4)	200	200	612 (4)	612	612

The data was cited from WHO guideline [2].

Table 21. Dosing schedule for artesunate + sulfadoxine–pyrimethamine (ASSP)

Age	Dose in mg (Number of tablets)					
	Artesunate (50 mg)			Sulfadoxine-pyrimethamine (500/25 mg)		
	Day 1	Day 2	Day 3	Day 1	Day 2	Day 3
5 – 11 months	25 (1/2)	25	25	250/12.5 (1/2)	-	-
≥ 1 – 6 years	50 (1)	50	50	500/25 (1)	-	-
≥ 7 – 13 years	100 (2)	100	100	1000/50 (2)	-	-
> 13 years	200 (4)	200	200	1500/75 (4)	-	-

The data was cited from WHO guideline [2].

Table 22. Dosing schedule for artesunate + mefloquine (ASMQ)

Age	Dose in mg (Number of tablets)					
	Artesunate (50 mg)			Sulfadoxine-pyrimethamine (500/25 mg)		
	Day 1	Day 2	Day 3	Day 1	Day 2	Day 3
5 – 11 months	25 (1/2)	25	25	-	125 (1/2)	-
≥ 1 – 6 years	50 (1)	50	50	-	250 (1)	-
≥ 7 – 13 years	100 (2)	100	100	-	500 (2)	-
> 13 years	200 (4)	200	200	-	1000 (4)	-

The data was cited from WHO guideline [2].

3.2.3.1. Clinical Outlooks of Treatment with Recommended ACTs

1) Artemether-lumefantrine (AL)

This was the first fixed dose combination of an artemisinin derivative with a second unrelated antimalarial compound. This is currently available as co-formulated tablets containing 20 mg of artemether and 120 mg of lumefantrine. The total recommended treatment is a 6-dose regimen of artemether-lumefantrine twice a day for 3 days (Table 19). Against multi-drug–resistant *falciparum* malaria the 6-dose regimen of AL is generally as effective as and better tolerated than artesunate–mefloquine [236]. AL is becoming increasingly available in tropical countries. The excellent adverse effects profile and recent price reductions (down to US $1 per adult treatment) make it an increasingly attractive treatment option. There is increasing evidence of safety for this combination in the second and third trimesters of pregnancy.

An advantage of this combination is that lumefantrine is not available as a monotherapy and has never been used by itself for the treatment of malaria. Recent evidence indicates that the therapeutic response and safety profile in young children of less than 10 kg is similar to that in older children, and artemether-lumefantrine is now recommended for patients is ≧ 5 kg. Lumefantrine absorption is enhanced by co-administration with fat. Low blood levels, with resultant treatment failure, could potentially result from inadequate fat intake, and so it is essential that patients or careers are informed of the need to take this ACT with milk or fat-containing food – particularly on the second and third days of treatment.

2) Artesunate + amodiaquine (AS + AQ)

In recent years resistance of amodiaquine has worsened considerably in parts of East and Southern Africa. After oral administration amodiaquine is largely converted to desethylamodiaquine, which contributes the majority of antimalarial activity. Amodiaquine is generally reasonably well tolerated and slightly more palatable than chloroquine, although in some areas it has not been a popular substitute. The combination of amodiaquine with artesunate is currently available as separate scored tablets containing 50 mg of artesunate and 153 mg base of amodiaquine, respectively. Co-formulated tablets are under development. The total recommended treatment is 4 mg/kg bodyweight of artesunate and 10 mg base/kg bodyweight of amodiaquine given once a day for 3 days (Table 20). This combination is sufficiently efficacious only where 28-day cure rates with amodiaquine monotherapy exceed

80%. Resistance is likely to worsen with continued availability of chloroquine and amodiaquine monotherapies. More information on the safety of artesunate + amodiaquine is needed from prospective pharmacovigilance programs.

3)   Artesunate + sulfadoxine–pyrimethamine (AS + SP)

Sulfadoxine-pyrimethamine (SP) is a fixed combination of a long-acting sulfonamide and the antifolate pyrimethamine. This ACT (AS + SP) is currently available as separate scored tablets containing 50 mg of artesunate, and tablets containing 500 mg of sulfadoxine and 25 mg of pyrimethamine. The total recommended treatment is 4 mg/kg bodyweight of artesunate given once a day for 3 days and a single administration sulfadoxine-pyrimethamine (25/1.25 mg base/kg bodyweight) on day 1 (Table 21). While a single dose of sulfadoxine–pyrimethamine is sufficient, it is necessary for artesunate to be given for 3 days for satisfactory efficacy. This combination is sufficiently efficacious only where 28-day cure rates with sulfadoxine- pyrimethamine alone exceed 80%. Resistance is likely to worsen with continued availability of sulfadoxine–pyrimethamine, sulfalene–pyrimethamine and cotrimoxazole (trimethoprim-sulfamethoxazole). This ACT is currently being used in parts of South America, the Middle East, and South Asia where SP susceptibility remains high.

4)   Artesunate + mefloquine (AS + MQ)

Combining artesunate or artemether (4 mg/kg/day for 3 days) with mefloquine has all the advantages of a combination treatment previously described [56], and the additional benefit that if mefloquine is split as 8.3 mg/kg/day for 3 days or not given until the second day of treatment then absorption is increased and gastrointestinal adverse effects are lessened. A fixed combination of mefloquine and artesunate has recently been developed. This is currently available as separate scored tablets containing 50 mg of artesunate and 250 mg base of mefloquine, respectively. Co-formulated tablets are under development but are not available at present. The total recommended treatment is 4 mg/kg bodyweight of artesunate given once a day for 3 days and 25 mg base/kg bodyweight of mefloquine usually split over 2 or 3 days (Table 22).

Two different doses of mefloquine have been evaluated, 15 mg base/kg bodyweight and 25 mg base/kg bodyweight. The lower dose is associated with inferior efficacy and is not recommended. To reduce acute vomiting and optimize absorption, the 25 mg/kg dose is usually split and given either as 15 mg/kg (usually on the second day) followed by 10 mg/kg one day later, or as 8.3 mg/kg per day for 3 days. Pending development of a co-formulated product, malaria control programs will have to decide on the optimum operational strategy of mefloquine dosing for their populations. Mefloquine is associated with an increased incidence of nausea, vomiting, dizziness, dysphoria, and sleep disturbance in clinical trials, but these are seldom debilitating and in general, where this ACT has been deployed, it has been well tolerated.

3.2.3.2. Efficacy of current ACTs Recommended

The International Artemisinin Study Group [11] has indicated higher cure rates, faster parasite clearance and decreased gametocyte carriage with ACT compared with monotherapy. WHO has stated that AL demonstrates cure rates above 95%, even in areas of multiple drug resistance [47, 64]. Preliminary results from Africa indicate that combinations of AS + AQ or AS + SP are highly effective, although efficacy may be compromised in areas with moderate to high levels of resistance to SP [206]. Hence, the combination AS + AQ is not only more effective than AS + SP at present but also may remain effective for an extended period of time [207]. Combinations with AS have proved very effective and could, in the right circumstances, protect drugs from the progressive development of resistance, as has been shown with the implementation of ACT elsewhere. AS + AQ has now been adopted as first line antimalarial treatment in 41 countries in Africa [55]. A number of studies [208, 209] in sub-Saharan Africa have shown the combination of AS + AQ to have a 28-day corrected efficacy of over 90%. However, in another study the efficacy of this combination was below 90% [210].

The safety and efficacy of three combinations, AQ + SP, AS + SP, and AS + AQ, were evaluated in children aged 6 - 59 months with axillary body temperature of $\geqslant$37.5 °C and non-complicated malaria [211]. Results indicated that the combinations were safe, had rapid fever reduction time, had 100% clinical efficacy and reduced gametocytaemia incidence during the 14-day follow-up time. Elsewhere in Congo, Swarthout et al. [212] showed that in children aged 0 - 59 months given 3 days observed treatment with AS + AQ (n = 90) or 3 days with AS + SP (n = 90) there was no significant difference between the two groups in time to parasite clearance, fever reduction and gametocyte clearance; however, AS + AQ had higher efficacy than AS + SP.

It was demonstrated that a six-dose course of AL administered to children with *P. falciparum* malaria reduced gametocyte prevalence, duration of gametocyte carriage and infectiousness to mosquitoes when compared to children treated with CQ + SP. None of the children treated with AL were infectious after day seven of treatment, compared with 37% of the children treated with CQ + SP; however, there was no significant advantage in terms of parasitological and clinical failure for the two regimens [213]. In Senegal, the efficacy and tolerability of AS + AQ, AS + MQ, AL (four doses and six doses) and AQ + SP in the treatment of uncomplicated *falciparum* malaria were evaluated in a multicenter study. Findings indicated an excellent clinical and parasitological response rate of 100%, except for the four-dose AL regimen (96.4%) [214]. The results from similar studies in Sudan [215] and Ghana [216] showed that ACT was very effective. More recently, one clinical trial demonstrated that AS + AQ and AL were still efficacious (94.2-98.2% curative rates) for treatment of children with uncomplicated malaria in Ghana and drug related adverse events were rare in treated subjects during one year of follow-up. However, the high prevalence of potentially AQ resistant parasites raises questions about the utility of AQ as a partner drug for ACT in Ghana. The efficacy of AS + AQ in Ghana requires, therefore, continuous monitoring and evaluation [161].

3.2.3.3. Mission for Selecting the Best ACT

Early and effective chemotherapy for malaria has a pivotal role in reducing morbidity and mortality especially since a vaccine is unlikely to emerge within the next decade. Multidrug resistance has been reported from most parts of the world and as a result, monotherapy or some of the available combination chemotherapies for malaria are either ineffective or less effective. New antimalarial regimens are, therefore, urgently needed and artemisinin-based combination therapy is widely advocated. ACTs can increase efficacy, shorten duration of treatment (and hence increase compliance), and decrease the risk of resistant parasites arising through mutation during therapy. However, the better treatment is required and has brought new challenges in the design and interpretation of efficacy assessments.

In the past few years, it has become accepted that antimalarial treatments must have high cure rates, which ideally should exceed 90% [2]. The corollary that malaria treatment recommendations should change if cure rates are below 90% requires further definition, but this is a considerable advance on the previous era when much lower rates were considered acceptable. The concept of combination therapy is based on the synergistic or additive potential of two or more drugs, to improve therapeutic efficacy as better treatments. If the cure rate of any ACT is below 90%, the treatment recommendation should change and the better ACT should be selected from a superiority clinical trial. In this earlier context of uncertainty it was reasonable to plan a randomized comparison to test if there was a difference between the regimens being tested in the "superiority" trials. The total effect of ACTs is to reduce the chance of parasite recrudescence, reduce the within-patient selection pressure, and prevent transmission.

There are many published trials comparing different ACTs for malaria. One difficulty in interpreting these studies is the different efficacy endpoints reported. Traditionally, the goal of treatment in low-transmission areas is to achieve a parasitological cure, i.e., the elimination of all parasites from the body, and the goal in high-transmission areas is a clinical cure. The WHO has recently updated its own endpoints to try to standardize these in a single scheme which includes both clinical and parasitological outcomes. Most studies in Africa report clinical outcome at 14 or 28 days as a primary endpoint [221]. From a public health point of view, documenting the effect of failure to achieve a parasitological cure is important. The continued presence of parasites in the blood after treatment may lead to anemia, clinical recrudescence, and severe disease. Longitudinal studies, where patients are followed over a longer period and given the same treatment at each malarial episode, may allow a better comparison of antimalarial therapies in high-transmission areas. The best ACT is selected based on various well-conducted and randomized comparative trials.

1) Six doses of AL is superior to four doses of AL

The data showed in one clinical trial that the study (238 adults and children, Thailand, 1996–1997) found a significantly higher rate of cure at day 28 with the 6-dose regimen given over 3 days than with the 4-dose regimen of artemether-lumefantrine (AL) also given over 3 days (PCR-unadjusted treatment cure rate for intention to treat population at day 28). The new oral fixed combination AL has proved to be an effective and well-tolerated treatment of multi-drug resistant *P. falciparum* malaria, although cure rates using the four-dose regimen have been lower than with the currently recommended alternative of artesunate-mefloquine (AS + MQ). Two six-dose schedules (total adult dose = 480 mg of artemether and 2,880 mg of lumefantrine) were therefore compared with the previously used four-dose regimen (320 mg of artemether and 1,920 mg of lumefantrine) in a double-blind trial involving 359 patients with uncomplicated multidrug-resistant *falciparum* malaria. There were no differences between the three treatment groups in parasite and fever clearance times, and reported adverse effects. The two six-dose regimens gave adjusted 28-day cure rates of 96.9% and 99.12%, respectively, compared with 83.3% for the four-dose regimen (P < 0.001). These six-dose regimens of AL provide a highly effective and very well-tolerated treatment when compared to four-dose regimens of AL [217].

2)   AL (6 doses) is not better than AS (3 days) + MQ

One clinical trial (537 participants) was studied the six-dose regimen of AL in comparison with AS + MQ. For the six-dose regimen, two studies compared AL with AS + MQ in comparative effects for day 28 parasitemia, and no difference in parasite or fever clearance time was detected. There were 11 parasitological failures with AL and none with AS + MQ [218]. The clinical trial suggested that AS + MQ therapy is more effective than that of AL.

3)   AL (6 doses) seems similar to AS (3 days) + AQ

To test the hypothesis that AS + AQ is as effective as AL in the treatment of acute uncomplicated malaria in Nigerian children. In an open label, randomized controlled clinical trial, children aged 6 months to 10 years were randomized to receive artesunate (4 mg/kg daily) plus amodiaquine (10 mg/kg daily) or AL (5-14 kg, one tablet; 15-24 kg, two tablets and 25-34 kg, three tablets twice daily). Both drug regimens were given for 3 days and follow-up was for 28 days. A total of 132 children (66 in each group) were randomized to receive either AS + AQ or AL. Day 28 cure rates in the per protocol population were 93% for AS + AQ and 95% for AL (OR = 0.71, 95% CI = 0.12-3.99, rho = 0.66). Using Kaplan-Meier product-limit estimates of failure, the median survival time for AS + AQ was 21 days and for AL 28 days (P = 0.294). PCR corrected day 28 cure rate for PP populations were 98.4% for AS + AQ and 100% for AL. Both drugs were well-tolerated. Therefore, AS + AQ is as effective as AL and both combinations were efficacious and safe [219].

Another trial was conducted with a 28-day follow-up period but one with a 14-day follow-up period [220]. A total of 295 children under 5 years were included; 153 children were treated with AS + AQ, and 142 children with the combination of AL. Among the 295 children, 290 were followed up to 14 days. In the group of 149 children treated with artesunate and amodiaquine, 142 (95.3%, 95% CI: 91.9-98.7%) presented with adequate clinical and parasitological response, five (3.3%) with late parasitological failure, one (0.7%) with late clinical failure and one (0.7%) with early treatment failure. Among the 141 children treated with AL, 140 (99.3%, 95% CI: 97.9-100%) presented with adequate clinical and parasitological response and one (0.7%) with late parasitological failure at Buhiga. Side-effects were comparable in both groups except for the vomiting. Vomiting was more frequent in the AS + AQ on D1 and D2. Both treatments decreased the gametocyte carriage but without getting full clearance in all the patients. During a consensus workshop, the Ministry of Public Health agreed on the combination of AS + AQ as the first line drug for the treatment of uncomplicated *falciparum* malaria in Burundi including epidemic outbreak.

4)   DHA-PQ (3 days) is better than AS (3 days) + AQ

A series of studies to optimize the treatment have been compared to various ACTs. Dihydroartemisinin-piperaquine (DHA-PQ) with artesunate-amodiaquine (AS + AQ) were conducted in 334 patients (185 were infected with *P. falciparum*, 80 were infected with *P. vivax*, and 69 were infected with both species) with a 42-day follow-up period. The overall parasitological failure rate at day 42 was 45% (95% confidence interval [CI], 36%-53%) for AS + AQ and 13% (95% CI, 7.2%-19%) for DHA-PQ (hazard ratio [HR], 4.3; 95% CI, 2.5-7.2; P < .001). Rates of both recrudescence of *P. falciparum* infection and recurrence of *P. vivax* infection were significantly higher after receipt of AS + AQ than after receipt of DHA-PQ (HR, 3.4 [95% CI, 1.2-9.4] and 4.3 [95% CI, 2.2-8.2], respectively; P < .001). By the end of the study, AS + AQ recipients were 2.95-fold (95% CI, 1.2- to 4.9-fold) more likely to be anemic and 14.5-fold (95% CI, 3.4- to 61-fold) more likely to have carried *P. vivax* gametocytes. DHA-PQ was more effective and better tolerated than AAQ against multidrug-resistant *P. falciparum* and *P. vivax* infections. The prolonged therapeutic effect of piperaquine delayed the time to *P. falciparum* re-infection, decreased the rate of recurrence of *P. vivax* infection, and reduced the risk of *P. vivax* gametocyte carriage and anemia [222, 223].

5) DHA-PQ (3 days) is better than AL (6 doses)

Uganda recently adopted AL as the recommended first-line treatment for uncomplicated malaria. However, AL has several limitations, including a twice-daily dosing regimen, recommendation for administration with fatty food, and a high risk of re-infection soon after therapy in high transmission areas. DHA-PQ is a new alternative artemisinin-based combination therapy that is dosed once daily and has a long post-treatment prophylactic effect. The efficacy of AL with DHA-PQ was compared in Kanungu, an area of moderate malaria transmission. Patients aged 6 months to 10 years with uncomplicated *falciparum* malaria were randomized to therapy and followed for 42 days. Genotyping was used to distinguish recrudescence from new infection. Of 414 patients enrolled, 408 completed follow-up. Compared to patients treated with AL, patients treated with DHA-PQ had a significantly lower risk of recurrent parasitemia (33.2% vs. 12.2%; risk difference = 20.9%, 95% CI 13.0-28.8%) but no statistically significant difference in the risk of treatment failure due to recrudescence (5.8% vs. 2.0%; risk difference = 3.8%, 95% CI -0.2-7.8%). Patients treated with DHA-PQ also had a lower risk of developing gametocytaemia after therapy (4.2% vs. 10.6%, p = 0.01). Both drugs were safe and well tolerated. DHA-PQ is highly efficacious, and operationally preferable to AL because of a less intensive dosing schedule and requirements [224].

Similar studies were performed by other scientists in different regents, and indicated that DHA-PQ regimens were more efficacious than the AL regimen for the treatment of uncomplicated *P. falciparum* malaria. DHA-PQ was also superior to AL for reducing the risk of recurrent parasitemia and gametocytemia, and provided improved hemoglobin recovery [133, 225-227].

6) DHA-PQ (3 days) is similar to AS (3 days) + MQ

Multi-drug resistant *falciparum* malaria is an important health problem in the Peruvian Amazon region. A randomized open label clinical trial was carried out in comparing AS + MQ, the current first line treatment in this region of Peru, with DHA-PQ. Total 522 patients with *P. falciparum* uncomplicated malaria were recruited, randomized (260 with AS + MQ and 262 with DHA-PQ), treated and followed up for 63 days. PCR-adjusted adequate clinical and parasitological response, estimated by Kaplan Meier survival and Per Protocol analysis, was extremely high for both drugs (99.6% for AS + MQ and 98.4% and for DHA-PQ) (RR: 0.99, 95%CI [0.97-1.01], Fisher Exact p = 0.21). All recrudescences were late parasitological failures. Overall, gametocytes were cleared faster in the AS + MQ group (28 vs. 35 days) and new gametocytes tended to appear more frequently in patients treated with DHA-PQ (day 7: 8 (3.6%) vs. 2 (0.9%), RR: 3.84, 95%CI [0.82-17.87]). Adverse events such as anxiety and insomnia were significantly more frequent in AS + MQ group, both in adults and children. DHA-PQ is as effective as AS + MQ in treating uncomplicated *P. falciparum* malaria but it is better tolerated and more affordable than AS + MQ [228]. Supporting studies were conducted by another two clinical trials [229, 230].

7) Other clinical trials

A meta-analysis of data from 16 randomized trials (12 from Africa) compared the effect of adding 3 days of any artemisinin to one of the standard treatment regimes of CQ, SP, AQ, or mefloquine. The study concluded that, for all trials combined, the addition of 3 days of artemisinin to standard antimalarial treatment significantly reduce parasitological failure on days 14 and 28. Gametocyte carriage was also reduced in the artemisinin treated patients. AS + AQ, AL and DHA-PQ were as reference antimalarial treatments, with candidate regimens using 2-3 day courses of artemisinin-piperaquine (ART-PQ). Initially, patients receiving each of the regimens had a rapid clinical and parasitological response. All treatments were well tolerated and no serious adverse effects occurred. The 28-day cure rates were < 80% for the 2-day treatments with ART-PQ at 2.4 mg/kg and 14.4 mg/kg, respectively, in the first study period and artemisinin-piperaquine at 3.2 mg/kg and 16.0 mg/kg, respectively, but > 98% for the 3-day regimens. These results suggest that a 3-day course of ART-PQ at 3.2 mg/kg and 16.0 mg/kg, respectively, deserve further evaluation as an alternative treatment for multidrug-resistant *P. falciparum* malaria [231]. Similarly, chlorproguanil–dapsone–artesunate (CDA), which is about to start clinical Phase III trials, is not yet available [221, 223].

3.2.4. Efficacy Challenges of ACTs in Uncomplicated Malaria Therapy

All of the ACTs presently recommended by the WHO rely on partner drugs that are already compromised by resistance. As ACTs are rolled out, optimizing their effectiveness will require more such trials that help target the right drugs to the right settings and that consider the broader impacts of different treatment regimens. These will include longitudinal trials, which take a step beyond extending the duration of follow up after a single treatment and assess the longer-term health benefit of treatment regimens by assigning participants to receive the same treatment each time they get malaria over the course of a year. In addition to the standard assessment of efficacy, longitudinal trials measure any loss of efficacy or selection for resistance after repeated treatments of ACTs were major challenge in the future treatment of uncomplicated malaria. Some African studies have already indicated a significant failure rate [148, 212] due to the presence of amodiaquine resistance. Resistance is likely to spread more rapidly as the combination is being adopted as the main treatment in several African countries [67].

Because new policies are made on the basis of good clinical research the mission to replace failed malaria drugs with more efficacious ACTs is being accomplished. However, maximizing the benefit and prolonging the life of ACTs will also require learning from the past. New strategies are needed to prevent ACTs from following trajectory of chloroquine.

- Can better understanding of mechanisms of action, pharmacokinetics, and pharmacodynamics be used to design combination therapies that provide prolonged prophylactic efficacy while still deterring resistance?
- Can drugs be rotated to preserve or resuscitate their efficacy?
- Can selective pressure for resistance be reduced by using multiple ACTs in parallel, instead of serially replacing failing drugs?
- Can use of ACTs in combination with insecticide impregnated bednets, residual spraying, and, eventually, vaccines reduce malaria transmission and block the dissemination of resistance when it arises?

As the campaign to push ACTs to the front line of the war on malaria succeeds, answering some of these questions may help us to contain the insurgency of drug resistance. The major challenge to the deployment of artemisinin-based antimalarial drug combinations particularly includes:

- the choice of ACTs best suited for the different epidemiological situations,
- when to deploy ACTs, and
- operational obstacles to implementation.

The next line of effective antimalarial drugs is significantly more expensive than current monotherapy with chloroquine, amodiaquine, or SP. In order to reduce malaria mortality, combination therapies must constitute treatment policy in Asia and Africa, and be useable as first-line treatment, affordable to communities, and available through the private sector. Appropriate political and institutional actions for sustained financing are prerequisites to addressing this important challenge of providing access to these drugs. Safety and efficacy have been demonstrated with extensive use of ACTs in South-east Asia and Africa. There is, therefore, an urgent need to document efficacy and safety among young children, pregnant women, and breastfeeding mothers and their babies.

Currently, it is of more important to note that there have been no reports of clinical resistance to the artemisinin drugs so far, although artemisinin-resistant strains of *P. falciparum* [232] and *P. yoelii* [233] have been developed in the laboratory. Clinical isolates and laboratory stains have been shown to vary in their sensitivities to these drugs but there is no evidence that this is related to clinical failure [234]. Apparent drug failure following AS treatment of *P. falciparum* malaria was reported in West Africa [235]. In another study, AL was associated with more failures than AS + MQ [236]. Although PCR genotyping was reported in this study, it was unclear how complete the results were. Other AS combinations tested in the same population, such as AS + SP or AL gave more sustained reductions in the levels of gametocyte carriage, as the level of resistance to the accompanying drug(s) remained low [237].

Over the past 10 years multiple comparative studies on ACTs were published, highlighting the relative advantages or drawbacks of these combinations. However, because none of the studies compare all combinations and methodology varies between studies, it is not easy to see clear results in this vast amount of data produced. Therefore a multi-treatment random-effects meta-analysis was used to provide a tool for comparative assessment of various ACTs [238]. Using this method the clinical efficacy at day 28 was compared between commonly used combination therapies. Results show that the combination AL has a higher clinical efficacy regarding PCR corrected ACPR at day 28 than all other combinations, but is closely followed by AS + MQ. The combinations AS + AQ and AS + SP are not necessarily inferior but their percentages of success at day 28 PCR corrected are very close to the threshold of 90%. Totally, twenty-six trials were conducted in Africa, four in Asia, and two in South America. The antimalarial combination therapies used for comparison against the combination SP + AS (12 studies) were SP + AS (2 studies), AQ + SP (3 studies), AQ + AS (10 studies), MQ + AS (5 studies), CQ + AS (4 studies), CQ + SP (2 studies), and AL (6 studies). The percentage of success for each antimalarial combination therapy at day 28 PCR corrected was calculated and is presented in Table 23 [238]. The lowest percentage of success was obtained with the combination CQ + AS (45.8%) and the highest percentage of success was obtained with the combination AL (97.4%).

One report described different methods for assessing the efficacy of ACTs that resulted in different estimates, with implications for changes in treatment policy. Data from different *in vivo* studies of ACT treatment of uncomplicated *falciparum* malaria were combined in a single database. Efficacy at day 28 corrected by PCR genotyping was estimated using four methods. In the first two methods, failure rates were calculated as proportions with either (1a) re-infections excluded from the analysis (standard WHO per-protocol analysis) or (1b) re-infections considered as treatment successes. In the second two methods, failure rates were estimated using the Kaplan-Meier product limit formula using either (2a) WHO [47] definitions of failure, or (2b) failure defined using parasitological criteria only. Data analyzed represented 2926 patients from 17 studies in nine African countries. Three ACTs were studied: AS+AQ (n = 1702), AS + SP (n = 706) and AL (n = 518). Using method (1a), the day 28 failure rates ranged from 0% to 39.3% for AS+AQ treatment, from 1.0% to 33.3% for AS+SP treatment and from 0% to 3.3% for AL treatment. The median [range] difference in point estimates between method 1a (reference) and the others were: (i) method 1b = 1.3% [0 to24.8], (ii) method 2a = 1.1% [0 to21.5], and (iii) method 2b = 0% [-38 to19.3]. The standard per-protocol method (1a) tended to overestimate the risk of failure when compared to alternative methods using the same endpoint definitions (methods 1b and 2a). It either overestimated or underestimated the risk when endpoints based on parasitological rather than clinical criteria were applied. The standard method was also associated with a 34% reduction in the number of patients evaluated compared to the number of patients enrolled. Only 2% of the sample size was lost when failures were classified on the first day of parasite recurrence and survival analytical methods were used. The primary purpose of an *in vivo* study should be to provide a precise estimate of the risk of antimalarial treatment failure due to drug resistance. Use of survival analysis is the most appropriate way to estimate failure rates with parasitological recurrence classified as treatment failure on the day it occurs [160].

Table 23. Percentages of success for each antimalarial combination therapy at day 28 PCR corrected

Treatment combination(s)	Percentage success (%)
Amodiaquine + Artesunate (AQ + AS)	88.5
Amodiaquine + Sulfadoxine-pyrimethamine (AQ + SP)	85.7
Chloroquine + Artesunate (CQ + AS)	45.8
Chloroquine + Sulfadoxine-pyrimethamine (CQ + SP)	72.1
Mefloquine + Artesunate (MQ + AS)	96.9
Sulfadoxine-pyrimethamine + Artesunate (SP + AS)	82.6
Artemether-lumefantrine (Art-lumef)	97.4

The data was cited from WHO guideline [238].

Whatever methods using in clinical failure measurements, the prevention of the resistance by ACTs should be necessary. Stable, therapeutically significant resistance to the artemisinin derivatives has not yet been identified and cannot be induced yet in the laboratory, which suggests that it may be a very rare event. But it would be foolish to bank on whether or not it can happen and should it arise, it would be a global disaster. For mutual protection against the emergence of drug resistance, these drugs should be used only in combination with other antimalarials. Artemisinin derivatives are particularly effective in combinations because of their very high killing rates, lack of adverse effects, and absence of significant resistance [6]. The ideal pharmacokinetic properties for an antimalarial drug have been greatly debated. From a resistance prevention perspective, the combination partners should have similar pharmacokinetic properties. Rapid elimination ensures that the residual concentrations do not provide a selective filter for resistant parasites, but these drugs (if used alone) must be given for 7 days, and adherence to 7-day regimens is poor. Even 7-day regimens of artemisinin derivatives are associated with approximately 10% failure rates. In order to be highly effective in a 3-day regimen, terminal elimination half-lives of at least one drug component need to exceed 24 hours. If two drugs are used with different modes of action, and therefore different resistance mechanisms, then the per-parasite probability of developing resistance to both drugs is the product of their individual per-parasite probabilities.

This strategy would be expected to be effective at preventing the *de novo* emergence of resistance at higher levels of transmission, where high-biomass infections still constitute the major source of *de novo* resistance. The main obstacles to the success of combination treatment in preventing the emergence of resistance will be incomplete coverage, or inadequate treatment, and, as for antituberculous drugs, use of one of the combination partners alone. Drugs of poor quality are common in tropical areas of the world, adherence to antimalarial treatment regimens is often incomplete, and antimalarials are available widely in the market place. Resistance to the artemisinins may not have happened yet. If it does, it will most likely arise in a hyperparasitemic patient who received an inadequate dose of a single antimalarial drug, not in combination with another suitable antimalarial agent. Irrespective of the epidemiological setting, ensuring that patients with high parasitemias receive a full course of adequate doses of artemisinin combination treatment would be an effective method of slowing the emergence of antimalarial drug resistance.

3.3. Chemotherapy for Severe Malaria

3.3.1. Severe and Complicated Malaria

Malaria is caused by protozoan parasites of the genus *Plasmodium (P.)*. In humans malaria is caused by *P. falciparum*, *P. malariae*, *P. ovale*, and *P. vivax*. *P. vivax* is the most common cause of infection, responsible for about 80 % of all malaria cases. However, *P. falciparum* is the most important cause of disease, and responsible for about 15% of infections and 90% of deaths. Severe malaria occurs when *P. falciparum* infections are complicated by serious organ failures or abnormalities in the patient's blood or metabolism. The manifestations of severe malaria include: cerebral malaria, with abnormal behavior, impairment of consciousness, seizures, coma, or other neurologic abnormalities. Consequences of severe malaria result in death if untreated - young children and pregnant women are especially vulnerable. The severe anemia and hemoglobinuria (hemoglobin in the urine) are due to hemolysis (destruction of the red blood cells). In the most severe cases of the disease fatality rates can exceed 20%, even with intensive care and treatment. In endemic areas, treatment is often less satisfactory and the overall fatality rate for all cases of malaria can be as high as one in ten. Over the longer term, developmental impairments have been documented in children who have suffered episodes of severe malaria.

Other manifestations that should raise concern are: acute kidney failure, hyperparasitemia (more than 5% of the red blood cells are infected by malaria parasites), metabolic acidosis (excessive acidity in the blood and tissue fluids), often in association with hypoglycemia, and hypoglycemia (low blood glucose). The hypoglycaemia may also occur in pregnant women with uncomplicated malaria, or after treatment with quinine. Severe malaria occurs most often in persons who have no immunity to malaria or whose immunity has decreased. These include all residents of areas with low or no malaria transmission, and young children and pregnant women in areas with high transmission. In all areas, severe malaria is a medical emergency and should be treated urgently and aggressively.

Severe malaria remains a major cause of mortality accounting for 800000 deaths per year worldwide [239]. Most of these deaths occur due to cerebral malaria which has a mortality rate of 15-20% despite intensive treatment. It is essential that every primary care doctor should be trained to identify and treat severe malaria. Modified WHO Criteria for Severe malaria [2, 73] are: a patient with *P. falciparum* asexual parasitemia and no other obvious cause of their symptoms, the presence of one or more of the following clinical or laboratory features classifies the patient as suffering from severe malaria:

Clinical manifestation:

- Prostration: extreme weakness;
- Cerebral malaria: Impaired consciousness, unrousable coma;
- Respiratory distress: acidotic breathing, pulmonary edema, acute respiratory distress syndrome
- Multiple convulsions: two convulsions with 24 hours;
- Circulatory collapse or shock: Systolic blood pressure < 70 mmHg with cold clammy skin or core, rapid thready pulse;
- Pulmonary edema (radiological): ARDS features;

- Abnormal bleeding/DIC;
- Jaundice: Serum bilirubin > 3 mg/dL plus either parasite count > 1,000,000/cmm in a peripheral blood film or renal failure-serum creatinine > 1.5 mg/dL;
- Hemoglobinuria:
- Jaundice: bilirubin > 3 mg/dl

Laboratory test:

- Severe anemia: hematocrit < 20%, hemoglobin level < 5 g/dl;
- Hypoglycaemia: whole blood glucose < 40 mg/dL;
- Acidemia/Acidosis: pH < 7.25 or plasma bicarbonate 2 in 24 hours;
- Renal failure: urine output < 400ml/day plus a serum creatinine > 3mg/dL;
- Hyperlactatemia;
- Hyperparasitemia: > 5% of the erythrocytes infected by parasites

Other criteria include manifestations which do not by themselves satisfy WHO criteria but alert the physician to treat them by parenteral antimalarials. Physicians should not worry about definitions. They should treat any patient about whom they are doubtful as having severe malaria [240].

3.3.2. Therapeutic Principles of Severe Malaria

3.3.2.1. WHO Policy of Antimalarial Drugs for Severe Malaria

The primary objective of antimalarial treatment in severe malaria is to prevent death. Prevention of recrudescence and avoidance of minor adverse effects are secondary. In treating cerebral malaria, prevention of neurological deficit is also an important objective. In the treatment of severe malaria in pregnancy, saving the life of the mother is the primary objective. The following policies are recommended for treatment of severe and complicated *falciparum* malaria [113]:

- Any of the following antimalarial medicines are recommended for initial treatment.
    - artesunate* (i.v. or i.m.)
    - artemether (i.m.)
    - quinine (i.v. infusion or i.m. injection).
- Follow-on treatment
    Once the patient recovers enough and can tolerate oral treatment, the following options can be used to complete treatment:
    - full course of an ACT or
    - quinine + clindamycin or doxycycline

Over the last three years (2004-2007) around 65 countries (41 in Africa) have updated their treatment policies to include ACTs as 1st-line or 2nd-line treatment of malaria (uncomplicated and severe malaria, Table 5). This was based on WHO advice, and was made possible with the participation of RBM partners and increased mobilization of international funding [55]. Consistent with WHO recommendations, malaria endemic countries which are experiencing resistances to currently used antimalarial drug monotherapies (chloroquine, sulphadoxine/pyrimethamine or amodiaquine) should change treatment policies to the highly effective ACTs [2, 47].

3.3.2.2. Treatment of Severe Malaria Depends on the following Issues

The mortality of untreated severe malaria is thought to approach 100%. Using antimalarial treatment the mortality falls to 15-20% overall, although within the broad definition are syndromes associated with mortality rates that are lower (e.g., severe anemia) and higher (metabolic acidosis). Death from severe malaria often occurs within hours of admission to hospital or clinic, and so it is essential that therapeutic concentrations of antimalarial are achieved as soon as possible. Management of severe malaria comprises four main areas: clinical assessment of the patient, specific antimalarial treatment, adjunctive therapy and supportive care.

1) Key issues
   - Artesunate, artemether and quinine are all effective treatments for severe malaria. Intramuscular artemether is probably equivalent to quinine.
   - In the South East Asian Quinine Artesunate Malaria Trial (SEAQUAMAT), intravenous artesunate was shown convincingly to be more effective than quinine in reducing mortality from malaria in adults with severe malaria from Asia.
   - Artesunate is likely to be more effective than quinine for the treatment of severe malaria in adults from all areas and for severe malaria in children from Asia.
   - Artesunate seems most effective in patients with very high parasite levels (hyperparasitemia).
   - It is not currently clear whether artesunate is superior to quinine in treating malaria in children in Africa; ongoing trials should answer this question.
   - There are theoretical risks to artesunate in early pregnancy but, in pregnant women with severe malaria in Asia, these risks are outweighed by the need for treatment with an effective drug.
   - Securing a reliable supply of good manufacturing practice standard parenteral artesunate should be a priority.
2) New strategy in severe malaria treatment
   - Rectal AS, and then intravenous artesunate, or intramuscular artemether, or intravenous quinine (infusion).
3) Clinical assessments
   - The airway should be secured in unconscious patients and breathing and circulation assessed.
   - The patient should be weighed or body weight estimated so that drugs, including antimalarials and fluids can be given on a body weight basis.
   - An intravenous cannula should be inserted and immediate measurements of blood glucose (stick test), hematocrit/hemoglobin, and parasitemia and, in adults, renal function should be taken.

- A detailed clinical examination should be conducted, with particular note of the level of consciousness and record of the coma score.
- Several coma scores have been advocated.
- The Glasgow coma scale is suitable for adults, and the simple Blantyre modification or children's Glasgow coma scale are easily performed in children.
- Unconscious patients should have a lumbar puncture for cerebrospinal fluid analysis to exclude bacterial meningitis.
- The degree of acidosis is an important determinant of outcome; the plasma bicarbonate or venous lactate level should therefore be measured if possible.
- If facilities are available, arterial or capillary blood pH and gases should be measured in patients who are unconscious, hyperventilating or in shock.
- Blood should be taken for cross-match, and (if possible) full blood count, platelet count, clotting studies, blood culture and full biochemistry should be conducted.
- The assessment of fluid balance is critical in severe malaria.
- Respiratory distress, in particular with acidotic breathing in severely anemic children, often indicates hypovolemia and requires prompt rehydration and, where indicated, blood transfusion.

3.3.2.3. Specific Antimalarial Therapies

It is essential that antimalarial treatment in full doses is given as soon as possible in severe malaria. Two classes of drugs are currently available for the parenteral treatment of severe malaria: the cinchona alkaloids (quinine and quinidine) and the artemisinin derivatives (artesunate, artemether, and artemotil). Although there are a few areas where chloroquine is still effective, parenteral chloroquine is no longer recommended for the treatment of severe malaria because of widespread resistance. Intramuscular sulfadoxine– pyrimethamine is also not recommended [2].

1) Quinine

Quinine treatment for severe malaria was established before modern trial methods were developed. Several salts of quinine have been formulated for parenteral use, but the dihydrochloride is the most widely used. Peak concentrations following intramuscular quinine in severe malaria are similar to those following intravenous infusion. Pharmacokinetic modeling studies suggest that a loading dose of quinine of twice the maintenance dose (i.e., 20 mg salt/kg bodyweight) reduces the time to reach therapeutic plasma concentrations. After the first day of treatment, the total daily maintenance dose of quinine is 30 mg salt/kg bodyweight (usually divided into three equal administrations at 8 h intervals).

WHO recommendation: 1) rate-controlled i.v. infusion is the preferred route of quinine administration, but if this cannot be given safely, then intramuscular injection is a satisfactory alternative; and 2) there is insufficient evidence to recommend rectal administration of quinine unless parenteral administration is not possible and no other effective options are available.

2) Artesunate

Various artemisinin derivatives have been used in the treatment of severe malaria including artemether, artemisinin (rectal), artemotil, and artesunate. The pharmacokinetic properties of artesunate are superior to those of artemether and artemotil as it is water soluble and can be given either by intravenous or intramuscular injection. Randomized trials comparing artesunate and quinine from South-East Asia show clear evidence of benefit with artesunate. In the largest multi-centre trial, which enrolled 1461 patients (including 202 children < 15 years old); mortality was reduced by 34.7% compared to the quinine group. The results of this and smaller trials are consistent and suggest that artesunate is the treatment of choice for adults with severe malaria. There are, however, still insufficient data for children, particularly from high transmission settings to make the same conclusion. An individual patient data meta-analysis of trials comparing artemether and quinine showed no difference in mortality in African children.

WHO recommendation: AS is the recommended first choice in areas of low-malaria transmission.

3) Artemether

Although artesunate has better pharmacokinetic properties than artemether and artemotil, there are relatively few published comparative clinical trials. Concerns have been raised, regarding the possibility that the intrinsic benefits of artemether as an antimalarial may have been negated by its erratic absorption following intramuscular injection. Artemotil is very similar to artemether, but very few trials have been conducted.

WHO recommendation: 1) artemether intramuscular is an acceptable alternative to quinine i.v. infusion; and 2) do not use artemotil unless alternatives are not available.

4) Quinidine

Quinidine commonly causes hypotension and concentration-dependent prolongation of ventricular repolarization (QT prolongation). Quinidine is thus considered more toxic than quinine and should only be used if none of the other effective parenteral drugs are available. Electrocardiographic monitoring and frequent assessment of vital signs are required if quinidine is used.

3.3.2.4. Management of Severe Malaria

1) Diagnosis

The differential diagnosis of fever in a severely ill patient is broad. Coma and fever may result from meningoencephalitis or malaria. Cerebral malaria is not associated with signs of meningeal irritation (neck stiffness, photophobia, and Kernig sign) but the patient may be opisthotonic. As untreated bacterial meningitis is almost invariably fatal, a diagnostic lumbar puncture should be performed to exclude this condition. There is also considerable clinical overlap between septicaemia, pneumonia and severe malaria – and these conditions may coexist. In malaria endemic areas particularly, where parasitemia is common in the young age group, it is often impossible to rule out septicaemia in a shocked or severely ill obtunded child. Where possible, blood should always be taken on admission for culture, and if there is any doubt, empirical antibiotic treatment should be started immediately along with antimalarial treatment.

2) Initial management

Severe malaria is a medical emergency. Initial management is based on that of any acutely and severely ill patient. The initial rapid clinical assessment should focus on the airway and circulation and include assessments of conscious level, respiratory status, and state of hydration. Hypoglycemia should be ruled out or, if the patient is comatose, treated empirically. Convulsions, which can present with subtle symptoms, especially in children, should be treated promptly. Intravenous rehydration should be commenced if indicated, oxygen given if there is clinical or blood gas evidence of respiratory distress or hypoxia, and an appropriate antimalarial drug administered. If the presence of severe malaria is suspected, the patient should be transferred to the highest level of care available (preferably an intensive care unit) [2, 73].

In areas of high transmission, peripheral parasitemia is common and relatively uninformative unless very high and other common infections may produce clinical pictures similar to the spectrum of syndromes produced by severe malaria. In cases with impaired consciousness, a lumbar puncture should be performed to exclude meningitis and the possibility of bacterial sepsis considered in all seriously ill individuals. Blood cultures are rarely available in endemic areas, but where they have been done systematically, bacteremias were found in a significant proportion of clinically severe patients with parasitemia. Clearly if there are focal signs suggesting a bacterial infection (such as pneumonia), broad-spectrum antibiotic cover should be given. However, at present there is no robust method of excluding bacterial sepsis in patients with parasitemia and "severe malaria" in high-transmission areas. In the absence of such a test, a strong case can be made for a broader use of empirical antibiotics (in addition to the antimalarials); although so far no clinical trials have addressed this issue [241].

3) Caution for quinine use
   - Infusion should be preferably given in dextrose saline to combat hypoglycemia and hypotension.
   - Tetracycline should be co-prescribed to decrease the duration of Quinine therapy to 5 days.
   - Loading dose of Quinine is to be avoided if the patient has already received Quinine, mefloquine, or chloroquine in the preceding 12-24 hours [242].
   - Quinine dose is to be reduced by 1/3 to ½ after 48 hours in all patients especially those with acute renal failure and / or hepatic failure [240].
   - Initial loading dose is to be avoided in those with prolonged QTc. Even subsequent doses can be reduced by 1/4 if QTc is prolonged.

4) Supportive care
   Patients with severe malaria require intensive nursing, in an intensive care unit if possible. Following the initial assessment and the start of antimalarial treatment, clinical observations should be made as frequently as possible. These should include recording of vital signs, with an accurate assessment of respiratory rate and pattern, coma score, and urine output. Blood glucose should be checked, using rapid stick tests every 4 h if possible, particularly in unconscious patients. Convulsions should be treated promptly with intravenous or rectal diazepam or intramuscular paraldehyde. It is necessary to try 5-7 mg/kg of phenobarbitone (IM to prevent convulsion) and diazepam IV 10 mg or paraldehyde 0.1 ml/Kg IM for convulsion. Heparin, prostacyclin, deferoxamine,

pentoxifylline, low molecular weight dextran, urea, high-dose corticosteroids, acetylsalicylic acid, deferoxamine, anti-tumor necrosis factor antibody, cyclosporin, dichloroacetate, adrenaline and hyperimmune serum have all been suggested – but none of these is recommended (Table 24). Other complications are listed for the supportive care.

a. Pulmonary oedema: 1. Nurse in 45° Position, 2. Oxygen, 3. Frusemide 40-20mg, 4.If not better PEEP Ventilation, 5. Venesection (last resort).

b. Renal Failure: Oliguria-hydrate well, Persistent oliguria: hemodialysis/Peritoneal dialysis

c. Shock: Adequate Saline, Blood transfusions, 400 mg Dopamine in 500 ml dextrose-7 rops/min, Antibiotics for gram negative septicemia

d. Hypoglycemia: 50ml 50% dextrose followed by 10% dextrose infusions

e. Severe Anemia: Blood transfusions.

f. Heavy Parasitemia: >5% exchange transfusion g. DIC: Vitamin K injections, blood transfusion

g. Hyperpyrexia: Tepid sponging, fanning, paracetamol.

In addition, constant monitoring, fluid and electrolyte balance are absolutely essential for favorable Outcome.

5) Follow-on treatment

Following initial parenteral treatment, once the patient can tolerate oral therapy, it is essential to continue and complete treatment with an effective oral antimalarial. Current practice is to continue the same medicine orally as given parenterally to complete a full 7 days of treatment. In non-pregnant adults, doxycycline is added to quinine, artesunate or artemether and should also be given for 7 days. Doxycycline is preferred to other tetracyclines because it can be given once daily, and does not accumulate in renal failure. But as treatment with doxycycline only starts when the patient has recovered sufficiently, the doxycycline course finishes after the quinine, artemether or artesunate course. Where available, clindamycin may be substituted in children and pregnant women, as doxycycline cannot be given to these groups. Although following parenteral treatment with a full course of oral ACT (artesunate + amodiaquine or artemether-lumefantrine) is theoretically a good alternative, this has not been evaluated in clinical trials.

The recommendation from experts' opinion is to complete treatment in severe malaria following parenteral drug administration by giving a full course of combination therapy, ACT or quinine + clindamycin or doxycycline. Regimens containing mefloquine should be avoided if the patient presented initially with impaired consciousness. This is because of an increased incidence of neuro-psychiatric complications associated with mefloquine following cerebral malaria.

6) Pre-referral treatment options

The risk of death from severe malaria is greatest in the first 24 h, yet in most malaria endemic countries, the transit time between referral and arrival at appropriate health facilities is usually prolonged thus delaying the commencement of appropriate antimalarial treatment, during which time the patient may deteriorate or die. It is recommended that patients are treated with the first dose of one of the recommended treatments by the parenteral route if possible or by the intra-rectal route before referral (unless the referral time is very short). This could be intramuscular artemether, artesunate or quinine, or a rectal formulation of artemisinin or artesunate.

Table 24. Immediate clinical management for severe manifestation and complication with *falciparum* malaria [2]

Manifestation/complication	Immediate management*
Coma (cerebral malaria)	Maintain airway, place patient on his or her side, exclude other treatable causes of coma (e.g., hypoglycaemia, bacterial meningitis); avoid harmful ancillary treatment such as corticosteroids, heparin and adrenaline; intubate if necessary.
Hyperpyrexia	Administer tepid sponging, fanning, cooling blanket and antipyretic drugs.
Convulsions	Maintain airways; treat promptly with intravenous or rectal diazepam or intramuscular paraldehyde.
Hypoglycaemia (blood glucose concentration of <2.2 mmol/l; <40 mg/100ml)	Check blood glucose, correct hypoglycemia and maintain with glucose-containing infusion.
Severe anaemia (hemoglobin <5 g/100ml or packed cell volume <15%)	Transfuse with screened fresh whole blood
Acute pulmonary oedema**	Prop patient up at an angle of 45°, give oxygen, give a diuretic, stop intravenous fluids, intubate and add positive end-expiratory pressure/continuous positive airway pressure in life-threatening hypoxemia.
Acute renal failure	Exclude pre-renal causes, check fluid balance and urinary sodium; if in established renal failure add hemofiltration or hemodialysis, or if unavailable, peritoneal dialysis. The benefits of diuretics/dopamine in acute renal failure are not proven.
Spontaneous bleeding and coagulopathy	Transfuse with screened fresh whole blood (cryoprecipitate, fresh frozen plasma and platelets if available); give vitamin K injection.
Metabolic acidosis	Exclude or treat hypoglycemia, hypovolemia and septicemia. If severe add hemofiltration or hemodialysis.
Shock	Suspect septicemia, take blood for cultures; give parenteral antimicrobials, correct hemodynamic disturbances.
Hyperparasitemia	Patients with high parasite counts are known to be at increased risk of dying, although the relationship between parasite counts and prognosis varies at different levels of malaria endemicity.

* It is assumed that appropriate antimalarial treatment will have been started in all cases.
** Prevent by avoiding excess hydration.

  7)   Adjustment of dosing in renal failure or hepatic dysfunction

The dosage of artemisinin derivatives does not need adjustment in vital organ dysfunction. Quinine (and quinidine) levels may accumulate in severe vital organ dysfunction. If there is no clinical improvement or the patient remains in acute renal failure the dose should be reduced by one-third after 48 h. Dosage adjustments are not necessary if patients are receiving either hemodialysis or hemofiltration. Dosage adjustment by one-third is necessary in patients with hepatic dysfunction.

Table 25. Extracts from current WHO guidelines for the treatment of severe malaria (reproduced with permission) [241]

Summary of recommendations on the treatment of severe malaria	Level of evidence
Severe malaria is a medical emergency. After rapid clinical assessment and confirmation of the diagnosis, full doses of parenteral antimalarial treatment should be started without delay with whichever effective antimalarial is first available.	E
Artesunate 2.4 mg/kg body weight (bw) i.v. or i.m. given on admission (time _ 0), then at 12 h and 24 h, then once a day is the recommended choice in low-transmission areas or outside malaria-endemic areas.	S
For children in high-transmission areas, the following antimalarial medicines are recommended as there is insufficient evidence to recommend any of these antimalarial medicines over another for severe malaria: ■ Artesunate 2.4 mg/kg bw i.v. or i.m. given on admission (time _ 0), then at 12 h and 24 h, then once a day; ■ Artemether 3.2 mg/kg bw i.m. given on admission then 1.6 mg/kg bw per day; ■ Quinine 20 mg salt/kg bw on admission (i.v. infusion or divided i.m. injection), then 10 mg/kg bw every 8 h; infusion rate should not exceed 5 mg salt/kg bw per hour.	S,T,O,E
Summary of recommendations on pre-referral treatment for severe falciparum *malaria* The following may be given: ■ Artesunate or artemisinin by rectal administration	T, E
■ Artesunate or artemether i.m.	E
■ Quinine i.m.	O, E
Recommendation for treatment for severe falciparum *malaria in pregnant women* Use the parenteral antimalarial treatment locally available for severe malaria in full doses. Where available, AS is the first and artemether the second option in the second and third trimesters. In the first trimester, until more evidence becomes available, both artesunate and quinine may be considered as options.	E

* Levels of evidence: S, formal systematic reviews, such as a Cochrane Review, including more than one randomized controlled trial; T, comparative trials without formal systematic review; O, observational studies (e.g., surveillance or pharmacological data); E, expert opinion/consensus.

The World Health Organization coordinated the production of guidelines for the management of severe and complicated malaria in 1990; and again in 2000, which also contain strict definitions of severe malaria, useful for standardizing clinical research [73]. More recently, in 2006, WHO published evidence-based guidelines for the treatment of malaria [2, 241], which include extensive advice on the management of severe malaria as well as a clinically useful distillation of the WHO severe malaria definitions (Table 25).

3.3.3. Efficacy of Three Drugs in Severe Malaria Therapy

3.3.3.1. Clinical Outlooks of Chief Drugs in Severe Malaria Therapies

Quinine

Since the advent of chloroquine resistance, quinine has been the drug of choice for the parenteral treatment of malaria. Whereas many antimalarials are prescribed in terms of base, for historical reasons quinine doses are often recommended in terms of salt (usually sulfate for oral use and dihydrochloride for parenteral use). Recommendations for doses of this and other antimalarials should state clearly whether the salt or base is being referred to (doses with different salts must have the same base equivalents). Quinine must never be given by intravenous injection, as lethal hypotension may result. Quinine dihydrochloride should be given by rate-controlled infusion in saline or dextrose solutions at a rate not exceeding 5 mg salt/kg per hour. If this is not possible then it should be given by intramuscular injection to the anterior thigh, not the buttock (to avoid sciatic nerve injury). The first dose should be split, 10 mg/kg to each thigh. Undiluted quinine dihydrochloride at a concentration of 300 mg/ml is acidic (pH 2) and painful when given by intramuscular injection, so it is best either formulated or diluted to concentrations of 60–100 mg/ml for intramuscular injection. Gluconate salts are less acidic and better tolerated than the dihydrochloride salt when given by the intramuscular and rectal routes. As the first dose (loading dose) is the most important in the treatment of severe malaria, this should be reduced only if there is clear evidence of adequate pretreatment before presentation. Although quinine can cause hypotension if administered rapidly, and overdose is associated with blindness and deafness, these adverse effects are rare in the treatment of severe malaria. The dangers of insufficient treatment (i.e., death from malaria) exceed those from excessive treatment initially. After the second day of parenteral treatment, if there is no clinical improvement or in acute renal failure, the maintenance doses of quinine given by infusion should be reduced by one-third to avoid accumulation.

Artesunate

Artesunate (AS) is available as an oral, rectal, intramuscular or intravenous formulation. The drug may be used rectally in children who have severe malaria and are unable to receive oral or intravenous therapy. The intravenous formulation is often used in emergency treatment of patients with severe *P. falciparum* malaria in Asia, and has been associated with a 35% reduction in mortality as compared with quinine [90]. Artesunate is rapidly hydrolyzed (within minutes) to its active metabolite, dihydroartemisinin (DHA), and is thus viewed as a prodrug. Although artesunate may retain some activity against *P. falciparum* parasites, most of the antimalarial activity is attributed to DHA. Like the naturally occurring artemisinin, both AS and DHA have short elimination half-lives ranging from 3 to 29 minutes and from 20 to 40 minutes for the parent and the metabolite, respectively. The short elimination half-lives of artemisinin and its derivatives highlight the need to administer these compounds with longer-acting drugs for the treatment of malaria.

Both AS and DHA are highly bound to plasma proteins [243]. DHA protein binding was noted to be as high as 93% in infected patients and to range from 88% to 91% in Vietnamese and Caucasian healthy volunteers, respectively, with binding being affected by the plasma pH and AAG concentration [244]. Variability in AAG plasma concentrations during malaria may impact on dihydroartemisinin free-fraction concentrations in infected patients. The dosing of artemisinin derivatives has been largely empirical. The doses recommended here are those that have been most widely studied. The only recent change is the higher maintenance dose of parenteral artesunate recommended (2.4 mg/kg), which is based on pharmacokinetic and pharmacodynamic studies and by extrapolation from studies with oral artesunate. Expert opinion is that the previously recommended maintenance dose of 1.2 mg/kg may have been insufficient in some patients.

Artesunate is dispensed as a powder of artesunic acid. This is dissolved in sodium phosphate salt (0.3 M PBS) to form sodium artesunate by intravenous injection or by intramuscular injection to the anterior thigh [243]. The solution should be prepared freshly for each administration and should not be stored. Artemether and artemotil are dispensed dissolved in oil (groundnut, sesame seed) and given by i.m. injection into the anterior thigh.

Artemether

Absorption of artemether is rapid, achieving the C_{max} approximately 2 hours after administration in human. In contrast to artesunate, artemether metabolism is largely dependent on the CYP system. Administration of artemether together with grapefruit juice resulted in increases in the drug's Cmax (by 155%) and AUC from 0 to 8 hours (by 90%) [245]. The drug is rapidly demethylated to DHA by CYP3A4/A5 which, in turn, is converted to inactive metabolites primarily by glucuronidation [246]. The AUC of DHA is 2–3 times greater than that of the parent compound following oral administration. This is in contrast to intramuscular and intrarectal administration, where the AUC of artemether is greater than that of DHA [244]. Artemether exhibits high protein binding *in vitro*, ranging from 95% to 98% [247]. The distribution in blood showed that 33% of artemether was bound to AAG, 17% to albumin, 12% to high dose density lipoproteins and 9.3% to low-density lipoproteins under physiological protein concentrations, whereas dihydroartemisinin binds predominantly to albumin [248]. The elimination half-life of artemether is longer than that of artemisinin or artesunate and amounts to 2–3 hours.

Artemether is formulated in oil and is given by intramuscular injection. It is absorbed erratically, particularly in very severely ill patients. There are also rectal formulations of artemether. Artemether exhibits time-dependent pharmacokinetics, as evidenced by changes in the DHA/artemether ratio after multiple doses in patients with multidrug-resistant *P. falciparum*. The changes in the metabolite: parent ratio does not impact on treatment efficacy, as both artemether and DHA are active against *P. falciparum* parasites [68].

3.3.3.2. Efficacy of Quinine, Artesunate and Artemether in Severe Malaria Therapy

Severe malaria develops when effective treatment is delayed, either because of poor health care and transport infrastructure, or because the drugs given are ineffective or counterfeit. Non-immune individuals are particularly susceptible. Once the disease takes a severe course, multiple organs may be affected, causing mortality as high as 10 - 50%. Prompt initiation of parenteral antimalarial treatment is pivotal to prevent death. Until recently, quinine was the mainstay of treatment for severe malaria worldwide, and was being replaced by artemisinins in Asia. To examine whether an advantage of artesunate over quinine could be translated into a reduction in mortality, 1461 patients (including 202 children) in four South and Southeast Asian countries were recruited between 2003 and 2005 into the SEAQUAMAT study, the largest ever clinical drug trial in severe malaria [90]. Patients randomized to receive the water-soluble artemisinin derivative artesunate intravenously had a 35% lower mortality than those receiving intravenous quinine. Neurological sequel were extremely rare, providing further reassurance of its safety. As a result of this study, parenteral artesunate is now recommended by WHO as the drug of choice for the treatment of severe malaria in low-transmission areas and in pregnancy in the second and third trimesters [2].

In high-transmission areas the current recommendation of WHO for treating severe malaria patients is either quinine or an artemisinin derivative. Artesunate use results in more rapid clearance of circulating malaria parasites compared with quinine, is active against almost all stages of the malaria parasites. Levels of quinine resistance have been reported to be increasing in Southeast Asia [12, 80, 249-251]. In addition, adverse effects resulting from quinine therapy are common. Cinchonism (symptoms of quinine overdose) often occurs at conventional dose regimens. This usually mild and reversible symptom complex consists of tinnitus, deafness, dizziness, and vomiting, and may affect adherence. Hypoglycemia is a less common but more serious adverse effect. Some people are allergic to quinine and develop skin rashes and edema even with small doses. In accordance with WHO policy of antimalarial drugs in severe malaria treatment [73], quinine remains the drug of choice, although it suffers from certain drawbacks. Quinine has narrow therapeutic ratio, causing hyperinsulinemic hypoglycemia (more frequent and severe in pregnancy) and prolongs the QTc interval when given parenterally, particularly if infused too rapidly. Intramuscular quinine is effective, but can cause local toxicity as well as hypoglycemia in patients who may not have intravenous access [79].

Not like the adverse effects, the uncommon resistance to quinine is being reported more and more often [249-251]. Development of artemisinins for treatment of severe malaria has been conducted since 1987 [83]. The most important question in antimalarial drug treatment of severe malaria is whether the artemisinin derivatives can reduce mortality better than quinine. Initially, intramuscular artemether has been the focus of many comparative studies versus parenteral quinine with over 3,000 cases having been reported in Asia and Africa. Artemether is a lipid-soluble agent that can only be administered intramuscularly and is now known to be subject to more erratic uptake and a slower time to peak plasma concentrations than artesunate [88]. For clearance of parasitemia current results document the greater efficacy of a five-day treatment using intramuscular artemether as compared to the standard seven-day treatment using intravenous quinine bichlorhydrate [84].

A number of trials in randomized comparisons of intramuscular artemether and quinine in Gambian children [85]; and Vietnamese adults [86]; and a meta-analysis of individual data from 1919 patients in 11 trials of parenteral therapy [77]; identified no significant difference in efficacy between these agents. Initially, a number of trials of parenteral artemether against quinine were undertaken in both children and adults with severe malaria [85–87]. In these trials, quinine was administered either intravenously or intramuscularly and artemether was administered intramuscularly. Although the results of these trials showed that artemether was safe and easier to use than quinine, none of these trials showed a survival benefit individually, although a survival benefit was demonstrated in subgroup analysis of a meta-analysis of adults in Southeast Asia. This evidence was not sufficiently convincing to prompt a change of policy from quinine to artemether. Most significantly, artemether may not have been the best choice of artemisinin to study in the first place, as the results of this study show that in terms of preventing deaths, artemether is as effective as quinine, which is comparable to findings of the 11 trials [77].

Compared with artesunate, artemether is less biotransformed to the more potent dihydroartemisinin and has slow, erratic absorption after intramuscular administration; in fact the ability of artemether to provide equivalent benefit to quinine is probably testament to the antimalarial potency of the artemisinin derivatives as a group. Attention has therefore switched to artesunate. The effective currently available artesunate formulation produced by Guilin Pharmaceutical factory in China is manufactured to Chinese GMP standards, but is not international GMP certified. Further certification is under review. A new GMP formulation of artesunate is currently being developed by Walter Reed Army Institute of Research (WRAIR), USA and should obtain FDA approval in the near future. Even if parenteral artesunate is widely available in hospitals, it will not be available to the majority of patients developing severe malaria living in rural tropics far from the nearest health clinic or hospital.

Severe malaria remains a major global health problem and, despite advances in prevention, it is likely to remain so for the foreseeable future. The SEAQUAMAT study demonstrates convincingly that treatment of severe adult malaria acquired in Asia with artesunate is associated with a reduction in mortality compared with quinine. Can the results be assumed to also apply to children in Asia, and also to adults and children in Africa, where the bulk of severe *falciparum* malaria occurs? Though most of the participants enrolled in this review were adults, there is no reason to believe that children should behave differently. It is possible that children might fare even better with artesunate as children tend to present with a more severe disease and hyperparasitemia. One of the included trials demonstrated the greater benefit of artesunate in patients with hyperparasitemia. Also, the complications and mortality of quinine induced hypoglycemia is expected to be higher in children. However, more clinical trials are needed to answer the question and in implication for practice and policy in children therapy [74].

3.3.3.3. Mission for Selecting the Best Drug to Treat Severe Malaria

1) Loading dose of quinine is not superior to unloading dose.

   The systematic review found no significant difference in mortality between a group receiving a high initial dose of quinine (20 mg salt/kg or 16 mg base/kg given by the intramuscular (i.m.) route or by intravenous infusion) followed by a standard dose of quinine, and one receiving the standard dose but no loading dose. Data are from two randomized controlled trials (RCTs); 2/35 (5.7%) died in the group receiving a high initial dose, 5/37 (13.5%) with no loading dose; RR: 0.43; 95% CI: 0.09–2.15). One of the RCTs (39 children) found no significant difference between the two groups in mean time to recover consciousness (14 h with a high initial dose, 13 h with no loading dose, weighted mean difference (WMD) 1.0 h; 95% CI: −8.8 h to +10.8 h). However, parasite and fever clearance times were reduced in the group of loading dose of quinine [2].

2) Intramuscular quinine is as effective as intravenous quinine.

Comparison trial was conducted in 59 children aged <12 years in Kenya following i.m. quinine (20 mg salt/kg loading immediately followed by 10 mg salt/kg every 12 h) with standard-dose quinine given by i.v. infusion (10mg salt/kg every 12 h) in severe *falciparum* malaria. The trial found no significant difference in mortality, mean parasite clearance time or recovery time to drinking or walking, but may have lacked the power to detect a clinically important difference (mortality: 3/20 deaths with i.m. quinine, 1/18 with i.v. quinine, RR: 2.7; 95% CI: 0.3–23.7) [252].

3) Intrarectal quinine is as effective as intravenous or intramuscular quinine.

One systematic review with eight RCTs was summarized from 1247 patients. Five trials compared intrarectal quinine with i.v quinine infusion, while 6 compared with intramuscular quinine. The systematic review found no significant difference between intrarectal with i.v. or i.m. routes for death, parasite clearance by 48 hours and 7 days, parasite clearance time, fever clearance time, coma recovery time, duration of hospitalization, and time to drinking. However trials reporting these outcomes were small, which resulted in large confidence intervals for all the outcomes [253].

4) Intravenous artesunate is superior to intravenous quinine.

One clinical trial was performed in 113 adults with severe malaria in Thailand by comparing i.v. artesunate (2.4 mg/kg initially, 1.2 mg/kg 12 h later, then 1.2 mg/kg daily) to i.v. quinine (20 mg/kg initially, then 10 mg/kg every 8 h). It found no significant difference between the treatments in mortality after 300 h (7/59 artesunate, 12/54 quinine, RR: 0.53; 95% CI: 0.23–1.26). It found that artesunate significantly improved parasite clearance time, but that there was no significant difference in fever clearance time or coma recovery time [89]. The second clinical trial is a large multi-center trial (SEAQUAMAT study group: Bangladesh, India, Indonesia, and Myanmar) with 1431 patients enrolled. It found that mortality of 15% (107/730 in the artesunate group was significantly lower than the 22% (164/731) in the quinine group. An absolute reduction of 34.7% (95% CI: 18.5–47.6%; *P* = 0.0002) in the artesunate group [90]. There are, however, still insufficient data for children, particularly from high transmission settings. The system review with 6 trials was summarized from 1938 patients. Five trials used intravenous artesunate and one trial intramuscular artesunate; all six used intravenous quinine. Treatment with artesunate significantly reduced the risk of death and reduced parasite clearance time. Intravenous artesunate is the drug of choice for adults with severe malaria, particularly if acquired in Asia. This review did not identify sufficient data to make firm conclusions about the treatment of children or the effectiveness of intramuscular artesunate. There is an urgent need to compare the effects of artesunate with quinine in African children with severe malaria. The applicability of these results to Asian children and the ethics of further research are points of debate [74].

5) Intramuscular artemether is as effective as intravenous quinine.

One review was reported no significant difference in mortality between i.m. artemether and quinine given by i.v. infusion or i.m injection in 1919 adults and children with severe *falciparum* malaria. The review found no significant difference in the speed of coma recovery, fever clearance time or neurological sequel between artemether and quinine [77]. The second review found a small significant reduction in mortality for i.m. artemether compared with i.v. quinine [254]. However, more rigorous analysis excluding three poorer quality trials found no significant difference in mortality. More clinical trial was performed in 41 children with severe malaria in Sudan by compared i.m. artemether with i.v. quinine [255]. There were no deaths in the artemether group, one child died with quinine (0/20 compared to 1/21). Another clinical trial was conducted in 77 comatose children aged 3 months–15 years with cerebral malaria) compared i.m. artemether with i.v. quinine [256]. It found no significant difference in death rates between the artemether and quinine groups. There was no significant difference between the two groups in mean fever clearance time, coma recovery time, and parasite clearance time.

6)   Similar outcomes of intramuscular artemotil to intravenous quinine

Two trials in 194 patients compared i.m. artemotil (arteether) with quinine given by i.v. infusion in children with cerebral malaria and reported on similar outcomes. There was no statistically significant difference in the number of deaths, neurological complications, or other outcomes including time to regain consciousness, parasite clearance time and fever clearance time [257]. The meta-analyses lack the statistical power to detect important differences.

7)   Rectal artemisinin is as effective as intravenous quinine.

Sixty-five adult patients of both sexes: 32 for artemisinin and 33 for quinine with complicated severe *falciparum* malaria. There was no significant different in the parasitological cure rates in both arms of treatment. Mortality rates were similar both in the artemisinin and quinine groups. The rectal artemisinin is more efficacious and safer than the intravenous quinine. Thus, artemisinin may be considered a potential drug which can replace quinine in the treatment of severe malaria [258].

Based on the missions of clinical trial, the WHO recommends intravenous artesunate as the treatment of choice for severe malaria in adults and children in areas of low-transmission. Data on children in high-transmission regions are limited, and the WHO recommends treatment with artesunate, artemether, or quinine. For severe malaria during pregnancy, additional data regarding the risks of artemisinins are needed. The WHO recommends artesunate or quinine during the first trimester and artesunate as the first-line therapy during the second and third trimesters [2].

3.3.3.4. Quinidine in Severe Malaria

Quinidine has a two-fold to three-fold greater antimalarial activity than quinine does, but it is also more cardiotoxic and mandates electrocardiographic monitoring. Therefore, quinidine is thus considered more toxic than quinine and should only be used if none of the other effective parenteral drugs are available for the treatment of severe malaria. In the United States, the only parenteral drug currently available is quinidine gluconate. Blood films should be examined every 12 hours until negative for malaria parasites. Parasite density typically decreases by 90% over the first 48 hours with quinine or quinidine therapy. Quinidine levels should be maintained in the range of 3 to 8 µg/ml because quinidine is more cardiotoxic than quinine and should be administered in an intensive care unit with continuous electrocardiographic and frequent blood pressure monitoring. Quinidine-related cardiovascular adverse effects are potentially serious and may be more frequent if the drug is administered rapidly. The risk of cardiotoxicity is increased with bradycardia, hypokalemia, and hypomagnesemia and if the patient has received other drugs that may prolong the QTc interval (e.g., quinine, mefloquine, or macrolide antibiotics) [259].

An initial loading dose of quinidine should be administered as soon as possible, followed by the maintenance dose. Drug should be given by a rate-controlled intravenous infusion (infusion pump), and never by a bolus injection, which may cause fatal hypotension or cardiac arrhythmia. Without a loading dose, it takes more than 24 hours to achieve therapeutic drug concentration. This delay may allow sequestration of trophozoites causing major organ dysfunction. The loading dose should not be administered to patients who received quinine, quinidine, halofantrine, or mefloquine within the preceding 12 hours. If intravenous treatment is continued past 48 hours, the maintenance dose should be reduced by 30-50%. In renal failure (clearance <10 ml/min) and in dialysis patients a normal loading dose should be administered, but the maintenance dose should be reduced by 30-50%.

If hemodialysis is performed, quinidine should be administered after dialysis. Electrocardiographic monitoring is mandatory with quinidine infusion and with quinine infusion if the patient has acute renal failure. If the QRS complex lengthens by more than 25% beyond baseline or the QTc interval increases to more than 500 ms, the infusion should be slowed or discontinued. Monitoring of plasma quinidine concentrations does not predict cardiotoxicity; electrocardiography may be a more accurate approach for monitoring cardiotoxicity. For quinidine the total therapeutic plasma concentration is between 4 to 8 µg/ml. Apart from lengthening of the QTc interval, common electrocardiographic abnormalities include supraventricular and ventricular ectopic beats, sinus bradycardia (<50 beats/min), and ventricular tachycardia [259].

3.3.3.5. Optimization of Follow-on (Sequential) Therapy

Following initial parenteral treatment, once the patient can tolerate oral therapy, it is essential to continue and complete treatment with an effective oral antimalarial. Current practice is to continue the same medicine orally as given parenterally to complete a full 7 days of treatment. In non-pregnant adults, doxycycline (7 days) or full course of an ACT or clindamycin is sequentially administered after the treatment with quinine, artesunate or artemether. Doxycycline is preferred to other tetracyclines because it can be given once daily, and does not accumulate in renal failure. But as treatment with doxycycline only starts when the patient has recovered sufficiently, the doxycycline course finishes after the quinine, artemether or artesunate course. Where available, clindamycin may be substituted in children and pregnant women, as doxycycline cannot be given to these groups. Although following parenteral treatment with a full course of oral ACT (artesunate + amodiaquine or artemether-lumefantrine) is theoretically a good alternative, this has not been evaluated in clinical trials.

The WHO recommendation from the opinion of experts is to complete treatment in severe malaria following parenteral drug administration by giving a full course of combination therapy, ACT or quinine + clindamycin or doxycycline. Regimens containing mefloquine should be avoided if the patient presented initially with impaired consciousness. This is because of an increased incidence of neuro-psychiatric complications associated with mefloquine following cerebral malaria.

3.3.4. New Strategy in Severe Malaria Treatment

In the SEAQUAMAT trial, parenteral artesunate was shown to be associated with a considerably lower mortality than quinine, and is now the recommended treatment for severe malaria in low-transmission areas and in the second and third trimesters of pregnancy. A trial is underway to establish its role in African children. The development of artesunate suppositories may provide the means to treat patients with severe disease in remote rural settings, potentially buying the time needed to reach a health care facility. The increasing availability of basic intensive care facilities in developing countries also has the potential to further reduce mortality.

Though parenteral artesunate is widely available in hospitals, it will not be available to the majority of patients developing severe malaria living in rural tropics far from the nearest health clinic or hospital. For these patients, artesunate suppositories offer the prospect of effective early treatment, preventing clinical deterioration and buying time to reach hospital. The WHO has sponsored the development of GMP-manufactured rectal artesunate capsules, which are well absorbed and effective for the treatment of moderately severe malaria [92]. Many children in Africa die of malaria before reaching formal healthcare services. One attractive method to administrater artesunate in severe disease is the rectal procedure. By using rectal artesunate, it is unlikely to challenge either parenteral artesunate or quinine in hospital practice, but it may have a significant role in pre-hospital settings; especially in rural parts of Africa where the bulk of malaria deaths occur. A trial in Uganda comparing rectal artemether with intravenous quinine showed a trend towards lower mortality, faster parasite clearance and faster time to reduce fever in the artemether compared with the quinine; however, these results failed to reach statistical significance [99] and rectal artesunate may well fare better [92].

One systematic review addresses the lack of any data directly comparing the therapeutic efficacy and safety of the different rectal preparations of artemisinin derivatives [100]. The pooled analysis of individual patient data suggests that artemisinin and artesunate suppositories rapidly eliminate parasites and are safe. There is far less evidence for artemether [99] and no studies of dihydroartemisinin suppositories were available to be included in this analysis. The results indicate that both artemisinin and artesunate, whether as single or multiple dose regimens, induce a superior parasitological response than parenteral quinine over the 24 hours following initiation of treatment. Regimens employing a higher single dose of rectal artesunate were five times as likely to result in > 90% parasite reductions at 24 hours as were multiple lower doses of rectal artesunate or than a single administration of artemether.

The evidence from this analysis shows rectal artesunate has superior effect in reducing parasite densities compared to quinine (intravenous or intramuscular) at 12 and 24 hours after administration. The result supports the WHO recommendation for the use of artesunate and artemisinin as initial pre-referral treatment [2]. Basis of systematic review, the WHO recommendation is to use artemisinins rectally for complete treatment only when parenteral antimalarial treatment is not possible [100]. Based on the risk of death from severe malaria is greatest in the first 24 h, yet in most malaria endemic countries, the transit time between referral and arrival at appropriate health facilities is usually prolonged thus delaying the commencement of appropriate antimalarial treatment, during which time the patient may deteriorate or die. It is recommended that patients are treated with the first dose of one of the recommended treatments by the parenteral route if possible or by the intra-rectal route before referral (unless the referral time is very short). This could be intramuscular artemether, artesunate or quinine, or a rectal formulation of artemisinin or artesunate.

Recommendations by WHO are to treat with artesunate or artemisinin by rectal administration first on pre-referral treatment for severe *falciparum* malaria children. The administration of an artemisinin by the rectal route as pre-referral treatment is feasible even at the community level. There is insufficient evidence to show whether rectal artesunate is as good as intravenous or intramuscular options in the management of severe malaria. The recommendation, therefore, is to use artesunate or artemisinin suppositories only as pre-referral treatment and to refer the patient to a facility where complete parenteral treatment with artesunate, quinine, or artemether can be instituted. If, however, referral is impossible, rectal treatment should be continued until the patient can tolerate oral medication, at which point a full course of the recommended ACT for uncomplicated malaria in the locality can be administered.

Dosing for antimalarials given by artesunate suppository is listed for adults (Table 26) and children (Table 27). The appropriate single dose of artesunate given by suppository should be administered rectally, as soon as the presumptive diagnosis of severe malaria is made. In the event that an artesunate suppository is expelled from the rectum within 30 min of insertion, a second suppository should be inserted and, especially in young children, the buttocks should be held together, for 10 min to ensure retention of the rectal dose of artesunate. Artemisinin suppositories are not widely available. Doses used have been variable and empiric: 10 - 40 mg/kg (at 0, 4 or 12, 24, 48 and 72 h). Some studies have given a maintenance dose of one- to two-thirds of the initial dose. The intrarectal dose used in treatment trials in Africa was either 12 mg/kg quinine base every 12 h without a loading dose, or 8 mg/kg every 8 h, also without a loading dose [2].

Table 26. Dosage of artesunate suppository for initial (pre-referral) treatment in adult patients (aged ≥ 16 years) [2]

Weight (kg)	Artesunate dose	Regimen (single dose)
< 40	10 mg/kg body weight	Use appropriate no. of 100-mg rectal suppositories
40 - 59	400 mg/kg body weight	One 400-mg suppository
60 - 80	800 mg/kg body weight	Two 400-mg suppositories
> 80	1200 mg/kg body weight	Three 400-mg suppositories

* It should be noted that the clinical trial data with rectal artesunate relate to a single suppository formulation and presentation17 which has well characterized absorption kinetics and so cannot necessarily be extrapolated to other rectal formulations of artesunate.

Table 27. Dosage of artesunate suppository for initial (pre-referral) treatment in children (aged 2-15 years) and weighing at least 5 kg [2]

Weight (kg)	Age	Artesunate dose (mg)	Regimen (single dose)
5 - 8.9	0-12 months	50	One 50-mg suppository
9 - 19	13-42 months	100	One 100-mg suppository
20 - 29	43-60 months	200	Two 100-mg suppositories
30 - 39	6-13 years	300	Three 100-mg suppositories
> 40	> 14 years	400	One 400-mg suppository

3.4. Treatments of Antimalarial Drugs in Special Populations

3.4.1. Malaria in Pregnancy and Children

3.4.1.1. Malaria in Children

Malaria accounts for one in twenty of all childhood deaths in Africa. Anemia, low birth-weight, epilepsy, and neurological problems, all frequent consequences of malaria, compromise the health and development of millions of children throughout the tropical world. Yet much of the impact of malaria on the world's children could be prevented with currently available interventions. In older children, malaria has a similar course as in adults. However, in children below the age of 5 years, particularly infants, the disease tends to be atypical and more severe. In the first two months of life, children may not contract malaria or the manifestations may be mild with low-grade parasitemia, due to the passive immunity offered by the maternal antibodies. In endemic and hyperendemic areas, the parasite rate increases with age from 0 to 10% during first three months of life to 80 to 90% by one year of age and the rate persists at a high level during early childhood.

The mortality rate is highest during the first two years of life. By school age, a considerable degree of immunity would have developed and asymptomatic parasitemia can be as high as 75% in primary school children. In Africa, on an average about 1 in 20 children die from malaria, and in worst affected areas, even 1 in 5 or 6 die from malaria and its related diseases (e.g., anemia). In areas of low endemicity, where the immunity is low, severe infection occurs in all age groups including adults. The morbidity and mortality due to malaria in children tends to be very high in these areas. Malnutrition does not increase susceptibility to severe *falciparum* malaria. In fact, it has been observed that well-nourished children are more likely to develop severe disease than those with malnutrition. However, when severe malaria does occur, malnourished children have a higher morbidity and mortality.

1) The unacceptable mortality from malaria

   Over 40% of the world's children live in malaria-endemic countries. Each year, approximately 300 to 500 million malaria infections lead to over one million deaths, of which over 75% occur in African children < 5 years infected with *Plasmodium falciparum*. The rapid spread of resistance to antimalarial drugs, coupled with widespread poverty, weak health infrastructure, and, in some countries, civil unrest, means that mortality from malaria in Africa continues to rise. The tragedy is that the vast majority of these deaths are preventable.

2) Low birth weight

   Malaria in pregnancy leads to low birth weight and premature delivery, both of which are associated with an increased risk of neonatal death and impaired cognitive development. In many parts of the developing world, specialist care for low birth weight babies is very limited, and untreated hypoglycemia (low blood glucose, a common problem in low birth weight babies) may cause brain damage.

3) Consequences of cerebral malaria

Approximately 7% of children who survive cerebral malaria (a severe form of the disease, characterized by coma and convulsions) are left with permanent neurological problems. These include weakness, spasticity, blindness, speech problems, and epilepsy. The limited availability of specialized educational provision and equipment for such children means that opportunities for subsequent learning, and for attainment of independence, are compromised even further. Epilepsy may be inadequately treated, or untreated, due to lack of appropriate drugs and expertise, and further injury or death may result from uncontrolled convulsions. Recent evidence suggests that some children who appear to have made a complete neurological recovery from cerebral malaria may develop significant cognitive problems (attention deficits, difficulty with planning and initiating tasks, speech and language problems), which can adversely affect school performance.

4) Anemia

Although nutritional deficiencies, hookworm infection, and HIV all predispose to anemia in children, evidence suggests that, in endemic countries, malaria is one of the most important factors. Antimalarial drug resistance exacerbates the situation, by increasing the proportion of children who fail to adequately clear parasitemia after treatment, and who consequently remain anemic. It has been estimated that severe malarial anemia causes between 190000 and 974000 deaths each year among children < 5 years. Although blood transfusion may be life-saving in this situation, it also exposes children to the risk of HIV and other blood-borne diseases.

5) Recurrent fever

It is estimated that African children have between 1.6 and 5.4 episodes of malarial fever each year, a figure that varies according to geographical and epidemiological circumstances. Children are vulnerable to malaria from about 4 months of age, and, in highly endemic areas during the peak transmission season, approximately 70% of one-year-olds have malaria parasites in their blood. Fever reduces appetite, and exacerbates malnutrition. Recurrent episodes of malaria in child, or in a family member (which may mean that the child is required to stay at home to help with domestic chores), has to the likely result in the loss of a substantial amount of time from school.

6) Severe *falciparum* malaria in children

Severe *falciparum* malaria is the commonest cause of death in infants and children in endemic and hyperendemic areas for malaria, and a discernable difference with adults (Table 28). Inadequate immunity results in rapid increase in the parasite count and development of complications. Delay in diagnosis and treatment also contributes to the mortality. Clinical features of severe disease should be given utmost priority. History of travel to malarious area, history of previous antimalarial therapy, history of vomiting, diarrhea, fluid intake, urine output, convulsions etc. should be obtained from parents. Physical examination should include assessment of hydration and of complications of *falciparum* malaria. Rectal temperature should be measured in infants and small children. All children should be weighed on admission. Thick and thin films for malaria, hematocrit, and hemoglobin, blood glucose (by finger prick) should be done in all cases. If the report is likely to be delayed, presumptive antimalarial treatment should be started. Parasite count should be done in all positive cases of *falciparum* malaria and a parasite count of > 2% indicates impending problems and > 5% should be considered as severe infection. All cases with severe *falciparum* malaria should be managed as medical emergency.

3.4.1.2. Malaria in Pregnancy

Malarial infection during pregnancy is a major public health problem in tropical and subtropical regions throughout the world. In most endemic areas of the world, pregnant women are the main adult risk group for malaria. Malaria during pregnancy has been most widely evaluated in Africa south of the Sahara where 90% of the global malaria burden occurs. The burden of malaria infection during pregnancy is caused chiefly by *P. falciparum*, the most common malaria species in Africa. The impact of the other three human malaria parasites (*P. vivax, P. malariae, and P. ovale*) is less clear. Every year at least 30 million pregnancies occur among women in malarious areas of Africa, most of whom reside in areas of relatively stable malaria transmission.

Table 28. Severe Malaria: Differences between Adults and Children [170]

Clinical manifestation	Adults	Children
Duration of illness prior to complications	5-7 days	1-2 days
Convulsions	Common	Very common; can be due to severe infection, hypoglycemia, febrile seizures, severe anemia etc.
Abnormal brain stem reflexes (oculovestibular, oculocervical)	Rare	More common
C.S.F. pressure	Usually normal	Variable, often raised
Resolution of coma	2-4 days	1-2 days
Neurological sequel	<5%	>10%
Cough	Uncommon	Common
Anemia	Common	More common and more severe; may be the presenting feature
Jaundice	Common	Uncommon
Pre-treatment hypoglycemia	Uncommon	Common
Pulmonary edema	Common	Rare
Renal failure	Common	Rare
Bleeding/clotting disturbances	Up to 10%	Rare

The symptoms and complications of malaria during pregnancy differ with the intensity of malaria transmission and thus with the level of immunity the pregnant woman has acquired. While these settings are presented as two distinct epidemiologic conditions, in reality the intensity of transmission and immunity in pregnant women occurs on a continuum, with potentially diverse conditions occurring within a country. In areas of epidemic or low (unstable) malaria transmission, adult women have not acquired any significant level of immunity and usually become ill when infected with *P. falciparum* malaria. Pregnant women resident in areas of low or unstable malaria transmission are at a two- or three-fold higher risk of developing severe disease as a result of malaria infection than are non-pregnant adults living in the same area. In these areas maternal death may result either directly from severe malaria or indirectly from malaria-related severe anemia. In addition, malaria infection of the mother may result in a range of adverse pregnancy outcomes, including spontaneous abortion, neonatal death, and low birth weight (LBW).

In areas of high and moderate (stable) malaria transmission, most adult women have developed enough immunity that, even during pregnancy, *P. falciparum* infection does not usually result in fever or other clinical symptoms. In these areas, the principal impact of malaria infection is associated with malaria-related anemia in the mother and with the presence of parasites in the placenta. The resultant impairment of fetal nutrition contributing to low birth weight is a leading cause of poor infant survival and development. In areas of Africa with stable malaria transmission, *P. falciparum* infection during pregnancy is estimated to cause as many as 10000 maternal deaths each year, 8% to 14% of all low birth weight babies, and 3% to 8% of all infant deaths.

Despite the toll that malaria exacts on pregnant women and their infants, until recently this was a relatively neglected problem, with less than 5% of pregnant women having access to effective interventions. The promising news is that during the past decade potentially more effective strategies for the prevention and control of malaria in pregnancy have been developed and demonstrated to have a remarkable impact on improving the health of mothers and infants. Malaria prevention and control during pregnancy has a three-pronged approach.

3.4.2. Management of *Falciparum* Malaria in Pregnancy

3.4.2.1. Treatment of Uncomplicated Malaria in Pregnant Women

Pregnant women with uncomplicated malaria infection should receive prompt treatment with effective antimalarial drugs to clear infection fast. Unlike severe malaria, where the aim is to save the pregnant woman's life, in the treatment of uncomplicated malaria a judgment must be made between the risks and benefits of a potentially toxic treatment and the consequences of infection. The options for treatment of pregnant women with malaria are few, and most endemic countries lack an official policy [2]. Some data are available on the safety of drugs when used in prevention regimens, but pregnant women have been systematically excluded from most malaria treatment trials for fear of toxicity to the fetus. Thus the options for treatment are limited by potential teratogenicity (the ability to cause defects in the developing fetus) coupled with a paucity of safety data for the mother and fetus [260].

Malaria in pregnancy is associated with low birth weight, increased anemia; and, in low-transmission areas, an increased risk of severe malaria. In high-transmission settings, despite the adverse effects on fetal growth, malaria is usually asymptomatic in pregnancy. There is insufficient information on the safety and efficacy of most antimalarials in pregnancy, particularly for exposure in the first trimester, and so treatment recommendations are different to those for non-pregnant adults. Organogenesis occurs mainly in the first trimester and this is therefore the time of greatest concern for potential teratogenicity, although nervous system development continues throughout pregnancy. The antimalarials considered safe in the first trimester of pregnancy are quinine, chloroquine, proguanil, pyrimethamine, and sulfadoxine–pyrimethamine. Of these, quinine remains the most effective and can be used in all trimesters of pregnancy including the first trimester. In reality women often do not declare their pregnancies in the first trimester and so, early pregnancies will often be exposed inadvertently to the available first line treatment. Inadvertent exposure to antimalarials is not an indication for termination of the pregnancy.

There is increasing experience with artemisinin derivatives in the second and third trimesters (over 1000 documented pregnancies). There have been no adverse effects on the mother or fetus. The current assessment of benefits compared with potential risks suggests that the artemisinin derivatives should be used to treat uncomplicated *falciparum* malaria in the second and third trimesters of pregnancy, but should not be used in the first trimester until more information becomes available. The choice of combination partner is difficult. Amodiaquine, chlorproguanil- dapsone, halofantrine, lumefantrine, and piperaquine have not been evaluated sufficiently to permit positive recommendations. Sulfadoxine–pyrimethamine is safe but may be ineffective in many areas because of increasing resistance. Clindamycin is also safe, but both medicines (clindamycin and the artemisinin partner) must be given for 7 days. Primaquine and tetracyclines should not be used in pregnancy. Despite these many uncertainties, effective treatment must not be delayed in pregnant women. Given the disadvantages of quinine, i.e., the long course of treatment, and the increased risk of hypoglycemia in the second and third trimesters, ACTs are considered suitable alternatives for these trimesters. In practice, if first-line treatment with an artemisinin combination is all that is immediately available to treat in the first trimester of pregnancy; pregnant women who have symptomatic malaria, then this should be given. Pharmacovigilance programs to document the outcome of pregnancies where there has been exposure to ACTs, and if possible documentation of the development of the infant, are encouraged so that future recommendations can stand on a firmer footing.

In addition, treatment options are reduced by the spread of drug-resistant *P. falciparum*. The WHO currently recommends treating uncomplicated malaria with chloroquine in chloroquine sensitive areas, sulfadoxine-pyrimethamine in areas with chloroquine resistance, and quinine in the first trimester (the first three months of pregnancy) and where both chloroquine or sulfadoxine-pyrimethamine are not effective [261]. However, chloroquine and sulfadoxine-pyrimethamine resistance is widespread, especially in Asia, and quinine sometimes causes hypoglycemia (low blood sugar) even in uncomplicated infections [262, 263]. New policy of WHO of antimalarial drugs in pregnancy with uncomplicated malaria are updated in these changes.

1)   WHO policy of antimalarial drugs in pregnancy with uncomplicated malaria [2]

      Systematic summaries of safety suggest that the artemisinin derivatives are safe in the second and third trimesters of pregnancy. One large observational study in Thailand suggests an increased risk of stillbirth associated with mefloquine but this was not found in a study conducted in Malawi. There are as yet insufficient safety data about the use of artemisinin derivatives in the first trimester of pregnancy. Summary of WHO recommendations on the treatment of uncomplicated *falciparum* malaria in pregnancy are follows:

    a.   *First trimester*: quinine + clindamycin to be given for 7 days. ACT should be used if it is the only effective treatment available.

    b.   *Second and third trimesters*: ACT known to be effective in the country/region or artesunate + clindamycin to be given for 7 days or quinine + clindamycin to be given for 7 days.

     c. *Antimalarial drugs that should not be used in pregnancy including:* (1) halofantrine, (2) tetracycline/doxycycline, and (3) primaquine.

2) Lactating women [2]

The amounts of antimalarials that enter breast milk and are therefore likely to be consumed by the breast-feeding infant are relatively small. The only exception to this is dapsone, relatively large amounts of which are excreted in breast milk (14% of the adult dose), and pending further data this should not be prescribed. Tetracyclines are also contra-indicated because of their effect on the infant's bones and teeth. Summary of recommendations on the treatment of uncomplicated *falciparum* malaria in lactating women are follows:

     a. Lactating women should receive standard antimalarial treatment (including ACTs) except for tetracyclines and dapsone, which should be withheld during lactation.

Although artemisinin derivatives and combinations have an excellent efficacy profile, there are very limited data on the safety of artemisinin use during pregnancy, particularly in sub-Saharan Africa. Amodiaquine + artesunate are a possible combination for use in pregnant women, although no trials have yet been conducted. In a retrospective study following a mass treatment campaign in Gambia using AS + SP, there was no evidence of teratogenic or otherwise harmful effects and no difference in the rate of abortion, still birth or infant death among those exposed or not exposed to the drug, including women exposed in their first trimester ($n = 80$) [264]. Treatment of a self-selected group of pregnant women with SP + AS during pregnancy was associated with a greater birth weight (0.48 kg), which may have resulted from clearance of malaria parasites [264]. Concerning the use of ACT in pregnancy, the WHO has recommended that, because of the limited experience with the artemisinins in pregnancy, ACT should only be used when other treatments are considered unsuitable and presently it cannot be recommended for the treatment of malaria in the first trimester. It should not, however, be withheld if it is considered lifesaving for the mother. To document further the safety of artemisinin compounds in pregnancy, careful follow up is required, with documentation of pregnancy outcome and the subsequent development of the child, whenever possible. To guide further development of policies on the use of artemisinin derivatives during pregnancy, alone or in combination, there is an urgent need for further research and documentation of their efficacy and safety for use as therapy for malaria [63].

Pharmacokinetic monitoring of artemisinin use during pregnancy is needed urgently as ACT is implemented in almost all countries in Africa. The pharmacokinetics of antimalarial drugs is altered in pregnancy. This is the consequence of multiple factors: expansion of the volume of distribution, increase in clearance, changes in the protein binding, lipid distribution, and absorption of drugs, as well as influence of hormonal changes on the drug metabolism. These changes can result in lower plasma concentrations and AUCs and this can lead to reduced efficacy. The pharmacokinetics of most antimalarial drugs is modified in pregnancy and dosages need to be adapted. Few studies have been published. Chloroquine in treatment [265], and as prophylaxis [266], but also proguanil [267], and atovaquone [268], as well as dihydroartemisinin [269]; all have altered kinetics in pregnancy, and plasma levels are significantly lower than in non-pregnant patients with malaria. This is likely to be due to increased clearance, larger volume of distribution and perhaps altered absorption. It is clear that dosage of these (and probably other) antimalarials need to be adapted when given to pregnant women [270].

3.4.2.2. Pregnant Women with Severe Malaria

Pregnant women who are infected with *P. falciparum* are at particular risk and need to be treated with effective antimalarials. The artemisinin derivatives are now recommended for the treatment of *P. falciparum* malaria in the second and third trimesters of pregnancy. In severe malaria at any time in gestation, intravenous artesunate is the drug of choice. The recommendations for dosing of artesunate used in monotherapy and artemisinin-based combination therapies (ACTs) have mainly been derived empirically. Artesunate is rapidly hydrolyzed *in vivo* to dihydroartemisinin, which has equivalent antimalarial activity. Thus, in terms of biological (i.e., antimalarial) effect kinetics, plasma concentrations of both compounds are assessed.

Pregnant women, particularly in the second and third trimesters of pregnancy are more likely to develop severe malaria than other adults, often complicated by pulmonary edema and hypoglycemia. Maternal mortality is approximately 50%, which is higher than in non-pregnant adults. Fetal death and premature labor are common. The role of early Cesarean section for the viable live fetus is unproven, but is recommended by many authorities. Obstetric advice should be sought at an early stage, the pediatricians alerted, and blood glucose checked frequently. Hypoglycemia should be expected and is often recurrent if the patient is receiving quinine. Antimalarials should be given in full doses. Severe malaria may also present immediately following delivery. Postpartum bacterial infection is a common complication in these cases. *Falciparum* malaria has also been associated with severe mid-trimester hemolytic anemia in Nigeria. This often requires transfusion, in addition to antimalarial treatment and folate supplementation.

Parenteral antimalarials should be given to pregnant women with severe malaria in full doses without delay. Artesunate or artemether are preferred over quinine in the second and third trimesters because quinine is associated with recurrent hypoglycemia. Recent evidence shows that in non pregnant adults with severe malaria in areas of low transmission, artesunate was superior to quinine, reducing mortality by 35% compared to quinine, which makes artesunate the preferred option in the second and third trimesters. In the first trimester, the risk of hypoglycemia associated with quinine is lower, and the uncertainties over the safety of the artemisinin derivatives are greater. However, weighing these risks against this evidence in favor of the efficacy of artesunate, and until more evidence becomes available, both artesunate and quinine may be considered as options. Treatment must not be delayed so if only one of the drugs artesunate, artemether or quinine is available it should be started immediately.

1) WHO policy of antimalarial drugs in pregnancy with severe malaria [2]

Severe and complicated malaria are more frequent in pregnant women than in non-pregnant adults. Artesunate reduces the mortality of severe malaria in non-pregnant adults compared with quinine in low transmission situations. The artemisinin derivatives (artesunate and artemether) may also have safety advantages compared with quinine in the second and third trimesters of pregnancy because they do not cause recurrent hypoglycemia. In the first trimester, the risk of hypoglycemia associated with quinine is lower, and the uncertainties over the safety of the

Table 29. Summary of the choice of drugs to treat malaria during pregnancy depend on the trimester and on the regional pattern of resistance to antimalarials [270]

Malaria	Treatments
Uncomplicated *falciparum* malaria	First trimester: • First episode: Quinine 10 mg/kg three times a day for 7 days with or without Clindamycin 5 mg/kg three times per day for 7 days. • Subsequent episodes: ACT locally effective or artesunate 2 mg/kg/d for 7 days with Clindamycin as treated in first episode. • Second and third trimesters: • First episode: ACT locally effective or artesunate plus clindamycin as treated in first episode. • Subsequent episodes: artesunate plus clindamycin as treated in first episode. • Or quinine plus clindamycin as treated in first episode. Prevention: IPT with SP where efficacy remains.
Severe malaria	Artesunate 2.4 mg/kg IV at hour 0, 12 and 24 and continued every 24 hours until the patient can tolerate oral artesunate 2 mg/kg/dose, for 7 days and clindamycin 5 mg/kg three times daily for 7 days. Or Quinine i.v: Loading dose (LD) 20 mg/kg given over four hours, then 10 mg given 8 hours after the LD was started, followed by 10 mg/kg every 8 hours for 7d. Once the patient has recovered sufficiently to tolerate oral medication both quinine 10 mg/kg and clindamycin 5 mg/kg, three times daily should be continued daily for 7 days.
Non-*falciparum* malaria	Chloroquine phosphate (1 tablet contains 250 mg salt, equivalent to 155.3 mg base). Dose is 10 mg/kg base once a day for 2 days followed by 5 mg/kg base on third day. For chloroquine resistant *P. vivax* amodiaquine, quinine, mefloquine, artemisinin derivatives can be used. Prevention: chloroquine phosphate 300 mg on admission followed by 150 mg per week.

artemisinin derivatives are greater. The WHO recommendations are follows:

a. To use the parenteral antimalarial treatment locally available for severe malaria in full doses.

b. Where available, AS is the first, and artemether the second option in the second and third trimesters.

c. In the first trimester, until more evidence becomes available, both artesunate and quinine may be considered as options.

2) Treatment during pregnancy in epidemic situations [2]

In epidemic settings where the need for simplicity is paramount, intramuscular artemether is the drug of choice for severe malaria in all trimesters of pregnancy. The summary of recommendation on treatment of severe *falciparum* malaria in pregnant women in epidemic situations is follows:

a. Artesunate by the intravenous route is the treatment of choice, but if not possible, artemether by the intramuscular route is the preferred alternative for severe malaria in pregnancy during a malaria epidemic.

P. falciparum severe malaria remains a potentially lethal yet treatable disease. However, we remain ignorant of the best treatments. Use of antimalarial drugs in pregnant women continues to be a problem in which the risks to the woman and fetus are not completely known. More information on the correct doses to be given to pregnant women is desperately needed. Large-scale trials and post-market surveillance systems to monitor drug safety in pregnancy are required. The choice of drugs to treat malaria during pregnancy will depend on the trimester and on the regional pattern of resistance to antimalarials. Despite the paucity of data, quinine and artemisinin-based combination treatment (ACT) are now recommended for therapeutic use in Africa, and elsewhere and general policy of antimalarial drugs in pregnant women are summarized in Table 29 [270].

3.4.3. Management of *Falciparum* Malaria in Children

3.4.3.1. Children with Uncomplicated Malaria

In children able to attend a health facility that is well staffed and with adequate supplies, most deaths occur within 24 hours after admission, underscoring the importance of early treatment for preventing deaths. It is therefore important to improve access to appropriate care. One way of tackling this problem is to simplify the treatment by using rectal quinine or rectal artemisinin or artesunate, which could be given promptly even at basic health facilities. A trial on prompt administration of rectal artesunate is ongoing and should provide some data on its usefulness in early treatment. Artemether is rightly classified among the interventions likely to be beneficial and has a marginal advantage over quinine. It is easier to use (intramuscularly) and is less likely to cause hypoglycemia, but the cost of injections for treating an adult is about three times that of quinine. In endemic countries, malaria is common in infants and children under 2 years of age. Immunity acquired from the mother wanes after 3–6 months of age, and the case-fatality rate of severe malaria in infants is higher than in older children. Furthermore, there are important differences between infants and older children in the pharmacokinetics of many medicines. Accurate dosing is particularly important in infants. Despite this, few clinical studies focus specifically on this age range, partly because of ethical considerations relating to the recruitment of very young children to clinical trials, and also because of the difficulty of repeated blood sampling.

In the majority of clinical studies, subgroup analysis is not used to distinguish between infants and older children. Infants are more likely to vomit or regurgitate antimalarial treatment than older children or adults. Taste, volume, consistency and gastrointestinal tolerability are important determinants of whether the child retains the treatment. Mothers often need advice on techniques of medicine administration and the importance of administering the medicine again if it is immediately regurgitated. The increased failure of chloroquine and sulfadoxine-pyrimethamine as front-line antimalarials emphasizes the challenge to find safe alternatives for this age group. Fortunately the artemisinin derivatives appear to be safe and well tolerated by young children, and so the choice of ACT will be determined largely by the safety and tolerability of the partner drug. The limited information available does not indicate particular problems with currently recommended ACTs in infancy.

The ACTs seem to be tolerated as well or better in children than in adults. There is no specific age-related toxicity. In younger children vomiting or regurgitation of the administered dose are always a concern but are no more common with ACTs than monotherapies. Recent pharmacokinetic studies indicate that the dose regimen advocated for SP in children for many years is probably too low. There are insufficient pharmacokinetic data on amodiaquine and more data on the pharmacokinetics of piperaquine in children are needed. Dose regimens for artesunate-mefloquine and artemether-lumefantrine in children are justified by pharmacokinetic studies. Further work is needed to optimize dosing in children based on weight or surface area and, where necessary, to introduce specific pediatric formulations [271].

Although dosing based on body area is recommended for many drugs in young children, for the sake of simplicity, dosing of antimalarials has been traditionally based on weight. The weight-adjusted doses of antimalarials in infants are similar to those used in adults. For the majority of antimalarials, however, the lack of an infant formulation necessitates the division of adult tablets, which leads to inaccurate dosing. There is a need to develop infant formulations for a range of antimalarials in order to improve the accuracy and reliability of dosing. Delay in treating *falciparum* malaria may have fatal consequences, particularly for more severe infections. Every effort should be made to give oral treatment and ensure that it is retained. In situations where it is not possible to give parenteral treatment, a sick infant who vomits antimalarial medicine treatment repeatedly, has a seizure or is too weak to swallow reliably should be given artesunate by the rectal method, pending transfer to a facility where parenteral treatment is possible. Large trials assessing the impact of this strategy on mortality have recently been undertaken in remote rural areas, but results are not yet available. Pharmacological and trial evidence concerning the rectal administration of artesunate and other antimalarial drugs is provided in section.

The summary of recommendations on treatment of uncomplicated *falciparum* malaria in infants and young children are given [2].

a.   The acutely ill child requires careful clinical monitoring as they may deteriorate rapidly.
b.   ACTs should be used as first-line treatment for infants and young children.
c.   Referral to a health center or hospital is indicated for young children who cannot swallow antimalarials reliably.

3.4.3.2. Children with Severe Malaria

The risk of death from severe malaria is greatest in the first 24 h. Yet, in most malaria endemic countries, the transit time between referral and arrival at appropriate health facilities is usually prolonged thus delaying the commencement of appropriate antimalarial treatment, during which time the patient may deteriorate or die. It is recommended that patients are treated with the first dose of one of the recommended treatments by the parenteral route if possible or by the intra-rectal route before referral (unless the referral time is very short). This could be intramuscular artemether, artesunate or quinine, or a rectal formulation of artemisinin or artesunate.

The administration of an artemisinin by the rectal route as pre-referral treatment is feasible even at the community level. There is insufficient evidence to show whether rectal artesunate is as good as intravenous or intramuscular options in the management of severe malaria. The recommendation, therefore, is to use artesunate or artemisinin suppositories only as pre-referral treatment and to refer the patient to a facility where complete parenteral treatment with artesunate, quinine, or artemether can be instituted. If, however, referral is impossible, rectal treatment should be continued until the patient can tolerate oral medication, at which point a full course of the recommended ACT for uncomplicated malaria in the locality can be administered.

The summary of recommendations on treatment of severe *falciparum* malaria in infants and young children are [2]:

a. One or more artesunate suppositories inserted in the rectum as indicated in Table 27. The dose should be given once and followed as soon as possible by definitive therapy for malaria.

b. The parenteral route is intramuscular artemether, artesunate, or quinine, or a rectal formulation of artemisinin or artesunate.

Rectal preparations have the advantage of being easy to administer in rural areas; therefore it is anticipated that rectal administration of an artemisinin derivative in remote settings might "buy time" by halting or slowing the progress of disease while a child is being transported to a health-care facility equipped to provide definitive treatment. Their utility would consequently be greatest in areas where access to injectable therapy is poor or does not exist. The clinical evidence accumulated in the initial phase of this development focused on measures of parasite reduction – a well-established indicator of clinical effect in the evaluation of antimalarial drugs.

In addition to artesunate, other artemisinin derivatives formulated for rectal administration now include artemisinin, dihydroartemisinin, and artemether. The evidence from this analysis shows rectal artesunate has superior effect in reducing parasite densities compared to quinine (intravenous or intramuscular) at 12 and 24 hours after administration. The result supports the WHO recommendation for the use of artesunate and artemisinin as initial pre-referral treatment [2]. Basis of systematic review, the WHO recommendation is to use artemisinins rectally for complete treatment only when parenteral antimalarial treatment is not possible [100]. Recommendations by WHO are to treat with artesunate or artemisinin by rectal administration first on pre-referral treatment for severe *falciparum* malaria children. The administration of an artemisinin by the rectal route as pre-referral treatment is feasible even at the community level. There is insufficient evidence to show whether rectal artesunate is as good as intravenous or intramuscular options in the management of severe malaria. The recommendation, therefore, is to use artesunate or artemisinin suppositories only as pre-referral treatment and to refer the patient to a facility where complete parenteral treatment with artesunate, quinine, or artemether can be instituted. If, however, referral is impossible, rectal treatment should be continued until the patient can tolerate oral medication, at which point a full course of the recommended ACT for uncomplicated malaria in the locality can be administered [101].

3.4.4. Treatments of Malaria in Traveler

Malaria is a protozoan disease transmitted by the bite of the blood-feeding female anopheline mosquito. Four species of the genus Plasmodium, *P. falciparum, P. vivax, P. malariae,* and *P. ovale,* are responsible for the vast majority of human infections. *P. falciparum* causes the most severe disease and is responsible for most malaria related deaths. From its influence on human genetic variation, and the toll it exacts in morbidity and mortality, malaria is considered to be the most important parasitic disease of man. Genetic mutations in hemoglobin genes are the commonest single gene disorders in man and their geographic coincidence with the malaria prone areas of the world illustrates their survival benefit for this disease. Although it has been eradicated from Europe, North America, and Russia; malaria is still found in the Middle East, China, and the Indian Subcontinent and, particularly, the tropics where the emergence of resistance to antimalarial drugs has partly contributed to a recent resurgence of the disease.

Travelers who acquire malaria are often non-immune adults either from cities with little or no transmission within endemic countries, or visitors from non-endemic countries. Both are likely to be at a higher risk of malaria and its consequences because they have no immunity to malaria. Within the malaria endemic country they should be treated in principle according to national policy. Travelers who return to a non-endemic country and then develop malaria present particular problems and, have a high case fatality rate. Doctors may be unfamiliar with malaria and the diagnosis may be delayed, relevant antimalarials may not be registered and/or therefore available. If the patient falls ill far from a major health facility, availability of antimalarials can be a life threatening issue despite registration. On the other hand prevention of emergence of resistance and transmission are of less relevance outside malaria endemic areas. Thus monotherapy may be given if it can be assured to be effective. Furthermore cost of treatment is usually not a limiting factor.

The principles underlying the recommendations given here are that effective medicines should be used to treat travelers; if the patient has taken chemoprophylaxis, and then the same medicine should not be used for treatment. The treatment for *P. vivax, P. ovale,* and *P. malariae* in travelers should be the same as for these infections in patients from endemic areas. In the management of severe malaria outside endemic areas, there may be delays in obtaining artesunate, artemether, or quinine. If parenteral quinidine is available but other parenteral drugs are not, then this should be given with careful clinical and electrocardiographic monitoring. (Chapter 3, section 3.3 Chemotherapy for severe malaria)

Rapid initiation of treatment for malaria after the diagnosis is established is necessary to avoid morbidity and mortality. Most symptomatic malaria infections are uncomplicated and the mortality is approximately 0.1% in the majority of cases of *P. falciparum* malaria, if effective treatment is initiated. In the minority of patients with severe malaria, the mortality can be 15-20% despite treatment. The patient should be admitted to an appropriate hospital setting, depending on the severity of presentation. Treatment should not be initiated before the diagnosis is established because a fever in a returning traveler from the tropics will be malaria in less than 50% of cases. Thick and thin blood smears should be examined at least every hour to monitor the efficacy of therapy until the parasitemia is below 1%. Smears should also be obtained at 3 days, one week and four weeks to exclude recurrence. Recommended treatments for malaria by WHO are included in travelers returning to non-endemic areas as follows. No medication kills all stages of the malaria parasite and the following antimalarials are suitable for use in travelers with uncomplicated malaria:

a.   artemether-lumefantrine (6-dose regimen),
b.   atovaquone–proguanil, and
c.   quinine + doxycycline or clindamycin.

The criteria for severe *falciparum* malaria include severe anemia, respiratory distress, cerebral malaria, renal failure, hypoglycemia, shock, coagulation failure, and for complicated malaria impaired consciousness of any degree plus prostration plus jaundice plus intractable vomiting plus parasitemia P 2%. These travelers should be admitted to the Intensive care unit and receive close monitoring and support, including hemofiltration, mechanical ventilation, and blood transfusions as necessary. Gram negative infections should always be excluded, since travelers with *P. falciparum* malaria are susceptible to other infections as well. Travelers with severe malaria, as well as those unable to tolerate oral therapy, should be given parenteral, intravenous, therapy with quinine or an artemisinin-based regimen. Alternatives are IM artemether or artesunate, artemisinin or artesunate suppository, or IM quinine in the thighs. When the patient improves, treatment can be changed to oral doxycycline or clindamycin.

Possible regimens include quinine dihydrochloride 10 mg (salt) per kg by infusion over 4 hrs in 500 ml in 5% dextrose, every 8 hours until parasites less than 1% and the patient can take by mouth, then quinine sulphate 600 mg three times a day orally until parasites have cleared, then doxycycline for seven days. Quinidine gluconate 10 mg/kg loading dose in normal saline (maximum 600 mg) over one to two hours, then continuous infusion of 0.02 mg/kg per minute until traveler can swallow. Travelers who have received other quinine derivatives or mefloquine within the last 12 hours should not receive a loading dose because of the risk of added toxicity. The dose of quinine or quinidine needs to be reduced by one-third to one-half after 48 hours. The traveler should be attached to a cardiac monitor and the glucose level followed because of the associated hyperinsulinemia caused by the two drugs. Artesunate 2.4 mg/kg intravenous as a first dose followed by 1.2 mg/kg at 12 and 24 hours followed by 1.2 mg/kg once daily for six days, or artemether 3.2 mg/kg intramuscular followed by 1.6 mg/kg daily for six days or artemisinin suppositories 40 mg/kg intrarectally followed by 20 mg/kg at 24, 48 and 72 hours followed by an oral drug.

Summary of WHO recommendations on the treatment of *falciparum* malaria in non-immune travelers returning to non-endemic countries are follows [2]:

a.   atovaquone-proguanil (15/6 mg/kg, usual adult dose, 4 tablets once a day for 3 days);
b.   artemether-lumefantrine (adult dose, 4 tablets twice a day for 3 days);
c.   quinine (10 mg salt/kg every 8 h) + doxycycline (3.5 mg/kg once a day) or
     clindamycin (10 mg/kg twice a day); all drugs to be given for 7 days

For severe malaria:
a.   The antimalarial treatment of severe malaria in travelers is the same as shown in section "3.3. Chemotherapy for severe malaria";
b.   Travelers with severe malaria should be managed in an intensive care unit;

c.   Hemofiltration or hemodialysis should be started early in acute renal failure or severe metabolic acidosis;

d.   Positive pressure ventilation should be started early if there is any breathing pattern abnormality, intractable seizure or acute respiratory distress syndrome.

In addition:

a.   Halofantrine is not recommended as first-line treatment for uncomplicated malaria because of cardiotoxicity.

b.   Doxycycline should not be used in children under 8 years of age.

3.5. Therapies of Malaria Caused by *P. Vivax, P. Ovale,* or *P. Malariae*

For the most part the interest of this resurgence has focused on infections with *P. falciparum*, which is associated with the greatest mortality and, in sub-Saharan Africa, intensity of transmission. *P. vivax*, the second most important species causing human malaria, accounts for about 40% of malaria cases worldwide and is the dominant malaria species outside Africa. It is prevalent in endemic areas in the Middle East, Asia, Oceania, Central and South America. In Africa, it is rare except in the Horn and it is almost absent in West Africa. In most areas where *P. vivax* is prevalent, malaria transmission rates are low, and the affected populations therefore achieve little immunity to this parasite. Consequently, people of all ages are at risk. The other two human malaria parasite species *P. malariae* and *P. ovale* are generally less prevalent but are distributed worldwide, especially in the tropical areas of Africa. Among the four species of *Plasmodium* that affect humans, only *P. vivax* and *P. ovale* form hypnozoites, parasite stages in the liver that can result in multiple relapses of infection, weeks to months after the primary infection. Thus a single infection causes repeated bouts of illness. This affects the development and schooling of children and debilitates adults, thereby impairing human and economic development in affected populations.

Although the emphasis on *P. falciparum* is appropriate, the burden of *vivax* malaria should not be underappreciated and exacts a significant toll on almost half of the world's population. The objective of treating malaria caused by *P. vivax* and *P. ovale* is to cure both the blood stage and the liver stage infections, and thereby prevent both relapse and recrudescence. This is called radical cure. Infection with *P. vivax* during pregnancy, as with *P. falciparum*, reduces birth weight. In primigravidae, the reduction is approximately two-thirds of that associated with *P. falciparum* (110 g compared to 170 g), but this adverse effect does not decline with successive pregnancies as with *P. falciparum* infections. Indeed, in the one large series in which this was studied, it increased. Reduction in birth weight (< 2500 g) increases the risk of neonatal death.

3.5.1. *P. Vivax* Transmission Areas and Diagnosis

There are between 72 and 80 million cases of malaria due to *P. vivax* each year with the greatest burden observed in South and East Asia (52%), Eastern Mediterranean (15%) and South America (13%) [272]. Recently these statistics have been challenged by an analysis using a combination of geographic information systems, malaria epidemiology, historical maps, and information on population densities, environment, and vector limits [273]. When the population density of areas endemic for *P. vivax* is taken into consideration, the number of people at risk of infection reaches 2.6 billion, slightly greater than that for *P. falciparum*. In South and Southeast Asia, where the majority of *vivax* malaria occurs, *P. vivax* accounts for up to 50% of malaria cases with prevalence rates between 1% and 6% of the population. The proportional burden of *vivax* is even greater in Central and South America, reaching 71–81% of all malaria cases. In eastern and southern Africa only 5% of malaria infections are attributable to *P. vivax*, although this still accounts for between 6 and 15 million cases per year. The emergence of chloroquine-resistant *P. falciparum* has tended to reduce the proportion of malaria cases due to *P. vivax*; nevertheless the absolute numbers of *P. vivax* remain high. Conversely, where successful malaria control strategies have been used the ratio of *P. falciparum* to *P. vivax* infections has fallen [274].

The clinical features of uncomplicated malaria are too non-specific for a clinical diagnosis of the species of malaria infection to be made. Diagnosis of *P. vivax* malaria is based on microscopy. Although microscopy remains the mainstay of diagnosis of *vivax* malaria, lack of access to good quality microscopy services in many endemic regions limits the reliability of diagnosis (as well as contributes to the underestimation of cases). *HRP-2-* based rapid diagnostic tests (RDTs) are usually sensitive for the diagnosis of *P. falciparum*, although aldolase-based and parasite lactic dehydrogenase-based RDTs have suboptimal sensitivity for *P. vivax*, limiting the utility of RDTs for *vivax* diagnosis [275]. The rapid diagnostic tests based on immune-chromatographic methods are available for the detection of non-*falciparum* malaria, but their sensitivities below parasite densities of 500/µl are low. Their relatively high cost is a further impediment to their wide use in endemic areas. Molecular markers for genotyping *P. vivax* parasites have been developed to assist epidemiological and treatment studies but these are still under evaluation.

3.5.2. Efficacy of Antimalarial Drugs in Single and Mixed Infections

P. vivax predominates in South America and parts of Asia and resistance of this parasite to chloroquine is geographically still limited [272]. In contrast to *P. falciparum*, asexual stages of *P. vivax* are susceptible to primaquine. Thus chloroquine + primaquine can be considered as a combination treatment. The only drugs with significant activity against the hypnozoites are the 8-aminoquinolines (bulaquine, primaquine, tafenoquine). There are very few recent data on the *in vivo* susceptibility of *P. ovale* and *P. malariae* to antimalarials. Both species are regarded as very sensitive to chloroquine, although there is a single recent report of chloroquine resistance in *P. malariae*. Experience indicates that *P. ovale* and *P. malariae* are also susceptible to amodiaquine, mefloquine and the artemisinin derivatives. Their susceptibility to antifolate antimalarials such as sulfadoxine-pyrimethamine is less certain. *P. vivax* susceptibility has been studied extensively, and now that short-term culture methodologies have been standardized, clinical studies have been supported by *in vitro* observations. *P. vivax* is still generally very sensitive to chloroquine, although resistance is prevalent and increasing in some areas, notably Oceania, Indonesia, and Peru (see Annex A.6.4). Resistance to pyrimethamine has increased rapidly in some areas, and sulfadoxine-pyrimethamine is consequently ineffective. There are insufficient data on current susceptibility to proguanil and chlorproguanil, although resistance to proguanil was selected rapidly when it was first used in *P. Vivax* endemic areas.

In malarious areas of Thailand, double infection with *P. falciparum* and *P. vivax* is common and usually manifests as sequential illnesses, commonly with *falciparum* malaria as the primary infection. Studies in artificial infections suggest suppression of *P. vivax* by *P. falciparum* and vice versa. In general, *P. vivax* is sensitive to all the other antimalarial drugs; it is more sensitive than *P. falciparum* to the artemisinin derivatives and slightly less sensitive to mefloquine (although mefloquine is still effective). There is no standardized *in vitro* method of drug assessment for hypnozoiticidal activity. *In vivo* assessment suggests that tolerance of *P. vivax* to primaquine in East Asia and Oceania is greater than elsewhere. For the mixed infection, the antimalarial treatment for *falciparum* malaria in patients is the same as shown in Chapter 3, see section 3.3 Chemotherapy for severe malaria.

3.5.3. Chemotherapies of *Vivax, Ovale, and Malariae* Malaria

3.5.3.1. Treatment of Uncomplicated Vivax Malaria

1) Blood stage infection

There have been fewer studies on the treatment of malaria caused by *P. vivax* than of *falciparum* malaria – only 11% of 435 published before 2004. For chloroquine-sensitive *vivax* malaria (i.e., in most places where *P. vivax* is prevalent) the conventional oral chloroquine dose of 25 mg base/kg is well tolerated and effective. Some have advocated lower total doses, but this is not recommended as it might encourage the emergence of resistance. Chloroquine is given in an initial dose of 10 mg base/kg followed by either 5 mg/kg at 6 h, 24 h and 48 h or, more commonly, by 10 mg/kg on the second day and 5 mg/kg on the third day. It is also clear that if ACT treatment is given, the response is as good as or better than in *falciparum* malaria. The exception to this is a regimen containing sulfadoxine-pyrimethamine. It appears that *P. vivax* has developed resistance to sulfadoxine-pyrimethamine more rapidly than has *P. falciparum*, so that artesunate + sulfadoxine-pyrimethamine may not be effective against *P. vivax* in many areas [2].

2) Chloroquine-resistant *vivax* malaria

There are relatively few data on treatment responses in chloroquine-resistant *vivax* malaria. Studies from Indonesia indicate that amodiaquine is efficacious, and there is some evidence that mefloquine and quinine can also be used. The artemisinin derivatives would also be expected to be highly effective, and artemether-lumefantrine could be an alternative treatment. However, there are insufficient clinical data to confirm this.

3) Liver stage infection

To achieve radical cure, relapses must be prevented by giving primaquine. The frequency and pattern of relapses varies geographically, however. It has become clear in recent years that whereas 50-60% of *P. vivax* infections in South- East Asia relapse, the frequency are lower in Indonesia (30%) and the Indian subcontinent (15-20%). Some *P. vivax* infections in the Korean peninsula (now the most northerly of human malarias) have an incubation period of nearly one year. Thus the preventive efficacy of primaquine must be set against the prevalent relapse frequency. It appears that the total dose of 8-aminoquinoline given is the main determinant of curative efficacy against liver-stage infection. There is no evidence that the short courses of primaquine widely recommended (such as 5-day regimens) have any efficacy. Primaquine should be given for 14 days. There has been debate as to whether primaquine should be given in endemic areas. Repeated *vivax* malaria relapses are debilitating at any age, but if reinfection is very frequent, then the risks of widespread use of primaquine may exceed the benefits. In low-transmission areas, the benefits of deploying primaquine are considered to exceed the risks, but in areas of sustained high-transmission (such as on the island of New Guinea), *P. vivax* infection is very frequent, immunity is acquired, and the risks of widespread deployment of primaquine are considered to outweigh the benefits [2].

4) WHO policy of antimalarial drugs for *vivax* malaria [2]

The summary of recommendations on the treatment of uncomplicated *vivax* malaria is shown:

c. Chloroquine 25 mg base/kg divided over 3 days, combined with primaquine 0.25 mg base/kg, taken with food once daily for 14 days is the treatment of choice for chloroquine-sensitive infections. In Oceania and South-East Asia the dose of primaquine should be 0.5 mg/kg.

d. Amodiaquine (30 mg base/kg divided over 3 days as 10 mg/kg single daily doses) combined with primaquine should be given for chloroquine-resistant *vivax* malaria.

e. In moderate G6PD deficiency, primaquine 0.75 mg base/kg should be given once a week for 8 weeks. In severe G6PD deficiency, primaquine should not be given.

f. Where ACT has been adopted as the first-line treatment for *P. falciparum* malaria, it may also be used for *P. vivax* malaria in combination with primaquine for radical cure. Artesunate +sulfadoxine-pyrimethamine are the exception as it will not be effective against *P. vivax* in many places.

3.5.3.2. Treatment of Severe Vivax Malaria

Although *P. vivax* malaria is considered to be benign malaria, with a very low case-fatality ratio, it may still cause a severe and debilitating febrile illness. It can also very occasionally result in severe disease as in *falciparum* malaria. Severe *vivax* malaria manifestations that have been reported are cerebral malaria, severe anemia, severe thrombocytopenia and pancytopenia, jaundice, spleen rupture, acute renal failure and acute respiratory distress syndrome. Severe anemia and acute pulmonary edema are not uncommon. The underlying mechanisms of severe manifestations are not well understood. Prompt and effective treatment and case management should be the same as for severe and complicated *falciparum* malaria. (Chapter 3, Section 3.3 Chemotherapy for severe malaria)

3.5.3.3. Treatment of Malaria Caused by P. Ovale *and* P. Malariae

Resistance of *P. ovale* and *P. malariae* to antimalarials is not well characterized and infections caused by these two species are considered to be generally sensitive to chloroquine. Only one study, conducted in Indonesia, has reported resistance to chloroquine in *P. malariae*. The recommended treatment for the relapsing malaria caused by *P. ovale* is the same as that given to achieve radical cure in *vivax* malaria, i.e., with chloroquine and primaquine. *P. malariae* should be treated with the standard regimen of chloroquine as for *vivax* malaria, but it does not require radical cure with primaquine as no hypnozoites are formed in infection with this species. *P. ovale* mainly occurs in areas of high stable transmission where the risk of re-infection is high. In such settings, primaquine treatment is not indicated.

3.5.3.4. Monitoring Therapeutic Efficacy for Vivax Malaria

The antimalarial sensitivity of *vivax* malaria needs monitoring, to track and respond to emerging resistance to chloroquine. The 28-day *in vivo* test for *P. vivax* is similar to that for *P. falciparum*, although the interpretation is slightly different. Genotyping may distinguish a re-infection from a recrudescence and from acquisition of a new infection, but it is not possible to distinguish reliably between a relapse and a recrudescence as they derive from the same infection. If parasitemia recurs within 16 days of administering treatment then relapse is unlikely, but after that time, relapse cannot be distinguished from a recrudescence. Any *P. vivax* infection that recurs within 28 days, whatever its origin must be resistant to chloroquine (or any other slowly eliminated antimalarial) provided adequate treatment has been given. In the case of chloroquine, adequate absorption can be confirmed by measurement of the whole blood concentration at the time of recurrence. Any *P. vivax* infection that has grown *in vivo* through a chloroquine blood concentration ≥ 100 ng/ml must be chloroquine resistant. Short-term *in vitro* culture allows assessment of *in vitro* susceptibility. There are no molecular markers yet identified for chloroquine resistance. Antifolate resistance can be monitored by molecular genotyping of the gene that encodes dihydrofolate reductase (*Pvdhfr*) [2].

3.5.4. Management of Resistant P. Vivax

In vivo drug studies on strains of *P. vivax* sensitive to chloroquine have demonstrated the efficacy of artemisinin derivatives, mefloquine, quinine, and halofantrine; however, few studies have addressed suitable treatment regimens for chloroquine resistance *vivax* malaria [276]. *P. vivax* has previously been regarded as intrinsically resistant to antifol. It is not; but resistance develops readily, through the sequential acquisition of mutations in *Pvdhfr* [277]. The prevalence of antifol resistance is similar to that of *P. falciparum* with high levels of resistance in much of Southeast Asia (Laos is a notable exception) and lesser degrees of resistance in the Indian subcontinent [278]. In Indonesia the efficacy of amodiaquine is superior to that of chloroquine, but in regions where high levels of chloroquine resistance predominate, day 28 failure rates exceed 25%, suggesting amodiaquine monotherapy is not a suitable alternative once high grade resistance has been reached [279]. Better success in the same region has been obtained with mefloquine and halofantrine, which retain almost 100% cure rates by day 28 [280].

Atovaquone-proguanil plus primaquine also showed excellent efficacy against chloroquine resistance *P. vivax*; however, the high cost of this regimen (~US$40) virtually precludes its use in endemic communities [274]. The artemisinin derivatives are highly potent antimalarials and are even more active against *P. vivax* than *P. falciparum.* Artemisinin-based combination therapy (ACT) is now widely advocated for the management of *P. falciparum* infections. In practice only a minority of infections has correct species identification prior to treatment and therefore in Asia and South America it is highly likely that other species will also be treated with ACT. Despite this, only 3 studies have addressed the efficacy of these combinations against *P. vivax*. In the first, a 3-day course of artesunate plus sulfadoxine-pyrimethamine resulted in a recurrence rate of 10.5% by day 28 [281]. In 2 subsequent studies of patients with pure *P. vivax* infections, recurrence rates by day 28 were significantly higher after artemether-lumefantrine (23%) and amodiaquine-artesunate (12%) than after dihydroartemisinin piperaquine (3.6%) [282]. None of these antimalarial drugs have any efficacy on the hypnozoite stages of *P. vivax* and hence the difference in cure rates is likely to be a consequence of the difference between the terminal elimination half life of lumefantrine (~4 days), amodiaquine (18 days) and that of piperaquine (28-35 days).

Piperaquine, like chloroquine, suppresses the first relapse whereas lumefantrine and amodiaquine do not. In areas where primaquine cannot be supervised, such post-treatment prophylaxis afforded by drugs with long half lives will not prevent subsequent relapses, but will provide the only practical means currently available of delaying the timing of clinical illness from these relapses. The latter affords patients a longer period free from symptomatic malaria, decreases the risk of anemia and reduces significantly gametocyte carriage and thus the transmission potential to the mosquito vector. Although longitudinal studies are needed to confirm these findings, when choosing an alternative to chloroquine in communities where chloroquine resistance *vivax* malaria is endemic, the relevance of post-treatment prophylaxis should not be overlooked [274].

Chapter IV

Antimalarial Drugs
in Chemoprophylaxis

Malaria chemoprophylaxis is the prevention of malaria disease by giving healthy pregnant woman, children and travelers medication prior to exposure to infective mosquitoes. Chemoprophylaxis is given to reduce the risk of malaria but is not perfectly effective, primarily because of non-compliance with medication regimens or parasite drug resistance. The principles of the use of antimalarial drugs for protection against malaria and the treatment of malaria were reviewed in International Travel and Health 2008 [283] and Guidelines for the Treatment of Malaria 2006 [2] by World Health Organization. This chapter considers the particular requirements for chemoprophylaxis in pregnant women and for chemoprophylaxis and stand-by treatment in travelers. It also comments on the management of severe malaria, the treatment of *vivax* malaria and the need for antimalarial formulations for pediatric use.

The aims of malaria chemoprophylaxis, in broad terms are to alleviate symptoms by the erythrocytic phase, to prevent relapses by the hypnozoites of *vivax* and *ovale*, and to prevent further transmission of the parasite. The principals of treatment depend on four factors, these being: 1) the type of infection, 2) the severity of infection, 3) the status of the host, and 4) any associated conditions or diseases present in the host. The new global and national antimalarial drug policies for chemotherapy and chemoprophylaxis updated in 2008 by World Health Organization are listed in Appendix 1. This chapter aims to present the current situation regarding malaria chemoprophylaxis, indicate what new drugs might be available in the future and discuss some of the common problems encountered when prescribing malaria chemoprophylaxis. There are approximately 14 antimalarials that are advised for use in the prevention and treatment of uncomplicated malaria. These will be discussed individually with particular attention to the specific recommended usage for each drug.

Chemoprophylaxis is recommended for (a) pregnant women, (b) children in high-risk areas, and (c) travelers including service personnel who temporarily go on duty to high malarious areas. Prior to the emergence of widespread chloroquine resistance, malaria prophylaxis was very simple: one chloroquine tablet a week was sufficient cover. In chloroquine sensitive areas chloroquine is to be given but in chloroquine resistant areas chloroquine is to be supplemented by proguanil. However chemoprophylaxis should not exceed 3 years due to the cumulative toxicity of chloroquine. Also considering the limitations of chemoprophylaxis personal protection measures should be encouraged. The efficacious new drugs and artemisinin-based combination therapy are available to prevent malaria, despite the rapid advance of drug resistance. The better tolerated regimens that require less compliance from the pregnancy, children and travelers may be the key elements for improved malaria chemoprophylaxis in the future.

4.1. Drugs in Current Malaria Chemoprophylaxis

4.1.1. Drug Resistance Affects Prophylactic Drugs

Resistance of *P. falciparum* to chloroquine, the cheapest and the most used drug has been reported in almost all the endemic countries, except in Central America and Hispaniola. Prior to the threat of widespread chloroquine resistance, malaria prophylaxis was straight forward and simple: one chloroquine tablet a week was sufficient coverage. However, today chloroquine resistance areas do exist and chloroquine is considered ineffective in its prophylactic effects. Resistance to the combination of sulfadoxine-pyrimethamine, which was already present in South America and in South-East Asia, is now becoming highly prevalent in Africa. Resistance to mefloquine is found mostly in Cambodia, Myanmar, Thailand, and Vietnam. Sporadic cases of prophylactic failure of mefloquine in travelers and therapeutic failure have been reported in Africa, South America, and in other Asian countries. These drugs (chloroquine, sulfadoxine-pyrimethamine, and mefloquine) were and still are first-line drugs in the chemoprophylaxis of malaria in many regions [283].

Prophylactic drug resistance results in enormous public health burden because of increase or recurrent illness with the infection and progression to severe malaria which is associated with increased hospitalization and death. Even lower levels of resistance tend to cause higher rates of infection that are associated with return of illness, prolonged or worsening anemia, and increased gametocyte carriage which fuels transmission; particularly for the resistant parasite and a higher risk of prophylaxis failure in primary infections. World efforts to monitor malaria with the goal towards controlling malaria are greatly challenged by the increase in the spread of antimalarial drug resistance. Use of ineffective antimalarials is thus considered partly responsible for the difficulties in reducing malaria infection, malaria morbidity and mortality. This is a problem particularly in sub-Saharan Africa, where susceptibility of *P. falciparum* to previously used cheap and commonplace antimalarials, such as chloroquine and sulfadoxine–pyrimethamine, has been declining rapidly in many high-transmission areas. To ensure that malaria control strategies and malaria treatment policies rely on the deployment of effective antimalarials, there is a need for systematic monitoring of antimalarial drug efficacy and drug resistance.

The mechanism underlying the development of resistance to chemoprophylactic agents originates from chromosomal mutations in the malaria parasite. Resistance to antimalarial drugs arises as a result of spontaneously-occurring mutations that affect the structure and activity at the molecular level of the drug target in the malaria parasite or affect the access of the drug to that target. Until the late 1950s, chloroquine was the mainstay of malaria chemoprophylaxis, but widespread resistance now limits its usefulness in almost all malaria endemic countries. Compelling evidence exists that multiple mutations in the *P. falciparum* chloroquine resistance transporter gene (*pfcrt*) are associated with chloroquine resistance [21]. These mutations alter chloroquine concentrations in the food vacuole, a major digestive organelle of the malaria parasite that harbors protein transporters and is responsible for the degradation of host-derived hemoglobin. The food vacuole is a proven chemotherapeutic target for chloroquine.

Similar to chloroquine, the site of action for the arylamino-alcohols, such as mefloquine, is within the digestive vesicle of the blood stage parasite where they complex with heme and prevent the detoxification of heme by crystallization into the malaria pigment. The mechanism of resistance to mefloquine is still unclear, but an increase in the copy number of the multidrug-resistance gene 1 (*pfmdr1*) may play an important role in modulating the sensitivity of the drug [22, 23]. Prophylactic failures observed with atovaquone-proguanil have been associated with a specific point mutation in the cytochrome b gene at codon 268, but other reported treatment failures with this drug were not associated with this mutation [24, 25].

Resistance to chloroquine took more than 20 years to eventuate, but resistance to the antifols, such as pyrimethamine and proguanil, and atovaquone, occurred shortly (< 1 year) after their release. For antifols and sulfonamides, point mutations in the dihydrofolate reductase (DHFR) and dihydropteroate synthase (DHPS) genes, respectively, have been identified as causing structural alteration in the binding site of the target enzymes which diminishes the affinity of these enzymes for the antifolate drugs. One to four mutations in DHFR confers a moderate to high level of resistance to DHFR inhibitors. The presence of mutations in DHFR appears to be more important in causing pyrimethamine/sulfadoxine failure than DHPS mutations [284].

As international travel becomes increasingly common and resistance to antimalarial drugs escalates, a growing number of travelers are at risk for contracting malaria. Parasite resistance to chloroquine and proguanil; and real or perceived intolerance among patients to standard prophylactic agents such as mefloquine have highlighted the need for new prophylactic drugs. Promising new regimens include atovaquone and proguanil, in combination; primaquine; and a related 8-aminoquinoline, tafenoquine. These agents are active against the liver stage of the malaria parasite and therefore can be discontinued shortly after the traveler leaves an area where malaria is endemic, which encourages adherence to the treatment regimen.

4.1.2. Current Drugs for Malaria Chemoprophylaxis

4.1.2.1. Current Drugs for Malaria Chemoprophylaxis

Use of antimalarial drugs to prevent the development of malaria is known as chemoprophylaxis. Currently, there are several substances used for treatment and, partially, for prevention (prophylaxis). Many drugs may be used for both purposes; larger doses are used to treat cases of malaria and low doses are used to prevent malaria infection. However, the spread of drug-resistant strains of *P. falciparum* since the 1960s has reduced the efficacy of chloroquine, which for several decades was a highly effective, convenient, and relatively safe drug for the chemoprophylaxis and treatment of malaria. At the beginning of the 1980s, however, chloroquine treatment became almost totally ineffective in Thailand and neighboring countries, and resistance to the treatment spread rapidly in South America. In 1993, Malawi was the first country in East Africa to change from chloroquine to the sulfadoxine–pyrimethamine combination as the first-line drug, and other African countries followed this example in the late 1990s. The combination of pyrimethamine and sulfadoxine and amodiaquine were introduced as alternative chemoprophylactic agents, but both proved to be too toxic to be widely recommended for this purpose. Because of extensive use of this combination, however, resistance also spread in East Africa. Subsequently, mefloquine became available as an effective chemoprophylactic drug against chloroquine-resistant *P. falciparum*, and doxycycline also proved useful for this purpose. The availability of these drugs simplified the recommendations for malaria chemoprophylaxis [285]. Nonetheless, some physicians remain concerned about the safety of mefloquine and favor other, less effective prophylactic regimens.

Pyrimethamine-dapsone was another combination drug used in Asia and the South Pacific, sometimes in combination with chloroquine, for prevention of chloroquine-resistant malaria. However, this regimen did not gain widespread use among international travelers from Western countries; in part, because access to pyrimethamine-dapsone was limited. In North America, currently recommended drugs for malaria chemoprophylaxis in chloroquine-resistant areas include mefloquine, doxycycline, and atovaquone/proguanil. Both mefloquine and doxycycline are associated with some well-known side-effects that may limit their use in some patients, and resistance to mefloquine has emerged in certain endemic areas. Atovaquone/proguanil is a fixed-dose combination drug that has been recently approved for both the prophylaxis and treatment of malaria, and is effective against malaria caused by chloroquine-resistant or multidrug-resistant *P. falciparum* strains. Although experience with the combination product is limited, proguanil has a long history of safety. Proguanil has been used alone and in combination with chloroquine for malaria chemoprophylaxis for many years following its development during World War II. The rapid emergence of resistance to the drug limited its use as a single agent, but the combination of chloroquine and proguanil appears to be more effective than either drug alone against chloroquine-resistant strains. Nonetheless, this combination is significantly less effective than the other drugs discussed here.

Therefore, their deployment depends mainly on the species and on the frequency of resistant parasites in the prevalent area where the drug is used. Presently available prophylactic drugs include:

- *Chloroquine* (Therapy and prophylaxis; usefulness now reduced due to resistance)
- Amodiaquine (Therapy and prophylaxis; usefulness now reduced due to resistance)
- *Sulfadoxine-pyrimethamine* (Therapy; prophylaxis for semi-immune pregnant women in endemic countries as "Intermittent Preventive Treatment" - IPT)
- *Proguanil* (Prophylaxis only)
- *Atovaquone-proguanil*, trade name *Malarone* (Therapy and prophylaxis)
- *Mefloquine*, trade name Lariam (Therapy and prophylaxis)
- Halofantrine (Therapy and prophylaxis)
- *Primaquine* (Therapy in *P. vivax* and *P. ovale* only; not for prophylaxis)
- Quinine (Therapy and prophylaxis)
- Artemisinin-based combination therapies (ACTs)
- *Doxycycline* (Therapy and prophylaxis)
- *Hydroxychloroquine*, trade name Plaquenil (Therapy and prophylaxis)

- *Cotrifazid* (Therapy and prophylaxis)

The individual information of development history and action mechanism for each drug is described in Chapter 3, section 3.1. Among prophylactic drugs that are well studied and are efficacious in suppressive and causal prophylaxis can guide use on disease prevalence and drug-resistance patterns.

- *Suppressive prophylaxis:* Chloroquine, proguanil, mefloquine, and doxycycline are suppressive prophylactics. This means that they are only effective at killing the malaria parasite once it has entered the erythrocytic stage (blood stage) of its life cycle, and therefore have no effect until the liver stage is complete. That is why these prophylactics must continue to be taken for four weeks after leaving the area of risk.
- *Causal prophylaxis:* Causal prophylactics target not only the blood stages of malaria, but the initial liver stage as well. This means that the user can stop taking the drug seven days after leaving the area of risk. Atovaquone-proguanil and primaquine are the only causal prophylactics in current use.

4.1.2.2. The Life Cycle of Plasmodium and Action Mode of Chemoprophylaxis

Four species of the protozoan parasite plasmodium infect humans. *P. falciparum* can cause a lethal infection, whereas *P. vivax*, *P. malariae*, and *P. ovale* cause milder but nonetheless debilitating acute disease. *P. vivax* and *P. falciparum* are the most abundant species; *P. ovale* is the rarest. Transmission occurs when an infected female anopheline mosquito injects sporozoites of plasmodium from its salivary glands into a human host, while taking a blood meal. The injected sporozoites rapidly circulate to the liver (30 minutes), where they invade hepatocytes (the exoerythrocytic stage) multiplying asexually and asymptomatically for a period of 6–15 days. After one to two weeks of incubation in the liver, the parasites transform and emerge as merozoites that then invade erythrocytes. The parasites replicate asexually within the erythrocytes and then burst out to infect other erythrocytes. This intraerythrocytic cycle gives rise to fever, the clinical hallmark of malaria [181]. Erythrocytes infected with *P. falciparum* adhere to post capillary venules. If such sequestration of infected erythrocytes is extensive, tissue blood flow and oxygenation may be severely reduced and cause potentially fatal organ dysfunction (such as renal failure and encephalopathy).

In *P. vivax* and *P. ovale* malaria, some of the liver-stage parasites remain dormant. Within the red blood cells the parasites multiply further, again asexually, periodically breaking out of their hosts to invade fresh red blood cells. Several such amplification cycles occur. Thus, classical descriptions of waves of fever arise from simultaneous waves of merozoites escaping and infecting red blood cells. The activation of these dormant forms (called hypnozoites) can give rise to relapses weeks, months, or years after the initial infection. *P. falciparum* and *P. malariae* do not form hypnozoites and do not cause relapses. The life cycle is completed when a vector anopheline mosquito ingests host erythrocytes that contain gametocytes, the sexual forms that arise from a subpopulation of merozoites. In the mosquito gut, male and female gametes emerge from the gametocytes and fuse to form the zygote. Subsequent stages occur in the gut wall of the mosquito. After about two weeks, sporozoites migrate from this site to the salivary glands, poised to infect a susceptible host.

Prophylactic drugs can prevent malaria infection and help bring about a reduction in malaria transmission by their effect on parasite infectivity. This can be a direct effect on the gametocytes, the infective stages found in human infections (gametocytocidal effect) or, when the drug is taken up in the blood meal of the mosquito, an effect on the parasite's development in the insect (sporonticidal effect) (Figure 9). Chloroquine acts against young gametocytes but has no suppressive effects on mature infective forms. Chloroquine has even been shown to be capable of enhancing the infectivity of gametocytes to the mosquito. In contrast, sulfadoxine-pyrimethamine increases gametocyte carriage but, provided there is no resistance, reduces the infectivity of gametocytes to mosquitoes. Artemisinins are the most potent gametocytocidal drugs among those currently being used to treat an asexual blood infection. They destroy immature gametocytes, preventing new infective gametocytes from entering the circulation, but their effects on mature gametocytes are less [170].

Antimalarial drugs act in tissue schizonticides for causal prophylaxis: these drugs act on the primary tissue forms of the plasmodia which after growth within the liver, initiate the erythrocytic stage. By blocking this stage, further development of the infection can be theoretically prevented. Pyrimethamine and Primaquine have this activity. However, since it is impossible to predict the infection before clinical symptoms begin, this mode of therapy is more theoretical than practical. Also these drugs affect tissue schizonticides for preventing relapse: these drugs act on the hypnozoites of *P. vivax* and *P. ovale* in the liver that cause relapse of symptoms on reactivation. Primaquine is the prototype drug; pyrimethamine also has such activity.

4.1.2.3. Chemoprophylaxis Regimen

No currently available regimen of chemoprophylaxis against malaria is completely effective, and drug resistance continues to evolve [121]. Accordingly, careful attention to avoiding contact with mosquitoes is an additional, essential facet of malaria prevention. The immediate challenges are to improve dissemination of health advice before travel begins, to enhance compliance, and to maintain vigilance for chemoprophylactic failures and to document them carefully. A clinician discussing malaria prevention should weigh the risk of malaria with the risk of adverse reactions to antimalarials. The risk depends on host factors, the place that will be visited, the duration of the visit, the degree of exposure and the level and type of drug resistance. The risk of acquiring malaria if prophylaxis is not used, extrapolated from morbidity data, after one month of travel in West Africa is estimated to be 1:40 and in East Africa is estimated to be 1:70 [286]. Business travelers may have a higher risk for malaria because of decreased compliance, despite being well informed about the malaria risk. Antimalarial drugs should be purchased before travel; drugs purchased overseas may not be manufactured according to developed countries standards and may not be effective [287]. They may also be dangerous, contain the wrong drug or an incorrect amount of active drug, be contaminated or simply be fake.

Medication should be started before going abroad (1–2 days for atovaquone/proguanil one week for chloroquine/proguanil). When mefloquine is used, it should be started two and a half weeks before travelling so that if adverse effects occur there will be time to switch to another medication. Medications should be continued throughout the stay in the malarious area and be discontinued four weeks after leaving it (atovaquone/proguanil should be continued for a week after leaving) [288]. The need for compliance with the regimen should be emphasized. Although lack of compliance appears to be only partly responsible for acquisition of malaria, it is more frequently noted in patients who acquired malaria compared to controls [289]. Formal controlled studies on the subject have not been performed. A study from Austria revealed that only 60% of patients completed the full course of prophylaxis [290]. Medications should be taken after meals, with water. Doxycycline should not be taken lying down. Oral typhoid and oral cholera vaccines and intradermal rabies vaccination are recommended to be given before starting antimalarial prophylaxis, although the subject warrants further research [291]. The usual regimens for chemoprophylaxis for adults (Table 30) and children (Table 31) are currently recommended [292].

Table 30. Malaria chemoprophylaxis for adults [292]

Drug	Usage	Adult dose
Atovaquone/proguanil	Areas with chloroquine-resistant or mefloquine-resistant *P. falciparum*	250 mg atovaquone and 100 mg proguanil hydrochloride. 1 tablet orally, daily
Chloroquine phosphate	Areas with chloroquine-sensitive *P. falciparum*	300 mg base (500 mg salt) orally, once/week. This is the equivalent of two tablets orally weekly (one tablet has 150 mg of base)
Doxycycline	Areas with chloroquine-resistant or mefloquine resistant *P. falciparum*	100 mg orally, daily (tablet or capsule)
Chloroquine PLUS proguanil	Areas with little chloroquine resistance.	2 tablets weekly (150 mg base/tablet) PLUS 2 tablets daily (100 mg/tablet)
Hydroxychloroquine sulfate	An alternative to chloroquine for primary prophylaxis only in areas with chloroquine-sensitive *P. falciparum*	310 mg base (400 mg salt) orally, once/week
Mefloquine	Areas with chloroquine-resistant *P. Falciparum*	228 mg base (250 mg salt) orally, once/week
Primaquine	An option for primary prophylaxis in special circumstances.	30 mg base (52.6 mg salt) orally, daily
Primaquine	Used to decrease risk of relapses of *P. vivax* and *P. ovale*	30 mg base (52.6 mg salt) orally, once/day for 14 days after departure from the malarious area.

Table 31. Malaria prophylaxis for children [292]

Drug	Pediatric dose
Atovaquone/proguanil	Pediatric tablets contain 62.5 mg atovaquone and 25 mg proguanil hydrochloride. 11–20 kg: 1 tablet; 21–30 kg: 2 tablets; 31–40 kg: 3 tablets; P40 kg: 1 **adult tablet daily **Adult tablets contain 250 mg atovaquone and 100 mg proguanil hydrochloride.
Chloroquine phosphate	5 mg/kg base (8.3 mg/kg salt) orally, once/week, up to maximum adult dose of 300 mg base. Doses if the chloroquine syrup is used are: under 4.5 kg, 2.5 ml, 4.5 to 7.9 kg 5.0 ml, 8.0 to 10.9 kg 7.5 ml, 11.0 to 14.9 kg 10 ml, 15.0 to 16.5 kg 12.5 ml.
Doxycycline	$\geq$ 8 years of age: 2 mg/kg up to adult dose of 100 mg/day (Health Protection Agency Advisory Committee on Malaria Prevention for UK Travellers recommends doxycycline starting at 12 years of age).
Hydroxychloroquine sulfate	5 mg/kg base (6.5 mg/kg salt) orally, once/week, up to maximum adult dose of 310 mg base.
Mefloquine	**5–10 kg: 1/8 tablet orally, once/week; **10–20 kg: 1/4 tablet once/week 20–30 kg: 1/2 tablet, once/week; 30–45 kg: 3/4 tablet once/week >45 kg: 1 tablet, once/week **The recommended dose of mefloquine is 5 mg/kg body weight once weekly. Approximate tablet fraction is based on this dosage. Exact doses for children weighing less than 10 kg should be prepared by a pharmacist.
Chloroquine PLUS Proguanil	Chloroquine phosphate 5 mg/kg base (8.3 mg/kg salt) orally, once/week, up to maximum adult dose of 300 mg base PLUS Proguanil (<6 kg, 0.125 dose = 1/4 tablet, 6.0 to 9.9 kg 0.25 dose = 1/2 tablet, 10.0 to 15.9 kg 0.375 dose = 3/4 tablet, 16 to 24.9 kg 0.5 dose = 1 tablet, 25.0 to 44.9 kg 0.75 dose = 1 1/2 tablets, weight 45 kg and over = adult dose).
Primaquine	0.6 mg/kg base (1.0 mg/kg salt) up to adult dose, orally, daily (for primary prophylaxis).
Primaquine	0.6 mg/kg base (1.0 mg/kg salt) up to adult dose orally, once/day for 14 days after departure from the malarious area.

The prophylactic regimen, preferably, should be started 1-2 weeks prior to travel to a malarious area. In addition to assuring adequate blood levels of the drug, this regimen allows for evaluation of any potential side-effects. Chemoprophylaxis should continue during the stay in malarious area and for 1-4 weeks after departure from the area. The following factors should be considered while choosing an appropriate chemoprophylactic regimen: 1) The travel itinerary should be reviewed in detail and compared with the information on areas of risk within a given country to determine whether the traveler will actually be at risk of acquiring malaria; 2) The risk of acquiring chloroquine resistant *P. falciparum* malaria

(CRPF) is another consideration, and 3) Any previous allergic or other reaction to the antimalarial drug of choice and the accessibility of medical care during travel must also be determined.

The following regimens are recommended by the WHO, UK HPA, and CDC:

- chloroquine 300 to 310 mg once weekly, and proguanil 200 mg once daily (started one week before travel, and continued for four weeks after returning);
- doxycycline 100 mg once daily (started one day before travel, and continued for four weeks after returning);
- mefloquine 228 to 250 mg once weekly (started two-and-a-half weeks before travel, and continued for four weeks after returning);
- Malarone® 1 tablet daily (started one day before travel, and continued for 1 week after returning).

What regimen is appropriate depends on the country or region travelled to. This information is available from the UK HPA, WHO or CDC (links are given later in this chapter). Doses depend also on what is available (e.g., in the US, mefloquine tablets contain 228 mg base, but 250 mg base in the UK). The data is constantly changing and no general advice is possible. The previously described dosage regimen given here are appropriate for adults and children aged 12 and over.

Other chemoprophylactic regimens that are also available:

- Dapsone 100 mg and pyrimethamine 12.5 mg once weekly (available as a combination tablet called Maloprim®;
- Primaquine 30 mg once daily (started the day before travel and continuing for seven days after returning): this regimen is not routinely recommended because of the need for G-6-PD testing prior to starting primaquine (see the article on primaquine for more information).
- Quinine sulphate 300 to 325 mg once daily: this regimen is effective but not routinely used because of the unpleasant side effects of quinine.

Notes on the prevention regimens [292]:

- Primaquine-induced gastrointestinal disturbances can be minimized if the drug is taken with food. Neither primaquine nor tafenoquine should be given to persons with glucose-6-phosphate dehydrogenase deficiency, to avoid the development of potentially severe drug-induced hemolysis. Tafenoquine is an analogue of primaquine that is more potent than the parent drug. Primaquine primary prophylaxis should begin 1–2 days before travel to the malaria-risk area. It should be continued once a day, at the same time each day, while in the malaria-risk area, and daily for 7 days after leaving the malaria-risk area.
- Chloroquine plus Proguanil for areas with little Chloroquine resistance: the regimen is recommended by the Health Protection Agency Advisory Committee on Malaria Prevention for UK Travelers. It is not recommended by the Centers of Disease Control because of concerns regarding reduced efficacy [289].

- There is less evidence for the effectiveness of hydroxychloroquine sulfate as an antimalarial drug. It cannot be used in areas where chloroquine resistance has been reported.
- Most of the medications taste bitter: parents should mix children's doses with something sweet like applesauce and give on a full stomach to minimize side-effects.
- Epileptic patients should not take mefloquine or chloroquine. Proguanil (200 mg daily) is recommended for areas without chloroquine resistance and doxycycline or atovaquone/proguanil is recommended for areas with chloroquine resistance.
- Patients with renal failure: proguanil pharmacokinetics is affected by renal failure. Proguanil can be substituted or the dose can be reduced. Dose regimen should be consulted by physician. In severe renal failure, there is a high risk for hematological toxicity.
- In patients with severe liver failure, antimalarials should be avoided. Doxycycline and mefloquine are excreted exclusively through the liver so, in practice, proguanil, chloroquine or atovaquone/proguanil are preferred in mild liver failure.
- Fansidar (500 mg of long-acting sulfadoxine and 25 mg of pyrimethamine in a single tablet once a week) has caused severe mucocutaneous reactions even after a few doses and is no longer recommended

Table 32. Medications that should not be used for malaria chemoprophylaxis [284]

Generic (trade) names	Severe adverse event or contraindication observed
Amodiaquine	May cause agranulocytosis and hepatitis
Pyrimethamine/sulfadoxine (Fansidar, Hoffman-LaRoche)Pyrimethamine/dapsone (Maloprim, Glaxo Wellcome)	Severe skin reactions (e.g., Stevens-Johnson syndrome) are seen in some persons who take the drug repeatedly. Agranulocytosis may occur if increased doses are given over time.
Quinine	Chronic (long-term) use quinine predisposes individuals to develop black-water fever
Azithromycin (Zithromax, Pfizer)	Efficacy of the drug in non-immune persons is insufficient to recommend its use as a single drug for prevention.
Artemisinin derivatives (i.e., arteether)	Severe CNS lesions in animal studies argues against any long-term use of these short-acting treatment drugs
Halofantrine	Cardiac dysrhythmias are seen when the drug is used in increased dosage or when increased drug absorption occurs.

4.1.3. Drugs that Should not be Used for Chemoprophylaxis

Chemoprophylaxis is not recommended for residents of endemic area.

Some very useful antimalarial treatment drugs have been tried for malaria prevention and have no rational use as malaria chemoprophylactic agents. Table 32 shows a brief description of antimalarial drugs that, because of adverse events, lack of efficacy, or unsuitable pharmacology, should not be used for malaria prevention [284]. Although azithromycin and doxycycline has been tested for malaria chemoprophylaxis, it has insufficient efficacy against *P. falciparum* to recommend its use for malaria prevention [293, 294].

4.1.4. Managements for Preventing Malaria

The primary consideration today in selecting a therapeutic agent to prevent malaria is the risk of exposure to chloroquine-resistant or multidrug-resistant *P. falciparum* in the traveler's itinerary. In previous years, when most *Plasmodium* strains were predictably sensitive, chloroquine was the drug of choice to both prevent and treat malaria. Chloroquine and other antimalarial drugs have been used in various ways to prevent malaria in the resident populations of endemic areas for a long history. The primary aim of most early studies was to interrupt transmission. This was rarely achieved, but administration of antimalarial drugs either through medication of salt or by mass administration frequently led to a marked reduction in the prevalence of malaria infection and in the incidence of clinical attacks. Chemoprophylaxis is highly effective in reducing mortality and morbidity from malaria in young children and pregnant women living in endemic areas, but is difficult to sustain and, in some studies, has impaired the development of naturally acquired immunity. Intermittent preventive treatment, in which full therapeutic doses of a drug are given at defined intervals, has the potential to provide some of the benefits of sustained chemoprophylaxis in pregnant women and young children without some of its drawbacks and is a promising new approach to malaria control (Table 9) [115].

4.1.4.1. Reduction of Malaria Transmission

1) Treatment of Clinical Cases

The idea that treatment of clinical cases with an effective antimalarial would reduce transmission and thus provide indirect protection to the population was in the early years of the 20th century. Quinine was used extensively in Italy and elsewhere during the first part of the 20th century, partly with this aim in mind. However, quinine has little or no effect on gametocytes, and then treatment of clinical cases with a drug, such as an artemisinin that kills gametocytes as well as asexual parasites, is more likely to be effective. Recent experience in Southeast Asia and in South Africa supports this view. On the Thai-Burmese border, replacement of mefloquine with mefloquine-artesunate as first line treatment of patients with symptomatic malaria was associated with a substantial reduction in the incidence of *P. falciparum* infection, and widespread use of artemisinins may have contributed to a marked decline in the overall incidence of malaria in Vietnam. In South Africa, replacement of ineffective sulfadoxine-pyrimethamine (SP) with an artemisinin-based combination therapy

(ACT) was again associated with a rapid decrease in the overall incidence of *falciparum* malaria. However, in Vietnam and in South Africa the introduction of artemisinins was associated with the introduction of other malaria control measures: insecticide treated nets (ITNs) in Vietnam and household residual spraying in South Africa. Thus, although these observations strongly suggest that introduction of ACTs for treatment of clinical cases had an effect on the overall level of malaria transmission in these low transmission areas, this has not been proven definitively [115].

2) Mass Drug Administration

In the highly endemic areas of Africa, in contrast to the situation in Southeast Asia and South Africa, most malaria infections are asymptomatic and go untreated, even when quite high levels of parasitemia are present. In such situations, a significant reduction in the transmission of malaria by administration of a gametocidal drug is likely to be achieved only if asymptomatic as well as symptomatic subjects are treated. Detecting the latter requires a major investment in surveillance so the concept of mass drug administration (MDA) was developed. During an MDA, the whole population of a community known to contain a number of asymptomatic subjects is treated without determining who is infected.

Antimalarials were distributed widely in endemic areas of Italy during the 1920s and the 1930s, partly as personal prophylaxis but also to reduce transmission. Increased consumption of quinine was accompanied by a marked decrease in the incidence of malaria, but it is not certain that this was cause and effect. Quinine was also used extensively by Gorgas during the construction of the Panama Canal, in part as personal prophylaxis for canal workers, but also to reduce transmission. The use of MDAs to reduce transmission became a more promising approach to malaria control following the discovery of the 8-aminoquinolines, which are gametocidal. During the 1960 s and 1970s, a number of trials of MDA were undertaken in Africa and Asia with the primary aim of interrupting transmission. The largest MDA program reported was conducted in Nicaragua in 1981. A single round of treatment with chloroquine plus primaquine was given to approximately eight million people [295]. There was a marked decrease in the incidence of malaria immediately after mass drug administration, but transmission was not interrupted and the incidence of malaria soon returned to its previous level.

A recent study of MDA was undertaken in The Gambia with the aim of delaying the onset of transmission in an area where malaria is highly seasonal and transmission limited to a few months of the year [296]. Sulfadoxine-pyrimethamine and artesunate were given to the whole of the population of 33 villages at the end of the dry season when gametocyte levels are at their lowest. A reduction in the incidence of malaria compared with that seen in nine control villages where placebo was distributed was observed during the month after the administration of the drug combination, but no overall effect on the incidence of malaria throughout the course of the malaria transmission season was seen.

One success for MDA in achieving the goal of interruption of transmission has been reported recently from the island of Aneityum, Vanuatu [297]. An MDA could also be useful in controlling an epidemic in a situation in which it is anticipated that transmission will decrease spontaneously within a short time; for example, after a period of flooding or a civil disturbance that leads to a short-term refugee situation. If an MDA is contemplated in such a situation, it should be given as soon as possible and the treatment regimen should incorporate a gametocidal drug, preferably an artemisinin combined with a longer acting partner.

3) Medication of Salt

The MDA studies in experimental systems have shown that exposure of malaria parasites to a sub-therapeutic dose of an antimalarial over a prolonged period of time is the optimum way of inducing resistance. Use of medicated salt is likely to have a similar effect and has no place in malaria control. Difficulties in delivery, problems with resistance, and the lack of a safe and highly effective anti-gametocidal drug limit the potential of MDAs in malaria control. The use of medicated salt and the unrestricted use of chloroquine and pyrimethamine, including their use for prophylaxis, in the 1960s are likely to have contributed to the initial emergence and spread of resistance to these drugs [298].

4.1.4.2. Prevention of Mortality and Morbidity from Malaria (Chemoprophylaxis)

Protection of the indigenous populations of malaria-endemic countries from malaria by regular administration of anti-malarial drugs has never found favor with the international malaria control community, although the operators of many mines and plantations have found it cost-effective to protect their work force and their families from malaria in this way. Nevertheless, when chemoprophylaxis has been evaluated in this kind of situation, it has almost always been found to be highly effective in reducing morbidity and mortality from malaria, especially when targeted at high risk groups such as young children and pregnant women. (Chapter 4, Sections 4.2 and 4.3.)

The use of targeted chemoprophylaxis is considered by following reasons.

1) Sustainability: Sustaining compliance with prophylaxis is a difficult but not an insurmountable challenge. In this community, chemoprophylaxis was given by village health workers who received no financial support from the public health services and limited help from their communities. In most malaria-endemic countries, a high proportion of pregnant women attend an antenatal clinic at least once during pregnancy, but few attend often enough to allow the distribution of weekly chemoprophylaxis.

2) Safety: Drugs given on a regular basis to large numbers of healthy children or pregnant women must be extremely safe. No serious adverse events attributable to drug administration have been reported during trials of chemoprophylaxis in children, but the possibility that occasional serious adverse effects may have been missed cannot be excluded.

3) Cost: Cost-effectiveness studies have shown that childhood chemoprophylaxis with a drug that costs approximately $0.1 per administration and chemoprophylaxis in pregnancy are both highly cost effective. However, donor funds might be needed to introduce chemoprophylaxis on a national scale.

4) Impairment of the development of natural immunity: A major concern over the use of chemoprophylaxis in young children has been a fear that this will impair the development of natural immunity.

5) Induction of drug resistance: It would be anticipated that widespread deployment of chemoprophylaxis in children and/or pregnant women would increase the rate of spread of drug resistant parasites, the extent of the effect being related to the relative proportion of drug used for prophylaxis compared with the overall use of the drug in

the community. Chemoprophylaxis will inevitably lead to an increase in drug pressure, but this may be an acceptable price to pay if the benefits are substantial and it may be possible to reduce this risk by the use of combination therapy.

6) Role of chemoprophylaxis in current malaria control programs: The effectiveness of chemoprophylaxis in reducing mortality and morbidity from malaria in young children and in pregnant women has been demonstrated convincingly on many occasions, and the difficulties of implementing chemoprophylaxis and its potential risks have probably been exaggerated. Constraints on the deployment of chemoprophylaxis could be overcome in most cases using the increased financial resources that are now available for malaria control. However, increasing drug pressure, impairment of the development of natural immunity and difficulty in identifying a drug that is safe and effective remain problems for this approach to malaria control. Intermittent preventive treatment, which provides some of the benefits of chemoprophylaxis while reducing some of its risks, may be a more promising way forward [115].

4.1.4.3. Intermittent Preventive Treatment

Intermittent preventive treatment (IPT) describes the administration of a full therapeutic course of an antimalarial to at risk subjects at specified times regardless of whether they are infected. IPT differs from chemoprophylaxis, which aims to sustain blood levels above the mean inhibitory concentration for a prolonged period, in producing protective drug concentrations for only short periods of time separated by periods when drug concentrations are below the level necessary to inhibit parasite growth. Because SP is the drug that has been used most widely for IPT in both pregnant women and children, it is not known whether IPT achieves its effect primarily by elimination of parasites or through the long-acting, prophylactic effect of SP. The IPT in pregnant women, in infants and children were reported in Chapter 4, Sections 4.2 and 4.3.

The use of drugs to prevent malaria in the population of malaria-endemic areas has a long history. It has progressed from mass drug administration involving the whole population through sustained chemoprophylaxis targeted at risk groups to IPT in which drug administration is reduced to the minimal level required to achieve a useful protective effect. The use of anti-malarial drugs to prevent malaria will always involve balancing the level of protection achieved with the cost of the drug, its side-effects, and its impact on the development of natural immunity and drug resistance. If the initial results of studies in infants and children are confirmed, the IPT approach may offer the optimum balance that can be achieved [115].

4.2. Chemoprophylaxis and IPT
of Malaria in Pregnancy

Malaria is a parasitic infection that causes significant morbidity and mortality worldwide, with more than 500 million people becoming severely ill every year [283]. For pregnant women, malarial infection can be severe, with high fevers, chills, and anemia leading to increased risk of poor maternal and fetal outcomes - including death. Pregnant women are also more likely to become infected and to develop more severe disease - they attract twice as many mosquitoes as non-pregnant women and have a relative immuno-suppression [260, 299].

4.2.1. Guiding Principles of Chemoprophylaxis and IPT during Pregnancy

The World Health Organization (WHO) recommends pregnant women avoid travel to malarial regions. If travel is required, WHO recommends chloroquine as first-line prophylaxis in pregnancy (plus proguanil if the region exhibits emerging chloroquine resistance). In areas with proven chloroquine resistance, mefloquine is the drug of choice. Other antimalarials - such as quinine, pyrimethamine, sulfadoxine, and artesunate - should not be withheld if the preferred drugs are not available, or if the infection is life-threatening [294].

The Centers for Disease Control and Prevention (CDC) also recommends avoiding travel to malaria-endemic regions during pregnancy, but if travel is necessary, the CDC advises use of chloroquine (or mefloquine in regions with chloroquine resistance). The CDC discourages the use of atovaquone/proguanil, doxycycline, and primaquine, due to known adverse fetal effects or inadequate experience in pregnancy [294]

Malaria infection in pregnancy poses a substantial risk to the mother, the fetus and the newborn infant. Pregnant women are less capable of coping with and clearing malaria infections. In areas of low transmission of *P. falciparum*, where levels of acquired immunity are low, women are susceptible to attacks of severe malaria, which may result in stillbirths or spontaneous abortions, or the death of the mother. In areas of high *P. falciparum* transmission, levels of acquired immunity tend to be high and women may have asymptomatic infections, which may result in maternal anemia and placental parasitemia. Both of these conditions can lead to low birth weight, an important contributor to neonatal mortality.

In programs for the prevention or treatment of malaria in pregnant women, two major issues are the safety and effectiveness of the antimalarial drug regimen. The programmatic effectiveness of a given drug is determined by the efficacy of that drug against the parasite and by the drug's characteristics, including affordability, availability, acceptability to the target population, and deliverability in terms of dosing requirements and incorporation into existing antenatal care delivery systems.

Weekly chloroquine chemoprophylaxis and preventive intermittent treatment with sulfadoxine-pyrimethamine during pregnancy have both been shown to reduce the rate of placental parasitemia and low birth weight [300]. Many national antimalarial treatment policies include a recommendation for chloroquine chemoprophylaxis. It is rarely implemented, however, because of problems with compliance; fears about adverse effects of the drug during pregnancy, and the concern of health workers that use of drugs for this purpose may deplete stocks needed for the treatment of acute infections.

An increasing number of countries, e.g., Malawi, are implementing intermittent treatment with 2- or 3-dose treatment regimens of sulfadoxine-pyrimethamine: once in the second and once or twice in the third trimester [301, 302] to prevent malaria in pregnancy. In HIV-positive women, the 3-dose treatment is significantly more efficacious than the 2-dose regimen [301]. It has been suggested that sulfadoxine-pyrimethamine should ideally be reserved for preventive intermittent treatment in pregnancy. In many endemic areas where sulfadoxine-pyrimethamine is one of the few replacement therapies for chloroquine, this may not be a viable option.

4.2.2. Chemoprophylaxis during Pregnancy and Lactation

Malaria infection in pregnant women may be more severe than in non-pregnant women. Women who live in areas with endemic malaria and who are pregnant for the first or second time are more likely to be infected with *P. falciparum* malaria than non-pregnant women of a similar age. In addition, the risk of adverse pregnancy outcomes, including prematurity, abortion, and stillbirth, may be increased. For these reasons, and because chloroquine has not been found to have any harmful effects on the fetus when used in the recommended doses for malaria prophylaxis, pregnancy is not a contra-indication to malaria prophylaxis with chloroquine or hydroxychloroquine (CRPF). However, because no chemoprophylactic regimen is completely effective in areas with CRPF, women who are pregnant or likely to become so should avoid travel to such areas.

Chloroquine and Proguanil are the preferred chemoprophylactic drugs against malaria in the first 3 months of pregnancy. Mefloquine can be given during the second and third trimesters if the situation demands. Mefloquine and doxycycline can be used in non-pregnant women with child bearing potential, but pregnancy should be avoided for 3 months after mefloquine use and for one week after doxycycline use. However, in case of unplanned pregnancy, malaria chemoprophylaxis is not considered an indication for termination of pregnancy. Doxycycline, a tetracycline, is generally contra-indicated for malaria prophylaxis during pregnancy. Adverse effects of tetracyclines on the fetus include discoloration and severe dysplasia of the teeth and inhibition of bone growth. In pregnancy, therefore, tetracyclines would be indicated only if required to treat life-threatening infections due to multi-drug resistant *P. falciparum*. Primaquine should not be used during pregnancy because the drug may be passed transplacentally to a G6PD-deficient fetus and cause life-threatening hemolytic anemia *in utero*. Whenever radical cure or terminal prophylaxis with primaquine is indicated, chloroquine should be given once a week until delivery, at which time the decision to give primaquine may be made.

This infection contributes to antenatal anemia and slows fetal growth, that may harm the mother and baby. Drugs have been widely used to prevent infection or its consequences. From the 1980s, prophylaxis to prevent, suppress, or eradicate malaria parasites with a variety of drugs has been tested. One of the main problems in evaluating the effects of routine antimalarial drugs on the mother and the infant is identifying meaningful outcomes in terms of the direct impact on mother or baby. This review has identified two pragmatic outcome measures:

- Severe antenatal anemia: there is a strong association between anemia and poor maternal outcomes, so preventing severe anemia is likely to be of benefit.
- Perinatal death: preventing death of the fetus and in the first week of life is one of the main reasons for giving malaria prophylaxis.

The first reported trial of chemoprophylaxis in pregnant women was undertaken in a mission hospital in western Nigeria in 1964. This showed that chemoprophylaxis with pyrimethamine led to a substantial increase in birth weight in primigravidae and in grand-multiparous women. Subsequent studies, conducted in several countries in Africa, have confirmed the effect of chemoprophylaxis on birth weight and shown that it also results in an increase in maternal hemoglobin levels. Most studies have shown that the protective effect of chemoprophylaxis is most marked in first and second pregnancies. Drugs that have been used for chemoprophylaxis in pregnancy include chloroquine, pyrimethamine, proguanil, Maloprim®, and mefloquine. On the basis of the result of trials of chemoprophylaxis conducted in several countries in Africa, the recommendation of the World Health Organization was, until recently, that all pregnant women resident in areas of moderate or high malaria transmission should receive chemoprophylaxis with chloroquine throughout the second and third trimesters of pregnancy. However, this intervention has rarely been implemented on any significant scale [115].

Although prophylaxis for pregnant patients traveling to malarial regions is a public concern, data for decision-making must be extrapolated from the available evidence, which is based primarily on women living in endemic areas. In a Cochrane systematic review, six trials (513 participants) met the inclusion criteria. Two were quasi-randomized, and none described allocation concealment. Data were scarce for the primary outcome, treatment failure. Antimalarial drugs were found to decrease the incidence of maternal infections (relative risk [RR] = 0.27; 95% confidence interval [CI], 0.17–0.44) and reduce maternal anemia (RR = 0.62; 95% CI, 0.50–0.78) in low-parity women - i.e., during a first or second pregnancy [260].

Another Cochrane systematic review indicated that sixteen trials (12,638 participants) met the inclusion criteria; two used adequate methods to conceal allocation. Antimalarials reduced antenatal parasitemia when given to all pregnant women (RR 0.53, 95% CI 0.33–0.86; 328 participants, 2 trials), placental malaria (RR 0.34, 95% CI 0.26–0.45; 1236 participants, 3 trials), but no effect was detected with perinatal deaths (2890 participants, 4 trials). Proguanil performed better than chloroquine in one trial of women of all parities in relation to maternal fever episodes. Sulfadoxine-pyrimethamine performed better than chloroquine in two trials of low-parity women. In low-parity women, these drugs were also found to decrease perinatal death (RR=0.73; 95% CI, 0.53–0.99) and low birth weight (RR = 0.57; 95% CI, 0.46–0.72) associated with malarial infection. When used in all parity groups, antimalarials were somewhat less effective, yet still reduced maternal infections (RR = 0.53; 95% CI, 0.33–0.86); the effects were similar with all antimalarials tested. In conclusion, the chemoprophylaxis reduces antenatal parasite prevalence and placental malaria when given to women in all parity groups. They also have positive effects on birth-weight and possibly on perinatal death in low-parity women [303].

Very small amounts of chloroquine and mefloquine are excreted in breast milk and the amount of drug is not sufficient to harm the infant nor is the quantity sufficient to protect the child from malaria. Breastfeeding infants should receive weight appropriate prophylaxis. Very limited data are available on the use of doxycycline in lactating women. It is generally thought to be safe. Primaquine should only be given to lactating women if both the woman and her infant have been tested for G6PD deficiency and have normal G6PD levels documented. Because safety data is not yet available, atovaquone/proguanil is not currently recommended for women breastfeeding infants <11 kg [292].

4.2.3. Intermittent Prevention Treatment in Pregnancy (IPTp)

From the 1990s, a modification called intermittent preventive treatment (IPT) had been used, where women are treated for malaria presumptively at fixed times during the pregnancy, usually drugs with a long half life, such as sulfadoxine-pyrimethamine. There remains debate as to whether the mechanism of IPT actually differs a lot from prophylaxis [304], but the regimens are very different. IPT requires just two or three doses during pregnancy, compared to prophylaxis regimens that may be daily (e.g., with proguanil) or weekly (e.g., with chloroquine). Prophylaxis and IPT are in addition to good care during pregnancy, which includes prompt treatment of women when they are clinically with fever or anemia [260].

The research and policy question addressed in this review is whether giving drugs on a regular basis (prophylaxis or IPT) to prevent malaria has additional advantages in terms of health outcomes over prompt, appropriate treatment. The advantages with women taking regular, routine antimalarial drugs are that the drug will treat or suppress parasites in the blood and reduce the chances of illness or anemia developing. In turn, this may ultimately benefit the fetus and impact on birth weight and long-term survival. The disadvantages of routine use of antimalarial drugs are that many drugs have adverse effects and that widespread use may contribute to the malaria parasite developing resistance to these drugs; and it requires health systems to help ensure their implementation. The question is first addressed in terms of all pregnant women, and then secondly in women of low parity, where the effects of malaria are more marked.

Because of the difficulty in sustaining chemoprophylaxis or IPT in pregnant women, Schultz and others working in Malawi compared the effects of treatment with a full course of SP given twice during pregnancy with the effects of weekly chemoprophylaxis with chloroquine [305]. The prevalence of low birth weight in the infants of women who received IPT with SP was significantly lower than that in women who received chemoprophylaxis with chloroquine. The protective effect against low birth weight of IPT with SP has subsequently been confirmed in larger effectiveness studies in Malawi, even when only one dose was given [306]. In Kenya, a controlled trial showed that IPT with SP given two or three times during pregnancy reduced the prevalence of severe anemia in primigravidae and secundigravidae [302]. However, a study in another area of Kenya showed that women infected with human immunodeficiency virus (HIV) required more than two or three doses of IPT to prevent infection of the placenta [301].

On the basis of these rather limited results, the World Health Organization now recommends that IPT with SP, given at each antenatal clinic attendance after quickening, should replace chemoprophylaxis as the preferred chemotherapeutic method for the prevention of malaria in pregnancy. However, there are still a number of uncertainties over the use of IPTp. There are no data on the effectiveness of IPTp in areas of low endemicity or in areas where malaria transmission is highly seasonal. Because the mode of action of SP when used for IPTp is uncertain, it is not known whether in the increasing number of areas of Africa where SP is losing its efficacy, SP should be replaced with a long-acting drug such as mefloquine or whether a short acting drug such as Lapdap® (proguanil plus dapsone, GlaxoSmithKline) would be as effective. The optimum approach to IPTp in women who are HIV-positive is not known. Thus, more research is needed if the full potential of this approach to the control of malaria in pregnancy is to be obtained [115].

Cochrane systematic review showed that sixteen trials (12,638 participants) met the inclusion criteria; two used adequate methods to conceal allocation. Antimalarials reduced antenatal parasitemia when given to all pregnant women (RR 0.53, 95% CI 0.33–0.86; 328 participants, 2 trials), placental malaria (RR 0.34, 95% CI 0.26–0.45; 1236 participants, 3 trials), but no effect was detected with perinatal deaths (2890 participants, 4 trials). In women in their first or second pregnancy, antimalarial drugs reduced severe antenatal anemia (RR 0.62, 95% CI 0.50–0.78; 2809 participants, 1 prophylaxis and 2 IPT trials), antenatal parasitemia (RR 0.27, 95% CI 0.17–0.44, random-effects model; 2906 participants, 6 trials), and perinatal deaths (RR 0.73, 95%CI 0.53–0.99; 1986 participants, 2 prophylaxis and 1 IPT trial; mean birth-weight was higher (WMD 126.70 g, 95% CI 88.64–164.75 g; 2648 participants, 8 trials), and low birth-weight less frequent (RR 0.57, 95% CI 0.46– 0.72; 2350 participants, 6 trials). The IPTs reduce antenatal parasite prevalence and placental malaria when given to women in all parity groups. They also have positive effects on birth-weight and possibly on perinatal death in low-parity women [303].

Falciparum malaria is an important cause of maternal, perinatal and neonatal morbidity in high transmission settings in Sub-Saharan Africa. IPT with sulphadoxine-pyrimethamine (SP-IPT) has proven efficacious in reducing the burden of pregnancy-associated malaria but increasing levels of parasite resistance mean that the benefits of national SP-IPT programs may soon be seriously undermined in much of the region. Hence, there is an urgent need to develop alternative drug regimens for IPT in pregnancy.

**Table 33. Antimalarials for prophylaxis: Chloroquine, mefloquine
are best choices during pregnancy [170]**

Drugs	Efficacy	Safety	Pregnancy Class	Availability
Chloroquine	Good	Excellent	C	Worldwide
Chloroquine/ proguanil	Good	Excellent	C	Worldwide
Mefloquine	Excellent	Good	C	Worldwide
Quinine	Excellent	Good	C[1]	Worldwide
Atovaquone/ proguanil	Excellent	Good	C (poorly studied)	Worldwide
Artesunate	Excellent	Good	N/A	Asia, Africa, limited in UK, not in US
Primaquine	Good	Fair	C[2]	Worldwide
Doxycycline	Excellent	Fair	D (teratogenic)	Worldwide
Sulfadoxine/ pyrimethamine	Fair	Poor	C	Worldwide, but restricted in US

* Prescribers and patients are urged to refer to the CDC reference about pregnancy in malaria (*wwwn.cdc.gov/travel/contentMalariaPregnantPublic.aspx*) and to specific country information regarding sensitivities of malaria (*wwwn.cdc.gov/travel/destinationList.aspx*).

** Pregnancy class C: Animal reproduction studies have shown an adverse effect on the fetus and there are no adequate and well-controlled studies in humans, but potential benefits may warrant use of the drug in pregnant women, despite potential risks.

[1] Monitor patients for maternal hypoglycemia.

[2] There is no evidence of teratogenicity, but primaquine is associated with fetal intravascular hemolysis.

4.2.4. The most Safe Malaria Prophylaxis during Pregnancy

Chloroquine and mefloquine have superior safety profiles in pregnancy, though all antimalarials are effective for prophylaxis (Table 33) even though their widespread resistance in the most regions of Asia and Africa. Antimalarials will decrease the severity of maternal malaria infection and malaria-associated anemia, while decreasing the incidence of low birth weight and perinatal death in women having their first or second baby. Chloroquine and mefloquine are the safest antimalarials for use in pregnant women, but personal protection measures are also critical.

While chemoprophylaxis in pregnancy appears efficacious, a major question remains - which agents are safest for both the woman and fetus? Some drugs routinely used in non-pregnant individuals should not be offered to pregnant women because of known direct effects on the fetus. Doxycycline is teratogenic, and primaquine poses a significant risk of fetal intravascular hemolysis in G6PD-deficient fetuses [307]. Other drugs, such as atovaquone/proguanil and artesunate, are not well studied in pregnancy, and therefore are not recommended for use unless other options are not available [283]. Among drugs that are well studied and without known direct fetal-damaging effects, adverse drug reaction profiles can guide use based on disease prevalence and drug-resistance patterns.

- *Chloroquine* is widely used because it is inexpensive and well tolerated, with only pruritus, mouth ulcers, and gastrointestinal upset as the most common adverse effects.
- *Mefloquine* is usually well tolerated, but can cause dose-related neuropsychiatric effects; it is contra-indicated in those with a history of epilepsy or psychiatric disease.
- *Sulfadoxine* and *pyrimethamine* are not normally used as prophylaxis for any patient, due to the risk of toxic epidermal necrolysis and Stevens-Johnson syndrome, and the possible risk of jaundice and kernicterus if used in the third trimester of pregnancy.
- *Quinine*, which can be used for treatment or prophylaxis, may cause hypoglycemia, an effect that is more pronounced during pregnancy and requires close monitoring of blood glucose levels [263, 307].

Therefore, given these reaction profiles, chloroquine or mefloquine are usually the best choice with their superior safety and efficacy. The recommendations from the World Health Organization (WHO) are consulting the pregnant women to avoid travel to malarial regions. If travel is required, WHO recommends chloroquine as first-line prophylaxis in pregnancy (plus proguanil if the region exhibits emerging chloroquine resistance). In areas with proven chloroquine resistance, mefloquine is the drug of choice. Other antimalarials - such as quinine, pyrimethamine, sulfadoxine, and artesunate - should not be withheld if the preferred drugs are not available, or if the infection is life-threatening [283].

The Centers for Disease Control and Prevention (CDC) also recommends avoiding travel to malaria-endemic regions during pregnancy, but if travel is necessary, the CDC advises use of chloroquine (or mefloquine in regions with chloroquine resistance). The CDC discourages the use of atovaquone/proguanil, doxycycline, and primaquine, due to known adverse fetal effects or inadequate experience in pregnancy [308].

Because of its safety, availability, and low cost, the drug gained widespread use throughout the world, becoming perhaps the second or third most widely consumed drug. Chloroquine continues to be widely used in developing countries despite the presence of chloroquine-resistant strains, a practice promoting drug pressure that selects for further resistance. Also, there is a growing public unease with the use of mefloquine for malaria chemoprophylaxis. Although rare severe neuropsychiatric adverse effects are well described with mefloquine [309], this controversy centered around the premise that mefloquine causes a great deal more serious psychiatric adverse effects than had been indicated by both the medical and pharmaceutical information available. The extent of the ongoing controversy, particularly via the Internet, makes it very difficult to untangle the various claims and counter-claims, especially since litigation is ongoing in several countries. This is important to the physician because of the increasing refusal and noncompliance rates associated with mefloquine prophylaxis. The recent safety and tolerability information on mefloquine is reviewed in Chapter 7, Section 7.1 with a view to inform current prescribing practice.

4.3. Chemoprophylaxis and IPT of Malaria in Children

4.3.1. Guiding Principles of Chemoprophylaxis and IPT on Children

There are no good guidelines for the use of antimalarial drugs in very young infants (< 5kg) because of a lack of relevant studies, although some public health authorities recommend the off-label use of mefloquine. It is important to distinguish between the chemoprophylaxis of children traveling temporarily to a malarious area and intermittent presumptive therapy (IPT) given to those indigenous children who grow up under the continuous threat of malaria infection. Although potentially a very important strategy for the protection of infants in highly malarious areas, IPT with drugs, such as sulfadoxine-pyrimethamine, should not be considered as chemoprophylaxis and is not appropriate for the children of travelers [284].

Children of any age can contract malaria. WHO advises against taking babies and young children to malarious areas, in particular where there is transmission of chloroquine-resistant *P. falciparum*. If travel cannot be avoided, children must be very carefully protected against mosquito bites and be given appropriate chemoprophylactic drugs. Long-term travelers and expatriates should adjust the chemoprophylaxis dosage according to the increasing weight of the growing child. Malaria can rapidly cause complications in children and therefore any child suffering from fever after returning from a malarious area should be considered to have malaria until proved otherwise. Since it may be difficult to administer drugs to children; where pediatric formulations and accurate dosage may not be available, it is best to protect babies and children against mosquito bites.

Antimalarial drugs are often not well adapted for infants. Mechanical prophylaxis is essential in childhood: devices as insecticide impregnated bed nets or repellents are the most efficient protection against mosquito bites and in fine against infection.

Chloroquine, proguanil, and mefloquine are considered compatible with breastfeeding. Breastfed, as well as bottle-fed, babies should be given chemoprophylaxis since they are not protected by the mother's prophylaxis. Dosage schedules for children should be based on body weight, and tablets should be crushed and ground as necessary. The bitter taste of the tablets can be disguised with jam or other foods. Chloroquine and proguanil are safe for babies and young children but their use is now very limited, due to the spread of chloroquine resistance. Mefloquine may be given to infants of more than 5 kg body weight. Atovaquone/proguanil is generally not recommended for prophylaxis in children who weigh less than 11 kg, because of limited data; in Belgium and the USA it is given for prophylaxis in infants of more than 5 kg body weight. Doxycycline is contra-indicated in children below 8 years of age. All antimalarial drugs should be kept out of the reach of children and stored in childproof containers: chloroquine is particularly toxic in case of overdose [283].

4.3.2. Chemoprophylaxis in Infants and Children

In 1956, McGregor and others reported the results of a trial in The Gambia in which children were given chloroquine weekly from birth until the age of two years [310]. Children who received chemoprophylaxis had fewer episodes of malaria, grew better, and had higher mean hemoglobin and a lower gamma globulin concentration than children in the control group. These findings were reproduced in studies conducted subsequently in several other countries in Africa [311]. A large trial of chemoprophylaxis in children less than five years of age was conducted in The Gambia in the early 1980s [312]. More than 700 children were given chemoprophylaxis with Maloprim® (pyrimethamine plus dapsone, GlaxoSmithKline) by village health workers throughout the rainy season over a period of five years. Overall mortality in children who received prophylaxis was reduced by approximately 35%. Protected children had fewer clinical attacks of malaria and a higher mean packed cell volume than control children. These results were sustained during several years of observation and are at least as impressive as those obtained with ITNs. A study conducted more recently in Tanzanian infants using the same antimalarial also showed a marked reduction in clinical attacks of malaria and in the incidence of severe anemia.

A review of Cochrane Database showed that Twenty-one trials (19,394 participants), including six cluster-randomized trials, met the inclusion criteria. Prophylaxis with antimalarial drugs resulted in fewer clinical malaria episodes (RR 0.53, 95% CI 0.38−0.74, REM; 7037 participants, 10 trials), less severe anemia (RR 0.70, 95% CI 0.52−0.94, REM; 5445 participants, 9 trials), and fewer hospital admissions for any cause (RR 0.64, 95% CI 0.49−0.82; 3722 participants, 5 trials). We did not detect a difference in the number of deaths from any cause (RR 0.90, 95% CI 0.65−1.23; 7369 participants, 10 trials), but the CI do not exclude a potentially important difference. One trial reported three serious adverse events with no statistically significant difference between study groups (1070 participants). Eight trials measured morbidity and mortality six months to two years after stopping regular antimalarial drugs; overall, there was no statistically significant difference, but participant numbers were small. In conclusion, the prophylaxis with antimalarial drugs reduces clinical malaria and severe anemia in preschool children [313].

Pediatric doses should be calculated carefully according to body weight. Pharmacists can pulverize tablets and prepare gelatin capsules with calculated pediatric doses. Mixing the powder in food or drink may facilitate the weekly administration of chloroquine to children. Alternatively, chloroquine in suspension is widely available overseas [283]. Overdose of antimalarial drugs can be fatal. The medication should be stored in childproof containers out of the reach of children.

In Africa, particularly in areas of high transmission, children under 5 years of age are the most affected by malaria, leading to a high case fatality rate in this age group. In spite of the importance of the disease in children and the fact that they are the major targets for antimalarial drugs, there are problems with existing pediatric formulations and regimens:

- Many tablet formulations are not scored, making it difficult to break the tablets into halves and quarters as required by current treatment regimens.
- Confusion arises from the availability on the market of some drugs at more than one strength (e.g., chloroquine as tablets of 100 mg and 150 mg base).
- Different spoon sizes lead to inconsistencies in the dosing of syrups.
- Syrups are often dispensed in volumes of 100-120 ml, tempting mothers to continue administering until finished, or to save the medication for administration to other children.
- Confusion arises because there are four different age groups for treatment with chloroquine in children under 5 years.

These and similar problems influence patient adherence and result in both under dosing and overdosing. Increasing attention to this area through operational research is beginning to suggest some solutions. Appropriate packaging and labeling improves compliance, enhances acceptability and greatly reduces the risk of overdosing. Adherence to prepackaged tablets is much better than to syrup, and the cost of prepackaged treatments is much lower. Training health facility workers and equipping them with packaged treatments have been shown to reduce case fatality rates. Prepackaging of drugs for specific age and weight ranges can improve home management of malaria. It now remains to translate these research findings into practice [2].

Very small amounts of antimalarial drugs are secreted in the breast milk of lactating women to infants. The amount of drug transferred is not thought to be harmful to the nursing infant; however, more information is needed. Because the quantity of antimalarials transferred in breast milk is insufficient to provide adequate protection against malaria, infants who require chemoprophylaxis should receive the recommended dosages of antimalarials.

4.3.3. IPT in Infants (IPTi) and in Children (IPTc)

Intermittent preventive treatment (IPT) involves giving a curative antimalarial treatment dose systematically at predefined intervals to asymptomatic individuals. Especially, those who are at special risk of malaria, regardless of whether or not they are parasitemic. Prophylaxis and IPT are widely used drug-based methods for preventing malaria. Prophylaxis refers to "the administration of a drug in such a way that its blood concentration is maintained above the level that inhibits parasite growth, at the pre-erythrocytic or erythrocytic stage of the parasite's life-cycle, for the duration of the period at risk." Drugs used for malaria prophylaxis are usually given in daily or weekly doses. IPT refers to "involves administration of a full therapeutic course of an antimalarial drug to the whole population at risk, whether or not they are known to be infected, at specific times with the aim of preventing mortality or morbidity" [314].

4.3.3.1. IPT in Infants (IPTi)

IPT administers a full therapeutic course of an anti-malarial drug at predetermined intervals, regardless of infection or disease status. It is recommended by the World Health Organization (WHO) for protecting pregnant women from the adverse effects of malaria (IPTp) and shows great potential as a strategy for reducing illness from malaria during infancy (IPTi). The IPT concept has recently been applied to the prevention of malaria in infants. Administered concurrently with standard immunizations, IPTi is expected to reduce the frequency of clinical disease, but to allow blood-stage infections to occur between treatments, thus allowing parasite-specific immunity to develop. While wide deployment of IPTi is being considered, it is important to assess other potential effects. Transmission conditions, drug choice and administration schedule will likely affect the possibility of post-treatment rebound in child morbidity and mortality and the increased spread of parasite drug resistance and should be considered when implementing IPTi.

A study undertaken in Tanzania by Schellenberg and others showed that administration of a full dose of SP to infants at the time that they received their second and third doses of DPT and measles vaccines resulted in a 59% reduction in the incidence of clinical attacks of malaria, and a 50% reduction in the incidence of severe anemia during the first year of life [315]. A second study undertaken in Tanzania in which amodiaquine was given in a full therapeutic dose three times during the first year of life at the time of regular infant check-up clinics achieved similar impressive results[316]. Importantly, there was no rebound in malaria attacks or in anemia during the second year of life in children who had received IPTi as infants. Trials of IPTi with SP are nearing completion in Ghana and Kenya, and further trials have started recently in Mozambique, Gabon, and Kenya. The results of these trials will help to define the epidemiologic conditions under which IPTi is effective. However, there are still many unanswered questions about the role of IPTi in malaria control. To address some of these important questions (Table 34), an IPTi consortium comprising several groups of investigators interested in this area of research. The protective efficacy against clinical attacks of malaria was 65% during the six months of treatment. Some protection persisted throughout the follow-up period; by the end of the study, 35% of treated infants remained free from a first attack of malaria compared to 13% in the control group.

Thus, the results of the two published IPTi studies are encouraging. Two other studies using strategies similar to IPTi failed to demonstrate statistically significant reductions in clinical episodes of malaria [317, 318]. Both administered SP at monthly intervals three times to anemic children in Kenya. These children were older than those in the two IPTi studies (mean age at enrollment = 11 [317] and 20 months [318], versus 0 [315] and three [316] months), but still within the age range most at risk for malaria-related morbidity and mortality in these regions [319].

The timing of IPTi administration has been recommended in accordance with drug pharmacokinetic data. Under conditions of intense, perennial transmission, infants may be susceptible to severe clinical attacks from a few months after birth, and antimalarials given during routine EPI visits at two, three and nine months of age are likely to be effective in reducing episodes of malaria. Where transmission is less intense or unstable, children are at great risk until 5 years of age or older. In low and moderate transmission areas, IPT may be most effective if the schedule is extended into early childhood. Highly seasonal malaria may require unique strategies which target the appropriate age groups during the transmission season. If mechanisms for delivery can be established that allow greater flexibility in the window of administration, IPTi could have a much broader impact. A study of IPT in children under five in an area of highly seasonal transmission administered doses of SP at monthly intervals exclusively during the high transmission season. This approach reduced the incidence of clinical malaria by more than 80% in the treated group compared to the control group, indicating that adapting the IPT schedule can be accomplished with good results [312]. Transmission intensity may vary considerably over very short distances and short time periods. While it will often be impractical to vary drug administration schedules in response, broader assessments of local transmission and disease patterns should guide IPTi schedules.

Table 34. Outstanding issues on the use of IPTi* [115]

Issues
• Lack of knowledge of the efficacy of IPTi in different epidemiologic situations
• Lack of knowledge of the mode of action of SP when used for IPTi
• The possible impact of co-administration of an anti-malarial drug at the time of vaccination on the immune response to that vaccine
• Whether administration of an anti-malarial drug at the time of vaccination would enhance or detract from compliance with routine immunization
• The identity of drugs that could be used to replace SP for IPTi in areas of high SP resistance
• The safety of SP and alternative drugs when used on a large scale for IPTi
• The effect of IPTi on the development of natural immunity
• The cost efficacy of IPTi in different epidemiologic situations
• Whether IPTi could be readily implemented on a programmatic scale

* IPTi = intermittent preventive treatment in infants; SP = sulfadoxine-pyrimethamine.

4.3.3.2. IPT in Children (IPTc)

Children living in areas where malaria is endemic will have acquired natural immunity to malaria by the time they are seven to 10 years old. Preschool children living in malarious areas have inadequate immunity to malaria; this explains why most of the one million malaria deaths that occur each year in endemic areas of sub-Saharan Africa occur in this age group [313].

In high transmission areas, where a major proportion of deaths and severe morbidity from malaria occur during the first year of life, IPTi could make a major contribution to reducing the burden of malaria as a whole. However, in many areas of Africa, perhaps covering as much as 50% of the population at risk, the major burden of malaria is not in infants but in older children. This is especially the case in countries of the Sahel and sub-Sahel where malaria transmission is intense but very seasonal. In such areas, IPTi, even if highly effective, would have only a limited impact on the overall burden of malaria in children. Thus, a pilot study in Senegal is exploring whether the IPT principle can be applied to younger children [320]. In this clinical trial, SP and one dose of artesunate have been given to all eligible children less than five years old in a community, where with a total population of approximately 12,000 was three times higher at one monthly interval throughout the peak period of malaria transmission. This intervention resulted in a reduction in the incidence of clinical attacks of malaria of approximately 86%. Observations are now under way to determine if this impressive effect is followed by a rebound. For the purpose of this trial, the anti-malarial drug was administered by project staff. Devising an effective method of delivery, which may be site specific, will be an important pre-requisite for this approach to malaria control and is likely to be most effective in areas of intense, seasonal transmission where drug administration may be required on only two to four occasions each year. The IPTi could be highly effective for prevention of malaria in children under 5 years of age living in areas of seasonal malaria [320].

A review of Cochrane Database showed that Twenty-one trials (19,394 participants), including six cluster-randomized trials, met the inclusion criteria. Prophylaxis or IPTs with antimalarial drugs resulted in fewer clinical malaria episodes (RR 0.53, 95% CI 0.38−0.74, REM; 7037 participants, 10 trials), less severe anemia (RR 0.70, 95% CI 0.52−0.94, REM; 5445 participants, 9 trials), and fewer hospital admissions for any cause (RR 0.64, 95% CI 0.49−0.82; 3722 participants, 5 trials). The review did not detect a difference in the number of deaths from any cause (RR 0.90, 95% CI 0.65−1.23; 7369 participants, 10 trials), but the CI do not exclude a potentially important difference. One trial reported three serious adverse events with no statistically significant difference between study groups (1070 participants). Eight trials measured morbidity and mortality six months to two years after stopping regular antimalarial drugs; overall, there was no statistically significant difference, but participant numbers were small. In conclusion, Prophylaxis and intermittent treatment with antimalarial drugs reduce clinical malaria and severe anemia in preschool children.

In this review, all 21 trials included in the review were conducted in areas in Africa where *P. falciparum* is the predominant cause of malaria. Transmission patterns were variable: seasonal in 10 trials and perennial in 11. The 10 IPT trials were more recent and contributed more data to meta-analyses than the prophylaxis trials. Several of the prophylaxis trials reported outcomes in different publications, but we have been careful to ensure the same participants are not included twice in each meta-analysis.

All 10 IPT trials and two of the 11 prophylaxis trials described adequate methods of generating the allocation sequence.

Overall, both prophylaxis and IPTs consistently reduced clinical malaria and admission to hospital. The effect of IPTs on the incidence of severe anemia appeared to be modest compared with the marked reduction observed in clinical malaria episodes and hospital admissions. Intermittent treatment did not appear to be effective in reducing the incidence of anemia in the three trials enrolling already moderately anemic children (secondary prevention), although it clearly was effective in preventing severe anemia in the primary prevention trials. The children in the primary prevention trials were mainly non-anemic and younger than those in the secondary prevention trials. This review did not provide convincing evidence that either prophylaxis or IPTs reduced the risk of death in preschool children, although the point estimate and confidence intervals are compatible with a potentially important effect. It is hoped that in the near future these trials would contribute data to confirm or disprove some of the inconclusive observations made in this review. In addition to providing more information on the effectiveness of these regimens, we expect that these trials will conduct long-term, follow-up studies to explore this and other research questions related to cost effectiveness, relative safety, and emergence of parasite resistance [313].

Almost all antimalarial drugs taste terrible and children object to taking them. Crushing tablets and placing them in food to disguise the taste does work but division of tablets is inexact, the role of the food in drug absorption is usually unknown and children are prone to spit out any undesired item, especially when the suspended drug still tastes foul. There are only two current choices for malaria chemoprophylaxis in small children: subdivided mefloquine tablets or the quarter-strength pediatric tablet of atovaquone/proguanil (Table 31). Some pediatricians recommend weekly mefloquine as it minimizes the number of times a parent has to induce an unwilling child to take medication, thus increasing the effectiveness of the chemoprophylaxis. Atovaquone/proguanil prophylaxis has been widely used in the pediatric population and has shown an excellent safety and efficacy profile [321]. Doxycycline is contra-indicated in young children because of the staining of teeth and bones. Chloroquine is worth mentioning, if only to warn parents about the risks to children. Despite the near universal presence of chloroquine resistance in malarious areas throughout the world, chloroquine is still very widely used. It is an all too common source of lethal poisonings in children who accidentally ingest only a few tablets and then die with intractable cardiac dysrrhythmias [322]. All medications should be kept out of the reach of children, but the ubiquity of chloroquine in some African settings makes it a risk which is particularly worth warning parents who are taking young children to the tropics.

4.3.4. Effect on Infants and Children after Stopping Intervention

Seven trials evaluated the impact after intervention was stopped: three prophylaxis trials; and four intermittent treatment trials [320]. These trialists reported outcomes assessed after intervention had been stopped for variable lengths of time: six months [323]; nine months [324]; four to 12 months corresponding to age 16 to 24 months [325]; 12 months [320]; and 18 months [326]. All four intermittent treatment trials contributed to the meta-analysis on clinical malaria, three on severe anaemia, and two on death rates, while one reported on the impact of IPT on measles immunization.

It has been widely speculated that giving prophylaxis to infants and preschool children resident in malaria-endemic areas would prevent natural immunity and result in (rebound) increase in morbidity and mortality after stopping prophylaxis. Data on possible rebound effect of prophylaxis in this review were scanty and showed inconsistent results. One prophylaxis trial detected a statistically significant increase in the incidence of clinical malaria and anemia among children who had previously received pyrimethamine-dapsone prophylaxis compared to the control group. The other two trials did not demonstrate any significant deleterious effects of taking the intervention, but sample sizes were small. The results of four trials of adequate methodological quality included in the meta-analysis demonstrated that intermittent treatment given over a short period during early childhood is unlikely to result in rebound effect malaria morbidity and mortality. While these trials are few with short follow-up periods, these results appear to support the hypothesis that IPT allows longer periods in between treatments for children to acquire protective malarial immunity and is therefore less likely to cause rebound morbidity and mortality than continuous prophylaxis [313].

There is an urgent need to deploy and develop new control tools that will reduce the intolerable burden of malaria. IPTi has the potential to become an effective tool for malaria control. A randomized, double-blind, placebo-controlled trial of sulfadoxine-pyrimethamine (SP) treatment was performed in 1503 Mozambican children. Doses of SP or placebo were given at 3, 4, and 9 months of age. The intervention was administered alongside routine vaccinations delivered through the Expanded Program on Immunization (EPI). Hematological and biochemical tests were done when infants were 5 months old. Morbidity monitoring through a hospital-based passive case-detection system was complemented by cross-sectional surveys when infants were 12 and 24 months old. IPTi was well tolerated, and no adverse events associated with SP were documented. During the first year of life, intermittent SP treatment reduced the incidence of clinical malaria by 22.2% (95% confidence interval [CI], 3.7%–37.0%; P=.020) and the rate of hospital admissions by 19% (95% CI, 4.0%–31.0%; P=.014). Although the incidence of severe anemia (packed cell volume of <25%) did not differ significantly between the 2 groups (protective effect, 12.7% [95% CI, -17.3% – 35.1%]; P=.36), there was a significant reduction in hospital admissions for anemia during the month after dosing for both the first and second dose. The serological responses to EPI vaccines were not modified by the intervention.

IPTi with SP has been shown to moderately reduce the incidence of clinical malaria in Mozambican infants without evidence of rebound after stopping the intervention or of interactions with EPI vaccines. Its recommendation as a malaria control strategy in Mozambique needs to be balanced against the scarcity of affordable control tools and the burden of malaria in children [327].

4.4. Prevention of Malaria in Travelers

4.4.1. Guiding Principles of Chemoprophylaxis and Stand-by Treatment in Travelers

The spread and intensification of drug resistance worldwide has greatly complicated recommendations for the prevention of malaria in travelers. Travel to malarious areas is on the increase, while many countries are experiencing a resurgence of malaria. As a short-term measure, chemoprophylaxis is recommended for international and national travelers from non-endemic areas, and for soldiers, police and labor forces serving or working in highly endemic areas. Detailed recommendations for the protection of travelers against malaria are updated and published annually by WHO in International travel and health: *vaccination requirements and health advice* [283]. All travelers to malarious areas should be clearly informed of: the risk of malaria; how they can best protect themselves against mosquito bites; the use of chemoprophylaxis wherever appropriate; and the need to seek early diagnosis and treatment if symptoms suggestive of malaria occur. Malaria must always be suspected if fever, with or without other symptoms, develops at any time between one week after the first possible exposure to malaria and two months, or even longer in exceptional cases, after the last possible exposure. Nearly all travelers who acquire a *P. falciparum* infection will have developed symptoms within 3 months of exposure. Medical attention should be sought and a blood sample examined for malaria parasites. If no parasites are found but symptoms persist, a series of blood samples should be taken and examined at appropriate intervals. Relapses of *vivax* and *ovale* malaria are not prevented by chemoprophylaxis with currently used prophylactic regimens [283].

Malaria chemoprophylaxis should be selected on the basis of an individual risk assessment for the traveler; an assessment of the safety and efficacy of potential chemoprophylactic regimens, drug resistance, and the extent of malaria transmission in the region to be visited. Weekly prophylactic antimalarial regimens should normally be started one week before travel. Daily drugs such as proguanil and doxycycline should be started the day before travel. Drugs should then be taken with unfailing regularity for the duration stay in the area of malaria risk, and continued for 4 weeks after leaving the endemic area. The exception is atovaquone/proguanil, which can be stopped one week after leaving the area with malaria risk. Mefloquine prophylaxis should preferably be started 2-3 weeks before departure, so that adequate blood levels are attained, and adverse reactions can be detected before travel, allowing consideration of alternative drug regimens. Antimalarial drugs should be taken with food and swallowed with plenty of water.

In general, travel areas are classified as:

- Areas with *P. vivax* transmission only.
- Chloroquine-sensitive--Malarious areas where chloroquine resistance has not been documented or is not widely present; these include Haiti, the Dominican Republic, Central America north-west of the Panama Canal and parts of the Middle East.
- Chloroquine-resistant--Most of Africa, South America, Oceania, and Asia.
- Chloroquine- and mefloquine-resistant--Resistance to both chloroquine and mefloquine is not a significant problem except in rural forested regions along the borders between Cambodia, Myanmar, and Thailand, and are infrequently visited by tourists; mefloquine resistance in varying degrees has been reported from Brazil, Cambodia, China (*in vitro*), French Guyana, Myanmar, Thailand, and Vietnam.

Data on the incidence of malaria and the effectiveness and tolerance of currently recommended chemoprophylaxis and self-treatment regimens for long-term travelers are limited. Chloroquine and mefloquine has been shown to be safe for at least 3 years. In chloroquine-resistant regions, mefloquine is more efficacious than the chloroquine plus proguanil combination [328]. In one study, chemoprophylaxis with atovaquone-proguanil for 20 weeks and with primaquine for 50 weeks had no significant adverse effects.

Children are at special risk of malaria since they may rapidly become seriously ill. Persuading young children to take antimalarial medications may be difficult because of the lack of pediatric formulations and the bitter taste of many drugs. Furthermore, some chemoprophylactic drugs are contra-indicated in children. Chloroquine remains the drug of choice in areas where malaria remains sensitive to this drug, while mefloquine is the preferred agent in chloroquine-resistant areas. Although the manufacturer recommends that mefloquine should not be given to children who weigh less than 5 kg, it should be considered for chemoprophylaxis of all children at high risk of acquiring chloroquine-resistant *P. falciparum* malaria. Atovaquone- proguanil may be a safe and effective chemoprophylactic alternative to doxycycline for children under 8 years of age who weigh more than 11 kg and are traveling to mefloquine-resistant areas. (Chapter 4, Section 4.3).

In non-immune pregnant travelers, malaria increases the risk of maternal and neonatal death, miscarriage, and stillbirth. When traveling to malarious areas, pregnant women should take special care to avoid mosquito bites. Chloroquine is known to be safe in pregnancy, but its usefulness is limited to a few areas with chloroquine-sensitive strains of *P. falciparum*. Available data indicate that mefloquine is safe as a chemoprophylactic agent after the first trimester. Becoming pregnant while taking mefloquine, chemoprophylaxis is not an indication for termination of the pregnancy. The combination of chloroquine plus proguanil is safe in pregnancy and provides more protection than chloroquine alone in areas with known chloroquine-resistant strains, but is significantly less efficacious than mefloquine. Doxycycline and primaquine are contra-indicated during pregnancy. Data on the safety of atovaquone-proguanil during pregnancy is insufficient, although studies with this combination are currently under way. (Chapter 4, Section 4.2)

Most travelers to urban or major tourist areas will be able to obtain prompt and reliable medical attention when malaria is suspected. However, a minority may be traveling to such isolated locations that they will have no access to competent medical attention within 24 h after the onset of symptoms, a week or more after first possible exposure. In such cases, stand-by emergency treatment should be prescribed to the traveler for self-administration should symptoms occur [329]. Precise instructions should be given on the recognition of malaria symptoms and the need to take the full treatment dose of the drug, and information should be provided on possible adverse reactions. Travelers must also be made aware that they should seek attention as soon as possible after symptoms appear and they begin their stand-by treatment, and that stand-by treatment is not a substitute for diagnosis and treatment by qualified medical personnel.

Because of the potential for additive toxicity and reduced efficacy, individuals who are on chemoprophylaxis should never attempt stand-by-treatment with the same drug. Following completion of the treatment, individuals should resume effective malaria chemoprophylaxis. As a general guide, chemoprophylaxis should be restarted one week after the first treatment dose. When the stand-by-treatment is quinine, however, mefloquine chemoprophylaxis should be restarted one week after the last dose [283].

Chloroquine, sulfadoxine-pyrimethamine, mefloquine, quinine plus tetracycline, atovaquone-proguanil, and artemether-lumefantrine can be prescribed as stand-by treatments, depending on the drug-resistance status of the parasites in the areas to be visited. Halofantrine is not recommended owing to the fact that it can result in ventricular dysrhythmias and prolongation of QTc intervals in susceptible individuals. While these antimalarial drugs described are all used for stand-by treatment, artemether plus lumefantrine is the only therapy registered by a national drug regulatory authority for this purpose. Efficacy, safety and ease of administration should be considered for the selection of stand-by treatment, but there is some concern about the use of drugs for this purpose in travelers who may have access to medical facilities [283].

4.4.2. Risk and Preventive Strategies for Travelers

Malaria is the most important parasitic disease in the world. Malaria incidence has increased during the last 2 decades, and it is now estimated that ~270 million people are infected annually, resulting in 1.5–2.7 million deaths. This increase can be attributed, in large part, to the appearance and spread of *P. falciparum* that is resistant to antimalarial drugs. Superimposed on this global resurgence in cases has been a major increase in travel to and emigration from countries where malaria is endemic. It is currently endemic in over 100 countries, which are visited by more than 125 million international travelers every year [283]. The risk for travel to areas where transmission is high, such as sub-Saharan Africa, has increased 120-fold since the 1960s. As a result of the combination of increased travel to areas where malaria is endemic and escalating drug resistance, a growing number of travelers are at risk for contracting malaria [330].

The risk of contracting malaria varies significantly from traveler to traveler, and this is the challenge of providing quality advice. These risk factors all have an impact on how significant the threat:

- specific areas visited within countries
- season of travel
- style of travel
- length of trip
- drug resistance patterns
- individual medical conditions, and
- the local medical infrastructure

General practitioners should be wary of databases providing generalized advice on malaria distribution and epidemiology since it is a changing phenomena. In general:

- Africa and Oceania present the highest risk
- Asia and South America moderate to low risk, and
- Central America very low risk.
- Many popular tourist destinations and urban centers in Asia have minimal or no risk.

Studies in travelers who have acquired malaria show that inappropriate or no prophylaxis is the greatest risk factor for contracting the disease. Factors contributing to this failure in prevention include the traveler's ignorance of the risk of contracting malaria, inadequate training in travel medicine for the prescriber, and fear of drug side-effects, cost issues, and cultural perceptions. The challenge is when to determine appropriately the need to prescribe. It is equally as inappropriate to prescribe medications when there is no risk, as it is to not prescribe when there is significant risk. Healthy travelers do not want to experience adverse reaction to a medication they did not require. The issue becomes more complex in those with low to moderate risk. If general practitioners are not confident in this area it is worth considering referral to a specialized travel medicine clinic.

The risk that a traveler will become infected depends on the overall rate of malaria transmission in the area visited and the extent of the traveler's contact with infected mosquitoes. Transmissions rates may vary greatly from region to region even within the same country; thus, the route and mode of travel and destination are important considerations. Furthermore, since the rate of transmission of malaria may vary from season to season in the same region, the timing of travel also may influence the risk. Finally, since female *anopheline* mosquitoes feed from dusk to dawn, the risk is influenced by a traveler's nighttime activities and the characteristics of his or her lodging. During the transmission season in malaria-endemic areas, all non-immune travelers exposed to mosquito bites, especially between dusk and dawn, are at risk of malaria. This includes previously semi-immune travelers who have lost or partially lost their immunity during stays of 6 months or more in non-endemic areas. Children of people who have migrated to non-endemic areas are particularly at risk when they return to malarious areas to visit friends and relatives.

During the period when chloroquine was widely effective as a chemoprophylactic agent, there was little motivation to analyze in detail the risk of acquisition of malaria among travelers to specific regions. When more toxic drugs had to be used to prevent infection with chloroquine-resistant strains of *P. falciparum*, however, it became necessary for health authorities, particularly the World Health Organization and the Centers for Disease Control and Prevention of the U.S. Public Health Service (CDC), to weigh risks and benefits in making specific recommendations. These authorities used the following imperfect sources to obtain crude estimates of the risk of malaria: the results of national surveillance programs in malarious regions [285]; data on travelers' worldwide destinations; selected surveys of the use of prophylaxis by travelers; reports of cases of imported malaria; and prospective cohort studies of populations at risk. From the study of such data has come the awareness that the risk of malaria for travelers is greatest in sub-Saharan Africa, Papua New Guinea, and the Solomon Islands; intermediate on the Indian subcontinent and in Haiti; and low in Southeast Asia and Latin America. Even in countries where the overall risk is relatively low, however, there may be foci of intense transmission. Regional variation in risk is exemplified by the situation in Kenya, a popular destination for tourists, where travelers' risk of malaria is highest near Lake Victoria, intermediate along the coast, and lowest in the game parks.

Behavioral factors are also influence risk. Travelers visiting friends and relatives in West Africa were found to have a much higher risk of acquiring malaria than other tourists. The relatively higher risk in this group - which included expatriate West Africans living in the United States or Europe - may have resulted from several factors. Having grown up in Africa, returning expatriates were more likely to have the misperception that they were immune to malaria and did not require chemoprophylaxis. They were also more likely to visit rural areas, where exposure to malaria is more intense, and they tended to stay in these areas for longer periods. Those who took any antimalarial chemoprophylactic drug were more apt to take those that were no longer effective. Moreover, their evening activities and lodging may have allowed greater contact with infected mosquitoes than is the case for most tourists [285].

The availability of chloroquine and of mefloquine and doxycycline (which are effective against chloroquine-resistant strains of *P. falciparum*) justifies the current general recommendation that one or another of these chemoprophylactic agents be used by all travelers to areas where there is any risk of malaria. However, the choice of drug needs to be tailored to the traveler's specific needs. No chemoprophylaxis may be required when travel is restricted to specific locations within countries where malaria is endemic but there is no risk of malaria for tourists (for example, certain tourist sites in Thailand and in Bali). On the other hand, there has been controversy about the best prophylactic regimen for persons who are traveling to regions where chloroquine-resistant organisms are widespread and in whom mefloquine and doxycycline are contra-indicated.

Therefore, travelers should have a clear understanding of the risk associated with the disease and myths such as 'the tablets are worse than the disease' should be vigorously dispelled with factual information. Education can lead to understanding and consequently to greater compliance.

4.4.3. Chemoprophylaxis for Long-term Travelers

4.4.3.1. Traveler and Chemoprophylaxis

Each year many international travelers fall ill with malaria while visiting countries where the disease is endemic, and well over 10,000 are reported to fall ill after returning home. Due to this lack of accurate reporting, the real figure may be as high as 30,000. International travelers from non-endemic areas are at high risk of malaria and its consequences because they lack immunity. Immigrants from endemic areas who now live in non-endemic areas and return to their home countries to visit friends and relatives are similarly at risk because of waning or absent immunity. Fever occurring in a traveler within three months of leaving a malaria-endemic area is a medical emergency and should be investigated urgently. Travelers who fall ill during travel may find it difficult to access reliable medical care. Travelers who develop malaria upon return to a non-endemic country present particular problems: doctors may be unfamiliar with malaria, the diagnosis may be delayed, and effective antimalarial medicines may not be registered and/or available, resulting in progression to severe and complicated malaria and, consequently, high case-fatality rates [283].

The overall case-fatality rate associated with imported *P. falciparum* malaria varies from 0.6% to 3.8%. The fatality rate may be ~20% among elderly patients or patients who have severe malaria, even when treatment is administered in modern intensive care units [330]. Yet malaria-associated deaths are largely preventable. Almost all fatal cases of imported malaria occur because the traveler did not use or did not comply with appropriate chemoprophylactic regimens, because of misdiagnosis by physicians or laboratories, or because the chosen prophylactic regimen was inappropriate. Recent reports of fatal cases of malaria in North America and Europe highlight these problems [330]. A number of prophylaxis failures and several deaths resulting from malaria have been reported in travelers to Africa who were taking chloroquine or a combination of chloroquine and proguanil, which highlights the need for more efficacious and acceptable drug regimens.

Travelers and their doctors should be aware that *no antimalarial prophylactic regimen gives complete protection,* but good chemoprophylaxis (adherence to the recommended drug regimen) does reduce the risk of fatal disease. The following should also be taken into account:

- Dosing schedules for children should be based on body weight.
- Antimalarials that have to be taken daily should be started the day before arrival in the risk area.
- Weekly chloroquine should be started 1 week before arrival.
- Weekly mefloquine should preferably be started 2–3 weeks before departure, to achieve higher pre-travel blood levels and to allow side-effects to be detected before travel so that possible alternatives can be considered.
- All prophylactic drugs should be taken with unfailing regularity for the duration of the stay in the malaria risk area, and should be continued for 4 weeks after the last

possible exposure to infection, since parasites may still emerge from the liver during this period. The single exception is atovaquone–proguanil, which can be stopped 1 week after return because of its effect on early liver-stage parasites ("liver schizonts"). Premature interruption of the daily atovaquone–proguanil prophylaxis regimen may lead to loss of this causal prophylactic effect, in which case atovaquone–proguanil prophylaxis should also be continued for 4 weeks upon return.

- Depending on the type of malaria at the destination, travellers should be advised about possible late-onset *P. vivax* and *P. ovale*.

Depending on the malaria risk in the area visited, the recommended prevention method may be mosquito bite prevention only, or mosquito bite prevention in combination with chemoprophylaxis, as follows in Table 35 [283].

4.4.3.2. Chemoprophylaxis for Long Term Travelers

Long-term travelers intending to stay for more than 1-3 months should seek the advice of local health care professionals familiar with the management of malaria in non-immune foreigners. The risk of serious side-effects associated with long term use of chloroquine and proguanil are low. However, twice yearly screening for the detection of early retinal changes should be performed in anyone who has taken 300 mg of chloroquine (as base) weekly for over five years and requires further prophylaxis. If changes are observed, an alternative regimen should be considered. Data indicate no increased risk of serious side-effects with long term use of mefloquine. The risk of long term use of doxycycline is not known. These two latter drugs should be reserved for those with greatest risk of infection.

Table 35. Type lists with recommended prevention methods [283]

Drugs	Efficacy	Safety
Type I	Very limited risk of malaria transmission	Mosquito bite prevention only
Type II	Risk of *P. vivax* malaria only; or fully chloroquine-sensitive *P. falciparum*	Mosquito bite prevention plus chloroquine chemoprophylaxis
Type III	Risk of *P. vivax* and *P. falciparum* malaria transmission, combined with emerging chloroquine resistance	Mosquito bite prevention plus chloroquine + proguanil chemoprophylaxis
Type IV	(1) High risk of *P. falciparum* malaria, in combination with reported antimalarial drug resistance; or (2) Moderate/low risk of *P. falciparum* malaria, in combination with reported high levels of drug resistance*	Mosquito bite prevention plus mefloquine, doxycycline or atovaquone–proguanil chemoprophylaxis (select according to reported resistance pattern)

* Alternatively, when traveling to rural areas with multidrug-resistant malaria and only a very low risk of *P. falciparum* infection, mosquito bite prevention can be combined with stand-by emergency treatment.

Adherence and tolerability are important aspects of chemoprophylaxis use in long-term travelers. There are few studies on chemoprophylaxis use in travel lasting more than 6 months. The risk of serious side-effects associated with long-term prophylactic use of chloroquine and proguanil is low, but retinal toxicity is of concern when a cumulative dose of 100 g of chloroquine is reached. Anyone who has taken 300 mg of chloroquine weekly for more than 5 years and requires further prophylaxis should be screened twice-yearly for early retinal changes. If daily doses of 100 mg chloroquine have been taken, screening should start after 3 years. Data indicate no increased risk of serious side-effects with long-term use of mefloquine if the drug is tolerated in the short-term.

Pharmacokinetic data indicate that mefloquine does not accumulate during long-term intake. Available data on long-term chemoprophylaxis with doxycycline (i.e., more than 12 months) is limited but reassuring. There are few data on long-term use of doxycycline in women, but use of this drug is associated with an increased frequency of *Candida* vaginitis. Atovaquone–proguanil is registered in European countries with a restriction on duration of use (varying from 5 weeks to 1 year); in the USA no such restrictions apply.

In type II areas, with exclusively *P. vivax* transmission or where *P. falciparum* can be expected to be fully sensitive to chloroquine, prophylaxis with chloroquine alone may be used. In type III areas, prophylaxis with chloroquine plus proguanil can be safely prescribed, including during the first 3 months of pregnancy. In type IV areas, mefloquine prophylaxis may be given during the second and third trimesters, but there is limited information on the safety of mefloquine during the first trimester. In light of the danger of malaria to mother and fetus, experts increasingly agree that travel to a chloroquine-resistant *P. falciparum* area during the first trimester of pregnancy should be avoided or delayed at all costs; if this is truly impossible, good preventive measures should be taken, including prophylaxis with mefloquine where this is indicated. Doxycycline is contra-indicated during pregnancy. Atovaquone-proguanil has not been sufficiently investigated to be prescribed in pregnancy.

The chemoprophylaxis for frequent travelers such as members of the aircraft crew may reserve chemoprophylaxis for high risk areas only. They should maintain rigorous self-protection against mosquito bites and be prepared for an attack of malaria and should carry a course of antimalarials as stand-by. In the multi-drug resistant malaria such in Thailand near the borders with Cambodia and Myanmar and in Western Cambodia, *P. falciparum* infections do not respond to chloroquine or sulfadoxine- pyrimethamine, and sensitivity to quinine is reduced. Treatment failures of over 50% are also being reported.

In these areas, chemoprophylaxis with doxycycline is recommended along with rigorous personal protection measures. Doxycycline is contra-indicated in pregnant women and children below the age of 8 years, and therefore they should avoid traveling to these areas. The WHO recommendation for using antimalarial drugs for prophylaxis in travelers is shown in Table 36a and Table 36b [283].

4.4.4. Chemoprophylaxis for Short-term Travelers

Short-term travelers are persons visiting an area where malaria is endemic for weeks or a few months. This group includes most tourists, and the main reason for travel is business and leisure. The travelers usually have little prior knowledge of disease, including malaria risk at their destination. Because of the incubation period of malaria, many short-term travelers who acquire the disease will be diagnosed after return. The actual profile of imported species varies among countries, reflecting the primarily geographic origin of the infection. Malaria deaths are due almost exclusively to infection with *P. falciparum* [120].

Table 36a. Use of antimalarial drugs for prophylaxis in travelers [283]

Generic Name	Dosage regimen	Duration of prophy-laxis	Use in special groups			Main contra-indications	Comments
			Preg-nancy	Breast-feeding	Children		
Atovaquone – proguanil combination tablet	One dose daily. *11–20 kg:* 62.5 mg atovaquone plus 25 mg proguanil daily (1 pediatric tablet) 21–30 kg: 2 pediatric tablets daily, 31–40 kg: 3 pediatric tablet daily, > 40 kg: 1 adult tablet.	Start 1 day before departure and continue for 7 days after return	No data, not recom-mended	No data not recom-mended	Not recom-mended under 11 kg because of limited data.	Hypersensitivity to atovaquone and/or proguanil; severe renal insufficiency (creatinine clearance < 30 ml/min)	Registered in European countries for chemoprophylactic use with a restriction on duration of use (varying from 5 weeks to 1 year). Plasma concentrations of atovaquone are reduced when it is co-administered with rifampicin, rifabutin, metoclopramide or tetracycline.
Chloroquine (CQ)	5 mg base/kg weekly in one dose, or 10 mg/kg weekly in 6 doses. Adult dose: 300 mg CQ base weekly in one dose, or 600 mg CQ base weekly in 6 doses of 100 mg base.	Start 1 week before departure and continue for 4 weeks after return. If daily doses; start 1 day before departure	Safe	Safe	Safe	Hypersensitivity to chloroquine; history of epilepsy; psoriasis.	Concurrent use of chloroquine may reduce the antibody response to intradermally administered human diploid-cell rabies vaccine
Chloroquine (CQ)/progua nil combination tablet	> 50 kg: 100 mg CQ base plus 200 mg proguanil (1 tablet) daily	Start 1 day before departure and continue for 4 weeks after return	Safe	Safe	Tablet size not suitable for persons of < 50 kg of body weight	Hypersensitivity to chloroquine and/or proguanil; liver or kidney insufficiency; history of epilepsy; psoriasis.	Concurrent use of CQ may reduce the antibody response to intradermally administered human diploid-cell rabies vaccine

The primary goal of malaria chemoprophylaxis is to prevent deaths from malaria, which are largely caused by *P. falciparum*. Another important goal is to prevent clinical malaria, which can lead to hospitalization with health and economic consequences. The currently recommended first-line drugs for chemoprophylaxis are 80–90% effective in preventing clinical episodes of primary malaria infection [331]. The antimalarial agents commonly used (e.g., chloroquine, mefloquine, doxycycline, atovaquone-proguanil) that are active against the blood stage parasites; however, do not prevent relapsing infection caused by *P. vivax* and *P. ovale*. These *Plasmodium* species are biologically different and can cause latent infection with hypnozoites in the liver that are not killed by the antimalarials active against the blood stages. The detail review of antimalarial drugs in causal prophylaxis for *vivax* and *ovale* malaria is described in Chapter 4, Section 4.5.

Table 36b. Use of antimalarial drugs for prophylaxis in travelers (*continued*) [283]

Generic Name	Dosage regimen	Duration of prophylaxis	Use in special groups			Main contra-indications	Comments
			Pregnancy	Breast-feeding	Children		
Doxycycline	1.5 mg salt/kg daily *Adult dose:* 1 tablet of 100 mg daily	Start 1 day before departure and continue for 4 weeks after return	Contra-indicated	Contra-indicated	Contra-indicated under 8 year of age	Hypersensitivity to tetracyclines; liver dysfunction.	Doxycycline makes the skin more susceptible to sunburn. People with sensitive skin should use a highly protective (UVA) sunscreen and avoid prolonged direct sunlight, or switch to another drug. Doxycycline should be taken with plenty of water to prevent oesophageal irritation.
Mefloquine	5 mg/kg weekly *Adult dose*: 1 tablet 250 mg weekly	Start at least 1 week (preferably 2–3 weeks) before departure and continue for 4 weeks after return	Not recom mended in first trimester because of lack of data	- Safe	Not recomme nded in under 5 kg because of lack of data	Hypersensitivity to mefloquine; psychiatric (including depression) or convulsive disorders; history of severe neuropsychiatric disease; concomitant halofantrine treatment; treatment with mefloquine in previous 4 weeks; not recommended in use of pilots, machine operators.	Do not give mefloquine within 12 hours of quinine treatment. Mefloquine and other cardioactive drugs may be given concomitantly only under close medical supervision. Ampicillin, tetracycline and metoclopramide may increase mefloquine blood levels.
Proguanil	3 mg/kg daily *Adult dose*: 2 tablets of 100 mg daily	Start 1 day before departure and continue for 4 weeks after return	Safe	Safe	Safe	Liver or kidney dysfunction.	Use only in combination with chloroquine. Proguanil may interfere with live typhoid vaccine.

At present, the key problem for travel health practitioners is when to advise the use of chemoprophylaxis for travelers to low-risk areas of endemicity. Based on the estimates of risk for malaria and adverse events, we propose that the prescription of chemoprophylaxis in areas with an endemicity level in the indigenous population below 10 cases per 1,000 populations per year is not justified (Table 37). To prevent one case at this level of transmission, 2,600 travelers should be prescribed chemoprophylaxis, and to prevent one fatality, assuming a 2% case fatality rate, 130,000 travelers should be prescribed chemoprophylaxis (Table 38). For comparison, the proper use of impregnated bed nets can reduce the risk of infection by 50% without causing adverse events in users. However, it should be noted that clinical malaria cases in the local population may reflect a lower attack rate than that for visitors if there is substantial clinical immunity in the local population. Therefore, the level of endemicity should be determined by the parasite rate and not the rate of reported clinical malaria cases in the indigenous population. Stand-by emergency treatment (SBET), where the traveler is prepared to self-treat a suspected case of malaria if he or she has symptoms suggestive of malaria and is out of reach of medical attention, is recommended by many authorities as a strategy for short-term travelers to low-risk destinations [283]. However, SBET is not indicated for travelers on trips shorter than the incubation period for malaria, which is a minimum of 6 days for *P. falciparum* malaria [120].

4.4.4.1. Treatment Abroad and Stand-by Emergency Treatment

An individual who experiences a fever 1 week or more after entering an area of malaria risk should consult a physician or qualified malaria laboratory immediately to obtain a correct diagnosis and safe and effective treatment. In principle, travelers can be treated with artemisinin-based combination therapy (ACT) according to the national policy in the country they will be visiting. National antimalarial drug policies for all endemic countries are listed at *http://www.who.int/malaria/treatmentpolicies.html*. In light of the spread of counterfeit drugs in some resource-poor settings, travelers may opt to buy a reserve antimalarial treatment before departure; so that they can be confident of drug quality should they become ill. Many travelers will be able to obtain proper medical attention within 24 hours of the onset of fever. For others, however, this may be impossible, particularly if they will be staying in remote locations. In such cases, travelers are advised to carry antimalarial drugs for self-administration ("stand-by emergency treatment").

Stand-by emergency treatment (SBET) may also be indicated for travelers in some occupational groups, such as aircraft crews, who make frequent short stops in endemic areas over a prolonged period of time. Such travelers may choose to reserve chemoprophylaxis for high-risk areas and seasons only. However, they should continue to take measures to protect against mosquito bites and be prepared for an attack of malaria: they should always carry a course of antimalarial drugs for SBET, seek immediate medical care in case of fever, and take SBET if prompt medical help is not available.

Furthermore, SBET – combined with protection against mosquito bites – may be indicated for those who travel for 1 week or more to remote rural areas where there is multidrug-resistant malaria but a very low risk of infection, and the risk of side-effects of prophylaxis may outweigh that of contracting malaria. This may be the case in certain border areas of Thailand and neighboring countries in South-East Asia, as well as parts of the Amazon basin. Studies on the use of rapid diagnostic tests ("dipsticks") have shown that untrained travelers experience major problems in the performance and interpretation of these tests, with an unacceptably high number of false-negative results. In addition, dipsticks can be degraded by extremes of heat and humidity, becoming less sensitive. Successful SBET depends crucially on travelers' behavior, and health advisers need to spend time explaining the strategy. Travelers provided with SBET should be given clear and precise written instructions on the recognition of symptoms, when and how to take the treatment, possible side-effects, and the possibility of drug failure. If several people travel together, the individual dosages for SBET should be specified. Weight-based dosages for children need to be clearly indicated.

Table 37. Risk of malaria in short-term travelers at different levels of endemicity in the indigenous population and mortality from malaria in short-term travelers* [120].

Annual incidence of malaria cases in local population	Example of area of endemicity**	Incidence per wk per 100000 travelers	Incidence per 2 wk per 100,000 travelers without prophylaxis	Incidence per 2 wk per 100000 travelers with prophylaxis assuming 90% efficacy of prophylaxis	Mortality per 100,000 travelers per 2 wk without prophylaxis	Mortality per 100,000 travelers per 2 wk with prophylaxis
1 per 1000	Mexico, parts of South America, Vietnam (except Binh Province)	1.9	3.8	0.4		
10 per 1000	Parts of Vietnam (Binh Phuoc province)	19.2	38.5	3.8	1	
20 per 1000	Parts of India (Assam, Gujarat, Orissa, Rajastan)	38.4	76.8	7.7	4	
50 per 1000	Parts of South Africa	96.1	192.3	19.2	8	
100 per 1000	Western Africa	192.3	384.6	38.5		0.5

* Assuming a case fatality rate of 2%. Short-term travel is considered to be travel in a region of endemicity lasting 2 weeks or less.

** The areas mentioned serve only as examples of areas with different levels of endemicity in the indigenous population. Risk in southeastern Asia is very unevenly distributed within each country and should be assessed at district levels based on the travelers' planned route and using malaria maps.

Table 38. Numbers of travelers needed to take chemoprophylaxis to prevent one case of malaria and to prevent one fatal case of malaria at different levels of endemicity [120]

Annual incidence of cases of malaria in local population	No. of travelers needed			
	Taking prophylaxis to prevent 1 case*	Taking prophylaxis to prevent 1 death**	With severe side- effects to prevent 1 malaria case***	With severe side- effects to prevent 1 death
1 per 1000	26,000	1,300,000	2,600	130,000
10 per 1000	2,600	130,000	260	13,000
20 per 1000	1,300	65,000	130	6,500
50 per 1000	520	26,000	52	2,600
100 per 1000	260	13,000	26	1,300

* Assuming 2 weeks of travel. ** Assuming 2 weeks of travel and assuming 2% mortality. *** Severe side-effects in 10% of cases.

Travelers should realize that self-treatment is a first-aid measure, and that they should still seek medical advice as soon as possible. In general, travelers carrying SBET should observe the following guidelines:

- Consult a physician immediately if fever occurs 1 week or more after entering an area with malaria risk.
- If it is impossible to consult a physician and/or establish a diagnosis within 24 hours of the onset of fever, start the stand-by emergency treatment and seek medical care as soon as possible for complete evaluation and to exclude other serious causes of fever.
- Do not treat suspected malaria with the same drugs used for prophylaxis.
- Vomiting of antimalarial drugs is less likely if fever is first lowered with antipyretics. A second full dose should be taken if vomiting occurs within 30 minutes of taking the drug. If vomiting occurs 30–60 minutes after a dose, an additional half-dose should be taken. Vomiting with diarrhea may lead to treatment failure because of poor drug absorption.
- Complete the stand-by treatment course and resume antimalarial prophylaxis 1 week after the *first* treatment dose. To reduce the risk of drug interactions, at least 12 hours should elapse between the *last* treatment dose of quinine and resumption of mefloquine prophylaxis.

The drug options for SBET are in principle the same as for treatment of uncomplicated malaria (Chapter 3, Section 3.2). The choice will depend on the type of malaria in the area visited and the chemoprophylaxis regimen taken. Artemether–lumefantrine has been registered for use as SBET for travelers. Chapter 3, Section 3.2 provides details on individual drugs.

4.4.4.2. Multidrug-Resistant Malaria

Multidrug-resistant malaria has been reported from South-East Asia (Cambodia, Myanmar, Thailand, and Vietnam) and the Amazon basin of South America, where it occurs in parts of Brazil, French Guiana, and Suriname. In border areas between Cambodia, Myanmar, and Thailand, *P. falciparum* infections do not respond to treatment with chloroquine or sulfadoxine–pyrimethamine, sensitivity to quinine is reduced, and treatment failures in excess of 50% with mefloquine are being reported. In these situations, malaria prevention consists of personal protection measures in combination with atovaquone-proguanil or doxycycline as chemoprophylaxis. SBET with atovaquone-proguanil or artemether-lumefantrine can be used in situations where the risk of infection is very low. However, these drugs cannot be given to pregnant women and young children. Since there is no prophylactic or SBET regimen that is both effective and safe for these groups in areas of multidrug-resistant malaria, pregnant women and young children should avoid traveling to these malarious areas.

4.5. Chemotherapies instead of Hemoprophylaxis

Poor compliance with chemoprophylaxis regimens has been identified as an important risk factor for malaria among travelers [332]. One US study found that only 23% of travelers to endemic areas had complied with recommended chemoprophylaxis [333]. A study of 547 Dutch travelers (most taking chloroquine/proguanil) showed an overall compliance rate of 60%, but this rate varied depending on the destination (South America 45%, West Africa 52%, Southeast Asia 53%, India 60%, and East Africa 78%) [332]. A study of 100 Spanish travelers showed that 44% did not use any malaria chemoprophylaxis, 27% defaulted from the recommended regimen, and 39% employed self-treatment. Several case-control studies have identified a 2- to 4-fold increased risk of malaria as a consequence of noncompliance with prescribed chemoprophylaxis [334].

Age, reason for travel, and length of travel also correlated with compliance rates. Young people traveling to "adventurous" destinations were the least likely to comply with prophylaxis in the Dutch study. In general, compliance tends to deteriorate with longer terms of exposure or travel, although short-stay travelers also tend to not use or misuse chemoprophylaxis. Immigrants returning to visit friends or relatives in their homelands rarely take chemoprophylaxis or use personal protection against mosquitoes. Lack of risk awareness and limited access to pre-travel health care are probable factors contributing to the suboptimal use of antimalarial measures among immigrant populations [335].

Other factors that might influence compliance with chemoprophylaxis include convenience of the prescribed regimen (daily or weekly, duration of pre-travel and post-travel medication), tolerability of side-effects, and even peer pressure [336]. Among travelers, the potential for serious toxicities is of great concern, and even minor side-effects sometimes cause travelers to discontinue their chemoprophylaxis. This happens even though the potential morbidity from malaria is clearly worse than the side-effects of prophylaxis. Some cases are in areas with severe, complicated, *vivax*, and *ovale* malaria, and the chemoprophylaxis may not good than treatment to reduce the adverse effects of this malaria.

4.5.1. Problems and Options of Presumptive Therapy

4.5.1.1. An Alternative Strategy: Intermittent Presumptive Therapy

An alternative control strategy that has proved effective in reducing the adverse effects of malaria in pregnancy and infancy is to give full treatment doses at intervals [304]. This intermittent presumptive therapy, also called intermittent prevention treatment, (IPT) has been evaluated mainly in areas of high stable transmission, and mainly with SP. IPT currently involves administering treatment doses in the second and third trimesters of pregnancy, or together with routine childhood immunizations at two, three, and nine months of age. (Chapter 4, Sections 4.2. and 4.3) To date, IPT deployment is limited and is largely confined to SP or amodiaquine as monotherapies, but deployment is likely to increase considerably. However, such extensive deployment of antimalarials to healthy pregnant women and infants would provide a selection pressure to the emergence of drug resistance. It might jeopardize the valuable new drugs now being introduced as treatments.

When the emergence of chloroquine-resistant *falciparum* malaria left many travelers who were taking chloroquine with inadequate protection, some physicians evolved a strategy of having travelers to malaria endemic areas carry treatment courses of antimalarial drugs, with these patients not actually taking any medication unless they developed a febrile disease thought to be malaria [329]. This approach has appealed to many travelers as a form of trip insurance against a presumably rare chance of a potentially lethal infection. As drug resistance has increased, selection of possible drug or drug combinations for such presumptive or stand-by therapy has become very limited. Use of presumptive stand-by therapy instead of malaria chemoprophylaxis is based on several assumptions and these need to be examined.

4.5.1.2. Problems of Intermittent Presumptive Therapy

For low malaria risk areas, presumptive or stand by therapy instead of chemoprophylaxis could be justified by the uncertain quality of antimalarial drugs at the traveler's destination, the distance from reliable medical care and the speed of developing life-threatening complications from *falciparum* malaria in a non-immune traveler. Unfortunately, these points in favor of presumptive therapy are often interpreted by the traveler as meaning that seeking medical evaluation for febrile disease is unnecessary and that the traveler could proceed with their planned trip without risk after taking the medication. Travelers' resistance to seeking appropriate medical care locally, despite pre-travel advice to the contrary, seemed to confirm that many travelers viewed presumptive therapy as trip insurance rather than a temporizing medical intervention.

Most febrile disease episodes, even in travelers to sub-Saharan Africa, are not caused by malaria, even if *falciparum* malaria is by far the most common potentially lethal infection. One approach for trying to focus presumptive therapy on those most likely to have malaria was to include a blood diagnostic antigen-capture card along with the malaria medication after instructing the traveler on the use of the test and when to take the medication. This refinement is limited by the ability of sick travelers to perform the test and make medically sound decisions when sick and in unfamiliar settings.

Given the practical limitations of presumptive or stand-by therapy, it is important to select such a strategy only for those most likely to benefit from it. There are exceptional travelers who work far from any medical care, such as some missionaries or field anthropologists, who truly need to be prepared to treat their own presumptive malaria attack before trekking out to find a physician. If these persons are in a very high malaria risk area, presumptive therapy needs to be coordinated with chemoprophylaxis and not used in its stead. This is particularly important when such long-term travelers are taking mefloquine, as subsequent presumptive treatment with halofantrine or quinine would be contra-indicated because of the risk of fatal cardiac dysrrhymias. Presumptive or stand-by therapy should only be chosen for very exceptional travelers, especially for those traveling to tropical Africa. If the risk of malaria is sufficient to be a serious concern for a particular traveler, then chemoprophylaxis is a better course of action than waiting for a potentially lethal infection when one has to rely on the traveler's own medical judgment and limited memory of what was discussed during pre-travel counseling [284].

4.5.1.3. Options for Intermittent Presumptive Therapy

1) Atovaquone/Proguanil

If it is decided that a traveler needs to carry presumptive malaria therapy in addition to other malaria prevention measures, then atovaquone/proguanil is probably the first option to be considered [337]. This does not apply to those already taking atovaquone/proguanil prophylaxis. Its excellent safety and tolerability profile makes it a drug combination that can be given to travelers to very isolated areas with confidence that the medication itself is very unlikely to cause a severe adverse effect, even if taken incorrectly. It is dispensed in a single treatment course blister pack that can easily be carried to very remote locations. Atovaquone absorption is limited if the ill traveler is unable to eat, since fatty food enhances the bioavailability of the drug. Despite scattered reports of drug resistance, atovaquone/proguanil continues to have a very high cure rate for *falciparum* malaria [338].

2) Artemisinin-based Combination Therapy (ACT)

Artemisinin combination therapy (ACT) has become increasingly popular in Southeast Asia for malaria chemotherapy that delays the development of drug resistance, and ACT is also expected to be used more frequently in tropical Africa. ACT can be a reasonable choice for an isolated traveler who needs presumptive therapy. However, two factors mitigate against the use of ACT in travelers, particularly in Southeast Asia. Rarely is any traveler in Southeast Asia more than several hours from adequate medical care, and seeking a well informed local physician for treatment of a febrile illness is a much superior option to carrying stand-by medication. Also, fake or counterfeit antimalarial drugs are increasingly common, particularly in Southeast Asia but also in Africa [339]. Although it is possible to have a European traveler obtain ACT therapy of a known quality by being prescribed a registered combination of artemether/lumefantrine, this is not currently possible in the US or Canada [340]. Mefloquine is often the partner drug in ACT combination therapy in Southeast Asia and the adverse events associated with therapeutic doses of mefloquine are much greater than those seen with lower prophylactic doses of mefloquine [341]. Although ACT is an important chemotherapeutic strategy in the era of drug-resistant *falciparum* malaria, it should only be a very rare choice for a traveler as presumptive or stand-by therapy.

4.5.2. Drug Treatment for Preventing Severe Malaria

Treatment with antimalarial drugs has a major role to play in preventing severe malaria and death. It reduces fever promptly and effectively to interrupt the progression of infection or mild illness to severe disease, and reduces fatality rates in severe malaria. Since the great majority of patients with fever or other symptoms suggestive of malaria receive their initial treatment at home, improving home management of fever is a critical component of this process. Patients and their families need up-to-date and practical guidance on when and how to use antimalarial drugs at home. How to recognize when a patient is not responding to therapy in order to seek medical attention, and the importance of correctly following, at home, the recommendations and treatments that are given in health centers and hospitals. This guidance should be complemented with more user-friendly treatment regimens, improved formulations, especially for the treatment of children, and pre-packaging of antimalarial tablets.

At the health post or health center level, availability of effective drugs is crucial. Health workers need clear guidelines on how antimalarial drugs should be used and how to deal with potential adverse reactions. In addition, there should be facilities to administer fluids, glucose, antibiotics, and anticonvulsants to severely ill patients. In suspected cases of severe malaria, rectal formulations of the artemisinin drugs and other preparations can be used as an emergency pre-referral treatment when parenteral antimalarial therapy is not available, and have the potential to reduce early mortality. In hospitals, prompt confirmation of diagnosis, rapid assessment of the severity of disease and the administration of prompt specific and supportive treatment, including safe blood for transfusion, are all critical. These issues are covered thoroughly elsewhere [2].

It is clear that IPT infants with an antimalarial drug significantly reduces severe malarial anemia and clinical malaria attacks [315-318]. Severe anemia in highly malarious areas is invariably a consequence of malaria and a major cause of infant death within malaria endemic regions in Africa. In recent years, TDR and partners have funded studies to answer the policy question of whether iron supplementation or intermittent chemoprophylaxis during infancy is the better strategy for preventing severe anemia and which should be recommended, by the United Nations Children's Fund (UNICEF) and WHO, on a large scale in high transmission areas. Iron supplementation has been shown to be safe and effective when given orally and daily during the infant's first year of life [342]. A recent follow up study shows that treatment with a single-dose antimalarial drug, sulphadoxine-pyrimethamine (SP), given to infants in an area of intense malaria transmission in Tanzania, reduced severe malarial anemia by 50% and clinical malaria attacks by 59% [315].

The findings of this recent study, therefore, open up the possibilities for an important new strategy for reducing childhood deaths in malaria endemic regions of the world – offering an inexpensive, high impact intervention against malaria in the context of the EPI program. The results of the ongoing studies are awaited with interest.

4.5.3. Therapy for Preventing Relapses of *P. Vivax* or *P. Ovale* Malaria

4.5.3.1. Therapy for Preventing Relapse

The primary goal of malaria chemoprophylaxis is to prevent *P. falciparum* infection, which is primarily responsible for malaria fatalities. However, both *P. vivax* and *P. ovale* can cause serious febrile illness in non-immune patients. The clinical presentation of *P. vivax* and *P. ovale* malaria cannot be distinguished from that of *P. falciparum* malaria [343]. The burden of infection is considerable in indigenous populations, and *P. vivax* is the most prevalent malaria type in Southeast Asia and South America and is found in eastern Africa. *P. vivax* is susceptible to most antimalarial drugs, although isolates with reduced sensitivity to chloroquine and primaquine have been found in parts of Indonesia and in Papua New Guinea, Iraq, and Afghanistan [344]. There is a decreased susceptibility of *P. vivax* to sulfadoxine-pyrimethamine in Thailand [345].

A study from Europe of 518 imported cases of *P. vivax* found that 60% of patients were admitted to the hospital on average 4 days after the start of symptoms, and seven patients had severe complications [346]. *P. vivax* and *P. ovale* develop hypnozoites when the host is infected, which relapse later and may cause malaria symptoms long after the traveler has returned home. Hypnozoites are susceptible only to primaquine. One study found that the first *P. vivax* attack was seen approximately 3 months after leaving the area where malaria is endemic regardless of whether the traveler had taken prophylaxis or not. The prevention of relapsing *P. vivax* infection can be achieved only by presumptive post-travel treatment with a course of primaquine or by using primaquine as a chemoprophylaxis during travel. As long as the absolute risk of infection by *P. vivax* and *P. ovale* is not known, the best strategy is still to inform travelers of the risk of a late-onset attack of *P. vivax* or *P. ovale* after return and cessation of chemoprophylaxis [120].

Relapsing malaria caused by *P. ovale* should be treated with chloroquine and primaquine. Malaria caused by *P. malariae* should be treated with the standard regimen of chloroquine as for *vivax* malaria, but it does not require radical cure with primaquine because no hypnozoites are formed in infection with this species. Returning travelers with severe *falciparum* malaria should be managed in an intensive care unit. Parenteral antimalarial treatment should be with artesunate (first choice), artemether or quinine. If these medicines are not available, use parenteral quinidine with careful clinical and electrocardiographic monitoring. The dosage regimens for the treatment of uncomplicated malaria are provided in Chapter 3, Section 3.2.

The main goal of malaria chemoprophylaxis is to prevent deaths from malaria, which are largely caused by *P. falciparum*. Another important goal is to prevent clinical malaria, which can lead to hospitalization with health and economic consequences. The currently recommended first-line drugs for chemoprophylaxis are 80−90% effective in preventing clinical episodes of primary malaria infection [347]. The antimalarial agents commonly used (e.g., chloroquine, mefloquine, doxycycline, atovaquone-proguanil) that are active against the blood stage parasites; however, do not prevent relapsing infection caused by *P. vivax* and *P. ovale*. These *Plasmodium* species are biologically different and can cause latent infection with hypnozoites in the liver that are not killed by the antimalarials active against the blood stages. Not only can hypnozoites survive despite the presence of a blood stage antimalarial during and after travel, they may emerge to cause symptoms months later. The late appearance of *vivax* or *ovale* malaria in a returned traveler who has taken antimalarials as prescribed reflects the different biological characteristics of *vivax* and *ovale* malaria. Resistance of *vivax* malaria to chloroquine has been well documented [348], although some reports of resistance may reflect relapses of the surviving liver forms. Use of agents active against the blood stage does not prevent infection but delays the first clinically apparent attack of malaria. In addition, atovaquone/proguanil may have some activity against the liver stage of *P. vivax*, it does not reliably prevent *vivax* malaria relapse [349].

In travelers, the percentage of infections caused by *P. vivax* varies widely depending on the common destinations for travelers from that geographic area. In the United States, among reported, the imported cases of malaria from 2001-2004 for which the species was identified, *P. vivax* caused from 22.9−27.8% and *P. ovale* caused from 2.0−3.6% [350]. In European centers, *P. vivax* caused from 9.3% in the Netherlands, 10.4% in France and 29.4% in Germany. *P. vivax* accounted for 12.9% of 4801 cases of imported malaria reported to TropNetEurop (European Network on Imported Infectious Disease Surveillance) between 1999 and 2003. *P. vivax* accounted for 44−51% of malaria in Canada, and 63−74% in

Australia. These studies demonstrate that *P. vivax* contributes significantly to the burden of malaria in travelers and that its relative importance varies by travel destination.

Diagnosis of infections caused by *P. vivax* manifest later may be problematic. The time of relapse also depends on the geographic origin of the parasite; those with parasites of tropical origin have a higher probability of relapse and a shorter interval to relapse than those with parasites of temperate origin and can occur more than a year after travel to a malarious area. The late appearance of *vivax* may delay diagnosis. In the United States, 12–15% of imported *vivax* malaria in 2003-2004 had onset more than 6 months after return [350]. In Europe, half of patients with *vivax* became ill more than 60 days after return from an endemic area [346]. Returned travelers and their health providers may not perceive the risk for *vivax* malaria when prophylaxis has been taken as prescribed. Among Israeli travelers with *vivax* malaria, 80% had taken suppressive prophylaxis [351]. In addition, parasitemia in *vivax* is typically lower than in *falciparum*, so laboratory personnel reading the malaria slides may not find the parasites. In Belgium, 9 (25%) of 36 of untreated patients with *vivax* malaria had parasitemia levels of less than 500 per µl, a level that may be difficult to detect in laboratories in which technicians have limited experience reading malaria smears. The rapid diagnostic tests, currently unavailable in the United States, also are less likely to be positive in *vivax* malaria [343].

4.5.3.2. Primaquine for Preventing Relapse

Presumptive anti-relapse therapy and radical cure to treat established infection with *P. vivax* or *P. ovale* have conducted by primaquine as primary prophylaxis. Primaquine is active against all malarial species [348]. Primaquine has protective efficacy of more than 85% for *falciparum* malaria and primary infections with *P. vivax*. In healthy non-immune Colombian soldiers going to a malarious area, primaquine 30 mg daily provided protective efficacy of 94% against *P. falciparum* and 85% against *P. vivax* compared with placebo, although follow-up was only for 3 weeks in base camp [352]. Primaquine can eliminate liver hypnozoites and is the only drug currently available with such activity. When given as presumptive anti-relapse therapy [353], it is administered to overlap with the blood stage active agent; though data demonstrating the efficacy of presumptive anti-relapse therapy are lacking. Studies done more than 50 years ago suggested that primaquine is more active against hypnozoites when given with chloroquine (or quinine). Individuals considered as possible candidates for primaquine must be screened first for glucose-6-phosphate dehydrogenase deficiency. Administration of primaquine to individuals with deficient glucose-6-phoshate dehydrogenase levels can have serious, even lethal, hemolysis. Primaquine should never be given as primary prophylaxis or as presumptive anti-relapse therapy to individuals with glucose-6-phospate dehydrogenase deficiency.

Although screening needs to be done only once, it involves the expense of testing and delays in decision making. Some strains of *P. vivax* are relatively resistant to primaquine, and high failure rates have been reported with the regimen used in the past. High rates of failure have occurred in persons with the Chesson strain of *P. vivax* from Papua New Guinea and in persons infected in Southeast Asia. A higher dose of 30-mg base daily is now recommended [349]. The total dose may be more important than the schedule of its delivery [348]. Weight-adjusted dosing of primaquine (0.5 mg/kg per day_14 days) has been recommended to prevent additional relapses in patients with *P. vivax* infection. Primaquine primary prophylaxis could be considered after screening to establish normal levels of glucose-6-

phosphate dehydrogenase for individuals with contra-indications to or intolerance of recommended first-line antimalarial agents; multiple, short exposures to malaria endemic areas; or travel to *vivax*-dominant areas [354]. Presumptive anti-relapse therapy should be considered for individuals who have had prolonged stays in malarious areas where *P. vivax*, *P. ovale*, or both are present; specific high-risk itineraries. In areas with high risk of exposure to *vivax* malaria, use of primaquine, either as primary prophylaxis or as post-exposure anti-relapse therapy, is an option.

Chapter V

Influence Factors
of Antimalarial Drug Efficacy

Many factors that might influence efficacy with chemotherapy and chemoprophylaxis include convenience of the prescribed regimen (daily or weekly, duration of malaria. The aim of antimalarial drug treatment in severe malaria is to save life of the patients. In uncomplicated malaria it is to reduce the parasite biomass to zero, or down to a level where host defenses can deal with the remainder. Treatment regimens with rapidly eliminated drugs must generally cover four asexual life-cycles to eradicate all the parasites in the blood. For slowly eliminated drugs, blood concentrations must exceed the parasites' minimum inhibitory concentration (and preferably the minimum parasiticidal concentration) until all parasites have been eradicated. Similar to other infections of HIV and tuberculosis, malaria does not always respond satisfactorily to treatment. It is easy to understand why a malarial infection would fail to respond to antimalarial treatment or prophylaxis if the drugs were not taken properly, or the parasites were intrinsically resistant to the drugs given.

The greatest problem is with drug resistance occurring with *P. falciparum*. Resistance means that there is a right shift in the concentration effect relationship. Although treatment failure in malaria usually results from poor compliance, inadequate dosing, pharmacokinetic and pharmacodynamic factors or resistance, some infections will recrudesce when none of these factors operates. How parasites persist despite apparently adequate antimalarial treatment remains unresolved. Unless the patient was observed during the administration of antimalarial treatment and for a minimum of 1 h after dosing, poor compliance or vomiting must always be considered as a possible contributor to treatment failure. In general, people do not like to take medicines or pills once they feel they have recovered from an illness. Thus treatment regimens longer than 3 days are often complied with poorly. Poor compliance results in low blood concentration and thus inadequate exposure of the infecting parasite population to therapeutic concentrations of drug [355]. Many limitations towards the satisfactory efficacy of antimalarial drugs are highlighted as being high malaria transmission rates, high reservoir of asymptomatic infections in semi-immune persons, inappropriate use of drugs, inadequate diagnostic facilities and capacities, ill-informed policy-makers and weak public health systems especially in Africa. Also, the limited knowledge on pharmacokinetics

(PK), pharmacodynamics (PD) and PK/PD relationship of antimalarial drugs are impacts of effective use of the drugs.

5.1. Efficacy Dependents on Antimalarial Drug Resistance

Antimalarial drug resistance is defined as the ability of a parasite strain to survive and/or multiply despite the proper administration and absorption of an antimalarial drug in the dose normally recommended. Drug resistance to an antimalarial compound results in a right shift in the concentration–effect (dose–response) relationship (Figure 10).

Drug resistance has arisen in most of the antimalarial drugs that have been used for more than 65 years, all except artemisinin and its derivatives. Widespread and indiscriminate use of antimalarials places a strong selective pressure on malaria parasites to develop high levels of resistance. Resistance of *P. falciparum* is of particular concern because of the enormous burden of disease caused by this species, its lethal potential, the propensity for epidemics, and the cost of candidate replacement drugs for areas with established drug resistance. Resistance to antimalarials has been documented for *P. falciparum, P. vivax,* and, recently, *P. malariae.* In *P. falciparum*, resistance has been observed to almost all currently used antimalarials (chloroquine, amodiaquine, mefloquine, quinine, and sulfadoxine-pyrimethamine) except for artemisinin and its derivatives. This has increased the global malaria burden and is a major threat to malaria control.

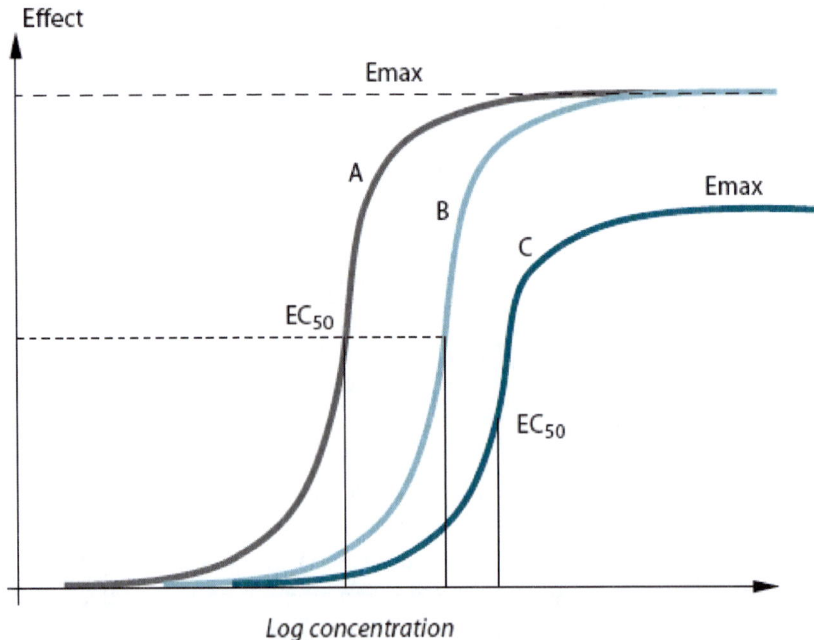

Figure 10. Resistance is a rightward shift in the concentration–effect relationship for a particular parasite population. This may be a parallel shift (B) from the "normal" profile (A) or, in some circumstances, the slope changes, and/or the maximum achievable effect is reduced (C). The effect is parasite killing [2, 3].

The risk of resistance varies according to species, strain, and drug. Estimates of risk are often confounded by the determinants already described. Methods of estimating risk also vary. The following estimates of the risk of treatment failure were distilled from many studies and should not be construed as quantitative region-specific risks, because patterns of resistance vary tremendously, even within nations. Protocols that evaluate efficacy generally follow the guidelines of the World Health Organization, which recommend a follow-up period of 7, 14, or 28 days [14]. The timing of recurrent parasitemia or disease reflects the degree of resistance - earlier failure represents higher-grade resistance. Late recurrence may be confounded by re-infection, especially in regions of intense transmission. Investigators in sub-Saharan Africa thus favor an *in vivo* test with duration of 7 or 14 days, whereas those in other regions tend to favor a 28-day test or even 63-day test. Comparison of genotypes of the strains causing original and recurrent parasitemias can be used to address confounding by re-infection, but relatively few reporting clinics or laboratories have the capacity to conduct such testing. Tests conducted in Africa tend to report both parasitological failure (i.e., recurrent parasitemia after treatment, independent of clinical presentation) and the clinical failure of treatment, whereas investigators elsewhere focus on parasitological failure.

The following methods are available for assessing resistance to antimalarials:

- *in vivo* assessment of therapeutic efficacy (see section 6.1),
- *in vitro* studies of parasite susceptibility to drugs in culture, and
- molecular genotyping.

Impacts of antimalarial drug resistance on health systems [4]:

- Prolonged or recurrent illness
- Increased outpatient cases
- More progression to severe malaria
- Increased cerebral malaria – neurological squeal
- Prolonged or worsening anemia and effect of anemia
- Increased hospital admissions and death
- Increased gametocyte carriage
- Increased demand on diagnosis
- Higher cost of combination treatment

Resistance can be monitored and prevented, or its onset slowed considerably, by combining antimalarials with different mechanisms of action and ensuring very high cure rates through full adherence to correct dose regimens. In general, with this class of compounds, it is the dose regimen, pharmacokinetic and pharmacodynamic properties of the drug used, and the number of parasites causing the infection that determines the risk of treatment failure. The development of resistance to this group of antimalarial drugs is best illustrated by the development of mefloquine resistance. This initially develops slowly, as treatment regimens are sufficient to eradicate the infecting parasite biomass, but the residual therapeutic blood concentrations that occur weeks or months after drug administration provide selective pressure on any newly acquired infections. At this early stage there is no selective pressure on the treated infection, as all the parasites are removed. Eventually some infections, particularly those with a high parasite biomass, or those in children or pregnant women who have less immunity, will survive the therapeutic onslaught and recrudescences occur. As mentioned previously, these tend to occur late (i.e., many weeks after treatment) at low levels of drug resistance and thus many parasite life-cycle s are exposed to concentrations of mefloquine below the MPC.

For bacteria, concentrations of antibiotics which exhibit between 20–80% of the maximum possible inhibitory activity provide the strongest selective pressures to the emergence of resistance [356]. The same probably pertains to malarial parasites. This selection process *in vivo* drives the emergence of resistance. The selective pressure is related to the reciprocal of the terminal elimination rate of antimalarial drugs constant and the reciprocal of the slope of the antimalarial concentration± effect relationship. The force that drives resistance is also related to the proportion of people harboring and capable of transmitting parasites who receive antimalarial drugs. The pressure that drives resistance is much higher. As recrudescent parasites are more likely to produce gametocytes, and therefore be transmitted, than those parasites causing primary infections, there is rapid selection of resistance in low-transmission areas. While it might be thought that chloroquine would provide the strongest selection pressure of all in this group of drugs, because of its extremely long terminal elimination half-life (1- 2 months), at intermediate levels of resistance the terminal elimination phase occurs with blood concentrations which may be below those providing the maximum selective pressure. This may partly explain why chloroquine resistance has been relatively slow to develop. In contrast, the terminal elimination phase for mefloquine covers a wider concentration range and contributes a significant part of the blood-concentration *vs.* time profile and is therefore the major determinant of selective pressure to resistance to this drug.

There is recent evidence that exposure of parasite populations to antimalarial drug pressure may select for resistance not only to the drug class providing the pressure but also to novel, unrelated, antimalarial drugs [357]. There is some epidemiological support for this, in that *P. falciparum* parasites from Southeast Asia tend to be resistant to all antimalarials, even those to which they have not been exposed [355].

5.1.1. The End of Chloroquine Age due to its Resistance

5.1.1.1. Resistance to Chloroquine

The growing problem of antimalarial drug resistance has greatly difficult in treatment for *falciparum* malaria. Whereas chloroquine and sulfadoxine-pyrimethamine could once cure most malaria, this is no longer possible and requires examination of alternative regimens. When chloroquine was introduced more than 65 years ago, after an analogue that had been developed in Germany was captured in 1943, chloroquine quickly came into universal use as therapy for and prophylaxis against malaria, it was considered a wonder drug. Successful treatment was then achieved with only two tablets, corresponding to a 300 mg chloroquine base. Chloroquine remained effective in Africa and in major parts of the other continents for nearly 30 years. Chloroquine was highly effective, easily administered, and inexpensive and had good safety and tolerability.

However, the first report of chloroquine resistance in *P. falciparum* appeared in the late 1950s in Thailand and Colombia and emerged in the 1970s in New Guinea and eastern sub-

Saharan Africa around 12 years after the drug's introduction. By 1980, all endemic areas in South America were affected, and by 1989, most of Asia and Oceania. In Africa, chloroquine resistance emerged in 1978 in the east, and gradually spread westwards through the 1980s. Resistance has now been documented in all *falciparum*-endemic areas except Central America and the Caribbean [12]. Recent molecular studies favor importation of chloroquine resistance to Africa from East Asia [358]. Chloroquine resistance has emerged independently less than ten times in the past 50 years (Figure 1).

Today, resistance to chloroquine in malaria caused by *P. falciparum* occurs everywhere except in Central America (and Hispaniola) and in some regions of southwestern Asia. In sub-Saharan Africa, the risk of parasitological treatment failure with the use of chloroquine was almost uniformly greater than 40%, whereas the risk of clinical treatment failure tended to be higher in the eastern and central portions of the African continent (typically >30 % and often >50%, respectively) than in the west (typically < 20 %). Elsewhere, the rates of parasitological failure at day 28 were 57% (of 209 evaluations) in southwestern Asia, percent (2280 evaluations) in southern Asia, 85% (223 evaluations) in Southeast Asia, and 66% (137 evaluations) in South America (Figure 1). Prescribing chloroquine monotherapy against this parasite in any setting, except one in which its effectiveness has recently been demonstrated, should be considered irresponsible [121].

Case management of malaria has relied largely on the use of single antimalarial agents, such as chloroquine, sulfadoxine-pyrimethamine and, more recently, mefloquine. Since chloroquine and sulfadoxine/pyrimethamine are inexpensive and widely available, their misuse has been extensive. Synthesized in 1943, chloroquine was considered too toxic to be used in humans and did not become the drug of choice for treatment of malaria until 1946 [359]. In 1957, the emergence of chloroquine-resistant *P. falciparum* was identified in Southeast Asia. Four years later, resistance to the drug was reported in Venezuela and Colombia [360]. In 1973, chloroquine was no longer recommended as a first-line therapy by Thailand, followed by South America and Asia. In 1993, single dose of chloroquine was ineffective in 35% *P. falciparum* and Malawi was the first African country to replace chloroquine with sulfadoxine/pyrimethamine [360]. Resistance to sulfadoxine/pyrimethamine was reported in 1967, the same year in which the drug was introduced, whereas mefloquine resistance was first identified in 1982, 5 years after its introduction (Figure 11) [53, 245, 360].

More recently, in order to improve the monitoring of the antimalarial drug resistance in Madagascar, a new national network based on eight sentinel sites was set up. A multisite randomized clinical trial was designed to assess the therapeutic efficacy of chloroquine (CQ), sulphadoxine-pyrimethamine (SP), amodiaquine (AQ), and artesunate plus amodiaquine combination (ASAQ). Children between six months and 15 years of age, with uncomplicated *falciparum* malaria, were enrolled. Primary endpoints were the day-14 and day-28 risks of parasitological failure, either unadjusted or adjusted by genotyping. A total of 1,347 of 1,434 patients (93.9%) completed treatment and follow-up to day 28. All treatment regimens, except for the CQ treatment group (32.0−44.0% follow-up to day 28) resulted in clinical cure rates above 97.6% by day-14 and 96.7% by day-28 (Figure 12). These findings show that antimalarial drug resistance remains low in Madagascar, except for CQ [361].

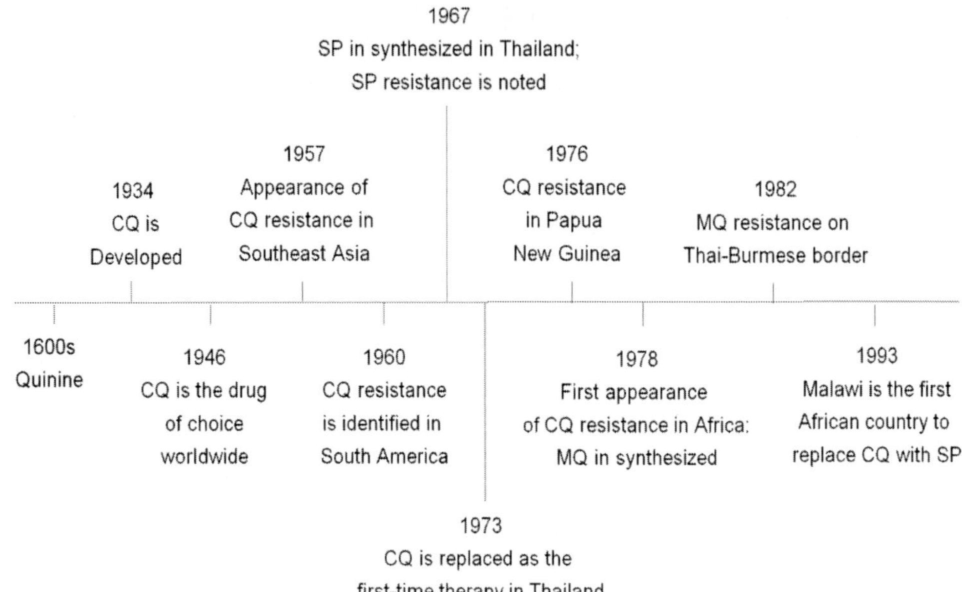

Figure 11. History of antimalarial drug resistance [245]. CQ = chloroquine; MQ = mefloquine; and SP = sulfadoxine-pyrimethamine.

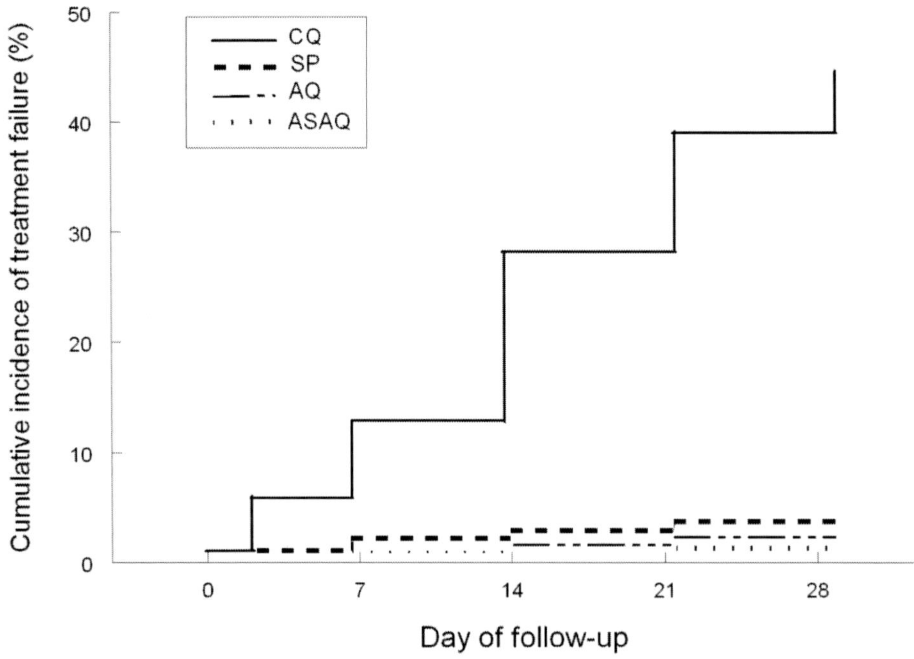

Figure 12. Kaplan-Meier curve of cumulative treatment failure over the 28-day follow-up period, adjusted by genotyping [361].

Widespread use of these agents contributed to selection pressure and emergence of *P. falciparum* resistance to almost all currently used antimalarials (i.e., amodiaquine, chloroquine, mefloquine, quinine, and sulfadoxine/pyrimethamine) with the exception of artemisinin and its derivatives. Resistance has also been observed with *P. vivax* and *P. malariae*, with *P. vivax* being resistant to sulfadoxine/pyrimethamine and chloroquine for many areas. Resistance of *P. vivax* is rare and generally limited to chloroquine resistance, which was first reported in the late 1980s in Papua New Guinea and Indonesia [2]. As a result, treatment including chloroquine or sulfadoxine-pyrimethamine is often ineffective [4]. Resistance to chloroquine by *P. vivax* has emerged, apparently having originated in New Guinea, where failure rates now approach 100% [362]. In contrast, surveys in Thailand reveal uniformly sensitive *vivax* malaria. Chloroquine-resistant *P. vivax* may be characterized as endemic to the Indonesian archipelago, especially in the east (including New Guinea), sporadic in the rest of Asia, and rare in South America [276].

5.1.1.2. Resistance to other Antimalarial Drugs

1) Sulfadoxine–pyrimethamine

Sulfadoxine–pyrimethamine has potent efficacy against chloroquine-resistant and pyrimethamine-resistant *P. falciparum*, and became available in 1971;to be used as the standard second-line therapy against chloroquine-resistant *falciparum* malaria. This combination acts synergistically against folate synthesis, inhibiting dihydropteroate synthase and dihydrofolate reductase. Although rare, idiosyncratic allergic reactions have occurred among users of sulfa drugs, sulfadoxine-pyrimethamine has otherwise offered superior safety and tolerability, along with the advantage of single-dose therapy. Resistance to pyrimethamine emerged rapidly after its deployment for treatment, prophylaxis and, in some areas; mass treatment in the 1950s. Resistance to both components of sulfadoxine-pyrimethamine was noted shortly after this drug was introduced over a decade later. However, resistance to sulfadoxine-pyrimethamine was recognized at the Thai-Cambodian border in the 1960s, and failures occurred in refugee camps in Thailand in the 1970s [12]. Resistance became an operational problem in the same area within the few years of the introduction of sulfadoxine-pyrimethamine to the malaria control program in 1975.

High-level resistance is found in many parts of South-East Asia, southern China and the Amazon basin, and lower levels of resistance are seen on the coast of South America and in southern Asia and Oceania. In eastern Africa, sulfadoxine-pyrimethamine sensitivity was observed to be declining in the 1980s and resistance has progressed westwards across Africa relentlessly over the last decade. Clinical failure rates of more than 25% have already been reported in Liberia [363], Guinea Bissau [364], and Malawi [365]. Many areas now have high-level resistance with high-treatment failure rates in children. Recent molecular evidence suggests a common South-East Asian origin of the resistant *P. falciparum* parasites now prevalent in much of southern and Central Africa (triple *dhfr* mutant) [2, 366] (Figure 1).

The rates of parasitological failure of treatment with sulfadoxine-pyrimethamine in sub-Saharan Africa were relatively high in southern regions of holoendemic infection (> 50 percent) and low elsewhere (< 5%, except in Cameroon, where the rate of failure was 10%). The rates of clinical treatment failure in sub-Saharan Africa were similarly distributed (< 5% in the west and 8−34% in the east and south) [42]. Studies in Southeast Asia indicated that the rates of parasitological failure at day 7 and day 28 were 36% and 49%, respectively. Good efficacy (80%) persisted elsewhere - in southwestern Asia and on the Horn of Africa, where no parasitological failures were reported among 362 evaluations; in southern Asia, where the failure rate was 18% by day 28; and in South America, where the failure rate was 9% by day 7, 14% by day 14, and 6% by day 28. The risk of resistance to sulfadoxine-pyrimethamine is relatively high in Southeast Asia and eastern Africa [120].

2) Mefloquine

Mefloquine emerged as a successor to chloroquine in 1983, but its resistance was first observed on the Thai-Cambodian and Thai-Burmese borders in 1991 [367] and the monotherapy are no longer effective there. Migrant gem miners returning from Cambodia may have been the means for the spread of mefloquine resistance to India and Bangladesh [368]. Resistance in that border region remains high [369]. Isolated cases of mefloquine resistance have also been reported from the Amazon basin, and *in vitro* studies in Africa have identified some *P. falciparum* strains with low mefloquine sensitivity. Overall, clinical mefloquine resistance outside South-East Asia is rare (Figure 1). The main determinant of mefloquine resistance is amplification of the gene (*Pfmdr*) that encodes the multidrug transporter [22]. Amplification occurs only for the "wild type" allele explaining the inverse relationship between sensitivity to chloroquine (the *Pfmdr* Tyr86 mutation is associated with reduced sensitivity) and to mefloquine (and to the structurally related drugs, quinine and halofantrine) [370].

In another study, of 79 patients hospitalized in Bangkok and treated with mefloquine (25 mg per kilogram), 68 (86%) remained free of parasitemia after 28 days [371]. Similar efficacy was observed in Bangladesh [372]. However, a group of Dutch marines in Cambodia who were receiving mefloquine prophylaxis during the 1990s had high attack rates of *falciparum* malaria [373]. In contrast, the protective efficacy of mefloquine among Indonesian soldiers in western New Guinea - where the efficacy of chloroquine against *P. falciparum* and *P. vivax* approaches zero - was 100% [374]. The risk of the prophylactic or therapeutic failure of mefloquine in Southeast Asia appears to be low outside the shared borders of Thailand, Myanmar, and Cambodia. Few studies have evaluated the effectiveness of mefloquine against *falciparum* malaria in Africa. In the mid-1990s, Lobel *et al.* examined mefloquine prophylaxis among 140 volunteers of Peace Corps that were infected by *P. falciparum* [375].

Poor adherence explained most of the infections, but in five cases the resistance appeared to be genuine. Despite sporadic, well-documented failures of mefloquine among travelers to Africa, the drug remains effective there. Studies conducted in coastal Peru and in the Amazon Basin among 153 subjects with *P. falciparum* malaria who were receiving mefloquine (at a dose of 15 mg per kilogram) revealed complete sensitivity [376]. Despite occasional reports of resistance in the Amazon basin, the available evidence shows a low risk of resistance throughout South America. Mefloquine has received adverse attention in the media in recent years. The drug has been blamed for suicides, homicides, and other personal tragedies on the basis of anecdotal accounts, which, by their nature, cannot establish cause and effect. Well-controlled trials consistently indicate that mefloquine given as prophylaxis is as well tolerated as other antimalarial drugs [377]. Nonetheless, the drug has been linked with a higher risk of insomnia, fatigue, and adverse neuropsychiatric effects (e.g., depression and anger) than other antimalarial drugs [378]. The risk appears highest among women, especially those taking the drug for the first time and those with a low body-mass index [379].

3) Quinine

The first reports of possible quinine resistance occurred in Brazil almost 100 years ago. Even today, however, clinical resistance to quinine monotherapy is reported only sporadically in South-East Asia and western Oceania, and resistance in Africa and South America is much less frequent. Zalis *et al.* suggested that there might be a link between poor *in vitro* responses to quinine and diminished clinical responsiveness in the Amazon basin, but without supporting clinical data [380]. Widespread use of quinine in Thailand in the 1980s led to significant reduction in its sensitivity [368]. Quinine is therefore now used in combination with an antibiotic, usually tetracycline, doxycycline or clindamycin, and is reserved for cases of severe malaria. *Pfmdr1* mutations associated with chloroquine resistance have been believed to be associated with reduced susceptibility to quinine [380]. Recent well-controlled trials of quinine monotherapy showed variable rates of efficacy against *P. falciparum*: 92% in a group of 49 patients in Bangladesh [372]; 67% among 54 patients in western Thailand [381]; and 80% among 30 patients in a clinic in Bangkok [382]. A seven-day regimen was more than 95% efficacious in Venezuela [383] and Equatorial Guinea [384].

Rare reports of failure of intravenous quinine for the treatment of severe or complicated malaria have appeared sporadically [385]. Oral quinine is used to treat uncomplicated malaria, generally over a period of three to seven days in combination with another blood-stage schizonticide, typically tetracycline or doxycycline. Although one study of 86 patients in Thailand showed the superiority of a seven-day regimen combined with tetracycline (a 100% efficacy rate) as compared with a five-day regimen combined with tetracycline (87%), studies performed elsewhere have shown complete efficacy with shorter regimens combining quinine, doxycycline, and primaquine [374, 386]. Poor adherence carries a high risk of treatment failure, particularly because quinine causes a syndrome of adverse effects known as cinchonism, including primarily tinnitus, nausea, and vertigo. A randomized trial in Thailand recorded a 71% rate of adherence [387]. However, in Cambodian villages, the rate of compliance with the same regimen was far lower — 11 to 20% — even after an intervention to raise awareness about the need for compliance [120].

4) Primaquine

In endemic regions, primaquine is widely used as a gametocytocide to prevent infection of mosquitoes. Despite more than 50 years of use in millions of people per year, primaquine is still shrouded in confusion and genuine mystery. At least four different regimens, some prescribed for only one week and others for as long as eight weeks, are aimed at the same objective of preventing relapse [353]. Three factors largely explain the lack of uniformity: first, the reluctance of health care programs to accept therapy that has a duration of two weeks; second, perceived issues of toxicity and tolerability; and third, the use of a total dose, rather than a dosing schedule, as the primary determinant of efficacy. How the same total dose of a rapidly eliminated drug kills organisms irrespective of whether it is delivered over a period of 7, 14, or 56 days defies explanation. Resistance is not fully understood either. Resistance in asexual blood stages of *P. vivax* has long been known but is of no clinical consequence [388]. Reports of resistance to tissue-stage schizonticidal activity fail to consider or describe patients' adherence, to exclude the possibility of reinfection, or to address the possible recrudescence of chloroquine resistant strains. Lack of evidence of resistance to primaquine in liver stages probably reflects a heavy burden of proof rather than an absence of resistance [353].

5) Multidrug Resistance

Multidrug resistance is generally defined as resistance to three or more antimalarial compounds from different chemical classes. Generally, the first two classes are 4-aminoquinolines (e.g., chloroquine) and antifolates (e.g., sulfadoxine-pyrimethamine). The precise amount of resistance needed in order for a drug to be considered as failing is not universally agreed. Some consider clinical cure rates of less than 75% to be the minimum required for classification as failure, while the current recommendations aim for cure rates over 90%. Established multidrug resistance occurs in South-East Asia (particularly along the borders of Thailand with Burma and Cambodia) and in the Amazon basin. In Thailand, mefloquine monotherapy was replaced with the combination of high-dose mefloquine and artesunate given for 3 days. Mefloquine resistance has been reduced by the use of this combination, as cure rates of more than 95% have been sustained for over 10 years, and susceptibility to mefloquine has actually improved despite extensive deployment of the combination. Several areas are at risk of multidrug resistance, as resistance to chloroquine and sulfadoxine-pyrimethamine is already widespread. Progressive loss of sulfadoxine-pyrimethamine efficacy should be taken as a warning sign.

5.1.2. Current only Free Resistance Drugs: Artemisinin Class

The artemisinin derivatives are currently the most rapidly acting and potent antimalarial drugs [6]. These pharmacodynamic effects are due to their rapid absorption and activity against many stages of the malaria life cycle, from young asexual forms (rings) to early sexual forms (gametocytes). The elimination half-life of drugs is short, which protects them from resistance. The reduce gametocyte carriage and infectivity and have been the main reason why transmission has been reduced on the Thai-Myanmar border [389]. Except in an animal model, there have been no confirmed reports of artemisinin resistance in malaria parasites that infect humans. The pharmacological characteristics of the drug, namely short elimination half-life, rapidity of action and ability to reduce gametocyte carriage, should delay the onset of significant resistance. Artemisinin derivatives are associated with high recrudescence rates (~10%) after monotherapy, so are usually combined with longer-acting antimalarials for clinical treatment. These recrudescences, however, are not a result of resistance.

Although a target for the artemisinins has recently been identified *(PfATPase6)*, preliminary studies have not so far associated polymorphisms in the gene encoding this enzyme with reduced susceptibility of malaria parasites. Amplification in *PfMDR* does reduce artemisinin susceptibility *in vitro*, but not to a degree that causes *in vivo* resistance. This has lead to erroneous claims that artemisinin resistance was being selected by widespread use of artemisinin derivatives, whereas in fact the selection pressure came from mefloquine use. The mutation frequencies derived from *in vitro* studies are often much higher than those derived from observations *in vivo* [357]. The absence of host defenses and differences in antimalarial concentration profiles contribute to this discrepancy. The highest rates for emergence of resistance *in vivo* are with pyrimethamine and atovaquone. In the case of atovaquone, it has been estimated that one in three patients with symptomatic *falciparum* "contained," at presentation have a spontaneously arising atovaquone-resistant mutant parasite [51]. For drugs such as chloroquine or artemisinin, the genetic events conferring resistance are much rarer. These genetic events may result in moderate changes in drug susceptibility, such that the drug still remains effective; or, less commonly, very large reductions in susceptibility such that achievable concentrations of the drug are completely ineffective [2].

The artemisinin compounds act by the heme-catalyzed intra-parasitic production of highly reactive carbon-centered free radicals [390]. Certain hemoglobinopathies are associated with reduced antimalarial activity of artemisinin, possibly because of reduced intra-erythrocytic availability of iron to catalyze opening of the peroxide bridge. Stable resistance to artemisinin or its derivatives has not occurred in clinical practice, and cannot yet be induced in the laboratory. Parasites reported as showing reduced susceptibility may simply represent the tail-end of a normal distribution of drug susceptibility [391], and remain well within the range of plasma concentrations obtained with clinical use of the drugs. It would be unwise to consider that resistance to these compounds cannot occur, although the information available to date is reassuring, and suggests that resistance will not emerge rapidly with appropriate prescribing in clinical practice. But this too depends on drug availability and use outside the medical system; for example, in some countries, misguided entrepreneurs have incorporated extracts of *A. annua* in herbal tonics and confectionery, providing a generally available and thus potentially powerful selective pressure for the emergence of resistance.

Stable, therapeutically significant resistance to the artemisinin derivatives has not yet been identified and cannot be induced yet in the laboratory, which suggests that it may be a very rare event [5]. A distinct set of *pfmdr1* mutation are not strongly associated with resistance to artemisinin derivatives in field isolates in Africa [392]. But it would be foolish to bank on this not happening, and should it arise, it would be a global disaster. Although clinically relevant resistance to artemisinin derivatives has not yet been reported and although *P. falciparum* isolates may vary two- to four-fold in their *in vitro* sensitivities to these drugs [393], artemisinin resistance is likely to develop in the near future. Recently, a strain of the murine malaria parasite *P. yoelii* with unstable resistance to artemisinin was obtained [395]. Further study demonstrated that although no DNA sequence difference was found, but the resistant strain was found to express 2.5-fold-more protein than the sensitive strain (P < 0.01). Thus, the phenotype of artemisinin resistance in *P. yoelii* appears to be multifactorial [233, 396]. For mutual protection against the emergence of drug resistance, these drugs should be used only in combination with other antimalarials. Artemisinin derivatives are particularly effective in combinations because of their very high killing rates (parasite reduction ratios 10,000-fold per cycle), lack of adverse effects, and absence of significant resistance [6].

5.1.3. Factors Contributing to the Spread of Resistance

The emergence of antimalarial drug resistance is dependent on the occurrence of a spontaneous genetic change (mutation or gene amplification) in a malaria parasite, and interferes with that parasite's susceptibility to a drug. A single mutation may be sufficient to confer almost complete resistance to some drugs (e.g., atovaquone) or more usually there is a series of mutations that confer increasing tolerance of the parasite to increasing drug concentrations, as in the cases of pyrimethamine and chloroquine [397]. However, for resistance to spread, the spontaneous occurrence of a mutation in itself is not sufficient. In the absence of the drug to which it is potentially resistant, a parasite with the resistant mutation does not have a survival advantage and therefore does not reproduce faster than the non-mutants. There may even be a survival disadvantage, a so-called fitness cost to having the mutation [398]. In the presence of the particular drug, the multiplication of the sensitive parasites is inhibited allowing the drug-resistant mutants to survive and multiply (i.e., selection), increasing the likelihood of transmission to the next host and therefore the spread of resistance.

Numerous factors contributing to the advent, spread, and intensification of drug resistance exist, although their relative contribution to resistance is unknown. Factors that have been associated with antimalarial drug resistance include such disparate issues as human behavior (dealt with in detail elsewhere), vector and parasite biology, pharmacokinetics, and economics. As mentioned previously, conditions leading to malaria treatment failure may also contribute to the development of resistance [399].

5.1.3.1. Molecular and Biological Influences on Resistance

The development of resistance can be considered in two parts: the initial genetic event, which produces the resistant mutant; and the subsequent selection process in which the survival advantage in the presence of the drug leads to preferential transmission of resistant mutants and thus the spread of resistance. A striking conclusion emerging from the molecular evolutionary studies described is that resistance genes may spread rapidly within countries and jump between continents. Both chloroquine resistant *pfcrt* and high-level *dhfr* alleles have invaded Africa from Southeast Asia, while Asian *pfcrt* alleles have also established in South America. The response of sensitive parasites to antimalarial drugs *in vitro* and the pharmacokinetic profiles of common antimalarial drugs are thought to always be a residuum of parasites with the resistance genes that are able to survive treatment. Under normal circumstances, these parasites are removed by the immune system [400]. Factors that decrease the effectiveness of the immune system in clearing parasite residuum after treatment also appear to increase survivorship of parasites and facilitate development and intensification of resistance. This mechanism has been suggested as a significant contributor to resistance in South-East Asia, where parasites are repeatedly cycled through populations of non-immune individuals; the nonspecific immune response of non-immune individuals is less effective at clearing parasite residuum than the specific immune response of semi-immune individuals. The same mechanism may also explain poorer treatment response among young children and pregnant women [6].

Some characteristics of recrudescent or drug resistant infections appear to provide a survival advantage or to facilitate the spread of resistance conferring genes in a population [401]. In one study, patients experiencing chloroquine treatment failure had recrudescent infections that tended to be less severe or even asymptomatic [402]. Schizont maturation may also be more efficient among resistant parasites [403]. There is some evidence that certain combinations of drug-resistant parasites and vector species enhance transmission of drug resistance, while other combinations inhibit transmission of resistant parasites. In South-East Asia, two important vectors, *Anopheles stephensi* and *A. dirus*, appear to be more susceptible to drug-resistant malaria than to drug sensitive malaria. In Sri Lanka, researchers found that patients with chloroquine-resistant malaria infections were more likely to have gametocytemia than those with sensitive infections and that the gametocytes from resistant infections were more infective to mosquitos [402]. The reverse is also true - some malaria vectors may be somewhat refractory to drug resistant malaria, which may partially explain the pockets of chloroquine sensitivity that remain in the world in spite of very similar human populations and drug pressure.

Many antimalarial drugs in current usage are closely related chemically and development of resistance to one can facilitate development of resistance to others. Chloroquine and amodiaquine are both 4-aminoquinolines and cross-resistance between these two drugs is well known [404]. Development of resistance to mefloquine may also lead to resistance to halofantrine and quinine. Antifolate combination drugs have similar action and widespread use of sulfadoxine-pyrimethamine for the treatment of malaria may lead to increased parasitological resistance to other antifolate combination drugs [405]. Development of high levels of SP resistance through continued accumulation of DHFR mutations may compromise the useful life span of newer antifolate combination drugs such as chlorproguanil/dapsone (LapDap) even before they are brought into use. This increased risk of resistance due to SP use may even affect non-malarial pathogens; use of SP for treatment of malaria increased resistance to trimethoprim/sufamethoxazole among respiratory pathogens [406].

The underlying mechanism of this plasticity is currently unknown, but this capacity may help explain the rapidity with which South-East Asian strains of *falciparum* develop resistance to new antimalarial drugs. The choice of using a long half-life drug (SP, MQ) in preference to one with a short half-life (artemisinins) has the benefit of simpler, single dose regimens which can greatly improve compliance or make directly observed therapy feasible. Unfortunately, that same property may increase the likelihood of resistance developing due to prolonged elimination periods. The relative contribution of low compliance versus use of long half-life drugs to development of resistance is not known. Parasites from new infections or recrudescent parasites from infections that did not fully clear will be exposed to drug blood levels that are high enough to exert selective pressure but are insufficient to provide prophylactic or suppressive protection [407].

When blood levels drop below the minimum inhibitory concentration (the level of drug that fully inhibits parasite growth), but remain above the EC5 (the concentration of drug that produces 5% inhibition of parasite growth), selection of resistant parasites occurs. This selection was illustrated in one study in Kenya that monitored drug sensitivity of parasites reappearing after SP treatment. Parasites reappearing during a period when blood levels were below the point required to clear pyrimethamine-resistant parasites, but still above that level required to clear pyrimethamine-sensitive parasites, were more likely to be pyrimethamine-resistant than those reappearing after levels had dropped below the level required to clear pyrimethamine-sensitive parasites [408]. This period of selective pressure lasts for approximately one month for mefloquine, whereas it is only 48 hours for quinine [407].

In areas of high malaria transmission, the probability of exposure of parasites to drug during this period of selective pressure is high. In Africa, for instance, people can be exposed to as many as 300 infective bites per year (in rare cases, even as much as 1000 infective bites per year), and during peak transmission, as many as five infective bites per night [409]. It is not clear what the relationship between transmission intensity and development of resistance is, although most researchers agree that there seems to be such an association.

It is apparent that there are more genetically distinct clones per person in areas of more intense transmission than in areas of lower transmission [410]. However, the interpretation of this and its implications for development of resistance has variously been described as resistance being more likely in low-transmission environments [411], high-transmission environments [410, 412], or either low- or high- but not intermediate-transmission environments [413]. This relationship between transmission intensity and parasite genetic structure is obviously complex and subject to other confounding/contributing factors [413, 414]. What is clear is that the rate at which resistance develops in a given area is sensitive to a number of factors beyond mere intensity of transmission (such as initial prevalence of mutations, intensity of drug pressure, population movement between areas, the nature of acquired immunity to the parasite or its strains, etc.), but that reducing the intensity of transmission will likely facilitate prolonging the useful life span of drugs [415].

5.1.3.2. Programmatic Influences on Resistance

The programmatic influences on development of antimalarial drug resistance include overall drug pressure, inadequate drug intake (poor compliance or inappropriate dosing regimens), pharmacokinetic and pharmacodynamic properties of the drug or drug combination, and drug interactions [368]. Additionally, reliance on presumptive treatment can facilitate the development of antimalarial drug resistance. Overall drug pressure, especially when exerted by programs utilizing mass drug administration, probably has the greatest impact on development of resistance. Studies have suggested that resistance rates are higher in urban and periurban areas than rural communities, where access to and use of drug is greater. Confusion over proper dosing regimen has been described. In Thailand, the malaria control program recommended 2 tablets (adult dose) of SP for treating malaria based on studies suggesting that this was effective. Within a few years, this was no longer effective and the program increased the regimen to 3 tablets. Although unproven, this may have contributed to the rapid loss of SP efficacy there. Similar confusion over the proper SP dosing regimen exists in Africa.

To simplify treatment, many programs dose children based on age rather than weight and, depending on the regimen being recommended, this has been shown to produce systematic under dosing among children of certain weight and age groups. The use of presumptive treatment for malaria has the potential for facilitating resistance by greatly increasing the number of people who are treated unnecessarily but will still be exerting selective pressure on the circulating parasite population. In some areas and at some times of the year, the number of patients being treated unnecessarily for malaria can be very large [416]. Concurrent treatment with other drugs can increase the likelihood of treatment failure and may contribute to development of drug resistance. Folate administration for treatment of anemia when used as a routine supplement can increase treatment failure rates during pregnancy [417].

Similarly, concurrent illness may have an influence, as was mentioned previously with regard to malnourishment. Drug quality has also been implicated in ineffective treatment and possibly drug resistance. Either through poor manufacturing practices, intentional counterfeiting, or deterioration due to inadequate handling and storage, drugs may not contain sufficient quantities of the active ingredients. In an analysis of chloroquine and antibiotics available in Nigeria and Thailand, between 37% and 40% of samples assayed had substandard content of active ingredients, mostly from poor manufacturing practices [418]. Another study in Africa found chloroquine stored under realistic tropical conditions lost at least 10% of its activity in a little over a year [399, 419].

5.1.3.3. Mismatched Pharmacokinetics Influence on Resistance

Multidrug resistance in parasites has forced the use of combination antimalarial regimens. The need for effective treatments has thrown together antimalarial combinations that have often worked better than monotherapies, though sometimes only temporarily. During the usage of combinations, mismatched pharmacokinetics can play a role in facilitating the development of resistance. The mismatched pharmacokinetics allow parasites to evolve resistance sequentially as the longer half-life partner persists as a vulnerable monotherapy (Figure 13), a result that can almost completely undermine the benefits of combination therapy [420]. Coartem (the Novartis brand name for the artemether-lumefantrine combination) was one of the first ACTs to be deployed, and field data provided by Sisowath *et al.* have provided the first indications of its likely fate [421]. They showed that parasites carrying the 86N form of the *pfmdr* gene are killed by treatment with Coartem but that their increased tolerance of residual drug levels allows them to re-infect a person more rapidly after his or her treatment, compared with parasites carrying the 86Y form of the gene. The 86N form therefore encodes increased tolerance of the drug, rather than clinical resistance, and spreads because of its ability to successfully infect people soon after treatment with the drug (Figure 13).

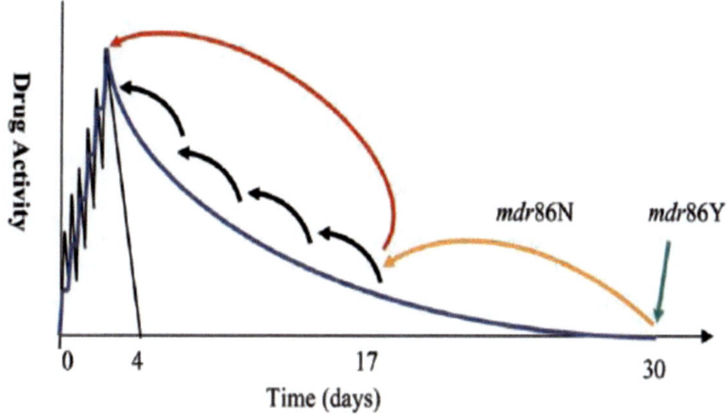

Figure 13. Likely evolution of resistance to Coartem (the Novartis brand name for the artemether-lumefantrine combination). The graph represents the drug levels in the serum of a patient starting the 3-day Coartem therapy regimen, a combination therapy (CT) comprising lumefantrine *(blue)* and artemether *(black)*. The drugs have decayed to non-therapeutic levels when they cross the X-axis (defined as the concentration at which the drug, if present as a monotherapy, would be unable to prevent infection by drug-sensitive parasites). The *pfmdr* 86N mutation's *(orange arrow)* increased tolerance of lumefantrine allows it to infect a person 17 days post-treatment, compared with 30 days post-treatment for the *mdr*86Y mutation. Two scenarios are given for subsequent mutational steps. Scenario 1 is a single large increase in drug tolerance *(red arrow),* in which the presence of artemether in the CT is powerless to stop the spread of this mutation, whose increased tolerance of lumefantrine allows it to infect a person 4 days post-treatment and to spread through the parasite population, displacing the 86N form, which can infect no earlier than 17 days post-treatment. Scenario 2 is a subsequent increase in tolerance, which occurs as a series of small increments *(black arrows);* the presence of artemether in the CT will slow this process when it reaches day 4, and Coartem may remain clinically effective for a longer period [420].

These observations raise a new additional point: can ACT accelerate the development of resistance to artemisinin derivatives? The presence of the *pfmdr1* 86N allele has also been shown to confer less sensitivity to artemisinin *in vitro* [23]. Although artemisinin or its derivative (short half-life) may not select *pfmdr1* 86N, lumefantrine (long half-life) pressure may provide the main selective force on the allele and, thus, co-drive a decrease in sensitivity to both compounds. The results of this study reinforce the idea that an understanding of the molecular basis of drug resistance of ACT components is fundamental for its maintained efficacy in malaria control. Since this strategy remains our best hope for a long-term treatment, the unambiguous message of these studies is that ACTs are not a panacea and that they may have a limited therapeutic life span. The first recommendation that needs to be considered is that ACTs with better-matched elimination rates should be developed, probably by using a partner drug that does not persist as long as lumefantrine. The second recommendation is that formalized mechanisms for the monitoring of resistance should be considered to be essential, to provide early indications of problems, thereby enabling the long-term strategic planning necessary to obtain the maximum benefit from ACT [420].

However, the wrong performances are still going on and the drug elimination rates are mismatched [67]. The elimination half-life of artemisinins are between 0.5-2.6 hours, but pyrimethamine is between 80 and 100 hours, is between 100 and 200 hours for sulfadoxine, and is between 336 and 432 hours for mefloquine, leaving an extended period when sulfadoxine is "unprotected" by synergy with pyrimethamine [408]. This sort of mismatched pharmacokinetics is even more apparent in the sulfadoxine-pyrimethamine, mefloquine + sulfadoxine-pyrimethamine, artesunate + sulfadoxine-pyrimethamine, artesunate + mefloquine, artesunate + amodiaquine, and proguanil + atovaquone combinations used in Asia and Africa [67, 422].

While the numbers of cases in the trials were small and uncertainties exist concerning the relevant tissue concentration of the agents, the trial does call attention to what may happen when there are mismatches in antimalarial drugs pharmacokinetics. The solution to the mismatch problem is conceptually straightforward: formulate and administer combination therapies in which pharmacokinetic profiles superimpose. Treatment failure due to other factors, such as lack of patient compliance with therapy regimens, will still occur. However, few new malaria strains should be generated as long as the equivalent of monotherapy is avoided. Such a result would strongly support the two-mutation idea for restricting the selection of resistance.

For the best pharmacokinetic match combination, there are still far from discovery of an ideal regimen. The best options available now include: artesunate + amodiaquine, which works well in Africa, although little is known about how this combination translates to high transmission areas; and clindamycin + quinine, which is effective in non-immune individuals, suggesting that clindamycin + artesunate should also be studied. Dihydroartemisinin-piperaquine and fosmidomycin-clindamycin are promising treatment candidates. Until better regimens become available there are enormous demands on policy makers to choose between existing combinations, many of which have not been sufficiently studied for informed judgments to be made.

5.1.3.4. Heavy Malaria Transmission and Selection on Resistance

The recrudescence and subsequent transmission of an infection that has generated a *de novo* resistant malaria parasite is essential for resistance to be propagated [51]. Gametocytes carrying the resistance genes will not reach transmissible densities until the resistant biomass has expanded to numbers close to those producing illness (>107 parasites). Thus to prevent resistance spreading from an infection that has generated *de novo* resistance, gametocyte production from the recrudescent resistant infection must be prevented. There has been debate as to whether resistance arises more rapidly in low or high-transmission settings [413], but aside from theoretical calculations, epidemiological studies clearly implicate low-transmission settings as the source of drug resistance. Chloroquine resistance and high-level sulfadoxine-pyrimethamine resistance in *P. falciparum* both originated in South-East Asia and subsequently spread to Africa [2].

In low-transmission areas, the majority of malaria infections are symptomatic and selection therefore takes place in the context of treatment. Relatively large numbers of parasites in an individual usually encounter antimalarials at concentrations that are maximally effective. But in a variable proportion of patients, for the reasons mentioned earlier, blood concentrations are low and may select for resistance. In high-transmission areas, the majority of infections are asymptomatic and infections are acquired repeatedly throughout life. Symptomatic and sometimes fatal malaria occurs in the first years of life, but thereafter it is increasingly likely to be asymptomatic. This reflects a state of imperfect immunity (premunition), where the infection is controlled, usually at levels below those causing symptoms. The rate at which premunition is acquired depends on the intensity of transmission. In the context of intense malaria transmission, people still receive antimalarial treatments throughout their lives (often inappropriately for other febrile infections), but these "treatments" are largely unrelated to the peaks of parasitemia, thereby reducing the probability of selection for resistance. Immunity considerably reduces the emergence of resistance [2].

Host defense contributes a major antiparasitic effect, and any spontaneously generated drug-resistant mutant malaria parasite must contend not only with the concentrations of antimalarial present, but also with host immunity. This kills parasites regardless of their antimalarial resistance, and reduces the probability of parasite survival (independently of drugs) at all stages of the transmission cycle. For the blood stage infection, immunity acts in a similar way to antimalarials both to eliminate the rare *de novo* resistant mutants and stop them being transmitted (i.e., like a combination therapy), and also to improve cure rates with failing drugs (i.e., drugs falling to resistance) thereby reducing the relative transmission advantage of resistant parasites. Even if a resistant mutant does survive the initial drug treatment and multiplies, the chance that this will result in sufficient gametocytes for transmission is reduced as a result of asexual stage immunity (this reduces the multiplication rate and lowers the density at which the infection is controlled) and transmission-blocking immunity. Furthermore, other parasite genotypes are likely to be present, competing with the resistant parasites for red cells, and increasing the possibility of out breeding of multigenic resistance mechanisms or competition in the feeding anopheline mosquito [423].

5.1.3.5. Other Factors Influences on Resistance

Moreover, the development to the spread of drug resistance by *P. falciparum* has been attributed to following a combination of factors:

- Human migration
- Severe infections, and
- Sustained and often haphazard use of drugs [408]

5.1.4. Preventing Resistance of New Antimalarial Drugs

Resistance has already developed to all the antimalarial drug classes with one notable exception - the artemisinins. These drugs are already an essential component of treatments for multidrug resistant *falciparum* malaria. If we lose artemisinins to resistance, we may be faced with untreatable malaria. The emergence of resistance to current antimalarial drugs is considered in two parts: first, the initial genetic event that produces the resistant mutant, and second, the subsequent selection process in which the survival advantage in the presence of the antimalarial drug leads to preferential transmission and the spread of resistance. Interventions aimed at preventing drug resistance, generally, focus on reducing overall drug pressure through more selective use of drugs; improving the way drugs are used like methods for prescribing, follow-up practices, and patient compliance; or using drugs or drug combinations that are inherently less likely to foster resistance or have properties that do not facilitate development or spread of resistant parasites.

5.1.4.1. Strategies for Preventing Spread and Emergence

Once a drug-resistant mutant has developed, preventing spread of resistance is difficult. Spread is facilitated by the exposure of malarial parasites to sub-therapeutic levels of antimalarial drugs, that kill sensitive parasites but allows parasites with a resistance mutation to survive and reproduce. Ensuring that drugs are taken in at a sufficient dose and for a sufficient duration reduces this risk.

Drug pressure is higher where a drug with a long half-life is taken because the drug remains in the patient's blood at low levels for weeks, exposing any newly introduced malarial parasites to sub-therapeutic levels [424]. This is particularly likely to occur in high transmission areas where people are not only infected more frequently, but also take antimalarial drugs frequently whether or not they are have malaria. Theoretically, this form of drug pressure can be reduced by using drugs with a shorter half-life and by restricting the use of the first line drug to patients with confirmed malaria: i.e., only treating those with a definitive diagnosis. There are drawbacks to both of these strategies that pit the long-term public health benefit against the benefit to the individual patient. Using drugs with short half-lives such as artesunate means that if they are used together with other rapidly eliminated drugs they need to be taken for a longer period resulting in poorer patient adherence and less likelihood of cure compared with drugs with longer half-lives such as mefloquine or SP, which can be taken over a three-day period or in a single dose. However, in combination with another effective drug, ACTs only require three days of treatment. Restricting usage of effective drugs to patients who have a definitive diagnosis of malaria would reduce access to cure to the most vulnerable communities because the availability and reach of diagnostic facilities is so poor.

This would be expected to lead to an increase in current morbidity and mortality, a trade-off between current and future burden of disease, which is clearly unacceptable. The use of rapid diagnostic tests that require minimal training and equipment may be a potential solution. However, their cost of US $0.7 per test and the high levels of asymptomatic parasitemia and therefore false-positive results in areas of high transmission, limit the appropriateness of this technology to lower transmission areas. The cost-effectiveness of the diagnostic tests depends on their cost relative to that of the ACT and the positive predictive value of clinical diagnosis.

For example, they are unlikely to be cost-effective when more than 50% of the patient group with a clinical diagnosis of malaria indeed does have an infection [134].

Because it is difficult to control antimalarial drug resistance once it has emerged, there is a need for strategies that prevent the rare event of initial emergence. Combinations of drugs which have different molecular targets delay the emergence of resistance. However, malaria control programs may be reluctant to adopt this strategy because until resistance emerges, there is no evident benefit to the more expensive combination treatment.

5.1.4.2. Reducing Overall Drug Pressure

Because overall drug pressure is thought to be the single most important factor in development of resistance, following more restrictive drug use and prescribing practices would be helpful, if not essential, for limiting the advent, spread, and intensification of drug resistance. This approach has gained support in North America and Europe for fighting antibacterial drug resistance [425]. The greatest decrease in antimalarial drug use could be achieved through improving the diagnosis of malaria. Although programs such as IMCI aim to improve clinical diagnosis through well designed clinical algorithms, a large number of patients will continue to receive unnecessary antimalarial therapy, especially in areas of relatively low malaria risk. Basing treatment on the results of a diagnostic test, such as microscopy or a rapid antigen test, however, would result in the greatest reduction of unnecessary malaria treatments and decrease the probability that parasites are exposed to subtherapeutic blood levels of drug.

There are notable exceptions to this, however. Presumptive therapy with SP during pregnancy has been shown to be an operationally sustainable intervention that offers significant protection from low birth weight associated with placental malaria [301]. There may be a role for presumptive treatment or even mass drug administration in response to an epidemic, although its cost-effectiveness has not been proven. Prophylaxis programs, however, should be used only among populations where compliance is likely to be high and where a highly effective prophylactic drug can be used.

5.1.4.3. Improving the Approach Drugs are Used

Other disease control programs, such as for TB, STDs, and HIV, have begun to rely heavily on directly observed therapy (DOT) as a way to ensure a high degree of compliance. While this has not yet received serious consideration for malaria, the use of drugs with single-dose regimens (SP, mefloquine) could potentially make DOT possible. The benefits of using single-dose DOT need to be weighed against the costs of using drugs with long half-lives. Another approach that has not been widely adopted is the close follow-up and re-treatment, if necessary, of patients. The success of this approach is dependent on availability of reliable microscopy (to diagnose the illness initially, as well as to confirm treatment failure), and either an infrastructure to locate patients in the community or a community willing to return on a given date, regardless of whether they feel ill or not. With this system, patients who fail initial treatment, for whatever reason, are identified quickly and re-treated until parasitologically cured, decreasing the potential for spread of resistant parasites [426].

5.1.4.4. Combination Therapy

A strategy that has received much attention recently is the combination of antimalarial drugs, such as mefloquine, SP, or amodiaquine, with an artemisinin derivative [43]. Artemisinin drugs are highly efficacious, rapidly active, and have action against a broader range of parasite developmental stages. This action apparently yields two notable results. First, artemisinin compounds, used in combination with a longer acting antimalarial, can rapidly reduce parasite densities to very low levels at a time when drug levels of the longer acting antimalarial drug are still maximal. This greatly reduces both the likelihood of parasites surviving initial treatment and the likelihood that parasites will be exposed to suboptimal levels of the longer acting drug [401]. Second, the use of artemisinins has been shown to reduce gametocytogenesis by 8- to 18-fold [46]. This reduces the likelihood that gametocytes carrying resistance genes are passed onwards and potentially may reduce malaria transmission rates. Use of combination therapy has been linked to slowing the development of mefloquine resistance and reductions in overall malaria transmission rates in some parts of Thailand and has been recommended for widespread use in sub- Saharan Africa [55].

The theory underlying combination treatment of tuberculosis, leprosy and HIV infection is well known, and has recently been applied to malaria. If two drugs with different modes of action, and therefore different resistance mechanisms, are used in combination, then the per-parasite probability of developing resistance to both drugs is the product of their individual per-parasite probabilities. Stable resistance to the artemisinin derivatives has not yet been identified, and cannot yet be induced in the laboratory, which suggests that it may be very rare indeed.

The ideal pharmacokinetic properties for an antimalarial have been much debated. Rapid elimination ensures that the residual concentrations do not provide a selective filter for resistant parasites, but drugs with this property (if used alone) must be given for at least 7 days, and adherence to 7-day regimens is poor. In order to be effective in a 3-day regimen, elimination half-lives usually need to exceed 24 h. Artemisinin derivatives are particularly effective in combinations with other antimalarials because of their very high killing rates (parasite reduction rate around 10 000-fold per cycle), lack of adverse effects and absence of significant resistance [51]. Combinations of artemisinin derivatives (which are eliminated very rapidly) given for 3 days, with a slowly eliminated drug such as mefloquine, provide complete protection against the emergence of resistance to the artemisinin derivatives if adherence is good, but they do leave the slowly eliminated "tail" of mefloquine unprotected. Perhaps resistance could arise within the residual parasites that have not yet been killed by the artemisinin derivative. However, the number of parasites exposed to mefloquine alone is a tiny fraction (less than 0.00001%) of those present in the acute symptomatic infection. Furthermore, these residual parasites "see" relatively high levels of mefloquine and, even if susceptibility was reduced, these levels may be sufficient to eradicate the infection (Figure 14). The long mefloquine tail does, however, provide a selective filter for resistant parasites acquired from elsewhere, and therefore contributes to the spread of resistance once it has developed. Yet on the north-west border of Thailand, an area of low transmission where mefloquine resistance had already developed, systematic deployment of the artesunate-mefloquine combination was dramatically effective in stopping resistance and also in reducing the incidence of malaria [53]. This strategy is thought to be effective at preventing the emergence of resistance at higher levels of transmission, where high-biomass infections still constitutes the major source of *de novo* resistance [2].

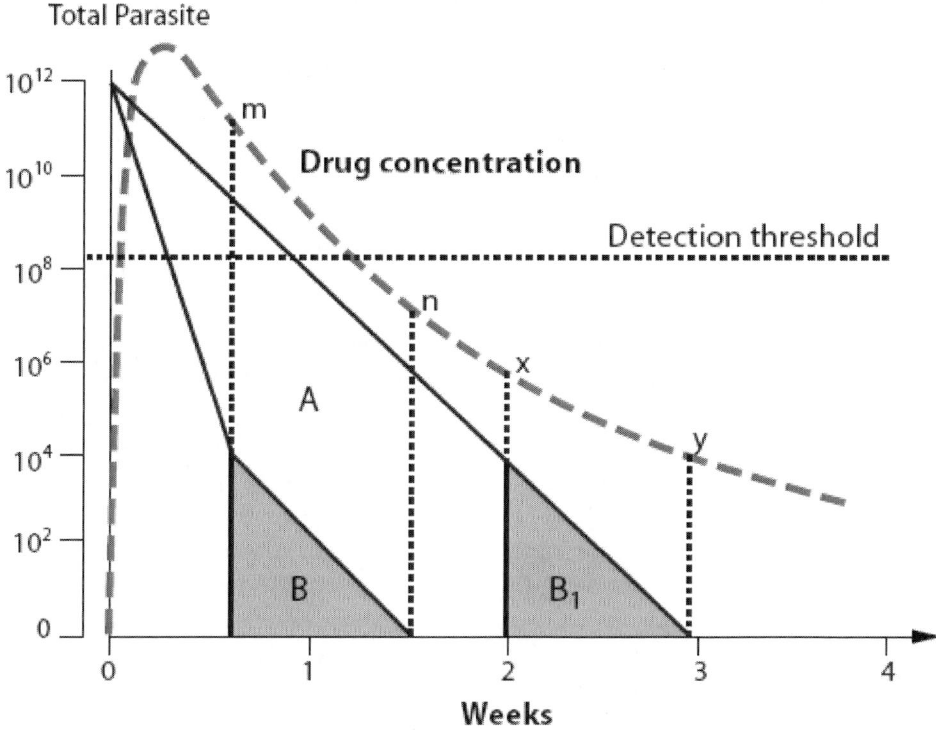

Figure 14. The artesunate + mefloquine combination. If no artesunate is given, then the number of parasites exposed to mefloquine alone is given by the area of A; with the combination administered for 3 days, the number of parasites exposed to mefloquine alone is given by the area of B (100 million times fewer). Furthermore, mefloquine levels are higher (m to n) when confronting B than when confronting the same number of parasites (B1) if no artesunate is given (x to y). If a parasite containing a *de novo* mefloquine-resistant mutation were to occur, then such a parasite should still be susceptible to artesunate. Thus the probability of selecting a resistant mutant is reduced by 100 million times, as only a maximum of 100,000 parasites are exposed to mefloquine alone after the fourth day (i.e., in the third cycle), and any artesunate-resistant parasite selected by artesunate initially would always be killed by the accompanying mefloquine. As a result, the combination is more effective, reduces transmission and prevents the emergence of resistance to both drugs. [2]

5.2. Therapeutic Efficacy Relates to Drug Pharmacokinetics (PK)

Most of the antimalarial drugs now in use to maximize their efficacy were introduced before the modern era of dosage design based on pharmacokinetic and pharmacodynamic principles. Antimalarial dose regimens have tended to be empirical. Then again, drug half-life, PK interindividual variation, drug exposure level and time have been inclined to be experiential or referred by data from the public trials.

5.2.1. Role of Antimalarial Drug Half-life to Resistance and Efficacy

5.2.1.1. Long Elimination Half-life Inducing Drug Resistance

In high transmission area (three infectious bites per person per day), a person who takes antimalarial treatment for symptomatic malaria exposes not only the parasites causing that infection to the drug, but also any newly acquired infections that emerge from the liver during the drug's elimination phase; the longer the terminal elimination half-life, the greater the exposure. The length of the terminal elimination half-life is an important determinant of the propensity for an antimalarial to select for resistance [408, 424]. In particular, the definition of optimal patterns of use for drugs with very different pharmacokinetic properties remains unresolved. Malaria is a blood infection. The response to treatment is determined by the blood concentrations of the drug. Drugs with a long elimination half-life have two valuable therapeutic properties.

First, they can provide long-term protection against re-infection (up to two months, in the case of pyrimethamine-sulphadoxine). Prolonged antimalarial drug activity *in vivo* is an advantage to the patient who is recovering from malaria in an area of moderate or high transmission, because further disease episodes are prevented. This assists recovery from anemia, a major cause of malaria morbidity and a contributor to mortality. Similarly, in epidemic malaria, where the population is highly susceptible and the clinical attack rate can approach 100%, but where the epidemic is of short duration and infrequent [427], long half-life drugs are of value. Second, long half-life drugs require a few (or only a single) administrations, which reduces the risk of under dosage and some of the problems of compliance. However, drugs that are eliminated slowly persist in the patient once the immediate therapeutic purpose has been achieved and the residual drug constitutes a potent selective force for the emergence of drug resistance [408].

However, there are also two related reasons for the expansion of resistance. First, there is an increased chance that a new, unrelated, infection will be acquired while drug concentrations following the treatment of an earlier infection have fallen below those sufficient to prevent the multiplication of the newly acquired malaria parasites. This risk depends on the frequency of infection and the shape of the drug's terminal elimination phase. In the case of mefloquine, with a terminal elimination half-life of between approximately 14 days in acute malaria and 21 days in healthy subjects, drug concentrations decline mono-exponentially and very slowly through the relatively wide concentration range associated with intermediate (20−80%) inhibitory activity for most *P. falciparum* parasites. In contrast to chloroquine, that has a terminal elimination half-life of between 1 and 2 months. This terminal phase occurs at concentrations that may already be below those exerting significant inhibitory effects.As such, although the drug has a long elimination half-life, the terminal elimination phase may exert relatively little selective pressure; paradoxically the more resistant the parasites, the less selective pressure this phase exerts. This may explain why chloroquine resistance has taken so long to develop despite pharmacokinetic properties which would appear, at first sight, to encourage the development of resistance.

The second factor that encourages the development of resistance in drugs which are slowly eliminated is within-patient *in vivo* selective pressure that occurs when an infection fails to be eradicated from the body. Multiple asexual cycles are exposed to progressively lower antimalarial drug concentrations. Given these conditions, if a high proportion of infections are treated and the mutations conferring resistance provide no survival disadvantage to the parasite, then resistance will develop rapidly. The importance of the elimination half-life as a determinant for the vulnerability of a drug is the development of resistance has been studied in detail by Watkins and Mosobo [408] in Kenya . They have provided persuasive evidence that short half-life dihydrofolate-reductase inhibitors are less vulnerable than the more slowly eliminated pyrimethamine/sulfadoxine and have evaluated the combination of chlorproguanil and dapsone as an alternative (half-life of1-2 days) [427].

A conflict therefore arises: drugs with a long half-life are beneficial at an individual therapeutic level but are disadvantageous at a population level. Several previous models have examined the spread of antimalarial drug resistance [423, 428,429]. These studies have clarified the general features underlying the process but contain two notable omissions [430]. First, they ignore the effects of drug elimination half-life: essentially they assume it to be zero and one study shows that the half-life may lead to serious underestimates of the rate of spread of resistance. Second, they assume that there are only two gene classes in the population: fully sensitive and fully resistant. Field evidence suggests that this second assumption is often invalid, an exception being atovaquone resistance where a single point mutation in the *cyt b* gene produces complete resistance. Resistance increases either gradually (e.g., mefloquine resistance) or in a stepwise manner from the baseline sensitivity, through a series of increasingly less sensitive forms, to the fully resistant form (e.g., pyrimethamine resistance).

These stringent definitions of resistance are necessary to make the mathematics soluble and have been employed in most previous population genetic models of antimalarial drug resistance [430]. There are assumed to be three general classes of susceptibility of the malaria parasite *P. falciparum* to a drug: *Res0*, the original, susceptible wild type; *Res1*, a group of intermediate levels of susceptibility that are more tolerant of the drug but still cleared by treatment; and *Res2*, which is completely resistant to the drug. *Res1* and *Res2* resistance both evolve much faster if the antimalarial drug has a long half-life. We show that previous models have significantly underestimated the rate of evolution of *Res2* resistance by omitting the effects of drug half-life. The methodology has been extended to investigate (i) the effects of using drugs in combination, particularly when the components have differing half-lives, and (ii) the specific example of the development of resistance to the antimalarial pyrimethamine-sulphadoxine [424].

An important detail of the model is the development of drug resistance in two separate phases. In phase A, *Res1* is spreading and replacing the original sensitive forms while *Res2* remains at a low level. Phase B starts once parasites are selected that can escape drug action (*Res1* genotypes with borderline chemosensitivity, and *Res2*): these parasites are rapidly selected, a process that leads to widespread clinical failure. Drug treatment is clinically successful during phase A, and health workers may be unaware of the substantial changes in parasite population genetic structure that predicate the onset of phase B. Surveillance programs are essential, following the introduction of a new drug, to monitor effectively changes in treatment efficacy and thus provide advance warning of drug failure. The model is also applicable to the evolution of antibiotic resistance in bacteria: in particular, the need for these models to incorporate drug pharmacokinetics to avoid potentially large errors in their predictions.

Several previous papers have discussed the effects of dosage and/or drug half-life in the context of drug resistance [431-433]. However, these studies all investigated their effects in determining whether an infection would actually be cleared by the therapeutic treatment. One study has investigated an entirely different aspect of drug half-life: it is assumed that therapeutic treatment always clears the current infection and focuses on the consequences of the drug's subsequent persistence, in chemoprophylactic concentrations, as a selective pressure for the evolution of drug resistance. The addition of this aspect of pharmacokinetics to models of drug resistance has resulted in several novel findings, as detailed earlier. Since this manuscript is long, we feel it appropriate to reiterate the most important findings, as follows. (i) A long elimination half-life at therapeutic concentrations results in long periods of chemoprophylaxis, so is a potent selective force increasing the rate at which resistance evolves. The absence of elimination half-lives from previous models is a significant omission resulting in underestimated rates of evolution of resistance. (ii) Resistance encoded by multiple mutations at a single locus may occur in two phases, phase A: where the drug is better tolerated by the parasites but therapeutic doses still usually clear the infection, and phase B, when clinical failures start to occur. (iii) Phase B is very rapid and it is essential that surveillance programs are in place and capable of monitoring the change from phase A to B. (iv) Phase A may occur more quickly, but the subsequent phase B more slowly, in areas of high transmission. (v) These general principles are strongly supported by field evidence for the antimalarial SP. (vi) Combination therapy significantly slows the rate of evolution of resistance but it should be instigated before significant resistance to either component is present in the population [424].

5.2.1.2. Short Elimination Half-life Avoiding Drug Resistance

Some rapidly eliminated antimalarials (e.g., the artemisinin derivatives) never present an intermediate drug concentration to infecting malaria parasites because they are eliminated completely within the two-day life-cycle of the asexual parasite. Other antimalarials (e.g., mefloquine, chloroquine) have elimination half-lives of weeks or months and present a lengthy selection opportunity. With the exception of the artemisinin derivatives, maximum antimalarial parasite reduction ratios (kill rates) do not exceed 1000-fold per cycle [6]. Following hepatic schizogony, exposure of at least two asexual cycles (4 days) to therapeutic drug concentrations is therefore required to eradicate the blood stage parasites emerging from the liver. Even with maximum kill rates in the sensitive parasites and maximum growth rates in the resistant parasites, the resistant parasites only "overtake" the sensitive parasites in the third asexual cycle. Thus rapidly eliminated drugs (such as the artemisinin derivatives or quinine) cannot select during the elimination phase. Obviously, the greater the degree of resistance conferred by the resistance mutation – i.e., the higher the $IC_{90}R$ relative to the IC_{90} for susceptible parasites ($IC_{90}S$) – the wider is the window of selection opportunity.

Patent gametocytemia is more likely in recrudescent than in primary infections. Therefore, if *de novo* resistance arose in an acute symptomatic treated infection, the transmission probability from the subsequent recrudescent infection (bearing the new resistance genes) would be higher than from an infection newly acquired during the elimination phase of the antimalarial given for a previous infection, even if it attained the same parasite densities [6].

The artemisinin derivatives and quinine are eliminated relatively rapidly (half-life, <16 h), so within hours or days of stopping treatment, drug concentrations in blood have fallen to sub-inhibitory levels. After quinine treatment is stopped, residual drug levels may still have some direct inhibitory effect on parasite multiplication for three to four half-lives (48 h), whereas artemisinin and its biologically active metabolites are eliminated in hours. If some viable parasites still survive when the concentrations of these drugs in blood are sub-therapeutic, their multiplication is then constrained only by any residual second-cycle or post-antimalarial effect or by immune defenses which may retard the initial phase of re-expansion of the parasite population. Recrudescences tend to occur earlier than those following mefloquine treatment, often occurring around 2 weeks after treatment are stopped. Thus, the selective pressure to drug resistance with artemisinin derivatives or quinine is minimal [359, 434]. Resistance therefore develops slowly or not at all. It is also evident that recrudescences that occur more than 4 weeks after antimalarial treatment with a drug that is eliminated rapidly is stopped are most unlikely. This result accords with clinical observations.

5.2.2. Dosing Regimens and Antimalarial Drug Efficacy

5.2.2.1. Various Half-lives Affecting Drug Regimens

As pathological processes in malaria result exclusively from the blood-stage infection, blood concentrations of the antimalarial drug determine the therapeutic effects directly, and it is not necessary to consider extravascular distribution and tissue penetration in assessing antimalarial drug responses. Nevertheless, many of the antimalarial drugs are distributed extensively in the body, by binding to tissues, leading to a large apparent volume of distribution, and they are cleared slowly. As a result residence times are long, and blood levels, exceeding those required to inhibit parasite multiplication, may persist for weeks or months following a course of treatment. The terminal elimination half-lives of chloroquine (one to two months), mefloquine (two to three weeks), halofantrine (one to three days), lumefantrine (three to four days), pyrimethamine (three days), sulphadoxine (ten days), and atovaquone (one to three days), are such that either a single dose or a short course (one to three days) of the drug provides a complete antimalarial treatment, whereas the considerably shorter half-life of lasting quinine (16 h) and artemisinin derivatives (1h) must be given at least once daily for seven days for complete treatment [6, 359]. Although the malarial parasites are intra-erythrocytic, the concentration of antimalarial to which they are exposed corresponds closest with the free (unbound) plasma concentration rather than the red-cell concentration (derived from measurements mainly in uninfected erythrocytes). Acute malaria may (i) affect the absorption of drugs because of reduced food intake, vomiting, or reduced visceral perfusion, and (ii) change the apparent volume of distribution by altering plasma and tissue binding, and also reduce clearance by impairment of liver and renal function.

5.2.2.2. Dosing Regimens and Antimalarial Drug Resistance and Efficacy

The contribution of under dosing to antimalarial treatment failure has been underappreciated. Most recommended dosage regimens are based on studies in non-pregnant adult patients. Young children and pregnant women, who bear the heaviest malaria burden, have the highest treatment failure rates [435]. This has been attributed previously to lower immunity, although blood concentrations of many antimalarial drugs are significantly lower in pregnant women and young children than in non-pregnant adults. Nevertheless, there have been no studies of higher dosages. Sub-therapeutic concentrations will certainly contribute to poorer responses to treatment and will fuel the emergence and spread of antimalarial drug resistance. In addition, in uncomplicated malaria it is the extent of the absorption of the antimalarial drug, rather than the rate of its absorption, which is important. In constrast to severe malaria, where the objective of treatment is to save life, the rate of drug absorption from an extra-vascular site can be a critical determinant of outcome. This is particularly relevant to the regimens for traditional drugs with long half-lives for inhibiting parasites completely for uncomplicated malaria, and of artemisinins with rapid effect for saving life.

Mature schizonts and tiny rings are the most drug-resistant stages of the blood infection [436]. Most of the available antimalarial drugs are active predominantly on the sequestered parasites that are not visible to the microscopist. The consequences of this are inhibition of trophozoite development and a considerable reduction in the parasite multiplication rate. As susceptibility is stage dependent, the timing of drug administration with respect to the stage and synchronicity of parasite development may determine the therapeutic response (chronotherapy). Antimalarial treatment with chloroquine and, possibly, some other compounds can be optimized if peak drug levels coincide with the most sensitive stage of parasite development [437]. This is rather difficult to arrange in practice. Artemisinin and its derivatives are peroxidic antimalarial agents which are particularly active against ring stage parasites. In the early Chinese studies with these drugs, treatment was started only when most of the infecting parasites were at the tiny-ring stage of development [438]. The advantages of chronotherapy remain to be established, although this is an important area for further research. The frequency of drug administration is usually determined by pharmacokinetic considerations.

Conventionally, a drug should be given at intervals roughly corresponding to its biological half-life (estimated in patients with the disease being treated, not extrapolated from studies with healthy volunteers). If the drug exerts its effects only while concentrations remain above a therapeutic level, then the biological half-life corresponds to the drug half-life. However, if there is concentration-dependent killing and biological effects persist well after the drug has declined below the therapeutic level, then the drug can be given less frequently than predicted from the drug half-life determined in pharmacokinetic studies. The dose and dosing interval should aim to provide antimalarial concentrations sufficient to ensure a clinical and parasitological cure in a non-immune patient, because antimalarial drug-treatment regimens are always more effective in semi-immune subjects than in the non-immune, and in some instances lower doses or shorter courses of antimalarial treatment than in non-immunes may be given [439].

Rapid clinical improvement and the prevention of transmission are important secondary objectives in uncomplicated malaria. Sub-therapeutic drug levels increase the risk of treatment failure and provide a selection pressure for resistant genotypes. Resistant parasites have a survival advantage because they are associated with increased gametocyte carriage both in the primary and subsequent recrudescent infection [440]. Increased gametocyte carriage in resistant compared with sensitive infections translates into a greater probability of transmission and thereby explains the spread of resistance.

Cure of uncomplicated malaria is ensured if antimalarial drug concentrations produce maximum effects until all malaria parasites are eliminated from the body. For drugs eliminated slowly, the area under the blood or plasma concentration time curve (AUC) is a useful pharmacokinetic predictor of therapeutic response because it captures both the drug concentration and duration of exposure [382, 439]. *P. falciparum* life cycles will reliably affect a cure, provided that the antimalarial drug achieves parasite reduction ratios (PRRs) of greater than 1000-fold per cycle at these concentrations [439, 441]. For relatively slow eliminated antimalarials (i.e., those with a terminal elimination half-life of > 24 h), the blood or plasma concentration on day 7 is correlated strongly with the AUC and, thus, treatment response [248, 442].

In severe malaria, the primary objective of treatment is to stop parasite development as rapidly as possible. Cure is a secondary concern that can be addressed once the complications of malaria have resolved. Thus, the key pharmacokinetic determinant of outcome is the interval from starting treatment to exceeding the minimum parasiticidal concentrations (MPC), for which the time to maximum concentration (Tmax) provides a useful correlate because the *in vivo* MPC is usually not defined precisely [439]. This underlies the use of a loading dose for quinine and the superiority of the more rapidly and reliably absorbed water-soluble artesunate, rather than the oil-based artemether following parenteral administration [87]. Suboptimal drug concentrations in an individual patient can result from the use of a substandard product, under dosing, inadequate drug absorption, an unusually large apparent volume of distribution or particularly rapid clearance. In addition to optimizing dosage regimens, urgent attention is needed to ensure both drug quality and full compliance and adherence.

The available studies suggest that despite rapid absorption and elimination half-lives for artesunate and the biologically active metabolite dihydroartemisinin that are less than 1 h, oral artesunate is equally effective when given once or twice daily [56]. This suggests that the biological effects of the artemisinin compounds extend beyond their presence at therapeutic concentrations in plasma. The broad stage specificity of action of these drugs ensures that only two exposures of a few hours in each asexual life cycle (i.e., once-daily administration) are sufficient for a maximal antimalarial effect. This challenges the concept that antimalarial drugs need to remain above parasiticidal levels throughout the dosing interval [6].

Although pharmacokinetic parameters are generally reported as means or medians, it is the extreme values that are often of greater interest because the lower blood concentrations are associated with treatment failure or a selection of resistant parasites, whereas the highest concentrations are more likely to be associated with toxicity [443]. It is therefore important to complete pharmacokinetic sampling when patients are withdrawn from a study following adverse events or treatment failure. The greater the inter-individual variability to narrow the therapeutic range, the greater will be the proportion of patients with sub-therapeutic or toxic drug levels. Low oral bioavailability is the main cause of large inter-individual variability in blood concentrations [248]. Inter-individual variability can be exacerbated by the use of use of concomitant medications that interact with the antimalarials.

Because therapeutic response is influenced by host factors, such as acquired immunity, it is necessary to define therapeutic drug-concentration ranges in non-immune populations; in stable malaria-transmission areas, this means young children. Pharmacokinetic studies have usually not been designed to describe adequately the relationship between drug concentrations over time (pharmacokinetics) and therapeutic response and drug safety (pharmacodynamics). As a result, most antimalarials have been introduced with dosage regimens derived from dose-finding studies in non-pregnant adult patients. These dosage regimens have often been too close to the minimum effective dose required, both for cost-saving and because of concerns over drug safety. The dosage regimens of amodiaquine, primaquine, mefloquine, halofantrine, artemether-lumefantrine, and the artemisinin derivatives have all been increased after initial deployment [3, 435].

5.2.2.3. Dosing Regimens and Efficacy in Special Risk Groups

The processes that determine a concentration profile of drug over time – namely absorption, distribution, metabolism and excretion (ADME) – can differ substantially between non-pregnant adults and the important target populations, such as young children or infants, pregnant women and those with HIV/AIDS co-infection. There are many reasons why pharmacokinetics might also be different in malnourished patients, although there are few studies in this group [2]. Thus, defining dosage regimens for these important groups based on studies in non-pregnant adult patients might be wrong [435]. These special risk groups are of particular concern because they are more vulnerable to malaria complications because of lower levels of malaria immunity.

1) Young Children

Dosage regimens in children have, generally, been derived from the same 'one dose fits all' milligram per kilogram dosage recommended for non-pregnant adults. The disposition of many drugs is altered significantly in children and infants. Despite extensive use for four decades, it has only been recognized recently that the currently recommended mg/kg doses of both SP and CQ achieve substantially lower plasma drug concentrations in young children than in older patients [444, 445]. Plasma lumefantrine concentrations achieved in children less than 5 years of age are also significantly lower than in older children and adults [446]. A decrease in the plasma lumefantrine concentrations is associated with increased risks of recrudescence and earlier re-infections [248, 485]. In some cases, the effect of age on drug disposition is complex. For example, the absorption and clearance of chlorproguanil, when co-administered with dapsone, is lower in young children than in adults, whereas clearance of its active metabolite chlorcycloguanil is higher in young children. The clearance and apparent volume of distribution of dapsone increases with body weight [191]. Concentrations of piperaquine, when co-administered with dihydroartemisinin, are lower at the beginning of the terminal elimination phase in children than in adults; low plasma piperaquine concentrations on day 7 were associated with treatment failure [282]. Young children with severe malaria might be more susceptible to quinine cardiotoxicity than older children [282]. There is no evidence currently that exposure to artemisinin derivatives is significantly reduced in young children, although marked inter-individual variability limits the power of small studies to define the effect of age.

2) Pregnant Patients

These physiological changes in pregnancy can cause a decrease in plasma concentrations and AUCs, resulting in reduced efficacy. Also, pregnant women are at greater risk of adverse outcomes from malaria than non-pregnant women. In areas of high-intensity malaria transmission, pregnant women with malaria are at particular risk of severe maternal anemia and of delivering low birth-weight infants [447]. In areas of lower-intensity transmission, malaria in pregnancy is associated with a higher risk of severe disease and maternal and fetal death [447]. Antimalarial drug concentrations are markedly lower in women in the second and third trimesters of pregnancy for artesunate, artemether, CQ, lumefantrine, mefloquine, atovaquone, proguanil (and cycloguanil) and sulfadoxine after oral administration [265, 268, 269, 448-453]. However, the pharmacokinetic parameters for the antimalarial agents show no significant difference between pregnant and non-pregnant women and animal species after intravenous or intramuscular injections [454-456]. As in young children, suboptimal dosing could partially explain the poorer treatment responses seen among pregnant women. Despite this, there has been no study of revised dosage regimens for malaria in pregnancy. The pharmacokinetic properties of quinine were not different between pregnant and non-pregnant women with uncomplicated malaria [457]. In severe malaria, pregnant women have a smaller Vd and more rapid elimination of quinine [458]. No data is published on the pharmacokinetics of amodiaquine, piperaquine or clindamycin in pregnant women, although all are used to treat malaria in pregnancy.

3) Intermittent Preventive Treatment

Intermittent preventive treatment (IPT) involves giving a curative antimalarial treatment dose systematically at predefined intervals to asymptomatic individuals, who are at special risk of malaria, regardless of whether or not they are parasitemic. This approach was taken initially for women in the second and third trimesters of pregnancy (IPTp); and, subsequently, during infancy (IPTi) and, most recently, in children in areas of high intensity, seasonal malaria transmission. SP has been by far the most widely used antimalarial for this indication. Only one of the studies published on IPT included a pharmacokinetic component [453]. SP is recommended for IPTp in Africa, despite the fact that it remains unknown whether curative blood concentrations, rather than merely prophylactic concentrations, are being achieved or for how long the increasingly prevalent resistant parasites are being suppressed [261]. This is surprising given the international consensus that SP is not indicated for prophylaxis and the considerable research investment evaluating SP for IPT. This knowledge gap is of particular concern because IPT is the only context in which antimalarial monotherapy is still advocated.

Use of slowly eliminated drugs, such as SP, will always result in a window of selection, when decreasing concentrations enable survival of resistant parasites while clearing sensitive parasites [66, 304]. If a substantial proportion of those receiving IPT achieve only sub-therapeutic concentrations that select for resistance, the benefits of these SP IPT policies could be short-lived and any future use of SP and the related antifolate, chlorproguanil-dapsone, could be threatened. These concerns are illustrated by a recent publication, which found that the AUC of sulfadoxine was approximately 40% lower in 30 pregnant women (second trimester) when compared with that in 10 of the same women who returned 2–3 months postpartum for another dose [453]. There are simply no data at all on SP pharmacokinetic parameters in infants, despite extensive evaluation of SP as IPTi. Amodiaquine has also been used for IPT without any published data on its pharmacokinetic properties in pregnant women or infants. Modeling the potential impact of IPT on resistance has been generally reassuring [459, 460], although these models have assumed – without any evidence – that IPT provides an effective treatment dose. The drug levels achieved in IPT programs cannot be extrapolated from those seen in young children or pregnant women being treated for uncomplicated malaria because acute disease can affect drug disposition. It is common sense that, when introducing drugs for new indications, consideration should be given to optimizing dosage in that particular target population and for that particular indication.

4) HIV/AIDS Co-morbidity

Both malaria and HIV/AIDS are prevalent in the tropics and recent evidence clearly supports the significant impact of each infection on the other. Inadequate clearance of parasites by IPTp has been shown in HIV-infected women; this led to a pragmatic recommendation of monthly SP treatment in areas with more than 10% HIV seroprevalence compared with two-dose SP regimens in areas with less than 10% HIV seroprevalence [461]. HIV-infected patients are more likely than HIV seronegative patients to develop severe and complicated malaria and to die from severe malaria, particularly in areas with lower-intensity malaria transmission [462]. Immune suppression associated with HIV co-infection increases the risk of SP treatment failure [463]. The potential for clinically significant drug interactions is greatest with the protease inhibitors, which inhibit the human cytochrome P450 enzymes (CYP) and, to a lesser degree, the non-nucleoside reverse transcriptase inhibitors, which can either induce or inhibit the CYP enzymes. As quinine, lumefantrine and halofantrine are all metabolized by CYP3A4, an increased risk of toxicity is possible with concomitant protease inhibitor use [464]. Concurrent administration of artesunate, amodiaquine, and efavirenz in healthy volunteers was associated with hepatotoxicity, resulting in early study discontinuation [465].

5.2.3. Drug Exposure Level and Time Involves Artemisinin Efficacy

Human and animal species studies have shown unambiguously that artemisinin derivatives are very potent drugs against *falciparum* and *vivax* malaria. These drugs initiate immediate decreases in parasitemia and, in most studies; clinical recovery of the patient is faster than with other antimalarials. However, a high recrudescence (about 30-50% for 3 days, 6-15% for 5 days, and 2-6% for 7 days regimens) is still a major disadvantage for all of these compounds after monotherapy [48, 466]. Although the clinical significance of the rapid killing advantage, in terms of mortality, has yet to be determined by using other artemisinins drugs [85, 86, 390], at least one published study has shown the potential in decreasing mortality. The SEAQUAMAT trial, a multicenter randomized trial conducted in Bangladesh, Thailand, Myanmar, Indonesia, India, and Vietnam, recently reported a 34.7% reduction in mortality associated with intravenous AS compared with intravenous quinine [90, 103, 467].

5.2.3.1. Effective Influences of Drug Exposure Level and Time in Animal Species

Injectable AS is being developed as a potential agent for the treatment of severe and complicated malaria. A risk assessment of the therapeutic index and related hematological changes of AS and AL following daily intravenous injection for 3 days has been conducted in *P. berghei* infected and uninfected rats. The minimum doses of AS and AL for parasitemia suppression were 2.3 and 2.5 mg/kg, respectively, and the suppressive doses for decreasing parasitemia by half (SD_{50}) were 7.4 and 8.6 mg/kg, respectively (Table 39). There seemed to be no significant difference between these two water-soluble compounds at suppressive dose levels [468].

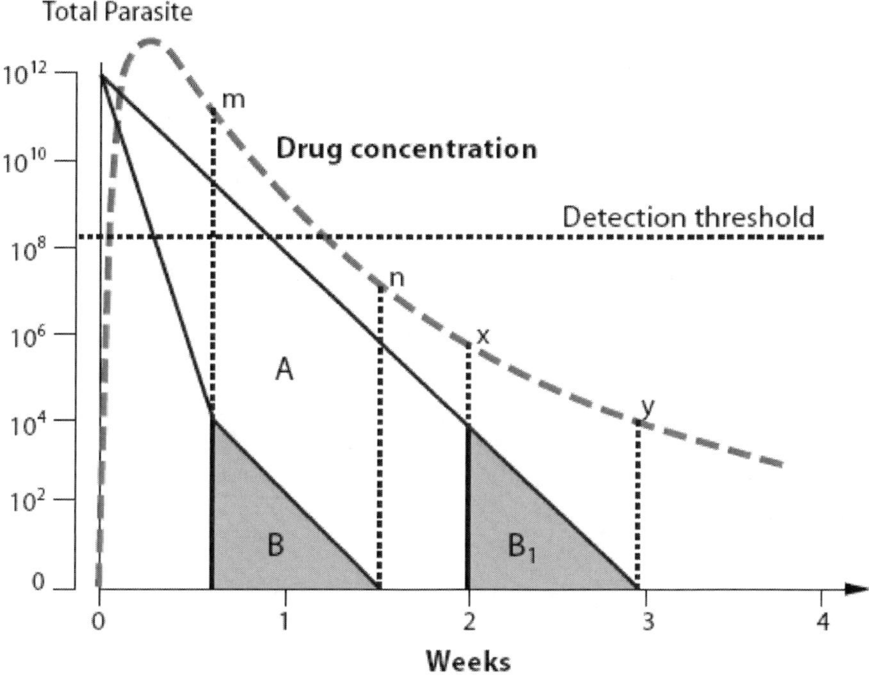

Figure 14. The artesunate + mefloquine combination. If no artesunate is given, then the number of parasites exposed to mefloquine alone is given by the area of A; with the combination administered for 3 days, the number of parasites exposed to mefloquine alone is given by the area of B (100 million times fewer). Furthermore, mefloquine levels are higher (m to n) when confronting B than when confronting the same number of parasites (B1) if no artesunate is given (x to y). If a parasite containing a *de novo* mefloquine-resistant mutation were to occur, then such a parasite should still be susceptible to artesunate. Thus the probability of selecting a resistant mutant is reduced by 100 million times, as only a maximum of 100,000 parasites are exposed to mefloquine alone after the fourth day (i.e., in the third cycle), and any artesunate-resistant parasite selected by artesunate initially would always be killed by the accompanying mefloquine. As a result, the combination is more effective, reduces transmission and prevents the emergence of resistance to both drugs. [2]

However, with regard to parasite clearance time (PCT), equimolar doses of 96 μmoles/kg, AS (36.7 mg/kg) cleared parasitemia in 7 out of 18 rats with a 39% clearance rate. At this dose, the PCT of AS was 52.8 hrs and clearance time lasted only 2.7 days with recrudescence back in the 7 animals. Accordingly, AS did not clear 0.339% (E_{max}) of parasitemia in rats treated at this dose level. The EC_{50} was 8.86 hrs, and the area under the effect curve (AUEC) was 536 % of parasitemia·hr for this dosage of AS/NaHCO$_3$ in the 7 rats. Using the same equimolar dose of 96 μmoles/kg, AL (40.6 mg/kg) cleared the parasitemia in all of 19 rats with 100% clearance rate. At this dose level, the PCT of AL was 57.2 hrs with clearance lasting for 3.2 days before recrudescence. Computer fitting revealed that E_0 was 6.2 ± 2.3%, whereas E_{max} was measured at 0.008 ± 0.01% of parasitemia. The EC_{50} was 10.9 hrs, and the area under the effect curve (AUEC) was 313 % of parasitemia·hr for this dosage of AL (Table 39) [468].

                    Qigui Li and Peter J. Weina

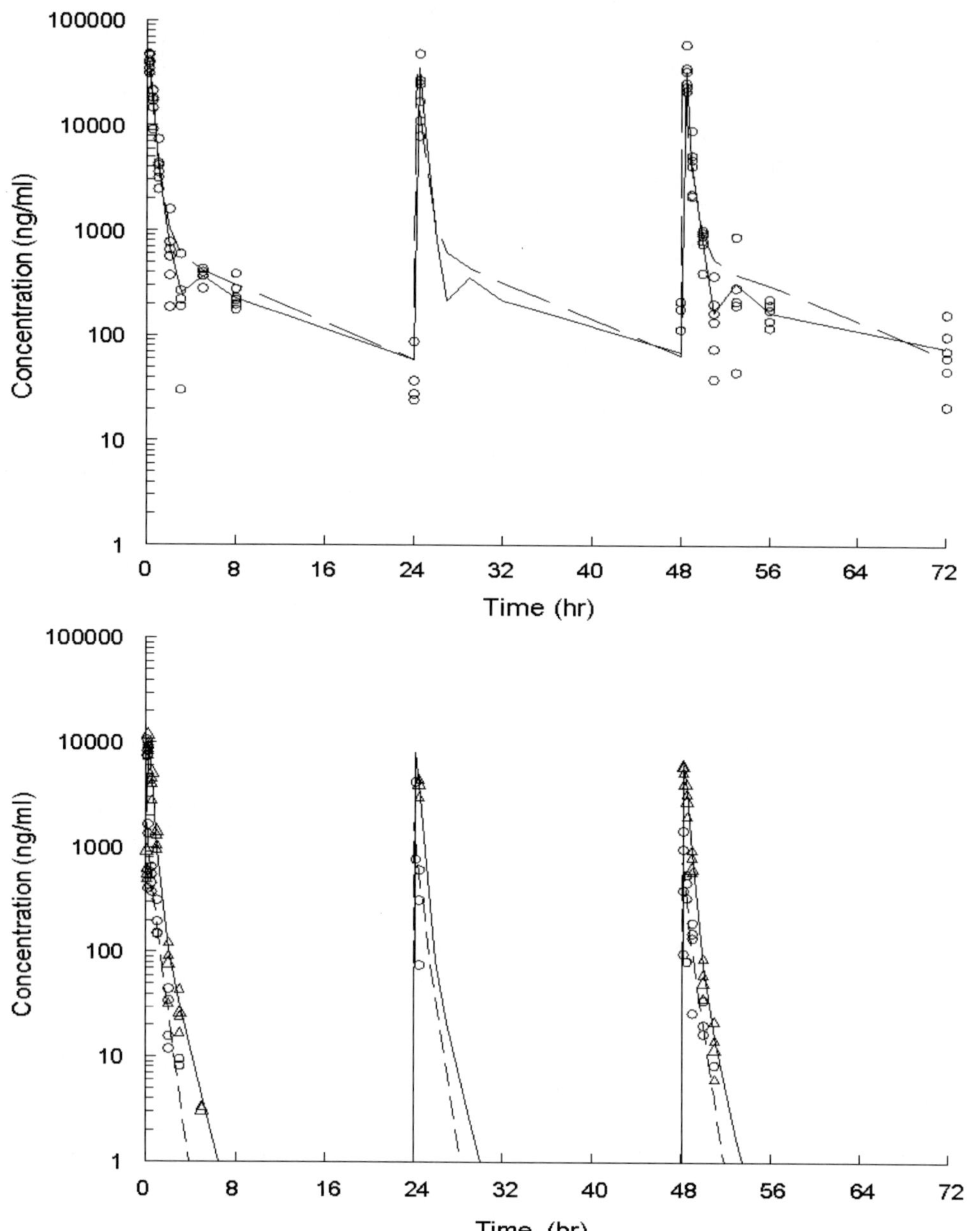

Figure 15. Individual plasma concentration-time profiles of artelinate (AL) measured by LC/MS/MS (markers) and mean concentration (solid line) as well as computer fitted curves (dashed line) by pharmacokinetic parameters at daily dosage of 40.6 mg/kg intravenously for three days in malaria-infected rats (top, n = 6). Individual plasma concentration-time profiles of artesunate (AS) measured by HPLC-ECD (markers) and computer fitted curves for parent drug, AS (dashed line), and its active metabolite, DHA (solid line), by pharmacokinetic parameters at dosage of 36.7 mg/kg once daily intravenously for 3 days in *P. berghei* infected male rats.(bottom, n = 4) [475].

Table 39. Definitive efficacy parameters of artelinate(AL)/lysine salt at 96 μmoles/kg (40 mg/kg) and artesunate(AS)/NaHCO$_3$ at 96 μmoles/kg (36.7 mg/kg) given as daily intravenous treatment for three days in *P. berghei* infected rats (2.0 x 10^7 parasitized erythrocytes per adult rat) [468]

Pharmacodynamic (PD) Parameters	AL/lysine 95 μmoles/kg (40.6 mg/kg) 95 μmoles/k (40.6 mg/kg) (36.7 mg/kg)	AS/NaHCO$_3$ 95 μmoles/kg (36.7 mg/kg)
Initial PD parameters		
Number of rats	19	18
Parasitemia on day 6 (%)	5.5 ± 1.1	6.7 ± 2.1
Rat number of clearance	19/19	7/18
Mean time for parasitemia to clear completely (PCT, hr)	57.2 ± 10.6	52.8 ± 14.9
Time parasitemia cleared (days)	3.2 ± 1.2	2.7 ± 0.7
Recrudescence rate (%)	100	100
Time recrudescence (days)	6.5 ± 1.0	6.0 ± 0
Mortality	0/19	0/18
Secondary PD parameters (computer fitting)		
Lag time (hr)	3.36 ± 1.64	4.01 ± 2.5
E_0 (%, drug effect on parasitemia at time 0)	6.2 ± 2.3	8.1 ± 2.7
E_{max} (%, maximum drug effect on parasitemia)	0.008 ± 0.01	0.339 ± 0.039**
EC_{50} (hr, time causes 50% of drug effect parasitemia)	10.85 ± 1.32	8.86 ± 2.87*
AUEC (%*hr, area under effect of parasitemia vs time curve)	313.3 ± 148.2	535.8 ± 252.4

Note. Values are means ± SD. Injection times are on day 6, 7, and 8 after infection day (day 0).

* $P = 0.0048$ (AS compared with AL).

** $P = 0.0007$ (AS compared with AL).

It is comprehensible from this data that at the equimolar dose of AL (with 100% clearance) has great efficacy than that of AS (with 39% clearance) in the malaria infected rat. However, the *in vitro* data indicate that, AS has a 5-8 fold greater antimalarial activity than AL [469]. Why do the results contrast between *in vitro* and *in vivo* data? A pharmacokinetic analysis may provide the answer to the question. The comparative pharmacokinetics and hydrolysis studies of intravenously administered AL and AS were performed in malaria-infected rats using 3 daily equimolar doses (96 μmoles/kg). The drug concentration of AS with daily doses of 36.7 mg/kg was one-third less on day 3 than on day 1, and the mean C_{max} of AS on day 1 was 16,492 ng/ml and 1,377 ng/ml on day 3. Similar to its active metabolite, DHA, it had peak plasma drug concentrations of 10,842 and 6,485 ng/ml on days 1 and 3, respectively, suggesting an auto-induction of hepatic drug-metabolizing enzymes for AS [470-474].

The auto-induction of the enzymes were similar for other artemisinin drugs (QHS, DHA, AE, and AM) [470-474], but not for AL [475]. Pharmacokinetic parameters of AL were very comparable from day 1 to day 3 with same AS molecular dose at 40.6 mg/kg. The mean C_{max} of AL on day 1 was 48,453 ng/ml and 44,101ng/ml on day 3. AS is the pro-drug of DHA with a DHA/AS ratio of 5.26 compared to a ratio of 0.01 for DHA/AL. Although total AUC_{1-3D} of 13,051 ng·h/ml DHA was produced by AS compared to 524 ng·h/ml DHA by AL, the peak level of AL was still 2-fold and 6-fold higher than that of AS plus DHA concentrations on day 1 and day 3, respectively. The increase in AL's peak concentration can explain why AL has a much great efficacy than AS *in vivo* (Figure 15). Other pharmacokinetic parameters revealed that the total AUC_{1-3} days (84.4 μg·h/ml) of AL was 5-fold higher than that of AS (15.7 μg·h/ml of AS plus DHA). The elimination half-life of AL (7.1 h) was much longer than that of AS (0.36 h) or DHA (0.72 h) (Table 40). The remarkable alteration of the pharmacokinetic shape of AL may be caused by poor conversion rates to DHA and an enterohepatic circulation, which has been confirmed by pharmacokinetics and tissue distribution studies [475, 476]. Compared to AS, higher drug peak and AUC values of AL in the rat blood may be the cause of its increased efficacy in malaria-rat.

Table 40. Toxicokinetic parameters of artesunate (AS) and artelinate (AL) at 96 μmoles/kg following multiple intravenous injections daily for three days in *P. berghei* infected rats by using Culex automated blood sampler [475]

PK Parameters	AS (36.7 mg/kg, n = 4)		AL (40.6 mg/kg, n = 6)	
	Parent drug (AS)	Metabolite (DHA)	Parent drug (AL)	Metabolite (DHA)
Day 1				
C_{max} (ng/ml)	16492 ± 4205	10842 ± 1950	48453 ± 7177	124.2 ± 36.2
T_{max} (h)	0	0.19 ± 0.03	0	0.33 ± 0.18
$AUC_{inf.}$ (ng·h/ml)	1324 ± 242	4893 ± 779	25857 ± 2114	160.4 ± 86.2
$t_{1/2}$ distribution (h)	0.04 ± 0.01	0.20 ± 0.02	0.29 ± 0.06	
$t_{1/2}$ elimination (h)	0.36 ± 0.03	0.72 ± 0.05	7.05 ± 2.01	
Vss (liter/kg)	8.10 ± 1.70	-	4.57 ± 2.01	
CL (ml/min/kg)	474.2 ± 89.8	-	25.93 ± 2.10	
MRT (h)	0.62 ± 0.33	0.43 ± 0.04	2.89 ± 1.11	
Day 3				
C_{max} (ng/ml)	1377 ± 898	6485 ± 1053	44101 ± 21452	116.6 ± 74.3
T_{max} (h)	0	0.18 ± 0.06	0	0.26 ± 0.12
$AUC_{inf.}$ (ng·h/ml)	490.6 ± 314.0	3327 ± 743	24650 ± 8663	204.2 ± 134
$t_{1/2}$ distribution (h)	0.12 ± 0.07	0.21 ± 0.02	0.30 ± 0.08	524 ± 277
$t_{1/2}$ elimination (h)	0.46 ± 0.06	0.65 ± 0.06	6.06 ± 0.64	0.01 ± 0.00
Vss (liter/kg)	76.87 ± 8.54	-	5.24 ± 3.50	
CL (ml/min/kg)	2172 ± 1274	-	29.40 ± 8.24	
MRT (h)	0.53 ± 0.01	0.60 ± 0.11	2.78 ± 1.34	
Total AUC_{1-3D} (ng·h/ml)	2717 ± 809	13051 ± 1926	84406 ± 19703	
Ratio of DHA/AS or AL		5.26 ± 2.22		

TK = toxicokinetics.
DHA = dihydroartemisinin.
MRT = Mean residence time.
D = days.

Table 41. Antimalarial activity (50% cure dose, CD_{50})* and corresponding 95% confidence interval (95% CI) of arteether and artemether given intramuscularly in daily dose for 3 days against *P. berghei* in female mice (n = 8 for each dose level) [44]

Drug formulation	Arteether (CD_{50})		Artemether (CD_{50})	
	Mg/kg	95%CI	Mg/kg	95%CI
Sesame oil formulation	34.1	12.3 – 66.9	35.6	7.6 - 63.6
Cremophore formulation	14.2	6.8 – 21.5	21.0	18.5 – 23.4

* When the mice have survived and showed negative smears at 60 days, they were considered cured.

Table 42. Pharmacokinetic parameters of arteether (AE) and dihydroartemisinin (DHA), a metabolite of AE, following single intravenous (25 mg/kg), and multiple intramuscular (25 mg/kg) administration with sesame oil and 1:2 cremophore (CRM)/0.9% saline daily for 7 days in male rats (n = 3-7) [44]

PK Parameters	Single IV CRM/Saline	Intramuscular Multiple			
		Sesame oil (AESO)		1:2 Cremophore/Saline (AECM)	
		Day 1	Day 7	Day 1	Day 7
Arteether					
C_{max} (ng/ml)	14482 ±1023	92.2 ± 19.9	311.9 ± 54.6	1227 ± 171	1826 ± 118
T_{max} (h)	0	1.42 ± 0.63	1.67 ± 0.58	1.33 ± 0.29	1.33 ± 0.58
$AUC_{inf.}$ (ng·h/ml)	5593 ± 591	1135 ± 277	8527 ± 1006	4165 ± 676	7312 ± 588
$AUC_{D1-8.}$ (ng·h/ml)			16922 ± 4038		46286 ± 2061
Vss (liter/kg)	0.98 ± 0.25	150.2 ± 33.7	54.1 ± 7.8	8.70 ± 1.21	5.06 ± 0.27
CL (ml/min/kg)	71.3 ± 9.11	73.7 ± 2.8	147.9 ± 6.19	76.3 ± 2.9	76.0 ± 1.9
$t_{1/2}$ distribution (h)	0.22 ± 0.04	0.71 ± 0.47	1.08 ± 0.49	1.00 ± 0.04	1.18 ± 0.14
$t_{1/2}$ elimination (h)	0.82 ± 0.05	17.66 ± 2.68	29.57 ± 2.07	6.96 ± 0.93	9.06 ± 1.69
MAT (h)	-	0.76 ± 0.40	0.87 ± 0.39	1.04 ± 0.03	1.07 ± 0.10
MRT (h)	0.74 ± 0.18	23.56 ± 3.42	41.51 ± 2.84	6.21 ± 0.66	5.83 ± 0.09
Bioavailability (%)	100	20.3 ± 5.0	152.4 ± 17.9	74.5 ± 12.1	130.7 ± 10.5
of single daily dose in		37.6 ± 8.2	91.4 ± 11.6	4.7 ± 1.8	17.3 ± 3.3
muscle					
Dihydroartemisinin					
C_{max} (ng/ml)	815.8 ±164.2	16.02 ± 4.05	123.8 ± 28.9	171.1 ± 79.4	188.4 ± 91.0
T_{max} (h)	0	0.92 ± 0.14	0.86 ± 0.35	0.92 ± 0.14	1.33 ± 0.29
$AUC_{inf.}$ (ng·h/ml)	350.0 ± 97.1	117.1 ± 33.0	790.3 ± 221	607.7 ± 102	708.2 ± 29.2
$AUC_{D1-8.}$ (ng·h/ml)			3427 ± 759		5346 ± 831
$t_{1/2}$ elimination (h)	1.08 ± 0.16	9.17 ± 2.54	14.47 ± 2.17	4.71 ± 0.71	6.64 ± 1.25
Ratio of DHA/AE	0.06 ± 0.02	0.09 ± 0.03	0.08 ± 0.03	0.14 ± 0.03	0.11 ± 0.01

* On day 1 the amount of dose in the muscle was measured at 24 h after first dosing, and on day 7 the amount of dose was measured at 24h after last dosing. IV = intravenous, CRM/saline = 1:2 cremophore/0.9% saline.

The efficacy and pharmacokinetic investigation of AE and AM with two different intramuscular vehicles, sesame oil and cremophore, have been performed in rats [44, 477]. AE and AM showed comparable efficacy against *P. berghei* when used cremophore vehicle to replace sesame oil following intramuscular treatment. The results of 50% cure dose (CD_{50})

were significantly different between the sesame oil and cremophore. AE and AM with sesame oil had the similar CD_{50} values of 34.1 (1.3-66.9) and 35.6 (7.6-63.6) mg/kg, respectively. However, the drugs with cremophore had 14.2 (6.8-21.5) and 21.0 (18.5-23.4) mg/kg for AE and AM, respectively, in malaria mice (Table 41). The results showed that AE and AM formulated in cremophore had a significantly higher efficacy than the sesame oil formulation following the three daily multiple intramuscular injections in mice [44, 477].

The mean plasma concentration-time profile (n = 3) of AE formulated with sesame oil (AESO) and 1:2 cremophore/0.9% saline (AECM) after single intravenous and intramuscular doses is shown in Figure 3 (bottom). AE with sesame oil and cremophore data showed a biphasic pattern of disposition and was fitted by a two-compartment open model after either intravenous or intramuscular injection. The concentration-time profiles of AE with the two vehicles are shown in Figure 7 after daily intramuscular for 7 days with the multiple administrations. The trough plasma concentrations increased significantly up to the last dosing in AESO treated rats, but not in AECM rats. Obviously, the drug accumulated (7.5-fold) in the plasma of animals administered with AESO (comparing the 7th dosing day to the first dosing day). In the AECM rats, the AE level increased slightly (1.8-fold) on day 7 over that on day 1 following the multiple dosing.

Significant differences in pharmacokinetic parameters between AECM and AESO are summarized in Table 42 after multiple intramuscular doses of 25 mg/kg. Plasma peak concentration on day 1 indicated that the C_{max} of AECM (1227 ± 171 ng/ml) was more than 13-fold higher than that of AESO (92.2 ± 19.9 ng/ml). The area under the curve (AUC) of AECM (4165 ± 676 ng·h/ml) was about 4-fold greater than that of AESO (1135 ± 277 ng·h/ml) on day 1. Total AUC of AECM (46286 ± 2061 ng·h/ml) during the 7-day dosing period was about 2.7-fold higher than that of AESO (16922 ± 4038 ng·h/ml). The elimination half-life was 7.0 ± 0.9 hr for AECM and 17.7 ± 2.7 hr for AESO on day 1. Based on the AUC data, the bioavailability of AECM was 74.5 ± 12.1% and 20.3 ± 5.0% for AESO after a single intramuscular injection. The data suggested that AECM, a less oily formulation, was more easily absorbed by the muscle with a higher C_{max} and AUC, which may be the explanation of its increased efficacy [44, 477].

The *in vivo* efficacy model shows that AE and AM formulated with cremophore had a significantly higher efficacy than that with sesame oil. It is apparent that the higher plasma concentration of AE produced by cremophore formulation has principle role in the increased efficacy. Plasma concentration showed that the C_{max} and AUC of AECM were about 13-fold and 5-fold, respectively, higher than that of AESO on day 1. Even the total AUC of AECM (46286 ng·h/ml) during the 7-day period was 2.7-fold higher than with AESO (16922 ng·h/ml). The much higher level of AE exposure in the rats using a cremophore vehicle truly contributed to the antimalarial efficacy, and without a doubt, the antimalarial efficacy is drug concentration dependent [436, 478-480], especially with the 13-fold higher peak concentration.

5.2.3.2. Effective Influences of Drug Exposure Level and Time in Human Studies

Karbwang *et al.* [481] reported on four patients, not cured of their malaria by the oral AM regimen used in his study, that had significantly lower plasma concentrations of DHA, an active metabolite of AM than others in the study. Using the same AM dose amount, a much lower C_{max} (196 ng/ml at 6 h and 54 ng/ml at 12 h) of DHA was measured in the four patients than the C_{max} (354 ng/ml at 6 h and 158 ng/ml at 12h) of DHA in 46 other subjects. Also, a lower AUC (834 ng·h/ml) of DHA was detected in these 4 patients but not in the remaining 46 patients (3684 ng·h/ml). Concentrations of the parent drug, AM, was not much different in these four patients (1044 ng·h/ml) when compared to the other subjects (1272 ng·h/ml). The results are unknown for those patientswith such low DHA concentrations, but these low levels of DHA may be the cause for the treatment failure. The four patients may have had more mature parasiteswhich could have required higher concentrations of AM/DHA than patients with similar numbers of parasites.

More recently, Sharma *et al.* [482] reported on two patients who died with a very high degree of parasitemia (16% and 50%) and cerebral malaria. The dose regimen of AM at 80 mg was intramuscularly injected twice on first day and then once daily for another four days. Intramuscular formulation of AM resulted in very low plasma concentration and low bioavailability [483] due to the prolonged absorption from the injection site [484]. This study showed that using a low dose regimen (total 480 mg) may not have enough drug concentration in blood to cure the heavy parasitemia in these patients.

During a phase III clinical trial of intramuscular Artemotil (AE), a patient had inadvertently obtained a full or fractional intravenous injection with the intramuscular formulation on the first day of treatment. The difference in concentration profiles vs. time of AE and DHA for patient number 151 from the high dose group is shown in Figure 9 (bottom). As first dosing, AE seemed to be accidentally injected in patient 151 through a vessel instead of the muscle. This resulted in an intravenous injection profile with very high C_{max} concentration right after injection and a rapid distribution phase. AE and DHA plasma concentrations decreased significantly after first dosing with an elimination $t_{1/2}$ of 1.1 hr in the first 6 hrs. After the second intramuscular injection at the 6th hour, the elimination $t_{1/2}$ of AE was immediately extended to 9.35 hr as a daily half-life. The C_{max} (2407 ng/ml) was shown to be about 22-fold higher in patient 151 than the mean value (110 ng/ml) reported in other patients in high dose group. AUC value (12259 ng·h/ml) was 2 times higher than the mean value (7403 ng·h/ml) shown in other subjects. The same results were also reported for increasing DHA levels in patient 151. To be expected, this patient was successfully cured at 36 hr (Parasitemia Clearance Time, PCT) and 4 hr (Fever Clearance Time, FCT) compared with other patients with 64.4 hr (PCT) and 70.1 hr (FCT) [485]. The clinical cure indicated that the intravenous formulation, like AS injection, is very important in the rapid treatment of malaria with high peak concentration (C_{max}), especially in the severe and complicated malaria. In the meantime, the C_{max} (22-fold) could have also been an important factor in the observed increase of drug efficacy.

Other cases of oral or intramuscular administration of AS and AM were compared in malaria patients [87, 486]. The first-dose pharmacokinetic properties of intramuscular AS at 2.4 mg/kg immediately followed by 1.2 mg/kg daily and AM at 3.2 mg/kg intramuscular start, followed by 1.6 mg/kg daily were compared in Vietnamese adults with severe *falciparum* malaria. A total of 19 patients were studied; 9 received AS, and 10 received AM. Intramuscular AS was rapidly absorbed. Concentrations of AS in plasma peaked between 523 and 3221 ng/ml (median, 2193 ng/ml) within 20 min of injection and then declined with a median (range) half-life of 30 (3 to 67) min. AS was hydrolyzed rapidly and completely to the biologically active metabolite DHA. Peak DHA concentrations in plasma ranged between 488 and 2011 ng/ml (median, 869 ng/ml) and declined with a $t_{1/2}$ of 52 (26 to 69) min. The median times for fever (FCT) and parasite clearance (PCT) were 65 h and 38 h, respectively, in the AS group. All patients survived.

In contrast, AM was slowly and erratically absorbed. The absorption profile appeared biphasic. Maximum AM concentrations in plasma ranged between 20 ng/ml and 487 ng/ml (median, 171 ng/ml) and occurred at a median (range) of 10 (1.5 to 24) h. There was relatively little conversion to DHA. After intramuscular injection in cases of severe malaria, absorption of the water-soluble AS is rapid and extensive, whereas the absorption of the oil-based AM is slow and erratic with relatively little conversion to the more active DHA. The median times to fever and parasite clearance were 56 h and 48 h, respectively, in the AM group. Two out of ten patients died in the process (20% mortality). Based on this pharmacological study, parenteral AS seems preferable to AM as an initial antimalarial therapy, particularly in seriously ill patients. Also, with the peak concentration (2193 ng/ml) of AS being 13-fold higher than that (171 ng/ml) of AM, it indicates that a high peak concentration contributes greatly in improving efficacy and reducing mortality [486].

After oral dose, despite a 29% lower molar dose, oral AS administration result in a significantly larger drug concentration. It also resulted in higher antimalarial activity than after oral AM ($P \leq 0.02$). The mean (95% CI) oral antimalarial bioavailability of AM, relative to oral AS, corrected for molar dose was 58% (40–76). The mean (95% CI) relative antimalarial bioavailability of AM was lower on the first day of treatment, 31% (17–100), compared to the second day, 72% (44–118) ($P = 0.018$). *In vivo* parasite clearance and time above the *in vitro* IC_{90} were similar for the two drugs, despite considerable differences in C_{max} and AUC. The oral antimalarial potency following AM was significantly lower than that after AS. AM oral antimalarial bioavailability was reduced in acute malaria. Oral AS provided greater antimalarial activity than AM at the same molar dose [486].

All of the evidence shows that higher plasma concentration (C_{max} and AUC), especially the peak concentration (C_{max}), will greatly increase the efficacy and clinical therapeutic potentials of artemisinin drugs. As previously discussed, from the severity of these complications, *P. falciparum* malaria constitutes a medical emergency, and appropriate treatment should be initiated as soon as infection is suspected. The appropriate treatment means that enough high dose regimens should be given as to provide a successful therapeutic efficacy for the malaria patients. It is known that 84% of all malaria related deaths occurs within 24 hr of hospital admission in African children [487]. Therefore, first exposure concentrations of artemisinin drugs are very clinically important.

5.2.4. PK Interindividual Variation Concerns Drug Efficacy

There is considerable inter-individual variation in plasma or blood concentrations for any given dose for many antimalarial drugs. Peak blood concentrations may vary by a factor of 10 or more. Several of the antimalarial drugs, such as atovaquone, halofantrine, benflumetol and, to a lesser extent, mefloquine, share common characteristics; they are hydrophobic and lipophilic and have very variable oral bioavailability [488, 489]. Absorption is often poor in the acutely ill patient who has not eaten for many hours. Co-administration with fat may increase absorption 10-fold or more [490]. Thus some patients will absorb these oral drugs inadequately despite having taken and retained a full treatment dose. Inadequate blood concentrations will result, and the infections may recrudesce.

If the minimum parasiticidal concentration (MPC) is defined as the minimum blood or plasma concentration which produces that maximum inhibitory activity of drug (i.e., IC_{99}), then, in general, blood levels of an antimalarial drug should exceed MPC values during at least four asexual-cycles (> 6 days) for maximum cure rates [6]. If the reduction in parasite biomass (parasite reduction ratio) is, 1000-fold per cycle, then the MPC must be maintained for more than four cycles to ensure high cure rates. For both halofantrine and benflumetol, the plasma concentrations 7 days after starting treatment have been shown to be a useful predictor of therapeutic outcome [491]. In the past it was generally assumed that antimalarial blood concentrations should remain within the therapeutic range throughout the dosing interval. This has been challenged by observations that once-daily administration of artemisinin derivatives, which have elimination half-lives of < 1 h, are not associated with inferior immediate responses compared with more frequent administration [56, 492]. In experimental animals, on the other hand, the timing of chloroquine administration to coincide with the greatest period of susceptibility in the parasite life-cycle has been associated with improved parasitological clearance [437].

Obviously, if the range of blood concentrations achieved in the patient considerably exceeds the concentrations giving 90% inhibition of multiplication (IC_{90} values) for the most resistant mutant ($IC_{90}R$), then resistance cannot be selected in the acute phase of treatment as even the resistant mutants are prevented from multiplying. Conversely, if the degree of resistance provided by the genetic event is very small, the window of opportunity for selection may be negligible. Provided that there is such a window of selection then the broader the range of peak antimalarial concentrations and the closer the median value approaches $IC_{90}R$, the greater the probability of selecting a resistant mutant in a patient. Peak drug concentrations are determined by absorption, distribution volume and dose. Several antimalarials (notably lumefantrine, halofantrine, atovaquone and, to a lesser extent, mefloquine) are lipophilic, hydrophobic and very variably absorbed (interindividual variation in bioavailability up to 20-fold) [248, 493]. Interindividual variation in distribution volumes tends to be lower (usually less than five-fold) but, taken together with variable absorption, the outcome is considerable interindividual variation in peak antimalarial blood concentrations.

The priority for treatment of complicated and severe disease is the parenteral administration of adequate, safe doses of an appropriate antimalarial, in the setting of the highest possible level of clinical care. Supportive management of complications such as coma, convulsions, metabolic acidosis, hypoglycemia, fluid and electrolyte disturbances, renal failure, secondary infections, bleeding disorders and anemia is also important. The most recent advance in decreasing mortality has been the use of intravenous AS instead of injection quinine [90, 103], which may very well revolutionize the management of severe disease [93]. However, recrudescence is still relatively high with monotherapy of intravenous AS alone [494]. Since recrudescence is not due to parasite resistance [495], one possible explanation of this parasitological failure may be a low peak concentration after first loading dose with the intravenous AS.

Table 43. The main pharmacokinetic parameters of artesunate (AS) and dihydroartemisinin (DHA), an active metabolite of AS, following oral, intrarectal (IR), intramuscular (IM) and intravenous (IV) administrations in malaria patients and volunteers* [44]

Patients	Where	Cases	Date	Route	Regimens	C_{max} ng/ml	AUC ng·h/ml	$t_{1/2}$ min	C_{max} ng/ml	AUC ng·h/ml	$t_{1/2}$ min	AUC_{DHA} /AUC_{AS}	References
						Artesunate (AS)			**Dihydroartemisinin (DHA)**			**Ratio of**	**References**
Children	Thailand	10	1998	IR	15.6 mg/kg	301	871	111.6	916	6145	40.2	7.06	[780]
Children	PNG	47	2004	IR	13 mg/kg	417	-	16.2	718	-	42.6		[718]
Children	Gabon	11	2002	IM	1.2 mg/kg	14 988	2000	11.5	2655	1585	32.0	0.79	[719]
Children	Gabon	10	2002	IV	2.4 mg/kg	29 683	3752	1.5	3012	3325	20.7	0.89	[719]
Children	Ghana	9	2001	IV	2.4 mg/kg				1593	1708	31.8		[709]
						198	210	-				6.35	
Adults	Vietnam	13	1998a	Oral	100 mg/person	-	161	-	737	1289	39.3	7.21	[771]
Adults	Thailand	3	2000	Oral	2 mg/kg	-	-	-	1052	1334	-	5.36	[773]
Adults	Vietnam	5	1998b	Oral	100 mg/person	-	-	-	851	1161	40.1		[772]
Volunteer	Cambodia	15	2005	Oral	4 mg/kg	112	154	-	748	1673	78.0	5.18	[722]
Volunteer	Thailand	10	2004	Oral	4 mg/kg	-	-	-	519	657	44.4	4.29	[778]
Volunteer	Caucasia	6	2002	Oral	150 mg/person	114	97	24.6	546	825	-	2.48	[720]
Volunteer	Vietnam	8	2001	Oral	150 mg/person	577	312	23.4	1309	2532	53.0	1.75	[620]
Adults	Thailand	6	2001	Oral	100 mg/person	887	999	41.0	339	502	52.2	2.08	[721]
Volunteer	Malaysia	1	1997	Oral	200 mg/person	2193	857	30.0	1141	1340	-	2.77	[737]
Adults	Vietnam	11	2002	IM	120 mg/person	11 343	1146	2.7	1166	2475	64.0	3.79	[788]
Adults	Vietnam	9	2004	IM	2.4 mg/kg	16 149	1038	3.2	869	1497	52.7	2.10	[44]
Adults	Vietnam	13	1998a	IV	120 mg/person	-	1056	-	2646	2378	40.2	1.61	[771]
Adults	Vietnam	11	2002	IV	120 mg/person	13 688	877	2.2	2760	2873	59.0	8.53	[788]
Adults	Thailand	2	2000	IV	2 mg/kg	-	846	2.6	-	3999	-		[773]
Adults	Vietnam	12	1998b	IV	120 mg/person	130	49	13.2	2193	1846	36.7		[772]
Volunteer	Vietnam	10	2001	IV	120 mg/person				1508	1365	53.0		[620]
Adults	Thailand	17	2006	IV	2.4 mg/kg				605	418	20.4		[711]

* The drug measurements are by using HPLC-ECD or LC-MS/MS methods, and some drug evaluations by using bioassay are not selected in this summary.
PNG = Papua New Guinea.

Table 44. Pharmacokinetics/pharmacodynamic (PK/PD) parameters of AS (IV, 120 mg and oral 100 mg at 8h, then oral 750 mg mefloquine at 24h), AS (oral, 100 mg and then 50 mg b.i.d x 4), DHA (oral, 200 mg and then 100 mg x 4), QHS (oral, 500 mg and 250 b.i.d x 4 and then 500 mg on D6), AM (IM, 3.2 mg/kg and 1.6 mg x 4), and AE (IM, 4.8 mg/kg at 0 h and 1.6 mg/kg at 6 h and then day 2-5 daily) in human treatment with uncomplicated and severe/complicated malaria on day 1* [44]

PK/PD Parameters	AS	AS	DHA	QHS	AM	AE
Administration Route	Intravenous	Oral	Oral	Oral	Intramuscular	Intramuscular
First loading dosage Maintaining	120 mg	100 mg	200 mg	500 mg	3.2 mg/kg	4.8 mg/kg
dosage Total dose	Oral 100 mg at 8 hr	50 mg b.i.d. x 4	100 mg x 4	250 x 2 x 5	1.6mg/kg x 4	1.6mg/kg x 5
	220 mg and mefloquine**	500 mg	600 mg	3000 mg	9.6 mg/kg	12.8 mg/kg
PK parameters (Day 1)						
C_{max} (ng/ml)	2646 (DHA); 11343(AS)	1052 (DHA); 198 (AS)	437.5	588.0	74.9	110.1
T_{max} (hr)	0.13	0.75	1.4	2.4	6.0	8.2
Tlag (hr)			0.2	0.45		
$AUC_{0-24\,hr}$ (ng·h/ml)	2378 (DHA); 1146 (AS)	1334 (DHA); 210 (AS)	1329	2601	1230	4702
$t_{1/2}$ (absorption, hr)		0.36 (DHA)	0.67	1.21	1.88	3.2
$t_{1/2}$ (elimination, hr)	0.67 (DHA); 0.05 (AS)	0.70 (DHA)	0.85	2.3	7.83	22.7
MRT (hr)	0.90 (DHA)	1.95 (DHA)	2.71	7.41	13.94	42.9
PD parameters (Day 1)*						
Time of lag phase (hr)	1.92	2.81	4.03	5.76	7.26	8.89
AUIC (%·h/µl)	397.3	921.2	1167.9	1464.4	1613.4	2463.5
PC_{50} (hr)	3.18	8.48	10.05	13.95	15.63	19.68
E_{max} (%) or MPC	0.0011	0.0016	0.2132	0.0100	0.0030	0.5504
Curative rate (%)	100**	81.3	76.0	74.3	86.7	48.0

* The data was fitted with WinNonlin (V5.0) by author. **Oral 750 mg mefloquine at 24 hr after IV injection. PK = pharmacokinetics; PD = pharmacodynamics; MRT = Mean residence time; PC_{50} = Mean time for parasitemia to fall by half; AUIC = area under inhibitory curve; QHS = Artemisinin; DHA = Dihydroartemisinin; AM = Artemether; AE = Arteether; AS = Artesunic acid; MPC = minimum parasiticidal concentration; IM = intramuscular.

Currently, the therapeutic dosages of injectable AS to treat severe and complicated malaria are in the range of 2-2.4 mg/kg alone. The dose was designed in accordance with prior experience of physicians concerned with neurotoxicity, to reduce dose-dependent hemotoxicity, to prevent a sub-maximal antimalarial effect, and to keep a minimum curative rate (> 90%) during over twenty years of treatments [466, 496]. However, as this report describes in previous sections, this regimen will only produce 735 - 18909 ng/ml of C_{max} with great variability (Table 43), which is a lower peak concentration than needed for patients with hyperparasitemia (i.e., the C_{max} in the malaria patients with a high variability). In theory, the patients with severe malaria require 7500-30000 ng/ml peak concentration to efficaciously cover each RBC (described in Section 6.3), including infected cells, because the infected cell can concentrate 100-300-fold more DHA than normal RBC during the first 1-2 hr after dosing. The large variability in pharmacokinetics and pharmacodynamics by using AS should also be considered. This means that the intravenous dose should be increased to a higher level (4-8 mg/kg) than current use for the loading dose. However, the question remains: is this intravenous dose safe for treatment of severe and complicated malaria?

Compared to oral AS, the injectable AS is superior in pharmacokinetics and pharmacodynamics to the oral AS and other four artemisinins in treating patients with severe malaria; injectable AS shows the highest peak concentration (Table 43), the fastest effective time, rapid elimination, and a very short exposure moment compared to oral AS [497]. As discussed in Section 3, the drug exposure time of artemisinins is the principle factor responsible for induction of neurotoxicity in animal species. Therefore, the active metabolites (DHA), the drug distributed in CNS, and/or the drug exposure level play relatively less important roles than the drug exposure time. The water-soluble artemisinins, AS in particular, have very short half-lives and produce limited exposure time in animal species and humans, indicating that once-daily oral or intravenous administration of AS is relatively safe. In fact, the animal experiments demonstrated that no neurotoxicity (pathologically or behaviorally) has been observed in animal species following intravenous administration at any repeated doses up to maximum tolerated doses (MTD). The MTD of AS was 240 mg/kg following intravenous injection daily in rats for 3 days, but no neurotoxicity was detected in these animals [94, 242]. In another study, intravenous AS had no effect on neurotoxic scores at 120 mg/kg in sodium carbonate daily for 7 days (our unpublished data). After one intramuscular injection, a high dose of 420 mg/kg of AS did not produce any neuronal necrosis in the rats. Also, up to 200 mg/kg of AS used orally daily for 5-7 days did not exhibit neuronal changes or any specific clinical signs [498]. Therefore, the 4-8 mg/kg dose level with intravenous formulation should be safe with regards to neurotoxicity [499, 500] and also safe with regards to hemotoxicity [468, 501]. One study used a high dose of oral AS, up to 6 mg/kg, with the AS being extremely well tolerated; to date, there is no convincing evidence that dose as high as this of oral AS have any significant toxicity [502]. In our current Phase I clinical trials, the 2-8 mg/kg doses have been exceedingly well tolerated following both single and multiple intravenous injections in volunteers with our new formulation (our unpublished data). In malaria patients, this dose offers a satisfactory margin of safety and would be unlikely to produce a sub-maximal antimalarial effect.

Angus *et al.* [502] characterized the *in vivo* dose-response relationship for AS (and thus, rationalize dosing), 47 adult patients with acute uncomplicated *falciparum* malaria and parasitemia >1% were randomized to receive a single oral dose of AS varying between 0 and 250 mg together with a curative dose of oral mefloquine. Acceleration of parasite clearance was used as the pharmacodynamic variable. An inhibitory sigmoidal maximum effect (E_{max}) pharmacodynamic model typical of a dose-response curve was fitted to the relationship between dose and shortening of parasite clearance time (PCT). The E_{max} was estimated at 28.6 h, and the 50% effective concentration was 1.6 mg/kg of body weight. These results imply that there is no reduction in PCTs with the use of single oral doses of artesunate higher than 2 mg/kg, effectively reflecting the average lower limit of the maximally effective dose [502]. It is suggested that the 2 mg/kg is a maximum effective dose in those selected patients. However, the author also pointed out that this dosage of 2 mg/kg would be the lowest to give maximum effects in an "average" patient. When taking into account the considerable and large variance between individuals in pharmacokinetics and pharmacodynamics [496, 502, 503], a larger dose would be required to ensure that a maximum effect was obtained in all patients. Therefore, the loading dose of injection AS is reasonable to increase to another higher level (4-8 mg/kg) for the consideration of WHO experts and investigators in the further development and treatment.

After the first drug exposure to reduce parasitemia below 1%, the maintaining dose could be significantly cut or switch (sequentially) to another slower acting and low dose antimalarial agents (not artemisinins) in order to protect AS from recrudescence or resistance and to optimize antimalarial therapy in severe and complicated malaria [401]. Because these artemisinins have excellent intrinsic activity, which produce reductions in asexual parasite biomass of up to 10,000-fold per cycle, these drugs still need to be present in therapeutic concentrations during four asexual life cycles, or 8 days [4]. Previous studies have shown that once-daily administration with the artemisinin derivatives provides equivalent cure rates relative to more frequent administration [492, 504]. This suggests that exposure of the infecting parasite population to therapeutic concentrations of drug twice each life cycle is sufficient for maximum effect [6].

5.2.5. The PK Matching in ACT Selection and Practice

Artemisinin derivatives are particularly effective in combinations because of their very high killing rates, lack of adverse effects, and absence of significant resistance [4]. The ideal pharmacokinetic (PK) properties for an antimalarial drug have been greatly debated. From a resistance prevention perspective, the combination partners should have similar pharmacokinetic properties (PK matching). Mutations that confer resistance occur spontaneously at a small but finite rate ($\sim10^{-6}$ per generation). When exposed to two drugs, parasites would need to develop mutations at two resistance loci simultaneously, and the likelihood of this occurring (i.e., $10^{-6} \times 10^{-6} = 10^{-12}$) becomes extremely small [4]. The probability of selecting a mutant parasite with resistance to both drugs is very unlikely.

Rapid elimination ensures that the residual concentrations do not provide a selective filter for resistant parasites, but these drugs (if used alone) must be given for 7 days, and adherence to 7-day regimens is poor. Even 7-day regimens of artemisinin derivatives are associated with approximately 10% failure rates [48, 466]. In order to be highly effective in a 3-day regimen, terminal elimination half-lives of at least one drug component need to exceed 24 hours. Combinations of artemisinin derivatives (which are eliminated very rapidly) given for 3 days, with a slowly eliminated drug such as mefloquine (20 day of half-life) provide complete protection for the artemisinin derivatives (< 2 h of half-lives) from selection of a *de novo* resistant mutant if adherence is good (i.e., no parasite is exposed to artemisinin during one asexual cycle without mefloquine being present). But this does leave the slowly eliminated "tail" of mefloquine unprotected by the artemisinin derivative (PK mismatching). The residual number of parasites exposed to mefloquine alone, following two asexual cycles, is a tiny fraction of those present at the peak of the acute symptomatic infection.

Furthermore, these residual parasites are exposed to relatively high levels of mefloquine, and, even if susceptibility was reduced, these levels are usually sufficient to eradicate infection. The long "tail" of the mefloquine elimination phase does, however, provide a selective filter for resistant parasites acquired from elsewhere; and, as described earlier, contributes to the spread of resistance once it has developed. Yet on the northwestern border of Thailand, an area of low transmission where mefloquine resistance had developed already, systematic deployment of artesunate-mefloquine combination therapy was dramatically effective, both in stopping resistance and also in reducing the incidence of malaria [53, 505]. This strategy would be expected to be effective at preventing the *de novo* emergence of resistance at higher levels of transmission, where high-biomass infections still constitute the major source of *de novo* resistance.

The main obstacles to the success of combination treatment in preventing the emergence of resistance will be incomplete coverage, or inadequate treatment; and, as for anti-tuberculous drugs, use of one of the combination partners alone. Drugs of poor quality are common in tropical areas of the world. Adherence to antimalarial treatment regimens is often incomplete, and antimalarials are available widely in the market place. Resistance to the artemisinins may not have happened yet. If it does, it will most likely arise in a hyperparasitemic patient who received an inadequate dose of a single antimalarial drug, not in combination with another suitable antimalarial agent. Irrespective of the epidemiological setting, ensuring that patients with high parasitemias receive a full course of adequate doses of artemisinin combination treatment would be an effective method of slowing the emergence of antimalarial drug resistance.

Rational choices of ideal partner drugs for antimalarial drug combinations will require defining and matching selective windows. Although in the past PK matching has been required by the US FDA for new anti-infective drug combinations, simply matching half-lives will not be sufficient. For example, the half-life of chloroquine is at least 1-2 months and probably longer, so in theory it should have protected against selection of resistance (and tolerance) to pyrimethamine, which has a half-life of only about 4 days. However, chloroquine is stored at high concentration in the liver and slowly leaches out for months after an initial steep decline in blood concentrations, so although its half-life is extremely long, its time-above-MIC, and thus its selective window, is probably relatively short but is undefined [506].

Artemisinins have been most prominently combined with the quinolines, amodiaquine, mefloquine, and lumefantrine. Pharmacokinetically the combination of artemether and lumefantrine (half-life: 3-6 days) [69] is better matched. However, it is also rather expensive and appears to be slightly less efficacious than artesunate-mefloquine in areas with a high prevalence of multidrug resistant parasites [70]. Artesunate-amodiaquine has been recently introduced as an inexpensive combination therapy (ASAQ) directed especially at Sub-Saharan Africa. ASAQ also suffers from a PK mismatch due to the long half-life (9-18 days) of its active metabolite (desethyl-amodiaquine) [71, 507]. Furthermore, some African studies have already indicated a significant failure rate [148, 212, 508] due to the presence of amodiaquine resistance. Resistance is likely to spread more rapidly as the combination is being adopted as the main treatment in several African countries. Artemisinins have also been combined with antibiotics possessing antimalarial properties that have close matching half-lives (2-4 h) which resulted in a cure rate of 87% [509].

There are major concerns that mismatched PK will allow parasites to evolve resistance to the longer half-life partner, undermining the benefits of ACTs. Mismatched drug combinations do not significantly impact the spread of resistance and may even risk jeopardizing the efficacy of components against which resistance has not yet been reported (such artemisinins). Accordingly, drug combination need to have similar half-lives in order to keep a constant dual pressure on current and reinvading parasites. Ideally, drugs with short half-lives should be preferred, with the intention of reduction of the exposure of reinvading parasites to suboptimal drug levels which may induce the selection for tolerance and eventually the development of resistance. Whether the observed phenomenon is labeled artemisinin 'resistance' or 'tolerance,' it is clear that parasites in some parts of the region have had reduced sensitivity to ACTs for several years [7]. PK/PD data that allow identification of *in vivo* minimum inhibitory or parasiticidal concentrations and definition of selective window times will provide the evidence needed for designing antimalarial drug combinations on a rational basis to prevent selection of resistance after clearance of the initial infection. Therefore, as with any medication, each combination has its own advantages and disadvantages based on expensiveness, tolerability, PK matching, and ease of administration.

5.3. Therapeutic Efficacy Concerns Drug Pharmacodynamics (PD)

Although more than 40% of the world population live in malaria endemic areas, there is only one class of drugs (artemisinins) available antimalarial drugs for the efficacious treatment of *P. falciparum* infection currently. Theoretical research and speculation into the dynamics determining the evaluation of drug usage have focused on five main areas: 1) the widespread of single- and multi-drug resistance; 2) the semi-immune and non-immune subjects; 3) the relationship between parasitemia and malaria diseases; 4) the large inter-individual variation of pharmacodynamics (PD); and 5) the dynamics in special populations. If the number of dynamics concerning an infection is known, then the pharmacodynamic properties required of an anti-infective drug to produce a cure can be defined. The assessment of the treatment response in malaria rests on the clinical outcome (mortality, speed of recovery from coma, fever clearance, etc.) and the parasitological outcome - the subjects of this section. The advantages of defining the pharmacodynamic properties of antimalarial drugs more precisely would be better methods of assessing the therapeutic response. This would also provide a firm basis for designing antimalarial drug regimens (dose, frequency, and duration), predicting treatment failure, and slowing the development of antimalarial drug resistance.

5.3.1. *In vivo* Pharmacodynamics Affects Use of Antimalarial Drugs

5.3.1.1. Antimalarial Effects on Parasite Clearance

Falciparum malaria may also exhibit rapid or slow declines in parasitemia following various treatment regimens as a result of sequestration of a synchronous infection. The parasite clearance times (PCT) is the time from the beginning of antimalarial treatment until parasites are no longer detectable in the peripheral blood film. This depends very much on the admission parasitemia and the frequency with which blood films are taken. PCTs will obviously be short if admission parasitemias are very low and will vary by ± 24 hr if parasite counts are taken only once a day. For these reasons, in small comparative clinical trials, parasitemias should be normalized and parasite counts should be taken at least every 12 h, preferably, every 4 to 6 h. As the sensitivity of microscopy depends on the quality of the microscope, the staining procedure, the slide, the duration of the examination, and the experience of the microscopist, the parasite clearance curve should be defined by more than one parameter [479]. The trial use PC_{50} (time until reduction by 50%), PC_{90} (time until reduction by 90%), and PCT. Others have used PC_{95} and PC_{99}. These measures are all misleading at low parasitemias because the intrinsic variability of the estimates of the parasite count increases disproportionally near the limits of detection. For drug comparisons, it may be better to exclude patients with admission parasite counts under 10,000/ml [6].

The dihydrofolate reductase inhibitors (cycloguanil and pyrimethamine) and the quinolines quinine, quinidine, and mefloquine appear to have relatively little effect on asexual malaria parasites in the first half (24 h) of the life cycle [436, 510-512]. As a result, most circulating *P. falciparum* parasites continue to mature, express adhesins on the infected erythrocyte surface, and begin to cytoadhere and form rosettes despite treatment. Thus, exposure of ring form parasites to these drugs does not appear to prevent sequestration significantly [513]. As a consequence, the early decline in parasitemia following treatment with these antimalarial agents is largely that which would have occurred in their absence. In contrast, chloroquine, halofantrine, and, to a greater extent, artemisinin and its derivatives do attenuate the growth of young parasites *in vitro* and increase the clearance of ring forms *in vivo* [436, 514, 515]. As a result, the early decline in parasitemia following treatment with these drugs is usually more rapid than that seen with treatment with those in the first category. These differences may not be appreciated from an individual patient's records but become evident when a series of patients with similar admission parasitemias and degrees of background immunity when treated with the different drugs are compared [6].

A population of patients with malaria will have infections different in magnitude (parasite numbers), stage, and synchronicity; and therefore have different parasitemia-time profiles. The population parasite clearance curve represents the average of the individual parasitemia-time profiles normalized to a baseline parasitemia of 100%. Following treatment with quinine or mefloquine, the average parasitemia-time profile is that of a lag phase, during which the population mean parasitemia remains unchanged, and then a decline. The population parasite clearance curves following treatment with chloroquine [6], halofantrine, and particularly the artemisinin derivatives [516] show a shorter lag phase and thus shorter PC_{50}, PC_{90}, and PCT, provided the infecting organisms are fully sensitive. These changes are too rapid to be accounted for by reduced merogony and can only represent accelerated ring form clearance

[517]. Accelerated cytoadherence (leading to enhanced sequestration) is unlikely as an alternative explanation, as the rapid decline in parasitemia coincides with morphological changes in the circulating parasites and a reduced ability to cytoadhere *in vitro* [6]. The stage specificity of drug action and the magnitude of the antimalarial effect can be deduced from population parasite clearance curves.

Antimalarial drugs are maximally active on the mature trophozoite stage of parasite development (roughly the middle third of the life cycle) [6, 436, 514, 515]. This is assessed *in vitro* in terms of inhibition of glucose, amino acid, and purine (hypoxanthine) uptake and also inhibition of morphological development and subsequent multiplication. Stage specificity varies between compounds and has not been defined precisely. Available evidence suggests that dihydrofolate reductase inhibitors have a relatively narrow time window of activity in the second half of the life cycle; quinine and mefloquine have a broader time window but still act mainly in the second half of the cycle. Chloroquine and halofantrine have a still wider time window of activity (hence their action on circulating forms), but artemisinin derivatives have the broadest range of all, with considerable effects on ring stage parasites. If the stages of parasite development are assumed to be distributed randomly in the patient population, then some of the patients, by chance, will have a predominance of mature meronts when drug treatment is given. Thus, the relative effects of two antimalarial drugs on meront development may be assessed by comparing the proportion of patients whose parasitemia shows an early rise (e.g., within 6 to 12 h) after the start of antimalarial treatment. Following treatment with oral artesunate, early increases in parasitemia are significantly less than those that occur following quinine or mefloquine treatment. This suggests that artesunate affects meront development at a later stage than these other antimalarial drugs [6].

5.3.1.2. Antimalarial Pharmacodynamics on Drug Regimens

Although a great deal remains to be learned about the pharmacodynamic properties of antimalarial drugs *in vivo*, sufficient information is already available to construct simple models which predict for how long antimalarial treatment should be given and the chances of treatment failure (i.e., recrudescence of the infection). Mature schizonts and tiny rings are the most drug-resistant stages of the blood infection [436]. Most of the available antimalarial drugs are active predominantly on the sequestered parasites that are not visible to the microscopist. The consequences of this are inhibition of trophozoite development and a considerable reduction in the parasite multiplication rate. As susceptibility is stage dependent, the timing of drug administration with respect to the stage and synchronicity of parasite development may determine the therapeutic response (chronotherapy).

Antimalarial treatment with chloroquine and, possibly, some other compounds can be optimized if peak drug levels coincide with the most sensitive stage of parasite development [437]. This is rather difficult to arrange in practice. Qinghaosu (artemisinin) and its derivatives are peroxidic antimalarial agents which are particularly active against ring stage parasites. In the early Chinese studies with these drugs, treatment was started only when most of the infecting parasites were at the tiny-ring stage of development [438]. The advantages of chronotherapy remain to be established, although this is an important area for further research. The frequency of drug administration is usually determined by pharmacokinetic considerations. Conventionally, a drug should be given at intervals roughly corresponding to its biological half-life (estimated in patients with the disease being treated, not extrapolated from studies with healthy volunteers). If the drug exerts its effects only while concentrations remain above a therapeutic level, then the biological half-life corresponds to the drug half-life. However, if there is concentration-dependent killing and biological effects persist well after the drug has declined below the therapeutic level, then the drug can be given less frequently than predicted from the drug half-life determined in pharmacokinetic studies [518]. It knows relatively little about antimalarial dose-response or concentration-effect relationships *in vivo*. *In vitro*, the relationships for antimalarial drug effects on hypoxanthine uptake or on schizont maturation display a sigmoid relationship [436].

These concentration effect relationships are probably different at different stages of parasite development. Although therapeutic concentrations are assumed to reflect drug activity at the flat (top) part of the sigmoid dose-response (concentration-effect) curve, i.e., E_{max}, there is little data to support this for any of the antimalarial drugs. These considerations may also be important for other short-half-life antimalarial agents, in particular, the artemisinin derivatives. The available studies suggest that despite rapid absorption and elimination half-lives for both the parent compound and the biologically active metabolite dihydroartemisinin that are less than 1 h, oral artesunate is equally effective when given once or twice daily [56, 517]. This suggests that the biological effects of the artemisinin compounds extend beyond their presence at therapeutic concentrations in plasma. The broad stage specificity of action of these drugs ensures that only two exposures of a few hours in each asexual life cycle are sufficient for a maximal antimalarial effect. This challenges the concept that antimalarial drugs need to remain above parasiticidal levels throughout the dosing interval.

5.3.1.3. Parasite Reduction Ratio Reflects Drug Efficacy

Parasitological recovery from malaria is assessed conventionally by the clearance of parasites from peripheral blood smears (68). In highly drug-resistant infections, parasites do not disappear from the peripheral blood or may increase following the administration of antimalarial drugs. Parasites with lower grades of resistance disappear from the peripheral blood (in fact, the concentration falls below the level of microscopic detection) but recur at a later time, usually in association with a return of symptoms. The efficacy of antimalarial drug treatment is assessed in terms of the speed at which symptoms and signs resolve and parasitemia declines (usually recorded as the PCT) and the proportion of patients in whom infections recur within a defined period (71).

In the management of individual patients, White [6] would like to credit a ratio of the parasitemia reduction at the time of treatment to the count 48 h later (the parasite reduction ratio [PRR]), representing the fractional reduction per asexual life cycle, to be a simple but useful predictive index.

The number of new young parasites that enter the circulation and invade erythrocytes depends on a number of factors, i.e., the sequestered biomass, its stage and synchronicity of parasite development, the number of viable merozoites released per schizont, the presence of merozoite agglutinating antibodies, the susceptibility of the remaining erythrocytes to invasion, the timing of antimalarial drug administration, the drug levels in blood achieved, and the drug's activity [479]. Antimalarial drugs differ in their concentration-effect relationships and also in their intrinsic activity (Figure 16). Thus, the E_{max} for one drug will be different from that of another, and so will the minimum concentration in blood or plasma required for a maximal effect (minimum parasiticidal concentration [MPC]). These concentrations will obviously be higher than those required to prevent parasite multiplication.

In the presence of effective antimalarial drug concentrations (i.e., concentrations greater than the MIC), parasitemia falls as a consequence of a reduced input of new young parasites, clearance of circulating rings, and sequestration.

In the non-sequestering benign human malarias (those caused by *P. vivax*, *P. ovale*, and *P. malariae*), the reduction in the total-body parasite burden following antimalarial treatment can be estimated by comparing the baseline and 48-h parasite counts and multiplying by the blood volume. Although the precise total burden of *P. falciparum* cannot be estimated, this baseline-to-48-h parasitemia ratio is still informative. The PRR or fractional reduction in parasitemia per asexual life cycle is analogous to the killing rate. The considerable differences in PRR between antimalarial agents reflect the differences in intrinsic activity (E_{max}) between the drugs as listed as follows [6]:

Antimalarial drugs	Estimated PRR *in vivo*
• Artemisinin, artesunate, artemether	10^3–10^5
• 4-Aminoquinolines, chloroquine, halofantrine	10^2–10^4
• Quinine, mefloquine, pyrimethamine-sulfadoxine	10^1–10^3
• Antimalarial antibiotics, Desferrioxamine	5–10

The therapeutic level of an antimalarial drug in plasma or blood *in vivo* and the parasite killing rate reflected by PRR are clearly related, and the minimum parasiticidal concentration (MPC) equals maximum PRR. Additionally, estimation of the PRR *in vivo* allows an assessment of the duration for which therapeutic concentrations of an antimalarial drug must be present (Figure 16), although this does make the considerable assumption that the reduction rate or killing rate remains constant with succeeding cycles. The simple PRR concept does not explain all aspects of the therapeutic response in malaria. The combination of quinine with tetracycline is consistently more effective than quinine alone, particularly in areas with quinine-resistant *falciparum* malaria, but mean PRRs are not significantly different whether or not tetracycline is added. To eradicate all of the parasites in the body, an effective antimalarial drug concentration (at least the MIC but preferably the MPC) needs to be present in the blood until either the last parasite has been removed or immune defenses are able to deal with the residuum. Figure 16 shows that courses of short-acting antimalarial agents of less than 7 days in nonimmune patients will not eradicate all of the malaria parasites in all patients (i.e., there will be significant recrudescence rates). Failure rates will be higher in patients with greater initial parasite burdens because they have more parasites to remove.

Antibacterial drugs with antimalarial properties, such as the sulfonamides, tetracycline, clindamycin, or rifampin, or other compounds such as desferrioxamine are much less active than the antifols, quinoline, or peroxidic antimalarial agents. They have estimated PRRs of only about 6 to 10 per cycle [6]. Therapeutic concentrations of these drugs alone need to be present for approximately 2 weeks in nonimmune patients; otherwise, the infection will recrudesce (this accords with the results of clinical trials). Chloroquine is more complex. Although it has an extremely long terminal elimination half-life (1 to 2 months), the blood concentration profile reflects a multiexponential decline after treatment. Concentrations in blood fall rapidly after absorption and the levels that persist in the blood for weeks after treatment are relatively low compared to those during the treatment phase [493]. If killing rates are low initially, then the infection cannot be eradicated.

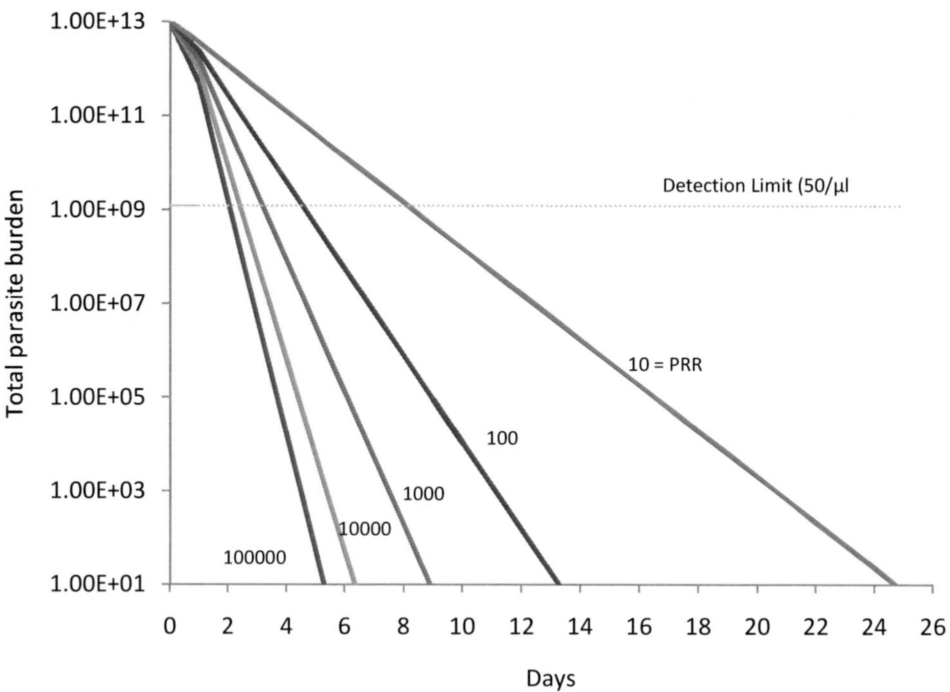

Figure 16. Parasite clearance following different antimalarial treatments. The initial parasite burden corresponds to a parasite count of approximately 100,000/ml, or 2% parasitemia, in an adult with *falciparum* malaria. The times taken to eradicate all parasites from the body are shown at different killing rates (PRRs), assuming constant fractional reduction and no contribution from induced immunity. It is evident that rapidly eliminated drugs with PRRs of #103 need to be given for .1 week. [156].

Resistance to the antifols usually develops more rapidly than resistance to quinoline antimalarial agents [359]. Resistance results from reductions in parasite DHFR affinity for these antimalarial agents and is usually caused by single or double base pair mutations in the DHFR gene. As only one or two steps are required to reduce affinity by several orders of magnitude, high-grade resistance develops rapidly. When proguanil (chloroguanide) was first introduced in peninsular Malaya, both *P. falciparum* and *P. vivax* were very sensitive to the drug, but over a 3-year period the mean PRR following administration of a single dose of 250 to 300 mg in uncomplicated *falciparum* malaria fell from 500 in 1947 to 100 in 1948 to 25 in 1949. By mid-1949, the early failure rate had risen to 25% [6]. The PCT and the PRR are obviously related closely, although the latter is much easier to measure and is more precise. Parasite counts taken at 2 days are already part of the standard assessment protocol for determining the level of 4-aminoquinoline drug resistance. PRRs of < 4 are classified as R_3 resistance. However, these observations suggest that the measurement may have a broader application. PRRs of <10 tend to be associated with R_2 resistance (i.e., parasite PCTs of over 7 days). The prognostic sensitivity and specificity of the PRR as a predictor of therapeutic failure should be defined for different drugs at different levels of antimalarial drug resistance.

5.3.1.4. Effects of Immunity on the Therapeutic Response

Antimalarial treatment responses in semi-immune or immune patients were always better than those in nonimmune patients. Even nonimmune patients eventually develop a strain-specific immunity to their malaria infection which at first limits and finally eradicates it. This is a lengthy process (several months) which approximates to the natural history of untreated *falciparum* malaria. Nonimmune patients with highly (multiple drug) resistant parasites may require three or more courses of antimalarial treatment before they are finally cured. In areas where malaria is endemic, the well-developed immune response acts together with the antimalarial drugs, allowing shorter treatment courses to be given and giving good therapeutic responses even when there is some degree of drug resistance. In these areas, the treatment response improves with age, coincident with the acquisition of immunity. Thus, age is an important factor determining the therapeutic response in such areas and must be taken into account in drug trials. Immunity is temporarily disrupted in pregnancy, and pregnant women are a particularly difficult group of patients to treat, since the choice of drugs is also limited.

5.3.1.5. PD of Antimalarial Drug Relevant to the Prevention of Resistance

To prevent the emergence of resistant mutants, combinations of antimalarial drugs with different loci of action have been employed. A similar strategy is employed in antituberculous chemotherapy, and also in antiretroviral and cancer chemotherapy. This works best if the drugs have well-matched pharmacokinetic properties and all have significant antimalarial activity against prevalent parasite strains. In a rodent model, the combination of sulfadoxine-pyrimethamine-mefloquine delayed the onset of mefloquine resistance in a strain of *P. berghei* (which was sensitive to all three drugs), but this did not work in *falciparum* malaria in Thailand, where the combination was introduced in 1984 [434], because the drugs were not well matched in terms of pharmacokinetic properties (mefloquine had a much longer terminal elimination half-life than the other two) and *P. falciparum* was already highly resistant to sulfadoxine and pyrimethamine [367]. The pharmacodynamic properties of antimalarial drugs are also relevant to the prevention of resistance.

If resistance potential is a function of viable parasite biomass (i.e., the more parasites there are, the more likely it is that a resistant mutant will emerge), then it follows that the risk of resistant isolates emerging from a primary infection will be greatly reduced if the parasite biomass is reduced. For example, artesunate is now increasingly used in combination with mefloquine for multidrug-resistant *falciparum* malaria. This is associated with a significant increase in cure rates, even in infections with highly mefloquine-resistant parasites [56, 359]. Artesunate given for only 3 days will reduce the parasite biomass by a factor of between 10^6 and 10^8. Thus, relatively few parasites (up to 10^4) remain for mefloquine to remove and the chances of a more mefloquine-resistant mutant emerging from this primary infection are reduced correspondingly. Therefore, the use of artesunate cannot protect mefloquine from any subsequent (new) infection which is acquired later, when mefloquine levels have declined to subtherapeutic values. In successfully treated patients, mefloquine also protects artesunate by removing all residual parasites originally exposed to the drug; thus, there is no pressure to artesunate resistance. The same principles should also apply to other combinations of artemisinin derivatives with more slowly eliminated antimalarial agents.

5.3.2. Poor Efficacy in Monotherapy of Artemisinin or its Derivatives

The therapeutic potentials for all artemisinin and its derivatives when using monotherapy within 54 clinical trials have been summarized in the treatment for malaria with various dose regiments (Table 4). Successful (or 100% cure) treatment is found neither in the 3-5 day treatments nor the high dose (4.8 mg/kg) regimens with monotherapy [44, 519]. The curative rate, fever (FCT), and parasite (PCT) clearances time are traditional evaluations for the efficacy of the antimalarial drugs in humans. These three parameters exhibit the therapeutic potential of intravenous, intramuscular, and oral agents in various regimens. Results have shown that arteether (AE) in sesame oil vehicle is not superior to that of the oral dihydroartemisinin (DHA), oral and intravenous artesunate (AS), intramuscular artemether (AM), or even intramuscular α/β-arteether formulated with peanut oil.

Indications from data, that show their known mode of rapid action, reveal the artemisinin group of antimalarials is considered to be highly and independently effective compared with other traditional malaria drugs. Although there is no evidence that drug resistance has occurred *in vivo*, short treatment courses (3-5 days) are associated with high recrudescence rates. The data that emerges from all available artemisinin drugs is that given a dose sufficient to induce initial parasite clearance and clinical recovery, the duration of therapy is the ultimate determinant for recrudescence. The rates of recrudescence (30−50% for 3 day regimen, 6−15% for 5 day regimen, and 2−6% for 7 day regimen) continues to be a big problem for all these compounds described previously [48, 466]. As the total parasite biomass in an adult with malaria may exceed 10^{12} parasites, then short-acting antimalarial drugs which cause <1000-fold reductions in parasite numbers per cycle must be given for longer than four asexual cycles. This simple numbers game explains most of the *in vivo* and *in vitro* results following antimalarial treatment [355]. It explains why short courses of treatment (< 7 days) with rapidly eliminated antimalarial drugs (artemisinins) cannot effect 100% cure rates, why infections with large biomass are more likely to recrudesce following antimalarial treatment (as the more parasites there are, the longer the duration of treatment required; why recrudescence's tend to occur 2 weeks after stopping treatment with short course antimalarials [355] and why recrudescence's following treatment with slowly eliminated antimalarial drugs tend to occur late (i.e., after 28 days) at low levels of drug resistance [56], but that the mean time to recrudescence shortens as antimalarial drug resistance worsens. It has been suggested that preventing recrudescence does not have the highest priority, but that a recrudescence should rather be treated with the same regimen again. Eventually, all patients would be radically cured with this strategy. The potential risk of development of resistance and recrudescence urges us to look for combination regimens that offer radical cure instead.

Reduction of parasitemia measured as the parasite clearance time (PCT); along with the fever clearance time (FCT) are primary measures of initial drug effect *in vivo*. Reviews of this therapeutic efficacy from multiple clinical studies demonstrate that when given alone (monotherapy) AM (oral and intramuscular administration), AS (oral and intravenous dosing), and DHA (oral) appear to have same effects for reducing PCT and FCT with various dose regimens (Table 4). The finding that therapeutic potentials of AM and AS in oral formulation could be equal to DHA may be contributed to the great conversion rate of AM or AS to DHA in human and high peak concentrations as previously described. AE after intramuscular injection appears to have fewer efficacies than the other 4 drugs, even with high dosage regimen or compared with AM intramuscular formulation (Table 4), which may have much lower blood concentration due to slow and prolong absorption [84, 87, 486].

Parasitemia count and clearance time (PCT) do not always accurately reflect the total burden of parasitized red blood cells, but better parameters are not available. All artemisinin compounds induce a remarkably rapid reduction of parasitemia, starting almost immediately after administration [520-522]. Fortunately the rapid parasite clearance induced by artemisinin drug treatment provides a simple pharmacodynamic measure that can be used in drug comparisons and assessment. Certainly, parasitological cure is the best end point measure of efficacy for patients with malaria. In this review, the available pharmacodynamic data of parasitemia reduction is evaluated with its associated PK result during the malaria treatment in human described as next.

5.3.3. Rapid Efficacy of Artemisinins Relates to Drug Peak Concentration

Pharmacokinetics and pharmacodynamics (PK/PD) evaluation indicates that the effect of parasitemia decrease rate appeared to be significantly slower (2.8-8.9 hours) than the decline of artemisinin drugs, reflecting a sum mean of the lag phase time, stages for effect time, and even body elimination time with the different compounds and dosage regimens, suggesting that PD effects are longer than PK outcomes. The elimination half lives in PK were in the range of 0.7-22.7 hr, whereas the PC_{50} in PD were in the range of 8.48-19.68 hr [44]. In this evaluation, AS is pharmacokinetically and pharmacodynamically superior to other four artemisinins for the treatment of malaria, demonstrating the most rapid killing activities with greater E_{max} in the PD analysis after oral administration. AS showed the highest peak concentration, the fastest and most reliable oral absorption rate, and the most limited drug exposure time than other four artemisinins. Although AM and AE showed the longest half lives and the highest exposure level (high AUC values), they had the worst efficacy (the longest lag time, the greatest AUEC, and the longest PC_{50}) due to the slowest absorption rate and the lowest peak concentration (Table 44). However, the inadmissible failure (48.0−86.8% curative rates) in those monotherapies led us to either re-assess their importance or seek alternative approaches (Table 4):

1) *In vitro*, at dose range 10^{-7}-10^{-5} M (30 −3000 ng/ml), synchronous culture study shows that artemisinin is rapidly effective against both rings and schizonts, demonstrating 100% growth inhibition of these stages within 4-6 h. The inhibitory effects of AM are more gradual, requiring 8-10 hr of exposure time to achieve 70% growth inhibition of the ring and 40−50% inhibition of schizonts stages. DHA is found to be highly effective against all stages of the parasite, in most cases achieving 100% growth inhibition by 2-4 hr [480]. Concentration of 10^{-6}-10^{-5} M (280-2800 ng/ml) artemisinin shows a parasitical effect after only 3 hr of drug exposure, whereas 10^{-7} M (28 ng/ml) artemisinin requires 24 hr of exposure for a cidal effect [478, 523]. It is clear that the low dose will increase the exposure time, and high dose will shorten the treatment period. Therefore, the low dose range cannot be selected for artemisinin drugs in human treatments in malaria because: i) the artemisinins

have very short half-lives, and ii) longer exposure time of artemisinins is principle factor to induce the fatal neurotoxicity [524].

In addition, in treating patients with severe and complicated malaria, where the ultimate survival of the patient is the key; low effective doses cannot be used without an undue delay in the necessary treatment time [479]. There is no evidence that rapid parasite killing with high sufficient doses is harmful and that the therapeutic blood drug concentration should be achieved as quickly as possible without danger of toxicity [493]. Therefore, the plasma concentration of the artemisinin derivatives in human against *P. falciparum* should at least be in the range of 10^{-6}-10^{-5} (300-3000 ng/ml) or even higher for AS or DHA as the more potent compounds. AE or AM should have higher plasma concentrations (more than that of AS or DHA). AL, being the lowest potency agent in regards to the first drug exposure [234, 469, 525, 526], should have the highest plasma concentration.

2) Artemisinins reliably reduce initial malaria parasitemia by a factor of 10^4 per 48 hour. Therefore, asexual cycle and modeling studies suggest that six days of treatment should cure parasite burdens of up to 10^{12} parasites [4]. This model is difficult to reconcile with the high recrudescence rates (10–15%) seen with artemisinin monotherapy and is also difficult to recognize with a higher potency agent (AS or DHA) and with higher concentrations. This poor efficacy of cure in the monotherapy is not due to resistance [527], but usually attributed to the intrinsically lower peak concentration of artemisinins, as demonstrated in this PK/PD evaluation.

The pharmacokinetic factors are not only focusing on the high peak concentration, but are also concentrating on tissue distribution in red blood cells of artemisinins [243]. More recently, the tissue distribution in rats and protein binding in human blood were conducted in WRAIR laboratory. Table 45 demonstrates the RBC/plasma ratios of five radiolabeled artemisinins in human whole blood at various concentrations [243, 528]. For equal volumes of RBC and plasma, the fraction of the five artemisinins found in the RBC is lower than that in the plasma in normal volunteer blood. The ratios are almost constant in human blood over a broad range of concentration. After correction for the 55:45 ratios of plasma and RBC volumes, the RBC/plasma ratios are 0.23 for ^{14}C-AE, 0.28 for ^{14}C-AM, 0.44 for ^{14}C-AL, 0.52 for ^{14}C-DHA, and 0.71 for ^{14}C-AS, indicating that the radioactivity remained in normal human RBC is the highest for AS and the lowest for AE.

The big advantage of DHA and AS is its readiness of entering human red blood cell (0.52 and 0.71 ratio of RBC/plasma, respectively) compared to AE and AM (0.23 and 0.28 ratio of RBC/plasma, respectively). It means that to reach same targets (RBC) the drug peak concentrations for AE and AM should be required to be two to three times higher than that of DHA and AS. In addition, the binding affinities are high for AE and AM radiolabels with 1945 and 1673 ng/ml of 50% bound concentration in human plasma (BC_{50}). Conversely, the binding affinities are low for AS and DHA radioactivities with 143 and 162 ng/ml of BC_{50}, indicating the free drug concentration can be much higher in human plasma with AS or DHA than that with AE and AM.

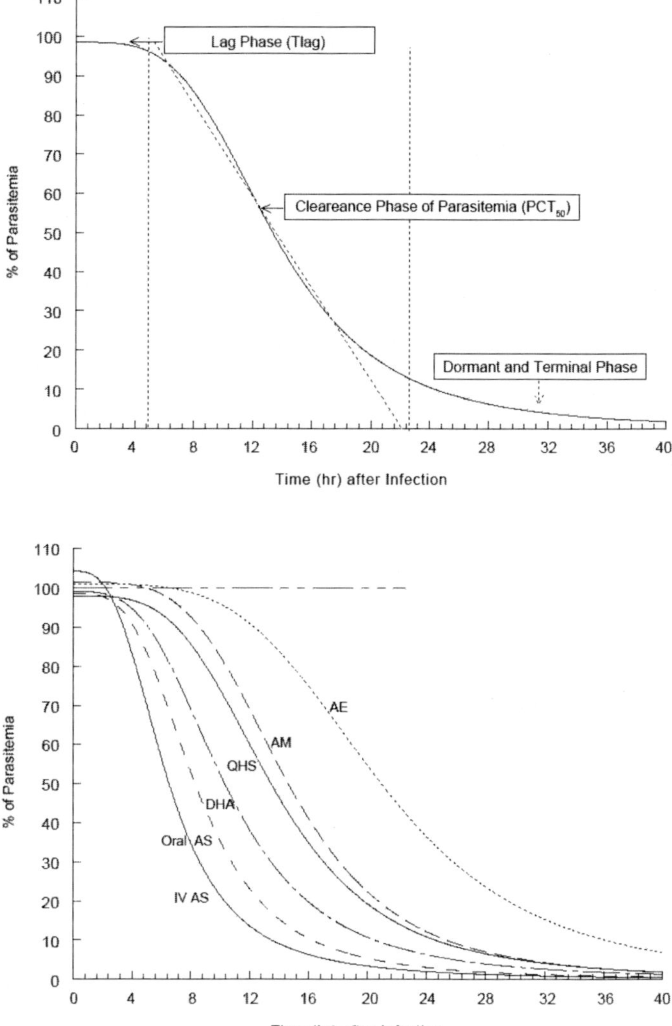

Figure 17. Standard parasite clearance curve in different stages (lag phase, clearance phase, dormant and terminal phase) and synchronicity of parasite development under antiparasitic drug received at 0 hour (top chart). The mean parasite clearance curves (bottom chart) of AS (solid line) at single intravenous 120 mg and oral 100 mg at 8 hr then oral 750 mg mefloquine at 24 hr [63, 64]; oral AS (dashed line) at 100 mg first day and 50 mg b.i.d 2-5 days orally [65, 69]; oral DHA (dot & dashed line) at 200 mg first day and 100 mg 2-5 days orally [17, 72, 73]; oral QHS (solid line) at 500 mg first day, 250 mg b.i.d 2-5 and 500 mg last day orally [65]; intramuscular AM (dashed line) at 3.2 mg/kg first day and 1.6 mg/kg 2-5 days intramuscularly [67, 70]; and intramuscular AE (dot line) of 4.8 mg/kg at 0 hour and 1.6 mg/kg at 6 and then 2-5 days intramuscularly [4, 66], in malaria patients (bottom chart) (Table 2). The parasitemia at 0 hour was set as 100% of parasitemia.

3) Although DHA tablets and AM intramuscular formulation have a similar AUC compared to AS tablets (Table 44), their efficacies are much slower than that of AS tablets (Figure 17, bottom). The initial killing of parasites do not appear to be done very well by the two drug regimens because they do not have enough high peak

concentration to efficaciously cover all red blood cells. The uptake of DHA in infected erythrocytes has been shown to be 100-300 times higher than in the uninfected erythrocytes, and the maximum uptake of the drug occurs after 1-2 hr [529]. In this case, as we discussed earlier, the plasma concentration of the artemisinin derivatives in human against *P. falciparum* should be much more than the range of 10^{-6}-10^{-5} (300-3000 ng/ml) for the first drug exposure in order to efficaciously cover all RBC targets of infected and uninfected cells [479, 523]. This report as compared to oral AS, the other 4 artemisinins with oral and intramuscular formulations have a lower peak level, could not stop the parasite and efficaciously cover all RBC in the initial several hours having greatly extended the lag and PC_{50} times.

4)   High peak concentration (C_{max}) is also more important than high drug concentration (AUC) to cure malaria because sufficient drug is needed to cover all RBCs (including infected cells) in order to kill as many parasites as possible in the initial dosing. AUC concentration is counted during this period of time and very often lacks the necessary high peak concentrations. For example, in the AE clinical trial, the AUCs are higher than other four artemisinin drugs (Table 44), but it has lower C_{max} value that results in poor efficacy. Therefore, the high concentration (AUC) without high peak level is not able to cover all RBCs; this could have given rise to some untreated RBCs, including infected cells. The untreated RBCs may be the reason for recrudescence in all monotherapy approaches with artemisinins. In addition, QHS is a low antimalarial potency agent [234, 469, 525, 526]. Even though oral doses 5-fold higher than AS have been given to patients in trials, QHS still demonstrates poor efficacy because its peak concentration is only a half of that of AS, suggesting that any artemisinin drug with lower antimalarial potency should not be used in malaria treatment.

In conclusion, the rapid efficacy of artemisinins is principally dependent on drug peak concentration (C_{max}) in the first drug exposure and drug relative antimalarial potency. The relative antimalarial potency is maintained by the capability for killing various stages of parasites, rapid oral absorption, high peak concentration, plasma protein binding, drug distribution in targets (RBCs), and membrane permeability of RBCs in infected subjects. Other factors of the pharmacokinetic parameters, such as drug exposure level (AUC) and drug exposure time tend to be of minor significance when considering efficacy.

5.3.4. PD Interindividual Variation Concerns Drug Efficacy

Antimalarial drugs are usually used in doses that give blood concentrations which exceed the maximum effect (i.e., levels greater than the MPC). However, the maximum effect differs between the drugs and response in interindividuals. The results of *in vivo* studies indicate that there are considerable large variations in the pharmacodynamics. Dose recommendations for relatively dynamic drugs have varied enormously (up to six-fold for chloroquine, and eight-fold for quinine!). Observations of minor drug efficacy in uncomplicated malaria have led to limitation of antimalarial doses in severe disease-where minor adverse effects are irrelevant, and the object of treatment is to save life.

Table 45. The bound ratios of human red blood cells (RBC) and plasma with ^{14}C-AE, ^{14}C-AM, ^{14}C-AL, ^{14}C-DQHS, and ^{14}C-AS in human whole blood from low to high concentration with incubated dialysis method at 37°C [44]

Concentration (MW)	^{14}C-AE (312.4)	^{14}C-AM (298.4)	^{14}C-AL (418.5)	^{14}C-DHA (284.9)	^{14}C-AS (384.4
0.125 nM	0.20 ± 0.01	0.28 ± 0.01	0.43 ± 0.01	0.52 ± 0.03	0.72 ± 0.05
0.25 nM	0.24 ± 0.03	0.27 ± 0.01	0.44 ± 0.02	0.53 ± 0.04	0.71 ± 0.03
0.5 nM	0.22 ± 0.02	0.28 ± 0.01	0.47 ± 0.02	0.52 ± 0.05	0.73 ± 0.05
1 nM	0.23 ± 0.02	0.27 ± 0.02	0.43 ± 0.03	0.51 ± 0.03	0.74 ± 0.04
2 nM	0.22 ± 0.01	0.26 ± 0.01	0.45 ± 0.02	0.50 ± 0.03	0.74 ± 0.06
4 nM	0.23 ± 0.01	0.29 ± 0.02	0.45 ± 0.01	0.51 ± 0.04	0.69 ± 0.06
8 nM	0.24 ± 0.03	0.27 ± 0.02	0.45 ± 0.02	0.53 ± 0.03	0.64 ± 0.04
16 nM	0.25 ± 0.03	0.29 ± 0.03	0.44 ± 0.02	0.51 ± 0.04	0.59 ± 0.08
Mean value	0.23 ± 0.02	0.28 ± 0.02	0.44 ± 0.02	0.52 ± 0.04	0.71 ± 0.08

All values represent mean ± SD for n = 4 individual samples, AE = arteether, AM = artemether, AL = artelinic acid, DHA = dihydroartemisinin; MW = molecular weight.

Table 46. Comparison of main pharmacokinetic parameters of artemisinin derivatives (DHA, AM, AE, AS, and AL) at 10 mg/kg in rats following single intravenous injection. Values are mean ± SD (n = 4) [497]

Parameters	DHA	AM	AE	AS	AL
MW =	284.9	298.4	312.4	384.4	418.5
Dose (mg/kg)	10	10	10	10	10
Dose (µM/kg)	35.1	33.5	32.0	26.0	23.9
IV Vehicles	CMP	1:3 CMP/sal	1:3 CMP/sal	Saline pH 8	Saline pH 8
C_{max} (ng/ml)	6683 ± 2405	6231 ± 1837	3724 ± 430	2106 ± 399	12706 ±1010
(µM)	23.5 ± 13.0	20.9 ± 6.2	11.9 ± 1.4	5.5 ± 1.0	30.4 ± 2.2
$AUC_{inf.}$ (ng·h/ml)	3184 ± 659	1857 ± 432	842 ± 154	738 ± 114	5481 ± 1754
(µM·h)	11.2 ± 2.3	6.2 ± 1.4	2.7 ± 0.5	1.9 ± 0.3	13.1 ± 3.8
Vss (liter/kg)	0.50 ± 0.19	0.67 ± 0.11	0.72 ± 0.09	0.87 ± 0.35	0.39 ± 0.12
CL (ml/min/kg)	55.4 ± 9.7	91.8 ± 21.0	200.0 ± 31.0	190.9 ± 30.3	32.1 ± 8.0
$t_{1/2}$ distribution (h)	0.22 ± 0.04	0.10 ± 0.01	0.10 ± 0.02	0.15 ± 0.02	0.19 ± 0.08
$t_{1/2}$ elimination (h)	0.95 ± 0.21	0.53 ± 0.14	0.45 ± 0.03	0.35 ± 0.08	1.35 ± 0.45

CMP = Cremophore EL, DHA = Dihydroartemisinin, AM = Artemether, AE = Arteether, AS = Artesunic acid, AL = Artelinic acid, MW = Molecular weight, IV = intravenous, sal = saline
*pH adjusted to 8.5 with 1N $NaHCO_3$.

Parasitemia falls as a result of clearance of circulating, drug-affected, ring form parasites, the killing of more mature stages, which in *P. falciparum* infections are sequestered and consequent prevention of multiplication [6]. Thus, providing that drug concentrations remain above the maximum effect value, there is a constant fractional decline in the parasite biomass with each asexual cycle. This log-linear decline continues until all the parasites have been removed, numbers are reduced low enough such that host defenses can contain the remainder, or until drug concentrations have fallen below the MPC. Then the rate of decline, or expansion, of the parasite population depends on the time profile of the blood concentration of the antimalarial drug, and the concentration effect relationships for the residual number of parasites [6]. In symptomatic malaria there are generally between 10^8 and 10^{13} parasites in the body [479]. Although the parasite population may be exposed to very high concentrations of antimalarial drug and up to 99.99% of the population killed during the first asexual cycle, large numbers of viable malarial parasites will still remain and must be killed in subsequent cycles.

If the minimum parasiticidal concentration (MPC) is defined as the minimum blood or plasma concentration which produces that maximum inhibitory activity of drug (i.e., IC_{99}); then, in general, blood levels of an antimalarial drug should exceed MPC values during at least four asexual-cycles (> 6 days) for maximum cure rates [6]. If the reduction in parasite biomass (parasite reduction ratio) is, 1000-fold per cycle, then the MPC must be maintained for more than four cycles to ensure high cure rates. For both halofantrine and benflumetol, the plasma concentrations 7 days after starting treatment have been shown to be a useful predictor of therapeutic outcome [491]. In the past it was generally assumed that antimalarial blood concentrations should remain within the therapeutic range throughout the dosing interval. This has been challenged by observations that once-daily administration of artemisinin derivatives, which have elimination half-lives of < 1 h, are not associated with inferior immediate responses compared with more frequent administration [56, 492]. In experimental animals, on the other hand, the timing of chloroquine administration to coincide with the greatest period of susceptibility in the parasite life-cycle has been associated with improved parasitological clearance [437].

To treat patients with severe malaria, physicians have to face many critical and variable malarial conditions such as cerebral infection, hyperparasitemia and renal failure. Also the parasitemia counts could be in the range from 0.1 – 20%, some cases with even more drastic changes. In an individual patient, several factors determine the profile of drug PD and PK. The stage and synchronicity of the infecting parasite population and the multiplication rate of sequestered mature stages (schizonts) determine the initial parasitemia time profile after drug administration [6]. If the patient is admitted to hospital at a time when a significant proportion of the parasites are undergoing schizogony, then parasitemia will rise following antimalarial drug treatment as schizonts rupture and release merozoites, which will then invade new erythrocytes to produce circulating ring stages. These stages can be seen by the microscopist and counted, whereas the sequestered stages are not seen and therefore are not counted. The implication is that parasitemia will fall. If the majority of the parasites in the blood are young rings and a drug which has little effect on the ring stage is used in treatment, then parasitemia will plateau and fall later as a result of sequestration [6, 436].

The overall individual parasitemia profile thus represents a hybrid of these different processes. A population of patients provides a series of very different parasitemia profiles. As a result there is considerable inter-individual variance in parasite clearance profiles. However, it is reasonable to assume that the distribution of parasite stages should be random in a patient population as there is no particular reason why a patient should present one level of parasite development – and not another – in the hospital. Thus, even before attempting to assess the effect of the drug *in vivo*, there is considerable inter-patient variance in the pharmacodynamic variables [502].

Although the high parasitemia and drug-RBC binding factors influence the drug concentrations that significantly affect the efficacy of AS in treatment of severe malaria [44, 97, 98, 454, 530, 531], there is now convincing evidence that for those who do develop severe malaria, intravenous AS will reduce the risk of death by approximately one-third [90, 103] in a milestone clinical trial in the treatment development for severe malaria. So far, this has been the largest trial (enrolling 1461 patients) ever performed in severe malaria and the first to demonstrate conclusively a mortality reduction over standard quinine therapy. The purpose to carry out large trials is to reduce the inaccuracy results due to big variation in pharmacodynamic responses. The World Health Organization now advocates the use of injectable artesunate for treatment of severe malaria is based on the SEAQUAMAT clinical trials [2, 90, 103]. In making the case for WHO and the international physician community to accept injectable AS in the role as the preferred treatment for severe malaria, two proposals are offered here:

1) The Monotherapy of Intravenous AS Should still be Used

To treat severe malaria, the monotherapy of injectable AS is needed in the first couple days in order to rapidly rescue lives and possibly reduce mortality. The TropNetEurop is recommending using intravenous AS to cure the patients with severe malaria [532] because they have been encouraged by the large and successful SEAQUAMAT trials in Asian. Unfortunately, there is currently no product available that is produced under good manufacturing practice (GMP) conditions. Patients with malaria in rural areas may be obtunded or be experiencing bouts of vomiting, making them unable to take oral medications and leading to significant delay in treatment if facilities for parenteral treatment are unavailable. In such circumstances, the intravenous or rectal route of administration is attractive because rural healthcare workers can be trained to identify moderate and severe malaria and administer intravenous or rectal drugs before transfer of patients to hospital; of course, this assumes that the intravenous or rectal route of administration is culturally acceptable in these areas.

There are some cases, highlighted especially in patients with uncomplicated hyperparasitemic *falciparum* malaria (> 4% infected RBC), who are at risk of severe malaria and treatment failure. These infections require a three to five day course of treatment with oral agents. The only alternative is parenteral quinine, which is less effective [90, 103], more expensive, and less tolerated [533]. In addition, in the second and third trimesters of pregnancy, injectable AS is also a better drug for uncomplicated malaria, but it needs to be given for long term oral treatment course with a dose adjustment because of the altered kinetics in pregnant women [269, 534, 535]. Physicians would like to prescribe appropriate antimalarial therapy without the need for oral tablets but this is unlikely to be possible in the near future, especially for injectable AS [93, 536]. Therefore, we still need AS monotherapy. They also hope that some manufacturers will continue to supply injectable AS alone to be used in specific circumstances [536]. Walter Reed Army Institute of Research (WRAIR) has being developed a novel cGMP injection of AS since 2003, which is in the process of US FDA approval by 2009 [94, 537].

2) Loading Dose of Intravenous AS Should be Pursued

All cases of *falciparum* malaria are potentially severe and life threatening, especially when managed inappropriately. A major reason for progression from mild through complicated to severe disease is missed or delayed diagnosis. Once diagnosed, the priority for treatment of complicated and severe disease is the parenteral administration of adequate, safe doses of an appropriate antimalarial, in the setting of the highest possible level of clinical care. Supportive management of complications such as coma, convulsions, metabolic acidosis, hypoglycemia, fluid and electrolyte disturbances, renal failure, secondary infections, bleeding disorders, and anemia is also important. The most recent advance in decreasing mortality has been the use of intravenous AS instead of injection quinine [90, 103], which may very well revolutionize the management of severe disease [93]. However, recrudescence is still relatively high with monotherapy of intravenous AS alone [494]. Since recrudescence is not due to parasite resistance [495], one possible explanation of this parasitological failure may be a low peak concentration after first loading dose with the intravenous AS.

Currently, the therapeutic dosages of injectable AS to treat severe and complicated malaria are in the range of 2-2.4 mg/kg alone. The dose was designed in accordance with prior experience of physicians concerned with neurotoxicity, to reduce dose-dependent hemotoxicity, to prevent a sub-maximal antimalarial effect, and to keep a minimum curative rate (> 90%) during over twenty years of treatments [466, 496]. However, as this report describes in previous sections, this regimen will only produce 735 - 18909 ng/ml of C_{max} with great variability, which is a lower peak concentration than needed for patients with hyperparasitemia (i.e., the C_{max} in the malaria patients with a high variability). In theory, the patients with severe malaria require 7500-30000 ng/ml peak concentration to efficaciously cover each RBC (as described in Section 6.3), including infected cells, because the infected cell can concentrate 100-300-fold more DHA than normal RBC during the first 1-2 hr after dosing. The large variability in pharmacokinetics and pharmacodynamics by using AS should also be considered. This means that the intravenous dose should be increased to a higher level (4-8 mg/kg) than current use for the loading dose. However, the question remains: is this intravenous dose safe for treatment of severe and complicated malaria?

Compared to oral AS, the injectable AS is superior in pharmacokinetics and pharmacodynamics to the oral AS and other four artemisinins in treating patients with severe malaria; injectable AS shows the highest peak concentration, the fastest effective time, rapid elimination, and a very short exposure moment compared to oral AS (Table 44 and Table 46) [497]. The drug exposure time of artemisinins is the principle factor responsible for induction of neurotoxicity in animal species. Therefore, the active metabolites (DHA), the drug distributed in CNS, and/or the drug exposure level play relatively less important roles than the drug exposure time. The water-soluble artemisinins, AS in particular, have very short half-lives and produce limited exposure time in animal species and humans, indicating that once-daily oral or intravenous administration of AS is relatively safe. In fact, the animal experiments demonstrated that no neurotoxicity (pathologically or behaviorally) has been observed in animal species following intravenous administration at any repeated doses up to maximum tolerated doses (MTD). The MTD of AS was 240 mg/kg following intravenous injection daily in rats for 3 days, but no neurotoxicity was detected in these animals [94]. In another study, intravenous AS had no effect on neurotoxic scores at 120 mg/kg in sodium carbonate daily for 7 days (our unpublished data). After one intramuscular injection, a high dose of 420 mg/kg of AS did not produce any neuronal necrosis in the rats. Also, up to 200 mg/kg of AS used orally daily for 5-7 days did not exhibit neuronal changes or any specific clinical signs [498]. Therefore, the 4-8 mg/kg dose level with intravenous formulation should be safe with regards to neurotoxicity [500] and also safe with regards to hemotoxicity [468, 501]. One study used a high dose of oral AS, up to 6 mg/kg, with the AS being extremely well tolerated; to date, there is no convincing evidence that dose as high as this of oral AS have any significant toxicity [495]. In our current Phase I clinical trials, the 2-8 mg/kg doses have been exceedingly well tolerated following both single and multiple intravenous injections in volunteers with our new formulation (our unpublished data). In malaria patients, this dose offers a satisfactory margin of safety and would be unlikely to produce a submaximal antimalarial effect.

Angus *et al.* [502] characterized the *in vivo* dose-response relationship for AS (and thus, rationalize dosing), 47 adult patients with acute uncomplicated *falciparum* malaria and parasitemia >1% were randomized to receive a single oral dose of AS varying between 0 and 250 mg together with a curative dose of oral mefloquine. Acceleration of parasite clearance was used as the pharmacodynamic variable. An inhibitory sigmoidal maximum effect (E_{max}) pharmacodynamic model typical of a dose-response curve was fitted to the relationship between dose and shortening of parasite clearance time (PCT). The E_{max} was estimated at 28.6 h, and the 50% effective concentration was 1.6 mg/kg of body weight. These results imply that there is no reduction in PCTs with the use of single oral doses of artesunate higher than 2 mg/kg, effectively reflecting the average lower limit of the maximally effective dose [502]. It is suggested that the 2 mg/kg is a maximum effective dose in those selected patients. However, the author also pointed out that this dosage of 2 mg/kg would be the lowest to give maximum effects in an "average" patient. When taking into account the considerable and large variance between individuals in pharmacokinetics and pharmacodynamics [496, 502, 503], a larger dose would be required to ensure that a maximum effect was obtained in all patients. Therefore, the loading dose of injection AS is reasonable to increase to another higher level (4-8 mg/kg) for the consideration of WHO experts and investigators in the further development and treatment.

After the first drug exposure to reduce parasitemia below 1%, the maintaining dose could be significantly cut or switch (sequentially) to another slower acting and low dose antimalarial agents (not artemisinins) in order to protect AS from recrudescence or resistance and to optimize antimalarial therapy in severe and complicated malaria [401]. Because these artemisinins have excellent intrinsic activity, which produce reductions in asexual parasite biomass of up to 10,000-fold per cycle, these drugs still need to be present in therapeutic concentrations during four asexual life cycles, or 8 days [4]. Previous studies have shown that once-daily administration with the artemisinin derivatives provides equivalent cure rates relative to more frequent administration [429, 504]. This suggests that exposure of the infecting parasite population to therapeutic concentrations of drug twice each life cycle is sufficient for maximum effect [6].

5.4. PK/PD Relationship and Antimalarial Drug Efficacy

In malaria many factors contribute towards the therapeutic response *in vivo*, and these are likely to be different in every patient. Antimalarial plasma drug concentrations are undoubtedly important, but as malaria is an intraerythrocytic parasite, it has been suggested that concentrations within the red cells may be more relevant. This is probably too simplistic. Most red cells are not parasitized, and measurements of red cell concentrations therefore reflect the majority uninfected population of cells. There are other important differences in comparison with antimalarial susceptibility testing. The asexual blood stage *falciparum* malaria parasite multiplies every 48 h. Infections tend to be synchronous from the outset, and there is a considerable variation in drug susceptibility during the life cycle. The middle third of the life cycle is most sensitive to the antimalarials, and the very young and very mature parasites are least affected [6]. Thus, circumstances where blood concentrations of the antimalarials fluctuate widely (e.g., chloroquine, artemisinin), the stage and synchronicity of the infecting parasite population become important determinants of immediate therapeutic response. Peak drug concentrations should theoretically coincide with the most sensitive stage of parasite development with the PK/PD relationship.

There may be up to 10^{12} parasites circulating or sequestered in the vasculature. It is asking a lot of any drug to kill all these parasites within one life cycle. Even if there was 99.99% kill/cycle, therapeutic concentrations would still be needed for four cycles (8 days) to eradicate all of the parasites in the body [479]. Host defense mechanisms, both nonspecific splenic clearance function and specific immune responses, are important allies in the therapeutic attack on the malaria infection. These processes can remove the residual burden of parasites after a brief period of antimalarial drug exposure. The therapeutic response is always better in semi-immune or immune subjects, and drug trials need to be interpreted with this in mind. In malaria endemic areas, adults therefore tend to respond better than children because they have acquired a more effective immune response [367]. Short course treatments are often effective in semi-immune patients. However, in non-immunes therapeutic antimalarial drug concentrations are probably needed for at least 3 to 4 parasite life cycles (6-8 days), unless the parasites are very drug sensitive.

The relationship between antimalarial drug concentrations and treatment failure has been studied on relatively few occasions. For quinine, recrudescent infections in Thai children were more likely if serum drug concentrations fell below 10 µg/ml in the latter half of the treatment course [538]. For chloroquine treatment, parasitemia rose when the median whole blood drug concentrations were 790 nmol/l in Tanzanian children with high-grade resistant (R^2) infections [539]. The corresponding estimated median *in vivo* M.I.C. for low grade resistant infections was 147 nmol/l. For mefloquine in Thailand, where resistance is also increasing rapidly, parasitemias in resistant infections rose through median serum mefloquine concentrations of 638 ng/ml [367]. In contrast in Malawi where the parasites are more sensitive, whole blood mefloquine concentrations of over 500 ng ml-' in children were associated with cures [540]. These examples obviously refer to particular populations whose infecting parasites were at a particular stage in the development of antimalarial resistance.

The MIC values derived cannot be extrapolated to other geographic locations or populations. In severe malaria the objective of treatment is to save life. Eradication of all the infecting parasites (i.e., prevention of recrudescence) is of less importance. Treatment aims to halt the pathological processes induced by the parasite as quickly as possible. This would be best achieved by switching off all the parasite's vital functions as soon as possible [479]. There is no evidence that rapid parasite killing is harmful; i.e., there is no evidence for a Jarisch-Herxheimer reaction in malaria. Thus, therapeutic blood drug concentrations should be achieved as soon as possible without risking toxicity. Where possible loading doses should be given by rate controlled intravenous infusion, although intramuscular administration of quinine or artemether is equally effective.

Stage specificity of drug action should also be considered. Antimalarial drugs with a broad stage specificity and rapid action such as the artemisinin derivatives, would be theoretically preferable to narrow spectrum drugs acting late in the parasite lifecycle such as pyrimethamine [479]. These theoretical predictions need now to be confirmed in clinical studies. Finally the practical realities of managing severe malaria in the rural tropics need to be faced. Complex intravenous drug regimens may give ideal blood drug concentration-time profiles, but most patients with severe malaria never reach a sophisticated hospital. The current strategies for primary health care assessment and administration of oral antimalarials, with referral of complicated malaria cases may lead to long delays in starting life-saving treatment. Simple methods of administering antimalarial treatment to unconscious patients in remote rural areas need to be developed. The pharmacokinetics, safety and efficacy of intramuscular injections, nasogastric instillation and suppository administration need to be evaluated prospectively. Treatment recommendations should be made only when this information is available.

5.4.1. PK/PD Evaluation of Artemisinin Antimalarial Drugs

The pharmacokinetic/pharmacodynamic (PK/PD) results for artemisinin drugs in monotherapy may give us more comprehension in dosage regimen investigation [541]. This section has analyzed the same subjects in one publication for the PK and PD analysis of intravenous AS and oral artemisinin (QHS). However, for the other four drugs with oral regimens (AS and DHA) and intramuscular dosing (AE and AM), the PK and PD with same drug and same dosage regimen are combined from two individual publications [454, 485, 542-546].

Pharmacokinetics (PK), in general, shows an absorption phase of artemisinin drugs following oral or intramuscular administration. It then reaches peak concentration at 1-3 hours followed by a sharp decline in elimination phase. However, no absorption phase shows after intravenous injection of AS. After multiple administrations, four drug concentrations (QHS, AS, AM, even DHA) has been reported to decline daily due to an auto-induction metabolism [470-474]. Following oral administration, AS and DHA has short mean residence time (MRT) with a range of 0.9 and 2.7 hr, and QHS has longer MRT with 7.4 hr. Due to prolonged absorption and accumulation in the injection sites, AE and AM have very long MRT at 13.9-42.9 hr in human treatment. For cases analysis, PK data of five artemisinin drugs (QHS, DHA, AE, AM, and AS) associated with parasitological consequences in malaria patients are shown in Table 44 [44].

Pharmacodynamics (PD) is similar to PK profiles but measures efficacy parameters. The mean parasitemia-time curve following treatment with any antimalarial compounds presents a lag phase. The lag phase is then followed by a sharp decrease in parasite counts (which presents a clearance phase), and is then slowly decreased to a very low parasitemia. This process is usually under the limitation of microscopic observation for both the dormant and terminal phases (Figure 17, top). The artemisinin drugs affect these profiles in several ways. First, they appear to prevent the continuation of merogony at a later stage of parasite development than either quinine or mefloquine. This prevents or attenuates the occasional alarming sharp rise in parasitemia immediately following treatment described earlier (lag phase). Second, the decline in parasitemia is accelerated. This can be explained only by enhancement of clearance of erythrocytes infected with ring forms (clearance phase) [517, 547]. After rapid termination, surviving and stage unaffected parasites have very low levels at a dormant status [529], which is a basis of the recrudescence (dormant and terminal phase). Different artemisinin drugs with different dosage regimens showed the different profiles of parasitemia-time (Figure 17, bottom). Although the PK/PD publication in artemisinin agents is very limited, the possible PK and PD evaluations for individual drug with same regimens in monotherapy for human-malaria are described here.

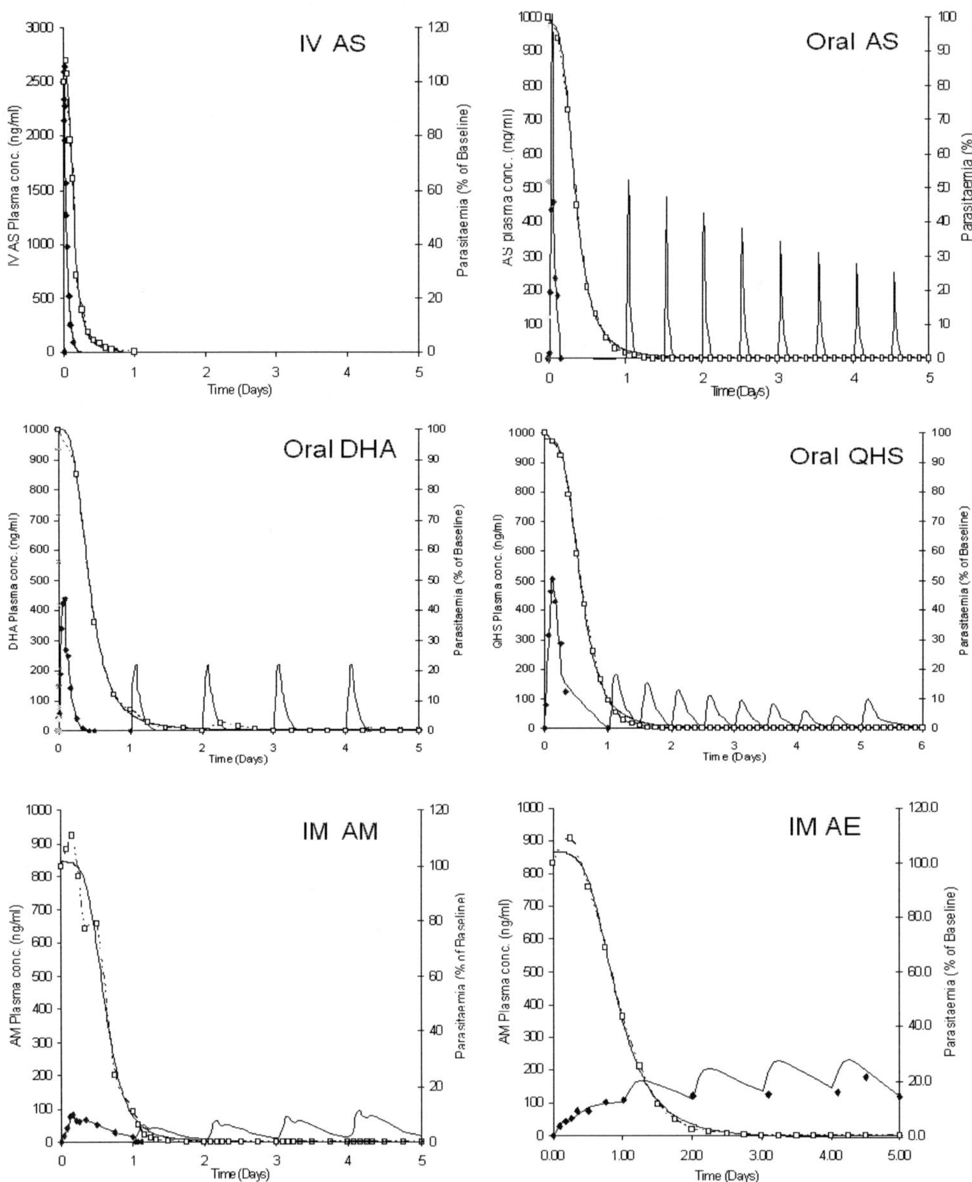

Figure 18. Pharmacokinetic/pharmacodynamic profiles are in plasma following intravenous and oral artesunate (AS, with DHA measurement) [63-65, 69], oral dihydroartemisinin (DHA) [17, 72, 73], oral artemisinin (QHS) [65], intramuscular artemether (AM) [67, 70] and arteether (AE) [4, 66] with various regimens (Table 3) measured by HPLC-ECD or LC-MS (solid markers) and computer fitted curves (solid-line) are from malaria human trials. Relative PK/PD parameters are given in Table 2, and the parasitemia was counted 100% at zero while the first dosing time.

5.4.1.1. Artesunate

1) Intravenous AS

PK/PD profiles in malarial patients have been studied following AS intravenous administration at dose of 120 mg/person on first day at 0 hr and then followed by oral 100 mg at 8 hr (Figure 18, IV AS). The mean peak level (C_{max}) of AS was achieved at 11343 ng/ml and then rapidly reduced with elimination half-life of 0.05 hr. The C_{max} of DHA, a metabolite of AS, was reached at 2646 ng/ml and then declined with elimination half-life calculated at 0.67 hr (Table 45). Mean AUC of AS and DHA was estimated at 2378 ng·h/ml, which was 2 times higher than that of its parent compound (AUC of 1146 ng·h/ml), AS, in this clinical trial [541].

At same dosage regimen, the time of lag phase in the parasitemia curve was 1.92 hr, which was the shortest lag time compared with oral AS and other 4 drugs (Figure 17, bottom), indicating that the effect of AS injection was very fast and efficient on killing parasites at a mean time of 3.18 hr for clearance of half the parasitemia (PC_{50}), the lowest PC_{50} and the lowest area under the inhibitory curve (AUIC) with 397.3%·h/μl in the evaluation (Table 44). After the computer fitting, the 0.00111% parasitemia of an E_{max} was estimated for the intravenous AS treatment in human on the first day for the computer modeling. The E_{max} was the principal pharmacodynamic parameter to determine the antimalarial response in maximum effect to the artemisinin derivatives, which had very steep concentration-effect relationships. The relationship may be caused by the time that blood concentrations exceed the minimum parasiticidal concentration (MPC), rather than the C_{max} or AUC [517]. For the data presented here, after 100 mg per patient of AS, the E_{max} was the lowest parasitemia remaining in blood (0.0011%) when compared to other four artemisinin derivatives (Table 44).

2) Oral AS

PK/PD profiles in patients with malaria has been studied following AS oral administration in tablets of 100 mg/person dose on first day and then followed by 50 mg twice daily for 2-6 days (Figure 18, Oral AS). The mean peak plasma concentration of DHA, a metabolite of AS, was reached at 1052 ng/ml and then declined with elimination half-life calculated at 0.75 hr (Table 46). Also, the AS concentration declined daily during the 5 days of dosing [474]. Mean AUC of DHA was estimated at 1334 ng·h/ml, which was 6.4 times higher than that of its parent compound (AUC of 210 ng·h/ml), AS, in this clinical trial [454]. At same dosage regimen, the time of lag phase in the parasitemia curve was 2.81 hr, which was the shortest lag time compared with other 4 drugs (Figure 12, bottom), indicating that the effect of AS tablets was very fast and efficient with 0.36 hr absorption half-life and 1052 ng/ml of C_{max} [544]. Similarly, very rapid parasite clearance produced a mean time of 8.48 hr for clearance of half the parasitemia (PC_{50}), the lowest PC_{50} and the lowest area under the inhibitory curve (AUIC) with 921.2%·h/μl (Table 44).

After the computer fitting, the 0.0016% parasitemia of an E_{max} was estimated for the oral AS treatment in human (Table 44) on the first day in (or for) computer modeling. The E_{max} was the principal pharmacodynamic parameter to determine the antimalarial response in maximum effect to the artemisinin derivatives, which had very steep concentration-effect relationships. The relationship may be caused by the time that blood concentrations exceed the MPC, rather than the C_{max} or AUC [517]. For the data presented here, after 100 mg per patient of AS, the E_{max} was still the lowest parasitemia (0.0016%) remaining in blood when compared to other four artemisinin derivatives (Figure 18, Oral AS).

5.4.1.2. Dihydroartemisinin (DHA)

PK/PD profiles in patients with malaria following DHA oral administration in tablets of 200 mg/person on day 1 and then 100 mg at day 2-5 are shown in Figure 18 (Oral DHA). After 200 mg of DHA by oral dosing, the mean peak plasma concentration of DHA was reached at 437 ng/ml and then declined with elimination half-life calculated at 0.85 hr. Mean AUC of DHA was estimated at 1329 ng·h/ml. At 200 mg dosage regimen orally, the time of lag phase in the parasitemia curve was 4.03 hr. The lag time was longer than AS (2.81 hr) and still shorter than AM (7.26 hr), AE (8.89 hr), and QHS (5.76 hr) (Table 44, Figure 17, bottom), indicating that fast effect of DHA tablets was not as good as AS, but was more rapid than AM, AE, and QHS regimens. In addition, the rapid parasite clearance resulted in a mean time of 10.05 hr of PC_{50} and the 1167.9 %·h/µl. of AUIC.

After first day computer modeling, the 0.213% parasitemia of an E_{max} was estimated for the oral DHA treatment in human. For the data presented here, after giving 200 mg per patient of DHA, more parasites remained in the blood when compared to the other three artemisinin derivatives (Table 44). The absorption of DHA tablets with 0.67 hr half-life was 50% slower than the AS (0.36 hr) trial too, suggesting that the slow reduction of parasitemia may have resulted from the slow absorption. In addition, the AUCs of the two treatments (AS and DHA with double dose) were the same, only the C_{max} of AS dosing was two times higher than DHA treatment, indicating that the high peak concentration (C_{max}) may be another reason for the decrease of parasitemia in human [542, 544]. Compared with the AS study, DHA appeared to have very low bioavailability (< 50%). If calculated with an equal dose level, the C_{max} and AUC_{DHA} estimated from AS treatment was 4-fold and 2-fold, respectively, higher than the DHA dosage regimen. Our data demonstrated that the DHA tablet (Cotexin) in beagle dogs had only 24.4% obsolete bioavailability [496]. Rebound parasites (3%) on day 3 and one-fourth recrudescence rate (24%) were found in patients treated with DHA tablets in clinical work [542]. The 3% of rebound parasites was only found in DHA oral treatment in this comparison (Figure 18, Oral DHA).

5.4.1.3. Artemether (AM)

PK and PD profiles in patients with malaria following AM intramuscular administration in sesame oil of 3.2 mg/kg on day 1 and then 1.6 mg/kg daily on day 2-5 can be seen in Figure 18 (IM AM). The mean peak plasma concentration was reached at 74.9 ng/ml and then declined with elimination half-life calculated at 7.83 hr (Table 44), indicating that this formulation had the lowest peak of drug concentration when compared to the other 4 artemisinin drugs. Mean AUC was estimated at 1230 ng·h/ml [545]. At the same dosage regimen the time of lag phase in the parasitemia curve was 7.26 hr, which was the second longest lag time compared with other 3 drugs except for AE, indicating that the effect of the AM intramuscular formulation had a very low and prolonged absorption from muscle injection sites [484]. Parasitemia counts in the beginning after dosing did not decrease due to the low peak concentration of AM, but increased markedly to 110% at 4 hours (parasitemia was counted 100% at zero while the first dosing time). That may be the reason why the lag time was longer at 7.26 hr with no parasites decrease after injection at hour 4. Similarly, there was very slow clearance to parasites produced at mean PC_{50} time of 15.6 hr, which was also the longest PC_{50} time compared to other 3 drugs (Table 44). The highest AUIC of 1613.4 %·h/µl also revealed that the drug with the intramuscular regimen was not efficacious in reducing the parasitemia [546].

Although the intramuscular AM was not fast enough to kill parasites in patients, the long lasting exposure level of AM (7.83 hr elimination half-life) could reduce parasitemia at a good rate. The computer modeling estimated that the 0.003% parasitemia was the E_{max} in this clinical trial. The parasitemia was two times more than oral AS in the first day modeling, but much less than oral DHA, AE and QHS (Table 44).

5.4.1.4. Arteether (AE)

PK and PD profiles in patients with malaria following AE intramuscular administration in sesame oil of 4.8 mg/kg at 0 h and 1.6 mg/kg at 6 h on day 1 and then daily 1.6 mg from day 2-5 are illustrated in Figure 18 (IM AE). The mean peak plasma concentration was reached at 110.1 ng/ml and then declined with elimination half-life calculated at 22.7 hr (Table 17), indicating that this formulation has a low peak of drug concentration when compared to AS, DHA, and QHS. Mean AUC was estimated at 4702 ng·h/ml [485]. At the same dosage regimen the lag time in the parasitemia curve was 8.89 hr, which was the longest lag time compared with other 4 drugs, indicating that the effect of the AE intramuscular formulation had a very low and prolonged absorption from muscle injection sites [484]. Parasitemia counts in the beginning after dosing did not decrease due to the low peak concentration of AE, but increased markedly to 108% at 6 hours (100% parasitemia was counted at zero hour during the first dosing time). The increase may be the reason why the lag time is longer at 8.89 hr with no parasites killing after injection at 6 hours. Similarly, a very slow parasite clearance was produced at mean PC_{50} time of 19.68 hr, which was the longest PC_{50} time compared to other 4 drugs (Table 17). The highest AUIC of 2463.5 %·h/µl revealed that the drug with the intramuscular regimen was not efficacious in reducing the parasitemia [546].

The intramuscular AE was neither able to exterminate parasites rapidly, nor to reduce parasitemia much in patients. The computer modeling estimated that the 0.550% parasitemia was the E_{max} in this clinical trial. In spite of two doses on day 1 (4.8 mg/kg at 0 hr and 1.6 mg/kg at 6 hr), the parasitemia was still the highest value compared to other four artemisinins (Table 44).

5.4.1.5. Artemisinin (QHS)

PK/PD evaluation in patients with malaria following artemisinin (QHS) oral administration of 500 mg/person on first day followed by 250 mg twice daily for 4 days and then another 500 mg on 6th day is seen in Figure 18 (Oral QHS). After 500 mg of QHS by oral dosing, the mean peak plasma concentration of QHS was reached at 588 ng/ml and then declined with elimination half-life calculated at 2.4 hr. Mean AUC of QHS was estimated at 2601 ng·h/ml, which was the highest AUC in comparison due to the highest oral dose taken. At 500 mg dosage regimen orally, the lag time in the parasitemia curve was 5.76 hr, which was longer than oral AS (2.81 hr) and DHA (4.03 hr) but was better than AM and AE trials (Figure 17, bottom), indicating that fast effect of QHS capsules was not as good as oral AS and DHA even given the higher doses [471, 472, 544]. The patients treated with oral QHS showed a 13.95 hr of PC_{50} and the 1464.4 %·h/µl. of AUIC (Table 44).

The absorption of QHS capsules with 1.21 hr half-life was slower than the absorptions of AS and DHA drugs, suggesting that the slow reduction of parasitemia may have resulted from the slow absorption. Compared with oral AS and DHA study, oral QHS appeared to have very low bioavailability of 30% [548, 549] and low antimalarial potency [234, 469, 525, 526].

The relationships of PK and PD have been previously evaluated for the five-artemisinin drugs, and the impressive inputs for each drug with the regimen could be summarized as follows:

1. Intravenous and oral AS is the fastest killer in the malaria treatments out of the other four artemisinins. Intravenous AS can provide the highest peak concentration and very short exposure time, while oral AS can also provide higher peak level and shorter exposure time; the high peak level as the role to eliminate parasites rapid (Figure 18, bottom) and the short exposure time to avoid fatal neurotoxicity [524] and to prevent resistance [543].

2. Although DHA has similar efficacy to AS *in vitro* [234, 469, 525, 526], the agent does not have a rapid effect like that of AS even with double oral doses. The explanations include lower bioavailability [531, 550], slower absorption phase, and long lag time on effect activity. However, the principle reason is due to the lower peak concentration in the treated humans when compared to oral AS.

3. Intramuscular AM or AE reveals a low efficacy and slow parasite killing. Due to the sesame oil formulation, AM and AE have a very low and prolonged absorption from muscle injection sites [484]. In the beginning, parasitemia counts after dosing did not decrease due to the low and delayed peak concentrations (75-110 ng/ml) of AM and AE. The low peak concentration may be the reason why the lag time is longer for 7.2-8.9 hr without parasites killing after injection. Similarly, there is very slow clearance rate on parasites with mean PC50 time of 15.63-19.68 hr, which may have resulted in inadmissible failures in AE trials [485]. Although the drug exposure times of AM and AE are much longer than AS, DHA and QHS, the long exposure time did not assist efficacy. Instead, it induced neurotoxicity, therefore, the formulation would not be encouraged to use in humans to treat acute and severe malarias.

4. Oral artemisinin (QHS) does not show well in this PK/PD comparison probably due to the highest oral dose (500 mg/person). Although the high dose produced a higher plasma concentration (2601 ng·h/ml of AUC) and high peak concentration (588 ng/ml), the low relative antimalarial potency [234, 469, 525, 526] and slow bioavailability [549] considerably limited the efficacy. It suggests that the artemisinin derivatives with lower antimalarial potency still cannot produce rapid effects even at an increased dose level.

5.4.2. Impact of PK/PD on Artemisinins Monotherapies

Drug peak concentration (C_{max}) is more important than plasma concentration (AUC) in producing the improved efficacy of antimalarial drugs that has been confirmed in the earlier PK/PD evaluation. The intravenous AS can provide sufficiently high peak concentration in the patients and can also provide the fastest efficacy in killing parasites, showing that injectable AS is a pharmacokinetic and pharmacodynamic superior when compared to other artemisinins with various regimens (Table 44). In current clinical trials, the C_{max} was shown to be from 605-18909 ng/ml (Table 43) from AS plus DHA resulting from a dose of AS 120 mg/person or 2.4 mg/kg in various malarias including severe malaria [551]. Although those dose regimens are still not enough to cure severe and complicated malaria in 100% [552, 553], there is now convincing evidence that for those who do develop severe malaria, intravenous AS will reduce the risk of death by approximately one-third [90, 103]. Moreover, it is the antimalarial activity with high killing capability in blood in the first few hours following the first administration of a parenteral drug that is critical from a therapeutic standpoint, particularly in patients with severe malaria, in whom mortality can reach between 15% and 20%, despite appropriate antimalarial and supportive treatment. It is also encouraging is that intravenous AS is also safer and easier to use than injectable quinine to treat severe malaria [467].

When the peak concentrations reached 198 ng/ml of AS and 1052 ng/ml of DHA after oral AS (100 mg/patient), and 11343 ng/ml of AS and 2646 ng/ml of DHA following intravenous AS (120 mg/patient), AS in the two clinical trials are superior pharmacokinetics and the fastest efficacy in malaria patients when compared to the other four-artemisinin drugs (Figure 17 and Table 44). The total drug concentrations (AS plus DHA) were calculated as 0.51×10^6 molecules/RBC in the oral patients and 4.67×10^6 molecules/RBC in the intravenous subjects treated with AS. This calculation was done by using Avogadro's number (mole = 6.024×10^{23} molecules) of equalmole drug divided by counts of RBC from humans samples (Table 47). In order to understand the drug molecules per RBC on antimalarial effects of artemisinins, we performed a culture incubation study of AS or DHA with *P. falciparum* clones (W2 and D6) with fixed parasitemia (0.8-1.0%) and RBC (5.0×10^7) by using flow cytometry measurement and ^{3}H-hypoxanthine incorporation assay [469].

In this experiment, the RBC infected with W2 and D6 parasites were incubated with AS and DHA at various concentrations [469], and the summary of the 50% and 99% inhibitory concentration (IC_{50} and IC_{99}) of the four compounds is shown in Table 47. The drug molecule concentrations showed that the IC_{99} had 0.31-0.49×10^6 molecules/RBC for AS and 0.39-0.73×10^6 molecules/RBC for DHA in both parasite strains (W2 and D6) infected RBCs. In other words, the minimum concentration of AS or DHA to kill 99% of culture parasites was in the range of 0.31-0.73×10^6 molecules/RBC in this study. When compared to clinical trial studies described earlier, the 0.51×10^6 molecules/RBC or 1250 ng/ml (AS + DHA) in patients following 100 mg oral administration of AS reached the exact minimum concentration needed to exterminate 99% parasites. Whereas the 4.67×10^6 molecules/RBC or 13989 ng/ml (AS + DHA) in malarial subjects after 120 mg intravenous injection of AS was about 10-fold higher than the IC_{99} level (0.31-0.73×10^6 molecules/RBC).

Although the *in vitro* culture data cannot closely reflect the results from *in vivo* clinical trials, the minimum drug molecules/RBC to clear parasites is still an important indicator when discussing the efficacy of antimalarials. In the present PK/PD analysis, the oral AS in humans has shown to be the best pharmacokinetic and pharmacodynamic role to treat uncomplicated malaria when compared to other four artemisinins (Table 44). However, the total peak concentrations (1250 ng/ml) of AS (198 ng/ml) plus DHA (1052 ng/ml) after oral administration only equaled to the minimal level for killing 99% parasites *in vitro* with the 0.8−1.0% parasitemia medium.

Table 47. Drug concentrations (ng/ml) and molecules per single red blood cell (molecules/RBC) needed to inhibit 50% and 99% of parasites (IC$_{50}$ and IC$_{99}$) in culture with 0.8% (A) and 1-5% (B) parasitemia and 5.0 x 10^7/ml RBC (1% of whole blood) to dihydroartemisinin (DHA) and artesunate (AS) against chloroquine-resistant clones of *P. falciparum* from Indochina (W2) and mefloquine-resistant clone of *P. falciparum* from Sierra Leone (D6). Two clinical trials (C) of AS after oral and intravenous administrations was compared to the *in vitro* results. [44]

A. Efficacy of AS and DHA against *P. falciparum* clones (n = 3) with 0.8 parasitemia [1]

Parasite Strain	Test Compound	IC$_{50}$ ng/ml	IC$_{99}$ ng/ml	IC$_{50}$Molecules/RBC*	IC$_{99}$Molecules/RBC*
W-2	AS	0.26 ± 0.04	7.22 ± 3.36	0.11×10^5	0.31×10^6
	DHA	0.36 ± 0.13	17.16 ± 6.17	0.15×10^5	0.73×10^6
D-6	AS	0.32 ± 0.12	9.69 ± 2.74	0.14×10^5	0.49×10^6
	DHA	0.24 ± 0.08	9.30 ± 0.85	0.10×10^5	0.39×10^6

B. IC$_{90}$ of AS and DHA against *P. falciparum* clones at different starting parasitemia

Parasite Strain	Test Compound	Starting Parasitemia (%)				
		0.2	0.5	1.0	2.0	5.0
W-2	AS	0.3	0.2	0.5	1.3	3.8
	DHA	0.2	0.2	0.4	0.8	2.8
D-6	AS	0.3	0.3	0.7	1.6	4.4
	DHA	0.1	0.2	0.5	1.0	2.3

C. Clinical trials after oral or intravenous AS on day 1 (n = 12-13)

Drug (dose)	AS C$_{max}$ ng/ml	DHA C$_{max}$ ng/ml	Drug molecules/RBC*		
			AS	DHA	AS + DHA
Oral AS (100 mg)	198	1052	0.06×10^6	0.44×10^6	0.51×10^6
IV AS (120 mg)	11343	2646	3.56×10^6	1.12×10^6	4.671×10^6

* Drug molecules were calculated in according to Avogadro's number (mole = 6.024 x 10^{23} molecules) of equalmole drug in divided by counts of RBC for samples of culture or humans. Values are mean or mean ± SD; W-2: Indochina clone, resistant to chloroquine, sensitive to mefloquine. D-6: Sierra Leone clone, sensitive to chloroquine, resistant to mefloquine. IC$_{50}$ = 50% inhibitory concentration; IC$_{99}$ = 99% inhibitory concentration. IV = intravenous.

It is obvious that the plasma concentrations after oral dose of AS are not enough to efficaciously cover all RBCs (infected and uninfected) and to completely kill 100% of the parasites, suggesting a reason for low curative rate (81.3%) and high recrudescence in those patients following the 5 daily oral doses treatments [454, 544]. The total plasma peak concentration (1250 ng/ml) should have drug balance and equilibrium among plasma as well as uninfected and infected RBCs in the blood of the patients treated with the oral treatment. It is also evident that the peak concentration after the oral AS dose proves to be a huge shortage to treat severe malaria, particularly if the parasitemia is over 0.8-1.0% parasitemia or hyperparasitemia (>5%) in patients. Even though the higher peak concentration of AS plus

DHA (11343 + 2646 ng/ml) after AS intravenous regimen [541] produced 10-fold higher level than the minimal level of IC_{99}, it is still a big clinical failure (7.5% recrudescence in average) in the severe malaria patients after being injected with AS by monotherapy [48, 554].

The high recrudescence (about 30−50% for 3 days, 6−15% for 5 days, and 2−6% for 7 days regimens) is a major disadvantage for all of these artemisinin compounds including injectable AS after the monotherapy [48, 466, 554]. However, it is still unclear how high the dosage or the blood concentration of artemisinins should be in order to be an efficacious chemotherapy for severe malaria. In order to have the efficacious blood concentration, there are three aspects should be considered.

5.4.2.1. Aspect 1: Effects of High Parasitemia on Drug Concentration

Some reports have shown that the uptake of DHA in infected erythrocytes has been 100-300 times more than the uninfected erythrocytes *in vitro* [529, 555, 556]. This is because the malaria parasite *P. falciparum* induces unselective actin conductance in the RBC membrane that changes the permeability [557]. Other mechanisms about highly concentrated artemisinins in infected erythrocytes have also been investigated by many scientists. For normal erythrocytes it is interesting that DHA is taken up by isolated red cell membranes but not intact red cells. The difference between intact red cells and isolated membranes is consistent with the fact that the proteins that bind DHA, alpha- and beta-spectrum, band 4.1, and 4.2. actin, and glyceraldehydes-3-phosphate dehydrogenase, is peripheral membrane protein that is located on the cytoplasmic face of the red cell [558]. Thus, binding of the drug to membranes may require its exposure to the membrane's inner surface. These observations may explain why radiolabeled artemisinin and DHA are taken up by infected red cells but not intact red cells [559]. The outer erythrocytic membrane of parasite-infected red cells has alterations in the distribution of phospholipids between inner and outer sheaths [560] and may contain clefts, pores [561-563], and parasite-produced proteins [564, 565]. These studies have shown that when *P. falciparum*-infected erythrocytes are treated with a radiolabeled DHA and AE, the radioactivities are bound with six malarial proteins with *P. falciparum* parasites. The bound occurs at physiological concentrations of the drug with parasite and no proteins are labeled when uninfected erythrocytes are treated with these drugs [566].

Thus, the level of parasitemia affects the drug concentration. In patients who have 1.0% of erythrocyte infected malaria with all drugs at C_{max} of 1250 ng/ml (AS + DHA) on average [454, 544], the amount entering infected RBCs can reach the minimal concentration to inhibit 99% parasites with efficacious equilibrium of drug between infected and uninfected erythrocytes. However, in 5% or 20% of the RBC infected patients, the uptake of DHA in infected RBCs can be significantly increased and the free plasma concentration can be decreased, indicating that the concentration (1250 ng/ml) in malaria patients after oral AS is not sufficient to distribute the drug to every erythrocyte (infected and uninfected) in each patient (from 1-20% infected RBC). Therefore, it is clear that the peak level at 1250 ng/ml in plasma is far from enough to cure the severe malaria with a hyperparasitemia.

In order to determine the concentration will be at the balance and equilibrium of drug among the plasma as well as the infected and uninfected erythrocytes in the 5% or 20% RBC infected patients, we performed an *in vitro* study on culture incubation of AS and DHA with *P. falciparum* clones (W2 and D6) in parasitemia (0.2, 0.5, 1.0, 2.0, and 5.0%) in 5.0 x 10^7 RBC/ml by using ^{3}H-hypoxanthine incorporation assay [567]. The results indicated that AS and DHA responses were more consistently altered by parasite burden against resistant and susceptible *P. falciparum*. For the D6 drug susceptible clone, increasing the parasitemia from 1.0 to 5.0% resulted in an increase of 7.6 and 7.0 folds in AS and DHA IC_{90}, respectively. For the W2 susceptible clone, increasing the parasitemia from 1.0 to 5.0% resulted in an increase of 6.3- and 4.6-folds in IC_{90} of AS and DHA, respectively (Table 47). Assuming an average increase of 6-fold from 1.0 to 5.0% derived from the *in vitro* study [567], the minimum plasma peak concentration for hyperparasitemia patients could be at least at a concentration of 7500 ng/ml to exterminate 5% parasitemia and 30000 ng/ml to execute 20% parasitemia. The calculation was based on the drug concentration in equilibrium after oral AS and it took 1250 ng/ml (AS plus DHA) for the 1.0 parasitemia patients to each IC_{99}. With this estimation, for patients with severe malaria, the concentration of AS plus DHA (11343 + 2646 ng/ml) following 120 mg intravenous injection [541] was not satisfactory to distribute the efficacious drug level to every erythrocyte (infected and uninfected) in each hyperparasitemia patient (from 5-20% infected RBC).

5.4.2.2. Aspect 2: Effects of the Protein and RBC Binding on Drug Concentration

The pharmacokinetic factors are not only focusing on the high peak concentration, but are also concentrating on tissue distribution in red blood cells and protein binding of artemisinins [243]. More recently, studies of tissue distribution in rats and protein binding in human blood have been conducted in our laboratory. Table 45 demonstrates the RBC/plasma ratios of five radiolabeled artemisinins in human whole blood at various concentrations [243, 528]. For equal volumes of RBC and plasma, the fraction of the five artemisinins found in the RBC is lower than in the plasma in normal volunteer blood. The ratios are almost constant in human blood over a broad range of concentration. After correction for the 55:45 ratios of plasma and RBC volumes, the RBC/plasma ratios are 0.23 for ^{14}C-arteether (^{14}C-AE), 0.28 for ^{14}C-artemether (^{14}C-AM), 0.44 for ^{14}C-artelinic acid (^{14}C-AL), 0.52 for ^{14}C-dihydroartemisinin (^{14}C-DHA), and 0.71 for ^{14}C-artesunate (^{14}C-AS), indicating that the radioactivity remained in human RBC is in the highest binding capacity for AS and in the lowest capacity for AE.

Comparison of 50% bound concentration (BC_{50}) of ^{14}C-AM, ^{14}C-AE, ^{14}C-AL, ^{14}C-AS and ^{14}C-DHA at varied concentration in human plasma samples is performed by microdialysis equilibrium method at 37°C for 6 hours. The calculations of BC_{50} are 1945 ng/ml with ^{14}C-AE, 1673 ng/ml with ^{14}C-AM, 482 ng/ml with ^{14}C-DHA, and 143 ng/ml with ^{14}C-AS (Figure 19) in human undiluted plasma [our unpublished data].

The big advantage of DHA and AS is its readiness of entering human red blood cell (0.52 and 0.71 ratio of RBC/plasma, respectively) compared to AE and AM (0.23 and 0.28 ratio of RBC/plasma, respectively) (Table 45). It means that to reach same targets (RBC) the drug peak concentrations for AE and AM will need to be two to three times higher than that of DHA and AS. In addition, the binding affinities are high for AE and AM radiolabels with 1945 and 1673 ng/ml of 50% bound concentration in human plasma (BC_{50}). Conversely, the binding affinities are low for AS and DHA radioactivities with 143 and 162 ng/ml of BC_{50}, indicating the free drug concentration can be much higher in human plasma with AS or DHA than that with AE and AM.

5.4.2.3. Aspect 3: Effects of High PK/PD Variability on Drug Concentration

1) PK variability in Humans

Fortunately, the intravenous AS can provide sufficiently high peak concentration in patients with a fast exposure time. In current clinical trials, the C_{max} is shown to be from 735 to 18909 ng/ml (total AS plus DHA levels) resulting from a dose of 120 mg/person or 2.4 mg/kg in various malaria including severe malaria (Table 43). However, the variability is 25-fold from different trials [454, 530, 551]. For clinical trials with oral formulations, the C_{max} is shown to be from 453 to 1718 ng/ml (total AS plus DHA levels) resulting from a dose of 100 mg/person or 2-4 mg/kg in patients with uncomplicated malaria [551, 568]. The variability is still about 4-fold from different trials. In addition, in the each trial, the large inter-individual variability has also been detected among the patients. It is obvious that such low AS and DHA concentrations presented in some patients may be the cause for the treatment failure [481].

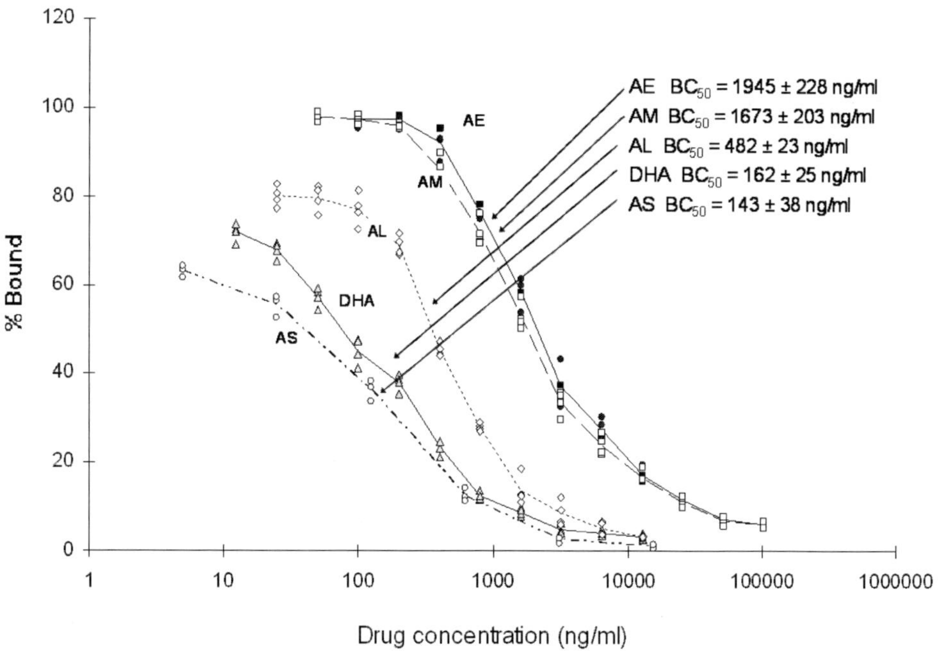

Figure 19. Comparison of 50% bound concentration (BC$_{50}$) of ^{14}C-artemether (AM), ^{14}C-arteether (AE), ^{14}C-artelinic acid (AL), ^{14}C-artesunate (AS) and ^{14}C-dihydroartemisinin (DHA) at varied concentration in human plasma samples by microdialysis equilibrium method at 37°C for 6 hours (n = 4).

2) PD Variability in Humans

Physicians have to face many critical and variable malarial conditions such as cerebral infection, hyperparasitemia and renal failure in patients with severe malaria. Also the parasitemia counts could be in the range from 0.1 − 20%, some cases with even more drastic changes. In an individual patient, several factors determine the profile of drug PD and PK.

The stage and synchronicity of the infecting parasite population and the multiplication rate of sequestered mature stages (schizonts) determine the initial parasitemia time profile after drug administration [6]. If the patient is admitted to hospital at a time when a significant proportion of the parasites are undergoing schizogony, then parasitemia will rise following antimalarial drug treatment as schizonts rupture and release merozoites, which will then invade new erythrocytes to produce circulating ring stages. These stages can be seen by the microscopist and counted, whereas the sequestered stages are not seen and therefore are not counted.

The implication is that parasitemia will fall. If the majority of the parasites in the blood are young rings and a drug which has little effect on the ring stage is used in treatment, then parasitemia will plateau and fall later as a result of sequestration [6, 436]. The overall individual parasitemia profile thus represents a hybrid of these different processes. A population of patients provides a series of very different parasitemia profiles. As a result there is considerable inter-individual variance in parasite clearance profiles. However it is reasonable to assume that the distribution of parasite stages should be random in a patient population as there is no particular reason why a patient should present one level of parasite development – and not another – in the hospital. Thus, even before attempting to assess the effect of the drug *in vivo*, there is considerable inter-patient variance in the pharmacodynamic variables [502].

5.4.3. Impact of PK/PD on Artemisinin-Based Combination Therapy

PK/PD interactions between dihydroartemisinin (DHA) and mefloquine were investigated in 10 healthy Thai males [569]. The study was of a three-way crossover design. Subjects were randomized to receive three drug regimens on three separate occasions as follows: regimen I: a single oral dose of 300 mg DHA; regimen II: a single oral dose of 750 mg mefloquine; regimen III: a single oral dose of 300 mg DHA, given concurrently with a single oral dose of 750 mg mefloquine. All regimens were well tolerated. Oral DHA was rapidly absorbed and disappeared from systemic circulation within 8-10 h. Mefloquine absorption and disposition were relatively slow processes. Pharmacokinetics of DHA and mefloquine when given concurrently were similar, except for the absorption rate of mefloquine which was faster in the presence of DHA. Pharmacodynamically, the combination of DHA and mefloquine resulted in a synergistic effect on *ex vivo* blood schizonticidal activity. Maximum activity (E_{max}) and area under effect-time curve (AUEC) of DHA and mefloquine were increased approximately two- and 20-fold (E_{max}), and four- and two-fold, respectively, compared with each individual drug alone. AUEC of mefloquine during the first 24 h ($AUEC_{0-24\,h}$) was increased approximately 50-fold in the presence of DHA (Table 48).

The dynamic profiles of both DHA and mefloquine generally coincide with their kinetic profiles in this report. The co-administration model provides more accurate estimation of the relative contributions of the beneficial to therapy without any enhancement of adverse effects of each individual drug. The PD measures (model-independent showed significant increase of the extent of response (AUEC) or the maximum response (E_{max}) of both DHA and mefloquine when they were given in combination, compared with each individual drug alone. Mefloquine did not have a significant effect on the maximal effect of DHA, while DHA dramatically increased the E_{max} of mefloquine [569].

Table 48. Safety and tolerability of available antimalarial drugs

Drug	Adverse effects	Contra-indications	Severe adverse events
Chloroquine	Gastrointestinal upset, itching, dizziness	Epilepsy	Death from overdose
Sulfadoxine-pyrimethamine	—	Pregnancy, renal disease	Stevens–Johnson syndrome
Quinine	Tinnitus, vertigo, headache, fever, syncope, delirium, nausea	G6PD deficiency, pregnancy, optic neuritis, tinnitus, thrombocytopenic, purpura, blackwater fever	Hemolytic anemia, coma, respiratory arrest, renal failure
Mefloquine	Vomiting, headache, insomnia, vivid dreams, anxiety, dizziness	Depression, schizophrenia, anxiety disorder, any psychosis, irregular heartbeat	Psychosis
Atovaquone–chloroguanide	Gastrointestinal upset, headache, stomatitis	Weight of <11 kg in children, pregnancy, breast-feeding, renal impairment	None known
Artemether–lumefantrine	Dizziness, palpitations	Pregnancy, severe malaria	Impaired hearing
Artesunate–mefloquine	Vomiting, anorexia, diarrhea	Depression, schizophrenia, anxiety disorder, any psychosis, irregular heartbeat	None known
Halofantrine	Gastrointestinal upset, prolonged QTc	Conduction abnormalities, pregnancy, breast-feeding, infancy, use of mefloquine	Cardiac arrest
Primaquine	Gastrointestinal upset, elevated levels of methemoglobin	Pregnancy, G6PD deficiency, breast-feeding	Hemolytic anemia

* Data were cited from [121].

Another combination of artemether and lumefantrine (benflumetol) is a new and very well tolerated oral antimalarial drug effective even against multidrug resistant *falciparum* malaria. The artemether component is absorbed rapidly and biotransformed to dihydroartemisinin and both are eliminated with terminal half-lives of around 1 hour. These are very active antimalarials which give a rapid reduction in parasite biomass and consequent rapid resolution of symptoms. The lumefantrine component is absorbed variably in malaria, and is eliminated more slowly (half-life of 3 to 6 days). Absorption is very dependent on co-administration with fat, and so improves markedly with recovery from malaria. Thus artemether clears most of the infection, and the lumefantrine concentrations that remain at the end of the 3- to 5-day treatment course are responsible for eliminating the residual 100 to 10,000 parasites. The area under the curve of plasma lumefantrine concentrations versus time, or its correlate the plasma concentration on day 7, has proved an important determinant of therapeutic response as follows [248, 570]:

1) Parasite Clearance

Characterization of these PK/PD relationships provided the basis for dosage optimization. Statistical evaluations were carried out to assess the effects of artemether and DHA pharmacokinetic parameters (AUC) on parasite clearance. The median parasite clearance time (PCT) was between 30 and 36 hours. As artemether is metabolized *in vivo* to DHA and both contribute to antimalarial effect, they were analyzed independently. A series of proportional hazards models were constructed using PCT as the response variable and any one, or a combination of the AUC values up to 24 and 36 hours for the 2 substances, together with parasite count at baseline, were used as the independent variables. These models showed that the AUC of both artemether and DHA contributed independently to the effect measure. However, when they were merged using a range of linear combinations, the fit of the model improved. The best fit was obtained with the AUC of DHA and artemether contributing equally to parasite clearance, suggesting equivalent *in vivo* activity. Lumefantrine pharmacokinetic parameters were not correlated significantly with PCT.

2) Curative Rates

The pharmacokinetics of artemether and DHA were not significantly associated with cure rates. The AUC of lumefantrine presumably reflects the AUC above the *in vivo* minimum parasiticidal concentration (MPC). The considerable intra-patient fluctuation in plasma concentrations of lumefantrine between individual doses meant that in some patient's plasma lumefantrine concentrations fell below the MPC between the fourth and eight day (3rd and 4th cycles) after treatment. These infections recrudesced subsequently if there was insufficient background immunity to affect a cure. For example, in an early Thai study [236], 75% of the patients with plasma lumefantrine concentrations above 280 ng/ml on day 7 were cured compared with only 51% of the patients with lower lumefantrine concentrations. The day 7 concentration giving the highest combined sensitivity and specificity for predicting subsequent cure was a concentration of 500 ng/ml: sensitivity 93.5%. As Thailand has the most drug resistant parasites in the world, this provides a 'worst-case' scenario. Thus, provided resistance does not develop, >90% of patients with day 7 plasma lumefantrine concentrations over 500 ng/ml would be expected to be cured. The parasite count at baseline reflects the total number of parasites in the body. Patients with a higher initial biomass will have a larger residuum of parasites remaining after the artemether has been eliminated for lumefantrine to remove. There is, therefore, a greater chance that plasma lumefantrine concentrations will fall below MPC and then minimum inhibitory concentration before these have been eliminated [6].

The clinical trial (238 adults and children) found a significantly higher rate of cure at day 28 with the 6-dose regimen given over 3 days than with the 4-dose regimen also given over 3 days (PCR-unadjusted treatment cure rate for intention to treat population at day 28: 96/118 (81%); 95% CI: 73.1–87.9% with the 6-dose regimen, 85/120 (71%); 95% CI: 61.8–78.8% with the 4-dose regimen, *P* < 0.001; PCR-adjusted treatment cure rate for evaluable population: 93/96 (97%); 95% CI: 91.1–99.4% with the 6-dose regimen, 85/102 (83%); 95%

CI: 74.7–90.0% with the 4-dose regimen, $P < 0.001$). The RCT reported all adverse events to be mild or moderate in severity and possibly attributable to malaria [217]. There is evidence from the data reported here that PK/PD measures to improve the oral bioavailability in short course regimens would augment therapeutic efficacy.

5.4.4. Stereoselectivity in PK/PD of the Chiral Antimalarial Drugs

Many synthetic drugs contain centers of asymmetry in their structure and exist as a pair of non-superimposable mirror images called enantiomers. Enantiomers may possess the same physicochemical properties, but often differ in their pharmacological properties. Several of the antimalarial drugs are chiral and administered as the racemate. These drugs include chloroquine, hydroxychloroquine, quinacrine, primaquine, mefloquine, halofantrine, lumefantrine, and tafenoquine. Quinine and quinidine are also stereoisomers, although they are given separately rather than in combination.

From the perspective of antimalarial activity, most of these agents demonstrate little stereoselectivity in their effects *in vitro*. Mefloquine, on the other hand, displays *in vitro* stereoselectivity against some strains of *P. falciparum,* with a eudismic ratio of almost 2:1 in favor of the (+)-enantiomer. Additionally, for some of these agents (e.g., halofantrine, primaquine, chloroquine), stereoselectivity has been noted in the ability of the enantiomers to cause certain adverse effects. These antimalarial drugs display stereoselectivity in their PK, leading to enantioselectivity in their plasma concentrations. Whereas the oral absorption of these agents appears to be non-stereoselective, stereoselectivity is often seen in their volume of distribution and/or clearance. When regarding distribution, plasma protein binding of some chiral antimalarial drugs exhibits a significant degree of stereoselectivity, leading to stereoselective distribution to blood cells and other tissues.

Over the past quarter-century there has been a significant interest in the stereoselective aspects of disposition and effects of chiral drugs, due to a generally established awareness that most marketed chiral drugs exhibit stereoselectivity in their PK and/or PD [571]. PK differences are present in the plasma or blood concentration-time profiles of the stereoisomers of these drugs, which may confer stereoselectivity in PD properties *in vivo*. However, because for most of the racemic antimalarial drugs the individual enantiomers are not available, or only small quantities are available, it is not possible to define a concentration-effect relationship for the enantiomers in their antimalarial efficacy in patients. This is in addition to the confounding variables of parasite sensitivity to the enantiomers of the chiral antimalarials. Attempts have been made, however, to delineate a difference in anti-inflammatory response between hydroxychloroquine enantiomers in their concentration-effect or concentration-toxicity profiles after administration of racemate.

Tett and colleagues examined plasma concentrations of the enantiomers of hydroxychloroquine after administration of racemate to 43 patients with rheumatoid arthritis [572]. A wide range of enantiomer concentrations was noted in whole blood, and it was not possible to identify a therapeutic range of concentrations for indices of arthritis such as duration and intensity of morning stiffness, rheumatoid factor and synovitis. It was of note, however, that patients with more severe symptoms had higher blood concentrations of both enantiomers. This paradoxical finding had been noted in a previous study that used non-stereo specific assay methodology [573]. This finding has been postulated to be caused by higher α1-acid glycoprotein levels as a result of more severe in flammation, causing increased binding, reduced clearance and, therefore, higher blood and plasma enantiomer concentrations in patients with more active disease [574].

(+)-Halofantrine was shown *in vitro* to be more potent than its antipode in inhibiting repolarizing currents in feline ventricular myocytes [575]. In a subsequent paper involving 15 subjects given (±)-halofantrine 500 mg/day for 42 days, the relationship between enantiomer concentration and QTc interval was assessed [576]. In 13 of the 15 subjects, significant positive correlations were found between (+)-enantiomer concentration and QTc interval. In contrast, in 10 of the subjects a significant ($p < 0.05$) relationship was found for (−)-enantiomer. As noted by the investigators, the plasma concentrations of the enantiomers after racemate covaried that makes it difficult to determine with certainty which of the enantiomers is more cardiotoxic. Nevertheless, because the regression coefficients were higher for (+)-enantiomer than for (−)-enantiomer in 80% of the subjects [576], the data seem to indicate a stronger association between the (+)-enantiomer and QTc interval prolongation.

To date, stereoselectivity in PK and/or PD of the chiral antimalarial drugs is incompletely understood. Noting with respect to antimalarial potency, with the exception of mefloquine, for most of the chiral antimalarial drugs stereoselectivity has not been demonstrated *in vitro*. Nevertheless, PK differences may be present between enantiomers, which in turn can impart stereoselectivity in the PD effects *in vivo*. Besides antimalarial effects *per se*, stereoselectivity is possible in other PD effects for some of these drugs. Only a limited number of antimalarial drugs are currently available, and treatment of the illness is complicated by an increasing incidence of drug resistance. In order to maximize clinical use of the existing drugs, it may be helpful to better understand their enantioselective properties so that optimal therapeutic benefit may be realized. The knowledge of the stereoselective aspects of these agents may be helpful in better understanding their mechanisms of action and possibly optimizing their clinical safety and/or effectiveness.

5.5. Well Compliance for Maximizing Efficacy of Antimalarial Drugs

The success or failure of any public health program is largely determined by the effective use of the services offered. This principle is especially relevant in treating malaria because prompt use of efficacious drugs greatly reduces the incidence of severe and complicated disease. Merely making an efficacious treatment available is not enough to reduce malaria mortality, however - the treatment also must be used optimally. Even the most highly efficacious treatment, if not used correctly, can fail to cure the illness, and may facilitate the development of drug resistance. In real life, many human factors thwart access and use of curative medications in resource-poor settings, including traditional beliefs, illiteracy, mistrust, and fatalistic attitudes.

In the case of malaria, economic barriers to purchase drugs; complex treatment and dosing regimens; limited understanding of how or why to adhere to recommended regimens; and adverse side-effects compound the problem. A false sense of security also affects choice and use of antimalarial treatments, particularly in sub-Saharan Africa. Most residents in malaria-endemic areas of Africa do not appreciate the rising toll of chloroquine-resistant *P.*

falciparum because, for many years, chloroquine worked so well. In some cases, loss of activity of chloroquine is also obscured by its partial antipyretic effects, which suggest to some patients that the drug is curing them as opposed to merely relieving symptoms. There is a tremendous public confidence in the old drug, chloroquine, even though the old drug is failing. That's part of the problem.

Another major problem is the inadequacy of public health infrastructure to serve basic health care needs in Africa and other resource-poor settings [577]. This inadequacy is reflected in a lack of electricity, clean water, or reliable medicines in many public health facilities as well as poorly trained and poorly motivated staff, leakage of drugs for private resale, unauthorized patient charges, mismanagement of patient user fees, distance to facilities, lack of drugs and other desired services, and lengthy waiting times. As a result, private-sector sources - ranging from licensed pharmacies to informal drug kiosks and itinerant drug sellers - have assumed an increasingly important role. In some studies, as many as 60% or more of patients seeking help for febrile illness received their medicines from the private sector. This section reviews specific barriers to access and use of antimalarial drugs in *falciparum*-endemic areas, as well as promising strategies to enhance appropriate drug use in the future [42].

5.5.1. Overcoming Economic Barriers

The high cost of treatment - real or perceived - is a major barrier to accessing malaria drugs and treatment. In addition to direct drug costs, indirect costs such as transportation and time affect utilization. When chloroquine was still effective, many mothers in Ghana who believed in its efficacy failed to use it simply because they lacked access to health services, pharmacies, or urban markets where they could buy the product. Similarly, in Papua New Guinea, patients' willingness to seek malaria care at primary health facilities was strongly influenced by the distance they had to travel to reach those facilities. Development of rational malaria treatment policies in endemic countries relies on consideration of a variety of data, including economic, behavioral, biomedical, and political. Among the relevant biomedical data, evidence from carefully conducted *in vivo* assessments of malaria therapy efficacy using established methodologies are critical for guiding appropriate drug choice, and for timing policy change [578].

Ultimately, however, the central purpose of formulating rational malaria treatment policies is to reduce malaria mortality. Making an appropriate drug available is merely the first step. The optimal use of that drug involves many additional steps. The patient (or their parent) must also:

- recognize the need for treatment with Western medicine;
- believe that the recommended treatment is useful, choose to use it, and know where to get it;
- perceive that treatment is safe, of reasonable quality, available, and affordable;
- at the time of dispensing, receive an appropriate amount of medicine from the provider - whether public sector- or private sector-based - along with adequate information on how to take the treatment correctly; and
- actually take the medicine completely and correctly.

This more holistic view of malaria therapy, i.e., one that takes into account all of the factors that determine treatment success in practice, underlies the concept of "programmatic effectiveness." Programmatic effectiveness recognizes that even the most highly efficacious treatment, if not used correctly, can fail to cure the illness, and possibly contribute to the development of resistance. In one of the few studies that have tried to measure programmatic effectiveness of malaria therapy, the authors identified a series of 7 critical factors leading to successful treatment [579]. They then measured the frequency with which each factor was satisfied in an actual health setting in Burkina Faso and found that the overall programmatic effectiveness of malaria treatment was only about 3% (i.e., only 3% of patients were successfully treated, according to their definitions). More important, increasing the efficacy of the malaria treatment from 85% to 100% increased overall programmatic effectiveness by less than 1%.

This example illustrates that optimizing a single factor while ignoring other factors may yield minimal improvements overall. In the case of malaria, simply optimizing drug efficacy by providing ACTs without attempting to optimize the environment in which treatment occurs may result in far less influence on overall morbidity and mortality rates than expected. Infusing ACTs in the same environment in which chloroquine, sulfadoxine/pyrimethamine, and other drugs have failed could even be counterproductive. An unprecedented investment in highly efficacious ACTs must be accompanied by investments aimed at improving the environments in which they will be used in order to maximize their benefits minimize the potential for harm resulting from misuse.

ACTs cost higher than most currently used monotherapies and alternative non-artemisinin-based combinations. Most cost up to 20 times that of conventional monotherapies and in view of the current high demand than supply, the costs are likely to remain high for several years to come. Approaches to facilitate program implementation include preferential reduction of market price for ACTs to developing countries and subsidizing implementation costs through bilateral financial assistance. In this consideration, negotiations between WHO and Norvatis Pharma has reduced the cost of artemether-lumefantrine (Coartem) to around US$ 0.9–1.4 for a treatment course of a child up to 7 years old and around US$ 2.4 per adult treatment dose [106].

Despite this reduction, prices are still high for most households. For example, in Tanzania where the mean monthly income of rural households is only US$ 13.4 and US$ 29.0 for urban households outside Dar-es-Salaam, one full treatment course for a child takes 1.4/13.4 (10%) of the monthly income. On average, one child may have four malaria episodes in a year and another four episodes that may be wrongly treated as malaria leading to a significant drain on the family income. Money is the main barrier when dealing with a malaria fight, as correctly pointed out in the Malaria World Report released by WHO and UNICEF on May 3, 2005. Unless cost is sufficiently addressed both locally and internationally, deploying ACTs could lead to adverse results ranging from an increase in potentially fatal delays in infected people presenting to medical services to systematically excluding poorest malaria sufferers from receiving treatment [122].

Price is an important factor in choosing both a treatment and a patient, but it interacts with a host of other variables. Even when drugs are free through public-sector sources, they are not necessarily accessible. Patients and their families may seek malaria treatment outside the public sector because private outlets are easier to reach, and their products are perceived to be of higher quality. Consumers also may obtain drugs from private facilities for the simple reason that antimalarial drugs are frequently out of stock at public clinics. A global subsidy to flood the public and private marketplace with highly effective artemisinin combination therapies (ACTs) in even the most peripheral areas where drug resistant *P. falciparum* is being transmitted would circumvent these economic barriers.

5.5.2. Making Antimalarial Drugs Safer and more Convenient

In many parts of the world, especially sub-Saharan Africa, antimalarial pills are still dispensed in unmarked paper envelopes, and syrups constitute the major form of treatment for children under five. Prepackaged tablets are one means by which to ensure proper dosing and adherence to a full course of antimalarial treatment. The Special Programme for Research and Training in Tropical Diseases (TDR) advocates a simple, dose-wise packaging of antimalarials to help patients take the correct drug dosage at the correct time [2]. Well designed drug packages increase adult compliance, on average, 20% [580], and the effect in children [581] may be even greater. In Ghana, a 91% adherence rate was found in children aged 0-5 years who were given prepacked tablets compared with a 42% adherence rate in users of antimalarial syrup. The tablets, in contrast to the syrup, did not require measuring, and the daily dose was clearly indicated. Most important, the cost of treating children with tablets was roughly 25% that of treating them with syrup [582]. Prepackaged tablets also have been associated with reduced drug wastage at health facilities and reduced patient waiting times at dispensaries [580]. The adherence is actually better for tablets in children.

Similar results have been reported elsewhere. In a comparison of blister packaging of antimalarials with other interventions (including public information campaigns and drug quality assessment), blister packaging was most effective at improving adherence [583]. Studies in Burma [584] and China [585] also suggest that blister packs are an effective way to improve adherence to short-course and combination treatments. A recent qualitative study among urban and rural women in western Uganda found pre-packed, unit-dosed malaria treatment for children widely accepted by mothers. Ninety-one percent of women said they preferred the pre-packed to the conventional type of treatment, and 94% were willing to pay between US$0.17 (rural women) and US$0.29 (urban women) more for this treatment. The main reasons for preferring pre-packs were safety and cleanliness; ease of application, dosing, and compliance were additional perceived benefits. In Burkina Faso, mothers who received training and packaged drugs recognized and treated malaria promptly and correctly. In sum, evidence that better packaging technology improves adherence is timely and relevant to the introduction of more efficacious antimalarial treatments such as ACTs, particularly in Africa. An emphasis on convenient dosing schedules, as well as forms developed in partnership with private industry will be essential to enhanced penetration of new-generation antimalarial drugs [42].

At the Abuja Summit in April 2000, African heads of state agreed that, by 2005, 60% of malaria sufferers should have prompt access to affordable, appropriate treatment within 24 hours of the onset of symptoms [2]. This recommendation coincided with increasing interest in extending malaria treatment closer to home, or even within the home. Although there is, at present, no single accepted definition of home-based malaria therapy, the general idea is to greatly improve access to efficacious medicines at the most peripheral levels, and to increase community members' knowledge about how to use antimalarial medicines properly [586].

In fact, unsupervised home treatment with antimalarials has been a common practice in certain African countries and settings for many years, especially where local residents are dissatisfied with formal health services. In a recent study in Kenya, home treatment with an antimalarial drug was given to 47% of children under 5 years (for 32%, this was their only antimalarial treatment), 43% were taken to a health facility, and 25% had no antimalarial treatment. In Mali and Nigeria, 76% and 71% of mothers, respectively, managed their child's illness at home with antimalarial drugs. Urban settings (where population density creates markets responsive to consumer demand) also promote home treatment. In rural Gambia, in contrast, low rates of home treatment with chloroquine (<10%) have been documented in several studies, possibly due to the restricted availability of antimalarials as well as lower costs for treatment at government health centers. A similar pattern has been reported from Zambia, where private providers also are few, and treatment at government facilities frees [587].

If the availability of effective antimalarials increased in rural African communities along with stakeholder education, many experts believe that residents could use drugs effectively. One community-based intervention in northern Ethiopia in which mother coordinators provided home treatment reduced under-5 malaria mortality by 40%. A smaller study in Guinea Bissau found similar treatment outcomes and day 7 chloroquine blood levels in children with symptomatic malaria whose treatment was either supervised in a health center or given at home following adequate caregiver education by health staff [588]. The main disadvantage of self-treatment is the lack of clinical evaluation of patients by trained health professionals, which can result in missed diagnoses and delays in appropriate treatment. In addition, presumptive treatment of febrile episodes at home is likely to result in overuse of antimalarial drugs and the attendant risk of escalating drug resistance.

5.5.3. Drug Quality Assurance

In May 2002, WHO and UNICEF have established mechanisms to pre-qualify manufacturers of artemisinin compounds and ACTs on the basis of compliance with internationally recommended standards of manufacturing and quality. Prequalified products and manufacturers are listed for procurement by UN agencies, and this could serve as a guide to Governments, NGOs and other partners procuring ACTs on behalf of endemic countries. Up to now one ACT has been prequalified, i.e., artemether/lumefantrine.

The quality of medicines in less developed countries is often poor due to a lack of quality assurance during manufacture, excessive decomposition of active ingredients under hot and humid conditions, and/or counterfeiting. Nonetheless, substandard and expired drugs continue to circulate because of simple economics, lack of adequate drug information, and weak country-level drug regulatory systems. In one study conducted by the Department of Pharmacy of the University of Nairobi, 46% of locally manufactured products were substandard when their active ingredient was measured by compendial methods. In another study, almost half of nearly 600 antimalarial, antibacterial, and anti-tuberculosis drugs purchased in Lagos or Abuja, Nigeria did not comply with set pharmacopoeial limits. The sample included chloroquine, sulfadoxine/pyrimethamine (SP), and quinine tablets, and chloroquine and SP syrups, some of which had less than 25% active ingredient. Failure to dissolve was found in 44% and 13%; respectively, of SP and amodiaquine tablets sold by private wholesale pharmacies in Dar es Salaam, Tanzania [589]. These findings validate consumers' concerns about antimalarial drug quality in many malaria-endemic countries.

In contrast to a substandard drug, a counterfeit drug is one that has been incorrectly packaged or constituted in a deliberate attempt to dupe sellers and consumers. Among cases in which pharmaceutical preparations contain no active ingredient, a drug other than that stated on the label, or consistently low concentrations of drug, fraud is the most likely cause. Counterfeit artesunate and mefloquine preparations are widely sold in Cambodia. Another recent study from Southeast Asia found that more than a third of tablets labeled "artesunate" purchased from shops in Cambodia, Laos, Burma, Thailand, and Vietnam contained no drug. Recognizable packaging is one means of increasing the likelihood that drugs are authentic, although it does not eliminate the need to verify quality using, for example, drug testing kits. In Southeast Asia where counterfeit packaging changes frequently, this type of testing must be conducted on a regular basis [339].

Labeling of pharmaceuticals, malaria drugs included, is woefully inadequate in many malaria-endemic settings. When drugs are accompanied by printed information, the information is often written in complicated or technical language, or, in some cases, a language entirely unknown to the consumer. And in many cases, the printed information is of no use whatsoever because the consumer is illiterate. Difficulties reading antimalarial instructions are well documented. In addition, verbal instructions are often inadequate. In Zambia, many children suffering from malaria did not receive the appropriate 3-day course of chloroquine because their caregivers did not receive adequate instruction on how to administer the drug. In Uganda, only 38% of children received chloroquine in compliance with instructions given by health workers or drug shop attendants. Obviously, consumers can only use antimalarials correctly if they receive a full explanation from a health care provider or drug seller, or appropriate written and graphic labeling accompanies the drug. Good labeling, in fact, is a minor added expense for producers that yield an excellent return on investment.

5.5.4. Appropriate Drug Use and Educating

1) Correct Approaches to Treatment

Inappropriate drug use is rampant in malaria endemic countries for varied reasons, such as poverty, ignorance and lack of effective drug regulatory bodies. Absence of effective regulatory bodies has led to importation of substandard drugs and unregistered drugs, uncontrolled prices of drugs, cross-boarder smuggling, purposeful under-dosing among prescribers/dispensers and illegal shift of drugs from the formal to the informal health sectors. Poor quality ACTs and artemisinin monotherapies do exist ahead of nationally agreed ACT policies. For example, prior to WHO recommendation on ACTs in 2001 [47], the informal health sector in a number of African countries already stocked artemisinins and incorrectly dispensed them as monotherapies thus jeopardizing its therapeutic longevity. Artemisinin monotherapy for less than 5 days is associated with high recrudescence rates, a gateway to resistance. This is a major ACT implementation issue requiring long-term solutions. Countries require well coordinated drugs regulatory mechanism, purposeful involvement of drug importers and distributors and well-informed prescribers/dispensers in both the formal and informal health sector. An International Consortium on ACTs implementation is being formed that will drive this agenda and other related issues forward in order to identify feasible and practicable solutions.

In endemic regions, some semi-immune malaria patients could be cured using partial treatment with effective medicines (i.e., use of regimens that would be unsatisfactory in patients with no immunity). This had led in the past to different recommendations for patients considered being semi-immune and those considered to be non-immune. Another potentially dangerous practice is to give only the first dose of the treatment course for patients with suspected but unconfirmed malaria, with the intention of giving full treatment if the diagnosis is eventually confirmed. Neither practice is recommended. If malaria is suspected and the decision to treat is made, then a full effective treatment is required whether or not the diagnosis is confirmed by a test. The exception to this is artemether-lumefantrine, the partner medicines of all other ACTs have been previously used as monotherapies, and still continue to be available as such in many countries. Their continued use as monotherapies can potentially compromise the value of ACTs by selecting for drug resistance. The withdrawal of artemisinins and other monotherapies is recommended.

Recommendations on treatment approaches that should be avoided

- Partial treatments should not be given even when patients are considered to be semi-immune or the diagnosis is uncertain. A full course of effective treatment should always be given once a decision to give antimalarial treatment has been reached.
- The artemisinins and partner medicines of ACTs should not be available as monotherapies.

2) *Education on Drug Usage*

Health education messages can provide information and address a variety of misconceptions regarding the use of antimalarials. Common formats include posters, video clips, radio, and other forms of mass media. Other methods include peer education, mothers' clubs, and school-based programs. In Tanzania, popular misconceptions about SP were undermining adherence to treatment. The Tanzanian health service responded by creating health education messages to counter the belief that SP is ineffective because it lacks immediate antipyretic plus anti-inflammatory effects of chloroquine. Other messages in Tanzania and elsewhere have focused on malarial symptoms and complications; correct dosing of antimalarial drugs, the use of oral versus injectable antimalarials, and persistent or recurrent symptoms following treatment. As with many health education messages, however, the Tanzanian malaria campaign was never formally evaluated.

One exception to the dearth of educational evaluations is a study conducted in Cambodia in which posters combined with video clips improved antimalarial adherence rates as much as 20%. The influence of posters alone was much more modest. Educational benefits were mainly reported among those who visited public and private health practitioners as opposed to consumers who purchased their drugs from vendors. While posters have been used widely in malaria education, this is the only study that has compared their effect on adherence with another form of mass media; namely, video.

Many consumers in Africa, Asia, and Latin America, seek medical treatment outside the formal heath care sector. Although close-to-home treatment is desirable in theory, in practice, private sector vendors are frequently unable to provide appropriate advice, complete doses, or even the correct category of drug for a specific complaint [586]. One way of improving quality of private-sector treatment is to train drug sellers and shopkeepers. In western Nigeria, primary health care training of patent medicine vendors improved knowledge about malaria as well as other infectious disorders and malnutrition. In Kenya, 500 private drug sellers were enrolled in a 2-4 day training workshop mounted by the Ministry of Health in conjunction with a public information campaign. The effort resulted in a short-term increase in the appropriate use of over-the-counter chloroquine of 62%. One year later, however, there was a 20% turnover in outlets and/or trained staff [590].

It should not be assumed that community-based health care providers - in particular midwives, traditional birth attendants, and health workers - are well informed about malaria. However, with regular training and updates, they can serve as potent advocates for proper case management and adherence to treatment guidelines. Medical assistants working in health posts are another target group for ongoing education and auditing. In Ghana, one study found a surprising discrepancy among clinic providers between knowledge and practice of malaria treatment as well as a tendency to yield to patient demands for injections and poly-pharmacy. In-service training did not alter these poor prescribing practices over the long term despite medical assistants' knowledge of correct treatment paradigms. The authors argue for an increased awareness of the sociocultural context of antimalarial drug use when designing educational programs for health providers and patients.

Understanding local knowledge, perceptions, and practices of malaria management has become the focus for increasing research during the last decade. In a recent study based on semi-structured interviews with community members in Mbarara, Uganda, causes of malaria − in addition to "fever due to mosquitoes" − were reported to be drinking dirty or unboiled water, breathing bad air, staying near somebody with malaria, witchcraft, avenging spirits, and eating fresh maize or sweet fruit such as mangoes, pineapples, and passion fruit. Even those who said that mosquitoes cause or transmit malaria had erroneous ideas about the route of transmission. Most believed that malaria was acquired by drinking mosquito eggs or larvae in dirty water as opposed to a mosquito bite [591].

Despite misconceptions about its cause, many people know malaria's common clinical features. Kenyan and Ghanaian mothers had sufficiently. Good knowledge to recognize mild to moderate malaria. In Uganda, not only were convulsions, anemia, and splenomegaly linked to malaria, interviewees also discussed complications of malaria in pregnancy, in particular abortions. However, many caregivers in Africa still blame convulsions—a prominent finding in severe malaria − on supernatural causes [591]. When spiritual causes are invoked, traditional healers, as opposed to medical professionals, are frequently consulted. Educating traditional healers about febrile convulsions and severe malaria is a particularly important means of rerouting children with seizures and impaired consciousness to effective antimalarials.

Another ethnographic study conducted in Tanzania found parents and traditional healers unanimous in their belief that febrile seizures require traditional treatments, at least initially, and that these treatments are effective. While traditional healers did refer some patients who were not improving to the District Hospital, the referral typically came late in the course of illness. In Uganda, traditional medicine also was endorsed for treatment of convulsions and splenomegaly [591]. Squeezing juice from herbs into the nose and mouth was a common way of treating a convulsing or unconscious child, while splenomegaly was often treated with therapeutic marks and herbs squeezed into cuts overlying the splenic area.

The education sector already knows the importance of health to schoolchildren as evidenced by the child-friendly schools of UNICEF (United Nations Children's Fund), the health promoting schools of the World Health Organization (WHO), and the International School Health Initiative of the World Bank. Malaria is a disease particularly well suited to public health partnerships with teachers, principals, and designers of school curricula. Although malaria in areas of stable transmission is most common and severe in children who are not yet attending school, it still remains an important cause of mortality (10−20% of all cases) and morbidity in schoolchildren [592]. In addition, malaria can significantly influence educational outcomes, accounting for roughly three to eight percent of all reasons for absenteeism, 13−50% of school days missed per year due to preventable medical causes, as well as residual neurologic sequel impairing cognitive and developmental potential in one to five percent of children infected early in life [42].

Chapter VI

Toxicity of Antimalarial Drugs

Various drugs are widely used in the prophylaxis and treatment of malaria. All antimalarial drug toxicity is viewed in the treatment and prophylaxis of *P. falciparum* malaria, which has a high mortality if untreated; a greater risk of adverse reactions to antimalarial drugs is inevitable. One of the most important current approaches to develop new drugs involves the synthesis of chemical libraries and their evaluation against most validated biochemical targets of malarial parasites. Avenues of research for the development of new antimalarials include lipid metabolism, degradation of hemoglobin and proteins, interaction with molecule transport, iron metabolism, apicoplasty, and signal transduction. Antimalarial drug toxicity is one side of the risk-benefit equation and is viewed differently depending upon whether the clinical indication for drug administration is malaria treatment or prophylaxis. Research that leads to drug registration tends to omit two important groups who are particularly vulnerable to malaria – very young children and pregnant women. Prescribing in pregnancy is a particular problem for clinicians because the risk-benefit ratio is often very unclear [263].

For the prevention of malaria in travelers, a careful risk-benefit analysis is required to balance the risk of acquiring potentially serious malaria against the risk of harm from the prophylactic agent. The therapeutic ratios for some antimalarials are narrow, and toxicity is frequent when recommended treatment dosages are exceeded; parenteral administration above the recommended dose range is especially associated with the hazards of cardiac and neurological toxicity [593]. The purpose of this chapter is to update investigators and physicians on the toxicity associated with antimalarial drugs. This summary's main focus is to provide knowledge on the incidence of major adverse effects and provide an outline of what is known about the minor side-effects of the antimalarial drugs.

All drugs with the potential to cause toxicity that was assessed and evaluated. Type A adverse effects result from excessive responses to a drug; these adverse effects are predictable from the known effects of that drug and are often dose or concentration related. In contrast, type B adverse effects are not predictable from the known effects of that particular drug; there may be an immunological basis to the adverse effect, and there is often no clear relationship with the dose or concentration of drug. Furthermore, certain patient groups are at particular risk of severe adverse effects – including the elderly, the very young, glucose-6-phosphate dehydrogenase (G6PD)-deficient people and HIV-positive people – and these may not be well represented in submissions to regulatory authorities [594]. Toxicity may range from mild to serious and from reversible to irreversible [594]. Adequate clinical response is defined as rare toxicities, e.g., those which occur in <1% of patients using the agent, uncommon in 1–10%, and common in > 10%.

6.1. General Toxicity

The toxicity of antimalarial drugs sets an unusual and interesting problem for the clinician. Unlike most clinical situations, antimalarial drugs are provided to healthy people to prevent malaria and to sick patients who are requesting treatment against ill health in malarial areas. Antimalarial drug toxicity is one side of the risk-benefit equation and is viewed differently depending upon whether the clinical indication for drug administration is malaria treatment or prophylaxis. Drug toxicity must be acceptable to patients and cause less harm than the disease itself. Prior to use these drugs, all adverse effects or toxicity of antimalarial drugs should be known.

6.1.1. Common Side-Effects of Antimalarial Drugs

Since chloroquine resistance has become widespread, alternative agents have been widely used in treatment and prophylaxis regimens, the toxicity of these antimalarial agents should be considered. Quinine is the mainstay for treating severe malaria due to its rare cardiovascular or CNS toxicity, but its hypoglycemic effect may be problematic. Mefloquine can cause dose-related serious neuropsychiatric toxicity and pyrimethamine-dapsone is associated with agranulocytosis, especially if the recommended dose is exceeded. Pyrimethamine-sulfadoxine and amodiaquine are associated with a relatively high incidence of potentially fatal reactions, and are no longer recommended for prophylaxis. Atovaquone/proguanil is an antimalarial combination with good efficacy and tolerability as prophylaxis and for treatment. The artemisinin derivatives have remarkable efficacy and an excellent safety record. Prescribing in pregnancy is a particular problem for clinicians because the risk-benefit ratio is often very unclear. High doses of the antimalarial drug pyrimethamine may cause blood problems that can interfere with healing and increase the risk of infection. People taking this drug should be careful not to injure their gums when brushing or flossing their teeth or using toothpicks. If possible, dental work should be postponed until treatment is complete and the blood has returned to normal. However, all drugs used for malaria therapy or prophylaxis have common adverse effects, in addition to rare, mild-to-severe and/or sometimes fatal adverse effects (Table 48).

1) Precautions

Antimalarial drugs may cause lightheadedness, dizziness, blurred vision and other vision changes. Anyone who takes these drugs should not drive, use machines or do anything else that might be dangerous until they have found out how the drugs affect them. The antimalarial drug mefloquine has received attention because of reports that it causes panic attacks, hallucinations, anxiety, depression, paranoia, and other mental and mood changes, sometimes lasting for months after the last dose. In fact, the U.S. Food and Drug Administration (FDA) began requiring warnings with mefloquine beginning in July 2003 because of serious psychiatric effects caused by the drug. Pharmacists are required to include a 2,000-word medication guide detailing the warnings. Anyone who has unexplained anxiety, depression, restlessness, confusion, or other troubling mental or mood changes after taking mefloquine should call a physician right away. Switching to a different antimalarial drug may be an alternative and can allow the side-effects to stop.

Anyone taking antimalarial drugs to prevent malaria who develops a fever or flu-like symptoms while taking the medicine or within 2-3 months after traveling to an area where malaria is common should call a physician immediately. If the medicine is being taken to treat malaria and symptoms stay the same or get worse. The patient should check with the physician who prescribed the medicine. Patients who take this medicine over a long period of time need to have a physician check them periodically for unwanted side-effects. Babies and children are especially sensitive to the antimalarial drug chloroquine. Not only are they more likely to have side-effects from the medicine, but they are also at greater risk of being harmed by an overdose. A single 300-mg tablet could kill a small child. This medicine should be kept out of the reach of children and safety vials should be used.

2) Allergies

Anyone who has had unusual reactions to antimalarial drugs or related medicines in the past should let his or her physician know before taking the drugs again. The physician should also be told about any allergies to foods, dyes, preservatives, or other substances.

3) Pregnancy

In laboratory animal studies, some antimalarial drugs cause birth defects. But it is also risky for a pregnant woman to get malaria. Untreated malaria can cause premature birth, stillbirth, and miscarriage. When given in low doses to prevent malaria, antimalarial drugs have not been reported to cause birth defects in humans. If possible, pregnant women should avoid traveling to areas where they could get malaria. If travel is necessary, women who are pregnant or who may become pregnant should check with their physicians about the use of antimalarial drugs.

4) Breastfeeding

Some antimalarial drugs pass into breast milk. Although no problems have been reported in nursing babies whose mothers took antimalarial drugs, babies and young children are particularly sensitive to some of these drugs. Women who are breastfeeding should check with their physicians before using antimalarial drugs.

5) Other Medical Conditions

Before using antimalarial drugs, people who have any of these medical problems (or have had them in the past) should make sure their physicians are aware of their conditions:

- Blood disease
- Liver disease
- Nerve or brain disease or disorder, including seizures (convulsions)

- Past or current mental disorder
- Stomach or intestinal disease
- Deficiency of the enzyme glucose-6-phosphate dehydrogenase (G6PD), which is important in the breakdown of sugar in the body
- Deficiency of the enzyme nicotinamide adenine dinucleotide (NADH) methemoglobin reductase
- Psoriasis
- Heart disease
- Family or personal history of the genetic condition favism (a hereditary allergic condition)
- Family or personal history of hemolytic anemia, a condition in which red blood cells are destroyed
- Purpura
- Hypoglycemia (low blood sugar)
- Blackwater fever (a serious complication of one type of malaria)
- Myasthenia gravis (a disease of the nerves and muscles).

In rural health care centers, staff with relatively little formal training may be responsible for taking care of patients with malaria. The patients often have no written medical history, and the centers have minimal capacity for laboratory screening for contra-indications to particular medications. Worse still, most of these people with malaria treat themselves with antimalarial agents acquired over the counter or passed from household to household. Any antimalarial drugs that are to be distributed in such regions need to be safe and well tolerated, and the risks associated both with the indicated use and with contra-indicated or inappropriate uses should be considered. Clinical studies supporting the licensing of new antimalarial drugs rarely include highly vulnerable groups such as infants, young children, and pregnant women [263]. Thus, safety issues often drive the decision to continue distributing chloroquine or sulfadoxine–pyrimethamine, despite substantial parasite resistance. Key data with regard to the safety and tolerability of the most commonly used antimalarial drugs are summarized in Table 48.

Research that leads to drug registration tends to omit two important groups who are particularly vulnerable to malaria - very young children and pregnant women. Prescribing in pregnancy is a particular problem for clinicians because the risk-benefit ratio is often very unclear. The number of antimalarial drugs in use is very small. Despite its decreasing efficacy against *P. falciparum,* chloroquine continues to be used widely because of its low cost and good tolerability. It remains the drug of first choice for treating *P. vivax* malaria. Pruritus is a common adverse effect in African patients. As prophylaxis, chloroquine is usually combined with proguanil. This combination has good overall tolerability but mouth ulcers and gastrointestinal upset are more common than with other prophylactic regimens. Sulfadoxine/pyrimethamine is well tolerated as treatment and when used as intermittent preventive treatment in pregnant African women. Sulfadoxine/pyrimethamine is no longer used as prophylaxis because it may cause toxic epidermal necrolysis and Stevens Johnson syndrome. Mefloquine remains a valuable drug for prophylaxis and treatment.

Additionally, tolerability is acceptable to most patients and travellers despite the impression given by the lay press. Dose-related serious neuropsychiatric toxicity can occur;

mefloquine is contra-indicated in individuals with a history of epilepsy or psychiatric disease. Quinine is the mainstay for treating severe malaria in many countries. Cardiovascular or CNS toxicity is rare, but hypoglycemia may be problematic and blood glucose levels should be monitored. Halofantrine is unsuitable for widespread use because of its potential for cardiotoxicity. There is renewed interest in two old drugs, primaquine and amodiaquine. Primaquine is being developed as prophylaxis, and amodiaquine, which was withdrawn from prophylactic use because of neutropenia and hepatitis, is a potentially good partner drug for artesunate against *falciparum* malaria, Atovaquone/proguanil is a new antimalarial combination with good efficacy and tolerability as prophylaxis and treatment. The most important class of drugs that could have a major impact on malaria control is the artemisinin derivatives. They have remarkable efficacy and an excellent safety record. They have no identifiable dose-related adverse effects in humans and only very rarely produce allergic reactions. Combining an artemisinin derivative with another efficacious antimalarial drug is increasingly being viewed as the optimal therapeutic strategy for malaria.

6.1.2. Adverse Effects of Drug-drug Interactions

Taking antimalarial drugs with certain other drugs may affect the way the drugs work or may increase the chance of side-effects. The most common side-effects of antimalarial drugs are diarrhea, nausea or vomiting, stomach cramps or pain, loss of appetite, headache, itching, difficulty concentrating, dizziness, lightheadedness, and sleep problems. These problems usually go away as the body adjusts to the drug and do not require medical treatment. Less common side-effects, such as hair loss or loss of color in the hair; skin rash; or blue-black discoloration of the skin, fingernails, or inside of the mouth also may occur and do not need medical attention unless they are long-lasting. More serious side-effects are not common, but may occur. If any of the following side-effects occur, the physician who prescribed the medicine should be contacted immediately:

- Blurred vision or any other vision changes
- Convulsions (seizures)
- Mood or mental changes
- Hallucinations
- Anxiety
- Confusion
- Weakness or unusual tiredness
- Unusual bruising or bleeding
- Hearing loss or ringing or buzzing in the ears
- Fever, with or without sore throat
- Slow heartbeat
- Pain in the back or legs
- Dark urine
- Pale skin
- Taste changes
- Soreness, swelling, or burning sensation in the tongue.

Other rare side-effects may occur. Anyone who has unusual symptoms after taking an antimalarial drug should get in touch with his or her physician. Some antimalarial drugs may interact with other medicines. When this happens, the effects of one or both of the drugs may change or the risk of side-effects may be greater. Anyone who takes antimalarial drugs should let the physician know all other medicines he or she is taking. Among the drugs that interact with some antimalarial drugs are:

- Beta blockers such as atenolol (Tenormin), propranolol (Inderal), and metoprolol (Lopressor)
- Calcium channel blockers such as diltiazem (Cardizem), nicardipene (Cardene), and nifedipine (Procardia)
- Other antimalarial drugs
- Quinidine, used to treat abnormal heart rhythms
- Antiseizure medicines such as vaproic acid derivatives (Depakote or Depakene)
- Oral typhoid vaccine
- Diabetes medicines taken by mouth
- Sulfonamides (sulfa drugs)
- Vitamin K
- Anticancer drugs
- Medicine for overactive thyroid
- Antiviral drugs such as zidovudine (Retrovir).

This list does not include every medicine that may interact with every antimalarial drug. It is advised to check with a physician or pharmacist before combining an antimalarial drug with any other prescription or nonprescription (over-the-counter) medicine.

6.1.3. Antimalarial Drug Toxicity on Glucose Metabolism

A number of the drugs used as prophylaxis against, or treatment for, malaria have clinically significant metabolic side-effects. Of these, hypoglycemia resulting from quinine administration is one of the most important [595]. Quinidine, the diastereoisomer of quinine, can also cause this complication [596]. The 4-aminoquinoline chloroquine is known to influence glucose metabolism in ways which could lead to low blood glucose concentrations [597], although reports of chloroquine-treated patients with acute malaria who have developed hypoglycemia are rare. Recently, the quinoline methanol mefloquine has been shown to reduce plasma glucose concentration during a conventional prophylactic course in healthy young adults [598]. In all these situations, raised serum or plasma insulin concentrations have been observed.

The maintenance of appropriate plasma glucose concentrations in both basal and fed states is largely a function of the action of insulin. Counter-regulatory hormones, especially glucagon, become important only when the plasma glucose starts to fail below normal. However, increased catecholamine and cortical secretion in the absence of hypoglycemia occurs in stress situations including infections. Other factors such as increased plasma free fatty acid concentrations and ketonemia may also contribute to raise basal plasma glucose.

However, continued host and parasite glucose demand, impaired hepatic glycogenolysis and gluconeogenesis, cytokine effects and the inability of the patient to take sufficient carbohydrate by mouth because of nausea and vomiting increase the propensity to hypoglycemia in malaria [595]. The added influence of hyperinsulinemia due to antimalarial treatment can result in severe, refractory hypoglycemia with serious neurological and other sequel if it remains undetected or inadequately treated.

Several questions arise concerning the effect of antimalarial drugs on plasma glucose and insulin concentrations. Which drugs have the greatest effect on insulin secretion and why? Is this effect related to the total or free (unbound) plasma concentration of the drug? Is there synergism between antimalarial drugs of the same or different classes? Do the drugs have other effects on glucose metabolism which could contribute to hypoglycemia? Are there special situations, during either prophylactic or treatment courses, in which drug effects on glucose metabolism are exaggerated or, conversely, attenuated? Do specific effects suggest novel ways of preventing or treating hypoglycemia? The published data relating to these questions will be assessed in the present review including, where possible, their reanalysis or reinterpretation in the light of the current understanding of mechanisms governing glucose homoeostasis in human [599].

1) Quinine and Quinidine

The drugs of cinchona alkaloid are known to increase insulin secretion by blocking ATP-sensitive potassium channels in pancreatic beta cells [600]. Though other host and parasite-specific factors contribute to hypoglycemia in *falciparum* malaria, quinine-associated hyperinsulinemia has been reported consistently in series of adults, pediatric, and pregnant patients from a variety of endemic areas [601]. Hypoglycemia during quinidine treatment has also been reported, though much less often. This could be because it has less antimalarial use than quinine on a world scale but, interestingly, reports of hypoglycemia in patients receiving quinidine as an antiarrhythmic drug have also been rare [602]. In a recently-published large-scale comparative trial of quinine and artemether in Vietnamese adults with severe *falciparum* malaria [86], hypoglycemia occurred in 25% of quinine-treated patients compared with 1% of those allocated artemether. These data confirm that hypoglycemia is a common complication of severe malaria and that quinine significantly increases its occurrence. The main result of the trial was that the two drugs were associated with similar mortality rates despite concerns of emerging resistance to quinine. This suggests that quinine will remain a first-line drug for malaria in tropical and other countries.

2) Chloroquine

There was a statistically significant ($P < 0.01$) reduction in fasting plasma glucose after chloroquine in the non-diabetic subjects. However, mean basal plasma insulin was greater after chloroquine than before even if this did not achieve statistical significance [598]. Chloroquine not only increased insulin secretion but that insulin clearance was reduced and peripheral glucose uptake accelerated [597]. Whether these three distinct effects operate in healthy travelers taking chloroquine prophylaxis or in patients with malaria during acute treatment is unknown but rare reports of serious chloroquine-associated hypoglycemia in situations including overdose [602], implies that they are of limited biological significance. Despite concerns over toxicity, chloroquine can be given parenterally with safety in severe malaria [86, 595]. However, chloroquine-resistance malaria is now widespread and the cinchona alkaloids and artemisinin derivatives are the drugs of choice in severely ill patients. Thus, one situation in which chloroquine metabolic effects might be important is now encountered rarely.

3) Mefloquine

Mefloquine is related chemically to the cinchona alkaloids. The data along with the finding where insulin secretion by rat islets of Langerhans is increased after exposure to mefloquine albeit at concentration higher than in human plasma [603], suggest that it also has the ability to close beta cell K^+ATP channels. Furthermore, total plasma mefloquine concentrations appear close to the same log-linear dose-beta cell response relationship as quinine and quinidine; though both prophylactic and treatment levels are usually less than those of the two cinchona alkaloids (<3 mg/l) [604]. However, mefloquine is at least 98% protein-bound [604] which means that enhanced beta cell function occurs at relatively low free drug concentrations. The divergent effects of free serum quinine, quinidine, and mefloquine on %B suggest that there are drug-specific differences in the nature of beta cell K^+ATP channel binding. Unfortunately, there are currently no *in vitro* data that allows an assessment of this. Principally because of poor solubility and irritant effects, mefloquine is not available as a parenteral preparation and so is not used in severe malaria. Analogous to the situation with chloroquine, this might have contributed to the apparently low rate of mefloquine-associated hypoglycemia.

4) Other Drugs

There have been no reported cases of hypoglycemia associated with the use of halofantrine, a phenanthrene methanol related chemically to mefloquine. As in the case of mefloquine, the drug cannot be given parenterally and so its clinical use is restricted largely to treatment of uncomplicated cases. However, halofantrine administration is associated with clinically significant prolongation of the electrocardiographic QT interval [605], an effect which is also seen with quinine. If beta cell K^+ATP channels effects parallel those in myocardial K^+ channels, it is possible that halofantrine also increase insulin secretion. This could contribute to or cause hypoglycemia but, as with mefloquine, this may still be an unrecognized phenomenon. Cardiotoxicity with halofantrine, especially when used in combination and other drugs influencing ventricular depolarization, could thus result from direct and indirect effects mediated through K^+ channels activity at two sites.

The artemisinin derivatives have not been associated with hypoglycemia; and the proportions of patients with hypoglycemia at study entry (8%) and during artemether treatment (11%) in published series of adult patients with severe malaria were similar. However, these compounds also prolong the QT interval in animals given high doses [606] and it is again possible that they also block pancreatic beta cell K^+ATP channels. However, these drugs and their active metabolites have relatively short half-lives [516, 607]. Any effect on insulin secretion may thus be short-lived and of limited clinical significance.

6.1.4. Safety Assessment of Antimalarial Endoperoxides

The endoperoxides are a promising class of antimalarial drugs which may meet the dual challenges posed by drug-resistant parasites and the rapid progression of malarial illness. The first generation endoperoxides include artemisinin (qinghaosu) and several semisynthetic derivatives (Figure 20A-F). Artemisinin, the prototype, is a sesquiterpene lactone. Its structure, which includes an endoperoxide bridge (C-O-O-C), is unique among antimalarial drugs. Dihydroartemisinin is the reduced lactol derivative of artemisinin, and the semisynthetic derivatives (artemether, arteether, artesunate, and artelinate) are ethers or esters of the lactol. The first-generation artemisinin drugs are being used widely in Thailand, Myanmar, Vietnam, and China, where multidrug resistant parasites are common [516]. The promise of these compounds lies in their effectiveness against chloroquine-resistant parasites and their rapid action against malaria. Moreover, they have been used for the treatment of malaria in three million patients with few reports of clinically significant toxicity [2]. The second-generation endoperoxides are synthetic compounds (Figure 20G-I). Hundreds of these have been made; many resemble artemisinin, but others (trioxanes, tetraoxanes) are relatively dissimilar. Only one of these compounds, arteflene (Figure 20G), has been taken beyond preclinical development.

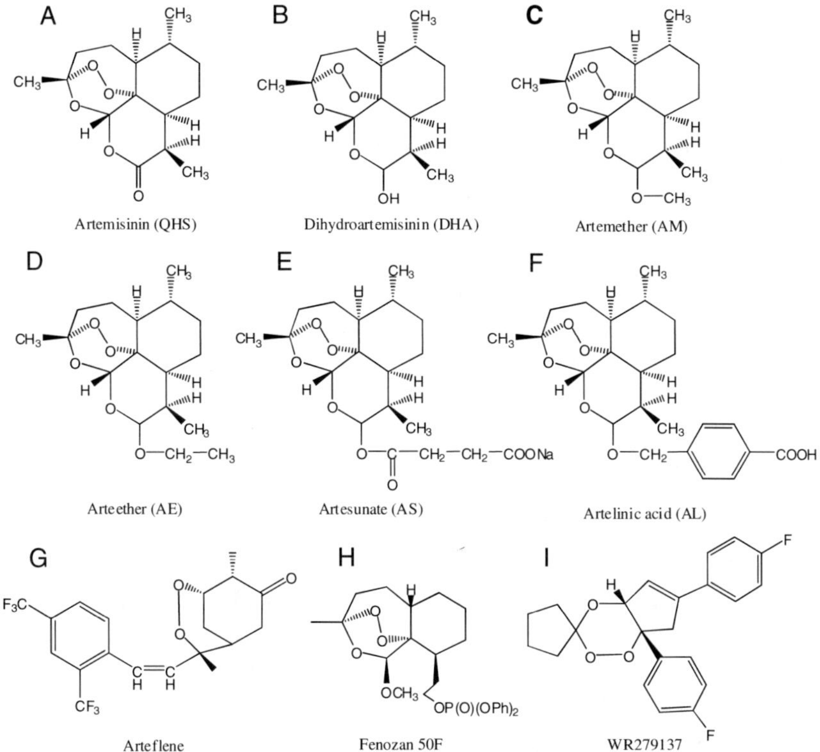

Figure 20. First-generation artemisinin derivatives including artemisinin (A), dihydroartemisinin (B), artemether (C), arteether (D), artesunate (E), and artelinate (F); and selected second-generation endoperoxides including arteflene (G), fenozan 50F (H), and WR279137 (I) [44, 390].

In general, the endoperoxides have several advantages over existing antimalarial drugs. First, there is little or no cross-resistance with other antimalarial drugs. Second, the endoperoxides clear the peripheral blood of parasites more rapidly than other available drugs do. Finally, resistance to the endoperoxides has not yet developed despite widespread clinical use [553]. However, there are also several disadvantages. The endoperoxides have short half-lives, and effective levels in plasma are sustained for only relatively brief periods. As a result, short-course treatment (less than 5 days) with the endoperoxides alone is generally associated with an unacceptably high rate (.10%) of recrudescent infections. Also, many of the drugs have poor oral bioavailability and the bioactivities of orally administered formulations are unpredictable.

Drug safety is an important consideration that has to be evaluated in concert with efficacy at the earliest possible stage in drug development. It is therefore essential to define the potential hazards associated with second or third generation peroxides that present greater and more prolonged systemic exposure and design appropriate test systems for screening for drug toxicity.

6.1.4.1. Pharmacological and Toxicological Mechanisms

In vitro studies have shown that all endoperoxide antimalarial compounds share common targets with respect to protein alkylation. It has been demonstrated that dihydroartemisinin associates with parasite membranes, hemozoin, and mitochondria; damage occurs in the mitochondria, rough endoplasmic reticulum and the nuclear envelope, and there are also changes in the nucleus, ribosomes, and food vacuole.

Antioxidants such as catalase, dithiothreitol and a–tocopherol have been shown to antagonize the antimalarial action of artemisinin. The role of iron was defined in experiments which showed that iron chelators antagonize the antimalarial activity of an artemisinin derivative, arteether [470].

Artemisinin and its derivatives are toxic to malaria parasites at nanomolar concentrations, whereas micromolar concentrations are required for toxicity to mammalian cells [608, 609]. One reason for this selectivity is the enhanced uptake of the drug by parasites. *P. falciparum*-infected erythrocytes take up and concentrate [^{3}H] dihydroartemisinin [529] and [^{14}C] artemisinin [555, 610] to more than 100-fold-higher concentrations than do uninfected erythrocytes. The uptake is rapid, reversible, and saturable and appears to be at least partially dependent on metabolic energy [529, 610], and artemisinins localize in specific parasite membranes [390].

Considerable evidence has accrued that the killing of parasites is mediated by free radicals. One piece of evidence is the observation that artemisinin derivatives lacking the endoperoxide bridge, a known source of oxygen free radicals, are devoid of antimalarial activity (13, 78). Further evidence includes observations that the *in vitro* antimalarial activities of artemisinin and artesunate are enhanced by high oxygen tension and by the addition of other free-radical-generating compounds. Most free-radical-generating drugs cause ''oxidant damage'' by producing oxygen free radicals, such as superoxide, which then cause indiscriminate damage to the cell . Artemisinin derivatives were at first thought to act in this manner (81, 112). Artesunate was shown to induce lipid peroxidation (112) in both infected and uninfected erythrocytes, as well as membrane protein thiol oxidation in isolated erythrocyte membranes (111). However, lipid peroxidation, membrane thiol oxidation, and most of the other effects could be observed only at very high drug concentrations (.100 mM), suggesting that these nonspecific oxidant reactions may not be mediating artemisinin killing of parasites.

In vitro, heme and iron catalyze the conversion of artemisinin and its derivatives into free radicals in a manner similar to the way that they catalyze the decomposition of hydrogen peroxide into free radicals . Not only does heme catalyze the breakdown of artemisinin, but also it forms a covalent complex with it. A reaction between artemisinin and heme could lead to parasite death via several different mechanisms. The artemisinin-heme adduct itself might be toxic to the parasite. However, *in vitro* data suggest that the artemisinin-heme adduct has little antimalarial activity (111). Another possibility is that free iron is released during the reaction with artemisinin, since other alkyl peroxides have been shown to degrade heme . The released iron might then be toxic to the parasite. However, free iron does not increase to detectable levels in artemisinin-treated parasites (110). Finally, the artemisinin-heme adduct might either inhibit hemozoin synthesis or cause its degradation. *In vitro* data suggest otherwise because there was no effect of artemisinin treatment on the hemozoin content of parasites in culture, even at concentrations which inhibit hypoxanthine incorporation (6).

Artemisinin, unlike other alkylating agents, does not react with DNA (184). There is a marked and reproducible reaction between artemisinin derivatives and several proteins when malaria-infected erythrocytes are exposed to artemisinin derivatives. When *P. falciparum* is incubated in the presence of a radiolabeled artemisinin derivative, as much as half of the parasite-associated radioactivity can be found in isolated hemozoin (53, 111), presumably in the form of drug-heme adducts. The association of drug-derived radioactivity with hemozoin has also been demonstrated by electron-microscopic autoradiography (104). Furthermore, mice infected with *P. berghei* RC, which does not synthesize visable hemozoin, require 50-fold more artemisinin for a cure than those infected with the N strain, which does make hemozoin (124).

The antimalarial endoperoxides appear to have a two-step mode of action. In the first step, artemisinin compounds are activated by heme or molecular iron to produce free radicals and electrophilic (alkylating) intermediates. In the second step, these reactive species react with and damage specific malaria membrane-associated proteins [611]. Thus, artemisinin derivatives are free-radical generators but are different from ''oxidant'' free-radical-generating drugs in two ways: (i) the resulting free radicals are carbon centered, and (ii) there are specific target proteins [612]. The identities of the artemisinin target proteins await elucidation.

The substantial neuropathology induced by artemether and arteether in animal models was predominantly distributed to the caudal brain stem, especially the nucleus trapezoideus and superior olive nucleus [606]. While the mechanism of neurotoxicity of the artemisinin derivatives is not fully understood, it appears related to its mechanism of antimalarial action. Cleavage of the endoperoxide bridge portion of the sesquiterpene lactone generates free-radical intermediates which cause oxidative damage to neuronal cells [390, 483, 613, 614].

6.1.4.2. Toxicities of Endoperoxide Drugs

1) Neurotoxicity

Although there is no clinical evidence for neurotoxicity in humans, high doses of artemisinin derivatives are neurotoxic *in vitro* and in experimental animals. Brewer *et al.* [606, 615] have shown that dogs administered six daily doses of arteether at 20 mg/kg/day or

28 daily doses of 15 mg/kg/day develop neurological symptoms, neuropathological findings, and electrocardiographic abnormalities. Similar neuropathological changes were obtained in rats by administering either arteether or artemether at doses as low as 25 mg/kg/day for 14 days. In both dogs and rats, lesions were most pronounced in the brain stem. Histopathologically, neuronal cells had undergone degeneration, showing a loss of *Nissl* substance and the presence of nuclear pyknosis. These results have recently been confirmed in studies of rats by Kamchonwongpaisan *et al.* [616], although no neuropathology could be detected at total doses less than 250 mg/kg.

Artemisinin derivatives are also toxic to neuronal cells *in vitro*. Wesche *et al.* demonstrated that a series of artemisinin derivatives inhibited protein synthesis by two neuroblastoma cell lines and by primary fetal rat neuronal cells [617]. Artesunate and dihydroartemisinin were the most toxic derivatives, with IC_{50}s of < 1 μM. Fishwick *et al.* also observed toxicity to neuroblastoma cells *in vitro* and found toxicity to glioma cells *in vitro* as well [618]. Surprisingly, these authors found artemether to be more toxic than dihydroartemisinin in inhibiting neuroblastoma cell proliferation. Several artemisinin derivatives significantly inhibited neurite formation in neuroblastoma cells at 100 nM. The toxicity of artemisinin derivatives to neuronal cells may also be iron dependent, since heme potentiates the toxicity of artemisinin derivatives to neuroblastoma cells *in vitro* [619]. An iron chelator also inhibits the acute toxicity of artemisinin to mice [614], although no effect of this chelator could be demonstrated on arteether-induced neuropathology in rats [616]. The toxic effect of artemisinin derivatives on neurons appears to involve protein alkylation [616]. Thus, the mechanism of neurotoxicity is similar in many respects to the mechanism of antimalarial action.

The recent reports of a prolonged length of time to recovery from coma (when compared with patients treated with quinine) in Vietnamese [86] and Gambian [85] patients, and an increased frequency of seizures in artemether-treated patients in Gambia [85], have raised concerns of the neurotoxic potential of these compounds. In addition, there has recently been one report of the occurrence of an acute cerebellar syndrome in a 36-year-old previously healthy man after five doses of artesunate, the features of which were still present (although improved) 4 months later [620]. Clearly, one such report does not constitute proof of a causal association [621], but the possibility that repeated administration of artemisinin compounds in malaria endemic areas will result in cumulative damage cannot be ruled out [483, 516], especially since these drugs are often freely available for use in tropical countries. Another possibility that cannot be discounted at this stage is that some patients may have suffered subclinical neurological toxicity which would have required sophisticated neurological and neuropsycho-logical tests for detection. The potential of the drugs to cause neurotoxicity will be discussed later in greater detail.

In summary, high doses of the artemisinins appear to have specific neurotoxic effects both *in vitro* and *in vivo*. However, since the drug has been used safely in millions of people, this neurotoxic effect is unlikely to be clinically relevant or drug exposure time related to [44, 524, 622].

2) Cardiovascular Toxicity

The question of drug toxicity, particularly with chronic or repeated use, becomes very important once a drug is released and is widely available. The safety of this class of drugs has been in question because the early preclinical and clinical studies did not meet the stringent requirements of Western and international drug-regulating agencies. Over two million patients are estimated to have been treated with artemisinin or its derivatives, and no adverse effects have been noted [553]: this admittedly anecdotal record is corroborated by the clinical trials conducted thus far in which toxicity has been limited to sporadic reports of mild, transient cardiac dysrhythmias [623-625].

Given that cardiotoxicity is a potential worry with many of the antimalarial drugs, and QT prolongation has been reported in animals treated with artemether [516], many of the clinical studies have actively monitored for potential cardiotoxicity. To date, most of the abnormalities detected have been minor and of little clinical significance. For example, transient first-degree heart block was noted in 1 of 82 patients receiving artesunate, and 3 of 39 patients receiving artemether [516]. In a recent study, it was demonstrated that 25% of 284 adults treated with intramuscular artemether for severe *falciparum* malaria had a QTc interval of more than 0.5 s [86]. However, it is possible that this was caused by the disease rather than the treatment [626].

3) Other Toxicities

Despite their extensive use, there have been very few reports of clinically significant toxicity with the peroxide antimalarials. Minor gastro-intestinal adverse effects such as nausea, diarrhea and abdominal pain have been reported in some patients; tenesmus occurs in 6% of patients administered artemisinin suppositories. Transient increases in serum transaminases, and decreases in the reticulocyte and neutrophil counts have also been encountered, but these have been of little clinical significance [516]. There have also been isolated reports of rash but a causal association was unlikely [466]. Interestingly, 25% of volunteers receiving artemisinin had an episode of drug-induced fever [516, 553] in some studies, though the significance of this is unclear.

Transient dose-related reductions in reticulocyte counts were noted in some early preclinical toxicology studies but have not been observed in human clinical trials. Fetal resorption was observed in some animal toxicology studies, so these drugs should be used during pregnancy only as a last resort [2]. There is little data on the cumulative toxicity of the artemisinin in humans, but given the neurotoxicity reported in animal models and *in vitro*, it is important to obtain such data. However, because oral-dosage forms are already in wide use, collecting this information will be challenging. Thailand has established a system for post marketing surveillance of the safety and efficacy of artesunate and artemether, and this may serve as a model for other countries in which these drugs are sanctioned for widespread use [390].

Fear of potential toxicity has limited the use of the artemisinin derivatives in pregnancy since animal studies have documented teratogenicity when the drugs are administered at high doses [307]. At present documentation of the safety of artemisinin derivatives in the treatment of malaria in pregnant women is limited and it has therefore been recommended that these drugs should be used in pregnancy only as a last resort [2]. Thus, there is a need to develop better peroxide compounds, the ideal properties of which would be (i) a pharmacological profile that allows for shorter durations of treatment with fewer doses, (ii) greater bioavailability to ensure sufficient parasite exposure, (iii) a predictable pharmacokinetic profile such that the drug is retained for sufficient lengths of time to ensure that it kills the parasites, but does not accumulate to any extent to cause damage to the host, (vi) low interindividual pharmacokinetic variability, and (v) lack of toxicity. Rational molecular design must be based on both structure-activity and structure-toxicity/metabolism relationships.

6.1.4.3. Operational Issues of Endoperoxide Drugs

The artemisinin drugs are being produced and used widely throughout Southeast Asia, and their use is increasingly promoted in Africa [553], but they cannot be manufactured or distributed by European or North American pharmaceutical companies until the regulatory requirements of the respective countries have been met. Despite having followed an unorthodox path of drug development, these drugs are already being used widely, particularly in parts of the world where multidrug-resistant *P. falciparum* is common. Expectations of toxicity based on findings in animal models have not been borne out in the clinical experience with these drugs. However, there is little experience with repetitive or chronic use, and as these drugs begin to penetrate areas where malaria is endemic it would be advisable to establish systems to detect serious adverse effects. The first-generation endoperoxides have their limitations, though, such as high recrudescence rates, poor oral bioavailability, and short half-lives. Second-generation endoperoxides may improve these features and, perhaps, be even less expensive.

The use of the artemisinin drugs in areas where multidrug resistant malaria is rare or absent is not recommended. If data from the large clinical trials recently completed suggest an advantage, this recommendation may be modified. At this time, however, there are effective oral drugs which are less expensive than the artemisinins, and quinine remains an effective treatment for severe disease. The availability of suppository formulations of the artemisinin allows home-based treatment of patients who are too ill to take antimalarial drugs by mouth. This might confer a significant advantage, because treatment could be started earlier, thus avoiding the delays incurred during travel to a health clinic. However, conclusive evidence for the effectiveness of these formulations in this situation has not been obtained.

The development of second-generation endoperoxides should be aided by our knowledge of their mechanism of action. The development of these new derivatives, however, is likely to be hindered more by economic and political issues than by scientific ones. How should the new derivatives be evaluated? Clinical measures of treatment efficacy should be determined in the context of the disease being treated; parasitological cure is the goal in patients with uncomplicated malaria, and survival is the important end point in patients with severe and complicated disease. Combinations of an artemisinin drug with another antimalarial agent are potentially more effective in patients with uncomplicated malaria because they are curative and can be administered over shorter periods. The duration of treatment is a less important consideration for patients with severe and complicated disease; in these patients, a rapidly effective, life-saving drug is required and radical cures can be accomplished at a relatively leisurely pace in the survivors.

In summary, the artemisinin derivatives are a highly promising new group of antimalarial agents. Second-generation derivatives may become widely used in the future. First-generation compounds are currently indicated in parts of the world where resistance to other antimalarial drugs is common. Premature and unregulated use of the artemisinins, especially oral formulations, should be avoided, because it might hasten the development of resistance.

6.2. Cardiotoxicity of Antimalarial Drugs

Chloroquine, artemisinin, sulphadoxine-pyrimethamine, tetracyclines, and primaquine can be safely used in these patients. Quinine can also be used carefully. Mefloquine and halofantrine are better avoided in patients with known cardiac illness. Oral chloroquine is safe in therapeutic or prophylactic doses. If the therapeutic dose or high dose is administered too rapidly by parenteral route, it can cause significant cardiotoxicity.

There are consistent differences in cardiovascular state between acute illness in malaria and recovery that prolong the electrocardiographic QT interval and have been misinterpreted as resulting from antimalarial cardiotoxicity. Of the different classes of antimalarial drugs, only the quinolines, and structurally related antimalarial drugs, have clinically significant cardiovascular effects. Drugs in this class can exacerbate malaria-associated orthostatic hypotension and several have been shown to delay ventricular depolarization slightly (class 1c effect), resulting in widening of the QRS complex, but only quinidine and halofantrine have clinically significant effects on ventricular repolarization (class 3 effect). Both drugs cause potentially dangerous QT prolongation, and halofantrine has been associated with sudden death. The parenteral quinoline formulations (chloroquine, quinine, and quinidine) are predictably hypotensive when injected rapidly, and cardiovascular collapse can occur with self-poisoning. Transiently hypotensive plasma concentrations of chloroquine can occur when doses of 5 mg base/kg or more are given by intramuscular or subcutaneous injection. At currently recommended doses, other antimalarial drugs do not have clinically significant cardiac effects.

6.2.1. Cardiovascular Toxicity

Antimalarial drugs are prescribed and self-medicated on a vast scale in the tropical areas of the world. In terms of human exposure, the 4-aminoquinoline chloroquine has been arguably the most widely used drug ever because of its long terminal elimination half-life (1–2 months). The first antimalarial drugs to be introduced into western medicine were the cinchona alkaloids, originally as bark or bark extracts containing a mixture of several different compounds. One of these alkaloids was quinidine - the prototype for agents causing prolongation of the electrocardiographic QT interval (often called the "quinidine effect," and the hallmark of class 3 antiarrhythmic drugs). Quinidine was used mainly as an anti-arrhythmic, and not as an antimalarial drug throughout most of the 20th century. It was also well known on occasions to cause tachyarrhythmias (quinidine syncope) and has recently been replaced by less toxic alternatives [627].

The lethality of chloroquine in overdose, the quinidine effect, and the belated discovery that halofantrine causes marked QT prolongation and sudden death, well after its registration by several regulatory authorities, have focused attention on the potential cardiotoxicity of the antimalarial drugs. Regulatory authorities and drug developers have become particularly concerned about QT prolongation. Several of the quinoline drugs and other compounds related to chloroquine, quinidine, and halofantrine are potentially hypotensive, and have small electrocardiographic effects, but have not been associated with significant cardiotoxicity in clinical practice. One should emphasize that their safety in patients with pre-existing cardiovascular disease, receiving other cardioactive drugs, has not been established because nearly all patients with malaria are young with otherwise normal hearts. Such patients in endemic areas form the bulk of the evidence base for antimalarial safety.

Chloroquine has three main cardiovascular effects: membrane stabilization, direct negative inotropic effects, and direct arterial vasodilation. The data also suggest a role for nitric oxide and histamine release in mediating this response leading to hypotension/postural hypotension. These effects are manifested as rhythm and conductance disturbances, myocardiopathy, or vasoplegic shocks. Quinine and halofantrine are capable of prolonging the QT interval. Quinine prolongs the QT interval at standard doses, similar to halofantrine. Halofantrine induces a dose-related prolongation of the QT interval whereas mefloquine has no effect on the QT interval.

However, the risk of significant QT prolongation was greater if halofantrine was given as a re-treatment following mefloquine failure than as primary treatment. Cardiotoxicity of antimalarials is increased in patients with acute renal failure, especially after 3 days of treatment. This is partly because the degree of QT prolongation is dependent on the plasma concentration of halofantrine. The frequency of QT interval prolongations following artemether-lumefantrine treatment was similar to or lower than that observed with chloroquine, mefloquine, or artesunate + mefloquine; these changes were considerably less frequent than with quinine or halofantrine [628, 629].

6.2.2. QT Prolongation with Drug Cardiotoxicity

Clinical assessment of drug-induced QT interval prolongation is crucially dependent on the quality of electrocardiographic data and the appropriateness of electrocardiographic analyses. To further evaluate the scope and mechanism of potential cardiotoxicity associated with the antimalarial drug, a significant number of adverse events related to the cardiovascular system were reported, including QT interval prolongation, life-threatening arrhythmias, and sudden death. Results indicate that some antimalarial drugs were able to prolong the QT interval that was primary measurement in evaluation of drug cardiotoxicity.

Prolongation of the electrocardiograph QT interval reflects prolongation of ventricular repolarization and thus the effective refractory period. Many factors affect the duration of ventricular repolarization. The QT interval varies from beat to beat, in daytime compared with night time, and from day to day. It is affected by age, sex, autonomic tone, myocardial ischemia, electrolyte concentrations, and importantly by several different classes of drugs. This pharmacological property is used to prevent ventricular arrhythmias, but it could also cause them. Heterogeneous prolongation of ventricular repolarization predisposes to intraventricular circuits of depolarization that manifest as potentially lethal polymorphic malignant ventricular tachyarrhythmias (torsades de pointes; TdP). Repolarization of cardiac ventricular myocytes is mainly caused by outward currents of potassium ions. One of the most important outward currents is the delayed rectifier current of potassium ions, IK, which has rapidly and slowly activating components (IKr and IKs, respectively). Activation of IKr leads to initiation of repolarization of the cardiac action potential. The human ether-related gene (hERG) on chromosome *7 q35-36* encodes the preforming subunits of the hERG channel, which carries the rapidly activating delayed rectifier current of potassium ions through IKr.9,10. Impaired IKr function leads to intracellular accumulation of potassium ions and consequent delay in ventricular repolarization (reflected electrocardiographically as prolongation of the QT interval).

Mutation of the hERG gene is a common cause of the inherited long QT syndrome, a disorder of cardiac repolarization that predisposes affected individuals to TdP and sudden death. hERG has been identified as the target for drugs (e.g., quinidine and other class 3 antiarrhythmics, phenothiazines, tricyclic antidepressants, some antihistamines, some antiemetics, macrolides, ketolides, fluoroquinolones, azole antifungals, pentamidine, etc.) that prolong the QT interval and can cause TdP. In general, drugs bind and then inhibit the channel only when the voltage-gated channels are open. Several unrelated drug classes are associated with QT prolongation and, especially in individuals with a hereditary long QT interval, can cause ventricular tachycardia, particularly TdP, and sudden death. Although many drugs predispose to TdP, in documented cases usually more than one predisposing factor is present—e.g., electrolyte abnormalities (hypokalemia, hypomagnesemia), myocardial ischemia, female sex, and inherited long QT syndrome, or there is a drug interaction (either a metabolic interaction or a second drug also causing QT prolongation).

6.2.3. Pre-existing Cardiovascular Disease

Several of the quinoline antimalarial drugs are vasodilators and affect myocardial depolarization, but only quinidine and halofantrine produce clinically significant prolongation of ventricular repolarization. Recovery from malaria is associated with significant lengthening of the electrocardiograph QT interval that has sometimes mistakenly been ascribed to the antimalarial drug treatment. In assessing iatrogenic effects of antimalarial drugs on the heart, one must also consider the underlying disease effects. Patients with severe valvular obstruction, compromised ventricular function and other conditions of cardiac decompensation may land in trouble on contracting malaria. In severe *falciparum* malaria, the myocardial function is remarkably maintained and most patients have an elevated cardiac index, with low systemic vascular resistance and low to normal right and left-sided filling pressures. However, in patients with decompensated heart, the high grade fever, tachycardia, hypoxemia, metabolic acidosis, etc., associated with malaria may add to the existing cardiac decompensation.

Orthostatic hypotension is common in febrile illnesses such as malaria, and can be exacerbated by quinoline antimalarials. Autonomic postural responses are blunted [630]. There are consistent differences in sympathetic tone when comparing the patient's acute admission to hospital with malaria, associated with arousal, stress, discomfort, anxiety, and usually fasting, and the recovery phase when the patient is relaxed, comfortable, supine in bed, and has often resumed eating. Stress, anxiety, discomfort, and the consequent increased sympathetic tone increases heart rate, and accelerate conduction and repolarization [631, 632] reflected electrocardiographically as QT shortening. The potentially arrhythmogenic asynchrony of ventricular repolarization is also increased, reflected electrocardiographically as QT dispersion [633]; although the relation between dispersion and arrhythmogenic potential is weak [634]. Thus, as the patient recovers from malaria, there will be a consistent reduction in heart rate and lengthening of the QT interval as a result of decreased autonomic (mainly sympathetic) tone which is independent of antimalarial treatment. Yet this effect has often been attributed to the direct cardiac effects of the antimalarial drugs.

Malaria is an acute illness associated with non-specific fever, sinus tachycardia, and a reduced systemic vascular resistance, largely attributable to release of proinflammatory cytokines. Fever increases the heart rate; for each 1°C increase in core temperature heart rate increases by 8.5 bpm. Small changes in electrocardiographic intervals have been noted when patients have received antimalarial treatments (e.g., sulfadoxine-pyrimethamine) which are very unlikely to affect the heart [65, 635]. Indeed, nearly all studies of antimalarial drugs report some prolongation of the QT interval in the days after the start of treatment. This observation suggests that malaria (or more precisely the difference between acute illness and recovery) can have small but significant differential effects on myocardial electrophysiology.

There are also methodological concerns in many studies; manual reading of electrocardiographic intervals at a paper speed of 25 mm/s is associated with substantial measurement errors. In addition to the paper speed, the leads chosen and the technique (especially for the definition of the end of the T wave) also affect the result. Automated methods have become more popular in recent years, but variance, especially for QT measurement, may be greater with these [636]. Finally, there is a tendency, particularly in regulatory studies, to record serial electrocardiographs and report the greatest deviation (e.g., the greatest prolongation of QT interval found among several recordings) from the baseline value [637]. Such reporting creates a random sampling error in that the more samples that are taken, the greater the maximum value is likely to be, on top of the systematic errors in comparing acute with convalescent values. The more the imprecision of the measurement, the greater will be the deviation.

6.2.4. Antimalarial Drugs Associated with Cardiotoxicity

In general, during the cardiotoxicity, hypotension, vasodilation, myocardial suppression, ECG abnormalities and cardiac arrest can occur. Treatment includes mechanical ventilation, adrenaline and diazepam. Concomitant use of chloroquine with amiodarone should be avoided. At therapeutic doses, quinine is relatively safe. Rapid intravenous administration may cause hypotension. Acute over dosage can cause fatal dysrhythmias such as sinus arrest, junctional rhythms, A-V block, ventricular tachycardia, and fibrillation. Quinine may delay the absorption and elevate the plasma levels of cardiac glycosides like digoxin. Quinine should not be used concomitantly with amiodarone. Concomitant use with astemizole and terfenadine can also increase the risk of ventricular arrhythmias. Halofantrine prolongs QT interval in a concentration dependent manner and it can result in ventricular arrhythmias and even death. It is therefore contra-indicated in patients with prolonged QT interval and with drugs known to cause prolongation of QT interval. Mefloquine should be used with extreme caution in patients suffering from cardiac conduction diseases. Mefloquine has been shown to cause asymptomatic sinus bradycardia and other conduction abnormalities; e.g., prolongation of the QT interval. Patients who are on either a beta-blocker or calcium channel blocker are at particular risk if there are signs of sinus bradycardia and/or atrioventricular block. Cardiac arrest has been reported in a patient receiving a single prophylactic dose of mefloquine while concomitantly taking propranolol. There appears to be no interaction between ACE inhibitors and mefloquine.

6.2.4.1. Quinine and Quinidine

Quinine is the levorotatory diastereomer of quinidine, but differs from it in many pharmacological respects. In terms of free drug concentration, quinine is about three to four times less active as an antimalarial drug [638]. Both drugs have alpha-blocking activity that can cause hypotension. This effect has been particularly studied for quinidine, which blocks the α_1ARs receptor subtype [639]. Quinine and quinidine both slow the rapid upstroke of the cardiac action potential by blocking the inward sodium current (I_{Na}), slowing depolarization, and thereby widening the QRS complex. This is the action of class 1 antiarrhythmic drugs. The cinchona alkaloids and related quinolines show little frequency dependence in this action, and so are classified as having class 1c activity. In common with other class 1 antiarrhythmic drugs, they are both negatively inotropic, although inhibition of I_{Na} does not correlate with the magnitude of negative inotropism. Both quinine and quinidine inhibit several different potassium channels [640].

QT prolongation with quinine is about four times less than with quinidine [641] and results mainly from prolongation of the QRS interval with little JTc prolongation. Quinine and quinidine are approximately equipotent in blocking the L-type calcium channel, and both cause weak vagal inhibition. No significant cardiotoxicity has been reported in large prospective studies of quinine both in uncomplicated and severe *falciparum* malaria, with total plasma concentrations up to 20 µg/ml. By contrast, at comparable concentrations of free drug quinidine is predictably hypotensive and causes marked QT prolongation. Although arrhythmias occur very rarely in severe malaria, there is no convincing evidence that the few that do occur result from quinine. QT prolongation by more than 25% occurs in less than 10% of patients receiving high dose intravenous quinine. Slight widening of QRS interval is usual, and is more pronounced in young children (mean 17%). Quinidine, by contrast, commonly causes hypotension when given by rate controlled intravenous infusion and predictably prolongs the QT interval with marked JT prolongation. Quinidine is therefore potentially dangerous, and needs careful monitoring when given intravenously. Hypotension should be treated with saline or volume expanders [75, 627, 642].

6.2.4.2. Halofantrine

Halofantrine is metabolized *in vivo* to desbutyl halofantrine, which is also biologically active. Halofantrine induces marked QT prolongation at therapeutic concentrations [643], and is associated with sudden death, presumably from malignant ventricular tachyarrhythmias [605, 644]. Halofantrine also prolongs the PR interval and can cause transient heart block at higher plasma concentrations [605]. The absorption of this lipophilic antimalarial drug is very variable and depends on co-administration with fats. Thus, the plasma concentrations vary considerably between individuals. Because cardiotoxicity is concentration dependent [605, 644], the cardiotoxic potential of halofantrine in an individual is unpredictable. Halofantrine and desbutyl halofantrine are both very potent inhibitors of the hERG potassium channel and readily produce TdP in experimental models [645]. Given that there are now several equally effective and much safer alternatives, halofantrine should not be used as an antimalarial drug.

6.2.4.3. Other seldom associated with Cardiotoxicity

1) Mefloquine

There is no convincing evidence for significant cardiotoxicity following mefloquine administration [598], despite initial reports that suggested mefloquine might cause significant bradycardia and hypotension, and very occasional case reports of cardiac dysrhythmias in people with pre-existing heart disease [646]. Sinus bradycardia does occur in some recipients of mefloquine, but is very seldom severe. Gap junctions are aggregates of cell–cell channels composed of connexins and these mediate direct intercellular diffusion of cytoplasmic ions, small metabolites, and signaling molecules. In the mouse heart the cell to cell conductance through one connexin (of the 20 identified) that is particularly abundant in the sinoatrial and atrioventricular nodes is blocked by mefloquine (IC_{50} 10 µmol/L), which could be relevant to the development of bradycardia [647].

Clinical and electrocardiographic studies have suggested either no or slight prolongation of the QT interval. Apart from mild sinus bradycardia there is no convincing evidence that mefloquine causes arrhythmias in the treatment of malaria [643]. Interaction studies with quinine do not suggest synergistic toxicity [648], although in one small volunteer study there was greater QTc prolongation with the two drugs combined than when given individually [649]. These data are generally reassuring with the caveat that, with halofantrine, there is a potentially dangerous interaction. Further interaction studies with other drugs that block the hERG channel are needed.

2) Chloroquine

Chloroquine rarely causes conduction disturbances and cardiomyopathy in chronic use in rheumatic diseases. Chloroquine in overdose (as in self-poisoning or when given by rapid intravenous injection) is certainly cardiotoxic and potentially lethal [650]. In some countries, chloroquine is an important cause of death from selfpoisoning.

Hypotension is common in self-poisoning; both tachycardias and bradycardia with atrioventricular block can occur, and there is consistent intraventricular conduction delay. Cardiovascular toxicity, and in particular hypotension, was the probable cause of sudden death that sometimes followed parenteral chloroquine administration for the treatment of malaria in children, and led WHO to recommend that its parenteral use should be discontinued in 1984 [2].

3) Lumefantrine

By contrast with halofantrine, lumefantrine is a very weak antagonist of the hERG cardiac potassium channel [644]. The drug has been assessed extensively and does not produce significant adverse cardiac effects *in vivo*, and has no significant effects on electrocardiograph [629]. In an overview of different trials, 6% of 291 children receiving the six-dose regimen of artemether-lumefantrine had a QT prolongation of longer than 60 ms (Fridericia's correction) [637]. In a large prospective assessment in Thailand there was no correlation between plasma concentrations of lumefantrine, over a 1000-fold range, and changes in electrocardiographic indices [651]. These studies indicate that lumefantrine has no significant cardiovascular toxicity.

4) Primaquine

The cardiovascular activity of the 8-aminoquinoline primaquine has not been studied extensively. Primaquine blocks the inward sodium current I_{Na}, slowing the upstroke of the action potential [652], so like many of the quinolines, it does have class 1 activity. Limited available evidence does not suggest significant cardiovascular toxicity.

5) Pyrimethamine-sulfadoxine

There is no evidence either from human or experimental animal studies that pyrimethamine, antimalarial biguanides, sulphones, or sulphonamides are directly cardiotoxic [653].

Atovaquone-proguanil

There is no evidence from clinical or laboratory studies that atovaquone, proguanil, or its biologically active metabolite cycloguanil, have significant cardiovascular toxicity in therapeutic use [654].

7) Artemisinin and Derivatives

These potent antimalarial drugs are remarkably well tolerated in therapeutic use. Artemisinin has now largely given way to the derivatives of the more potent dihydroartemisinin, artesunate, artemether, and artemotil. These are metabolized *in vivo* back to dihydroartemisinin. In clinical trials there have been no adverse cardiovascular effects noted in thousands of severe malaria patients and tens of thousands of uncomplicated malaria patients treated with these drugs. Where neurotoxicity of arteether and artemether was shown in beagle dogs and rats, prolongation of the QT interval was also noted [606, 615]. However, because neurotoxicity was coincident with cardiotoxicity, it was not clear whether these effects represented direct cardiotoxicity, or were an indirect result of central nervous system toxicity. Neurotoxicity has not been found in human beings. Nonetheless, these findings, and confusion between disease and possible drug-related effects, have left uncertainty over the cardiotoxic potential of this class of drugs.

There are no reported electrophysiological data on artesunate, artemether, or dihydroartemisinin. In a large randomized comparison of high-dose artemether and quinine in adults with severe malaria, serial electrocardiographs were recorded in 301 patients; slight QT prolongation was noted in both groups, 11 of 152 (7%) artemether recipients and 12 of 133 (9%) quinine recipients had an increase of more than 25% [86], but the contributions of drug and disease cannot be determined. In another comparative trial one of 31 patients with severe malaria treated with artemether had transient right bundle branch block, six had non-specific T-wave changes and two had QT prolongation (both had sinus bradycardia) [655]. There were no dysrhythmias or adverse cardiovascular effects in either study. However, when artemether is given orally to healthy individuals, despite plasma concentrations of artemether and dihydroartemisinin that were up to ten times higher than followed intramuscular administration, no electrocardiographic changes were seen [627]. Rapid intravenous administration of artesunate is not associated with cardiovascular effects despite transiently high plasma concentrations of artesunate and dihydroartemisinin. These concentrations are up to two orders of magnitude higher than with intramuscular artemether [87].

6.3. Neurotoxicity

Central Nervous System (CNS) adverse drug events are dramatic, and case reports have influenced clinical opinion on the use of antimalarials. Malaria also causes CNS symptoms, thus establishing causality is difficult. CNS events are associated with the quinoline and artemisinin derivatives. Chloroquine, once considered too toxic for humans, has been the antimalarial of choice for 40 years. While a range of serious CNS effects have been documented during chloroquine therapy, the incidence is unclear (extrapyramidal symptoms occur with an incidence of 1 in 5000). Amodiaquine has a higher incidence of mild CNS effects than chloroquine. Mefloquine therapy causes dose-related transient dizziness. Serious CNS events during mefloquine therapy occur in 1:1200 Asians and 1:200 Caucasians/Africans.

Risk factors include dosage, concomitant drug use/interactions, previous history of a CNS event and disease severity. Retreatment (within a month) increases the risk in Asians 7-fold. Studies indicate that the frequency of serious CNS events with mefloquine prophylaxis (1:10,000) is similar to that with chloroquine (1:13,600). Quinine causes cinchonism at standard therapeutic doses. High-tone hearing loss occurs, but irreversible auditory or ocular effects are very rare. The artemisinin derivatives are associated with dose-dependent brain lesions in rodent, canine, and nonhuman primates. At low doses, histological injury has been demonstrated, without clinical neurological signs. No significant toxicity has been reported in humans. Other antimalarial drugs are seldom associated with CNS adverse events. Data do not suggest a need to diminish the correct use of the quinoline derivatives. Irreversible effects are extremely rare and usually associated with overdosing or prior history of a serious CNS event. Concomitant therapeutic use of 2 drugs from the same family, or retreatment with the same drug should be avoided. Onset of drug-associated serious CNS events requires drug discontinuation and future avoidance of the drug [656].

6.3.1. Neurotoxicity Definitions and Risk Factors

Adverse reactions refer to disorders judged to be caused by the suspect drug. The section examines adverse events, that include all events reported in association with the use of an antimalarial, regardless of causality rating. These were classified as 'serious' if they constituted an apparent threat to life, required (or prolonged) hospitalization or resulted in severe disability. CNS adverse events and defined according to the Council of International Organizations of Medical Sciences [1] criteria. They are classified into 10 major categories according to the standardized dictionary of adverse drug reactions. These are:

1. major psychiatric disorders and symptoms
2. disorders of affect
3. neurosis
4. other psychiatric symptoms
5. seizures
6. disturbances in level of consciousness
7. dizziness/vertigo/ataxia
8. neuropathies and sensory disturbances

9. headache
10. other neurological disorders

Seizures were reported in almost 40% of children admitted with malaria compared with 12% of children with other acute medical illnesses. The proportion of admitted children with neurological involvement (in particular, seizures) increased until 1997. Possible reasons for this include an increasing awareness of the problem, patient selection, and a growing confidence of the community in the ability of the hospital to manage seizures. Seizures were associated with high parasitemia but not fever or hyponatremia, supporting the hypothesis that *falciparum* malaria might be epileptogenic *per se*. The presence of additional neurological features supports the hypothesis of direct cerebral involvement. However, it is also likely that some seizures in malaria are simple febrile seizures similar to those seen in non-malarial endemic areas. It is also notable that not all children with cerebral malaria reported seizures and those without seizures but with coma had a worse outcome, suggesting there are other mechanisms by which coma might develop.

The following factors may contribute to the pathogenesis of neurological involvement in childhood malaria: biochemical perturbations (hypoglycemia, acidosis), impaired perfusion, high parasitemia, and some children may have a predisposition to seizures (higher frequency of history of seizures). Although peripheral parasitemia correlates poorly with vascular sequestration, high parasite density does predict poor outcome in cerebral malaria. This may be due to systemic derangements in immunologic, metabolic, and cardio-respiratory functions. These changes may be associated with altered consciousness and may precipitate or lower the threshold for seizures. We propose that neurological involvement in *falciparum* malaria may arise from (1) a direct effect of sequestered parasites possibly through parasite-induced toxins or immune responses to sequestered parasites on neuronal/blood-brain barrier function or mechanical vascular blockage or (2) an indirect effect from parasite-induced local and systemic metabolic derangements or impaired perfusion (impaired delivery of substrates). Genetic susceptibility to seizures, including febrile seizures, may be important. Acute and long-term imaging studies, especially magnetic resonance imaging, will greatly assist in determining pathogenesis [657].

In the evaluation of the risk factor, it is unclear why a higher incidence of CNS events has been reported in Caucasian and Africans (1 in 100 to 1 in 200) compared with the Asian ethnic groups (1 in 1200) studied. Since under-reporting was unlikely in the latter groups, other factors need to be examined, including social and genetic factors [656].

6.3.2. Antimalarials often Associated with Neurotoxicity

Neurotoxic events are often associated with the quinoline, mefloquine and artemisinin derivatives. Chloroquine, once considered too toxic for humans, has been the antimalarial of choice for 40 years. While a range of serious CNS effects have been documented during chloroquine therapy, the incidence is unclear (extrapyramidal symptoms occur with an incidence of 1 in 5000). Amodiaquine has a higher incidence of mild CNS effects than chloroquine. Mefloquine therapy causes dose-related transient dizziness. Serious CNS events during mefloquine therapy occur in 1:1200 Asians and 1:200 Caucasians/Africans. However, no significant neurotoxicity of artemisinin and its derivatives was found in humans [44, 524].

6.3.2.1. Chloroquine

Chloroquine has a small therapeutic window and is extremely toxic at high doses, with rapid onset of neurological, respiratory, and cardiovascular effects, and a mortality rate of 35% in overdose. Neurological symptoms begin with dizziness, vomiting, and headache, and progress rapidly to CNS depression and visual disturbance. Hyperexcitability, restlessness and seizures precede circulatory arrest. Historically, when chloroquine was first discovered by Bayer in Germany in 1934, it was considered too toxic for therapeutic use in humans on the basis of studies in bird malaria. The cutting of quinine supplies necessitated its wartime use and it subsequently became the drug of choice in the treatment of malaria. In the treatment of chronic rheumatoid disease at dosage of 250 mg/day, neurological effects, including headache, neuropathies and sensory disorders, and psychiatric symptoms such as confusion, apathy and anxiety, were identified in 15 of 100 patients monitored. While the frequency of adverse effects was not correlated with cumulative dose, there was a relationship with serum chloroquine concentrations. During the past 40 years, chloroquine has seldom been associated with serious neuropsychiatric effects when used as an antimalarial. Only recently has its safety, in relation of CNS effects, been contested. CNS effects resulting from standard dosage used for malaria treatment (25 mg/kg) and prophylaxis (300mg weekly or 100mg daily) are reviewed below [656].

A range of neurological and psychiatric events have been reported with therapeutic use of chloroquine for malaria. Psychotic episodes were reported in 10 young adults, 6 of whom were administered chloroquine at standard therapeutic dose (1500mg total dose), and 4 at doses of 1800 to 2000mg [658]. The onset of events occurred between the third and tenth day after the start of chloroquine therapy, and resolved within a week of drug discontinuation. In this small series, age and sex were not predisposing factors and no patients had personal or family histories of mental illness. The frequency of these events was not calculated. During a 15-month period, toxic psychosis was reported in 4 out of 73 young Canadians treated for malaria while working in Nigeria. None had a psychiatric history; all worked in separate locations and had received standard therapeutic doses (25 mg/kg) [659]. Two experienced recurrences several months later, without further chloroquine exposure. Four additional less severe psychiatric reactions were retrospectively detected in the 69 other volunteers in the country. Reasons for the outbreak were not elucidated.

Neurological disorders have been reported in 13 young adults and 5 children. Notably, 2 to 96 hours after chloroquine administration for malaria therapy. Two of the patients also exhibited major psychiatric symptoms. Chloroquine-induced seizures have not been reported following the standard 25 mg/kg dose. Four patients developed seizures when administered chloroquine, between 500 and 1000 mg/day for 14 days, for treatment of amoebiasis. Complaints of increasing headache, nausea, blurred vision and dizziness in the first 3 to 7 days of treatment preceded seizures occurring on days 3, 5, 6, and 12 after starting chloroquine therapy. Therapy was in combination with paromomycin or hydroxyquinoline known to cause a clinical neurological syndrome in approximately 1% of patients taking 750 to 1500mg daily for less than 2 weeks, with toxicity evident within 24 hours at higher dosages.

Tonic-clonic seizures have been described in 4 patients exposed to prophylactic dosage of chloroquine. All patients had good responses to standard antiepileptic drugs, and none experienced further episodes. Three more incidents of seizures were possibly related to chloroquine prophylaxis have been reported. Two other persons in this prospectively monitored population had psychotic episodes. The frequency of serous CNS adverse events associated with chloroquine prophylaxis was estimated to be 1 in 13,600. The estimated frequency of non-serious effects such as headache, dizziness, insomnia and depression during chloroquine prophylaxis were 6.4, 5.3, 4.5, and 1.4%, respectively. The incidence of CNS effects was 1.5- to 2-fold higher than in pyrimethamine-sulfadoxine users, and 5-fold higher than in people who took no chemoprophylaxis [656].

6.3.2.2. Amodiaquine

Amodiaquine, a 4-aminoquinoline similar in structure to chloroquine, has also been associated with extrapyramidal effects. In 86 patients with rheumatoid arthritis, amodiaquine was shown to cause a significantly higher incidence of CNS effects than chloroquine. Dizziness and headache were reported in 22 and 6% of amodiaquine users, respectively, compared with 5 and 0% of chloroquine users. No effects were reported in 33 controls. Visual impairment was described by 20% of patients using amodiaquine and 7.5% using chloroquine. During malaria therapy, neurological disorders (protruding tongue, speech defects, and excessive salivation) occurred in a woman receiving amodiaquine 600mg daily for 3 days. Later re-challenge with a single dose resulted in similar symptoms. Akindele and Odejide described 3 other cases of amodiaquine-induced extrapyramidal symptoms: 1 adult had standard chloroquine therapy followed by 1 dose of amodiaquine 600mg; another developed symptoms 3 hours after ingestion of 1 dose of amodiaquine 600mg; and a 7-year-old child developed a protruding tongue after receiving amodiaquine 400 mg/day for 2 days. No reports of neurotoxic events associated with amodiaquine prophylaxis were found in the literature [656].

6.3.2.3. Mefloquine

CNS adverse events in association with mefloquine, a 4-methanolquinoline derivative structurally related to quinine, have been prominent in the literature. In an overview of the tolerability of mefloquine, adverse were reported to ranges from 47–90% in adults and 57–61% in children [263, 442, 660]. CNS events commonly encountered were headache, dizziness, and insomnia. Dizziness is recognized as a frequent but transient adverse effect of mefloquine therapy. Seven healthy Caucasian volunteers administered mefloquine 25 mg/kg all experienced lightheadedness, 4 of whom were severely incapacitated for 3 to 4 days [661]. This study had a maker influence on policies regarding the therapeutic use of mefloquine. In a more controlled setting on the Thai-Myanmar (Burma) border, in more than 3500 patients with uncomplicated malaria, dizziness occurred in 83% of adults and 59% of children under 15 years of age in the first 3 days post-treatment [491]. Dizziness just prior to treatment was found to occur in 63% of adults and 38% of children.

Following treatment with mefloquine 25 mg/kg, 5% of adults and 3% of children were severely dizzy and unable to walk for 1 to 3 days, compared with none prior to treatment; in 76% of these patients dizziness resolved within 24 hours, and in all patients within 72 hours. Both pre-and post-treatment, females were noted to report CNS evens more frequently than males. Administration of mefloquine 25 mg/kg, compared with 15 mg/kg, significantly increased the frequency of dizziness. In this large study, headache was associated with acute malaria but nor with mefloquine. In a similar study population, rates for dizziness were 31% in adults and 35% in children treated with halofantrine 24 mg/kg. None of the halofantrine-treated patients were severely dizzy [491].

Between 1987 and 1989, a spectrum of serious neurological and psychiatric adverse events attributed to mefloquine therapy, including affective and anxiety disorders, psychosis, toxic encephalopathy and acute brain syndrome, were brought to the notice of authorities. Since then, a number of pharmacoepidemiological studies have been conducted to elucidate causality and determine risk factors. In a review of controlled clinical trials between 1977 and 1989, severe CNS events occurred in 7 of 735 adults (1%) administered mefloquine 500 to 1250mg total dose and 5 of 500 patients (1%) administered mefloquine 500 to 750mg in combination with pyrimethamine-sulfadoxine. Weinke and colleagues estimated that 1 in 215 European malaria patients had mefloquine associated CNS events [662]. Monitoring for CNS effects in such studies was not standardized, and events were only ascribed retrospectively.

In a prospective study of 317 Nigerian patients treated with mefloquine, serious CNS events were recorded in 2 (1:159) [663]. In a closely monitored community-based study in northern Thailand, 10,446 patients with uncomplicated malaria were treated with mefloquine 15 mg/kg (10036 were given mefloquine in combination with pyrimethamine-sulfadoxine); 5 severe CNS events (i.e., 1 in 2089) were documented. A history of CNS disorders was a predisposing factor, but age and gender were not. A higher dosage (25 mg/kg) did not significantly increase the frequency of serious CNS toxicity (1 in 1217); however, retreatment with mefloquine 25mg/kg within 1 month was demonstrated to increase the risk of serious events 7-fold, rising from 1 in 1217 to 1 in 173. A priming of the CNS was suggested since serious neuropsychiatric events were not experienced during the first treatment. An analysis was conducted of 217 CNS adverse events in 98 patients receiving mefloquine therapy, reported to the manufacturer and the World Health Organization between 1985 and 1990. No patients died as a result of the adverse events. Symptoms occurred within 3 days in 72% of patients, with only 9% reporting onset 10 days or more after treatment. The majority (78%) reported symptoms resolution within 3 weeks [656].

It is unclear why a higher incidence of CNS events has been reported in Caucasian and Africans (1 in 100 to 1 in 200) compared with the Asian ethnic groups (I in 1200) studied. Since under-reporting was unlikely in the latter groups, other factors need to be examined, including social and genetic factors. The therapeutic dose taken by patients with serious events in the international database (mean 1637mg) was significantly higher than those with non-serious events (mean 1085mg). Besides dose-dependency of CNS effects, drug interaction or differences in disease severity may have contributed to the higher rates in the former group. Concomitant drug use is postulated, since over one-half of the Europeans experiencing CNS events during mefloquine therapy had received quinine or chloroquine, drugs which are also associated with such events. Quinine appeared to increase the severity of mefloquine-associated CNS effects; of 53 patients with serious CNS events during mefloquine therapy reported to Roche, 25% received concomitant quinine, compared with only 4% of 45 patients reporting non-serious events during mefloquine therapy (p=0.006). In a separate study of 18 patients (1 with cerebral malaria) with clinical conditions necessitating intravenous quinine *prior* to mefloquine 25 mg/kg, no CNS events were observed [664].

6.3.2.4. Quinine and Quinidine

Cinchonism is a well recognized syndrome associated with quinine. Nausea, vomiting, tinnitus, deafness, headache, disturbed vision and vasodilation are relatively mild at standard therapeutic doses. Quinine intoxication following doses of 4 to 12 g is characterized by seizures and coma. Early symptoms are mild visual and hearing complains. A principle sign is the sudden onset of bilateral pupil dilatation. Lethal doses may be around 8 g. Irreversible hearing impairment and blindness are very rare at standard dosages of quinine. Reversible high tone hearing loss was shown using serial autiometry in 10 patients receiving quinine for acute malaria. Hearing loss was unnoticed, although 7 had tinnitus, and recovery was complete. Three patients had received standard oral dosages (quinine sulphate 10 mg/kg 3 times daily for 7 days), and all had plasma concentrations in usual therapeutic range.

A recent pharmacokinetic study estimated 43% higher peak plasma quinine concentrations in very young children (< 24 months) compared with adults after rapid intravenous administration. The authors postulated that young children given the standard dosage regiment could, theoretically, be at greater risk of quinine-induced visual impairment. Quinidine is more potent as an antimalarial and more toxic than quinine, although apparent volume of distribution and systemic clearance are greater for Quinidine and the elimination half-life is shorter. Quinidine toxicity is associated with psychiatric disorders. Two cases of acute psychosis after standard quinine therapy were described as idiosyncratic drug reaction [656].

6.3.2.5. Artemisinin and its Derivatives

In animal toxicity studies of the oil soluble compounds, artemether and arteether, all dogs that received 20 mg/kg/day for 8 days developed clinical neurological defects, with progressive cardiorespiratory collapse leading to death in 5 of 6 animals [615]. Neuropathies and sensory disturbances included gait disturbances, loss of spinal and pain response reflexes, and prominent loss of brain stem and eye reflexes. In a dose-ranging study using between 5 and 20 mg/kg/day for 28 days, dose-related, region-specific neurophathic lesions, most pronounced in the popns, were identified. Similar dose-related lesions were confirmed in rats treated with 12.5 to 50 mg/kg/day for 28 days. Potential neutotoxicity of arteether in *malaca mulatta* monkeys was recently investigated, at dosages of between 6 and 24 mg/kg/day for 14 days [665]. Fine motor tremor with nystagmus and diffuse piloerection were the only neurological signs demonstrated in the animals given the highest dosages. Dose-dependent brain lesions were seen, confirming the previous pattern of neuropathology seen in rodent and canine brains. Despite histological injury, neurological signs were absent in monkeys receiving low dosages.

The dose-related CNS effects demonstrated in animals are warring, but difficult to interpret clinically. These studies provide little insight into the early signs of CNS toxicity or the reversibility of neurological lesions. Dosages used for malaria (total dose 10 mg/kg, given over 3 to 7 days) are lower than those in these preclinical studies, but the safety margin has yet to be determined. To date no significant toxicity has been reported in humans, despite the use of artemisinin and its derivatives in at least 2 million patients [516]. However, the majority of these patients were no reports of single episodes of CNS events with the oil soluble derivative, artemether, were in severely ill patients, making it difficult to distinguish between CNS events related to the drug and the malaria (poast-malaria neurological syndrome). No CNS events were identified in over 1000 patients with uncomplicated malaria administered oral doses of artesunate of artemether [666].

6.3.3. Antimalarials seldom Associated with Neurotoxicity

6.3.3.1. Proguanil and pyrimethamine

Pyrimethamine and proguanil are both dihydrofolate reductase inhibitors and have similar actions. Proguanil has not been associated with CNS adverse events. While standard therapeutic dosages of pyrimethamine (i.e., ± 1.5 mg/kg, combined with sulfadoxine) appear to be safe, CNS events have been reported in young children taking overdoses. Within 1 hour of ingesting 450mg pyrimethamine, a 14-month-old lost consciousness and convulsed every few minutes. Continuous generalized convulsions were controlled with diazepam but he remained comatose. EEG showed slow amplitude waves and no response to visual or auditory stimuli. Symptoms gradually resolved over 3 months. The speed of onset of symptoms suggests that the drug had a direct toxic action on the CNS, rather than metabolic effects due to folate antagonism. Seven infants with pyrimethamine at dosages ranging from 5.5 to 40 mg/kg per day (standard dosage 1 mg/kg/day for 3 weeks) experienced vomiting, tremors, and seizures [656].

6.3.3.2. Sulphonamines

Hadache, drowsiness and reduced acuity have rarely been encountered with sulphonamides utilized in the past 20 years. A causal relationship has not been established. In a review of 8339 adverse events associated with sulphur drugs reported to Swedish and UK drug authorities between 1965 and 1988, 6% were described with cotrimoxazole and 3% or less with each of the other sulphonamides. None of these were classified as serious. Occasional reports of peripheral neurophathytemporally associated with sulphones and with a dose-response effect have been published. Dapsone overdose can result in severe neurological symptoms, including disturbances in leval of consciousness and seizures. Of a total of 1438 adverse events reported in association with pyrimethamine-sulfadoxine, 145 (10%) were CNS disorders, of which 46 were classified as serous. Frequent concomitant use of chloroquine or mefloquine was reported, with only 1 patient with seizures having no other risk factors. These events are estimated to have occurred in a population of 300 million persons exposed to pyrimethamine-sulfadoxine for chemoprophylaxis and treatment of malaria. In view of this, measurement of rates of CNS adverse events associated with pyrimethamine-sulfadoxine are extraneous [656].

6.3.3.3. Primaquine

Primaquine, an 8-aminoquinoline, has rarely been associated with CNS adverse events. Severe depression and confusion were reported in 1 patient immediately following the onset of Primaquine therapy. Prior chloroquine use may have caused the event. Headache occurs when large dosages of 60 to 240 mg/day are given.

6.3.3.4. Tetracyclines

Tetracyclines are rarely associated with CNS effects in humans. Signs include headache, nausea, vomiting, dizziness, tinnitus, blurred vision and occasionally double vision. Tetracycline-induced benign intracranial hypertension in infants (bulging fontanelle syndrome) was a complication of the former use of this antibiotic in young children. It has rarely been seen in older children and adults. Symptoms develop 12 hours to 4 days after the onset of therapy and regress days or weeks after stopping the drug. Papilloedema persisted for months in 2 of 4 women with benign intracranial hypertension who took tetracycline 250 mg/day for acne. No CNS events have been described during malaria therapy. No case reported of CNS toxicity was associated with doxycycline malaria chemoprophylaxis are apparent in the literature.

6.3.3.5. Halofantrine

Halofantrine, a phenanthrene-methanol, has rarely been associated with CNS effects. Of 1973 patients treated with halofantrine 1,500 mg, the incidence of treatment-related adverse events was estimated to be 0.81 per 100 patients [488, 667]. CNS events occurred with equal frequency before and after treatment with halofantrine, at either 24 or 72 mg/kg. From 1988 to April 1995, 47 neurological events in 26 patients, and reported to the manufacturer (Horton J, SmithKline Beecham, personal communication). The most frequent serious CNS event was seizures (13 cases), half of which were associated with cardiac or concomitant risk factors. An estimated 3 million people were exposed to halofantrine, and the incidence of CNS adverse events resulting from treatment with this drug appears to be negligible.

6.3.3.6. Atovaquone

Preliminary clinical trial data suggest that Atovaquone, a hydroxyl-naphthoquinone, dose not produce CNS effects. Of 50 patients treated with 750mg 8 hours for either 4 doses or for 7 days, a 7 day post-treatment follow-up identified minimal CNS events of headache (4%0, insomnia (6%) and dizziness (4%) (Hutchinson D, personal communication). Of 163 patients treated with Atovaquone 750mg 8 hourly for 4 doses in combination with proguanil 200mg daily, headache, insomnia, and dizziness were reported in 1, 2, 5 and 2.5%, respectively. In combination with other antimalarials (doxycycline, tetracycline, and pyrimethamine) in a total of 86 patients, headache, insomnia, and dizziness were reported in 3.5, 0 and 1%, respectively. Atovaquone, most commonly at a dosage of 750mg 3 times daily, demonstrated an acceptable tolerability profile in patients with AIDS and Pneumocystis carinil pneumonia enrolled in 5 clinical trials [668].

6.3.4. Avoiding Neurotoxicity from Mefloquine with well Compliance

Mefloquine is predominantly used in a weekly dose of 250 mg. The most frequently encountered adverse reactions are nausea, abdominal pain, and dizziness. After therapeutic use of mefloquine, often as 1.5 g in three administrations, neuropsychiatric adverse effects have been noted [341]. Similar cases after both therapeutic and prophylactic treatment with mefloquine have been reported in the medical literature on several occasions, but few data exist on the frequency of neuropsychiatric effects to mefloquine and other antimalarials. In this prospective cohort study, we investigated the occurrence of adverse events to antimalarials in travelers.

Neurological and psychiatric adverse events have also been attributed to mefloquine prophylaxis [328]. Between 1985 and 1989, 138 patients with 278 CHS effects associated with mefloquine chemoprophylaxis were reported to the international database. Of the 138 patients, 45% experienced events classified as serious [669]. Symptoms developed in 41% of patients during the first week, and in 78% by the third week. Nearly one-half of the patients reported symptoms continued for over 2 months, with less than 2% experiencing long term sequel [669]. A higher proportion of female were documented with events during prophylaxis (63%), which could not be explained by user rates. This was similar to the observation with treatment doses in Thailand. Furthermore, significantly more females experienced serious events compared with males [328]. Higher reporting rates, greater compliance with prescriptions or a true increased risk were postulated. Regardless of weight, most females received the standard prophylactic dose of 250 mg and susceptible individuals could, therefore, be prone to dose-related toxicity.

The mechanism of psychic effects to mefloquine is unclear. Like the other antimalarials with a quinoline structure, mefloquine inhibits the enzyme acetylcholinesterase. The symptoms in the aforementioned travelers, such as agitation, concentration impairment, anxiety, insomnia, and blurred vision, are well-known effects of inhibitors of acetylcholinesterase such as physostigmine, carbamates, and organophosphates. Because the inhibitory action of mefloquine on acetylcholinesterase *in vitro* is less than that of chloroquine and amodiaquine, an additional pharmacokinetic factor might be important. Mefloquine has an elimination half-life of 2 - 3 weeks and reaches a steady state in the blood after approximately 6 - 11 weeks following the beginning of prophylactic treatment with 250 mg weekly. The fact that 80% of neuropsychiatric adverse events seem to appear in the first 3 weeks of treatment is compatible with this time interval, and could be explained by the accumulation of a relatively high concentration of mefloquine or one or more of its metabolites at receptor sites in the central nervous system. Concomitant symptoms such as dizziness, headache, paraesthesia, blurred vision, tremor, coordination impairment, and convulsions support to the hypothesis that the psychic effects have a neurotoxic pathogenesis [309, 328, 341].

The fact that blood concentrations of the drug give no straightforward indication of the concentration at receptor sites in the central nervous system, and the fact that many travelers had already stopped taking the drug before presenting themselves, makes it difficult to study the toxicology in these travelers in more detail. In those cases in which blood levels have been assessed, they were not abnormally elevated [1]. In conclusion, the literature and results from our study demonstrate that mefloquine can cause insomnia after prophylactic use of weekly doses of 250 mg. It should be emphasized that despite its neuropsychiatric effects, mefloquine remains a useful antimalarial drug. Even so, the guidelines for prophylaxis in the Netherlands have recently been reconsidered because of an increase in reports of serious neuropsychiatric adverse effects [328, 341]. In these guidelines, it is advised to start mefloquine 3 weeks before departure. As 80% of the serious neuropsychiatric adverse events appear within 3 weeks, this facilitates a timely discontinuation or change of antimalarial prophylaxis. A second reason for starting treatment 3 weeks before departure is that it increases the likelihood of adequate antimalarial blood levels. The neurological adverse effects could be avoided by only to use mefloquine within 3 weeks at 250 mg weekly dose [669].

6.3.5. Avoiding Neurotoxicity from Artemisinins with Appropriate Formulations

6.3.5.1. Neurotoxicity of Artemisinins Minor Concerning with Dihydroartemisinin

The effectiveness of AS has been attributed to its rapid and extensive hydrolysis to DHA. The efficacy of AM and AE has also been credited to the conversion to DHA, but their conversion rates were significantly lower and less complete than in AS [501]. DHA is obtained by sodium borohydride reduction of artemisinin, an endoperoxide containing sesquiterpene lactone; this was isolated by Chinese researchers and characterized as the antimalarial principle of the plant *Artemisia annua*. Through *in vitro* bioassay tests, DHA has been shown to be 1.4-5.2 folds more active than other artemisinins [161, 469, 525] but has also been shown to be 2-4 folds more toxic than other artemisinins *in vitro* and *in vivo* [44, 500, 618, 670]. Due to its poor solubility in water or oils, DHA has only been formulated as an oral preparation and has been used primarily as a semisynthetic compound for derivatization to other oil soluble and water-soluble artemisinins [33].

Table 49. Mean ratios of AUC_{DHA} (an active metabolite) /$AUC_{AM\ or\ AE,\ AS,\ and\ AL}$, and the concentration values (ng·h/ml) of $AUC_{DHA}/AUC_{AM\ or\ AE,\ AS,\ and\ AL}$ in ng·h/ml as the conversion evaluation of artemisinin derivatives (AM, AE, AS, and AL) to DHA in animal species and humans [44]

Drugs and administrations	Animal Species		Humans	
	Rats	Dogs	Volunteers	Patients
Artemether (AM)	0.04 (65/1857)			
Intravenous	0.07 (71/1007)	-	-	-
Intramuscular	0.10 (38/366)	0.14 (312/2240)	0.41–1.62	1.06
oral		2.96 (5592/1887)	1.12-7.69	1.42–8.55
Arteether (AE)				
Intravenous	0.04 (29/842)	-	-	-
Intramuscular	0.15 (43/286)	0.23 (181/804)	-	-
oral	0.17 (50/298	-	-	-
Artesunate (AS)				
Intravenous	0.64 (474/738)	0.50 (778/1533)	-	2.81–8.12
Intramuscular	0.31 (236/773)	0.73 (1109/1521)	-	-
Oral	2.74 (595/217)	0.45 (299/660)	4.29–5.36	6.35–9.71
Rectal	-	-	2.31	-
Artelinic acid (AL)				
Intravenous	0.01 (63/12706)	0.001 (7/11262)	-	-
Intramuscular	0.01 (44/5023)	0.003 (23/10195)	-	-
oral	0.03 (62/1650)	0.002 (14/8849)	-	-

DHA = dihydroartemisinin.

Thus, evaluating the conversion rates of various artemisinin drugs is very important for assessing the risk of neurotoxicity for the four compounds that are the focus of this article. A conversion evaluation of AM, AE, AS, or AL to DHA is summarized in Table 49 [44]. The summary shown illustrates that the conversion of artemisinin drugs to DHA in humans, which seem to be much more efficient than that in other animal species. Compared to the data from animal species and humans treated with AM or AS in various regimens, the ratios of AUC_{DHA} to $AUC_{AM\ or\ AS}$ are significantly higher in humans (0.41-9.71) than in animal species (0.04-2.96), suggesting that the human hydrolysis system is more active and complete.

In the animal species, the conversion rates are different with the four artemisinin agents. The highest hydrolysis found for AS are in rats and dogs with the DHA/AS ratio in the range of 0.31-2.74, whereas less than 0.23 of the ratio of DHA to their parent drugs have been measured for AM and AE. Moreover, the ratio in the range of 0.001-0.03 for AL conversion to DHA calculated in rats and dogs, suggests that DHA should not contribute as much toxicity for the 3 drugs (AM, AE, and AL) as that for AS (Table 50) [44]. The concentration with AUC data has demonstrated that DHA hydrolyzed from AS is in the range of 236–596 ng·h/ml compared to that from AM and AE in the range of 29-71 ng·h/ml at same dose level in rats [497]. However, while AM and AE induce neurotoxicity in the animals, AS does not [671]. Similar results are observed in dogs, 1109 ng·h/ml of DHA is produced from AS at a dose of 10 mg/kg without neurotoxicity symptoms [672], whereas 312 and 181 ng·h/ml of DHA from AM and AE, respectively, at a dose of 20 (AM) and 15 (AE) mg/kg has resulted in fatal neurotoxicity following intramuscular injection for the drugs [673, 674].

Table 50. Neurotoxicity findings in pathological and behavioral evaluation in animal species treated with dihydroartemisinin (DHA), artemether (AM), arteether (AE), artesunate (AS) and artelinate (AL) following multiple intramuscular (IM) and intragastric or oral administrations (n = 4-20)

Drugs	Species	Pathological Neurotoxicity			Behavioral Neurotoxicity		
		Dose mg/kg	Route	Regimen	Dose mg/kg	Route	Regimen
AE	Rats	12.5	IM	7 doses daily	50	IM	7 doses daily
	Dogs	5 – 7	IM	28 doses daily	20	IM	7 doses daily
	Monkeys	8	IM	14 doses daily	24	IM	14 doses daily
AM	Dogs	20	IM	8 doses daily			
	Mice				150	IM	28 doses daily
	Mice				300	Oral	28 doses daily
	Mice	300	Oral	28 twice daily			
AS	Mice				250	IM	28 doses daily
	Mice				250	Oral	28 doses daily
	Mice	300	Oral	28 twice daily			
AL	Rats	160	Oral	9 doses daily			
DHA	Mice	200	Oral	28 doses daily	200	Oral	28 doses daily

DHA = dihydroartemisinin, AM = artemether, AE = arteether, AS = artesunic acid, AL = artelinic acid; IM = intramuscular. *Data were cited from [44].

Data from our recent work found the maximum tolerant dose (MTD) for AS to be 240 mg/kg daily for 3 days after intravenous injection [94]. The animals were sacrificed at 24 hrs (1 day) and 192 hrs (8 days) after last dose. Histopathological evaluation demonstrated mild and moderate tubular necrosis in uninfected rats treated with AS 240 mg/kg; interestingly, fewer pathological lesions (minimal tubular necrosis) were observed in malaria infected rats.

Renal injury was reversible in all cases by day 8 after cessation of dosing and no neurotoxicity was found in any case with all intravenous regimens [66]. In this case, the AUC of 6984–15768 ng·h/ml of DHA, which is converted from AS, can be simulated daily in this regimen of rats according to our previous study [497]. Although DHA is 2-4 folds more active and more toxic than other artemisinin derivatives [44, 500, 618, 670], the high level of DHA generated from AS does not seem to have a role in neurotoxic induction in animal species.

In other comparative studies of toxicokinetics and hydrolysis, AL and AS were administered to malaria-infected rats using 3 daily equimolar doses (96 μmoles/kg) via intravenous administration. Plasma concentration of AS with a daily dose of 36.7 mg/kg was one-third less on day 3 than on day 1, resembling that of its active metabolite, DHA. The results were similar for other artemisinin drugs, but not for AL. Toxicokinetic parameters of AL were very comparable from day 1 to day 3 at the same molecular doses of AS at 40.6 mg/kg. AS was the pro-drug of DHA with a DHA/AS ratio of 5.26 compared to the ratio of 0.01 for DHA/AL. Other toxicokinetic parameters revealed that the total $AUC_{1-3\ days}$ (84.4 μg·h ml^{-1}) of AL was 5-fold higher than that of AS (15.7 μg·h/ml of AS plus DHA). In this study, the two drugs did not produce any CNS toxicity [475], but the general toxicity of AL was 3-fold greater than that of AS in malaria-infected rats [468]. The minimum doses of AS and AL producing parasitemia suppression were 2.3 and 2.5 mg/kg, respectively. The suppressive doses for decreasing the parasitemia by half (SD_{50}) were 7.4 and 8.6 mg/kg, respectively. The MTD for artesunate was 240 mg/kg with a therapeutic index of 32.6, whereas the MTD for AL was 80 mg/kg with a therapeutic index of 9.3 [468]. It is barely comprehensible that AS with 4893 ng·h/ml of DHA was not only significantly safer than AL with only 160 ng·h/ml of DHA in general toxicity, but it also did not produce neurotoxicity [94, 475].

Therefore, the review believes that DHA is not an important factor in neurotoxicity induction in animal species even though DHA is 2-4 times more toxic than other artemisinin derivatives in direct *in vitro* and *in vivo* studies [44, 500, 524, 618, 670].

6.3.5.2. Neurotoxicity of Artemisinins Minor Concerning with CNS Distribution

Data on the distribution of artemisinin analogues in the brain have few available studies, and of the published data, their methods all differ significantly. So to date, there are no comparative published results for quantitative analysis of the artemisinin drugs in brain tissue.

Using the thin-layer chromatography densitometry method, artemisinin (QHS), AM, and AS are found to cross the blood-brain and blood-placenta barriers with intravenous injection in rats [675]. QHS seems easier to detect in the brain after oral administration than after intravenous injection [676]. By radioimmunoassay, the highest level of AS was found in the rat intestine 10 minutes after intravenous administration, followed by the brain, liver, kidney, and other organs, in decreasing order. After one hour, the drug levels dropped significantly but not uniformly in all tissues, with high levels still remaining in the brain, fat, intestine, and serum [677]. The highest levels were found in rat brains 5 minutes after intravenous injection. Moderate levels of AM were also found in the heart, lung and skeletal muscles, while liver and kidney levels were low [678].

At 24 h after intramuscular injection of ^{14}C-arteether, the tissue homogenate method was used to identify that the highest levels of radioactivity were present in the intestinal tracts, liver, kidneys, and spleen. Lower amounts of radioactivity were found in brown fat and brain. The lowest levels of radioactivity were present in the heart, testes, muscle and residual carcass. High-performance liquid chromatographic data from this work indicated that three metabolites of arteether, but not arteether itself, could cross the blood-brain barrier (BBB). The calculated apparent concentration in the whole brain was 0.89% of the total radioactivity after intramuscular injection of 25mg/kg arteether [our unpublished data].

Autoradiographs of rats sacrificed 1 h following administration of ^{14}C-AL showed widespread distribution of the radiolabel. The relative activity of tissues (as density per unit area) were intestinal tract > liver > kidney > bone marrow > spleen > brown fat & salivary glands > heart & testes > brain. At 48 h, the relative activities were as follows: intestinal content > spleen > kidney > liver > salivary glands > brown fat & bone marrow > testes > brain. The low residual activity found in the brain appeared uniformly distributed. Autoradiographic evidence confirmed the presence of ^{14}C-AL in the brain from 1 to 192 h. However, the presence of ^{14}C-AL in the brain was at a lower level than the presence of ^{14}C-AL in most other tissues. There was no evidence of a label concentration in any area of the brain that was visible in the whole body sections. It was found that ^{14}C-labeled AL could not easily penetrate through the blood-brain barrier. An apparent concentration of 0.1% of total radioactivity/brain was calculated from an intravenous and oral administration of 10 mg/kg AL [475].

More recently, a neurotoxic dose range study showed that neuronal damage occurred in all beagle dogs at 40 and 80 mg/kg of AM following multiple intramuscular daily treatments. At 20 mg/kg, minimal neuronal effects occurred in 5 out of 8 dogs. Two hours after the 8th administration (on day 8), only low levels of AM were found in cerebrospinal fluid (CSF) with 25, 60, and 71 ng/ml for 20, 40, and 80 mg/kg dosing, respectively [674]. AM levels in the CSF were < 10% of plasma concentration.

These limited and incomparable data indicates that the oil-soluble derivatives, AE (0.89% of total dose in brain) and AM (10% of plasma level in CSF) seem to have the ability to cross the BBB more readily than water-soluble artemisinin compounds, like AL (0.1% of total dose). If this is true, then the differential neurotoxicity of AE and AM with the water soluble forms is easily understood. However, this deduction is not supported by the observation that AS, a water-soluble drug, has the ability to easily cross the BBB. The largest fraction of AS has been found in rat brain after intravenous administration by radioimmunoassay [677]. Recently it was found that the levels of ^{14}C-AS in rat brain were higher (>2-fold) than that in plasma; the AUCs of radioactivity in brain and plasma were 70.7 and 29.4 µg·h/g, respectively. During a 192 hour period, 1.27% of total radioactivity dose was detected in the rat brain. The half-life of ^{14}C-AS in brain tissue was 94.2 hours, significantly longer than that in plasma with only a 63.7 hour half-life [184]. These results indicated that the resident time of ^{14}C-AS was longer in the brain than in plasma. This may reflect a "sink effect" of DHA uptake transfer by lipid-rich brain structures [679].

If the absence of AS in CSF is consistent with its low lipid solubility and a decrease of concentration in plasma after intravenous injection [680], then the types of metabolites present in the brain tissue are unknown [243]. Whether these metabolites in the brain are associated with neurotoxicity is also unknown. Artesunate is converted stoichiometrically to DHA, concentrations of which peak at 5-10 min [530, 680] in plasma. DHA is highly lipid soluble and has a low molecular mass (284 Da) - both factors favoring penetration of CSF [679]. Since DHA has relatively low solubility in water, it should be able to cross cell membranes and the BBB. After treatment with AS in patients, the parent drug (AS) cannot be detected in the CSF. However, DHA levels in the CSF increase with time while DHA levels in plasma decrease. These finding suggest a continuing influx but a slower efflux of DHA in which DHA might accumulate in CSF during frequent AS dosing [679, 680].

Although (i) radiolabeled AS has been shown to easily cross the BBB and have more accumulation in brain tissue when compared to other artemisinins, as described earlier [243, 675-679], and (ii) other investigators has found increasing DHA levels in CSF (1100-1450 ng/ml) than in plasma (104-120 ng/ml) after intravenous injection of 120 mg AS in malaria patients [680], but no neurotoxicity has been observed. Therefore, while the artemisinin agents distribute in brain tissue and CSF, they do not appear to be a major factor in neurotoxicity.

6.3.5.3. Neurotoxicity of Artemisinins Minor Concerning with Drug Exposure Level

Toxicokinetic (TK) data demonstrate that AE and AM accumulation is observed in the plasma of rats [484], beagle dogs [674], rhesus monkeys (our unpublished data), and humans [681] following multiple intramuscular injections. Data collected from rat studies have demonstrated that the accumulation of AE in plasma is due to a slow and prolonged absorption from the injection sites [484]. The drug formulation, administration route, and plasma accumulation all appear to be associated with both general toxicity and neurotoxicity [484, 682, 683].

1) Drug Exposure Level in Rats

An accumulation and prolonged exposure of AE in plasma was first reported in rats with severe neurotoxicity due to slow and prolonged absorption in the muscle injection site following daily intramuscular injections of 25 mg/kg for 7 days [484]. The absorption of AE from muscle (at the injection site) was incomplete in the first 48 hours after a single injection. At 24 and 48 hours after dosing, 38% and 22% of the total dose of AE still remained in the injection site, respectively. The amount of AE in muscle rapidly decreased for 1-2 hours and then exhibited a much slower rate of decrease. Half-lives for the fast and slow absorption phases in the muscle were 1.0 hour and 26.3 hours, respectively. It is expected that the remaining dose amount would be absorbed later because an intramuscular total dose of AE with sesame oil should be 100% absorbed in the absence of decomposition or metabolism at the injection site.

The analysis of TK parameters on day 1 was similar to that of the single intramuscular dose. However, after day 1, the kinetic data estimates from days 2-7 showed a notable change in the PK parameter estimated for day 1 (Figure 21, top). The elimination $t\frac{1}{2}$ of AE after 7 daily doses was prolonged from 13.7 hours on day 1 to 31.2 hours on day 7, suggesting that the exposure time of AE had been greatly extended. C_{max} of AE (410 ng/ml) on day 7 was found to be three times higher than that on day 1 (130 ng/ml). AUC intramuscular dose was found to be 4.5-fold higher on day 7 at steady state (4367 ng·h/ml) than on day 1 (905 ng·h/ml) [484]. This indicates that the exposure concentration of AE has been greatly increased and the system of excretion and biotransformation in the rats may not have been normal [684]. Only a pathophysiological factor could change the slopes of the distribution and elimination due to biotransformation, excretion, and protein or tissue binding [685, 686].

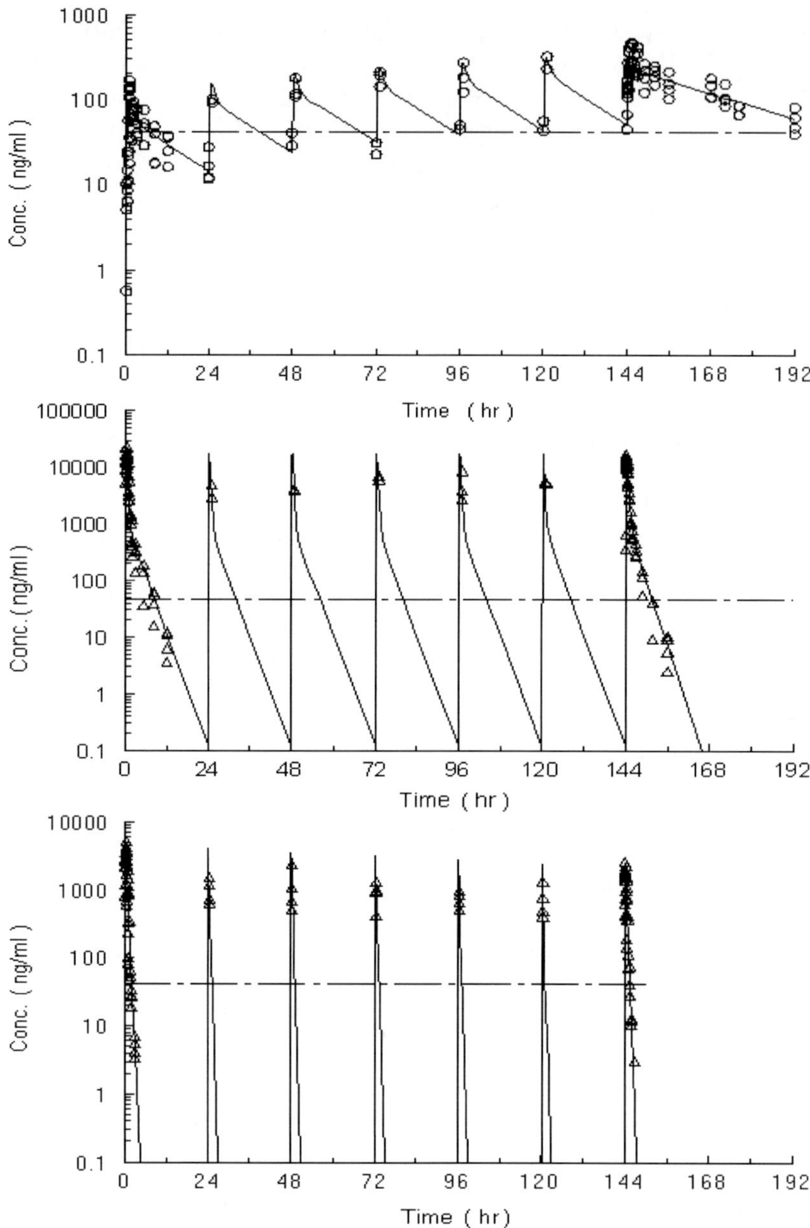

Figure 21. Pharmacokinetic profiles measured by HPLC-ECD (markers) and computer fitted curves (solid line) of 25 mg/kg of arteether with sesame oil (top, n = 4, Reference 484). Artelinate (middle) and artesunate (bottom) with 5% NaHCO$_3$/0.9% saline following daily intramuscular injection 25 mg/kg daily for 7 days in male rats. (n = 3, Reference 44). The minimal detected pathological effect level (MDPEL) from arteether measurement was estimated with 41.32 ng/ml (dashed line) in rats as described in Section 3.4.1.

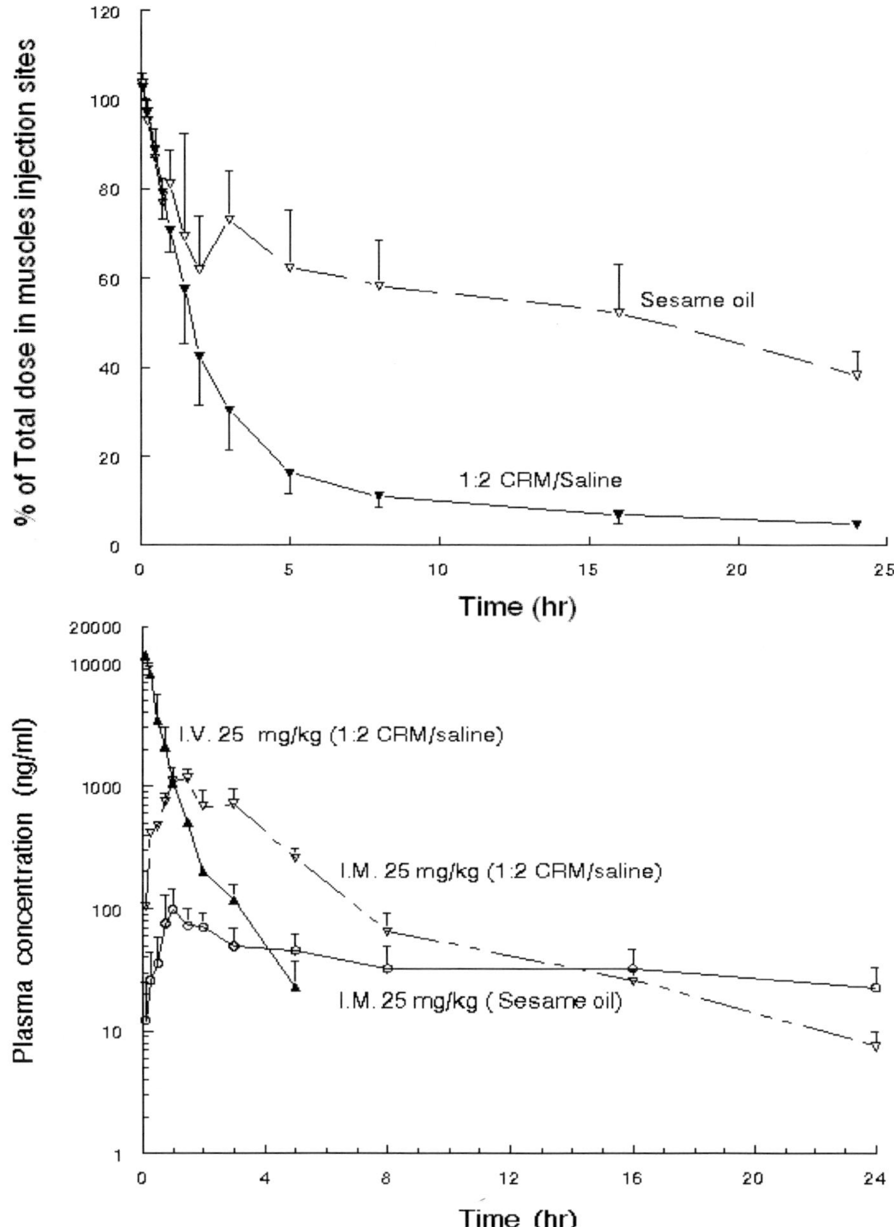

Figure 22. Mean amounts of arteether (AE) in muscle (injection site) measured by HPLC-ECD (top chart, n = 3) and mean plasma concentration-time profiles (bottom chart, n = 3) and by using 25 mg/kg of AE with sesame oil (AESO, dashed-line) and with 1:2 cremophore/0.9% saline (AECM, sold-line) after 7 days intramuscular injection in male rats [484, 500].

Drug exposure levels of 25 mg/kg AE after a 7 day intramuscular treatment was 16.92 ± 4.04 µg·h/ml for the AUC (a total of 7 days of AUC data), which induced severe neurotoxicity and animal death [484]. However, this exposure concentration level of AE was just one-fifth of the total AUC of AL (86.84 ± 5.59) with the same regimen of AE following intramuscular injection [44] (Table 51). The clinical observation did not show any neurotoxicity effect in those animals treated with AL (Figure 21, middle) and AS (Figure 21, bottom), even at high AL doses of 100 mg/kg after multiple intramuscular doses in rats [44]. Computer simulation indicated that the total AUC for the high dose treatments (100 mg/kg) throughout the 7 days was 341.88 μg·h/ml and was 21 times higher than that of AE treated rats. This result suggests that the 5-21 times higher exposure concentrations of AL did not seem to be a principal independent factor in causing neurotoxicity (Table 51).

Table 51. Neurotoxicity effects with exposure concentration (total AUC) and neurotoxic exposure time of artemisinin derivatives at an over 41.32 (rats) and 40.92 (beagle dogs) ng/ml of the minimal detected pathological effect level (MDPEL) of arteether, respectively, after intramuscular or oral administrations in rats (7 days dosing) and dogs (14 days dosing) [44]

Rats	Arteether Intramuscular (Sesame oil) AESO	Arteether Intramuscular (1:2CRM/saline) AECM	Artelinate Intramuscular (5%NaHCO$_3$)	Artesunate Intramuscular (5%NaHCO$_3$)
Dose	25 mg/kg x 7	25 mg/kg x 7	25 mg/kg x 7	25 mg/kg x 7
Neurotoxicity	Severe & death	Moderate	No	No
AUC $_{total\ 7D}$ (μg·h/ml)	16.92 ± 4.04	46.29 ± 2.06	86.84 ± 5.59	8.57 ± 3.34
AUC $_{total\ 7D}$ (DHA)	3.43 ± 0.76	5.35 ± 0.83	0.56 ± 0.18	5.06 ± 1.21
AUC $_{Day\ 1}$ (μg·h/ml)	1.31 ± 0.34	4.17 ± 0.68	12.46 ± 1.85	1.89 ± 0.80
AUC $_{Day\ 7}$ (μg·h/ml)	11.37 ± 2.05	7.31 ± 0.59	10.56 ± 0.76	0.99 ± 0.34
C$_{max\ Day\ 1}$ (ng/ml)	130.4 ± 39.0	1227 ± 171	20044 ± 3459	3981 ± 992
C$_{max\ Day\ 7}$ (ng/ml)	410.0 ± 90.9	1826 ± 118	15585 ± 1476	2078 ± 416
Neurotoxic Exposure time (hr)	164.28 ± 7.91	102.96 ± 5.26	42.48 ± 6.80	22.26 ± 4.87

Dogs	Arteether Intramuscular (Peanut oil)	Arteether Intramuscular (Sesame oil)	Artelinic acid* Oral (Suspension)	Artelinic acid Oral (Capsules)
Dose	20 mg/kg x 7	15 mg/kg x 14	20 mg/kg x 14	25 mg/kg x 14
Neurotoxicity	Moderate 32.13	Severe & death	No	No
AUC $_{7\ or\ 14D}$ (μg·h/ml)	2.39	24.39 ± 18.04	231.58 ± 115.81	191.14 ± 90.80
AUC $_{Day\ 1}$ (μg·h/ml)	5.11	0.81 ± 0.32	13.97 ± 3.11	10.46 ± 4.89
AUC $_{Day\ 7\ or\ 14}$ (μg·h/ml)	190	7.09 ± 3.74	12.78 ± 3.37	11.35 ± 5.78
C$_{max\ Day\ 1}$ (ng/ml)	346	90.11 ± 14.74	12360 ± 2706	6279 ± 2995
C$_{maxDay\ 7}$ (ng/ml)		352.7 ± 202.6	9780 ± 2885	7177 ± 6248
Neurotoxic Exposure time (hr)	174	277.59 ± 10.67	57.98 ± 14.55	47.41 ± 11.78

*Report from our contract company (Southern Research Institute, Birmingham, Alabama); DHA = Dihydroartemisinin, CRM = cremophore, AUC = area under the curve.

Table 52. Average effects in rat brain areas of repeated arteether (AE) injection with vehicle sesame oil (AESO) and cremophore (AECM) by daily intramuscular dosing of 25 mg/kg for 7 days (n = 3)* [484]

Date	Findings	Vehicle control	AESO	AECM	Student Test
Day 7**	Neuronal chromatolysis	0	3.00 ± 1.00	2.33 ± 1.15	P = 0.092
Day 10	Neuronal chromatolysis	0	4.67 ± 0.58	3.00 ± 1.00	P = 0.019

* Finding were grading as 0 = no significant lesions, 1 = minimal, 2 = mild, 3 = moderate, 4 = marked, and 5 = severe. ** The date is given as following dosing day as day 0.

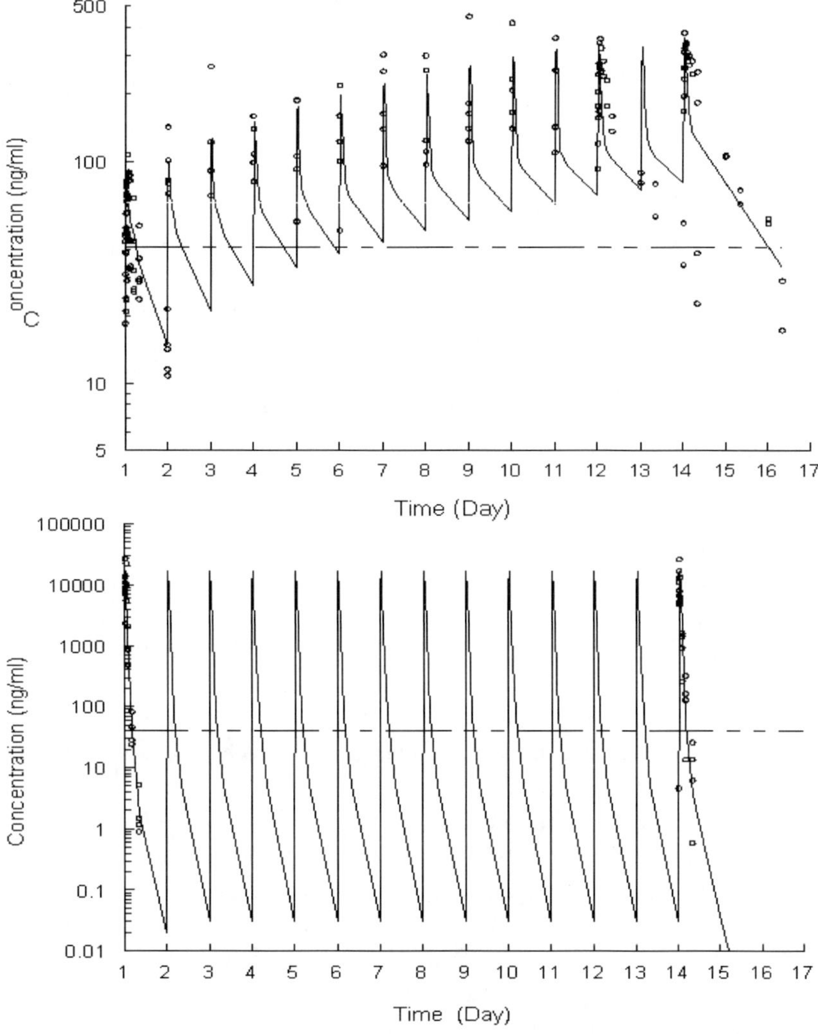

Figure 23. Pharmacokinetic profiles measured by HPLC-ECD (open circle) and computer fitted curves (solid line) of 15 mg/kg of arteether [44, 673] with sesame oil after intramuscular injection (top, n = 5); 20 mg/kg suspension of artelinic acid with carboxymethyl cellulose (CMC) and water (bottom, n = 4) following intragastric administration daily for 14 days in beagle dogs [44]. The minimal detected pathological effect level (MDPEL) of AE was estimated as 40.92 ng/ml (dashed line) in male dogs as described in Section 3.4.2.

In order to demonstrate that the neurotoxicity of AE is related to drug accumulation in blood and is due to the slow and prolonged absorption from the intramuscular injection sites, a study was designed to decrease the accumulation and toxicity of AE through the replacement of traditional sesame oil with a cremophore vehicle [500]. When administered at a daily dose of 25 mg/kg for 7 days, AE blood accumulation with sesame oil (AESO) had a 7.5-fold higher AUC (on first versus last day dosing), while AE with cremophore (AECM) had only a 1.8-fold higher AUC. Although the accumulation of AECM was greatly reduced, its total exposure level (46.29 µg·h/ml) was still 2.7-fold higher than AESO (16.92 µg·h/ml) due to a higher bioavailability of AECM (74.5%) compared with AESO (20.3%) (Table 51, Figure 22). While histopathological examinations of the brain demonstrated neurotoxic changes in both groups, the AESO group was significantly more severe than the AECM group. Brain injury scores with AECM were mild to moderate (2.3 to 3.0), but with AESO they were moderate to severe (3.0 to 4.7) on day 7 and day 10, respectively (Table 52). This study further demonstrated that the toxicity of AE was not dependent on its exposure level in rats.

2) Drug Exposure Level in Dogs

AE accumulation was also observed in beagle dogs' plasma with severe neurotoxicity and death after daily intramuscular administration of 15 mg/kg for 14 days [673]. C_{max} of AE (352 ng/ml) was found to be four times higher on the last day (day 14) compared to the C_{max} on day 1 (90 ng/ml). AUC (7.09 µg·h/ml) on the last day was 8.7-fold higher than that on the first day (0.81 µg·h/ml) indicating that the exposure concentration of AE had been greatly increased (Figure 23, top). The total AUC during the 14 days at 15 mg/kg AE daily intramuscular repeated injection was 24.39 µg·h/ml (Table 51),

Similar accumulation results for AM were reported in dog plasma after daily intramuscular administration of 20, 40, and 80 mg/kg for 7 days [674]. The concentration-time profile of 20 mg/kg AM in the beagle dog is shown in Table 51. The analysis of the toxicokinetic (TK) parameters on day 2-7 was changed, with the parameter estimated on day 1. The elimination t½ of AM after 7 daily dosing was prolonged from 9 hours on day 1 to 37 hours on day 7, suggesting that the exposure time of AM had been significantly increased (Figure 24, top). C_{max} of AM (340 ng/ml) doubled on day 7 from that on day 1 (190 ng/ml). AUC (6.79 µg·h/ml) on day 7 was 2.8-fold higher than AUC of the IM dose on day 1 (2.39 µg·h/ml), also indicating that the exposure concentration of AM had been increased [674].

Compared to drug exposure levels, much higher plasma concentrations were found in dogs treated with AL following oral administration. An AUC value produced by AL at 20 mg/kg in a suspension formulation was 231.6 ± 115.8 µg·h/ml following daily oral doses for 14 days (Figure 23 bottom). However, the value dropped to 191.14 ± 90.80 µg·h/ml for oral AL in a daily 25 mg/kg capsule formulation for 14 days (Table 51). The animals in both groups did not show any clinical toxicity [687]. The two dose regimens in dogs treated with AL showed at least 7 times higher exposure concentrations than those in dogs treated with AE. Neurotoxicity was neither detected in those animals with 20 or 25 mg/kg oral dose levels for 14 days [687], nor in animals treated at even very high AL doses of 320 and 300 mg/kg after multiple oral doses [688]. The drug exposure levels of these treatments suggest that a much higher exposure concentration of AL in the dogs does not produce neurotoxicity compared to AE and AM, which induce neurotoxicity with much lower exposure concentrations (Table 51) in dogs.

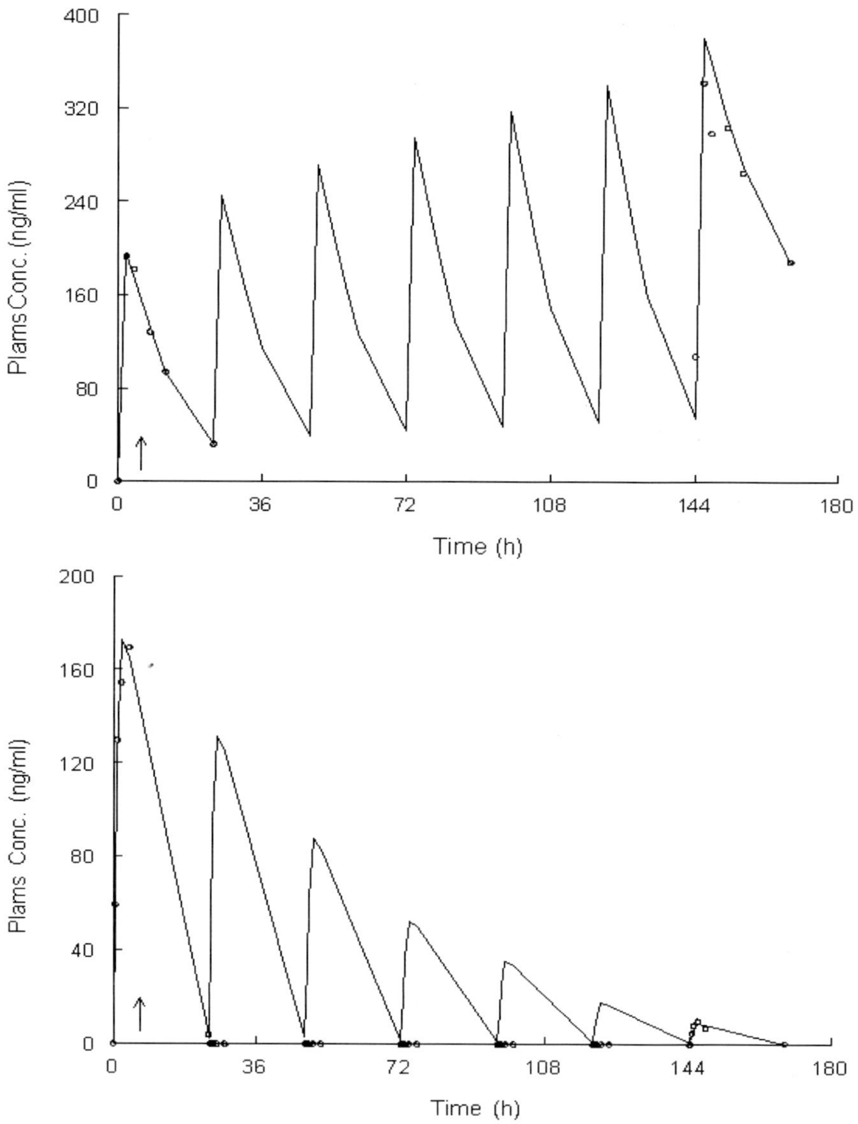

Figure 24. Pharmacokinetic profiles measured by HPLC-ECD (open circle) and computer fitted curves (solid line) of 20 mg/kg of artemether with sesame oil (top, showing drug accumulation) after intramuscular injection and 600 mg/kg of artemether in capsule (bottom, showing drug decline) following oral administration daily for 7 days in beagle dogs. [674]

3) Drug exposure Level in Monkeys

Similarly, great plasma accumulation of AE was shown in rhesus monkeys' plasma after daily intramuscular administration of 16 mg/kg for 14 days (Figure 25, top). The concentration-time profile of 16 mg/kg AE in the monkeys showed that the C_{max} of AE (1230 ± 577 ng/ml) on day 14 was 19-fold higher than that on day 1 (63.36 ± 9.08 ng/ml). AUC (53101 ng·h/ml) on the last dosing day was 60-fold higher than the AUC of the intramuscular dose on the first dosing day (877 ng·h/ml), indicating that the exposure concentration of AE was greatly increased [44].

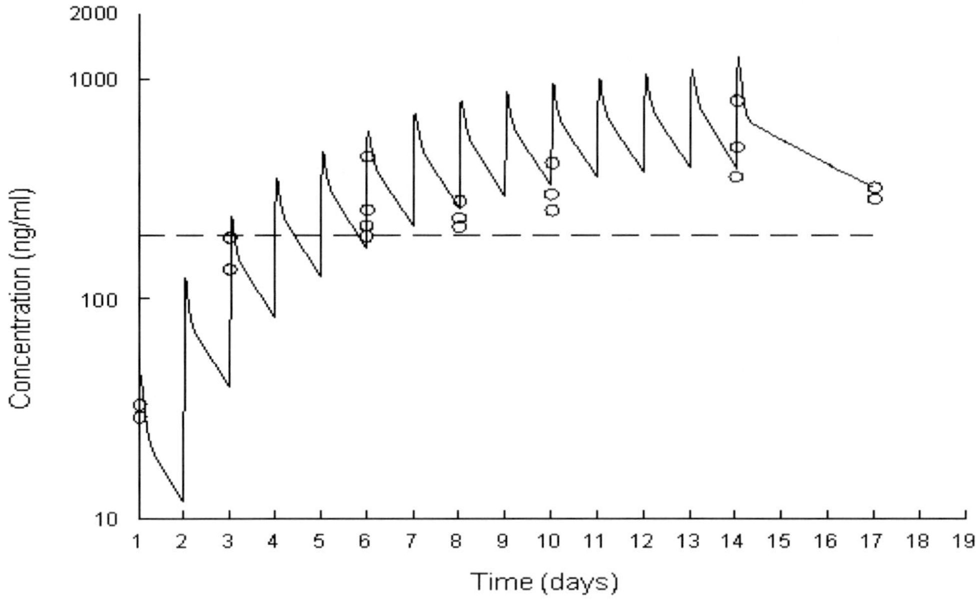

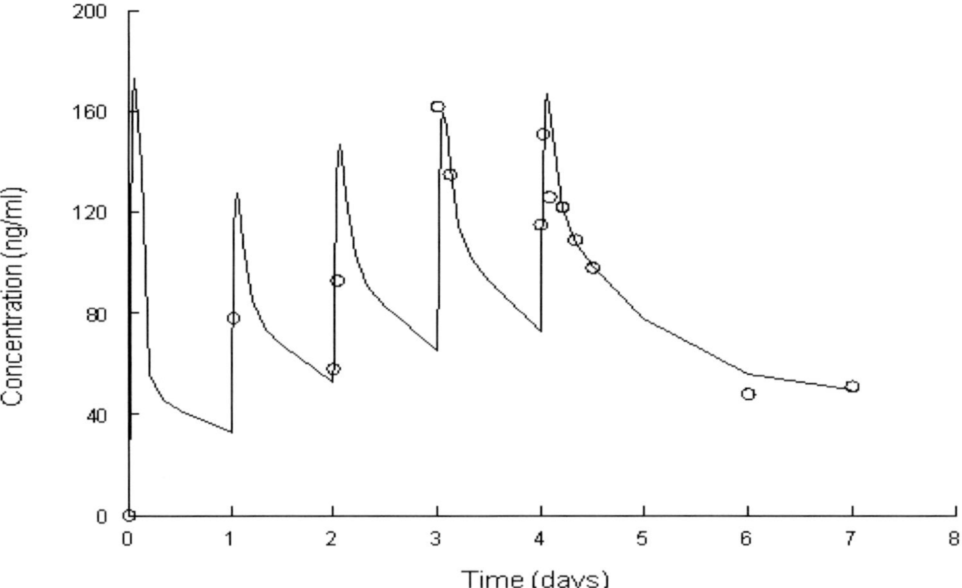

Figure 25. Pharmacokinetic profiles measured by HPLC-ECD (empty diamonds) and computer fitted curves (solid-line) of 16 mg/kg daily for 14 days in rhesus monkeys (top, n = 2) [44], and the minimal detected pathological effect level (MDPEL) of AE was estimated as 193.8 ng/ml (dashed line) in male dogs as described in Section 3.4.3. In humans, 3.2 mg/kg in first day and then 1.6 mg/kg for another 4 days of arteether with sesame oil from health subjects (bottom, n = 27) [681]. Both data showed the drug accumulation in monkeys and humans after the 5 days treatments.

4) Drug Exposure Level in Humans

Except for AE and AM intramuscular injections with oils, no plasma accumulation is found for all artemisinin drugs by other administration routes in human. On the contrary, four artemisinin drugs (QHS, AS, AM, even DHA) have reported a declining concentration in plasma during multiple oral treatments in malaria patients and health subjects. The C_{max} and AUC values are markedly reduced from one-third to one-seventh on last dose day compared with first day. The decrease in drug exposure levels during treatment is not disease-related, since the artemisinin drug PK on day 1 is similar to that reported in healthy subjects [470-474, 485]. Similar time-dependent declining of oral AM was also found in beagle dogs (Figure 24, bottom) [674]. One possible explanation for the decrease in plasma concentration-time during treatment is an increase in metabolic capacity due to auto-induction of hepatic drug-metabolizing enzymes. However, the drug decrease did not occur for AE in sesame oil after intramuscular administration in humans [689].

Table 53. Main pharmacokinetic parameters of ß-arteether with sesame oil (Artemotil) following multiple intramuscular administration of low dosage regimen (3.2 mg/kg on day 0 and 1.6 mg/kg on day 1-4, n = 11) and high dosage regimen (4.8 mg/kg on day 0 and 1.6 mg/kg on time 6, 24, 48, 72, and 96 hr, n = 44) in Thai patients with severe *P. falciparum* malaria* [690]

Parameters	Day 1	Day 5	Day 1-5	Ratio of Day 5/ Day 1
Low dosage regimen (n =11)				
Dose (mg/kg)	3.2	1.6	-	-
C_{max} (ng/ml)	63.6 ± 39.9	140.8 ± 74.2	-	2.9 ± 2.4
$AUC_{inf.\ Artemotil}$ (ng·h/ml)	2521 ± 1950	8409 ± 4730	16242 ± 9113	4.5 ± 2.9
$AUC_{inf.\ DHA}$ (ng·h/ml)	-	-	2842 ± 1119	-
$t_{1/2}$ absorption (h)	3.20 ± 0.64	2.13 ± 1.44	-	-
$t_{1/2}$ elimination (h)	12.53 ± 3.91	40.66 ± 14.82	-	3.4 ± 1.1
High dosage regimen (n = 44)				
Dose (mg/kg)	6.4	1.6	-	-
C_{max} (ng/ml)	110.1 ± 56.9	241.1 ± 121.4	-	2.7 ± 1.5
$AUC_{inf.\ Artemotil}$ (ng·h/ml)	4702 ± 2964	9484 ± 4365	19782 ± 7725	2.9 ± 1.9
$AUC_{inf.\ DHA}$ (ng·h/ml)	-	-	3209 ± 1058	-
$t_{1/2}$ absorption (h)	3.18 ± 2.90	3.46 ± 1.96	-	-
$t_{1/2}$ elimination (h)	22.36 ± 9.24	31.62 ± 12.73	-	2.0 ± 1.3
Patient 151 (who likely got first dosing intravenously, n =1)				
Dose (mg/kg)	6.4	1.6	-	-
C_{max} (ng/ml)	2407.5	317.3	-	0.1
$AUC_{inf.\ Artemotil}$ (ng·h/ml)	12259	7403	24529	0.6
$AUC_{inf.\ DHA}$ (ng·h/ml)	-	-	2432	-
$t_{1/2}$ elimination (h)	9.4	28.9	-	3.1

*Values were indicated are the mean ± SD. Min = minutes; AUC = area under the curve; Vss = volume of distribution at steady state; F = fraction of drug absorbed into blood; CL = clearance rate; MRT = mean residence time; DHA = dihydroartemisinin. – No data available.

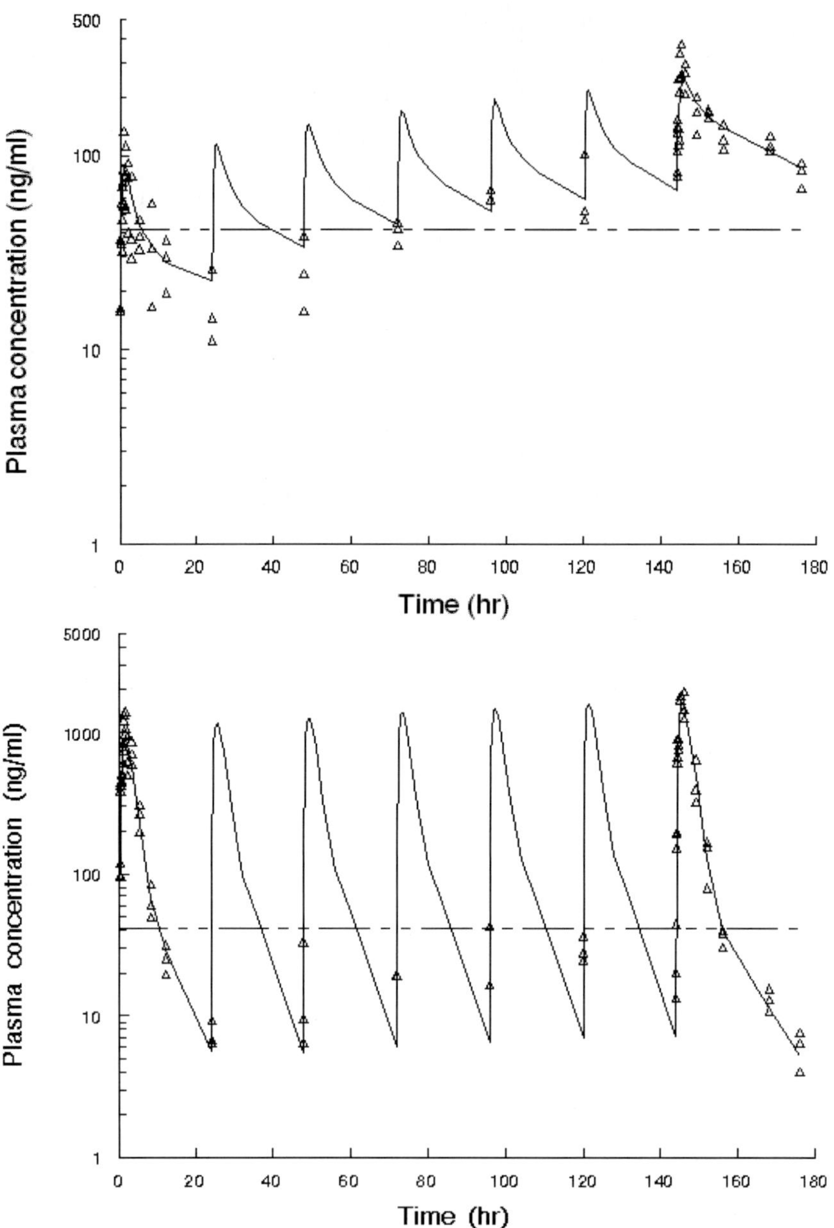

Figure 26. Toxicokinetic profiles in plasma measured by HPLC-ECD (open circle) and computer fitted curves (solid-line) of 25 mg/kg of arteether with sesame oil (AESO, top, n = 3) and arteether with cremophore (AECM, bottom, n = 3) following daily intramuscular injection for 7 days in male rats [500]. The minimal detected pathological effect level (MDPEL) was estimated as 41.32 ng/ml (dashed line) [section 3.4.1].

Kager *et al.* [681] reported on the first use of AE in humans on the studies of safety, tolerance, and preliminary pharmacokinetics. An initial AE intramuscular dose of 3.2 mg/kg was used in 27 healthy subjects on day 1. The dosage was then decreased to 1.6 mg/kg for four additional days. The PK data showed a long elimination half-life of 25-72 hr, explaining the marked accumulation in plasma observed in the multiple dose injection (Figure 25 bottom). Recently, another study reported administering AE by intramuscular injection to human subjects in multiple doses in a clinical treatment trial. Two dose regimens of AE (AE with 3.2 mg/kg intramuscular start and then 1.6 mg/kg for 4 days as low dose regimen, and high dose of AE regimen with 4.8 mg/kg at 0 hr and 1.6 mg/kg at 6, 24, 48, 72, and 96 hr) were administered for the patients with severe and complicated falciparum malaria [485, 690]. Although these dose regimens have been designed as loading (high) and maintaining (low) doses, AE accumulation was still observed in the human plasma for both of those two dose regimens (Table 53).

In the low dose regimen, the elimination t½ of the low dose of AE regimen in human after 5 daily dosing was prolonged from 12.5 hr (day 1) to 40.7 hr (day 5), suggesting that the exposure time of AE had been greatly extended. The C_{max} of AE was two times higher on day 5 (140.8 ng/ml) than that on day 1 (63.6mg/ml). $AUC_{96-120h}$ (8409 ng·h/ml) on day 5 was 3.5-fold higher than AUC_{0-24h} of the intramuscular dose on day 1 (2521 ng·h/ml), indicating that the exposure concentration of AE had been obviously accumulated. Similar profiles of the prolonged half-lives and increased C_{max} and AUC were also confirmed in the high dose regimen of AE in humans (Table 53). These dose regimens proved to be safe and had potent antimalarial efficacy in reducing parasitemias from the peripheral blood.

5) Prolonged Absorption Effect on the Accumulation of AE and AM

Many scientists believed that the intramuscular administration of AM and AE was associated with the slowest absorption because the drugs were dissolved in oil (sesame or peanut oil) and, when injected, formed a depot from which drug was slowly released [390, 497, 681, 683]. The inference of slow elimination was recently demonstrated in a rat study, which also found a great deal of accumulation in plasma at injection sites [484]. Following 25 mg/kg of AE in sesame oil daily intramuscular injections for 7 days, the study confirmed and extended the results of earlier studies [497] by demonstrating that the absorption of AE from muscle (injection site) was incomplete. This study also indicated that up to 38% of the total single dose of AE remained in the injection site 24 hr after dosing and 22% of the total single dose still remained in the muscle after 48 hr. The dose amounts of AE remaining in the injection site vs. time are shown in Figure 22 (top). The amount of AE rapidly decreased for 2-3 hours and then exhibited a slower decrease. The half-lives for the fast and slow absorption phases were 0.97 ± 0.48 hr and 26.3 ± 1.2 hr, respectively. Following 7 days of multiple dosing of AE (25 mg/kg), 91.4% of a single dose (25 mg/kg) was still left in the muscles from the 7 injection sites 24 hrs after last dose [484].

After intramuscular multiple dosing, accumulation of AE levels in plasma and muscle were evaluated. Plasma concentration after 7 days of dosing increased 3-4 folds, from an increase in C_{max} of 130.4 ng/ml to 410.0 ng/ml to an increase in $AUC_{inf.}$ from 1309 ng·h/ml to 11366 ng·h/ml. When compared with single dosing, AE accumulations in the muscle at injection sites increased 2.4-fold after multiple injections from 38% (day 1) to 91.4%, (day 7) of the total single doses following the 7 daily doses (Table 51). When muscle absorption data from the single dose study was used to simulate the amount of dose remaining in the muscle after 7 days of dosing, 89% of a single dose was calculated to be present, which is similar to the actual measured amount [484]. This suggested that the change in TK observed during day 1 was not due to a change in absorption, but rather a change in body distribution or/and elimination. Therefore, the volume of distribution at a steady state on day 7 (41.8 L) was 40%

of the volume on day 1 (104.3 L). Clearance value was increased by 89% from day 1 to day 7, 0.98 L/h to 1.85 L, respectively. The elimination t½ of AE after 7 daily dosing was also prolonged from 13.7 hr (day 1) to 31.2 hr (day 7).

Furthermore, the histopathological study demonstrates that the slow and prolonged absorption of AE in rat muscle is due to a slow release of AE from the sesame oil phase and therefore associates the cytotoxicity of the drug on the muscle cell at the injection site [WRAIR unpublished data].

6.3.5.4. Neurotoxicity of Artemisinins Major Concerning with Drug Exposure Time

1) Drug Exposure Time in Rats

Genovese *et al.* [692] found that a minimal daily dose of AE at 6.25 mg/kg for 7 days has no detectable pathological effect dose (NDPED), which does not produce any neurotoxicity (clinical observation and/or neuropathological detection) in rats. Statistically significant neuropathology in brain stem nuclei was observed in the group of rats (6/6 affected) treated with 12.5 mg/kg. These results demonstrate that AE-induced brainstem neuropathology in rats could occur at the relatively high dose of 12.5 mg/kg (double the NDPED dose) for 7 days (Table 50). The pharmacokinetic parameters and the minimal plasma AE concentration of 41.32 ng/ml at 12.5 mg/kg dose has been determined by our previous studies [500], as the minimal detected pathological effect level (MDPEL) in plasma that should result in a neuropathological effect [498, 524, 691].

When calculating the exposure time of AE with sesame oil spent over the MDPEL (41.32 ng/ml) period in rats, neurotoxic exposure time was 164.28 ± 7.91 hours in the 7 day treatment after daily 25 mg/kg intramuscular injections (Figure 26 top; Table 51). This exposure time affected neurotoxicity is seen at 85.6% of the total dose treatment period (192 hours) with AE [500]. The total exposure time of AE with cremophore was 102.96 hours during the treatment period (Figure 26, bottom), which was about one-third (37%) less than the exposure time of AE with sesame oil (164.28 hours). Neurotoxicity effects in those rats were reduced from severe to moderate, and the drug exposure time seems to be related to the effects of neurotoxicity (Table 52) [484].

Compared with AL treatment, drug accumulation and neurotoxicity were not found in animals treated at the same dose (25 mg/kg), same period (7 days) and same conditions as the AE study [44, 484, 497]. The drug exposure period of AL spent over the MDPEL was 42.48 ± 6.80 hours (Figure 21, middle, Table 51) in rats, which was significantly shorter (4-fold) than the AE study (164.28 hours). Therefore, the animals treated with AL would have avoided the risk and the neurotoxic outcome. An even wider safety margin is seen in animals treated with AS. Using the same conditions, 25 mg/kg of AS daily for 7 days generated 22.26 ± 4.87 hours of neurotoxic exposure time in rats (Figure 21, bottom; Table 51), is 7-fold shorter than that in animals treated with AE (Figure 21, top; Table 51). This result suggests that the 4-7 time shorter exposure times of AL and AS over the MDPEL compared to the exposure times of AE seem to play a major role in avoiding clinical neurotoxicity [524, 693].

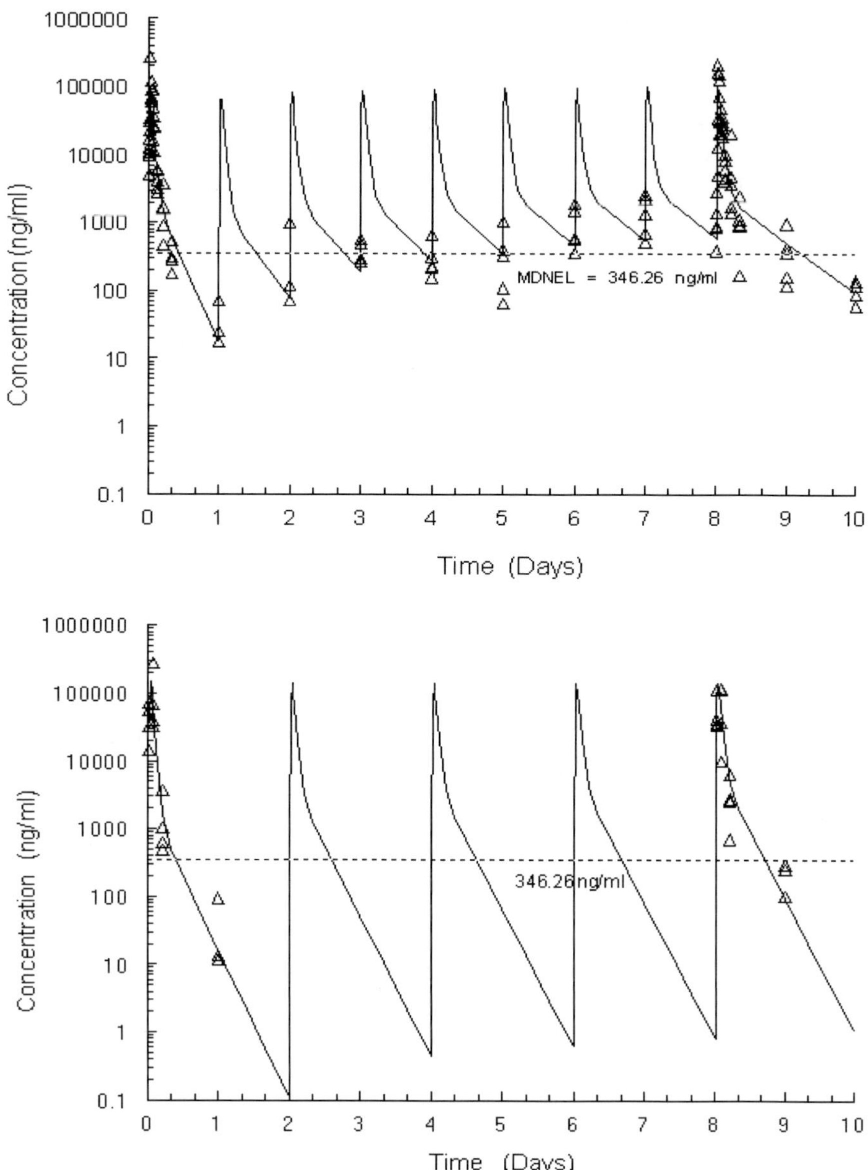

Figure 27. Pharmacokinetic profiles measured by HPLC-ECD (markers) and computer fitted curves (solid line) of artelinic acid (AL) in suspension at 160 mg/kg daily for 9 dosages (top, n = 5), and artelinic acid in suspension at 288 mg/kg every other daily for 5 dosages in rats (bottom, n = 4). The two regimens have same total dose with 1440 mg/kg and same treatment period for 9 days [524, 694]. The minimal detected neurotoxic effect level (MDNEL) was defined as the plasma level (346.26 ng/ml) at which anorectic toxicity was first noted in daily dosing cohort of rats.

Changing administration route from intramuscular to intragastric, AL showed moderate neurotoxicity in rats treated with 160 mg/kg daily in 9 multiple doses for 9 days but not in animals with 288 mg/kg every other day in 5 multiple doses for 9 days [694]. When calculating the neurotoxic exposure time we cannot have the MDPEL data from animals treated with oral AL. However, there was available with a minimum detectable neurotoxic effect level (MDNEL), defined as the minimal inhibition of gastric emptying on day 5 [694], as this seems to relate to a neurotoxic effect induced by various artemisinins, including: AM, AE, and DHA [500]. In this study, the stomach contents were examined in the rats belonging to the daily 160 mg/kg dosing regimen for AL at 8 and 24 hours post dosing at various time points in the study. The quantity of AL detected in the stomach was expressed as both the amount of drug per gram of stomach contents and as a percentage of a single dose. On days 3, 5, 7, and 9 we detected 0, 46.52, 178.59, and 486.21 μg, respectively, of AL per gram of stomach contents 8 hours after the last dose. The increasing drug remaining in stomach over the course of dosing suggests that gastric emptying is inhibited. This inhibition of gastric emptying progresses from mild inhibition on day 5 to severe inhibition on day 9. This inhibition persisted even at 24 hours post dosing. Twenty-four hours post dosing on days 5, 7, and 9 we found 5.74, 25.50, and 29.11 μg of AL/g, respectively, contents remained in the stomach. It is postulated that this decrease in GI motility could result from a decrease in vagal tone from a decrease in sympathetic outflow [670].

Therefore, on the day 5 post dosing the beginning inhibition of gastric emptying was defined as an initial neurotoxic effect in the rats with 160 mg/kg regimen. The corresponding plasma concentration of AL on day 5 in minimum was 346.26 ng/ml, which was the minimal anorectic toxicity in the rats. The neurotoxic exposure time with AL's MDNEL (346.26 ng/ml), 186 hours was measured as the exposure period threshold to affect neurotoxicity (Figure 27, top). These results indicate a correlation between neuropathology of rats and a prolonged AL exposure. The AL exposure was related to drug accumulation in the blood resulting from a delayed gastric emptying-induced prolonged absorption of the drugs from the stomach [694]. When rats were treated intermittently with AL at 288 mg/kg every other day in 5 multiple for 9 days, the neuronal degeneration was not identified (not significantly different from that of vehicle control group) until day 7 after the last treatment. The neurotoxic exposure time was calculated as 75 hours in the animals (Figure 27, bottom). The data further indicated that by shortening the drug exposure time at MDNEL level, it was possible to reduce the neurotoxic risk in the animal treated with the artemisinins [524, 694].

The drug retained in the stomach appears to be a reservoir which delays and prolongs the absorption of AL and results in drug accumulation (high AUC) and prolongs elimination (great Vss and long $t_{1/2}$). This reservoir is very similar to the depot effect seen with the oil-soluble artemisinins (AM and AE) intramuscularly in the injection sites. This accumulation increases the time that the drug levels are above a toxicity threshold and can thereby induce neurotoxicity. In this study, we attempt to correlate the neurotoxic exposure time to the observed histopathological findings to the two dosing groups. We found that the neurotoxic exposure time to be significantly extended for animals in the daily AL treatment cohort (186 hrs) compared to those in the every other day dosing cohort (75 hrs). To determine the neurotoxic exposure time for the two regimens, we calculated the time that the AL plasma level was above the minimum detectable neurotoxic effect level (MDNEL), which defined in this study as the plasma level at the time of onset of the anorectic effect. Although the MDNEL is not a minimum detectable pathological effect level (MDPEL), which defined as the first observation of neuronal injury, as was used in a previous study where AE was shown to produce neurotoxicity when given intramuscularly at 12.5 mg/kg daily for 7 days [500]; its estimation can give us an initial estimation of neurotoxic exposure time and allow us to correlate this exposure time with observed histopathology from different AL dosing regimens.

It is clear that in the present study the delayed gastric emptying resulted in AL accumulation in blood and prolonged a neurotoxic exposure time (186 hr) in the 160 mg/kg x 9 rats when compared to 75 hr in 288 mg/kg x 5 animals. It is obvious that the neurotoxic exposure time is a key factor in the neurotoxicity induced by oral AL in rats.

2) Drug Exposure Time in Dogs

Davidson [695] showed that a minimal daily dose of AE at 3 mg/kg for 28 days has no detectable pathological effect dose (NDPED) and does not cause neurotoxicity in beagle dogs; this was confirmed by histopathological examinations. AE-induced brainstem neuropathology has been detected though in some dogs with a dose as low as 5 mg/kg/day from Brewer [606, 615], 6.25 mg/kg from Dayan [498], and 6.75 mg/kg from Davidson [695], based on a daily dose for 28 days. Thus, the average of the minimal dose to produce neurotoxicity in dogs should be 6 mg/kg (calculated from the three findings.) The minimal plasma concentration of AE with this low dose (6.0 mg/kg) has been simulated by previous publications in Figure 23 (top). The concentration level of 40.92 ng/ml could be calculated as the MDPEL in plasma that should be the first risk for causing neurotoxicity in dogs [473].

AE accumulation was also observed in plasma of the beagle dogs after daily intramuscular administration of 15 mg/kg for 14 days [673]. The mean concentration-time profile of AE in the beagle dog is shown in (Table 51). The kinetic data from toxicokinetic parameters indicated notable changes on the last dosing day from the pharmacokinetic parameters estimated on day 1. The important change was the elimination t½ of AE during the two weeks of daily dosing. The elimination was prolonged from 11.9 hours on the first dosing day to 22.5 hours on the last dosing day, suggesting that the exposure time of AE had been doubly extended. Also, a MDPEL (40.92 ng/ml) of AE in beagle dogs first appeared in that study on day 6-7. Therefore, day 6-7 would be the earliest time that neurotoxicity occurred in the beagle dogs, based on the toxicities and the toxicokinetic data analysis.

When estimating the exposure time of AE over the concentration level of 40.92 ng/ml (MDPEL) in dogs with severe neurotoxicity and death, the exposure period was 277.59 ± 10.67 hours in a 14-day study after daily intramuscular injections of 15 mg/kg (Table 51). The neurotoxic exposure time of AE was 82.6% of the total dosing treatment period (336 hours). Compared with AL dosing by suspension or capsules orally, the neurotoxic exposure times were calculated to be 57.98 ± 14.55 hours for AL suspension (Figure 23, bottom), and 47.41 ± 11.78 hours for AL capsules (Table 51). All animals treated with AL were not observed to have any neurotoxicity [688].

These evaluations strongly support the findings that neurotoxicity is exposure time dependent in the animal species regardless of formulation and administration route. As a result, the artemisinins exposure time in animal species has an imperative role in the induction of neurotoxicity [524]. Other factors of the pharmacokinetic parameters, such as drug distribution in the brain, toxicity of active metabolite (DHA), or drug exposure level, tend to be of minor importance.

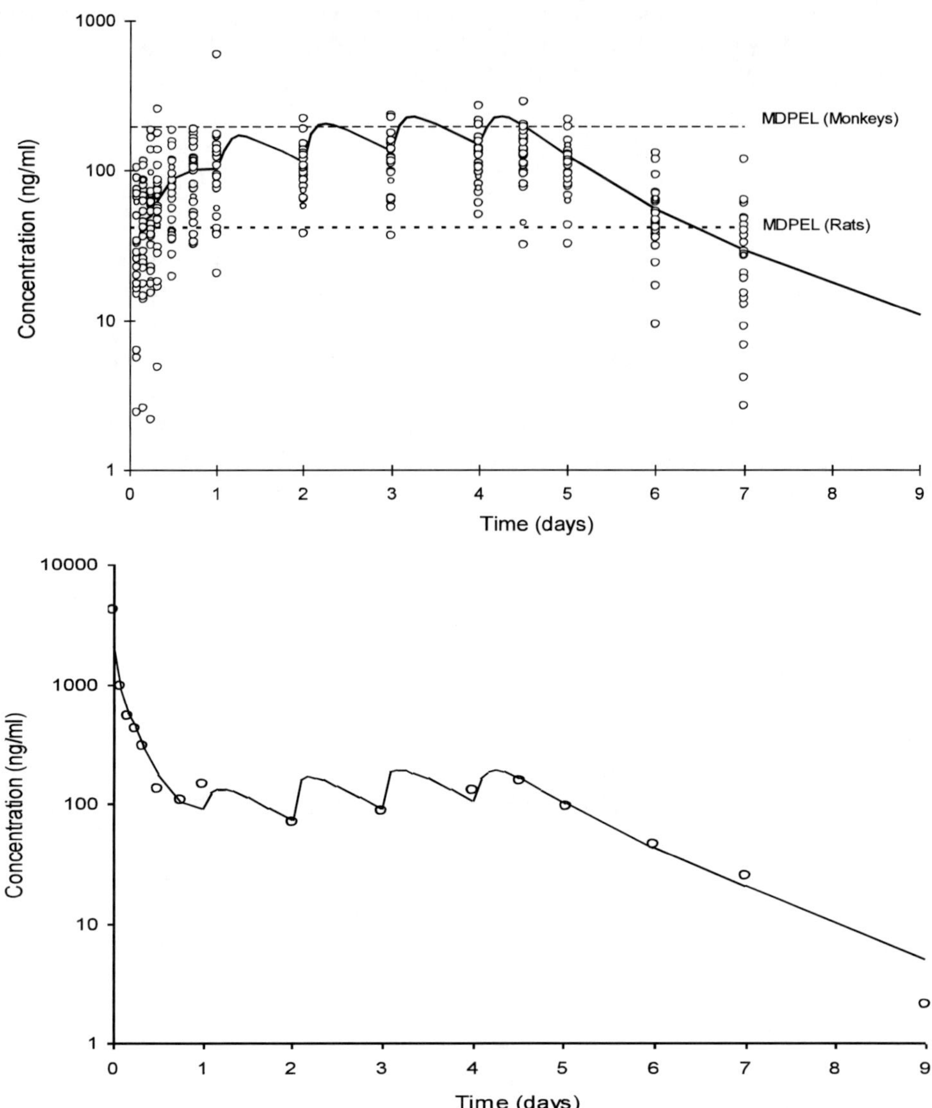

Figure 28. Pharmacokinetic profiles in plasma measured by HPLC-ECD (open circle) and computer fitted curves (solid-line) of 4.8 mg/kg at 0 hour and 1.6 mg/kg at 6, 24, 48, 72, and 96 hours of arteether (AE) with sesame oil in 42 patients summary as intramuscular administration (top chart), and in a single patient (bottom chart), who seemed to be injected into vessel as intravenous injection at first dosing and appeared a great efficacy [690].

3) Drug Exposure Time in Monkeys

AE was present in rhesus monkeys plasma after daily intramuscular administrations of 16 mg/kg for 14 days [our unpublished data]. The concentration-time profile of 16 mg/kg AE in the monkeys showed that the C_{max} of AE (1230 ± 577 ng/ml) on day 14 was 19-fold higher than that on day 1 (63.36 ± 9.08 ng/ml). AUC (53101 ng·h/ml) on the last dosing day was 60-fold higher than the AUC of the intramuscular dose on the first dosing day (877 ng·h/ml), indicating that the exposure concentration of AE had been evidently increased (Figure 25, top) [44]. Since AE-induced brainstem neuropathology in monkeys occurred at a minimal dose of 8 mg/kg daily doses for 14 days [665], the minimal plasma concentration of AE 193.8 ng/ml with this low dose (8.0 mg/kg) had been simulated by our laboratory, as the minimal detectable pathological effect level (MDPEL) in plasma that should be at first to risk causing neurotoxicity in monkeys. The MDPEL value (193.8 ng/ml) of monkeys is 4-fold greater than that of rats (41.32 ng/ml) and dogs (40.92 ng/ml), indicating that the rat and dog seems to be more sensitive than monkey species to suffer neurotoxicity.

4) In Conclusion: Drug Exposure Time Resulting in Neurotoxicity

To date, there have been no neurological side-effects noted in humans due to DHA, QHS, AM, AE, and AS treatments. After intramuscular administration, AE and AM provide a longer exposure time in humans the same way they did in animal species (rats, dogs, and monkeys) [524, 681, 690, 696]. The conversion of AM and AS to DHA in human is much higher than that in animal species [44], but why humans are able to avert the fatal neurotoxicity described in animal species is unknown. This review finds that the drug exposure time is the principle factor in the induction of neurotoxicity in rats, dogs, and monkeys. The minimal detected pathological effect level (MDPEL) for AE and minimal detected neurotoxic effect level (MDNEL) for AL have been used to measure the exposure period at neurotoxic effect levels in the animals in the present discussion. When using the rats (41.32 ng/ml) or monkeys (193.8 ng/ml) MDPEL to estimate human samples treated with AE with regimen of 4.8 mg/kg at 0 hr and 1.6 mg/kg at 6 hr and then daily in 2-5 day following intramuscular injection, the 162.4 or 33.2 hrs (Figure 28, top), respectively, of drug exposure time is estimated during the 5 days treatment in the human subjects [690]. However, in the 42 patients treated with this regimen there was a failure to detect any neurotoxicity even under the very carefully monitored design in those clinical trials. Regarding acute toxicity, monkeys have fewer responsive than rats, whereas humans appear to be less sensitive than animals [39, 40] and humans have much better repair capabilities than animals do [41].

Other possible explanations for the neurotoxicity discrepancy observed between animal and human can be:

i. The therapeutic dose (2-4 mg/kg) in humans is never higher than the minimal dose that exhibited neuropathologic findings in rats (12.5 mg/kg for 7 days), dogs (6 mg/kg for 28 days), or monkeys (8 mg/kg for 14 days). Also, it is never higher than the minimal doses that showed clinical signs in rats (50 mg/kg for 7 days), in dogs (20 mg/kg for 7 days), or in monkeys (24 mg/kg for 14 days), all of which are 3-4 times higher than the doses resulting in neuropathologic findings (Table 50);

ii. The treatment period (3-7 days) in humans is never longer than the minimal time for the development of neurotoxicity in rats (7 days), dogs (28 days), or monkeys (14 days);

iii. Many of the patients followed had been seriously ill with cerebral malaria due to severe *falciparum* infection. The resulting evidence may lead to future studies on whether the pattern of neurotoxicity occurred from the illness or from the drug(s);

iv. A number of early reports did not pay sufficient attention to the feasibility of associating brain damage and neurological disorders in patients with such treatments of malaria; and

v. Post-marketing surveillance is limited in developing countries and the potential neurological side-effects of the drug are not recorded.

In conclusion, the drug exposure time of artemisinins is the principle factor responsible for the induction of neurotoxicity in animal species. Therefore, the active metabolites (DHA), the drug distributed in CNS, and/or the drug exposure levels do not play important roles in inducing neurotoxicity. The oil-soluble artemisinins, AM and AE, possess the ability to induce fatal neurotoxicity because their sesame oil vehicle delays and prolongs the absorption from the intramuscular injection sites leading to an enormous drug accumulation, notably extending the exposure time to the toxic effect level. This would indicate that relinquishment of sesame oil as a vehicle should be considered. Oral AL induced the moderate neurotoxicity in rats also because the AL long exposure time was related to drug accumulation in the blood resulting from a delayed gastric emptying-induced prolonged absorption of the AL from the rat stomach. However, the water-soluble artemisinins, injectable AS and AL, have very short half-lives and produce limited exposure time in animal species and without any neurotoxicity observation with various dosages and regimens, indicating that once-daily intravenous administrations of artemisinins are relatively safe [94, 524].

6.4. Reprotoxicity

Malaria in pregnancy reduces birth weight (and thereby infant survival). In low-transmission areas there is an increased risk of severe *falciparum* malaria and consequent death of both mother and fetus. The health community has finally realized that even if the mother is asymptomatic, the presence of malaria parasites in the blood is always harmful to the fetus and must be treated promptly and effectively. There have been remarkably few studies of antimalarial drugs in pregnancy. Of over 500 antimalarial drug trials conducted between 1966 and December 2006, only 31 evaluated antimalarial treatments (including intermittent preventive treatments) specifically in pregnant women (and 14 of these were from a single centre) [700]. Only primaquine and tetracyclines are considered contra-indicated in pregnancy, although the evidence base for the safety of widely used antimalarials such as amodiaquine is weak [701], and recommended dose regimens for pregnant women are all derived from studies in non-pregnant adults.

Chloroquine is safe throughout pregnancy, and mefloquine and artemisinins are safe in the second and third trimesters, with limited data suggesting safety in the first trimester [593]. Malaria often occurs in chloroquine-resistant regions, thus the pregnant traveler cannot generally choose chloroquine. Effectively, she has the choice of mefloquine in the second and third trimester, and nothing for the first trimester. The data suggest that mefloquine may lead to stillbirths if administered in the first trimester [381]. Published data on 607 pregnancies in which artemisinin compounds were given during the 2nd or 3rd trimesters indicate no evidence of treatment-related, adverse pregnancy outcomes. Similar data show normal outcomes in 124 pregnancies exposed to artemisinin compounds in the 1st trimester. Artemisinin compounds cannot be recommended for treatment of malaria in the first trimester. Because the safety data is limited, artemisinin compounds should only be used in the second and third trimester. Artesunate-atovaquone-proguanil is a well-tolerated, effective, practical, but expensive treatment for multidrug resistant *P. falciparum* malaria during the second or third trimester of pregnancy [702]. Very small amounts of antimalarial drugs have been discovered in the breast milk, but the amount of drug transferred is not thought to be harmful to the infant. Because the quantity of antimalarials transferred in breast milk is insufficient to provide adequate protection against malaria, infants who require treatment or chemoprophylaxis should obtain the recommended antimalarial dosages.

6.4.1. Malaria in Pregnancy and Safe Antimalarial Drugs

Malaria is an enormous global health problem and most of the disease burden affects young children and pregnant women. The adverse impact of malaria in pregnant women is largely caused by *P. falciparum*; approximately 90% of *P. falciparum* clinical cases globally occur in sub-Saharan Africa. Malarial infection during pregnancy poses substantial risk to the mother, her fetus, and the neonate. In areas of low transmission of *P. falciparum*, women do not acquire substantial antimalarial immunity and are susceptible to episodes of severe malaria, which may result in stillbirths, spontaneous abortions, or maternal death. In areas of high transmission of *P. falciparum*, where adult women have considerable acquired immunity, women may have asymptomatic infections or be minimally symptomatic, but such infections can contribute to maternal anemia and cause placental parasitemia, both of which may subsequently lead to low birth weight (LBW). Prevalence of parasitemia is greatest in the second trimester, and susceptibility to clinical malaria appears higher in both second and third trimesters. Although there is less data about the role of *P. vivax*, there is evidence that it may also lead to anemia and LBW. LBW is an important contributor to neonatal mortality. It is estimated that malaria in pregnancy is responsible for 5–12% of all LBW, 35% of LBW that is preventable during pregnancy, and contributes to 75,000–200,000 infant deaths each year.

For decades the prophylaxis and treatment of malaria during pregnancy have relied on chloroquine. Resistance in *P. falciparum* to chloroquine and increasingly to sulfadoxine-pyrimethamine in Africa, means that other antimalarials will have to be used in pregnancy on this continent, the worst affected by malaria. In South East Asia, multidrug resistance in *P. falciparum* emerged more than 30 years ago and drugs such as quinine, mefloquine, artemisinin derivatives, and atovaquone-proguanil have been used to treat pregnant women with malaria. For an antimalarial to be considered for use in pregnancy, it must be safe for the mother, the fetus, and later for the breast-feeding infant. However, safety is a relative term and it is difficult to unequivocally prove safety.

There are inherent limitations within this drug safety data when evaluating risks during pregnancy. For some drugs, there are years of programmatic experience to suggest that a drug is generally safe for use during pregnancy. Sometimes no data exists at all from use by pregnant women. At other times, a drug may be a new combination of component drugs where the individual drugs are considered to be safe in pregnancy, but for which there is no data on the combination. Ultimately, we have widely varying amounts of information on antimalarial drugs that allow us to broadly categorize them as (i) useful for a pregnant woman, as there are extensive data from programmatic use during pregnancy, (ii) possibly useful for a pregnant woman, but more data is needed, or (iii) not useful for a pregnant woman because they have known adverse events associated with their use in pregnancy, safe and efficacious alternatives exist. Drugs in this third category are not considered further in the paper, as efficacy and effectiveness are irrelevant in the face of poor or questionable safety. Finally, at the end of this section, we briefly review safety concerning antimalarial use during lactation [144].

6.4.2. Safety of Antimalarial Drugs in Pregnancy

6.4.2.1. Safe Antimalarial Drugs in Pregnancy

Drug regimens should be recommended on the basis of pharmacokinetic and pharmacodynamic information in the target population. Where blood or plasma concentrations of the antimalarial drugs have been measured in late pregnancy, they have usually been found to be reduced [435, 458]. For artemether, dihydroartemisinin, atovaquone, proguanil, and lumefantrine, the reductions have been substantial, and likely to have contributed to poor therapeutic responses, yet there have been no studies of higher dose regimens in pregnancy. In recent years, sulfadoxine-pyrimethamine (SP) replaced chloroquine as the most widely used antimalarial drug in pregnancy. However, the pharmacokinetic properties of sulfadoxine and pyrimethamine in pregnancy have only just been reported, years after the policy recommendations and widespread deployment of SP first as a treatment and later as intermittent preventive treatment in pregnancy. Blood concentrations of sulfadoxine were found to be 40% lower during pregnancy compared with after pregnancy [453], suggesting that the dose of SP in pregnancy may have been too low.

Antimalarial drugs have received widespread programmatic use during pregnancy, and are thought to be safe, although overdoses and idiosyncratic reactions could lead to detrimental effects on the mother and/or the fetus. Chloroquine (CQ), perhaps the most widely used antimalarial, is generally considered safe in all trimesters of pregnancy [307]. Children born to a cohort of 169 non-immune women who took CQ chemoprophylaxis throughout pregnancy had no more birth defects than 454 births to women who had not received CQ [703]. Among >2500 women who received CQ, there was no reported increase in abortions, stillbirths, or congenital abnormalities, although there were frequent non-severe side-effects such as itching, dizziness, and gastrointestinal complaints. There are reports of increased spontaneous abortions, particularly in patients with systemic lupus erythematosus treated with high doses of CQ over prolonged periods, and in some settings, CQ has gained a reputation as an abortifacient at higher doses. CQ overdoses have been responsible for numerous deaths, but its abortifacient effects appear to be limited to these very high doses that are life-threatening to the mother [144].

In many region areas, quinine remains the principal treatment for severe malaria. It has been associated with teratogenic effects and damage to fetal optic and auditory nerves when taken at very high (abortifacient) doses, but should be considered safe in pregnancy when taken at normal therapeutic doses. There is, however, a risk of hyperinsulinemia and subsequent hypoglycemia in women who take quinine. The related compound, quinidine, is also considered safe in pregnancy. There are no reports of congenital abnormalities associated with its use during pregnancy, although there have been reports of neonatal thrombocytopenia after maternal use. Proguanil is generally considered safe during pregnancy. A cohort of pregnant travelers who took proguanil in combination with CQ had no higher rates of spontaneous abortions or congenital anomalies than the expected background rate. A study in

Nigeria found no increase in adverse outcomes among pregnant women receiving daily proguanil (100 mg) in addition to weekly CQ. Another study in Tanzania found no increase in adverse outcomes among women receiving proguanil alone or in combination with weekly CQ compared with those receiving CQ alone. Sulphonamides are also generally considered safe in the second and third trimesters of pregnancy.

Although there is very limited evidence that sulpha drugs may be associated with kernicterus when given to premature neonates, this problem has not been noted in studies of IPT where sulphadoxine-pyrimethamine (SP) was administered to the mother. Studies examining the risk to the fetus from *in utero* exposure to SP combination have generally not found any increased risk in spontaneous abortions or congenital defects. A retrospective study of antifolate drugs given before and during pregnancy found that there was an increased risk of birth defects when such drugs were taken during the first trimester, but not during the second or third trimester. When given weekly as prophylaxis, SP has been associated with rare and severe cutaneous reactions such as toxic epidermal necrolysis and Stevens–Johnson syndrome; there is no evidence that this risk is any greater in pregnant women. In summary, these drugs are considered safe in the second and third trimesters of pregnancy [144].

Pyrimethamine is usually given in combination with sulphadoxine (see previous section). However, studies in which pyrimethamine has been given alone have also found no increase in adverse pregnancy outcomes. Dapsone has been used extensively in pregnant women with leprosy, without reported adverse effects. Studies in which dapsone was given in combination with pyrimethamine for malaria chemoprophylaxis during pregnancy had no increase in adverse pregnancy outcomes. However, a severe hypersensitivity syndrome has been described in a non-pregnant woman receiving weekly Maloprim. Additionally, occasional reports have been received of non-pregnant patients developing agranulocytosis following Maloprim chemoprophylaxis; although usually after twice weekly dosing, which is no longer recommended. A study using dapsone in combination with SP in children found no serious adverse effects. Dapsone is most often used currently with chlorproguanil.

Clindamycin has been used routinely as an antibiotic in pregnancy without any evidence of adverse effects. Although the drug does cross the placenta and does accumulate in fetal tissues perhaps to the point of therapeutic levels, there is no evidence that this accumulation is harmful. A recent trial in quinine comparing quinine-clindamycin with artesunate for the treatment of *falciparum* malaria during pregnancy found no serious adverse events, no increase in stillbirths or congenital anomalies above expected levels, and no negative impact on infant development [144].

6.4.2.2. Antimalarials seldom Associated with Reprotoxicity

For a number of other drugs there are some concerns of safety in pregnancy, or there are limited data about their use during pregnancy. Amodiaquine (AQ), a 4-aminoquinoline related to CQ, falls under the category of antimalarials about which there are insufficient data to be certain about their use in pregnancy [704]. In non-pregnant women taking AQ for chemoprophylaxis, there have been reports of agranulocytosis and granulocytopenia, hepatitis, and increases in serum aspartate aminotransferase (AST) levels. A study comparing AQ with atovaquone + proguanil found no serious adverse events, but found that pruritis, weakness, insomnia, and dizziness were more common in the AQ group. However, a recent systematic review of studies of AQ for the treatment of malaria found no increase in adverse events when compared with CQ or SP, and found that all adverse events were minor or moderate, and not life threatening. A recently completed three-country trial in Africa found no cases of clinical hepatitis among children >10 years receiving AQ or AQ + artesunate, but did find that 6% of children developed neutropenia (neutrophil count < 1000/ll) and that 60% of children experienced a decline in serial neutrophil counts. In a smaller study involving younger (6–59 month) Tanzanian children, no serious adverse effects of AQ were noted, and no increase in neutropenia was noted when compared with children receiving SP. In Burma, AQ and quinine were compared for treatment of malaria in pregnant women. A high overall rate of spontaneous abortion was noted, but was not stratified by treatment type [701].

Chlorproguanil (Lapudrine), in combination with dapsone, has only been evaluated in a single published trial among pregnant women [705]. However, the study did not specifically address whether any adverse fetal outcomes were observed. There are two studies in children that have demonstrated good safety when using the combination [706], although one trial did find a higher rate of severe anemia among children treated with the combination than among those treated with SP [706]. Given that the component drugs are both considered safe in pregnancy, it is expected that the combination will also be safe for use in pregnancy.

Mefloquine (MQ) is a quinoline methanol compound that has been used extensively, particularly in Asia for the treatment of *P. falciparum*. It has also been used extensively for malaria chemoprophylaxis among travelers to areas with CQ-resistant *P. falciparum* malaria. Safety of MQ during pregnancy has been evaluated through post-marketing surveillance, retrospective and prospective studies. An early dose-finding study in Thailand found no increase in adverse pregnancy outcomes among women who took either 125 or 250 mg/week for prophylaxis during the third trimester. Post-marketing data collected by the manufacturers of MQ showed that among 1627 women exposed to the drug during pregnancy (95% for chemoprophylaxis), there was no increase in congenital malformations over the expected background rate. In one prospective study of pregnant women, MQ chemoprophylaxis was more frequently associated with stillbirths than SP chemoprophylaxis, (9.1% vs. 2.6%), but this rate did not differ from the background rate of stillbirths among the population studied (7–10%). Among 451 women in Malawi who took MQ treatment (750 mg) followed by weekly MQ chemoprophylaxis (250 mg), no increase in the incidence of stillbirths or spontaneous abortions was observed [705].

Among US soldiers in Somalia, 72 women used MQ for chemoprophylaxis before learning of their pregnancies, and a greater expected percentage (16.7%) had spontaneous abortions. Data from two studies in Thailand have showed no difference in infant development between infants born to mothers given MQ chemoprophylaxis during pregnancy and those given placebo, or between women treated with MQ + artesunate and those treated with quinine [270]. Data from a retrospective study in Thailand suggest a significantly increased risk of stillbirth among women exposed to treatment doses of MQ during pregnancy compared with those exposed to quinine, other treatments or women who had no malaria. However, no general patterns of physical abnormalities or specific defects were observed in the stillbirths from MQ-using women. Although most of the data from a variety of sources suggest that MQ is safe for use in pregnancy, recent data from Thailand highlight the need for continued vigilance in monitoring adverse events of women treated with MQ in pregnancy [270].

Malarone, a fixed combination of atovaquone (250 mg) and proguanil (100 mg), is currently being evaluated in pregnant women. (See earlier section on proguanil). In clinical trials among non-pregnant people, Malarone appears to have an excellent safety profile when used at the dosage recommended for prophylaxis of *P. falciparum*. The most commonly reported adverse effects include gastrointestinal disturbances and headache. There were no significant differences in moderate to severe adverse effects in the randomized, placebo-controlled, double-blind studies. Generally, atovaquone/proguanil is as well or better tolerated than most drugs at doses necessary for treatment of malaria, although there are not yet any published data on the use of Malarone for treatment of pregnant women with malaria. Treatment limiting adverse events occurs in <1% of patients receiving treatment doses, and serious adverse effects attributable to treatment doses are rare. The most common adverse effects at treatment doses include vomiting, nausea, and abdominal pain. Although the clinical significance of these elevations is unknown, studies have not shown liver enzyme elevations to be treatment limiting.

Azithromycin is another compound for which there is limited data about safety in pregnancy. There are several trials in which azithromycin was used in the treatment of sexually transmitted diseases and other genital infections during pregnancy; no adverse neonatal outcomes were noted. Azithromycin does not readily cross the placenta, and the drug appears to be generally safe for use in pregnancy. There are no published data evaluating the safety of azithromycin for the treatment or prevention of malaria during pregnancy.

6.4.2.3. Contra-indicated Antimalarial Drugs in Pregnancy

Certain antimalarials should not be used in pregnancy because of their effects on the fetus. Halofantrine is neither mutagenic nor teratogenic but is embryotoxicity in animals. Pregnant rabbits exposed to high doses (60-120 mg/kg/d) between day 7 and 19 of gestation gave birth to pups with skeletal abnormalities. Body-weight was decreased in female rats exposed to 25-100 mg/kg/d of halofantrine from day 15 of gestation and day 21 postpartum, and the pups also showed a loss of weight. No data exists on the use of halofantrine in pregnant women, but the cardiotoxicity of the drug has compromised its role in the treatment of uncomplicated *falciparum* malaria [270].

Two antibacterials, tetracycline and doxycycline, fall into this category. Tetracycline easily crosses the placenta, and can lead to disturbances of skeletal growth, permanent discoloration of teeth, corneas and lenses. Additionally, tetracycline has been associated with an increased hepatotoxicity among pregnant women, especially in the last trimester. Doxycycline, a closely related compound, has been presumed to be capable of causing similar tooth discoloration, although recent data suggest that perhaps because of its lower binding affinity for calcium than tetracycline, doxycycline does not cause clinically significant staining of permanent teeth following use in childhood.

Primaquine (PQ), an 8-aminoquinolone used primarily for the radical cure of *P. vivax* and *P. ovale* infections, but also used as chemoprophylaxis against *P. falciparum*, is also generally considered contra-indicated in pregnancy. Persons who have a deficiency of glucose-6-phosphate dehydrogenase (G6PD) are at increased risk of acute hemolytic events associated with the administration of PQ. There is a theoretical increased risk of hemolysis and subsequent jaundice among G6PD deficient infants whose mothers received PQ. This risk would be greatest in Africa, where the gene responsible for G6PD deficiency is most common. Because of the theoretical risk, the recommendation for laboratory screening for G6PD deficiency, the ability to delay radical cure with PQ until after pregnancy, and the availability of other antimalarials, PQ is not recommended for use in pregnancy.

Tafenoquine (WR 238605) is, like PQ, an 8-aminoquinolone. It has been shown to be effective prophylaxis against *P. falciparum* infections in semi-immune teenagers and young adults, and effective in preventing relapse with *P. vivax*. There are currently no published data about its safety in pregnancy, as trials to date have specifically excluded pregnant or lactating women. However, it is more likely to have properties similar to other 8-aminoquinolone, and therefore pose a theoretical risk to women and neonates with G6PD deficiency, and therefore be inappropriate for use during pregnancy.

Other antimalarials have been associated with serious side-effects in non-pregnant adults and should therefore also be used with caution in pregnant women. Halofantrine has been associated with lengthening of the QT interval, and with fatal arrhythmias in some persons. Current recommendations suggest that this drug be used for treatment only in those who have a documented normal electrocardiogram, which makes its use in many developing world settings impractical. The questionable safety of these drugs (tetracycline, doxycycline, PQ, tafenoquine, and halofantrine) in pregnant women does not imply that they are completely contra-indicated. In the face of serious illness and in settings where a limited number of drugs are available, it is necessary to balance the risk to the life of the mother (whose death could also lead to fetal demise), with hypothetical risks to the fetus [144].

6.4.2.4. Antimalarial Drugs Safety during Lactation

A number of antimalarials are considered safe during lactation, including: CQ, quinine, SP, pyrimethamine, dapsone, clindamycin, mefloquine, doxycycline, and tetracycline [8]. Six of these (CQ, quinine, pyrimethamine, dapsone, clindamycin, and tetracycline) are specifically mentioned as compatible with breastfeeding by the American Academy of Pediatrics (AAP) [8]. There is some evidence that caution should be exercised when using sulphonamides in infants who are sick, premature, or have G6PD deficiency; because these drugs are present in breast milk, lactation under these conditions should take place with caution [707]. For other drugs, there are no data about secretion into breast milk or use during lactation. However, as proguanil is considered safe for use in pregnancy, and AQ, chlorproguanil-dapsone, azithromycin, and artemether-lumefantrine are considered safe for use in infancy; it is more likely that these drugs would not pose serious harm to the infant through lactation.

6.4.3. Safety of Artemisinin and its Derivatives in Pregnancy

6.4.3.1. No Adverse Effects of Artemisinins in Pregnancy

Artemisinins, a group of related compounds (sesquiterpene lactones) also known by their Chinese name, Qinghaosu, are derived from the medicinal herb *Artemisia annua*, which is also known as annual or sweet wormwood. Its antimalarial properties appear to have been known in ancient times, but were rediscovered in China in the early 1970s [516]. Artemisinins are available in a variety of oral (artemisinin, artesunate, artemether, and dihydroartemisinin), parenteral (artemether, arteether, and artesunate), and rectal (artesunate and dihydroartemisinin) formulations. These compounds have received widespread attention in recent years in the treatment of severe malaria, and in the treatment of multidrug-resistant *falciparum* malaria, particularly in Southeast Asia.

The safety of the artemisinin derivatives in early pregnancy remains uncertain, as these drugs are embryotoxic in animals [708, 709]. They are not recommended for uncomplicated malaria in the first trimester, unless effective alternatives are unavailable. There is now a reasonable body of evidence for safety from clinical trials in nearly 1,000 women in the second and third trimesters of pregnancy, and artemisinin-based combination treatments are now recommended treatments in the second and third trimesters [458]. Severe malaria in late pregnancy carries a mortality approaching 50%—approximately 2.5 times higher than in non-pregnant patients. Artesunate has been shown to reduce the mortality of severe malaria by 35% compared with quinine, and, unlike quinine [90], it does not cause severe recurrent hyperinsulinemic hypoglycemia, for which pregnant women are at considerably increased risk. But quinine is still the most widely used treatment for severe malaria in pregnancy. Somehow, the unknown, but certainly very small, risk to the fetus posed by artesunate is considered to exceed the 35% increased risk of both mother and baby dying from malaria. Although artesunate is clearly now the drug of choice in severe malaria, there has been no pharmacokinetic studies in pregnant women to guide dosage. By extrapolation from studies in uncomplicated malaria, the currently recommended dose could again be too low [710].

A small case series from China found that among six pregnant women treated for malaria (*P. falciparum* or *P. vivax*) at a mean of 21.7 weeks of gestation none had adverse outcomes [711]. In another trial from China, seven children exposed *in utero* between 17 and 27 weeks of gestation were tracked after birth from 3 to 10 years; no adverse outcomes were found. In another clinical trial in China, 21 pregnant women were included and were given a variety of artemisinins; no adverse outcomes were recorded [552]. A study in 83 women in Thailand treated with either artesunate or artemether found no increase in adverse outcomes (4% spontaneous abortion and 3% stillbirth – all of which were explainable by other events). Sixteen of the women in this study were accidentally exposed to artemisinins during the first trimester. Follow up of the live born children from this cohort found no developmental delay [712]. Further work in Thailand has shown that women treated with MQ + artesunate or 7 days of artesunate alone had no increase in adverse effects or adverse birth outcomes, and no negative developmental impact when compared with women treated with quinine. A further 461 women in Thailand treated with either artesunate or artemether for *P. falciparum* malaria had no increase in rates of abortion, stillbirth, congenital abnormality, or mean gestation at delivery. A total of 287 pregnant women in the Gambia were exposed to artesunate in combination with SP during a mass drug administration; no difference was noted in the rates of abortions, stillbirths, or infant deaths among those exposed or not exposed to the drugs [144].

A recent consultation held at WHO has recommended that: (1) because of the limited experience with the artemisinins in pregnancy they should only be used when other treatments are considered unsuitable, (2) presently, they cannot be recommended for treatment of malaria in the first trimester. They should not, however, be withheld if they are considered lifesaving for the mother, (3) to further document the safety of artemisinin compounds in pregnancy, careful follow-up is required, with documentation of pregnancy outcome and the subsequent development of the child whenever possible, and (4) to guide further development of policies on the use of artemisinin derivatives during pregnancy, alone or in combination, there is an urgent need for further research/documentation of their efficacy and safety for use as therapy for malaria and for IPT.

6.4.3.2. Major Embryotocixity Associated with Artemisinins in Animals

1) Embryotoxicity of Artemisinin and its Derivatives

Artemisinin derivatives are not currently recommended for use during the first trimester of pregnancy in women because the derivatives cause embryo death and some abnormalities in early pregnancy in animals [708-719]. In order to better understand the mechanism of the developmental toxicity in animals, the main animal experiments have been listed here for the discussion.

Use of big doses (40 mg/kg subcutaneously on day 6-10 of gestation), Lou and Zhou observed that AS significantly decreased the concentration of serum progesterone in early pregnant rats; decidual cells and fetuses of treated groups were found to be degenerated at day 11. Drug was shown to directly damage the decidual cells. Similar results were found in human decidual cell cultures with AS for 48 h. The damage of AS on the decidua and placenta may be the mechanism of its contra-gestational action [720]. These results also suggested that in order to examine the effect and mechanism of AS on embryo development, rat embryo and placental glutathione peroxidase (GSH-Px) and malondialdehyde (MDA) must be identified by using dithionitrobenzene (DTNB) direct method and thiobarbituric acid (TBA). They found that AS induced developmental toxicity in rat embryos and placenta by neutralizing the antioxidant defense mechanism [721].

Chen *et al.* reported that AS [722] and DHA [723] were found to inhibit angiogenesis *in vivo* and *in vitro*. The anti-angiogenic effect *in vivo* was evaluated in nude mice by means of human ovarian cancer HO-8910 implantation and immunohistochemical staining, vascular endothelial growth factor (VEGF) and VEGF receptor *KDR/flk-1*. The *in vitro* effect of AS was tested on models of angiogenesis, namely, proliferation, migration and tube formation of human umbilical vein endothelial cells (HUVEC). The results showed that AS and DHA significantly inhibit angiogenesis in a dose-dependent form. Additionally, the inhibitory effect of AS and DHA on HVUEC proliferation was stronger than that on Hela, JAR, HO-8910 cancer cells, NIH-3T3 fibroblast cells and human endometrial cells, indicating that AS and DHA were selectively against HUVEC. These findings and the known low toxicity of AS and DHA are clues that artemisinins may be a promising angiogenesis inhibitor [722, 723].

The effects of DHA in rat's whole embryo cultures (WEC) were recently conducted by Longo *et al.* [709]. DHA was added to the culture medium for the entire 48-hour culture, 1.5 hours after the beginning or 1.5 hours before the end of the culture dose of 0.01–2 µg/ml. DHA affected primarily red blood cells during yolk sac hematopoiesis. Higher concentrations and longer exposure inhibited angiogenesis. More studies the inhibition of angiogenesis by artemisinins were conducted by various investigators [724-729]. AS strongly reduced angiogenesis *in vivo* in terms of vascularization of Matrigel plugs injected subcutaneously into syngenic mice. Based on these studies, AS represents a promising candidate drug for the treatment of the highly angiogenic Kaposi's sarcoma. As a low-cost drug, it might be of particular interest to areas of Kaposi's sarcoma endemics. These results and the known low toxicity are clues that AS and DHA may be promising novel candidates for cancer chemotherapy [726, 729].

It is clear that the embryotoxicity and contra-gestational effect of artemisinins (AS and DHA) in pregnant animal species is mainly due to the angiogenesis and the expressions of vascular endothelial growth factor on embryos and fetuses. Further research has demonstrated that the embryotoxicity of artemisinins is drug exposure level dependent [709, 723].

2) Effects of Drug Exposure Level and Time on Inhibition of Angiogenesis

To study the effects of DHA on embryonic development, a rat whole embryo culture (WEC) study was conducted by Longo *et al.* [709]. This model allowed the investigation of early embryonic developmental stages, which reportedly were highly susceptible to artemisinin compounds *in vivo*, in the absence of maternal interaction. Ultimately, the WEC study was conducted in order to better understand the mechanism of the developmental toxicity and to identify the initial target of action of DHA.

On gestation day 9, pregnant female rats were euthanized to extract embryos (9.5 days of age, 1–3 somites). The extracted embryos were randomly distributed to experimental groups and cultured in a 25-ml glass bottle (5 embryos/bottle) containing 5 ml of heat-inactivated sterile rat serum. DHA was dissolved in DMSO and added to the culture medium (5 μl/bottle) in order to obtain a range of final concentrations of 10 – 2000 ng/ml. Group 1: control, DMSO 0.1% from GD 9.5 to 11.5; Group 2: DHA 10, 50, 100, 500, 1000, and 2000 ng/ml from GD 9.5 to 11.5; Group 3: DHA 50, 100, 500, 1000, and 2000 ng/ml at GD 9.5 for 1.5 h, thereafter the medium was changed and embryos were cultured in normal serum until GD 11.5; Group 4: DHA 1000 and 2000 ng/ml at GD 11, 1.5 hr before the end of the culture. At the end of the culture period, the embryos (11.5 days old) were transferred into Tyrode's salt solution (Sigma) and examined under the stereomicroscope.

All embryos exposed to DHA for 48 hr at all concentrations were alive at the end of the culture period. At DHA concentrations ≥ 50 ng/ml, the visceral yolk sac was well vascularized but pale; yolk sac vessel formation and circulation appeared normal but blood was visibly paler than in controls. This effect increased in proportion to increasing DHA concentrations. At concentrations ≥ 500 ng/ml, the vasculature was also affected. Vessel diameter was reduced, small anastomotic vessels were not visible, and the number of circulating cells was further reduced. At 1 μg/ml, yolk sac vasculature and circulating cells were markedly reduced and at 2000 ng/ml only a few poorly organized yolk sac vessels were present.

The allantois, although well developed up to the DHA concentration of 500 ng/ml, showed pale umbilical vessels starting from 50 ng/ml. At 1 μg/ml, the allantois was present but attached ectopically to the yolk sac (fusion with the chorion did not occur near the ectoplacental cone but in other regions of the yolk sac). At 2000 ng/ml, the allantois was lying free in the exocelom. At DHA ≥ 1000 ng/ml, the heart-rate was reduced and irregular. Analysis of cytospin blood preparations showed a reduction in the number of RBCs starting at concentrations ≥ 100 ng/ml, and deformed or damaged RBCs at concentrations ≥ 500 ng/ml. At 2000 ng/ml, DHA induced complete embryonic disruption with multiple non-specific abnormalities. Histological examination showed areas of cell death in the branchial arches, in the mesenchyme of the caudal region and near the dorsal aorta at 100, 500, and 1000 ng/ml. Complete tissue disorganization was detected at 2000 ng/ml. Histological examination confirmed the reduced presence of RBCs in the heart and aorta of embryos exposed to DHA at 100 ng/ml concentrations. For embryos exposed to DHA for 48 hr, no observed adverse effect level (NOAEL) was 10 ng/ml.

Similar observation of the embryotoxicity was also found at 1.5 hr after culture with DHA. The embryos exposed to DHA at all concentrations tested were alive at the end of the culture period and had normal heart rate. At DHA concentrations ≥ 100 ng/ml, the visceral yolk sac, although well vascularized, was pale; yolk sac vessel formation and circulation appeared normal, but blood was visibly paler than in control embryos. This effect was more obvious with increasing concentrations. Starting from 1μg/ml, the circulation was also affected, and small anastomotic vessels were not visible. The allantois, although well developed at all concentrations tested, showed pale umbilical vessels starting from 100 ng/ml and cases of ectopy from 1000 ng/ml. At DHA concentrations ≥ 2000 ng/ml, the heart rate was reduced and irregular in the majority of embryos. The analysis of cytospin blood preparations showed a clear reduction in the number of RBCs starting from 500 ng/ml and deformed or damaged RBCs from 1000 ng/ml. For embryos exposed to DHA for 1.5 h at the start of the culture, the NOAEL was 50 ng/ml.

The results evidently indicated that the drug exposure level is more important to induce the embryotoxicity than drug exposure time; this was evident because the AS and DHA have very short half lives [497] in clinical therapy. When concentrations of DHA were increased to 100 ng/ml for 1.5 hr, the initial damage could be seen on the embryos. At 500 ng/ml, the analysis of cytospin blood preparations showed a clear reduction in the number of RBCs. At 1000 ng/ml level, the significant ectopy, irregular heart rate, and deformed or damaged RBCs were observed and detected. Those indicators would be great references for the clinical monitoring on the embryotoxicity induced by artemisinins. The literature on human trials and animal experiments concluded that a range of 1000-2000 ng/ml was an acceptable estimation of median AS or DHA maximum concentration [44].

3) Drug Sensitivity Period in Pregnant Animals and Women

Although the drug exposure level is a major factor in the development of embryotoxicity in animals, the sensitive duration period of the embryos during the gestation is another important factor for the toxic investigation. In most studies in the rat, treatment usually started around implantation on day 6 and continued thereafter. A common effect seen was a dose-related increase in post-implantation loss with total resorptions at higher dose levels. Some studies show instances of morphological abnormalities at doses that cause fetal resorption, without maternal toxicity. All studies using AS, AM, or AE show a steep dose–response curve and some have demonstrated marked sensitivity to the day of dosing between Days 9–14 of gestation in the rat. In particular, in Chinese studies on rats, with AS, a dose of 0.4 mg/kg produced no increase in resorptions (5.8%), but 0.8 mg/kg caused 98% resorption [15]. Total resorption was also induced by AS at 25 mg/kg SC, when administered during GD 9–11 or 12–14, while no effect was seen when treatment was given prior to GD 8 or after GD 14. Further comparative *in vivo* experiments with AS, DHA, AM, and AE indicate that the critical period for induction of resorption is GD 9–14; GD 10 is identified as the most sensitive day [709].

From day 9.5 of gestation in rats (day 7.5 in mice) and day 15 in humans, simple diffusion is no longer sufficient to meet the increasing demand for oxygen supply of the growing embryo. At that point, undifferentiated mesenchyme on the yolk sac condenses to form angiogenetic cell clusters, the blood islands that give rise to hematopoietic stem cells (the precursors of the blood cells) and angioblasts (the precursors of blood vessels). As soon as a capillary network is created, mature erythroid cells formed within the yolk sac enter the embryos and reach the vitelline vessels. Hematopoiesis within the yolk sac occurs in rats during GD 9–14 (GD 7–12 in mice) and in humans between GD 15 and 6 weeks of gestation, when it is replaced by late embryonic, fetal, and subsequently definitive adult sites of hematopoiesis (the liver, spleen, and bone marrow) [730]. The effects of exposure to artemisinins during this special period of organogenesis can be noticed especially on the embryotoxicity.

During the development period embraced by this experiment, it becomes more apparent that a transition occurs from predominantly anerobic to aerobic glycolysis, with active Krebs cycle and oxidative phosphorylation; this requires significantly higher oxygen concentrations, as well as an adequate circulation. Accordingly, culture conditions are changed from 5 to 20% O_2. Under the latter conditions, the probability of reactive oxygen species (ROS) formation is increased. The visceral yolk sac has a protective role for the embryo proper by removing ROS and antioxidant defenses increase several times in the yolk sac from GD 9 to 13 [731].

Longo's results are compatible with the hypothesis that DHA also affects RBC precursors (hemoangioblasts) and the Wolffian blood islands, thus explaining why hematopoiesis as well as vasculogenesis and angiogenesis are affected, although with different sensitivities. Some of the features observed in embryos seem to point to similar mechanisms as those described earlier for the malaria parasite. Embryonic RBCs derived from the yolk sac have transferrin receptors as they react with anti-rat CD71 monoclonal antibody. Moreover, some of the *in vivo* studies reported fetal growth retardation, delayed skeletal development and cardiovascular malformations in the animals [14-21]. The damage to the hematopoietic system and angiogenetic process may all be induced by exposure to those artemisinins. Finally, the days of marked sensitivity correspond to the period of yolk sac hematopoiesis, which occurs during GD 9–14 in rats and, correspondingly, GD 15 to week 6 in humans [732].

4) Drug Peak Level concerning the Reprotoxicity

The adverse impact of malaria in pregnant women is largely caused by *P. falciparum*; approximately 90% of clinical cases globally occur in sub-Saharan Africa. Every year there are approximately 50 million pregnancies in women living in malarious areas [455]. Artemisinins have been used to treat pregnant women since 1987 [711]. Data from limited clinical trials in pregnant women (2101 cases) exposed to artemisinin compounds, including a small number (108 cases) in the first trimester, do not show an increase in the rates of abortion or stillbirth or evidence of abnormalities. There were no clinically significant adverse effects of the drug, neither in the outcomes of the pregnancies, nor in the development (neurological and physical) of the infants, including 44 infants exposed during the first trimester [206, 260, 268, 307]. Are artemisinins really not toxic to either the women, the fetuses during pregnancy, or to the infants during lactation? Without other relevant human pharmacokinetic data, it is difficult to quantify the risk of possible embryonic death or teratogenicity with exposure to artemisinin compounds in the first trimester in women. However, the two factors of the pharmacokinetic characteristics (including tissue distribution) and the drug sensitive period of the embryos during the gestation (previously summarized) may assist us in avoiding the low birth weight, abortion, and even potentially fetal death for pregnant women requiring malaria therapy.

Table 54. Embryotoxic effects (NOAEL and ED$_{50}$) of artemisinin (QHS), dihydroartemisinin (DHA), artemether (AM), artesunate (AS) given intragastrically, intramuscularly, subcutaneously, and intravenously on pregnant animal species (female mice, rats, hamster, guinea pig, and rabbits) [44]

Animal (drugs)	Dose duration	Dose regimens (daily)	Dosing route	No-observed-adverse-effect-level (NOAEL) on fetus resorption (mg/kg)				ED$_{50}$ (95% CL) (mg/kg)
				Oral	IM	SC	IV	
Mice (AM)	GD 6-15	Multiple x 10	IM		5.4			11.3 (10.6 - 12.0)
(DHA)	GD 7	Single	SC			10		32.8 (27.7 – 38.9)
Rats (QHS)	GD 1-6	Multiple x 6	Oral	5.6				11.5 (10.5 – 12.2)
(DHA)	GD 9.5, 10.5	Single	Oral	7.5				NA
(AM)	GD 6-15	Multiple x 10	IM		2.7			6.13 (4.52 – 8.26)
(AM)	GD 6-15	Multiple x 10	Oral	2.5				14.4 (10.4 – 17.8)
(AS)	GD 6-15	Multiple x 10	SC			0.2		0.58 (0.55 – 0.61)*
(AS)	GD 6-18	Multiple x 11	IM/IV		0.5		0.4	0.61 (0.59 – 0.62)*
(AS)	GD 6-17	Multiple x 12	Oral	5-7				7.74 (6.92 – 8.57)
Hamster								
(DHA)	GD 7	Single	SC			4.2		6.06 (5.90 – 6.21)
(DHA)	GD 7	Single	Oral	20				51.0 (37.9 – 68.7)
(AS)	GD 5	Single	SC			0.35		1.0 (0.9 – 1.2)*
Guinea Pig								
(DHA)	GD 18	Single	IM		2.5			18.3 (13.9 – 24.2)
Rabbits (AM)	GD 7-18	Multiple x 12	IM		0.7			NA
(AS)	GD -7-19	Multiple x 13	Oral	5-7				NA
(DHA)	GD 9	Single	IM		5.0			7.57 (7.48 – 7.67)

DHA = Dihydroartemisinin; AM = Artemether; AS = Artesunate; IM = intramuscular; SC = subcutaneous; IV = intravenous; ED$_{50}$ = drug concentration induces 50% fetus resorbed; GD = gestation day (The day of mating was defined as day 0 of gestation.)

* The severe toxic effects were detected in the animals treated with AS after single or multiple intramuscular, intravenous or subcutaneous injections. Values of ED$_{50}$ are given as median (95% confidence limits). NA = not available.

In the summary of embryotoxicity in animals (Table 54), the severe toxic effects were detected in the animals treated with AS after single or multiple intravenous, intramuscular or subcutaneous injections [44, 307, 733,734]. The three ED$_{50}$ doses (0.58-1.0 mg/kg) are all below human therapeutic dose (2-4 mg/kg). Other dosage regimens of AS, AM, and DHA produced the high ED$_{50}$ doses with values of 6.1-32.8 mg/kg in those animal species, which seems to be safer than injectable AS in clinical usage because those dosages are higher than the human therapeutic doses with wide safety margin [735]. Why is the situation so severe such that there is AS only following the single or multiple injections intravenously, intramuscularly, and subcutaneously? The possible explanations could be:

1. The administration is during the days of marked sensitivity corresponding to the period of yolk sac hematopoiesis, which occurs during GD 9–14 in rats;
2. The bioequivalence of AS by intramuscular and subcutaneous administration are fulfilled for AS by intravenous injection in rats [44], and the 3 administrations of AS can produce a significantly higher peak concentration than AS, AM, and DHA after oral administration or AM with oil vehicle after intramuscular injection in rats [497].

Again, the data revealed that the peak concentration is a major issue that causes the toxicity;

3. As with multiple administrations, a single subcutaneous injection of AS also induced severe embryotoxicity in hamsters, demonstrating that the drug exposure level (even with just one exposure) is the principal role for persuading the toxicity when compared to longer drug exposure time with multiple doses;

4. When the peak concentrations are high in the blood, the concentration could be even higher in the location of the feto-placental unit because the drug distribution of AS in feto-placental tissues was always higher (2-4 folds) than that in the blood of the pregnant rats [736].

To treat women with artemisinins, a recent WHO consultation has recommended that: i) because of the limited experience with artemisinins in pregnancy they should only be used when other treatments are considered unsuitable; ii) presently, they cannot be recommended for treatment of malaria in the first trimester. They should not, however, be withheld if they are considered lifesaving for the mother; iii) to further document the safety of artemisinin compounds in pregnancy, careful follow-up is required, with documentation of pregnancy outcome and the subsequent development of the child whenever possible; and iv) to guide further development of policies on the use of artemisinin derivatives during pregnancy, alone or in combination, there is an urgent need for further research/documentation of their efficacy and safety for use as therapy for malaria [737].

In accordance with WHO recommendations and the new research described here, the two major issues for considering artemisinin drug use in a program for prevention or management of malaria in pregnant women are safety and effectiveness. First, the exposure to artemisinins should be avoided during the early sensitive days (GD 15 to week 6 in humans), which is the critical period for induction of embryo damage and resorption, to protect pregnant women from malaria with treatment of artemisinin derivatives. This is essentially the same recommendation as WHO consult declaring that the artemisinin drugs should not be used in the first trimester of pregnancy in women. Secondly, WHO recommends that artemisinin compounds should only be used in the second and third trimester when other treatments are considered unsuited. However, we feel that oral doses or oral doses in combination regimens could be used to treat pregnant women in all trimesters when other treatments are considered unsuited, because the oral regiments provides lower peak concentrations, and that can make the agents safer than intravenous or intramuscular injection of AS. In fact, reprotoxicity was not found in humans treated with artemisinins during the past twenty years; this conclusion resulted from the more than 90% of pregnant women treated orally or intramuscularly with AM in sesame oil over that period whose tests resulted in much low plasma concentration and prolonged absorption from muscle [206, 260, 268, 307, 733,734].

In addition, further studies to define the precise mechanism of damage in animal models are warranted. Reliable pharmacovigilance on the use of these drugs in pregnancy and the careful monitoring of safety after exposure in the first trimester of pregnancy, when treatment may occur inadvertently or be necessary to save life even with injection AS, are needed.

To date, there have been no convincingly documented cases of embryotoxicity or other reprotoxicity side-effects noted in humans due to QHS, DHA, AM, and AS treatments [206, 260, 268, 307, 733, 734]. After the four oral drugs and one intramuscular drug (AM) administration, the conversion of AM and AS to DHA in humans was been much higher than that in animal species although they produced lower peak concentrations in plasma [44]. It remains unknown how humans have been able to avert the death embryotoxicity described in animal species. Data from limited clinical trials in pregnant women (2,101 cases) exposed to artemisinin compounds, including a small number (108 cases) in the first trimester, have not shown an increase in the rates of abortion or stillbirth; they have also not shown evidence of abnormalities. Regarding acute toxicity, humans appear to be less sensitive than animals [697, 698] and humans have much better repair capabilities than animals do [699]. Other possible considerations for the reprotoxicity discrepancy observed between animal and human [44]:

i. The animal data revealed that only intramuscular or subcutaneous (including intravenous) AS causes reprotoxicity at lower dose than the therapeutic dose (2-4 mg/kg) in humans. Other doses in different regimens (oral or intramuscular AM) are safe at higher levels than the therapeutic doses. Since more than 90% of pregnant patients have been treated with oral or intramuscular AM in our counted trials (2101 cases), it may be the reason for the lack of toxicity observed;

ii. Many of the pregnant patients followed have been seriously ill with malaria, which is responsible for 5-12% low birth weight (LBW), 35% of LBW that is preventable during pregnancy [738], and contributes to 70,000-200,000 infant deaths each year [739]. The resulting evidence may lead to future studies on whether the pattern of reprotoxicity has occurred from the illness or from the drug(s) [144, 307];

iii. A number of early reports have not paid much attention to the feasibility of associating low birth weight, abortion, and/or infant disorders in patients with such treatment of malaria with artemisinins; and

iv. Post-marketing surveillance has been limited in developing countries, and the potential reproductive side-effects of the drug have not been well recorded.

6.4.4. Pregnant Pharmacovigilance of Antimalarial Drugs

WHO defines pharmacovigilance as the science and activities relating to the detection, assessment, understanding, and prevention of adverse reactions or any other possible drug-related problems [740, 741]. The major aims of pharmacovigilance studies are the early detection of unknown safety problems, the detection of increases in frequency of known adverse drug reactions, the identification and quantification of risk factors for adverse reactions, and the prevention of unnecessary risk to the patient by promoting the rational and safe use of medicines. Preclinical drug studies and formal phase I, II, and III clinical trials are generally accepted to have serious limitations in terms of establishing safety.

Spontaneous reporting is a relatively new phenomenon in the history of medicine. The first national reporting schemes for adverse drug reactions were set up in the 1960s in ten countries after the thalidomide disaster. However, only a few African countries have implemented a reporting mechanism. Implementation of spontaneous reporting in low income countries is particularly problematic because of other pressing health-care priorities and specific challenges; such as geographical remoteness of many of the health facilities, poor telecommunication systems, and inadequate education of health professionals and patients. Additionally, problems with availability of drugs, caused by lack of funds and failures in systems and markets, could also interfere with the reporting process. Since most of these countries have well established (although often overstretched) public-health programs, which operate according to standard guidelines and are supported at both national and international level, there is an opportunity for these structures to interact with pharmacovigilance initiatives. WHO is promoting the introduction of pharmacovigilance into public-health programs, and some countries have started to implement their monitoring systems in collaboration with malaria control programs [458].

Recently, a new policy in the treatment of malaria, with ACTs, was adopted by several countries. The therapeutic profile of these artemisinin-based drugs seems to be good under well-conducted clinical trials, but their efficacy and safety have not been adequately monitored in large-scale use in populations outside Southeast Asia [263]. Safety monitoring is important in all countries, but especially in African populations in which the presence of comorbid conditions such as HIV/AIDS, malnutrition, and tuberculosis could be important issues. As mentioned previously, there are concerns related to the safety of ACTs during the first trimester of pregnancy [737]. Use of artemisinin combined with amodiaquine or sulfadoxine-pyrimethamine has raised different concerns over dermatological, hematological, and hepatic toxicity [263, 704].

To address the issue, five African countries (Burundi, Democratic Republic of Congo, Mozambique, Zambia, and Zanzibar), supported by WHO Roll Back Malaria, participated in a training course in which they designed action plans to introduce pharmacovigilance systems, together with the implementation of new antimalarial therapy [742]. In each of these countries, mechanisms are being established to collect information about adverse reactions, and some of them have become members of a WHO monitoring program [740]. In addition, Ghana and South Africa are reinforcing established pharmacovigilance systems to better monitor the safe use of antimalarial drugs. Nevertheless, none of these activities have focused on safety monitoring of drugs used during pregnancy. However, some have included safety monitoring of antimalarial drugs used during the implementation of intermittent preventive treatment during pregnancy. These studies were designed to monitor adverse events in the mother rather than the unborn child, and have not produced any signals of significant risk [458].

During implementation of the pharmacovigilance systems, special attention should be given to specific risk groups, particularly pregnant women. The first trimester of pregnancy carries the highest risk of fetal adverse reactions, and some women are exposed to medicines during this period because they are unaware that they are pregnant or do not declare their pregnancy. Recently, studies have described drug exposure prevalence of 86–97% [743] with an average of 2.9–4.2 drugs per woman. The most commonly prescribed medicines are antimicrobials, analgesics, anti-emetics, tranquilizers, vitamins, mineral salts, and vaccines. In areas of high malaria prevalence, this list also includes antimalarial drugs.

Indiscriminate use of medicines in pregnancy is not recommended because of the risk of adverse reactions in the mother and fetus, and the possibility of irreversible effects. The decision to give drugs to pregnant woman must be made based on a balance between risk and benefits. In particular, the potential benefits must outweigh the potential risk to the fetus. The adverse consequences of malaria in pregnancy are well described: untreated malaria poses a far greater risk than treatment, although the mechanism for monitoring pregnant women exposed to these drugs is limited. Causality assessment is difficult in pregnant woman because some adverse reactions can only be identified after delivery. Date of the last menstrual period, or some other reliable method of gestational age, is notoriously difficult to obtain in low-income countries, but is crucial in determining first trimester exposures. Different factors should be considered to estimate the strength of the association between the drug and the reaction, including specific and possibly unique pathognomonic defects, plausible temporal exposure, consistency of the observed evidence, dose-response relations, duration of exposure, and confounding factors (e.g., drugs, environmental factors, chemicals, and traditional medicines).

Pregnancy registries are recognized as one method for detecting major risks associated with a drug or biological exposure during pregnancy. At the time of pregnancy registration, information is collected on drug exposure, maternal disease status, gestation, and other factors that may affect pregnancy outcome. An active follow-up of these pregnancies including outcome of the pregnancy and the infant are done using various approaches, including maternal interviews, medical record abstraction, or a combination of these methods to avoid recall bias. From this system, accurate data should be recorded to calculate the prevalence of adverse reactions, identify risk factors, better estimate the magnitude of exposure risk, and detect long-term reactions such as delayed development, neurological impairment, or any effects that might be detected in older children of at least1 year who might have been exposed to antimalarial drugs in the uterus.

These surveillance mechanisms are susceptible to under-reporting, selection bias (some pregnancies will not be registered, and some defects will not be diagnosed at birth), and loss to follow-up, and there may be difficulties linking specific maternal exposures to fetal anomalies. Despite limitations of these methods, they have been used to supplement animal toxicology studies and clinical trials, and to generate signals of risk to help health-care providers assess the risk of antimalarial drug use in pregnant woman. Single methods should not be used alone to identify increases in the prevalence of adverse events, particularly in pregnant women. Combinations of different methods are needed for the early detection of any safety issue [458].

6.5. Other Toxicities and Drugs Withdrawn due to Adverse Effects

6.5.1. Other Adverse Effects of Antimalarial Drugs

1) Ocular Toxicity

Ocular toxicity caused by antimalarials was first described in the literature as early as 1957. As antimalarials were also found to be effective in the treatment of rheumatoid diseases apart from the treatment and prophylaxis of malaria, the risk of ocular toxicity is increased. The incidence of early retinopathy in ophthalmologically unmonitored patients was estimated to be 10% for chloroquine and 3-4% for hydroxychloroquine. Advanced retinopathy had an incidence of 0.5%. These risks might be reduced substantially by regular observation and testing. The major toxicity of antimalarial agents is retinal damage (rare), which can lead to visual impairment. The major risk factor for retinal toxicity appears to be the combination of cumulative doses > 800 g and age > 70 years (presumably due to the increased prevalence of macular disease in the elderly). In the absence of risk factors, it is recommended that an ophthalmologic examination and central field testing be performed every 6-12 months. The central 10° of the visual field is the initial site of antimalarial retinal toxicity. There is a higher risk of visual loss when plasma concentrations of quinine exceed 15 mg/l at any stage of over dosage. Blurred vision may proceed to complete blindness within a few hours. As vision is lost, the pupils become dilated and unresponsive to light. Initially, only narrowing of the retinal arterioles may be seen on fundoscopy but after 3 days retinal edema may appear. Another study reported that a 34-year-old man treated once with 250 g of amodiaquine hydrochloride during 1 year was noted to have diffuse conjunctival and corneal changes and also demonstrated abnormal results in retinal function tests [744].

2) Myopathy

Factors increasing the risk of muscle disorders may depend on concomitant disease (diabetes, hypothyroidism, renal, and hepatic disease), advanced age and dose.

Myopathy has rarely been reported with these agents [745]. Clinicians should be aware that treatment may lead to neuromyopathy, as well as irreversible retinopathy with chronic use. Usually patients complain of muscle weakness with or without muscle pain. Peripheral sensory abnormalities, such as lack of deep tendon reflexes, may be noted on examination. Muscle enzymes are normal or slightly elevated. In cases suspected of drug-induced myopathy, plasma concentrations of cellular contents released from damaged muscle are assessed. These laboratory parameters include creatine kinase, lactate dehydrogenase, aspartate aminotransferase, alanine aminotransferase, aldolase myoglobin; potassium and phosphorus both of which increase with muscle injury. Serum creatine kinase is considered to be the most sensitive indicator, but its lack of specificity is a major limitation. In the presence of drug-induced myopathy, serum creatine kinase may be normal, slightly elevated, or as high as 10–20 times the upper normal limit [746]. Myopathy and cardiomyopathy are thought to be caused by damage to the mitochondria in muscle cells.

3) Hepatotoxicity

Amodiaquine can cause adverse effects including liver damage. The observed drug toxicity is believed to involve the formation of an electrophilic metabolite, amodiaquine-quinoneimine, which can bind to cellular macromolecules and initiate hypersensitivity reactions. Since hepatitis and agranulocytosis occurred in prophylactically treated patients, it is no longer recommended as prophylactic treatment of malaria. Repeated exposure to the quinoneimine-generated antigen may be important in the generation of organ damage. One study reported on a 24-year-old woman with severe liver failure following trimethoprim-sulfamethoxazole treatment, demonstrating the possible severity of the drug hypersensitivity syndrome associated with trimethoprim-sulfamethoxazole. The abrupt onset of hepatic failure, 5 days after the start of doxycycline, and the rapid normalization after drug discontinuation leads to suspect a causal relationship between doxycycline and liver insufficiency. Another study found significant changes in the proportions of plasma proteins in malarious birds 8 days after infection; albumin and α_2-globulin were reduced, while γ_1- and γ_2-globulin were increased. Those changes coincided with significant increases in the plasma concentrations of total protein and enzymes, and a decrease in creatinine. A prospective study done in 216 children with complicated *P. falciparum* malaria showed hepatopathy in 33.3% of cases, with a higher incidence in children aged > 5 years. Bilirubin and alanine aminotransferase were moderately raised in most cases [744].

4) Hemotoxicity

Artesunate hemotoxicity in the rats, and to a greater degree in rhesus monkeys, parallels the effects reported in human (0.6% reticulocytopenia in 4062 patients at low doses of 2-3 mg/kg) [747]. In humans artesunate caused rapid inhibition of hematopoiesis and lower peripheral reticulocyte counts by day 5 of treatment compared to the quinine group (p = 0.011) [96, 748]. These reductions in reticulocyte counts and anemia have also been confirmed by other investigators [749-751]. Hematological changes induced by artesunate administered at therapeutic doses in humans are not completely clear. In another study, both artesunate and artelinate-treated rats, dose-dependent and rapidly reversible hematological changes (significant reductions in RBC, HCT, Hb, and reticulocyte levels) were seen in the peripheral blood. Bone marrow evaluation revealed a statistically significant reduction in the myeloid/erythroid ratio only at the highest dose of AS (240 mg/kg), albeit still within the normal ratio range (1.0-1.5:1.0). Looking at the respective therapeutic indices we have concluded that AS is much safer than AL. Both drugs induced hematological changes in rats that parallel the dose-dependent, reversible anemia and reticulocytopenia previously reported in animals and humans. However, no significant bone marrow depression was seen for either agent [468].

6.5.2. Chemoprophylactic Drugs with Unknown Toxicity

1) Doxycycline

The tetracyclines have been in clinical use for many years and have been recently suggested as potential chemoprophylactic drugs. In one randomized study, minor adverse events were reported to be more common than with chloroquine alone: for instance, abdominal symptoms occurred in 40% of patients compared with 15% in the chloroquine group. These suggestions have been made in the absence of reliable data on the incidence of fatal toxicity with this group of drugs. The theoretical risks are great: doxycycline can produce photosensitivity, allergic skin reactions, and skeletal deposition with dental staining, oesophagitis, candida infections, pseudomembranous colitis, and perhaps enhancement of shigella and salmonella enteritis. The use of this drug in young children and in pregnancy is contra-indicated because of discoloration of the teeth and possible adverse effects on development [752].

2) Mefloquine

Mefloquine was first used experimentally in human beings in 1972, and in recent years has been increasingly used for treatment and prophylaxis of malaria. Unfortunately, there is insufficient post-marketing experience to assess the drug's toxicity. Early studies have already shown that both bradycardia and a prolonged QT interval are common(9%) and, alarmingly, in 1% of patients neuropsychiatric changes occurred. There have been reported instances of acute brain syndrome presenting with convulsions, psychosis, and depression, and the significance of these reports is, as yet, unclear. No new cases of neutropenia have, so far, been reported. Minor side-effects include gastrointestinal disturbances and dizziness. Mefloquine has a very long half-life which means that the drug must be taken every two weeks after the first three or four weeks of use. This is recommended because of the increased frequency of adverse side-effects on a weekly regimen [752].

6.5.3. Antimalarial Drugs with Non Fatal Adverse Events

1) Chloroquine

Chloroquine was first used in 1945, and since then has been very widely employed throughout the world. During this time there have been few, if any, reports of severe or fatal adverse effects attributed to the use of the drug at the normal prophylactic dose; thus, it is reasonable to assume, in view of its huge consumption, that instances of fatal adverse effects to chloroquine are substantially less than 1 in a 100 000. This is equivalent to it being safe. Chloroquine causes a short term and reversible effect on optical accommodation which can potentially affect eyesight during performance of operators of high performance machinery or cars. The true incidence of this effect has not been determined. Chloroquine binds irreversibly to melanin and long term use of high dose daily chloroquine in patients with rheumatoid arthritis may lead to the accumulation of chloroquine in retinal melanin. There are only a few reports of retinopathy which have occurred in patients taking weekly chloroquine for malarial suppression. In these cases the total dose of chloroquine has not been properly assessed.

The experience of rheumatologists with higher (500 mg) daily doses of chloroquine suggests that retinopathy, lens and corneal changes can occur after total doses of 100 g; experience with lower (250 mg) daily doses suggests that retinopathy does not occur until over 1000 g have been given. Hydroxychloroquine appears to be better tolerated than chloroquine. There is some suggestion that sunlight aggravates the effect of chloroquine. It would, therefore, appear prudent to suggest that long term users of chloroquine should switch to hydroxychloroquine and have a careful ophthalmological examination after every 100 g of chloroquine (that is every 4-6 years of use). The effect of chloroquine on the skin may also be due to the binding of chloroquine to skin melanin.

Chloroquine is associated with reversible pruritus in about 10% of black Africans; the incidence in white Europeans appears to be much lower. A few reports suggest that chloroquine can aggravate psoriasis, particularly in the subgroup of patients with light intolerant disease, although there are very few cases following malarial prophylaxis. As psoriasis affects at least 1% of the population, the paucity of exacerbations of psoriasis suggests that this adverse effect is very rare. In general, chloroquine can be considered to be a safe drug. It has been widely used and even at the much higher doses used in the treatment of arthritis, side-effects are relatively uncommon [753].

2) Proguanil

Proguanil was first used in 1948 at a high dose for the treatment of malaria; it was then widely used at low doses for chemoprophylaxis throughout the former British colonies. Until 1985, proguanil was taken at a dose of 100 mg daily with remarkably few adverse effects. No severe toxic effects were reported. Since 1985, the dose has been increased to 200 mg per day and prescribed in combination with chloroquine. To date no serious adverse events have been reported. Similarly, the other anti-folate drugs, pyrimethamine and trimethoprim, which have been used both for malarial prophylaxis and for other diseases are also free of serious adverse events at low doses. It is reasonable to presume that proguanil is safe at doses of 200 mg a day, although the full weight of experience is less than that for chloroquine, since proguanil has been used less widely.

Since the dose of proguanil has been increased to 200 mg there have been an increasing number of reports of reversible aphthous ulceration. It is unclear what the incidence of this effect is, for it has varied from different reports; it is also unclear whether chloroquine taken in combination with proguanil aggravates and is responsible for the increasing incidence of this effect reported since 1986. Other adverse effects reported recently include reversible alopecia and nausea and gastric irritation, although these occur just as frequently with other antimalarial agents [752].

3) Artemisinins

The artemisinin and its derivatives were discovered in China. A crude extract of the wormwood plant *Artemisia annua* (qinghao) was first used as an antipyretic 2000 years ago, and its specific effect on the fever of malaria was reported in the 16th century [43]. The active constituent of the extract was identified and purified in the 1970s, and named qinghaosu, or artemisinin. Although artemisinin proved effective in clinical trials in the 1980s, a number of semi-synthetic derivatives were developed to improve the drug's pharmacological properties and antimalarial potency [44]. The pharmacological and clinical evaluations of artemisinin group of drugs have been taken place for 30 years and four advantages have been evaluated.

The therapeutic index of the artemisinin derivatives is wide. The most common reported adverse effects include nausea, vomiting, bowel disturbance, abdominal pain, headache and dizziness - symptoms that can result from malaria infection itself. Mild and reversible hematological and electrocardiographic abnormalities, such as neutropenia and first-degree heart block, are observed infrequently.

Artemisinin and its derivatives are safe and remarkably well tolerated [65]. There have been reports of mild gastrointestinal disturbances, dizziness, tinnitus, reticulocytopenia, neutropenia, elevated liver enzyme values, and electrocardiographic abnormalities, including bradycardia and prolongation of the QT interval, although most studies have not found any electrocardiographic abnormalities. The only potentially serious adverse effect reported with this class of drugs is type 1 hypersensitivity reactions in approximately 1 in 3,000 patients [2]. Neurotoxicity has been reported in animal studies, particularly with very high doses of intramuscular artemotil and artemether, but has not been substantiated in humans [2]. Similarly, evidence of death of embryos and morphological abnormalities in early pregnancy has been demonstrated in animal studies [741]. Artemisinin has not been evaluated in the first trimester of pregnancy; so, should be avoided in first trimester patients with uncomplicated malaria until more information is available.

6.5.4. Antimalarial Drugs Withdrawn from Use due to Adverse Effects

This theoretical view has been followed in practice whenever toxicity has been measured. In 1985, Fansidar was withdrawn as a recommended drug for routine prophylaxis on the basis of an estimated fatal adverse reaction rate of about 1 in 20,000. A year later amodiaquine was withdrawn because of an incidence of fatal neutropenia of about 1 in 2,000. Unfortunately, there are few good techniques available to measure rates of severe adverse effects which are lower than 1 in 10,000. Prospective trials are not large enough to reliably detect side-effects or adverse effects of such frequencies and the much less reliable techniques of post-marketing surveillance must be used. This depends on using isolated case reports, reports to government agencies and to the pharmaceutical industry. These reports have to be assessed in the context of estimates of overall drug usage. Clearly, such estimates are very imprecise and drugs often have to be used for several years before even this imperfect information can be obtained. In contrast, frequent but mild side-effects are much easier to determine. Care is needed in interpreting the nature of mild side-effects because placebo controlled trials have shown that patients often suffer from non-specific side-effects such as nausea, dizziness, and headaches [752].

1) Mepacrine

Mepacrine was first used in 1935, and was widely employed throughout the Second World War. Severe cases of aplastic anemia, transient psychotic reactions and exfoliative dermatitis have been described, together with more minor adverse events including yellow skin pigmentation and gastrointestinal disturbances. The incidence of adverse events is unknown. It is likely that the drug was withdrawn because of the high frequency of minor adverse events, rather than the high frequency of life-threatening events. Also, at the time of withdrawal, the non-toxic drugs chloroquine and proguanil became widely available.

2) Sulphonamides

The use of sulphonamides was started in the 1930s. The problems of severe skin reactions and neutropenia were well described. Nevertheless, a combination of pyrimethamine and sulfadoxine (Fansidar) was introduced in 1965. Twenty-two cases of Stevens-Johnson syndrome were observed with three deaths. In 1985, 19 reports of severe skin reactions with six fatalities were reported in the United States and a corresponding number of nine cases (four fatal) in UK were also reported [754]. From an estimation of the frequency of the reported reactions and the number of tablets sold within the US, an incidence of fatal reactions of a frequency of 1 in 18 to 1 in 24,000 (with 95% confidence limits about 1 in 10-50,000) has been reported [755]. It is unlikely that this toxicity is due to combination treatment as similar frequencies were observed in Beira when single doses of sulphadoxine were given to 150,000 people. Examples of neutropenia have also been recorded with Fansidar, although the frequency of this has not been properly measured. Many studies suggest that this occurs approximately as frequently as severe skin reactions. One dissenting Swiss study shows a much lower (1 in 150,000) incidence of severe adverse effects. The reason for this difference remains obscure, although it may simply reflect over-estimates of drug usage [756]. It is unclear whether different formulations of sulphonamides have a significantly different incidence of severe adverse effects but, as these effects are so rare, it is unlikely that any high quality data will ever be produced that can be used to disprove this hypothesis.

3) Dapsone

Dapsone had been used since 1965 as prophylaxis against malaria and, ever since, its use has been associated with neutropenia. Originally, it was used in combination with chloroquine and primaquine at doses of 25 mg a day, and neutropenia occurred in 1 in 10,000 cases, 40% of which died. Since then the combination of 12.5 mg pyrimethamine and 100 mg dapsone (Maloprim) at a dose of two tablets a week has been shown to be associated with agranulocytosis. A dose of one Maloprim tablet a day has also been associated with four cases of neutropenia, including two deaths. Dapsone is also associated with specific minor side-effects, in particular methemoglobinemia. The toxicity of low dose Maloprim (one a week) is still contentious as only a few reports of neutropenia have been associated with low dose use. Maloprim is not licensed in the US and is only used by a minority of travelers who are advised in Britain and Australia. Thus, in spite of the few cases reported, the frequency of fatal adverse effects is likely to lie in the grey area of 1 in 20 to 1 in 50,000, where the benefits of prophylaxis may not out weight the toxicity [752].

4) Amodiaquine

Amodiaquine was introduced in 1948 and widely used as an analogue of chloroquine for the treatment and prophylaxis of malaria, as well as for the treatment of rheumatoid arthritis. It was mainly used in Africa and was reintroduced to the UK in 1986 following the development of chloroquine resistance. Observations were made that chloroquine resistant strains of malaria were sensitive to amodiaquine. It is interesting that, at that time, there were a number of case reports of neutropenia associated with amodiaquine at different doses; these included four cases of neutropenia following low dose amodiaquine use. In spite of these anecdotal reports, both WHO and MMWR [757], described the drug as safe, presumably on the basis that anecdotal reports were not in themselves sufficient evidence to class the drug as toxic. After reintroduction in the UK there was a cluster of case reports of neutropenia, and estimates of the frequency of the adverse effects based on the national case reports of neutropenia, together with an assessment of the number of UK prescriptions, suggested that the frequency of adverse effects was 1 in 2,000 [758]. It should be noted that the errors in such estimates are so great that it is impossible to be certain that the frequency of adverse events with amodiaquine is much greater than that with Fansidar or dapsone. No minor adverse events specific to amodiaquine were reported.

Chapter VII

Antimalarial Drug Toxicity

Malaria affects 300-500 million and kills 1-3 million individuals annually, and has an enormous economic impact in the developing world, especially in sub-Saharan Africa [759]. There are various antimalarial drugs in use for a long time either for the treatment for malaria or, to a lesser extent, as prophylactic agents against malaria infection. Malaria, caused mostly by *P. falciparum* and *P. vivax*, remains one of the most important infectious diseases in the world. Antimalarial drug toxicity is one side of the risk-benefit equation and is viewed differently depending upon whether the clinical indication for drug administration is malaria treatment or prophylaxis. In recent years, treatment of malaria using combination of available antimalarial drugs raises the hope that it will have enhanced efficacy along with reduction in rate of spread of resistance to antimalarial drugs [401]. However, the risk of combination-drug-related toxicity must be assessed against the potential outcome and benefit of the treatment or prophylaxis.

Antimalarial drug toxicity must be acceptable to patients and cause less harm than the disease itself. Research that leads to drug registration tends to omit two important groups who are particularly vulnerable to malaria - very young children and pregnant women. Prescribing during pregnancy is a particular problem for clinicians because the risk-benefit ratio is often very unclear. The number of antimalarial drugs in use is very small. The physician based in the temperate zone is orientated more towards malaria prevention than treatment and may be unfamiliar with the clinical assessment of malaria. Cost and access to treatment often limit therapeutic choices in the tropics; these limitations do not generally apply in temperate zones. Drug-related toxicity and its risk must be balanced against the likely outcome of malaria treatment or prophylaxis and the circumstances of clinical practice. In this regard, there are no contra-indications to prescribing an effective drug that will save a life.

Considering the risk-benefit ratio is not primarily the task of busy clinicians, who rely on recommendations by national or international authorities that have reviewed all the pertinent research data. Nevertheless, clinicians should have an appreciation of how these guidelines are produced and of the quality of the research data that has been evaluated. Estimating the risk of drug toxicity is also important but no standard system exists to quantify and describe risk. In addition, acute malaria is associated with malaise, fever, nausea, vomiting, abdominal pain, anemia, and sometimes diarrhea - all of which could be iatrogenic. These symptoms are commonly mentioned in reports as being possible drug-related adverse effects, even though they resolve coincident with recovery from the infection and are therefore in all probability malaria related.

Pregnant patients with malaria are a group for whom the risk-benefit ratio is often unclear. Consequently, clinicians are wary of prescribing during pregnancy. Reluctance by drug manufacturers to test drugs in pregnant women is well known. As a result, data in pregnancy often become available because of accidental exposure or because public health necessity leads to the use of drugs. Malaria in pregnancy poses a significant risk to mother and fetus because of high maternal and fetal morbidity and mortality. Currently, antimalarial drug resistance is a global challenge that compromises drug efficacy. *P. falciparum* is resistant to virtually all of the standard antimalarial drugs and chloroquine-resistant *P. vivax* is emerging. Recommending optimal treatment or prophylaxis becomes more challenging. Patient compliance is another important determinant of drug effectiveness and may be compromised if drug toxicity is unacceptable [263].

Antimalarial drug treatment policy in most malaria-endemic countries centers on the use of single agents used in sequence once the first-line drug line is failing. This policy will become unsustainable because of the limited number of affordable antimalarial drugs, and the small number of drugs in development. A new therapeutic strategy has been proposed recently, namely, combining standard or new antimalarial drugs with an artemisinin derivative with the aim of increasing drug efficacy, reducing transmission, and retarding the development of resistance [401]. This approach has a similar rationale to the use of multiple drugs in the treatment of tuberculosis and HIV infection. It has two major implications. Firstly, for drug-related toxicity, will two drugs be more toxic than one? Secondly, older drugs whose efficacy has been eroded by resistance may be given a new lease of life. This chapter will review the adverse effects of the widely used traditional antimalarial drugs, the artemisinin derivatives, and drug combinations newly introduced.

7.1. Antimalarial Drug Toxicity in Monotherapy

7.1.1. Traditional Antimalarial Drugs

Despite its decreasing efficacy against *P. falciparum,* chloroquine, a traditional antimalarial drug, continues to be used widely because of its low cost and good tolerability. It remains the drug of first choice for treating *P. vivax* malaria. Pruritus is a common adverse effect in African patients. As prophylaxis, chloroquine is usually combined with proguanil. This combination has good overall tolerability but mouth ulcers and gastrointestinal upset are more common than with other prophylactic regimens. Sulfadoxine/pyrimethamine is well tolerated as treatment and when used as intermittent preventive treatment in pregnant African women. Mefloquine remains a valuable drug for prophylaxis and treatment. Tolerability is acceptable to most patients and travelers despite the impression given by the lay press. Dose-related serious neuropsychiatric toxicity can occur; mefloquine is contra-indicated in individuals with a history of epilepsy or psychiatric disease. Quinine is the mainstay for treating severe malaria in many countries. Cardiovascular or CNS toxicity is rare, but hypoglycemia may be problematic and blood glucose levels should be monitored. Halofantrine is unsuitable for widespread use because of its potential for cardiotoxicity. There is renewed interest in two old drugs, primaquine and amodiaquine. Primaquine is being developed as prophylaxis, and amodiaquine, which was withdrawn from prophylactic use because of neutropenia and hepatitis, is a potentially good partner drug for artesunate against *falciparum* malaria, Atovaquone/proguanil is a new antimalarial combination with good efficacy and tolerability as prophylaxis and treatment. The most important class of drugs that could have a major impact on malaria control is the artemisinin derivatives. They have remarkable efficacy and an excellent safety record. They have no identifiable dose-related adverse effects in humans and only very rarely produce allergic reactions. Combining an artemisinin derivative with another efficacious antimalarial drug is increasingly being viewed as the optimal therapeutic strategy for malaria [263].

7.1.1.1. Chloroquine

Chloroquine (CQ), a 4-aminoquinoline, has been the standard antimalarial drug of choice for half a century for its rapid schizonticidal activity against all malarial parasite infections. CQ is given once daily for treatment of all species of *Plasmodium* infection. Children and adults should receive a full treatment dose of CQ base 25 mg/kg given over 3 days. The most practical regimen used in many areas is CQ base 10 mg/kg on days 1 and 2, and 5 mg/kg on day 3 [2]. The use of CQ as a single first-line drug treatment is now limited due to high number of CQ-resistant *P. falciparum* cases and, as a result, its use is limited only to certain countries and populations [759]. Resistance of *P. vivax* to CQ was also reported [760]. In the areas where *P. falciparum* and *P. vivax* are still sensitive to CQ, the recommended weekly prophylactic dose is CQ base 5 mg/kg along or in combination with proguanil 200 mg/day [2].

1) CQ with Mild and Moderate Adverse Effects

CQ is one of the safest antimalarial drugs ever discovered. Serious adverse reaction to CQ is rare at the regular dose and it is generally well tolerated when used for treatment or prophylaxis. Commonly reported symptoms include headache, malaise, dizziness, blurred vision, difficulty focusing, mild gastrointestinal upset, and itching. Oral antihistamine treatment is not usually very effective. By contrast, in a retrospective study of itching in Thai and Burmese patients with *vivax* malaria, only 1.9% reported itching that was rated as mild and responded favorably to antihistamine treatment. Less common CQ adverse effects include skin depigmentation, whitening of scalp hair and eyebrows, hair loss, various rashes, and eye pathology.

CQ prophylaxis, either alone or in combination with proguanil, is well tolerated. In one open-label, prospective study in 384 Scandinavian travelers, adverse effects associated with CQ plus proguanil were low: nausea (3%), diarrhea (2%), and dizziness (1 %); all were considered mild. However, larger, retrospective, post-travel studies have recorded considerably higher rates of reported symptoms. One such study collected data from 145,003 travelers who used several regimens while on holiday in Kenya. Non-serious adverse effects were assessed in 89,902 travelers who used no prophylaxis (n = 4,026), CQ 300mg weekly (n = 3,354), CQ 600mg weekly (n = 3,646), CQ + proguanil (n = 20,150), sulfadoxine/pyrimethamine (n = 8,673), and mefloquine (n = 50,053). Adverse effects were reported frequently by recipients of all prophylactic regimens and by 5.3% of travelers not on prophylaxis. The corrected rates (reported rate less 5.3%) were 18.5% (CQ 300 mg), 17.2% (CQ 600 mg), 30,1% (CQ + proguanil), 11,6% (sulfadoxine-pyrimethamine) and 18.7% for mefloquine. Overall, some 50% of adverse effects were mild, 36–40% was moderate, and 10–14% was severe.

Several studies indicated that mild symptoms like nausea, dizziness, and diarrhea are reported among the users of prophylactic dose of CQ, either alone or in combination with proguanil, sulfadoxine/pyrimethamine, and mefloquine. Nausea was the most common adverse effect for all regimens, reported by 10.8–18.8% of travelers. The CQ + proguanil group had the highest proportion of travelers who reported symptoms. The rates for nausea (18.8%) and for mouth ulcers (7.9%), a known adverse effect of proguanil, were the reasons for the overall excess of adverse effects with this regimen. Dizziness was reported by more people taking mefloquine (7.6%) than those taking CQ (5–6%), whereas the opposite trend was noted for insomnia. No differences were found between regimens for depression and headaches [263].

2) CQ with Severe Adverse Effects

Association of long-term CQ use and severe toxicity, such as neuromyopathy and retinopathy, has been reported in rare occasions. The neuromyopathy leads to progressive weakness and atrophy of proximal muscles. It is reported that after several years of treatment with chloroquine, its discontinuation results in slow return of sensory functions. Long-term, high-dose CQ therapy has been reported to be associated with CQ-induced retinopathy, which results in irreversible visual impairment. In addition, CQ-induced bull's eye maculopathy in rheumatoid arthritis patients has been documented [759].

CQ has a low safety margin and is very dangerous in overdosage. Larger doses of CQ are used for the treatment of rheumatoid arthritis than for malaria, so adverse effects are seen more frequently in patients with arthritis. The drug is generally well tolerated. The principle limiting adverse effects in practice are the unpleasant taste, that may upset children; and pruritus, which may be severe in dark-skinned patients. Other less common side-effects include headache, various skin eruptions, and gastrointestinal disturbances, such as nausea, vomiting, and diarrhea. More rarely central nervous system toxicity including, convulsions and mental changes may occur. Chronic use (> 5 years continuous use as prophylaxis) may lead to eye disorders, including keratopathy and retinopathy. Other uncommon effects include myopathy, reduced hearing, photosensitivity and loss of hair. Blood disorders, such as aplastic anemia, are extremely uncommon.

Reversible corneal opacities have been reported in 30–70% of rheumatology patients within a few weeks of treatment. Although usually asymptomatic, some patients report photophobia, visual halos around lights, and blurred vision. CQ recipients who report impaired vision should be assessed by an ophthalmologist. Generalized convulsions have rarely been reported in travelers who took weekly CQ prophylaxis. In a case series of four women who developed clonic-tonic fits, all had either a previous history of fits and/or electroencephalogram evidence of a lowered seizure threshold [761].

3) Toxicity after CQ Overdose

Acute overdosage is extremely dangerous and death can occur within a few hours. The patient may progress from feeling dizzy and drowsy with headache and gastrointestinal upset, to developing sudden visual disturbance, convulsions, hypokalemia, hypotension, and cardiac arrhythmias. There is no specific treatment, although diazepam and epinephrine (adrenaline) administered together are beneficial. A one-time dose of 20 mg/kg is considered toxic and doses as low as 30 mg/kg have resulted in fatalities [762]. A CQ dose of 5 g was an accurate predictor of a fatal outcome in adults [763]. Severe clinical features are of rapid onset, occurring between1-3 hours, and include visual disturbances, vomiting, hypokalemia (< 3 mmol/L) convulsions, drowsiness, shock, cardiorespiratory arrest, and death. The case fatality rate is 10–30%. Diazepam is a specific antidote. Parenteral CQ may cause potentially lethal hypotension. It must never be given by intravenous injection. It should be administered either as a constant-rate infusion or as small (< 5mg base/kg) frequent (4- to 6-hourly) intramuscular or subcutaneous injections [764].

4) Drug Use in Pregnant and Lactation Women

CQ is considered safe for treatment or chemoprophylaxis during pregnancy as no abortifacient or teratogenic effects have been reported so far. CQ has teratogenic effects only at high doses in animals (250 to 1500 mg/kg) resulting in a 25% fetal mortality rate [307]. It is considered suitable for use in all three trimesters of pregnancy [2]. CQ is excreted in human milk in small quantities with a mean half-life of 8.8 days and a breastfeeding baby would receive only 4.2% over 9 days of the daily, maternal dose. No congenital abnormalities at birth or developmental delay in the first year after birth in the children were observed whose mothers used hydroxychloroquine throughout their pregnancy. However, CQ has a low safety margin and can be extremely toxic when ingested at higher than recommended dose. A single dose of 20 mg/kg (i.e., 2-3 times the recommended daily treatment dose) is considered toxic and a 30 mg/kg dose has resulted in a fatal outcome. Death may occur within a few hours due to acute CQ poisoning. The main effect of overdose is cardio-toxicity that leads to cardiac and respiratory arrest and death [759]. The future role of CQ as an antimalarial in pregnancy will be determined by the prevalence and degree of CQ resistance. This limits CQ use for prevention of *P. falciparum* but *P. vivax* remains generally sensitive in most areas.

However, at prospective monitoring of pregnant women with autoimmune disease taking large daily doses of CQ did not identify rates of abnormalities significantly higher than in diseased women not exposed to antimalarials. Out of 14 women taking chloroquine 200 to 500 mg/day throughout pregnancy in one study, there were 6 (43%) full term babies, and no congenital defects. In another study, of 17 pregnancies in 14 women taking chloroquine 125 to 400 mg/day during the first trimester, 11 (65%) resulted in full term babies, 4 (23%) spontaneous abortions and 2 (12%) stillbirths, again with no congenital defects. Fetal loss in the non-diseased control populations was between 8 and 14%, and in SLE patients ranged between 12 and 29%. Preterm delivery occurred in 45% of prospectively monitored diseased patients [307].

CQ is excreted into breast milk in small quantities with a mean half-life of 8.8 days. The dose that a breast-feeding baby would receive was estimated at 0.7% of the daily, maternal dose, and 4.2% over 9 days. CQ can be given to breast-feeding mothers.

7.1.1.2. Mefloquine

Minor adverse effects are common following mefloquine (MQ) treatment, most frequently nausea, vomiting, abdominal pain, anorexia, diarrhea, headache, dizziness, loss of balance, dysphoria, somnolence and sleep disorders, notably insomnia and abnormal dreams. Neuropsychiatric disturbances (seizures, encephalopathy, psychosis) occur in approximately 1 in 10,000 travelers receiving mefloquine prophylaxis, 1 in 1000 patients treated in Asia, 1 in 200 patients treated in Africa, and 1 in 20 patients following severe malaria. Other side-effects reported rarely include skin rashes, pruritus and urticaria, hair loss, muscle weakness, liver function disturbances and very rarely thrombocytopenia and leukopenia. Cardiovascular effects have included postural hypotension, bradycardia and, rarely, hypertension, tachycardia or palpitations and minor changes in the electrocardiogram. Fatalities have not been reported following overdosage, although cardiac, hepatic and neurological symptoms may be seen. Mefloquine should not be given with halofantrine because it exacerbates QT prolongation. There is no evidence of an adverse interaction with quinine [2].

1) MQ with Mild and Moderate Adverse Effects

MQ is generally well tolerated by malaria patients. However, in some treatment trials, MQ recipients reported a higher rate of certain adverse effects compared with other antimalarials. Thus, of those reporting significant gastrointestinal or CNS symptoms, approximately two-thirds can be ascribed to MQ and one-third to the malaria. Early vomiting was associated with reduced MQ absorption on day 2 and an increased risk of persistent parasitemia on day 7. MQ- induced early vomiting is a significant clinical problem when treating malaria in young children. Studies also found that high-dose (25 mg/kg) MQ produced a significant increase in moderate dizziness (feeling of swaying) and severe dizziness in children and adults. Between day 1 and day 3, moderate and severe dizziness increased from 1.6% to 6%, and from 0% to 3%, in children, respectively; corresponding figures for adults were 2.6% to 15%, and 0% to 5%. Severe dizziness usually lasted 1 day and had resolved completely within 3 days. Insomnia was present in 1% (n = 336) on day 0; this proportion rose to 4.3% after MQ 25 mg/kg and resolved by day 7. General malaise, headache, abdominal pain, diarrhea, myalgia, and arthralgia were also reported but these symptoms were closely related to the malaria and not caused by the MQ.

2) MQ Induced Neuropsychiatric toxicity

Much attention has focused on the neuropsychiatric effects of MQ. Several reviews have summarized the data [309, 656]. A neuropsychiatric event includes any CNS or psychiatric symptom (e.g., headache, dizziness, insomnia, nightmares, and anxiety) or illness and a serious neuropsychiatric event encompasses the following principal diagnoses: convulsions, disturbance of consciousness, acute confusion, inability to walk unaided because of vertigo or ataxia, psychosis, disorder of affect, and acute neurosis. Dizziness and anxiety are the most frequently reported neuropsychiatric adverse effects [669]. Predisposing factors are a past history of neuropsychiatric disorders, recent (within 2 months) MQ exposure, previous MQ-related neuropsychiatric adverse effects, and previous treatment with psychotropic drugs. It is unclear if a family history is a risk factor. The data on neuropsychiatric adverse effects during MQ prophylaxis have been collected in tourists and soldiers using different study designs. Essential points from these studies are:

- many (40%) neuropsychiatric adverse effects occur soon after the first dose and 75% are manifest by the third dose
- most are of mild or moderate severity and resolve
- some require medical management including hospitalization
- <2% of patients have sequel
- women are more likely to experience neuropsychiatric adverse effects

- MQ in soldiers (fit, young men) is well tolerated, including the loading dose of MQ (250 mg/day for 3 days)
- in prospective studies, MQ-related neuropsychiatric adverse effects were broadly similar to other antimalarials or placebo
- long-term MQ prophylaxis is well tolerated
- the risk of serious neuropsychiatric adverse effects during prophylaxis is estimated to be 1/10,600 for MQ and 1/13,600 for CQ.
- In a meta-analysis of controlled trials, MQ recipients were more likely to withdraw from studies compared with placebo but not when compared with other prophylactic regimens [765].

A number of studies have examined psychomotor function, motor function, action requiring fine coordination, balance, and hearing. This regimen was well tolerated but one trainee who reported dizziness, diarrhea, and flu-like symptoms during the loading-dose phase was withdrawn from the study. There were no significant differences in flying performance, psychomotor functions, and postural sway between the two arms. Poorer sleep quality was reported by the MQ recipients, who had a lower mean reduction in sleep times of 34 minutes that was not statistically significant compared with placebo. Similar findings were reported in another placebo-controlled trial of MQ prophylaxis and the effect of alcohol when driving a car. These data are evidence of the lack of interference of higher mental function in these cohorts. These data show that MQ tolerability was acceptable over the medium to long term.

3) Other Toxicity and Overdose

Cutaneous toxicity with MQ is rare. In a review of published data from 1983-1997, only 74 cases of any skin reaction were reported. These reactions included itching, red maculopapular rashes, urticaria, cutaneous vasculitis, exfoliative dermatitis, Stevens Johnson syndrome, toxic epidermal necrolysis and severe facial lesions. Pruritus (4−10%) and maculopapular rashes (30%) were the most commonly reported reactions [766]. MQ is not generally associated with cardiovascular toxicity. Reports of sinus bradycardia and sinus arrhythmia are more likely to be related to resolution of malaria, especially in young fit individuals. There have been reports of complete atrioventricular block (with MQ treatment), J atrial flutter with 1 to 1 conduction (MQ prophylaxis), and aberrant atrioventricular conduction (MQ prophylaxis). Acute intravascular hemolysis has been reported with malaria treatment in expatriate Europeans Agranulocytosis has also been reported. MQ causes transient elevation of transaminases but is rarely associated with hepatitis [263]. Published data confined to a small number of those taking prophylaxis, some of whom also took CQ and sulfadoxine/pyrimethamine. An acute encephalopathy was a prominent feature [767].

4) Drug Use in Pregnant and Lactation Women

MQ has been used both for prophylaxis and treatment during pregnancy. Prophylaxis after the first trimester of pregnancy was found to be highly effective and well tolerated in Malawi and the Thai-Myanmar border. A retrospective study from the latter setting showed that MQ treatment at any time during pregnancy has a higher risk of stillbirth compared to treatment with quinine alone or other antimalarial drugs. In contrast, a prospective study on prophylaxis of weekly MQ after week 24 of gestation found non-significant increased risk of stillbirth. Post-market surveillance in 1627 pregnant women from Western countries, who had been exposed to MQ chemoprophylaxis, reported 32 cases of congenital malformation and 79 cases spontaneous abortions among those who used MQ before becoming pregnant or within the first trimester of pregnancy. Although this report could not establish a definitive cause and effect relationship between MQ use and development of congenital malformations, abortions or stillbirth, it raises the possibility of increased risk of using MQ in the early pregnancy. Thus, although MQ may be given for both prophylaxis and treatment during the second and third trimester of pregnancy, it should be avoided in the first trimester. MQ, is excreted in small amount in breast milk, the activity of the excreted drug is unknown. Absence of adverse effects on breast-fed infants whose mothers are taking MQ has been suggested by circumstantial evidence. Thus, based on available data, MQ use is contra-indicated in persons with known allergy to MQ, with a history of neuropsychiatric disorder, receiving halofantrine treatment, performing activities that require high degree of fine coordination and spatial discrimination, and women in first trimester of pregnancy [759].

Women MQ is excreted into breast milk in small amounts, the activity of which is unknown [768]. Circumstantial evidence suggests that adverse effects do not occur in breast-fed infants whose mothers are taking MQ. MQ use on the Thai-Burmese border has not been associated with apparent ill effects in breast-fed infants [263].

7.1.1.3. Quinine

Administration of quinine or its salts regularly causes a complex of symptoms known as cinchonism, which is characterized in its mild form by tinnitus, impaired high tone hearing, headache, nausea, dizziness and dysphoria, and sometimes disturbed vision. More severe manifestations include vomiting, abdominal pain, diarrhea, and severe vertigo. Hypersensitivity reactions to quinine range from urticaria, bronchospasm, flushing of the skin and fever, through antibody-mediated thrombocytopenia and hemolytic anaemia, to lifethreatening hemolytic-uremic syndrome. Massive hemolysis with renal failure (black water fever) has been linked epidemiologically and historically to quinine, but its etiology remains uncertain. The most important adverse effect in the treatment of severe malaria is hyperinsulinemic hypoglycemia. This is particularly common in pregnancy (50% of quinine-treated women with severe malaria in late pregnancy). Intramuscular injections of quinine dihydrochloride are acidic (pH 2) and cause pain, focal necrosis and in some cases abscess formation, and in endemic areas are a common cause of sciatic nerve palsy. Hypotension and cardiac arrest may result from rapid intravenous injection. Intravenous quinine should be given only by infusion, never injection. Quinine causes an approximately 10% prolongation of the electrocardiograph QT interval – mainly as a result of slight QRS widening. The effect on ventricular repolarization is much less than that with quinidine. Quinine has been used as an abortifacient, but there is no evidence that it causes abortion, premature labor or fetal abnormalities in therapeutic use [2].

1) Quinine with Mild and Moderate Adverse Effects

Ototoxicity is a well-recognized adverse effect of quinine. Hearing loss in Thai adults affected the high-tone range, was rapid in onset, not always clinically apparent, and resolved spontaneously. Tinnitus, detected after quinine plasma levels exceeded 5 mg/L, was not related to the degree of audiogram-measured hearing loss. In a study comparing healthy volunteers and patients with malaria, hearing loss was audiometrically documented in nine healthy volunteers and all patients. Hearing loss in the volunteers, who received quinine 300 mg, was more often in the high-frequency range, occurred at a mean total peak concentration of 2 mg/L, and was not always clinically manifest. Hearing loss was in the high- and standard-frequency range, and was maximal on the third day of treatment. Full recovery of hearing was achieved for both groups. Serious toxicity has been reported very rarely. Transient and permanent blindness has been documented in a small number of patients with uncomplicated or severe malaria, most of whom were overdosed. The incidence in severe malaria is <0.1%.

Hypoglycemia is a significant adverse effect, affecting up to 10% of patients with severe malaria given quinine, and is mediated through a quinine induced increase in insulin secretion. Quinine-induced thrombocytopenia was reported in 43 patients in Sweden and 11 from the US over 11 and 9 years, respectively, making this adverse effect very rare. Acute intravascular hemolysis, caused by various mechanisms, has often been labeled black-water fever. However, this clinical picture in the tropics is more often caused by infections or hemolytic drugs (e.g., primaquine) given to patients with G6PD deficiency. Quinine can cause the hemolytic-uremic syndrome, disseminated intravascular coagulation (DIC), and Die with renal failure [263]. Various skin reactions are associated with quinine. Pruritus, skin flushing, and urticaria are the most common manifestations of quinine hypersensitivity. Other rashes, reported rarely, have included photosensitivity, cutaneous vasculitis, lichen planus, and lichenoid photosensitivity.

2) Quinine Associated with Cardiotoxicity

Quinine should never be given by intravenous injection as potentially lethal hypotension may result from transiently high plasma concentrations in the distribution phase. Iatrogenic hypotension does not occur with carefully rate-controlled infusions or following intramuscular injection. Quinine may cause lengthening of the QTc interval that is mainly a result of QRS widening; JT intervals are usually normal, making quinine less arrhythmogenic than quinidine. Two studies have examined the cardiac interaction of mefloquine and quinine. In seven human normal volunteers, the increase in QTc was similar between quinine and quinine combined with mefloquine. There was a positive correlation between the increase in QTc and the total and free quinine concentrations but no correlation between the QTc interval increase and the mefloquine concentrations. There has been a report of an elderly Caucasian woman who died of ventricular fibrillation while receiving intravenous quinine. She had a slightly prolonged QTc interval on admission that increased during treatment and was caused by unexpectedly high free quinine levels [77].

3) Toxicity after Overdose

Overdosage of quinine may cause oculotoxicity, including blindness from direct retinal toxicity, and cardiotoxicity, and can be fatal. Cardiotoxic effects are less frequent than those of quinidine and include conduction disturbances, arrhythmias, angina, hypotension leading to cardiac arrest, and circulatory failure. Treatment is largely supportive, with attention being given to maintenance of blood pressure, glucose and renal function, and to treating arrhythmias.

4) Drug Use in Pregnant and Lactation Women

The standard treatment of 10 mg/kg 3 times daily over 10 days has evolved to include a 20 mg/kg loading dose in order to raise plasma quinine levels to enhance a rapid therapeutic response during early treatment [307]. High dose of quinine stimulate the pregnant uterus and have previously been used as an abortifacient. Deafness and hyperplasia of the optic nerve have been described in children born after unsuccessful attempts to induce abortion in women taking quinine overdose [307]. Quinine has been used in the past to induce abortions but this resulted in significant maternal morbidity and mortality, and possible congenital abnormalities of the auditory and optic nerves. However, evidence for the latter has been strongly contested. The weight of evidence suggests that therapeutic doses of quinine are relatively safe in pregnancy, apart from its propensity to cause hypoglycemia. In the treatment of malaria in late pregnancy, quinine actually reduces uterine contractions by treating the infection which increased uterine. Quinine crosses the placenta and is excreted into breast milk. The mean quinine concentration in umbilical cord blood and breast milk is about 30% of the maternal concentration. Quinine at therapeutic doses is suitable for use in pregnant and breastfeeding women.

7.1.1.4. Halofantrine

Adverse effects of halofantrine include nausea, abdominal pain, diarrhea, pruritis and skin rashes. Prolongations of the QTc interval and rare cases of serious ventricular dysrhythmias, sometimes fatal have also been reported. The latter have usually occurred in patients receiving higher than recommended doses that had also received recent or concomitant treatment with mefloquine or were known to have pre-existing prolongation of the QTc interval. Convulsive seizures, intra-vascular hemolysis that compromises renal function and elevation of serum transaminases have also been observed. The relation of the elevation of serum transaminases to medication is unclear since such changes are commonly seen in acute malaria. Values return to normal within one week after treatment.

1) Halofantrine with Mild and Moderate Adverse Effects

Clinical experience has shown that halofantrine is generally well tolerated. Rare reports of nausea, vomiting, abdominal pain, diarrhea, and transient rise in liver transaminases, convulsions, pruritus, skin reactions and acute intravascular hemolysis are documented. Gastrointestinal adverse effects, e.g., nausea, vomiting, abdominal pain and diarrhea, are rare. Hepatic toxicity has been documented in the form of transient rises in liver transaminases. A variety of symptoms attributable to halofantrine have included convulsions, cough, pruritus, skin rash, and acute intravascular hemolysis; all have been reported rarely.

2) Halofantrine with Severe Adverse Effects

Cardiac toxicity is the most severe adverse effect of halofantrine, which include prolongation of PR, QRS and QTc intervals, and serious cases of ventricular dysrythmias, which may be fatal. Cardiac toxicity is unpredictable because of the unpredictable absorption of halofantrine, which can be increased considerably by fatty food. Significant QTc prolongation occurs at the standard and higher dosages (24 mg/kg/day for 3 days) and is directly correlated with halofantrine serum levels. It is more likely to occur with mefloquine failures that are treated with high-dose halofantrine. Over 20 sudden, unexpected fatalities have occurred. First- and second-degree heart blockage has also been documented. In a clinical series of 42 halofantrine-treated Nigerian children aged between 1 and 11 years of age with uncomplicated *falciparum* malaria, two developed first-degree heart blockage and one developed Mobitz type I second-degree heart blockage. These ECG changes occurred within 48 hours, resolved after 72 hours, and were not associated with clinical symptoms. Interestingly, desbutyl-halofantrine, the main metabolite, retains antimalarial activity but may lack the cardiotoxicity of halofantrine [263].

3) Drug Use in Pregnant and Lactation Women

Preclinical studies demonstrated toxicity in terms of increased frequency of post-implantation embryonic death and reduced fetal bodyweight at dosages >15 mg/kg/day in rodents. No teratogenic effects have been reported. Data in pregnant and breast-feeding women are lacking and, because of these limitations. Halofantrine is advised to be avoided in pregnant and breastfeeding women.

7.1.1.5. Amodiaquine

The adverse effects of amodiaquine (AQ) are similar to those of chloroquine. Amodiaquine is associated with less pruritus and is more palatable than chloroquine, but is associated with a much higher risk of agranulocytosis and, to a lesser degree, of hepatitis when used for prophylaxis. The risk of a serious adverse reaction with prophylactic use (which is no longer recommended) appears to be between 1 in 1000 and 1 in 5000. It is not clear whether the risks are lower when amodiaquine is used to treat malaria. Following overdose cardiotoxicity appears to be less frequent than with chloroquine. Large doses of amodiaquine have been reported to cause syncope, spasticity, convulsions, and involuntary movements.

1) AQ with Mild and Moderate Adverse Effects

The systematic collection of tolerability data on AQ has been somewhat thin until recently. AQ alone or combined with oral artesunate for 3 days was used in 941 children with uncomplicated *falciparum* malaria. Tolerability was good. Drug-related symptoms were reported by a small number of children, e.g., 11 (1.2%) had drug induced vomiting necessitating alternative treatment, and 8 (0.9%) reported itching. No clinical or biochemical evidence of hepatic reactions were detected; indeed, liver function tests improved with disease resolution. Case reports in the literature have documented rare neurological problems such as protruding tongue, intention tremor, excess salivation, and dysarthria in four African patients who were treated with AQ. In two patients, these signs occurred on re-exposure to AQ. Yellow pigmentation of skin and mucose, the development of comeal and eonjunctival inclusion bodies, and retinopathy have been reported in one patient following AQ use for 1 year [263].

2) AQ with Severe Adverse Effects

AQ has caused serious and, in some cases, fatal liver and bone marrow toxicity in European travelers when used as prophylaxis. Agranulocytosis usually developed between 5 and 14 weeks of prophylaxis and was associated with hepatitis in some travelers. Based on UK data, the risk of developing agranulocytosis was estimated at 1 in 2000 to 2200, with a risk of death of 1 in 31,300, and 1 in 15,650 for a serious hepatic reaction. AQ had been known to cause neutropenia on rare occasions when used for the treatment of rheumatoid arthritis but data from this era are limited. The pathogenesis of AQ-induced hepatitis is unclear but AQ may exert a direct toxic effect on the liver through production of a quinine imino intermediate or may act via IgG anti-AQ antibodies. Hepatitis has occurred from as early as 3 weeks to as long as 10 months of AQ prophylaxis. The clinical features of reported cases have ranged from a mild transient elevation of liver enzymes with few symptoms, to fulminant hepatitis resulting in slow recovery of liver function or death [263].

3) Drug Use in Pregnant and Lactation Women

Published data is lacking. AQ cannot be recommended in breast-feeding women at current therapies.

7.1.1.6. Primaquine

The most important adverse effects of primaquine (PQ) are hemolytic anemia in patients with G6PD deficiency, other defects of the erythrocytic pentose phosphate pathway of glucose metabolism, or some other types of hemoglobinopathy. In patients with the African variant of G6PD deficiency, the standard course of primaquine generally produces a benign self-limiting anemia. In the Mediterranean and Asian variants, hemolysis may be much more severe. Therapeutic doses may also cause abdominal pain if administered on an empty stomach. Larger doses can cause nausea and vomiting. Methemoglobinemia may occur. Other uncommon effects include mild anemia and leukocytosis.

1) PQ with Mild and Moderate Adverse Effects

Gastrointestinal toxicity is dose related and improved by taking PQ with food. Abdominal pain and/or cramps are commonly reported when taken on an empty stomach. PQ at a dosage of 15 mg/day resulted in mild abdominal pain in 3% of American Caucasian individuals, a rate similar to placebo. This increased with an increase in dosage to 12% (22.5 mg/day), 10% (30 mg/day), including 3% who developed epigastric distress necessitating the administration of antacids, 71% (60 mg/day), and 100% (120 mg/day). These higher doses also caused nausea, cramping abdominal pain, and occasional vomiting. Mild diarrhea has also been reported occasionally. Increased and decreased white cell counts have been associated with PQ. PQ is oxidant and converts hemoglobin to methemoglobin, an action which is dose dependent.

2) PQ Associated with Hemolysis

Acute hemolysis is a well recognized side-effect of PQ in individuals with G6PD deficiency, other enzyme deficiencies that counter oxidant stress, and several hemoglobinopathies. Although first recognized in the 1920s with pamaquine, it was not until the early 1950s and the development of PQ that hemolysis was noted as a significant problem amongst American soldiers of African descent. Work in the 1950s, using 30 mg of PQ base administered daily to 110 African Americans, found that hemolysis usually appeared on the second or third day of drug administration and continued for a further 5-7 days. Hemolysis ceased within 4 days of discontinuing PQ. In another cohort of 50 African Americans, daily administered PQ (15 mg base) resulted in mild anemia in 12; the mean drop in Hb was 2 g/dL. PQ base at a dose of 15mg for 14 days can be given to these patients without individual monitoring because the hemolysis is mild and self-limiting. If higher doses are required e.g., 30 mg daily (14 days) to treat the Chesson strain of *P. vivax,* then monitoring is required because of the risk of severe, acute hemolysis.

Individuals with these forms of G6PD deficiency have low enzyme levels in old and young red cells and are prone to develop severe anemia. PQ base, at doses of 15 mg/day or 45 mg/week, has produced severe hemolysis, progressive hemoglobinemia, hemoglobinuria, acute renal failure, and death. In a study of Thai patients with *vivax* malaria (n = 13), who were given PQ daily (15mg base) after standard dose chloroquine, hemolysis occurred in all 13. The degree of hemolysis, graded only as < 20% vs. > 20% hemolysis, was not related to PQ pharmacokinetics, which were similar between the two groups and to 13 control subjects without G6PD deficiency 1336] pj Lanka, a hospital based study of PQ induced hemolysis in 21 children with partial or gross G6PD deficiency, was associated significant morbidity and a CFR of 19% (4/21). On admission, nine children had congestive cardiac failure, five acute renal failures, and 14 required blood transfusion. Two children who died had also been given aspirin, a known hemolytic drug in G6PD deficiency. Half of the 21 children had been given higher than the recommended doses of PQ, emphasizing the potential severe toxicity of PQ in such patients [263].

3) Toxicity after Overdose

Overdosage may result in leucopenia, agranulocytosis, gastrointestinal symptoms, hemolytic anemia and methemoglobinemia with cyanosis.

4) Drug Use in Pregnant and Lactation Women

Neither high nor standard doses are administered to pregnant women; the fetus is relatively G6PD-deficient and, thus, even standard doses of PQ are avoided in pregnancy to minimize the chance of hemolytic effects [307]. Case reports of exposure during pregnancy are evident in the literature. Pregnant women who require radical treatment for *vivax* malaria should receive PQ after delivery. There are no data on PQ excretion into breast milk.

7.1.2. Artemisinin and its Derivatives

Artemisinin and its derivatives are safe and remarkably well tolerated. There have been reports of mild gastrointestinal disturbances, dizziness, tinnitus, reticulocytopenia, neutropenia, elevated liver enzyme values, and electrocardiographic abnormalities, including bradycardia and prolongation of the QT interval; although most studies have not found any electrocardiographic abnormalities. The only potentially serious adverse effect reported with this class of drugs is type 1 hypersensitivity reactions in approximately 1 in 3000 patients. Neurotoxicity has been reported in animal studies, particularly with very high doses of intramuscular artemotil and artemether, but has not been substantiated in humans. Similarly, evidence of death of embryos and morphological abnormalities in early pregnancy has been demonstrated in animal studies. Artemisinin has not been evaluated in the first trimester of pregnancy so should be avoided in first trimester patients with uncomplicated malaria until more information is available [2].

1) Artemisinins with Mild Adverse Effects

Adverse effects of artemisinins may include headache, nausea, vomiting, abdominal pain, itching, drug fever [666], abnormal bleeding, and dark urine. Minor cardiac changes (mainly non-specific ST changes and first degree atrioventricular blockage) have been noted during clinical trials. These returned to normal after improvement of malaria symptoms. Experience indicates that artemisinin and its derivatives are less toxic than the quinoline antimalarial drugs, few adverse effects being associated with their use. Several clinical trials involving artemisinin and its derivatives conducted in different parts of the world suggest that these drugs are much less toxic than most of the antimalarial drug families. A review of published and unpublished studies of the artemisinin derivatives (n = 8844) has confirmed the earlier Chinese findings of excellent tolerability. No serious adverse drug reactions were reported. Adverse events were independent of the artemisinin and the route of administration. Hematological changes, assessed in 4062 patients, were neutropenia (but not agranulocytosis) in 52 patients (1.3%), reduced reticulocyte count in 25 patients (0.6%), anemia in eight patients (0.2%), and eosinophilia in 40 patients (1.0%). Acute hemolysis occurred in seven patients treated with artemether. An elevated aspartate aminotransferase occurred in 36/3893 patients (0.9%). ECG abnormalities without clinical effect were reported in <1.3% of 2638 patients: transient bradycardia (1.1%); prolongation of the QTc interval (1.2%); and very small numbers of patients with prolonged PR interval (1st degree atrioventricular blockage), atrial extrasystoles, and nonspecific T-wave changes [666].

In these prospective studies of uncomplicated *falciparum* malaria, the oral artemisinins were used either as monotherapy (artesunate, n = 630; artemether, n = 206), or in combination with mefloquine (15 or 25 mg/kg) in 2826 patients. The artemisinin derivatives were associated with substantially fewer adverse effects than the mefloquine-containing regimens: acute nausea (16% vs. 31%), vomiting (11% vs. 24%), anorexia (34% vs. 51%), and dizziness (15% vs. 47%). Oral artesunate or artemether alone were very well tolerated. There was no difference in the incidence of possible adverse effects between the two drugs, and no evidence that either derivative caused allergic reactions, neurological or psychiatric reactions, or cardiovascular or dermatological toxicity. Acute intravascular hemolysis occurred in three patients treated with mefloquine and artesunate. Since these data were published, there have been two cases of acute urticaria and anaphylaxis following oral artesunate alone. In total, six patients have experienced these reactions out of a total of some 17,000, giving an estimated risk for developing an allergic reaction as 1 in 2833 [263].

2) On the Question of Artemisinins Neurotoxicity in Humans

Neurotoxicity has been demonstrated in mice, rats, dogs, and rhesus monkeys that have been repeatedly administered with oil-soluble analogues of AM or AE at multiple doses of 5-24 mg/kg/day [606, 616, 665, 769]. Studies show that AM and AE are toxic to the central nervous system and induce neuropathologic changes in those animal species. The neurotoxic findings produced by artemisinins in animal species are summarized in Table 50. Neuropathology has been detected in dogs given an AE dose of 5.0 – 6.75 mg/kg daily for 28 days [498, 615, 695]. The same neuropathology was found in rats at a dose of 12.5 mg/kg of AE given daily for 7 days [692] and in monkeys at 8 mg/kg of AE for 14 days [665]. However, clinical signs (ataxia and shaking) of neurotoxicity appeared at significantly higher doses in dogs (20 mg/kg, 7 days), in rats (50 mg/kg, 7 days), and in monkeys (24 mg/kg, 14 days) [498, 615, 616, 770]. This suggests that by the time neurological signs are observed in those animals, the drug concentrations are already three or four times higher than the doses known to cause neuropathologic changes. In other words, the doses to produce clinical and neurotoxic signs are much higher (3- to 4-fold) than that to generate the neuropathological confirmations in those animal species.

Despite extensive studies with AM and AE neurotoxicity, there is limited evidence of neurotoxicity in animals related to the water soluble artemisinin derivatives, for example AS, the most widely used artemisinin in humans. AL, another water-soluble artemisinin derivative, has also been reported to have some pathological neurotoxicity in rats following intragastric dose at 160 mg/kg daily for 9 days [694]. These studies help us see that the neurotoxicity of oil-soluble artemisinins is more potent than that of water-soluble analogues (Table 50). The neurotoxicity of AL has been investigated in beagle dogs following daily oral administration at 25, 100, and 300 mg/kg dose levels for 14 days. No behavioral neurotoxicity has been observed in these beagle dogs at these dose levels. The considerable lack of toxicity detected in these dogs very well may be caused by very low oral bioavailability at these doses. Actually, in the highest dose level (300 mg/kg), the animals only received one-fifth of total dose due to the very low bioavailability (22%, our unpublished data). A recent study looked at AM, AS, and DHA orally administered daily for 28 days has helped us compare the two groups. There were little clinical or neuropathological evidence of the neurotoxicity found at high dose levels of over 200 mg/kg/day in mice for the 3 drugs [683, 771, 772]. At the highest doses of intramuscular AM and AS, there was neurotoxicity and behavioral deficits of potentially irreversible abnormalities in balance and equilibrium [772]. AM was determined to have much greater neurotoxicity than intramuscular AS though [674, 772].

The water-soluble artemisinin derivatives, AS and AL, were designed for intravenous injection. To date, no neurotoxicity (pathologic or behavioral) has been observed in animal species following intravenous administration at any repeated doses up to maximum tolerated doses (MTD). The MTDs of AS and AL are 240 and 80 mg/kg, respectively, following intravenous injection daily in rats for 3 days, but no neurotoxicity was detected in these animals [94]. In another study, intravenous AS had no effect on the neurotoxic scores at 120 mg/kg in sodium carbonate daily for 7 days (our unpublished data). After one intramuscular injection, a high dose of 420 mg/kg of AS and AL did not produce any neuronal necrosis in the rats [498]. Also, up to 200 mg/kg of AS used orally daily for 5-7 days did not exhibit neuronal changes or any specific clinical signs [498].

Although artemisinin agents, mainly oil-soluble drugs, induced fatal neurotoxicity in animal models [524], a striking fact in treating patients with the artemisinins is the lack of any serious adverse events, despite careful monitoring in several clinical studies. Severe malarial infection itself, particularly cerebral malaria in multi-drug resistant strains in the developing world, often leaves patients with decreases in neural function. This development has made characterization of potential neurotoxic effects of the artemisinin derivatives very difficult to ascertain, even though these drugs are clinically utilized in ever growing numbers [502, 773]. In a study of 3500 patients in Thailand, evidence of serious toxicity was not noticed [65]. From other reports, rare neurological side-effects have been noted in humans receiving artemisinin drugs [390, 552, 774-775], and these findings were difficult to dissect from effects due to the concomitant malarial infection [776, 777].

Does the neurotoxicity of artemisinins in humans? Animal studies have documented CNS toxicity with artemisinin derivatives; however, such adverse effects were absent in well-designed clinical trials, which has raised the question of whether neurotoxicity might occur in humans.

A number of studies indicate that no neuralgic side-effects are noted in humans from all artemisinin and its derivatives treatments. Even AE was safely used from phase I to III clinical trial studies [257, 495, 553]. However, a 36-year-old American geologist had neurological symptoms two days after completing AS treatment with worsening ataxia and slurred speech nine days after completing AS treatment [620]. Two months later, the patient still had slurred speech; a wide-based, ataxic gait; and impaired heel-to-shin and rapid alternating movement. He was re-treated with oral quinine and doxycycline after testing positive again for *Plasmodium* species (too scant to speciate) in a smear. Four months later, his ataxia was moderately improved, and his speech was nearly normal.

In another case report, a 39-year-old man who returned from Sierra Leone with probable extrapyramidal syndrome that was self-limiting. He had an undetermined febrile illness for which he had received inappropriately long and repeated courses of both AS and chloroquine [778, 779]. Franco-Paredes *et al.* concluded that the patient's condition "suggests neurotoxicity due to multiple courses of AS therapy, perhaps compounded by chloroquine therapy [780]. However, Newton *et al.* pointed out that the patient was most likely suffering from the adverse effects of excessive chloroquine [781].

Certainly, these two reports do not constitute proof of any causal association, but the possibility that repeated administration of artemisinin drugs in malaria endemic areas will result in cumulative damage cannot be ruled out [483], especially since these drugs are often freely available for use in tropical countries [782]. However, another four patients experienced neuropsychiatric adverse events. One patient with history of epilepsy, seizure-free for 10 years, had an episode of tonic-clonic seizures; one patient suffered from anxiety, palpitations and sleep disturbance; one developed frank psychosis and another, depression. These patients were also receiving mefloquine though, a known cause of neuropsychiatric disturbances [747].

In the Gambian children with cerebral malaria described by van Hensbroek [85], intramuscular AM was associated with more convulsions (39 vs. 28%, P = 0.01) and more prolonged coma than was intramuscular quinine (26 vs. 20 hr, P = 0.05). Similar results were found in Vietnamese adults [86]. Survival analysis indicated a significant divergence in the times to recovery from coma among patients who remained unconscious for more than two days. In these patients, treatment with AM was associated with slower recovery than that of quinine dosing (66 vs. 48 hr, P = 0.003). Although in the current reports there was no increase in neurological sequel in patients after multiple treatments with AM, significantly prolonged coma in both studies and increased incidence of convulsions in the AM group of Gamnian children indicates the need for further attentiveness and active investigation of the neurological side-effects of artemisinin drugs. A recent report from Mozambique describes a small but significant and irreversible hearing loss in patients exposed to AM-lumefantrine [783, 784]. The results of this trial are controversial though as the treatment group had both malaria and treatment with AM-lumefantrine and the control group had neither. In another case control study, there were no differences in the test results between cases and controls. There was no neurophysiologic evidence of auditory brainstem toxicity that could be attributed to AM-lumefantrine in this study population [785].

Either intravenous followed by oral AS, or oral followed by intravenous AS were given at a dose of 2 mg/kg for 2 days to patients with uncomplicated acute *P. falciparum* malaria

[454]. At the time of presentation the patients had been ill for a median of 3 days (range, 1 to 10 days) with fever, headache, anorexia, nausea, and vomiting. No adverse effects of the study drug were observed. In another trial, quinine plus tretracycline versus oral AS given at a total dose of 700 mg over 5 days was studied in patients with acute uncomplicated *falciparum* malaria [623]. The occurrence of adverse effects, such as tinnitus, was significantly higher in the quinine-tetracycline group. Nausea and dizziness were rather commonly associated with the AS group in this study. Toxicity has also been observed after 5 days of multiple intramuscular doses with AE. One person out of 14 subjects developed shiverers, clammy hands and feet, dizziness, headache, nausea and a 'metallic taste' 1 hr after the injection; the symptoms lasted about one hour and then were resolved spontaneously [681]. In view of their potential for neurological adverse effects, artemisinin analogues after multiple treatments should be borne in mind. Karbwang [623], Kager [681], Hoffeman [786], and Park [470] expressed concern about the potential neurotoxicity of all artemisinin derivatives, which are the drugs of choice for treating severe malaria caused by parasites with suspected resistance to chloroquine and quinine.

3) On the Question of Artemisinins Reprotoxicity

The artemisinin derivatives (mainly AM and AS) have been extensively studied in animals and in humans, and the experience in human pregnancy are encouraging. However, artemisinins have been shown to be very toxic to animal embryos in repeated preclinical studies. This preclinical data included *in vivo* embryo-fetal developmental toxicity studies carried out on mice, rats, hamsters, guinea pig, and rabbits, by different laboratories using for the most part ICH-compliant protocols [708, 713-719].

Early study of QHS and AM were conducted in mice, rats and rabbits that were intramuscularly from gestation day (GD) 6 to 15 to observe for reproductive toxicity (Table 54). The no observed adverse effect level (NOAEL) was 5.4 mg/kg in mice, 2.7 mg/kg in rats and 0.7 mg/kg/day in rabbits; above these doses, there were increased resorptions of fetuses: 100% resorption at 21.4 mg/kg/day in mice, > 90% resorption at 10.7 mg/kg/day in rats and > 90% resorption at 2.7 mg/kg/day in rabbits [708, 713]. By subcutaneous administration, the study of AS was dosed in rats from GD 6 to 15, and the developmental retardation of sterna and ribs was noted in surviving fetuses in groups treated with 0.54 mg/kg of AS, respectively. The dose that induced 50% fetuses resorbed (ED_{50}) was 0.58 mg/kg [714]. When rats were given a dose of 26 mg/kg/day at GD 1–6, 82.3% of fetuses survived. When the same dose was given in early organogenesis, GD 6–8, only 17.1% of fetuses survived, and treatment occurring after GD 8 resulted in 100% fetal resorption [714]. Recently, our study of AS was also performed in rats following intravenous and intramuscular injection from GD 5 to 15. The ED_{50} of AS after intravenous dosing was 0.61 mg/kg [715].

In experiments carried out in mice, hamsters, guinea pigs and rabbits both DHA and AS intragastrically, intramuscularly, or subcutaneously showed from mild to severe embryolethality effect. In mice and rabbits they caused embryo absorption whereas in hamsters and guinea pigs they induced abortion [716]. In rats, oral AS was administered from GD 6–19. The primary reported effect was embryo-fetal loss manifested as abortion in rabbits and resorption in both rats and rabbits. No or low adverse effect levels were in the range of 5 to 7 mg/kg/day AS [717, 787]. All preclinical studies of the reproductive toxicity in animal species are summarized in Table 7. AS has the greatest toxicity to animal embryos by routes of subcutaneous, intramuscular and intravenous injections with less than 1 mg/kg of the ED_{50}, which is lower than therapeutic dose of 2-4 mg/kg in humans. [719]

Some recent studies found that tissue damage in embryos exposed to DHA affect primarily red blood cells during yolk sac hematopoiesis and higher concentrations and/or longer exposure inhibited angiogenesis [709, 718, 722]. Another study in rats showed that AS subcutaneous injection at a dose of 40 mg/kg/d during days 6-10 of gestation significantly decreased the concentration of serum progesterone in early pregnant rats and discontinued the degeneration of decidual cells and fetuses (720). In cell culture, AS significantly decreased rat embryonic and placental glutathione peroxidase (GSH-Px) and increased malondialdehyde (MDA) level in these tissues (721).

The WHO expert panel judged from their review of these and 28 other experimental laboratory reports that in animals "all the artemisinin analogues have been associated with embryolethality over a narrow dose range and that there are instances of effects on development of the cardiovascular system, the axial skeleton and the limbs," but that "these do not suggest the nature of the toxic action." They considered that "because the safety data are limited, artemisinin compounds should only be used in the second and third trimester when other treatments are considered unsuited" [737].

Thus, an important population at high risk for malaria will continue to be excluded from treatment with artemisinin based combination therapy unless these treatments are shown to be effective and safe in pregnancy. Two early reports in China on pregnant women described that the effects of QHS, AM or AS were not found evident of drug toxicity [711, 788]. There were 94 cases of first trimester exposure to AS identified in the published literature [264, 534] and no unexpected adverse effects were reported. In areas of multi-drug resistance, only treatments based on artemisinin compounds were effective, and pregnant women have been treated for malaria with these drugs throughout pregnancy. Data from limited series of pregnant women (2045 cases) exposed to artemisinin compounds, including a small number (108 cases) in the first trimester, did not show an increase in the rates of abortion or stillbirth or evidence of abnormalities. There were no clinically significant adverse effect of the drugs, on the outcomes of the pregnancies, or the development (neurological and physical) of the infants, including 44 infants exposed during the first trimester [534, 733, 734, 789]. Treatment with AS during pregnancy is associated with higher birth weight, which may have resulted from clearance of malaria parasites, as placental malaria is associated with reduced birth weight [790]. However, the influence of confounding factors should not be excluded.

4) Drug Use in Pregnant and Lactation Women

Pregnant rats given artemisinin in doses up to 3000 mg/kg within 6 days of gestation and normal fetuses; however, the drug caused fetal death and resorption in rodents even at relatively low doses when given after the sixth day of gestation. Fetal resorption occurred in 30% of mice administered artemisinin 10.7 mg/kg/day and it occurred in 100% of those given 21 mg/kg/day. No resorption occurred if artemisinin was administered on days 1 or 2 of gestation. There was no evidence of retatogenicity in the offspring that survived organogenesis. Reports on the use of artemisinin and its derivatives during pregnancy are very limited. No adverse effects were found during long term follow-up in 6 children whose mothers were exposed to artemisinin or artemether, a derivative of artemisinin, between 17 and 27 weeks' gestation. In another report of 17 children exposed between 16 and 38 weeks' gestation, who were followed for periods of between 3 months to 10 years, no abnormalities were seen. The same group recently reported that a total of 21 pregnant women have now been exposed to artemisinin and monitored to outcome with no adverse outcomes [307].

The safety of the artemisinin derivatives in early pregnancy remains uncertain, as these drugs are embryotoxic in animals [708, 709]. They are not recommended for uncomplicated malaria in the first trimester, unless effective alternatives are unavailable. A small case series from China found that among six pregnant women treated for malaria (*P. falciparum* or *P. vivax*) at a mean of 21.7 weeks of gestation none had adverse outcomes [711]. In another clinical trial in China, 21 pregnant women were included and were given a variety of artemisinins; no adverse outcomes were recorded [552]. A study in 83 women in Thailand treated with either artesunate or artemether found no increase in adverse outcomes. Sixteen of the women in this study were accidentally exposed to artemisinins during the first trimester. Follow up of the live born children from this cohort found no developmental delay [712]. Further work in Thailand has shown that women treated with MQ + artesunate or 7 days of artesunate alone had no increase in adverse effects or adverse birth outcomes, and no negative developmental impact when compared with women treated with quinine. A further 461 women in Thailand treated with either artesunate or artemether for *P. falciparum* malaria had no increase in rates of abortion, stillbirth, congenital abnormality, or mean gestation at delivery [44].

Published data on the use of the artemisinin derivatives in breast-feeding women is lacking but experience in Thailand has not shown any apparent infant toxicity [263].

7.1.3. Antibiotic Class Drugs

Based on the recent evidence from current practice and the consensus opinion of the guidelines development group, the use of some antibiotics in combination with artesunate as a second-line of treatment for the uncomplicated *falciparum* malaria by WHO [2]. The antibiotics in use for malaria are tetracycline (4 mg/kg q.i.d.) or doxycycline (3.5 mg/kg/day) or clindamycin (10 mg/kg b.i.d.) along with artesunate (2 mg/kg/day). All these combination are given for 7 days. Among these, tetracycline and doxycycline should not be used in pregnancy [759] and children < 8 years old. The reported side-effects of tetracycline and doxycycline include nausea, vomiting, diarrhea, abdominal pain, dizziness, photosensitivity, headache, esophagitis, odynophagia and rarely may also cause hepatotoxicity, pancreatitis and benign intracranial hypertension seen with the tetracycline class of drugs. Tetracycline accumulates in patients with renal impairment and may cause renal failure. In contrast, doxycycline is less prone to accumulation and preferred in patients with renal impairment [2]. As with tetracycline, oesophageal ulceration can be prevented if the oral dose is washed down with copious amounts of water. Other gastrointestinal symptoms can be reduced if doxycycline is taken with a meal. Milk products must be avoided since they reduce absorption.

In contrast to tetracycline, doxycycline can also be used for chemoprophylaxis. Doxycycline prophylaxis is recommended in areas of mefloquine-resistant *falciparum* malaria and for those visiting high-risk areas and unable to take mefloquine. It has been used successfully for this purpose by United Nations forces in Cambodia and Somalia. In a tolerability trail, doxycycline, mefloquine, the fixed combination chloroquine and proguanil, or fixed combination stovaquone and proguanil were compared for use in malaria chemoprophylaxis, in non-immune travelers to sub-Saharan Africa. The chloroquine and proguanil group had the highest proportion of mild-to-moderate adverse events (45%), followed by mefloquine (42%), doxycycline (33%), and Atovaquone and proguanil (32%). In another trail in Australian soldiers, tolerability of mefloquine as malaria prophylaxis was compared to doxycycline in East Timor. Approximately 57% of soldiers using mefloquine prophylaxis reported at least one adverse event, compared with 56% using doxycycline. The most commonly reported adverse effects of both drugs were sleep disturbance, headache, tiredness and nausea [759, 791].

1) Tetracycline

All the tetracyclines have similar adverse effect profiles. Gastrointestinal effects, such as nausea, vomiting and diarrhea, are common, especially with higher doses, and are due to mucosal irritation. Dry mouth, glossitis, stomatitis, dysphagia and oesophageal ulceration have also been reported. Overgrowth of *Candida* and other bacteria occurs, presumably due to disturbances in gastrointestinal flora as a result of incomplete absorption of the drug. This effect is seen less frequently with doxycycline, which is better absorbed. Pseudomembranous colitis, hepatotoxicity, and pancreatitis have also been reported.

Tetracyclines accumulate in patients with renal impairment and this may cause renal failure. In contrast doxycycline accumulates less and is preferred in patient with renal impairment. The use of out-of-date tetracycline can result in the development of a reversible Fanconi-type syndrome characterized by polyuria and polydipsia with nausea, glycosuria, aminoaciduria, hypophosphatemia, hypokalemia, and hyperuricemia with acidosis and proteinuria. These effects have been attributed to the presence of degradation products, in particular anhydroepitetracycline. Tetracyclines are deposited in deciduous and permanent teeth during their formation and cause discoloration and enamel hypoplasia. They are also deposited in calcifying areas in bone and the nails and interfere with bone growth in young infants or pregnant women. Raised intracranial pressure in adults and infants has also been documented. Tetracyclines use in pregnancy has also been associated with acute fatty liver. Tetracyclines should therefore not be given to pregnant or lactating women, or children of less than 8 years of age.

Hypersensitivity reactions occur, although they are less common than for ßlactam antibiotics. Rashes, fixed drug reactions, drug fever, angioedema, urticaria, pericarditis, and asthma have all been reported. Photosensitivity may develop and, rarely, hemolytic anemia, eosinophilia, neutropenia and thrombocytopenia. Pre-existing systemic lupus erythematosus may be worsened and tetracyclines are contra-indicated in patients with the established disease [2].

2) Clindamycin

Diarrhea occurs in 2–20% of patients. In some, pseudomembranous colitis may develop during or after treatment, which can be fatal. Other reported gastrointestinal effects include nausea, vomiting, abdominal pain and an unpleasant taste in the mouth. Around 10% of patients develop a hypersensitivity reaction. This may take the form of skin rash, urticaria or anaphylaxis. Clindamycin is a promising drug with a short elimination half-life and a good safety and tolerability profile for antimalarial therapy. Usually combined with quinine, it has been used extensively in South America and has also proven effective in adults and children with acute malaria in Africa. In a trial in Thailand, clindamycin in combination with quinine was found safe and effective treatment for multi-drug resistant *P. falciparum* malaria. This combination may be of particular value in children and pregnant women, in whom tetracycline are contra-indicated. The principle adverse effect of clindamycin is diarrhea, which occurs in 2 – 20% of the patients [2].

In some cases, pseudomembranous colitis caused by *clostridium difficile* may develop in during and after treatment that can prove fatal. Other reported gastrointestinal effects include nausea, vomiting abdominal pain and an unpleasant taste in mouth. Approximately 10% of the patients develop a hypersensitivity reaction. This may take the form of urticaria or anaphylaxis. Some other adverse affects include leucopenia, agranulocytosis, eosinophilia, thrombocytopenia, erythema, multiforme, polyarthritis, jaundice, and hepatic damage. Some parenteral drug formulations contain benzyl alcohol, which may cause fatal 'gasping syndrome' in neonates [759, 792].

3) Fosmidomycin

Fosmidomycin is a new class of antimalarial drugs that inhibits the synthesis of isoprenoid by *P. falciparum* and suppresses the growth of multi-drug resistant strains *in vitro*. The combination of fosmidomycin plus clindamycin (FC) is promising as an antimalarial treatment; in a randomized, controlled, open-label study to evaluate the efficacy and safety of fosmidomycin combined with clindamycin among Africa children in Gabon, the FC combination was found to be well tolerated and superior to either agent on its own with respect to the rapid and radical clearance of *P. falciparum* infections.

In a randomized trail carried out with 105 Gabonese children aged 3 – 14 years with uncomplicated malaria, FC given every 12 h for 3 days was compared with a standard single oral dose of SP. The two treatments showed equally well tolerability and had an identical 94% efficacy. In another study conducted in Thailand, fosmidimycin was accessed when given alone and in combination with clindamycin in patients with acute uncomplicated *falciparum* malaria. Both mono and combination therapy were well tolerated but combination therapy gave 100% cure rate [759, 793].

4) Doxycycline

Gastrointestinal effects are fewer than with tetracycline, although oesophageal ulceration can still be a problem if insufficient water is taken with tablets or capsules. There is less accumulation in patients with renal impairment. Doxycycline should not be given to pregnant or lactating women, or children aged up to 8 years. Adverse effects include gastrointestinal irritation, phototoxic reactions (increased vulnerability to sunburn), transient depression of bone growth (largely reversible) and discoloration of teeth and enamel hypoplasia (permanent). Aggravation of renal impairment may occur but is less likely than with tetracyclines.

7.1.4. Other Antimalarial Drugs

1) Chlorproguanil

The adverse effects are similar to those of proguanil. Apart from mild gastric intolerance, diarrhea, occasional aphthous ulceration and hair loss, there are few adverse effects associated with usual doses of proguanil hydrochloride. Hematological changes (megaloblastic anemia and pancytopenia) have been reported in patients with severe renal impairment. Overdosage may produce epigastric discomfort, vomiting, and hematuria. Proguanil should be used cautiously in patients with renal impairment and the dose reduced according to the degree of impairment.

2) Dapsone

Varying degrees of hemolysis and methemoglobinemia are the most frequently reported adverse effects and occur in most patients given more than 200 mg of dapsone daily. Doses of up to 100mg daily do not cause significant hemolysis but patients deficient in G6PD are affected by doses of >50 mg daily. Hemolytic anemia has been reported following ingestion of dapsone in breast milk. Agranulocytosis has been reported following use of dapsone and pyrimethamine together as malaria prophylaxis – particularly when used twice weekly. Aplastic anemia has also been reported. Rashes, including pruritus and fixed-drug reactions may occur but serious cutaneous hypersensitivity is rare. "Dapsone syndrome" consists of rash, fever, jaundice and eosinophilia, and has been reported in a few patients using dapsone as malaria prophylaxis, but mainly in leprosy patients on long treatment courses. Other rare adverse effects include anorexia, nausea, vomiting, headache, hepatitis, hypoalbuminemia and psychosis.

7.2. Antimalarial Drug Toxicity in Combination Therapy (CT)

7.2.1. Sulfadoxine-Pyrimethamine (SP)

1) Adverse Effects of Sulfadoxine

Sulfadoxine shares the adverse effect profile of other sulfonamides, although allergic reactions can be severe because of its slow elimination. Nausea, vomiting, anorexia and diarrhea may occur. Crystalluria causing lumbar pain, hematuria and oliguria is rare compared with more rapidly eliminated sulphonamides. Hypersensitivity reactions may affect different organ system. Cutaneous manifestations can be severe and include pruritus, photosensitivity reactions, exfoliative dermatitis, erythema nodosum, toxic epidermal necrolysis, and Stevens-Johnson syndrome. Treatment with sulfadoxine should be stopped in any patient developing a rash because of the risk of severe allergic reactions. Hypersensitivity to sulfadoxine may also cause interstitial nephritis, lumbar pain, hematuria, and oliguria. This is due to crystal formation in the urine (crystalluria) and may be avoided by keeping the patient well hydrated to maintain a high urine output. Alkalinization of the urine will also make the crystals more soluble. Blood disorders that have been reported include agranulocytosis, aplastic anemia, thrombocytopenia, leucopenia and hypoprothrombinemia. Acute hemolytic anemia is a rare complication, which may be antibody mediated or associated with glucose-6-phosphate dehydrogenase (G6PD) deficiency. Other adverse effects, which may be manifestations of a generalized hypersensitivity reaction, include fever, interstitial nephritis, and a syndrome resembling serum sickness, hepatitis, myocarditis, pulmonary eosinophilia, fibrosing alveolitis, peripheral neuropathy and systemic vasculitis, including polyarteritis nodosa. Anaphylaxis has been reported only rarely. Other adverse reactions that have been reported include hypoglycemia, jaundice in neonates, aseptic meningitis, drowsiness, fatigue, headache, ataxia, dizziness, drowsiness, convulsions, neuropathies, psychosis and pseudomembranous colitis [2].

2) Adverse Effects of Pyrimethamine

Pyrimethamine is generally very well tolerated. Administration for prolonged periods may cause depression of hematopoiesis due to interference with folic acid metabolism. Skin rashes and hypersensitivity reactions also occur. Larger doses may cause gastrointestinal symptoms such as atrophic glossitis, abdominal pain and vomiting, hematological effects including megaloblastic anaemia, leukopenia, thrombocytopenia and pancytopenia, and central nervous system effects such as headache and dizziness. Acute overdosage of pyrimethamine can cause gastrointestinal effects and stimulation of the central nervous system with vomiting, excitability, and convulsions. Tachycardia, respiratory depression, circulatory collapse and death may follow. Treatment of overdosage is supportive.

3) Sulfadoxine-Pyrimethamine

Some consider sulfadoxine-pyrimethamine a single drug rather than a combination in which each component acts independently, since, although pyrimethamine inhibits dihydrofolate reductase (DHFR) and sulfadoxine inhibits dihydropteroate synthase (DHPS), inhibition of both enzymes prevents synthesis of folic acid in parasites. Pyrimethamine is also structurally related to proguanil, and to cycloguanil and chlorproguanil, other drugs that inhibit DHFR. After chloroquine, sulfadoxine-pyrimethamine was the most commonly used antimalarial treatment in almost all endemic areas between the 1960s and the 1980s. However, because of drug resistance, the combination is now rarely used outside Africa. As efficacy falters, the risk-benefit ratio for use of sulfadoxine-pyrimethamine is also falling, as a result of reports of rare occurrences of severe and fatal adverse events—mainly Stevens-Johnson and Lyell syndromes—as well as even rarer instances of hepatotoxicity and agranulocytosis almost exclusively described after use in chemoprophylaxis in expatriates or travelers to the tropics.

4) SP with Mild and Moderate Adverse Effects

Generally, SP is well tolerated with infrequent and mild adverse reactions when used at the recommended doses. SP is a fixed-dose combination that has the advantage of being a one-dose treatment (25 mg/kg based on sulfadoxine) for *P. falciparum* malaria. Fever resolution is slower than with chloroquine or amodiaquine, and resistance has eroded its efficacy. It is now largely ineffective in many parts of Latin America, and South East Asia, where *P. vivax* is also highly resistant. In general, adverse reactions are infrequent and mild when SP is used for malaria treatment. Commonly reported adverse effects include gastrointestinal upset (abdominal pain, nausea, and diarrhea) and headache. Itching was reported by 12% of patients in one treatment trial in Africa and by 2.1% of Indonesians. In clinical trials sponsored by the WHO/Tropical Disease Research (TDR), SP was well tolerated in 2400 patients, mostly African children. There was no serious toxicity attributed to SP [263].

5) SP with Severe Adverse Effects

The most serious events are associated with hypersensitivity to the sulfa component, involving the skin and mucous membranes and normally occurring after repeated administration. Serious cutaneous reactions following single-dose treatment with sulfadoxine-pyrimethamine are rare. Of 12 cases of cutaneous events reported to the manufacturer following therapeutic use of the combination, none had received the recommended single dose (Hoffmann-La Roche, personal communication, 1995). However, such events, including life-threatening erythema multiforme (Stevens-Johnson syndrome) and toxic epidermal necrolysis, have been reported in 1 in 5,000 to 1 in 8,000 people taking the drug for weekly chemoprophylaxis. The combination is therefore no longer recommended for prophylactic use. Data on the incidence of serious cutaneous events following sulfalene-pyrimethamine use are lacking.

Cutaneous drug reactions are more common in patients who are HIV positive (83). There is therefore concern that the high prevalence of HIV infection in parts of Africa may result in an increased frequency of sulfa drug-associated toxicity in HIV-positive people treated with sulfa drug-pyrimethamine combinations for a concomitant malaria infection. There have been isolated reports of a transient increase in liver enzymes as well as hepatitis occurring after administration of sulfadoxine-pyrimethamine. Hematological changes including thrombocytopenia, megaloblastic anemia and leukopenia have also been observed. These conditions have usually been asymptomatic but, in very rare cases, agranulocytosis and purpura have occurred. As a rule, these changes regress after withdrawal of the drug. Concomitant or consecutive administration of sulfa drug-pyrimethamine combinations with trimethoprim or sulfa drug-trimethoprim combinations such as cotrimoxazole may intensify the impairment of folic acid metabolism and related hematological adverse reactions, as well as increasing the risk of severe adverse skin reactions. It should, therefore, be avoided [794].

6) Toxicity after Overdose

High doses of the combinations are potentially fatal. Symptoms include headache, anorexia, nausea, vomiting, excitation and possibly convulsions and hematological changes. In cases of acute intoxication, induction of emesis or gastric lavage is useful if undertaken within a few hours of ingestion. Convulsions can be controlled with diazepam and blood dyscrasias treated with intramuscular folinic acid. Renal function and routine hematology should be checked weekly for 4 weeks after the overdose.

7) Drug Use in Pregnant and Lactation Women

Administration of a full adult treatment dose of SP given in the second trimester and one dose in the third trimester of pregnancy in HIV-negative primigravidae in Kenya and Malawi proved to be effective in clearance or prevention of placental malaria infection and reduction in the risk of low birth weight. Clinical evidence of effect of SP on the fetus is lacking, although there is a presumptive risk of jaundice among premature babies born to mothers who are treated with SP late in the third trimester of pregnancy. No apparent risk of kernicterus development has been found. Folic acid antaginists, as a group, have been reported to increase the risk of neural tube defects, cardiovascular defects, oral clefts and urinary tract defects. Like CQ, both sulfadoxine and pyrimethamine are excreted in small amounts in breast milk and are deemed to be harmless to the breastfeeding infants [759]. Where it is still effective, SP has an important role in the prevention of the deleterious effects of malaria in pregnancy in areas of high malaria transmission, where maternal anemia and low birth weight babies are the two main problems.

Published data on SP use in the first trimester of pregnancy are limited. Animal toxicology studies show dose-related toxicity of pyrimethamine in the form of fetal resorption and growth retardation that were prevented by folic acid administration. There is a case report of a French woman who delivered a child with severe congenital abnormalities who took chloroquine before and after conception and three doses of dapsone/pyrimethamine after conception as prophylaxis. The authors thought that the congenital malformation was related to the dapsone/ pyrimethamine. Sulphonamides are not teratogenic in humans. Pyrimethamine use in the past for mass prophylaxis was not apparently associated with congenital abnormalities. In a study of travelers who had been accidentally exposed to SP or mefloquine as prophylaxis in the first trimester of pregnancy, the rates of congenital abnormalities were 7.0% (12/172), and 3.8% (16/421), respectively. These reported rates of congenital abnormalities were comparable to background rates and provide some reassurance. Although not directly applicable to malaria in pregnancy, the limited published data on the use of SP for the treatment of toxoplasmosis in pregnancy indicate a favorable outcome for children in terms of a reduction in the severe forms of congenital toxoplasmosis.

Both pyrimethamine and sulfadoxine are excreted into the breast milk. In a study of three women given concurrent chloroquine and dapsone/pyrimethamine, the secretion of all three drugs was low and deemed harmless to the infant. Although sulfadoxine secretion is low it carries the theoretical risk of precipitating kernictems in jaundiced and/or premature neonates, and hemolysis in G6PD-deficient neonates. Substantial experience of prescribing SP to breast-feeding women in Malawi has shown that SP is well tolerated by both mothers and breast-fed infants. SP is still given to mothers even if the infant is < 2 months old or is jaundiced. Acute hemolysis has not been observed in breast-fed infants. Clinical experience from western Kenya also suggests that SP use in breast-feeding women has caused no ill effects in breast-fed infants [263].

7.2.2. Atovaquone/Proguanil

1) Adverse Effects of Atovaquone

Atovaquone is generally very well tolerated. Skin rashes, headache, fever, insomnia, nausea, diarrhea, vomiting, raised liver enzymes, hyponatremia and, very rarely, hematological disturbances, such as anemia and neutropenia, have all been reported.

2) Adverse Effects of Proguanil

Apart from mild gastric intolerance, diarrhea, occasional aphthous ulceration and hair loss, there are few adverse effects associated with usual doses of proguanil hydrochloride. Hematological changes (megaloblastic anemia and pancytopenia) have been reported in patients with severe renal impairment. Overdosage may produce epigastric discomfort, vomiting and hematuria. Proguanil should be used cautiously in patients with renal impairment and the dose reduced according to the degree of impairment.

3) Atovaquone/proguanil

Atovaquone acts by inhibiting the parasite's mitochondrial complex bc_1 and disrupting the membrane potential. Because of the high frequency of resistant mutants associated with atovaquone monotherapy it was combined with proguanil (a biguanide acting on DHFR) after promising results of *in vitro* experiments, showing synergistic activity. Atovaquone is erratically absorbed since it is fat soluble, and should therefore be taken with food (not always possible when patients are in the throes of an attack of malaria). Atovaquone has an elimination half-life of about 73 h in African patients. The elimination half-lives of proguanil and its metabolite cycloguanil are estimated at 14 h. Atovaquone-proguanil has been developed for therapy and prophylaxis of malaria, and shows good safety and tolerability in children and adults, with high efficacy against *P. falciparum* malaria when used in a 3-day regimen. However, its use in the tropics is severely curtailed by its high price.

4) Atovaquone/proguanil with Mild Adverse Effects

Atovaquone has been reported to be well tolerated in placebo-controlled double-blind chemoprophylaxis trails. Percentages of patients reporting symptoms varied widely, between 10% and 65%. Adverse effects include abdominal pain, nausea, vomiting, and diarrhea. Vomiting is the most commonly described side-effect reported in several trails. Anaphylaxis following treatment with Atovaquone-proguanil has been reported in one case. Due to lack of data on safety efficacy, Atovaquone-proguanil is not recommended for use in pregnancy. During clinical trials, one case of anaphylaxis following treatment with atovaquone-proguanil was observed. In double-blind placebo-controlled chemoprophylaxis trials, the frequency of adverse events was similar to that in the placebo group, indicating that the combination is very well tolerated.

The tolerability of atovaquone/proguanil as treatment has been similar to that of other drugs. However, in one treatment trial of Thai adults, vomiting occurred in 14 (15.4%) of 91 atovaquone/proguanil recipients of whom 13 (14.3%) required re-treatment; this compared with 2 of 91 mefloquine recipients, none of whom required re-treatment ($p < 0.001$). Vomiting was also a frequent symptom in Gabonese adults (33%) and adult travelers (44%) with *falciparum* malaria. Overall, the rates of early vomiting (with- in 1 hour) from pooled data were 8% in adults and 11% in children. Vomiting after re-treatment was low (< 1%) [263].

5) Atovaquone/Proguanil with Severe Adverse Effects

Serious toxicity is rare. Anaphylaxis was reported in one patient during clinical trials.

6) Drug use in Pregnant and Lactation Women

The manufacturer stated that animal studies revealed no teratogenic or embryotoxic effects. Atovaquone/proguanil cannot be recommended yet; however, it may be used if no other suitable antimalarial drug is available. It is not known whether atovaquone is excreted into human milk. In a rat study, atovaquone concentrations in the milk were 30% of the concurrent atovaquone concentrations in the maternal plasma. Proguanil is excreted into human milk in small quantities. Caution should be exercised when atovaquone/proguanil is administered to nursing

7.2.3. Other Combination Drugs

1) Quinine + Tetracycline

Tetracycline is an antibiotic that probably acts by inhibiting the binding of aminoacyl tRNA to the ribosome. It acts fairly slowly (over days rather than hours), is well absorbed orally, and has an elimination half-time of 8 h, which is similar to that of quinine. Tetracyline maintains activity against multidrug resistant parasites. In Thailand, a 7-day course of quinine+ tetracycline, in which the drugs need to be taken three-to-four times a day, has been used to treat *falciparum* malaria successfully for decades. A major limiting factor in compliance has been symptoms of cinchonism, leading to replacement of this combination by artemisinin-based regimens. Because the use of tetracyclines is contra-indicated in children and pregnant women, use of this combination has always been restricted. Emergence of parasite resistance over the past 10 years in areas where the combination has been extensively used has further hampered its use.

Doxycycline is related to tetracycline but has a longer elimination half-life (about 20 h), allowing once-daily dosing. However, doxycycline, like tetracycline, cannot be used by children and pregnant women. Although dosing of doxycycline is more convenient than dosing of tetracycline and therefore compliance might be improved, quinine-doxycycline has rarely been advocated or used in clinical trials for malaria. Recently, quinine + tetracycline have been recommended as second-line drug in uncomplicated malaria treatment [2].

2) Quinine + Clindamycin

Quinine + clindamycin have never been widely used, although numerous studies show both good efficacy and sound safety profiles in adults, children, and pregnant women. Results of studies in South America, Africa, Southeast Asia, and in non-immune travelers who acquired malaria in different endemic areas, support the efficacy of this combination. Both drugs have fairly short plasma half-lives (clindamycin about 2–4 h), perhaps reducing the risk of selecting for resistant parasites. A clear disadvantage is cinchonism.

Although clindamycin is generally well tolerated, it can lead to the development of *C. difficile*-associated diarrhea, a common complication of antibiotic treatment. The risk of developing colitis with clindamycin treatment is similar to the risks with treatment with expanded-spectrum cephalosporins and broad-spectrum penicillins, and the rate of colitis is about 5% among hospitalized patients. The risk increases by use of combinations of quinine (see adverse effects of quinine). The strongest risk factor for the disease is hospitalization, with outpatients having a fraction of the risk compared to that for inpatients. Other important factors are concomitant disease, advanced age, and the duration of treatment [795]. Recently, quinine + clindamycin have been recommended as first-line drug in first trimester treatment of pregnancy [2].

3) Sulfadoxine-pyrimethamine + Chloroquine

Chloroquine is an aminoquinoline, which interferes with the parasite's heme degradative pathway and thereby prevents detoxification of harmful products of metabolism. It was used as the antimalarial treatment of choice for decades in all malaria-endemic areas and is still used as first-line treatment in Central America and parts of Africa for *P. falciparum* malaria. Today, *P. falciparum* is resistant to chloroquine in all endemic areas except Central America.

The combination of chloroquine with sulfadoxine-pyrimethamine has been sporadically tried in the past. The results of a review, summarized the findings from five trials of sulfadoxine-pyrimethamine +4-aminoquinolines showed a quicker resolution of symptoms with the combination than with sulfadoxine-pyrimethamine alone, as being well tolerated. However, other data do not encourage the use of this combination on grounds of poor effectiveness [796].

4) Sulfadoxine-pyrimethamine + Amodiaquine

Amodiaquine, like chloroquine, leads to annoying pruritus, especially in Africans. Additionally, it causes agranulocytosis and liver damage in about one in every2000 white people taking the drug for chemoprophylaxis. This ad hoc triple combination regimen is used sporadically in some endemic areas, despite the fact that it combines the potential of rare but fatal adverse events associated with amodiaquine and with sulfadoxine-pyrimethamine, as observed when the drugs were used as chemoprophylactic agents for travelers in malaria endemic areas. These observations have led to the withdrawal of both amodiaquine and sulfadoxine-pyrimethamine in almost all European countries. However, when used to treat malaria, similar serious (and perhaps allergic) adverse events have not been observed so often, perhaps because such events only arise after repeated exposure, particularly during chemoprophylaxis. If repeated exposure is indeed a prerequisite for adverse events, then newer intermittent prophylactic regimens in infants and pregnant women might carry similar risks. Nevertheless, renewed interest in this combination for treating African children has been generated by favorable results from Tanzania and elsewhere [796].

5) Sulfadoxine-pyrimethamine + Quinine

Quinine needs to be administered on 7 consecutive days when given as a monotherapy, which is not always achievable because of cinchonism (dysphoria, tinnitus, and nausea) and the very bitter flavor of most salts. However, apart from artemisinins, quinine is the only drug used to treat severe malaria and, despite widespread use, it (with a combination partner such as clindamycin or tetracycline in Thailand) retains excellent efficacy. Quinine + SP have been used in several trials to improve efficacy, but with unconvincing results in Brazil and Bangladesh, where parasite resistance to sulfadoxine-pyrimethamine was already established. In an area where most parasites are still sensitive to sulfadoxine-pyrimethamine, this drug combined with a 3-day regimen of quinine might be more effective than either drug alone, as suggested by the results of a study from South Africa. The adverse effects were reported similar to SP and quinine individually.

6) Sulfadoxine-Pyrimethamine-Mefloquine

This single dose, fixed, triple combination formulation has never been extensively used in most countries, and was introduced onto the market when the efficacy of sulfadoxine-pyrimethamine against parasites was already impaired. It is not an additively or synergistically acting combination, and each drug has a different pharmacokinetic profile. Mefloquine has a long (about 3 weeks) half-life. Other disadvantages are its high price and moderate safety profile.

7) Chlorproguanil-dapsone

Dapsone is a sulfa drug, with an elimination half-life of about 30 h, which has been combined with chlorproguanil (with an elimination half-life of about 20 h). Chlorproguanil-dapsonc was developed by a private-public partnership as an affordable combination treatment for African children. A phase III multicenter study compared the efficacy and safety of chlorproguanil-dapsone with sulfadoxine-pyrimethamine in children in Gabon, Kenya, Malawi, Nigeria, and Tanzania. Results indicate that the combination is well tolerated and efficacious. However, there was a higher frequency of serious hematological adverse events after chlorproguanil-dapsone treatment than after treatment with sulfadoxine-pyrimethamine. The safety of this combination is, therefore, a major concern, particularly since dapsone can cause methemoglobinemia and hemolysis in individuals with glucose-6-phosphate dehydrogenase deficiency. To test for glucose-6-phosphate dehydrogenase routinely in malaria-endemic areas before giving antimalarials is impracticable. There are also concerns about the potential for rapid emergence of cross-resistance to this combination in areas where resistance to sulfadoxine-pyrimethamine is established or becoming so. In Thailand, for example, efficacy of this combination has been severely limited for some time [796].

8) Chloroquine + Proguanil

There has been some experience with the fixed combination of CQ + proguanil in travelers. The combination was well tolerated in 194 French travelers. Gastrointestinal adverse effects were the most commonly reported, affecting about 10% of travelers. Non-serious neuropsychiatric effects were reported by 2%, a significantly lower rate compared with those taking mefloquine (11.5%). Proguanil alone is used occasionally as prophylaxis. Two studies have assessed its tolerability. Mouth ulcers were a particular problem in a cohort study of 470 British soldiers in Belize, affecting 37% of those who took CQ + proguanil and 24% of those who took proguanil alone (relative risk [RR] = 1.56). In a series of Dutch travelers, dizziness and nausea were notable adverse effects compared with travelers who took no prophylaxis; interestingly, there were no reports of mouth ulcers in the 103 proguanil.

In a telephone survey of returning British travelers, the reporting of any adverse effect was similar for those taking CQ + proguanil and mefloquine. Most adverse effects were considered 'trivial' and did not interfere with daily activities. Approximately 0.7% of those taking mefloquine had disabling neuropsychiatric adverse effects, compared with 0.09% of those taking CQ + proguanil (p = 0,021). Mild gastrointestinal upset was reported more frequently in the CQ + proguanil group: 16.3% versus 12.5% (p = 0.009). These findings were broadly similar to a postal survey of visitors to South Africa. People taking CQ H-proguanil reported more gastrointestinal upset and mouth ulcers, and those taking mefloquine reported more neuropsychiatric effects [263].

9) Fosmidomycin-clindamycin

Fosmidomycin-clindamycin is a new antimalarial combination where both drugs act on the parasite's apicoplast, and not on targets of conventional antimalarials. In initial clinical trials, fosmidomycin rapidly eliminated malaria parasites, but did not lead to a sufficiently high final cure rate when used alone. However, when fosmidomycin was combined with the synergistically acting clindamycin, rapid parasite clearance and 100% cure rates were achieved in Gabonese children at 28 day follow-up [796].

7.3. Antimalarial Drug Toxicity in ACTs

Artemisinin-based combination therapies with traditional antimalarial drugs are the most preferred treatment available for malaria in present time for their enhanced efficacy, lower malaria incidence with their potential to lower the emergence and spread of drug resistance rate. A WHO expert consultative group recommended four artemisinin-based combination therapies (ACTs) in 2001 [47]: artemether/lumefantrine (AL); artesunate + mefloquine (AM); artesunate + amodiaquine (ASAQ); and artesunate + SP (ASP).

AL is reported to be effective both in sub-Saharan Africa and in areas with multi-drug resistant *P. falciparum* in Southeast Asia; and, therefore, is recommended as first-line treatment for uncomplicated malaria in several countries [797]. Furthermore, of the abovementioned ACTs, WHO recommends pre-qualified AL as the only fixed-dose ACT for the treatment of uncomplicated *falciparum* malaria. The current recommendation for AL is six doses, twice daily for 3 days, a dose regimen that is known to offer a higher efficacy [2]. So far, no report on adverse effects of AL has been documented. However, a recent report from Mozambique described a small but significant and irreversible hearing loss in patients exposed to AL [783, 798], an adverse reaction not seen in another study from the Thailand-Myanmar border [785].

The human safety data on AM is very limited, although it has been tried as a combination therapy, especially in Southeast Asia. It has recently been reported that in Laos, patients treated with AM had a significantly higher incidence of post-treatment headache, weakness, dizziness, nausea, vomiting, insomnia, palpitations, dyspnoea, nightmares, and pruritus; in comparison to the dihydroartemisinin-piperaquine, recipients [230]. In another study in Kenya, there was no significant derangement in the hematological, biochemical, and electrocardiogram parameters in the recipients of AM, and AM combination was found to be highly effective and safe in the treatment of uncomplicated malaria [521]. A study from southern Papua, Indonesia, indicated that DHP was more effective and better tolerated than ASAQ against multi-drug resistant *P. falciparum* and *P. vivax* infections [222]. Extensive use of AM on the Thai-Burmese border has produced consistently high cure rates (>95%) for uncomplicated malaria, and reduced the transmission of *P. falciparum*. The toxicity of AM is the same as mefloquine monotherapy [263].

ASAQ is an efficacious and affordable antimalarial combination drug. Recently, artesunate amodiaquine was launched as a fixed dose tablet. Under this formulation, all individuals in age groups will receive treatment regimens of once daily dosing for 3 days: 1 tablet/day for children up to 13 years of age ($\geq$ 35 kg) or 2 tablets/day for adolescents aged 14 years or older, and adults ($\geq$ 36 kg). In a large study with ASAQ in Senegal in > 3000 patients, no adverse events were reported. The most commonly reported adverse events possibly related to treatment were vomiting (5.6%), pruritus (3.8%), and dizziness (3.1%). It is important to note that some patients had been complaining of vomiting and/or dizziness prior to the treatment, as often seen in malaria attacks.

Another ACT drug, ASP, was found to be well tolerated, the parasitemia was cleared and the patients were symptom-free within 2 days in a study from Sudan involving 32 pregnant women. All the patients delivered full-term live babies. One of the babies died on the fourth day; none of the women died and there was no miscarriage, stillbirth, or congenital abnormalities in the newborn babies. Studies from Kenya and Uganda also indicated that this combination is effective and safe [759]. This has also been found with ASP or ASAQ. There are several other artesunate combination studies that have been conducted and as more safety data will be reported.

7.3.1. Artemether/Lumefantrine (AL)

AL is the only fixed-dose artemisinin-containing formulation registered after internationally recognized guidelines. It seemed safe and well tolerated in children, as well as in adults [248, 629]; however, a study showed irreversible hearing impairment [783]. Like atovaquone, lumefantrine absorption is enhanced with food, causing problems in children with malaria. It is also expensive. Lumefantrine is structurally related to halofantrine and has a half-life of several days but, unlike halofantrine, does not cause cardiotoxicity.

1) AL with Mild and Moderate Adverse Effects

Experience with AL is increasing rapidly. In large scale clinical trials AL was very well tolerated whether used as four or six doses. AL has been shown to have better tolerability than mefloquine alone or artesunate plus mefloquine. Reported adverse effects have generally been mild and have included gastrointestinal upset (anorexia, nausea, vomiting, abdominal pain, and diarrhea), headache, dizziness, fatigue, sleep disturbance, palpitations, myalgia, arthralgia (all of which could be disease related), and rash. Because it has a similar structure to mefloquine, halofantrine, and quinine, it might be expected that cardiotoxicity would be a problem. However, detailed prospective studies have not demonstrated any cardiotoxicity. In healthy human volunteers and patients, there were no clinically significant changes in the QTc interval and there was no correlation between the QTc intervals and plasma lumefantrine concentrations. Furthermore, in the volunteers, there was no clinically relevant difference in the QTc intervals comparing mefloquine or artemether/lumefantrine alone and when given together [263].

AL was well tolerated in two treatment clinical trials [799]. The proportion of reporting any symptom or sign within the first week after treatment was statistically similar between the arms ($P > 0.05$), except for vomiting reported and the total number of symptoms or signs, which were significantly higher in artesunate + mefloquine arm (5.1% and 27.8%) compared with the AL arm (1.7% and 18.4%). One (1.5%) of 68 participants in the AL arm had mild and transient elevation of the ALT level on day 14. This abnormal value was not associated with clinical illness and resolved spontaneously by day 28 [799]. The excellent adverse effects profile and recent price reductions (down to US $1 per adult treatment) make it an increasingly attractive treatment option.

2) AL Associated with Ototoxicity

The controversy about artemisinins and ototoxicity in humans has only recently been investigated by Toovey and Jamieson [783]. The authors compared audiometric data from 150 adult construction site employees who have been treated with AL for uncomplicated malaria with 150 matched controls who neither suffered malaria nor received artemether. Significant hearing loss over the term of their employment was found in frequencies between 1 and 8 kHz. This was judged to be irreversible, because the time between treatment and exit audiogram (mean = 163 days, range 3-392 days) did not correlate with the degree of hearing loss [784]. However, possible confounding factors like the influence of noise exposure in these construction site workers or the lack of a control group of malaria patients treated with other antimalarials make it difficult to establish a causal relationship between hearing loss and AL therapy from this retrospective evaluation.

Detailed neurological data provided by Price *et al.* had more than 1,000 patients above five years of age treated with artemether or artesunate (alone or in combination with mefloquine) and examined on days 2, 7, and 28 post-treatment, no patient developed deafness (assessed by tuning-fork test) or permanent neurological abnormalities. A retrospective study carried out in Vietnam [800] compared 337 subjects who had received from two to 21 courses (median = 2) of either artemisinin or artesunate with 108 controls from the same village. Even though 20% of the subjects had received cumulative doses of $\geq$ 500 mg/kg artemisinin (or the adjusted equivalent of artesunate), which might be more than in any other group of people in the world; the authors found no evidence of a drug effect on screening audiometry (testing for hearing loss $\geq$ 40 dB), brainstem evoked auditory potential latencies or neurological examination. Similar results were obtained in two case-control studies from Thailand in 79 subjects treated at least twice with oral artesunate or artemether [801] and in 68 patients who had been treated with AL [785]. A recent study in 15 adult volunteers with experimental *falciparum* malaria treated with AL could not detect any ototoxicity by using conventional and evoked response audiometry, but did not compare artemisinins to other antimalarials [802].

In more recent clinical trial, no vomiting occurred after ingestion of the antimalarial drugs, and no serious adverse events were reported during treatment and follow-up. Most symptoms present at the time of diagnosis resolved until day 7. However, hearing problems and tinnitus were more common on day 7 with nine of thirty patients complaining of hearing problems in the Q group. In seven of these, audiometry and OAE testing confirmed significant hearing loss. Patients reporting subjective hearing impairment in the AL group did not have abnormal hearing test results. In the AP group, only the reported hearing loss by one patient on day 90 corresponded to significantly impaired audiometry and OAE results; in this patient malaria re-infection was diagnosed.

Air conduction hearing thresholds were compared for a standard range of frequencies from 0.125 to 8 kHz. Bone conduction thresholds were also measured at all time points in order to exclude a possible conductive hearing loss. In the Q group, a hearing loss affecting all frequencies is evident on day 7 and has disappeared by day 28. Otherwise, no significant changes of the mean hearing thresholds compared to day 0 were evident, except for some slight general improvement in all groups [803].

Brainstem evoked response audiometry was tested. Interpeak latencies (IPL) were calculated for the I–V, I–III, and III–V intervals. In all groups, IPL I–V was shorter on day 0 than on later time points. The difference in IPL I–III between the AL group and the other two groups on day 28 was limited to the right ear. However, only one patient in the AL group had a potentially clinically relevant interaural difference of IPL I–III greater than 10% on day 28, which disappeared by day 90. No permanent drug-related prolongation of interpeak latencies occurred. This study is the first randomized clinical trial directly comparing ototoxicity of AL with other antimalarial drugs. In conclusion, this study could not detect any detrimental effect of a standard oral regimen of AL on peripheral hearing or brainstem auditory pathways in patients with uncomplicated *falciparum* malaria as assessed by pure tone audiometry, otoacoustic emission recording and brainstem evoked response audiometry. In contrast, this study clearly detects the transient quinine induced hearing loss due to temporary outer hair cell dysfunction. These results therefore support the continued use of oral artemisinin based combination therapy for uncomplicated malaria [803].

3) AL with Severe Adverse Effects

According to the manufacturer, there were 20 reported serious adverse in 1869 patients. Nineteen of these adverse events could be explained on the basis of the malarial episode or represented concurrent illnesses (including severe anemia in children, unsatisfactory response, pneumonia, hepatitis). AL may have contributed to only one of the reported serious adverse events – the development of hemolytic anaemia in a 35-year-old patient after the drug had been discontinued for 13 days [629].

4) Drug Use in Pregnant and Lactation Women

In animal studies (Novartis data-on-file), lumefantrine was shown to be neither mutagenic, nor embryotoxic. Doses up to 1000 mg/kg have been tested and shown to be well tolerated, with no evidence of maternal or embryofetal toxicity. Additional studies were conducted with the combination (ratio 1: 6) of artemether and lumefantrine. In rats receiving doses of 0, 30, 100, and 300 mg/kg of artemether/lumefantrine, reduced maternal body weight and food consumption and a high level embryotoxicity occurred with doses greater than 100 mg/kg. Doses of the combination of artemether/lumefantrine of 30 mg/kg were neither embryotoxic or teratogenic. In rabbit studies, a more favorable profile was seen. Doses greater than 210 mg/kg of this combination were associated with embryo-fetal lethality. Viable fetuses, however, showed no abnormalities. Doses up to 105 mg/kg were free of treatment-induced effects. Artemether, however, showed maternal effects and embryo-fetal toxicity at 10 mg/kg in rats but no teratogenicity, with a no adverse effect level of 3 mg/kg. Similarly, embryo-fetal toxicity was documented in rabbits at 30 mg/kg, with no evidence of teratogenicity. The no adverse effect level in rabbits was 25 mg/kg, which suggests a steep dose-response curve for embryotoxicity for this species. In summary, artemether, like all artemisinin derivatives, but not lumefantrine can induce fetal toxicity/re-sorption in the early days of gestation in rodents.

There are no published data on the use of AL in pregnant or breast-feeding women, and early gestational fetal toxicity/re-sorption would be very difficult to document in human pregnancy. More studies are needed. Until more data becomes available, clinicians should only use AL for treating drug-resistant *falciparum* malaria if no other suitable drugs are available.

7.3.2. Artesunate + Amodiaquine

Amodiaquine, like chloroquine, is a 4-aminoquinoline; it is effective against chloroquine-resistant strains of *P. falciparum*, although there is some cross-resistance. In recent years resistance has worsened considerably in parts of East and Southern Africa. After oral administration amodiaquine was largely converted to desethylamodiaquine, and contributes to the majority of antimalarial activity. Amodiaquine is generally reasonably well tolerated and slightly more palatable than chloroquine, although in some areas it has not been a popular substitute. The serious adverse effects that have been associated with its prophylactic use (agranulocytosis and severe liver toxicity) are considered rare when amodiaquine is used in malaria treatment, although more data are needed to characterize the risks. More data is also required for comparison during pregnancy.

The effectiveness of 3 days of treatment (rapid clearance of fever and malaria parasites) in western and central Africa, where resistance to amodiaquine is low, and the combination of artesunate plus amodiaquine may delay or prevent the emergence of resistance to both drugs. An important step is the recent registration in Morocco (the country where the drug is manufactured) of a fixed combination of artesunate plus amodiaquine by the Drugs for Neglected Diseases initiative with sanofi-aventis as the industrial partner. A prequalification dossier of this fixed combination has been submitted to the WHO. This new co-formulation will almost certainly increase its effectiveness by improving drug compliance [804].

In Africa, artesunate + amodiaquine (ASAQ) is one of the recommended ACT for the treatment of malaria where amodiaquine (AQ) monotherapy remains effective. Using this combination, the artesunate (AS) acts quickly to kill most of the parasites and then the longer-acting AQ clears the residual parasites (the elimination half-life of dihydroartemisinin, the active metabolite of AS, is only about 1 h whereas that of desethylamodiaquine, AQ's active metabolite, lies between 6 and 18 days). Although there is recent evidence indicating that the total drug exposure to both AS and AQ is significantly reduced when the two drugs are given in combination [805], the clinical implications of this reduction remain unclear.

Many African countries have already adopted ASAQ as the first-line treatment for uncomplicated malaria [805]. In the last few years, the recommended first-line treatment in Ghana, for example, was changed from chloroquine to AS–AQ, based on the high frequencies of cure seen with the combination and the growing capacity for the local production of both AS and AQ. Also important was the expectation that with the increasing use of insecticide-treated bed nets; the high efficacy, low cost, and simplicity of ASAQ treatments would, in the long run, help to reduce malarial transmission in Ghana. Unfortunately, in the minds of many Ghanaians, the ASAQ combination treatment is associated with several adverse events (apparently because adverse reactions were initially reported in a few patients treated with some brands of the combination) and this perception, although largely unfounded, has probably reduced compliance. To facilitate the change from monotherapy to ACT, policy-makers in Ghana (and elsewhere) need site-specific, evidence-based data from trials of the effectiveness of ACT when used under 'everyday' conditions rather than the near-optimal conditions, with supervised treatments and perfect compliance, seen in many clinical trials [806].

1) ASAQ with Mild and Moderate Adverse Effects

Within the first three days of treatment, complaints of dizziness were reported by 14 (12%) subjects in the ASAQ patients. Complaints of fatigue and excessive sleepiness were reported by five (4.3%) subjects in the ASAQ groups. Pruritus was reported by three subjects. A history of vomiting and nausea was part of the presenting complaint by 44.8% (52/116) subjects respectively in the ASAQ groups. Between the first three days of treatment, 27.5% (32/116) subjects respectively complained of vomiting and or nausea in the AS+AQ groups [161]. In another study, the ASAQ treatment was found to be fairly well tolerated; none of the patients in the two arms developed severe adverse effects and no patient was withdrawn for reasons of safety or tolerability. All the treatment-emergent signs and symptoms documented in the study were related to malarial symptoms and resolved during the follow-up. They were similar in the two arms and similar to those effects reported previously in trials of ASAQ [806].

Of the 301 children who completed the study with ASAQ, 70 (23.2%) suffered weakness, 56 (18.5%) itching, 53 (17.5%) abdominal pains, 46 (15.2%) body pains, 26 (8.6%) nausea, 25 (8.3%) vomiting, 15 (5.0%) shaking chills, and nine (3.0%) fatigue. The incidences of these adverse effects, which were all mild or moderate, were similar in the two treatment arms [807, 808].

2) ASAQ with Severe Adverse Effects

Although the limited data available suggest that the combination of ASAQ is well tolerated as short-course treatment, there are two safety concerns that have not yet been fully addressed. Neutropenia and hepatitis curtailed the use of amodiaquine as prophylaxis in the 1980s [807]. A recent randomized, controlled trial comparing treatment with amodiaquine alone or in combination with artesunate found, asymptomatic neutropenia on day 28 in 6% of 153 children with uncomplicated *falciparum* malaria, all of whom had normal baseline absolute neutrophil counts. We have previously reported a case of hepatitis in this group of volunteers that developed after the second dose of amodiaquine and were considered probably related to amodiaquine. A similar report of delayed-onset asymptomatic hepatitis has been described in healthy American volunteers who had received artesunate plus amodiaquine alone followed by artesunate plus amodiaquine plus efavirenz, resulting in early study discontinuation. In that study, efavirenz was thought to have caused an increase in amodiaquine concentrations by competitive inhibition of the P450 cytochrome enzymes that metabolize amodiaquine. Widespread use of artesunate plus amodiaquine treatment in Africa support the need for laboratory monitoring of the treatment's safety, particularly if antiretrovirals are used concomitantly [2, 805].

In healthy volunteers, twenty-nine adverse events were reported over the course of the study by 10 (67%) of the 15 volunteers. The frequency of adverse events was similar across all three treatment arms: 4/15 (27%) following artesunate alone, 8/15 (53%) following amodiaquine alone, and 5/15 (33%) following the combination. All adverse events were consistent with the product information available, resolved spontaneously and, except for the transaminitis were mild or moderate in intensity. A case of asymptomatic, prolonged, severe transaminitis that developed after phase 3 of the study has been previously published. This was considered to be probably related to amodiaquine, rather than artesunate, before the pharmacokinetic results were available. The subsequent finding that this volunteer had the highest measured desethylamodiaquine AUC following the administration of artesunate plus amodiaquine strengthened this assessment.

There were no liver function abnormalities detected in any other volunteers. No renal or electrolyte abnormalities were detected in any volunteers. Significant hematological changes were confined to the white cell counts. During phase 2 of the study, two volunteers (one following amodiaquine alone, one following the combination) developed asymptomatic, NCI grade 1 leucopoenia and either a grade 1 or grade 2 neutropenia. None of the other observed changes in hematological parameters were outside the normal range. There were no significant changes following treatment seen on the ECGs; the mean (95% CI) QTc interval was 398 (390–406) ms at screening, 401 (393–409) ms following artesunate alone, 400 (385–

414) ms following amodiaquine alone and 412 (400–424) ms following the combination. In two patients, the prolongation following treatment with amodiaquine or artesunate plus amodiaquine was considered of borderline clinical significance.

Despite our small sample size, we detected the well known amodiaquine-related adverse drug reactions. Clinically important adverse events were reported by a quarter of these volunteers, even though all were healthy and only two doses were administered 3 weeks apart (rather than the daily administration for 3 days recommended for malaria treatment). When the artesunate plus amodiaquine combination becomes widely used, monitoring of liver function tests and hematological parameters are warranted to define these risks in malaria patients, particularly in patients at special risk such as those at each age extreme, with HIV co-infection or malnutrition. However, the feasibility of this recommendation outside of the research setting is of concern.

To conclude, the total exposure to the main active metabolites of both artesunate and amodiaquine was significantly reduced when administered in combination to healthy African volunteers; this might be important clinically. However, because cure rates with this combination are generally higher than with amodiaquine monotherapy, artesunate and amodiaquine could remain in the armamentarium of drugs used to combat *falciparum* malaria, provided efficacy continues to be monitored and adequate safety precautions can be taken. Further pharmacokinetic research on artesunate plus amodiaquine, when administered concurrently or as a fixed dose combination, is urgently required in patients with malaria to establish the extent and clinical significance of these pharmacokinetic interactions [805].

In another clinical trial, severe anemia (hemoglobin 4.8 g/dl) occurred on day 14 in a 1.5 year old subject (ASAQ). Severe anemia and dizziness were considered unlikely to be drug related, but pruritus and excessive sleepiness were considered possibly drug-related. Other reported adverse events were mild in intensity, and overlapped with known malarial symptomatology. These were mostly classified as unrelated to study medications [161].

3) Malaria-Associated Neutropenia

Neurological examination could be done for 92 children treated with ASAQ. Nystagmus was observed in two subjects (n = 1) between days one and seven. A detailed history indicated that one of these two subjects (7-year-old male) was admitted as a neonate into neonatal intensive care (NICU) because of probable birth asphyxia. This subject was still attendant at pre-school at seven years of age (which is unusual in Ghana), and demonstrated excessive emotional liability. An electroencephalography (EEG) done for this subject because of a history suggesting the possibility of a seizure disorder, reported the presence of epileptiform loci and generalized cerebral dysfunction. This subject had poor academic performance (had repeated first and second grades in school) and possible cognitive impairment. A routine EEG done for this subject was normal. No other abnormal neurological findings were observed in the remaining children during the 28-days follow-up, monthly follow-up visits, or for subjects who received multiple treatments.

There were no differences in the mean total WBC or absolute neutrophil counts between the two groups on days 0, 3, 7, 14, or 28. Neutrophil counts on follow-up days 3, 7, 14, and 28 were significantly lower (p < 0.01) compared with day 0 counts in both treatment arms. Pre-treatment neutropenia was observed in three subjects (ASAQ, n = 1) on admission. Between days 3 and 28, neutropenia was observed in 13 (11.7%) subjects in the ASAQ groups. The median pre-treatment neutrophil count of subjects who developed neutropenia was lower than subjects who did not develop neutropenia, but the mean age was similar. Neutropenia was severe (<500/µl) in three subjects (ASAQ, n = 1) between days 3 and 28. Neutrophil counts did not show any pattern of difference between the initial and subsequent episodes of uncomplicated malaria in subjects who were treated more than once. In two subjects, each of which was treated for five episodes of uncomplicated malaria (with ASAQ) during the one-year follow-up; no clinical signs suggestive of neutropenia was noticed, and all neutrophil count measurements were higher than 1,000/µl in these two subjects. The neutrophil count profile of one of these individuals is shown [161].

There was a drop in neutrophil counts in ASAQ arms after treatment. This pattern of neutrophil count reduction, as well as the absence of differences in neutrophil count profiles between children who received single course and those who received multiple courses, suggests the possibility of a disease-rather than drug-related effect. Malaria-associated neutropenia has been reported for subjects treated with AQ, either as monotherapy, or in combination with AS [809, 810]. Neutrophil count reduction during acute malaria may be due to altered intravascular granulocyte distribution, or to enhanced phagocytosis. AQ-associated agranulocytosis on the other hand, may be due to inhibition of bone marrow precursor cells, or to immune-mediated hypersensitivity. The finding that neutrophil counts were not increasingly reduced in subjects who received more than one treatment course in this study lends support to the conclusion that, brief three-day administration of AQ, as used for treatment, is unlikely to be associated with the described side-effects observed during prophylaxis. However, there was a large random variability around the confidence interval of the difference in neutropenia incidence between the groups, and the sample size of the study was not sufficient to detect rare events; therefore, this potential side-effect should be continuously monitored.

This is one of the few studies to conduct systematic longitudinal neurological examinations for children treated with artemisinin-based regimens. A total follow-up lasting 12 months did not reveal abnormal neurological signs that could be ascribed to artemisinin-based treatment in this study. The two subjects with observed nystagmus had underlying conditions that could easily explain the noted abnormalities. Furthermore, the observed nystagmus was not associated with other cerebellar signs, suggesting these were unlikely to be of clinical significance, or could be false positive due, for instance, to excessive lateral eye abduction. However, nystagmus has been reported in artemisinin-treated subjects without reported underlying neurological abnormality, as well as in an experimental study of artemisinin-treated rhesus monkeys. Because available data from animal studies do not provide much insight into the early signs of possible artemisinin-related neurotoxicity, and because of the lack of a well-defined neurological syndrome to characterize potential late artemisinin toxicity during clinical use, detailed focused studies on this subject are still required.

ASAQ was efficacious for the treatment of Ghanaian children with uncomplicated malaria. After 12 months of detailed systematic follow-up, no overt clinical evidence of neurotoxicity or neutropenia was observed. However, the sample size was limited for conclusive evidence on safety; therefore, post-marketing surveillance, with particular emphasis on neurological and hematological changes, is still indicated. In view of the high prevalence of potentially AQ-resistant mutant parasites in the study area, the efficacy of the ASAQ combination, the first-line treatment for uncomplicated malaria in Ghana, requires continuous monitoring and evaluation [161].

4) Drug Use in Pregnant and Lactation Women

WHO recommends this regimen for treatment of uncomplicated *falciparum* malaria in African children and women. Indeed, in some African countries ASAQ is considered as first-line treatment for children and women with uncomplicated malaria. In one study of healthy volunteers, one woman developed an asymptomatic rise in liver function tests following two sequential doses of AS (day 0), AQ (day 7), and both drugs together on day 28 [270, 809]. There are no other published data on the use of ASAQ in pregnant or breast-feeding women.

7.3.3. Artesunate + Mefloquine

Mefloquine is a quinoline methanol compound related to quinine. Several different mefloquine formulations are now available with different oral bioavailability. Combining artesunate with mefloquine has all the advantages of a combination treatment previously described [56], and the additional benefit that if mefloquine is split as 8.3 mg/kg/day for 3 days or not given until the second day of treatment then absorption is increased and gastrointestinal adverse effects are lessened. A fixed combination of mefloquine and artesunate has recently been developed. This is dispensed as tablets containing 200 mg of artesunate and 400 mg mefloquine (base). Recent trials in Asia indicated that the tolerability of this new regimen (mefloquine dose 8 mg/kg/d for 3 days) is better than that of the standard regimen. This combination has been evaluated and used mainly in Southeast Asia and South America. More information on tolerability, safety, and efficacy is needed in African children so that its potential utility in Africa can be assessed objectively.

A 3-day regimen of artesunate + mefloquine (AM) has been the preferred treatment for malaria in Thailand for almost a decade. It is safe, well tolerated, and highly effective, and has also been investigated in South America and Africa. In Thailand, malaria incidence and *in vitro* mefloquine resistance of parasites has decreased since this combination has come into use. Disadvantages of this regimen include its price and the pharmacokinetic mismatch of each drug. In a hyperendemic area, widespread treatment with this combination would lead to long-term exposure of parasites to low doses of mefloquine [796].

1) AM with Mild and Moderate Adverse Effects

The human safety data on AM is very limited, although it has been tried as a combination therapy, especially in Southeast Asia. It has recently been reported that in Laos, patients treated with AM had a significantly higher incidence of post-treatment headache, weakness, dizziness, nausea, vomiting, insomnia, palpitations, dyspnoea, nightmares and pruritus, in comparison to the dihydroartemisinin-piperaquine, recipients [230]. In another study in Kenya, there was no significant derangement in the hematological, biochemical and electrocardiogram parameters in the recipients of AM, and AM combination was found to be highly effective and safe in the treatment of uncomplicated malaria [521]. A study from southern Papua, Indonesia, indicated that DHP was more effective and better tolerated than ASAQ against multi-drug resistant *P. falciparum* and *P. vivax* infections [222]. Extensive use of AM on the Thai-Burmese border has produced consistently high cure rates (>95%) for uncomplicated malaria, and reduced the transmission of *P. falciparum*. The toxicity of AM is the same as mefloquine monotherapy [263].

No serious adverse event was reported during the study in all the four trial sites. Many adverse events of the antimalarial drugs were most likely related to the underlying malaria disease. Of the 39 reports, 10 patients (2.3% of total patients treated) reported vomiting. The others were as follows: headache (9, 2.1%), dizziness (12, 2.8%), and abdominal discomfort (4, 0.9%). There was one report (0.2%) each of sleeplessness, fast breathing, weakness, and back pain. Another report showed that more than half of the patients (54.6%) reported suffering from dizziness. Headache (28.1%), nausea (27.8%), insomnia (24.3%) and vomiting (19.5%) were also common. The other side-effects, such as abdominal pain (8%), tinnitus (7.9%), diarrhea (3.4%), itching (0.6%), tachycardia or palpitation (1.5%) were reported by a few patients. All these symptoms were mild and transient, patients recovered without any medical intervention. Based on the experience of this study, AM is safe and well-tolerated. The laboratory values were not significantly different pre- and post-treatment. The marginal variations in liver function test results may be related to stabilization of the liver following successful treatment. The same result was observed with mean hemoglobin values which returned to normal after recovery. The slight reduction in mean platelet count (data not presented) was consistent with the reported findings of relative thrombocytopenia in 50 to 75% of patients with acute malaria [245, 271, 810].

2) AM with Severe Adverse Effects

Potential disadvantages of AM include the neuropsychiatric adverse effects of mefloquine and the current lack of a co-formulated product, both of which may reduce adherence [811].

3) Drug Use in Pregnant and Lactation Women

McGready 2000 [381] reported fewer treatment failures (excludes new infections) at day 63 with artesunate plus mefloquine (RR 0.09, 95% CI 0.02 to 0.38; 106 participants). There was also a trend towards better performance at day 28 in this trial (approximately 97% of the artesunate plus mefloquine group and 88% of the quinine group were without parasite recrudescence; data estimated from figure in original article). Bounyasong 2001 [789] reported no treatment failures in 57 participants at day 28. Bounyasong 2001 also reported that mean hematocrit was not statistically significantly different between treatment groups on admission (artesunate plus mefloquine 34.29% versus quinine 35.17%). By the end of treatment hematocrit was slightly reduced in both treatment groups, but it was higher with artesunate plus mefloquine (33.2%) compared with quinine (28.4%); trial authors' P test was statistically significant [789].

Anemia on admission was similar in both treatment groups in McGready 2000, but by day seven more women in the artesunate plus mefloquine group had anemia. This did not persist at further time points (days 28, 42, and 63) and by day 63 the number of participants with anemia, in both treatment groups, was about half that on admission. The trial authors also examined the number of women who developed anemia during the 63 days of follow up (overall 35/53) and found no difference between the groups.

Both trials reported on adverse events experienced by the women. Those affecting the nervous system, including tinnitus and vertigo, were generally more common with quinine. In addition, gastrointestinal adverse events were more commonly reported with quinine. The trials reported other adverse events − hypoglycemia, muscle and joint pain, and palpitation, but there was no statistically significant difference between the groups except for hypoglycemia, which was more common in the quinine group in Bounyasong 2001. McGready (2000) did not test for hypoglycemia. And healso reported one death, unrelated to malaria [381].

Both trials reported on mean birth weight. McGready (2000) reported no statistically significant difference in the mean birth weight and the number with low birth weight [381]. Bounyasong (2001) reported that infants born by mothers in the mefloquine group were 140 grams heavier overall compared with quinine (trial authors' P = 0.041) [789]. McGready 2000 reported no stillbirths in either group and no statistically significant difference in the number of abortions, but there were few participants [381]. There were two neonatal deaths in the artesunate plus mefloquine group compared with one in the quinine group. Bounyasong 2001 did not demonstrate a statistically significant difference in gestational age (trial authors' data) or neonatal jaundice. Neither trial reported congenital abnormalities [260].

7.3.4. Artesunate + Sulfadoxine-Pyrimethamine

Sulfadoxine-pyrimethamine (SP) is a fixed combination of a long-acting sulfonamide and the antifolate pyrimethamine. These are synergistic against sensitive parasites. Minor adverse effects are unusual. Serious sulfonamide toxicity is unusual with a single-dose treatment of malaria. The anti-folate properties of pyrimethamine rarely produce toxicity. The combination with artesunate is available as separate scored tablets containing 50 mg of artesunate, and tablets containing 500 mg of sulfadoxine, and 25 mg of pyrimethamine. There are no plans for developing a fixed dose combination. The total recommended treatment is 4 mg/kg BW of artesunate, given once a day for 3 days and a single administration of SP 1.25/25 mg base/ kg BW on admission. This SP dose was developed in adults but in the main target group (children aged 2–5 years) the weight adjusted dose produced blood concentrations of both components that are approximately half those in adults. Thus the standard dose may be sub-optimal in younger children. The combination has been evaluated extensively in adults and children with uncomplicated malaria and is sufficiently efficacious in areas where 28-day cure rates with SP alone exceed 80%. This ACT is currently being used in parts of South America, the Middle East, and South Asia where SP susceptibility remains high. Because SP, sulfalene–pyrimethamine, and trimethoprim–sulfamethoxazole are still widely used as "monotherapies," resistance is likely to worsen.

Artesunate + sulfadoxine-pyrimethamine (ASP) has also been assessed in African children. It gave promising results in The Gambia, where it was as well tolerated and as efficacious as sulfadoxine-pyrimethamine alone. This combination is currently available as separate tablets containing artesunate 50 mg and tablets containing both sulfadoxine 500 mg and pyrimethamine 25 mg (20 parts : 1 part). As with other artesunate-based regimens, the recommended dose of artesunate for the treatment of uncomplicated *P. falciparum* malaria is 200 mg for 3 days. The long-acting SP combination is administered at 1500 mg/75 mg (three tablets) on treatment day 1. SP is available as an intramuscular injection at the same doses as the oral form. Prophylaxis with sulfadoxine-pyrimethamine is no longer recommended because of its ability to cause Stevens-Johnson syndrome and toxic epidermal necrolysis [245].

However, in further WHO-led trials in African children, the combination was disappointing. Studies from China, the Thai-Myanmar border and one study of accidental exposure to artesunate and SP in Gambian women in all trimesters of pregnancy showed that the artemisinins are well tolerated in pregnancy without any adverse effects, and newborns who were followed up for 1 year after birth had a normal rate of development. However, there is no sufficient data to definitely determine any potential risks during pregnancy, particularly in the early first trimester. WHO recommends that artemisinins should not be used for treatment for malaria in the first trimester (but should not be withheld if they are lifesaving for the mother) and should be used in later pregnancy only when other treatments are considered unsuitable [759].

1) ASP with Mild and Moderate Adverse Effects

In the recent clinical trial, no serious or severe adverse events were reported. Four of the participants (3.8%) vomited after the first treatment dose of ASP; 1 on ASP and 3 on ASP+PQ arm. One of these vomited immediately, refused to take the drug again and withdrew consent. The other three cases (2.9%) vomited after more than eight hours after taking the drug and the dose was therefore, not repeated for them. Two people complained of insomnia and another two complained of itching [812].

Most ASP clinical trials did not clearly describe the methods for reporting adverse events or the precise numbers of events. Abacassamo 2004 reported "no severe adverse reactions attributable to treatment"; Rwagacondo 2003 reported "no major drug related adverse effects"; Dorsey 2002 reported "no severe adverse reactions to trial drugs" and that "mild adverse reactions did not differ between the three treatment groups." Mockenhaupt 2005 did not mention adverse events. There are presently insufficient data to determine the effects of those treatments on outcomes such as symptom resolution and adverse events. Common adverse effects (nausea, vomiting, diarrhea, itching, dizziness, and higher hemoglobin levels) were observed with similar frequency in those clinical trials and all of them were mild and resolved spontaneously [812-815].

In a clinical trial with children and women, both drugs of SP and ASP were well tolerated. No child vomited during the 1-h observation period. There were no clinically significant adverse experiences in either group during the 28 days of follow-up that were attributed to the study medication. A total of 10 children, 4 of whom received SP alone and 6 of whom received the ASP combination, experienced an adverse experience during follow-up, namely acute respiratory infection (4), cellulites (2), and gastro-enteritis, otitis media, clinical jaundice, and facial puffiness (1 each), all of which resolved completely.

There were no clinically significant adverse experiences in terms of either the hematological or the biochemical indices. The proportions of children with PCV < 20% on Days 0, 7, and 28 were 26%, 17% and 0% for SP alone and 16%, 32%, and 0% for the combination, respectively. Three children had an ALT concentration > 42 IU/L at admission, 1 of whom received SP alone and the other 2 received the ASP combination; the values for 2 of these children had fallen within the normal range by Day 7. The 3rd child, treated with the combination, had an ALT concentration of 368 IU/L at enrollment which was still elevated at 135 IU/L by Day 7, but, clinically, the child made an uneventful recovery. No child had a serum creatinine concentration > 130 μmol/L at any time point. Total WBC, neutrophil and lymphocyte counts on Days 0 (n = 38) and 7 (n = 37) were unremarkable.

In addition, a total 'symptom score,' the sum of the severity of each listed symptom divided by the number of children seen at each time point, and the proportion of children who did not complain of any symptom. By Day 28, 2 children, both originally treated with SP alone, complained of mild weakness and mild headache, respectively. All the other children were asymptomatic. Although the sample size was small, these data suggest that the ASP combination of a 3-day course of artesunate given with a single dose of SP is safe and well tolerated as a treatment for uncomplicated *P. falciparum* malaria among Gambian children. There were no adverse experiences attributed to the study medication nor was there any evidence of toxicity in terms of either the hematological or the biochemical indices [816].

2) ASP with Severe Adverse Effects

There was no report of serious adverse events and none of the patients followed-up reported any drug-related side-effects. Some mild adverse events such as weakness were noted, but these events resolved spontaneously and gave no reason to stop the treatment. Thus, both drugs were effective in treatment of uncomplicated malaria in children. The difference in recrudescence indicates that ASP may be the preferred treatment.

3) Drug Use in Pregnant and Lactation Women

The ASP combination regimen was relatively well tolerated. The minor side-effects reported by the women, were difficult to distinguish from symptoms of uncomplicated malaria. There was no difference in the perinatal mortality in the three treatment groups. Although spontaneous abortions only occurred in the SP-azithromycin group, three of the four abortions occurred long after drug ingestion in women with HIV and/or malaria infections. This single unexplained abortion could have been a chance occurrence. A previous study reported 6 abortions out of the 123 women who were exposed to azithromycin during pregnancy [817]. Therefore, this outcome needs to be closely monitored in future studies that use azithromycin to rule out an association. Three still births occurred in the SP-artesunate group, but, again, a long time had elapsed between the time of drug administration and the occurrence of the still births. The short half-life of artesunate would suggest that it was unlikely these still births were associated with the treatment. Additionally, other obstetric explanations could be found for all 3 still births [813].

In this trial,106 (89.8%) of the 118 babies were weighed and gestational ages assessed within 24 hours of delivery. There was no difference in the gestation age between the groups, and no visible physical abnormalities were detected in any of the newborn babies. There was also no significant difference in the mean birth weights of infants born to mothers in the SP (2,868 ± 625 g) and SP-artesunate (2,836 ± 482 g), probably because of the small sample size[813]. A total of 287 pregnant women in the Gambia were exposed to artesunate in combination with SP during a mass drug administration; no difference was noted in the rates of abortions, stillbirths, or infant deaths among those exposed or not exposed to the drugs [816].

In another trial, among the 459 known pregnancy outcomes, there were 39 (8.5%) fetal and infant deaths within 9 months of the MDA. There was no difference in the proportion of abortions, stillbirths, and infant deaths among those exposed or not exposed to ASP during gestation. Findings during the first week of life 200 live infants were examined within 6 h to 7 days of delivery; the mean age at the time of examination was 3 days. Six (5.0%) of the 119 infants who were exposed to ASP during gestation, and 2 (2.5 %) of the 81 that was not exposed had an abnormal physical examination finding (P = 0.48). Among the exposed infants the findings included an umbilical hernia (1), undescended testis (l), skin infection (1), dry and parched skin (1), acute respiratory infection (1), and purulent eye discharge (1). Among the unexposed infants the findings included a hemangioma (1) and jaundice (1).

The weights of 195 singleton infants were measured within the first week of life. The mean weight of infants born to all mothers who had received ASP regardless of gravidity was 140 g greater than that of infants born to untreated mothers (P = 0.05). After adjusting for the number of previous deliveries, gestational month of exposure to ASP, exposure to other antimalarials during pregnancy, age of the infant when weighed and sex, there was no significant difference between the mean weight of infants born to treated and untreated mothers (P = 0.10). Of the 195 infants exposed to ASP, 77 were exposed during the first trimester, 90 during the second trimester, and 28 during the third trimester. The mean weight of infants born to mothers who had received ASP during the third trimester was 480 g greater than that of infants born to untreated mothers (P = 0.01). The effect of ASP given during the third trimester on the weight remained statistically significant after adjustment (P = 0.05).

The mean weight of infants born to primi- and secundigravidae who received the drug combination was 190 g greater than that of infants born to untreated primi- and secundigravidae (P = 0.05, adjusted P = 0.18). Of the 71 infants born to primi- and secundigravidae, 30 had gestational exposure to ASP during the first trimester, 31 during the second trimester, and 10 during the third trimester. The mean weight of infants born to primi- and secundigravidae who had received ASP during the third trimester was 530 g greater than that of infants born to untreated primi- and secundigravidae (P = 0.03, adjusted P = 0.03). On interview 9 months after the MDA, the parents stated that 34 (8.1%) of 420 surviving infants were not well. Twenty-four (9.1%) of 264 infants who were exposed to ASP during gestation and 10 (6.4%) of 156 infants who were not exposed were ill on follow-up (P = 0.4). The problems reported by the parents were diarrheal disease (14), acute respiratory infection (13), malaria (3), scabies (2), conjunctivitis (1), and trauma (1). There was no report of any abnormality that could be related to exposure *in utero* to a teratogenic agent.

This trial recorded 3 deaths among the 517 women who had a pregnancy outcome within 9 months of the MDA. One had received ASP during pregnancy, 1 had not and for 1 mother the exposure status could not be confirmed. Verbal autopsies indicate that all 3 deaths were due to postpartum hemorrhage. Two of the women had symptoms of anemia before the onset of labor. Among the 287 pregnancy outcomes with gestational exposure to the combination ASP, there was no evidence of increased fetal loss or infant death. Fetal losses were probably under-reported since pregnancy is often denied until the mother feels movement of the fetus. The rate of under-reporting is likely to have been similar between the exposed and unexposed groups.

Despite these limitations, the differential effect in weight seen with gravidity and trimester of exposure may suggest a true beneficial effect. Intermittent treatment with as little as a single dose of ASP could be a cheap and safe alternative for primi- and secundigravidae where the efficacy of SP and chloroquine has been reduced by drug resistance. To investigate this hypothesis randomized, controlled trials are required. The progressive, resistance-related loss of efficacy of traditional antimalarial drugs increases the urgency to undertake such trials [264, 818].

7.3.5. Other ACT Drugs in Developments

1) Dihydroartemisinin + Piperaquine

New combination of dihydroartemisinin-piperaquine (DHAPQ) has undergone clinical trials with well tolerated, and compares favorably in efficacy (99% cure rate, 164/166 patients) with mefloquine combined with artesunate (99% cure rate, 76/77 patients) in Vietnamese patients assessed 56 days after treatment [151]. Although dosing may need optimization, DHAPQ might prove to be more affordable than other combinations of comparable efficacy.

Piperaquine is a bisquinoline compound related to chloroquine. It was discovered in France and developed as an antimalarial by Chinese scientists over 30 years ago. Piperaquine replaced chloroquine as the first-line treatment of *falciparum* malaria in China in 1978. After over 200 metric tons were used, including in mass treatment, resistance to piperaquine developed in *P. falciparum* in the late 1980s. Piperaquine was not used outside of China. The mechanism of action of the drug and mechanism of resistance have not been well characterized but are likely to be related to those of the other drugs in this general class. Piperaquine has a large apparent volume of distribution of > 500 L/kg and a terminal elimination half-life estimated at 2 to 3 weeks. Since there was increasing sensitivity of the assay, the true terminal half-life is probably similar to that of chloroquine; 1–2 months. Oral bioavailability increases with co-administration with fat. The fixed dose combination formulated in tablets containing dihydroartemisinin (40 mg) and piperaquine (320 mg) is commercially available in many countries in Asia, and also more recently in Africa.

Recent clinical trials have shown that the fixed combination given once daily for 3 days was effective and well tolerated. The most common adverse effects are gastrointestinal (nausea, vomiting, abdominal pain, and diarrhea), but they are usually mild and self limiting. The main determinant of the parasitological efficacy is the slow elimination of piperaquine. This also determines the "post-treatment prophylactic effect," which is important if the drug is going to be used as Intermittent Preventive Treatment (IPT). The simplicity of administration, the excellent efficacy even against multi-drug–resistant strains and the favorable toxicity profile, make the DHAPQ combination one of the more promising of the currently available ACTs [271].

2) Artesunate + Chlorproguanil-Dapsone

Chlorproguanil-dapsone is an antifol-sulfonamide combination with a similar mode of action and synergistic properties to SP. Chlorproguanil can be considered as a prodrug for the active antifol chlorcycloguanil to which it is metabolized by the polymorphic CYP_{450} 2C19. Activity of this enzyme is reduced in approximately 20% of Orientals and is also reduced by estrogens (e.g., pregnancy, oral contraceptive). The advantages of this combination are good tolerability and rapid elimination providing less selective pressure on the spread of resistance and greater activity than SP against moderately resistant *P. falciparum* (although both compounds are ineffective against *P. falciparum* with the Ile164Leu mutation common in Asia and South America, and recently identified in Africa). Disadvantages are resistance selected already by SP, some concerns over the safety of dapsone (hemolytic anemia), and lack of a post-treatment prophylactic effect. This ACT is in the late stage development and should be registered in the near future. [271].

3) Artesunate + Pyronaridine

Pyronaridine is one of many synthetic antimalarials developed in China and used originally as a monotherapy. It bears closest structural similarity to amodiaquine, although pyronaridine is much more active against resistant parasites. Pyronaridine's pharmacokinetic properties have not been fully characterized yet but like several other drugs in this general class of antimalarials, it is extensively distributed and eliminated slowly. The mechanism of action of the drug and mechanisms of potential resistance have not been well characterized but are likely to be related to those of the other drugs in this general class. Pyronaridine and the ACT combination are well tolerated and effective. The fixed dose ACT is in late-stage development.

4) Artesunate Plus Atovaquone-Proguanil

This fixed dose combination of 2 established drugs has not been developed as a 3-drug fixed dose ACT although it has been evaluated and found to be well tolerated, safe, and effective. Atovaquone-proguanil has a different (and synergistic) mode of action to other antimalarials affecting parasite respiration at the level of the cytochrome chain. Proguanil is acting itself in the combination, and not via the antifol triazine metabolite cycloguanil − so it remains effective against antifol-resistant parasites. Mutations in the gene encoding cytochrome b confer high level atovaquone resistance. Atovaquone absorption (like that of lumefantrine and halofantrine) is augmented by co-administration with fats. Its elimination half-life of 1–2 days provides for an effective 3-day treatment regimen. As with many antimalarials, plasma concentrations of both drugs are reduced in pregnancy. It is remarkably well tolerated with no serious adverse effects. The main impediment to its use is the high cost of atovaquone manufacture. The drug is essentially unaffordable in malaria endemic areas [819].

McGready 2005 reported fewer treatment failures at day 63 with artesunate plus atovaquone-proguanil in 81 pregnant Karen women with uncomplicated *falciparum* malaria with a restricted sequential trial design of 7 days of supervised quinine (SQ7) versus 3 days of artesunate-atovaquone-proguanil (AAP). Eighty-one pregnant women entered the study; 42 were treated with SQ7 and 39 were treated with AAP. There were no significant differences in birth weight, duration of gestation, or congenital abnormality rates in newborns or in growth and developmental parameters of infants monitored for 1 year. AAP is a well-tolerated, effective, practical, but expensive treatment for multidrug-resistant *falciparum* malaria during the second or third trimesters of pregnancy. Despite the small number of subjects, these results add to the growing body of evidence that AAP is safe for the mother and the fetus [702, 820].

7.4. Differences in Toxicity State Between Diseases and Drugs

Malaria is probably the only infection that can be treated in just three days, yet that kills millions every year. Without prompt and appropriate treatment, malaria may become a medical emergency by rapidly progressing to complications and death. Most cases of severe malaria are caused by *P. falciparum* infection. Rarely, *P. vivax* or *P. ovale* produce serious complications, debilitating relapses, and even death. One can also come across problems related to the use of antimalarial drugs. Malaria can also aggravate certain pre-existing illnesses and may even prove fatal for patients with end stage organ disease.

7.4.1. Pre-existing Toxicities in Patients with Complicated Malaria

Severe *falciparum* malaria is defined by the demonstration of asexual forms of *P. falciparum* in a patient with a potentially fatal manifestation or complication of malaria in which other diagnoses have been excluded. Even though the complications are almost unique to *P. falciparum* infection that does not mean that all cases of *P. falciparum* malaria invariably develop complications. The case fatality of *P. falciparum* malaria is around 1% and this accounts from 1 to 3 million deaths per year all over the world. An estimated 80% of these deaths are caused by cerebral malaria.

Malaria can exacerbate/complicate pre-existing clinical conditions, adding to its morbidity and mortality. In this regard, the following points should be kept in mind [170]:

- Patients with these conditions should be managed energetically to avoid any potential problems.
- Even *P. vivax* (or other milder types) can lead to deterioration in the condition, even causing death.
- Since *P. vivax* and other milder types generally do not cause any complications or death by themselves, one should be careful in filling up the death certificates in these patients. In these cases, malaria can at the most be sited as a contributory cause and not as a primary cause.
- Pre-existing problems may influence treatment of malaria, especially the choice of antimalarials.

1) Pre-existing Cardiovascular Disease

Patients with severe valvular obstruction, compromised ventricular function and other conditions of cardiac decompensation may be in serious trouble contracting malaria. In severe *falciparum* malaria, the myocardial function is remarkably maintained and most patients have an elevated cardiac index, with low systemic vascular resistance and low to normal right and left-sided filling pressures. However in patients with decompenzated heart, the high grade fever, tachycardia, hypoxemia, metabolic acidosis, etc. associated with malaria may add to the existing cardiac decompensation.

Antimalarials may relate to these cardiovascular diseases. Chloroquine, artemisinin, pyrimethamine-sulphadoxine, tetracyclines and primaquine can be safely used in these patients. Quinine can also be used carefully. Mefloquine and halofantrine are better avoided in patients with known cardiac illness.

Oral chloroquine is safe in therapeutic or prophylactic doses. If the therapeutic dose or a high dose is administered too rapidly by parenteral route, it can cause significant cardiotoxicity. Hypotension, vasodilation, myocardial suppression, ECG abnormalities, and cardiac arrest can occur. Treatment includes mechanical ventilation, adrenaline, and diazepam. Concomitant use of chloroquine with amiodarone should be avoided.

At therapeutic doses, quinine is relatively safe. Rapid intravenous administration may cause hypotension. Acute over-dosage can cause fatal dysrhythmias such as sinus arrest, junctional rhythms, A-V block, ventricular tachycardia, and fibrillation. Quinine may delay the absorption and elevate the plasma levels of cardiac glycosides like digoxin. Quinine should not be used concomitantly with amiodarone. Concomitant use with astemizole and terfenadine can also increase the risk of ventricular arrhythmias.

Mefloquine should be used with extreme caution in patients suffering from cardiac conduction diseases. Mefloquine has been shown to cause asymptomatic sinus bradycardia and other conduction abnormalities, e.g., prolongation of the QT interval. Patients who are on either a beta-blocker or calcium channel blocker are at particular risk if there are signs of sinus bradycardia and/or atrioventricular block. Cardiac arrest has been reported in a patient receiving a single prophylactic dose of mefloquine while concomitantly taking propranolol. There appears to be no interaction between ACE inhibitors and mefloquine.

Halofantrine prolongs QT interval in a concentration dependent manner and it can result in ventricular arrhythmias and even death. It is therefore contra-indicated in patients with prolonged QT interval and with drugs known to cause prolongation of QT interval.

2) Pre-existing CNS Disease

Malaria and antimalarial drugs may pose problems in patients with pre-existing CNS disorders like dementia, epilepsy, etc. Severe *P. falciparum* infection, dehydration, hyponatremia, high grade fever can lead to deterioration of patients having pre-existing dementia. Elderly patients and patients with dementia contracting malaria may be prone for secondary infections like aspiration bronchopneumonia. Chloroquine, quinine, and mefloquine can cause neuropsychiatric side-effects. Chloroquine and mefloquine are better avoided in patients with significant neuropsychiatric disorders. To date, the CNS adverse effects of artemisinins were only found in animal species, and no confirmed case was reported in humans.

Chloroquine has a small therapeutic window and is extremely toxic at high doses, with rapid onset of neurological, respiratory, and cardiovascular effects, and a mortality rate of 35% in overdose. Neurological symptoms begin with dizziness, vomiting, and headache, and progress rapidly to CNS depression and visual disturbance. Amodiaquine, a 4-aminoquinoline similar in structure to chloroquine, has also been associated with extrapyramidal effects. In 86 patients with rheumatoid arthritis, amodiaquine was shown to cause a significantly higher incidence of CNS effects than chloroquine. Dizziness and headache were reported in 22 and 6% of amodiaquine users, respectively, compared with 5 and 0% of chloroquine users [656].

CNS adverse events in association with mefloquine, a 4-methanolquinoline derivative structurally related to quinine, have been prominent in the literature. In an overview of the tolerability of mefloquine, adverse were reported to ranges from 47 to 90% in adults and 57 to 61% in children [422]. CNS events commonly encountered were headache, dizziness, and insomnia. Dizziness is recognized as a frequent but transient adverse effect of mefloquine therapy. Seven healthy Caucasian volunteers administered mefloquine 25 mg/kg all experienced lightheadedness, 4 of whom were severely incapacitated for 3 to 4 days [661]. This study had a manufacturer influence on policies regarding the therapeutic use of mefloquine. In a more controlled setting on the Thai-Myanmar (Burma) border, in more than 3500 patients with uncomplicated malaria, dizziness occurred in 83% of adults and 59% of children under 15 years of age in the first 3 days post-treatment [491]. Dizziness just prior to treatment was found to occur in 63% of adults and 38% of children.

Quinine intoxication following doses of 4 to 12 g is characterized by seizures and coma. Cinchonism is a well recognized syndrome associated with quinine. Nausea, vomiting, tinnitus, deafness, headache, disturbed vision, and vasodilation are relatively mild at standard therapeutic doses. Early symptoms are mild visual and hearing complains. A principle sign is the sudden onset of bilateral pupil dilatation. Lethal doses may be around 8 g. Irreversible hearing impairment and blindness is very rare at standard dosages of quinine. Reversible high tone hearing loss was shown using serial audiometry in 10 patients receiving quinine for acute malaria. Hearing loss was unnoticed, although 7 had tinnitus, and recovery was complete. Three patients had received standard oral dosages, and all had plasma concentrations in usual therapeutic range.

The dose-related CNS effects demonstrated in animals are warring, but difficult to interpret clinically. These studies provide little insight into the early signs of CNS toxicity or the reversibility of neurological lesions. Dosages used for malaria (total dose 10 mg/kg, given over 3 to 7 days) are lower than those in these preclinical studies, but the safety margin has yet to be determined. To date no significant toxicity has been reported in humans, despite the use of artemisinin and its derivatives in at least 2 million patients [516]. However, the majority of these patients were no reports of single episodes of CNS events with the oil soluble derivative, artemether, were in severely ill patients, making it difficult to distinguish between CNS events related to the drug and the malaria (post-malaria neurological syndrome). No CNS events were identified in over 1000 patients with uncomplicated malaria administered oral doses of artesunate or artemether [666].

3) Pre-existing Anemia

Anemia is a common problem in developing countries of Africa and Asia and it is commonly due to helminthiasis and malnutrition. Malaria is also common in these areas. Both *vivax* and *falciparum* malaria can exacerbate the anemia, specially causing problems in pregnancy and in children. Also, blood transfusion for anemia may transmit malaria.

Artesunate hemotoxicity in the rats, and to a greater degree in rhesus monkeys, parallels the effects reported in human (0.6% reticulocytopenia in 4062 patients at low doses of 2-3 mg/kg) [747]. In humans artesunate caused rapid inhibition of hematopoiesis [748] and lower peripheral reticulocyte counts by day 5 of treatment compared to the quinine group (p = 0.011). These reductions in reticulocyte counts and anemia have also been confirmed by other investigators [749-751]. Hematological changes induced by artesunate administered at therapeutic doses in humans are not completely clear. In another study, both artesunate and artelinate-treated rats, dose-dependent and rapidly reversible hematological changes (significant reductions in RBC, HCT, Hb, and reticulocyte levels) were seen in the peripheral blood. Bone marrow evaluation revealed a statistically significant reduction in the myeloid/erythroid ratio only at the highest dose of AS (240 mg/kg), albeit still within the normal ratio range (1.0-1.5:1.0). Looking at the respective therapeutic indices we have concluded that artesunate is much safer than artelinate. Both drugs induced hematological changes in rats that parallel the dose-dependent, reversible anemia and reticulocytopenia previously reported in animals and humans. However, no significant bone marrow depression was seen for either agent [468].

4) Pre-existing Renal Disease

Severe *falciparum* malaria can compromise renal blood flow by sequestration and obstruction to the microcirculation, by hemolysis, by dehydration and hypovolumia, by acidosis etc. Any of this could prove detrimental to patients with pre-existing renal disease. Acute intrinsic renal impairment occurs during apparently 'uncomplicated' *falciparum* malaria in children. Malaria has been reported in renal transplant recipients. Malaria should be considered in the differential diagnosis of fever in transplant recipients who have received organs or blood products from an area of endemic malaria. The dose of quinine needs modification in renal failure whenever S. creatinine is > 3 mg %. Chloroquine increases plasma cyclosporine concentration and may increase the risk of toxicity.

5) Pre-existing Liver Disease

Patients with hepatocellular failure due to cirrhosis, etc., may deteriorate if they contract malaria. A case of malaria in a recipient of orthotopic liver transplantation has been reported. The patient was found to have *Plasmodium ovale* malaria during evaluation of a severe febrile illness. The infection was traced to a platelet transfusion and responded to treatment with chloroquine. Risk factors associated with the development of malaria infection are identifiable and should be reviewed from the recipient and donor when possible. Routes of infection in the liver transplant patient would include blood products, the organ itself, and resurgence of latent infection. None of the antimalarial drugs have any direct hepatotoxic effect. However, chloroquine is not advisable in patients with severe hepatic insufficiency.

Amodiaquine can cause adverse effects including liver damage. The observed drug toxicity is believed to involve the formation of an electrophilic metabolite, amodiaquine-quinoneimine, which can bind to cellular macromolecules and initiate hypersensitivity reactions. Since hepatitis and agranulocytosis occurred in prophylactically treated patients, it is no longer recommended as prophylactic treatment of malaria. Repeated exposure to the quinoneimine-generated antigen may be important in the generation of organ damage. One study reported on a 24-year-old woman with severe liver failure following trimethoprim-sulfamethoxazole treatment, demonstrating the possible severity of the drug hypersensitivity syndrome associated with trimethoprim-sulfamethoxazole. The abrupt onset of hepatic failure, 5 days after the start of doxycycline, and the rapid normalization after drug discontinuation leads to suspect a causal relationship between doxycycline and liver insufficiency. Another study found significant changes in the proportions of plasma proteins in malarious birds 8 days after infection; albumin and α_2 -globulin were reduced, while γ_1 - and γ_2 -globulin were increased. Those changes coincided with significant increases in the plasma concentrations of total protein and enzymes, and a decrease in creatinine. A prospective study done in 216 children with complicated *P. falciparum* malaria showed hepatopathy in 33.3% of cases, with a higher incidence in children aged > 5 years. Bilirubin and alanine aminotransferase were moderately raised in most cases [744].

6) Epilepsy

There are some reports of chloroquine causing convulsions even in previously healthy patients. Mefloquine is contra-indicated in patients with a history of convulsions. Several case reports of first-time seizures in patients taking mefloquine in prophylactic doses have been reported. There have also been reports of mefloquine reducing the half-life and lowering the blood levels of sodium valproate. This may be due to mefloquine accelerating the hepatic metabolism of sodium valproate, because they are both metabolized by the same hepatic enzyme system.

Doxycyline does not affect epilepsy, but may interact with some of the anti-convulsants. Carbamazepine, phenytoin, and barbiturates may shorten the half-life of doxycycline by up to 50% and lower mean serum levels by liver enzyme induction, thus possibly compromising its therapeutic efficacy. The degree to which the levels are affected is not clear. In theory, this means that the usual recommended prophylactic dose of 100mg daily could be taken more frequently, probably twice daily. However an exact recommendation cannot be made because there is limited experience with an increased incidence of side-effects. Therefore, epileptic patients not taking carbamazepine, phenytoin, and barbiturates can safely use doxycycline as malarial prophylaxis. Patients taking these drugs must be aware of the fact that the normal dose of doxycycline may not provide adequate protection and increasing the dose may result in an increased incidence of side-effects [170].

7) Diabetes Mellitus

Severe *P. falciparum* malaria can cause hypoglycemia and this fact should be considerednwith diabetics receiving insulin and/or oral hypoglycemic agents. Suitable dosage adjustments may be needed. Quinine has stimulatory effects on the pancreatic beta cells and is known to cause severe hypoglycemia. Thereby it may potentiate the effects of sulfonylureas. In normal patients and in normal doses chloroquine does not appear to cause increased pancreatic secretion of insulin and has no effect on plasma glucose concentrations. Some studies suggest that in non-insulin-dependent diabetes mellitus chloroquine may improve glucose tolerance, possibly by decreased metabolic degradation of insulin rather than increased pancreatic secretion.

There is very limited evidence that doxycycline occasionally increases the hypoglycemic effects of insulin and sulphonylureas. Although there is no need to avoid concomitant use, patients must be aware of signs of hypoglycemia and, if needed, the dose of hypoglycemic agent should be adjusted. It is unknown whether mefloquine interacts with oral antidiabetic agents. Treatment doses of mefloquine have caused hypoglycemia especially in children and pregnant women, but mefloquine apparently does not stimulate the release of insulin. Patients should be made aware of the possibility and should be able to reduce the hypoglycemic dose if necessary. The impact on the control of diabetes is unknown; it is therefore suggested that blood glucose be monitored even more closely and that medication adjustments are made as required.

8) Myasthenia Gravis

Quinine decreases the excitability of the motor end-plate region so that responses to repetitive nerve stimulation and to acetyl chorine are reduced. Quinine may produce alarming respiratory distress and dysphasia in patients with myasthenia gravis. Chloroquine also may increase the symptoms of myasthenia gravis and reduce the effect of neostigmine and pyridostigmine.

9) Dermatitis

Concomitant use of chloroquine with gold salts and phenyl butazone should be avoided because all this class of drugs can cause dermatitis.

7.4.2. Neurological Involvement in Children with Severe Malaria

P. falciparum appears to have a particular propensity to involve the brain but the burden, risk factors, and full extent of neurological involvement have not been systematically described. Objective of this section is to determine the incidence and describe the clinical phenotypes and outcomes of neurological involvement in African children with acute *falciparum* malaria.

1) Burden of Malaria with Neurological Involvement

A total 58,239 children were admitted to Kilifi District Hospital in Kenyan during the study period. Of these, 22,441 had malaria parasites detected on blood test results and 19,560 (33.6%) had malaria as the primary clinical diagnosis. Of the 19,560 admissions with malaria, neurological involvement was observed in 9313 children (47.6%) at admission compared with 7794 (21.8%) of the 35,798 admissions without malaria (P < 001). The mean annual incidence of admissions with malaria among children younger than 5 years during the study period was 2694 (range, 1506-3744) per 100,000 persons. The mean annual incidence of malaria with neurological involvement among children younger than 5 years was 1156 (range, 474-2075) per 100,000 persons. The peak incidences of malaria and malaria with neurological involvement (3794 and 1181 per 100,000 persons, respectively) were in the first year of life. A marked decline in incidence was observed in children older than 5 years. The incidence of admission in children aged 5 to 9 years was 344 per 100 000 persons for malaria and 120 per 100 000 persons for malaria with neurological involvement; for children aged 10 to 14 years, the incidence of admission was 39 per 100,000 persons and 9 per 100,000 persons, respectively.

2) Features of Neurological Involvement on Admission

Seizures were the most common neurological feature of acute *falciparum* malaria and were reported or observed at admission in 37.5% of children. Seizures were not common before the age of 6 months. After 12 months, the age-specific prevalence increased rapidly reaching a peak prevalence of 48.6% among patients aged 27 to 33 months. Patients with seizures had a shorter duration of illness and higher parasitemia. Multiple seizures were common with 56% of those with a history of seizures reporting 2 or more episodes during the illness. In 22% of children, seizures lasted longer than 30 minutes, fulfilling the definition of status epilepticus. During the course of inpatient stay, seizures were observed in 13.3% of children with neurological involvement. These were not associated with hypoglycemia or hyponatremia.

Temperature appeared to influence seizure manifestation; 63% of generalized seizures were reported in febrile children compared with 54% secondarily generalized seizures and 44% focal seizures. The recurrence of seizures in the ward was associated with increased mortality. Among 6212 children about whom the presence or absence of seizures during previous illnesses was known, a history of seizures was more common among those admitted with seizures (41.5%) compared with those admitted without seizures (15.1%). The majority of these past seizures were associated with febrile illnesses including respiratory tract infections and malaria. Sixty percent of patients with prostration reported a seizure. The clinical risk factors for prostration were examined among 2967 consecutive children assessed for both seizures and prostration. Comatose patients had been ill for a longer period prior to admission compared with those with agitation or prostration. Overlap between the features of neurological involvement was common, particularly seizures and prostration or seizures and impaired consciousness. All patients with agitation had at least 1 other feature of neurological involvement (i.e., seizures, prostration, or impaired consciousness).

Deterioration in consciousness during admission was observed in 219 (14.3%) of 1533 children with neurological involvement admitted to the high-dependency unit. It was associated with recurrence of seizures and abnormal motor posturing but not duration of illness, admission temperature, parasite density, hypoglycemia, metabolic acidosis, hyponatremia, or severe anemia. However, secondary deterioration in consciousness was associated with higher mortality than established impaired consciousness (38.6 vs. 12.8%). Some children developed agitation during the course of the admission. This was associated with worsening level of consciousness and increased mortality.

3) Factors Associated with Neurological Involvement in Falciparum *Malaria*

Children with neurological involvement were older, had a shorter duration of illness, and a higher geometric mean parasite density. The majority of children admitted to the hospital with fever lasting less than 2 days had neurological involvement (65.4% vs. 34.6%). Apart from seizures, medical history in the 2 groups was similar. Features of shock or impaired perfusion and metabolic acidosis were associated with neurological involvement. Vomiting, diarrhea, and cough were less common in patients with neurological involvement. Although neurological involvement was observed at low levels of parasitemia, the proportion with neurological involvement increased with rising parasitemia: neurological involvement was present in 40% of patients with parasite densities lower than $100 \times 10^3/\mu l$, 50% of patients with densities between $100 \times 10^3/\mu l$ and $500 \times 10^3/\mu l$, and in more than 60% of those with densities higher than $1000 \times 10^3/\mu l$.

Among patients with life-threatening features, the most common biochemical derangements were hyponatremia, acidosis, hyperkalemia, hypoglycemia, and elevated plasma creatinine level. Only metabolic acidosis, hypoglycemia, and hyperkalemia were significantly associated with neurological involvement.

4) Factors Independently Associated with Neurological Involvement

Clinical and laboratory features associated with neurological involvement on univariate analysis with a *P* value of less than .10 were entered in a logistic regression model to identify those features independently associated with neurological involvement. Factors independently associated with neurological involvement included past history of seizures, fever lasting 2 days or less, delayed capillary refill time, acidosis, and hypoglycemia.

Neurological involvement is common in children in Kenya with acute *falciparum* malaria, and is associated with metabolic derangements, impaired perfusion, parasitemia, and increased mortality and neurological sequel. This study suggests that *falciparum* malaria exposes many African children to brain insults [657].

7.4.3. Adverse Effects in Patients with Severe Malaria and Antimalarials

Although these artemisinins have proved highly efficacious in clinical trials with very few reported adverse effects, there are few systematic large detailed clinical studies of the toxicity of artemisinin derivatives in clinical practice. The general toxicity profile in experimental animals has been good, but in all mammal species tested to date, these compounds have produced an unusual selective pattern of damage to certain brain stem nuclei, particularly those involved in auditory processing. Although there have been no reported neurotoxic reactions to artemisinin or its derivatives in humans, the relevance of these observations in animals to toxicity in humans is unresolved. The extensive use of the artemisinin derivatives in this area provided us with the opportunity to review the human toxicity of these compounds in a cumulative experience of more than 3500 prospectively studied treatment courses.

In one reported study [65], of the 4965 patients who completed a full course of treatment, 2593 (52%) were recruited into randomized prospective studies (1041 were treated with artemisinin derivatives plus mefloquine [AM], 461 with artemisinin derivatives alone [A], and 1,091 with mefloquine alone [M]), and an additional 2372 (48%) patients presented to the clinics. Although the latter group were not part of comparative studies, they were treated in the same way (1785 AM, 375 A, and 212 M), and therefore included in this analysis. Overall, 3,461 (70%) patients were recruited into studies of primary infections (1,043 M, 2,412 MA, and 6 A), 1,201 (24%) into studies of recrudescent infections (260 M, 197 MA, 744 A), and 303 (6%) into studies of hyperparasitemic infections (217 MA and 86 A). Of the 3662 patients who received an artemisinin derivative, 3276 (89%) received a treatment regimen containing artesunate, and 386 (11%) received artemether. Mefloquine, when given alone, was given as a split dose (15 mg/kg followed by 10 mg/kg 8–24 hr later) in 353 patients (27%), and as a single dose in the remainder (950; 73%). Of the 2,826 patients of patients who received combination therapy, 4% (102) had mefloquine administered as a split dose. In the remaining patients, mefloquine was given as a single dose either on admission (587, 21%), day 1 (204, 7%), or thereafter (1933; 68%).

1) Early Vomiting (within 1 hr)

Data on drug vomiting were available for artesunate, artemether, and mefloquine treatments in > 95% of patients on all observed days. Overall, vomiting of artesunate or artemether within 1 hr of administration occurred in 3% (93 of 3561) of the patients on the day of admission, 1.5% (50 of 3339) on day 1, and 2% (72 of 3022) on day 2, and 0.3% (13 of 3782) of the drug administration's were vomited on subsequent days. After stratifying by day of administration, co-administration of mefloquine increased the risk of vomiting the artemisinin derivative by a factor of 2.8%. In those not receiving mefloquine, 2.2% (65 of 2972) of patients vomited on the day of admission, 1.3% (40 of 3145) on day 1, and 1.1% (12 of 1131) on day 2, with 0.2% (9 of 3674) of the drug administration vomited on subsequent days.

On admission, the following were found to be independent risk factors associated with vomiting artesunate or artemether: age # 14 years, co-administration of mefloquine, a history of vomiting, or nausea. The population attributable risks (PAR) for vomiting an artemisinin derivative (artesunate and artemether pooled) within 1 hr were 65% for an age # 14 years, 69% for co-administration of mefloquine, 29% for a history of vomiting, and 30% for nausea. The overall PAR, calculated as one minus the product of one minus each individual PAR, was 94%. Although fever on admission was a univariate risk factor for vomiting, it was not a significant factor in a multivariate model. The presenting parasitemia, type of derivative (artesunate or artemether), or dosage of artemisinin derivative given (2 mg/kg vs. 4 mg/kg) did not affect vomiting. Co-administration of an artemisinin derivative did not increase the risk of vomiting mefloquine on admission.

2) *Late Side-Effects (Onset. 1 hr)*

Both the malaria infection itself and mefloquine cause adverse effects on the neurologic and gastrointestinal systems are difficult to separate and therefore confound assessment of possible artemisinin derivative side-effects.

In confounding factors, on admission, 65% (3160 of 4889) of the patients reported anorexia, 46% (1795 of 3871) dizziness, 38% (1525 of 3979) nausea, 25% (1250 of 4943) vomiting, and 2% (110 of 4949) diarrhea. An admission parasitemia greater than 10,000/µl and fever were taken as markers of disease severity, and were found to be significant independent risk factors for the following signs/symptoms on admission: dizziness, anorexia, nausea, and vomiting. Admission parasitemia and fever were independent risk factors for anorexia on days 1 and 2, but patients had largely recovered by day 7 and initial disease severity was no longer a confounder at that time. Primary infections were significantly associated with nausea, vomiting, and anorexia on admission and on days 1 and 2. They were also associated with dizziness on days 1–2, but with none of the symptoms on day 7.

The administration of mefloquine was a major independent risk factor for the following symptoms/side-effects in the first two days of treatment: dizziness, vomiting, nausea, anorexia, and diarrhea. On day 7, only dizziness was influenced by prior mefloquine administration. There were significant differences in the frequency of reported side-effects between adults and children. Vomiting and diarrhea were more common in children than adults on admission. This remained apparent for diarrhea on days 1 and 2, and at day 7. Conversely, dizziness and nausea (both only assessed reliably in those over 5 years of age) and anorexia was more common in adults from admission until day 7. On admission and days 1 and 2, females were significantly more likely than adult males to report dizziness, anorexia, and nausea, but by day 7 this was no longer apparent. Patients who failed to defervesce within 24 hr were more likely to report the following associated symptoms on days 1 and 2: dizziness, vomiting, nausea, and anorexia. By day 7, only anorexia was more common in those still febrile after 24 hr. The reporting of symptoms or side-effects was not influenced by the parasite clearance time.

On effect of dose and derivative, the administration of mefloquine was associated significantly with all symptoms and side-effects reported on days 1 or 2. Thus, all potentially iatrogenic adverse effects in patients given both antimalarial drugs could be attributed to mefloquine. After correcting for confounding baseline characteristics, there was no effect of the type of artemisinin derivative used (artesunate or artemether) on dizziness, diarrhea, nausea, vomiting, or anorexia. Of the 2801 patients treated with two or more days of artesunate therapy, 460 received (16%) a low dose of artesunate in the first 48 hr (2 mg/kg/day) and 1972 (70%) received 4 mg/kg/day. After correcting for confounding factors, dosage did not independently influence any of the symptoms/side-effects. Overall, 2814 (77%) patients treated with an artemisinin derivative (2428 with artesunate and 386 with artemether) either received mefloquine on or after day 2 or not at all. In these patients, mefloquine toxicity could not be a confounding factor; 65% (1528 of 2363) reported anorexia on days 1–2, 42% (750 of 1765) dizziness, 29% (493 of 1710) nausea, 17% (361 of 2137) vomiting, and 2% (42 of 2107) diarrhea (after treatment). By day 7, 12% (399 of 3462) of all patients studied complained of anorexia, 3% (80 of 2,834) of nausea, 0.5% (19 of 3500) of vomiting, and 1% (34 of 3859) of diarrhea.

There was no difference between short (less than four days) and longer courses of drug treatment or the derivative used in these rates. Furthermore, the incidence of reported diarrhea, vomiting, nausea, and dizziness on day 7 did not differ significantly from the incidence of these complaints on day 14 or day 28. This suggests that these symptoms were not attributable to the artemisinin derivatives. Anorexia was more common on day 7 than on day 14, but thereafter the incidence did not change significantly. Overall, 2,151 (43%) patients received mefloquine in the first two days of treatment: 1,303 were treated with mefloquine alone and 848 received mefloquine plus artesunate or artemether. On days 1–2, the incidence of nausea was significantly lower in that receiving combination treatment with artesunate. This apparently beneficial effect was no longer significant after controlling for the more rapid fever clearance associated with combination treatment. The addition of an artemisinin derivative did not affect the reported incidence of dizziness, vomiting, anorexia, or diarrhea on days 1–2, or any of the symptoms on day 7 [65].

3) Neurologic Side-Effects

Neurologic examinations could be performed reliably only in patients who were 5 years old. Examinations were carried out in 66% (1971 of 3003) of the patients receiving an artemisinin derivative (307 following treatment with artemether and 1664 with artesunate). Of these, 1512 (77%) also received mefloquine, and in 181 (9%) this was on admission or day 1. An additional 134 (12%) patients treated with mefloquine alone were also tested using similar procedures. Data were recorded in 1327 (63%) patients on admission, 1099 (52%) on day 2, 1503 (71%) on day 7, and 1438 (68%) on or after day 28. Of the patients in whom neurologic examinations were not possible on admission, 45% (353 of 778) were tested instead on day 2. On day 2, reporting of any neurologic disturbance was associated with dizziness. Of the 746 (35%) patients tested on both admission and day 2, six of the 733 patients (0.8%) without neurologic deficit on admission developed neurologic signs on day 2. All six patients had disturbed balance. One patient also developed nystagmus and another had impaired finger dexterity.

Although all patients were aparasitemic at the time of testing, three reported headache and fever before testing and the other three reported feeling dizzy. The signs had resolved in all cases by day 7. A total of 955 (45%) patients were tested on both admission and on day 7. Of the 938 patients without neurologic deficit on admission, four (0.4%) had developed neurologic signs on day 7 (two had heel toe ataxia, one was unable to stand with his feet together and one had both signs). All had received prior mefloquine as part of their treatment regimen. One of these patients had also developed nystagmus, and another reported dizziness, headache, and weakness. In all but one case the neurologic disturbance had resolved by day 14: in the remaining patient the disturbance persisted until day 42. There was no apparent association between neurologic side-effects and type or dose of derivative. No patient developed deafness or permanent neurologic abnormalities.

4) Rare Adverse Effects

Rare adverse effects that could have been related to antimalarial medication were reported in 17 patients: 15 following treatment with mefloquine plus artesunate/artemether, none following artesunate monotherapy, and two following mefloquine monotherapy. Five patients had generalized tonic seizures. In two of these cases the seizure occurred in children, 6 years old on the day of admission 6 and 10 hr after starting antimalarial treatment with artesunate and prior to receiving mefloquine. At the time of the seizures, both were febrile. One patient had an episode of hypertonia, generalized shaking, and brief unresponsiveness associated with a high fever on day 1, 24 hr after receiving mefloquine plus artesunate. This could have been a rigor. A six-year-old boy with a history of epilepsy was treated with artemether plus mefloquine for three days and had a seizure two weeks later. All 12 patients with either seizures or neuropsychiatric adverse effects made full recoveries. Three children treated with mefloquine plus artesunate regimens developed frank hemoglobinuria within 24 hr of the first dose of artesunate.

Resolution of hemoglobinuria occurred in all three within 48 hr. Two children, 5 years old developed a generalized urticarial rash 3 hr after their second dose of artesunate. They responded to treatment with chlorpheniramine and did not develop a further episode following treatment with subsequent doses of artesunate. Two patients treated with a seven-day course of artesunate died. One of these patients, a 17-year-old man, had had a three-month history of swollen ankles, and peripheral paresthesia. He was seen on day 14 of follow-up apparently well, but two days later had a sudden onset of dyspnea and palpitations while digging in the forest. The other patient was a 35-year-old woman who made a rapid and uneventful recovery from her malaria and was followed until day 35 of follow up with no adverse sequel. She was reported as having returned to Burma and died two weeks later. No further details were available.

5) Hematology

Serial white blood cell counts and platelet counts were recorded in 154 patients (132 of whom received artesunate and 22 artemether). Platelet counts were significantly lower on admission than on day 28. Platelet counts decreased below 100,000/µl on day 7 in two patients receiving mefloquine plus artesunate treatment (to 27,000/µl and 92,000/µl), but in both cases these returned to the normal range by day 14. Overall, there was no change in the mean white blood cell or absolute neutrophil counts during follow-up.

6) Biochemical Investigations

Samples for biochemical studies were taken from 180 patients (150 treated with mefloquine plus artesunate and 30 with mefloquine plus artemether). Liver function was assessed by analysis of alanine transaminase (ALT) and gamma-glutamyl transpeptidase (GGT) levels and renal function was assessed by plasma creatinine and blood urea nitrogen levels. On admission, nine (5.5%) patients had increased serum levels of liver enzymes. Of the 156 patients with normal biochemical findings on admission, seven (4.5%) developed increased serum levels of liver enzymes during follow-up. Three adults had small increases in their ALT levels on day 7 (60–70 IU/L) that had returned to normal by day 14. These had returned to normal by the subsequent week and were not associated with a concomitant increase in levels of GGT. A 10-year-old boy treated with artesunate + mefloquine developed increased ALT levels (to 137 IU/L) on day 7 that had decreased to 100 IU/L by day 14, but these was not measured subsequently. A 35-year-old woman treated with artesunate + mefloquine had a normal GGT level and a high ALT level on admission (104 IU/L) and had an increase in her ALT and GGT levels by day 28 (to 331 and 190 IU/L, respectively). Both patients remained asymptomatic with no clinical evidence of hepatic dysfunction during their 63-day follow-up. There was no evidence of renal dysfunction in any patient tested.

7) Electrocardiographic Findings

Electrocardiograms were recorded in 216 patients during the course of treatment (88 were treated with mefloquine plus artesunate, 56 with mefloquine plus artemether, 53 with artesunate monotherapy, and 19 with artemether monotherapy). Of these patients 113 patients also had ECGs recorded immediately prior to receiving each dose of artesunate (n = 94) or artemether (n = 19) and again 1 hr later. There were no significant changes in the serial measures of the RR, QRS, PR, QT, or QTc intervals between day 28 and day 63. Day 63 values were therefore taken as being representative of the healthy population. Apart from changes in heart rate associated with illness and fever, there were no significant differences in any of the other ECG parameters on days 3 or 7 compared with day 63. One hour after taking an artemisinin derivative (when peak blood concentrations would be anticipated), there was a small but significant decrease in the heart rate, a small increase in the PR interval, and a very small decrease in the rate corrected QT interval (QTc). Overall, 2% (5 of 216) of the cases had borderline first-degree heart block (PR 5 0.21-0.22 sec) on admission that was unchanged by treatment. Neither the small increase in QTc or PR was influenced by the co-administration of mefloquine, the type of artemisinin derivative, or the dose given. The QRS interval did not change significantly during the study. Although the rate-corrected QT interval or QTc is considered to be a measure of the QT interval independent of heart rate, both the QTc and the PR interval were in fact significantly correlated with the RR interval. Therefore, the changes observed after day 2 may be attributable to changes in the heart rate alone.

In conclusion, the combined regimens of mefloquine plus an artemisinin derivative were associated with more side-effects than those with an artemisinin derivative alone; acute nausea (31% versus 16%), vomiting (24% versus 11%), anorexia (51% versus 34%), and dizziness (47% versus 15%), (respectively, $P < 0.001$). Oral artesunate and artemether alone were very well tolerated. There was no difference in the incidence of possible adverse effects between the two drugs, and no evidence that either derivative caused allergic reactions, neurologic or psychiatric reactions, or cardiovascular or dermatologic toxicity. Blackwater fever occurred in three patients treated with mefloquine plus artesunate regimens. Oral artesunate and artemether are safe and well tolerated antimalarial drugs [65].

7.4.4. Drug Toxicity in Differentiation from Complicated Malaria

The toxicity of antimalarial drugs sets an unusual and interesting problem for the clinician. Unlike most clinical situations, antimalarial drugs are provided to healthy people to prevent malaria and to sick patients who are requesting treatment against ill health in malarial areas. Antimalarial drug toxicity is one side of the risk-benefit equation and is viewed differently depending upon whether the clinical indication for drug administration is malaria treatment or prophylaxis. Drug toxicity must be acceptable to patients and cause less harm than the disease itself. Prior to use these drugs, all adverse effects or toxicity of antimalarial drugs should be known.

Use of these drugs in the treatment of both severe and uncomplicated malaria is increasing in many tropical areas. The artemisinin derivatives appear to be remarkably safe in clinical practice, and have been used to treat several million patients without reported serious adverse effects.16–18 Worldwide, artesunate is the most widely used of these antimalarials, but in some countries the cheaper but 5–10 times less active parent compound artemisinin is used. *In vivo* artesunate and artemether are both biotransformed rapidly to the active metabolite dihydroartemisinin, which is eliminated with a half-life of approximately 1 hr. Artemisinin is also eliminated rapidly, although it is not biotransformed to dihydroartemisinin. Artesunate and artemether were remarkably well tolerated with no evidence of attributable neurotoxicity, cardiotoxicity, or allergic reactions. Since acute malaria is associated with symptoms of lassitude, nausea, vomiting, abdominal pain, dizziness, headache, muscle pain, and sometimes diarrhea, it is often difficult in the acute phase of the disease to distinguish disease effects from drug effects. In this large series, these symptoms and signs all resolved with recovery from malaria, and in those patients who received treatment with the artemisinin derivatives for longer periods, there was no relationship between duration of treatment and duration of symptoms. This suggested that the symptoms resulted from malaria and not its treatment.

In experimental studies in mice, rats, dogs, and monkeys, artemether, arteether, artesunate, and the common metabolite dihydroartemisinin have all produced an unusual selective pattern of damage to certain brain-stem nuclei [606]. Two patterns of damage have been observed: irreversible neuronal loss leading to permanent neurologic deficit or death, and transient neurologic abnormalities with full recovery of function. The brain stem nuclei most affected include those involved in the auditory and vestibular relays. There is some interspecies variability in susceptibility; for example, mice appear to be more resistant than rats, but in all species tested this unusual anatomic pattern of damage is observed. Neurotoxicity in experimental animals has been associated particularly with the use of intramuscular injections of artemether or arteether. These two oil-based compounds, which are released relatively slowly from the intramuscular injection site, are more toxic than the rapidly absorbed water-soluble artesunate [772]. Intramuscular administration of artemether or arteether is also more neurotoxic than oral administration of the same drugs. Neurotoxicity has been observed with parenteral doses close to those used in the treatment of malaria and has given rise to concern that similar effects could occur in humans [620]. The principal objective of this large study was to identify whether there was any evidence of clinically apparent neurotoxicity with the routine use of these oral artemisinin derivatives in the treatment of malaria.

Although five patients had seizures and five additional patients had neuropsychiatric reactions, these could be attributed to malaria and mefloquine exposure. There was no difference in reported side-effects between artesunate or artemether or any observed dose-related phenomenon. Artemisinin and its derivatives are known to suppress reticulocyte production. There was no difference in mean hematocrit or the prevalence of anemia in patients who received antimalarial treatment with artemisinin derivative and those who did not. In the early Chinese studies, neutropenia was reported, but no significant changes in leukocyte counts were noted this study. Hemoglobinuria had been reported in previous studies with artemether and artesunate and occurred in three patients in this study [86, 821]. All had high parasite counts (104,000–790,000/ml). The relative contributions of the drug, deficiency of glucose-6-phosphate dehydrogenase, and the infection cannot be assessed. Although two children had generalized urticarial reactions, subsequent treatment with artesunate did not cause the reactions to recur, and their relationship to the drug is again uncertain.

This largely negative study in more than 3,500 treated patients is reassuring. If clinically detectable neurologic deficit follows administration of oral artemether or oral artesunate, there is a less than a 5% chance that it occurs with a frequency greater than one per 1,000 treatments, if it occurs at all. These drugs are safe and remarkably well tolerated in the treatment of malaria.

The malaria infection itself and artemisinins cause adverse effects on the neurologic and gastrointestinal systems are difficult to separate and therefore confound assessment of possible artemisinin derivative and other antimalarials side-effects. However, the possible differentiations from drug toxicity to malaria infection were carefully described by Kakkilaya (Table 55) [170] and also summarized here by present authors to evaluate and confirm the differentiations:

1) Antimalarial drug related toxicities should be confirmed by
   - Specific drug has been used (IV quinine for cardiotoxicity; high dose mefloquine for neurotoxicity; or primaquine for anemia) (Table 55);
   - The adverse effects detected in malaria patients should be also found in healthy volunteers;
   - The adverse effects are not in dose-dependent;
   - The adverse effects are not found during the admissions;
   - Historical record with same antimalarial drugs administrated;
   - Overdose is confirmed before admission or during the hospitalization;
   - Pharmacokinetic results indicated an extremely higher peak concentration or/and longer drug exposure time than other normal cases;
   - Drug accumulation is happened after multiple administrations;
   - A number of early reports did not pay sufficient attention to the feasibility of associating brain damage and neurological disorders in patients with such treatments of malaria.
2) Malaria diseases related adverse effects should be with
   - The adverse effects were diagnosed during admissions of patients;
   - Patients with *P. falciparum* or *P. vivax* malaria;
   - Patients are with severe and complicated malaria (Table 55);

- Misdiagnosis in the patients without right treatments;
- Under dose therapy was confirmed;
- Drug resistance was confirmed by PCR test;
- Pharmacokinetic results indicated an extremely lower peak concentration or/and shorter drug exposure time than other normal case.

**Table 55. Drug induced problems and in differentiations
from severe malaria [170]**

Problem	Drugs	Differentiation from severe malaria	Treatment
Vomiting	Chloroquine, Quinine, Mefloquine, Halofantrine, Tetracyclines, Primaquine	Vomiting is even otherwise common in malaria, usually at the height of fever.	Antiemetics like domperidone and metaclopramide. In the young, metaclopramide can cause extra-pyramidal signs.
Dizziness	Chloroquine, Quinine, Mefloquine, Halofantrine	Could be due to high fever, dehydration and postural hypotension.	Usually mild; if bothersome, drugs like cinnarazine, Betahistine etc. can be used
Itching	Chloroquine		Antihistamines can be tried
Pain abdomen (very rarely abdominal cramps)	Chloroquine, Quinine, Mefloquine, Primaquine	In malaria, particularly falciparum, there may be pain over upper or right lower abdomen, mimicking acute abdominal syndromes.	Drug induced pain can be managed with antacids or H_2 receptor blockers.
Altered behavior, confusion, delirium, hallucinations, etc.	Chloroquine, Quinine, Mefloquine	These symptoms can be due to severe *falciparum* infection or due to high grade fever in any type of malaria.	Watchful expectancy; if needed, tranquilizers like Haloperidol can be used.
Convulsions: Some antimalarials can induce convulsions.	Chloroquine, Quinine, Mefloquine	In severe malaria, convulsions may be recurrent and may lead to unarousable coma.	Anticonvulsants like phenobarbitone for recurrent convulsions. Mefloquine is better avoided in known epileptics.
Coma	Quinine can cause hypoglycemia, which may present as coma.	In cerebral malaria, coma persists even after infusion of 50% dextrose	25-50% dextrose, 50-100 ml intravenously
Hypoglycemia	Quinine	In severe *falciparum* malaria, especially in pregnancy and children, hypoglycemia can occur even without quinine therapy	25-50% dextrose, 50-100 ml intravenously

Problem	Drugs	Differentiation from severe malaria	Treatment
Anemia	Primaquine: It can cause massive hemolysis with G6PD deficiency	Anemia is a common feature in malaria, especially in children.	Usually self-limiting; withdraw the drug; blood or packed cell transfusion if needed.
Jaundice	Primaquine may cause hemolytic jaundice with G6PD deficiency	Severe malaria can cause hemolytic jaundice, or rarely malarial hepatitis	Withdraw the drug
Hemoglobin-uria	Primaquine (same group as above)		
Fever	Artemisinin derivatives	In cases of resistant malaria, with the continuation of fever, the general condition deteriorates and parasitemia increases.	Self-limiting, disappears after the drug is stopped.

Chapter VIII

Antimalarial Drugs Withdrawn and in the Pipeline

There are a number of antimalarial drugs withdrawn from current use due to drug resistance, severe adverse effects, and inappropriate dose regimens. Because of widespread and unsupervised use of malaria drugs, chloroquine-resistant *P. falciparum* emerged in the early 1960s and rapidly spread around the world. Today there are reported cases of *Plasmodium* parasite resistance to most of the currently available antimalarial therapies, thus necessitating the development of new antimalarial treatments. Only class of antimalarials - artemisinin derivatives - has been shown to be highly efficacious against parasites resistant to other antimalarial drugs, and currently becomes to the last line of defense against malaria.

Faced with malaria as the greatest scourges to threaten humanity, there can be no doubt that the need for new antimalarial drugs is being taken very seriously both by the scientific community and the pharmaceutical industry. Even a cursory glance at the primary scientific literature reveals that a wide range of novel chemical structures are being assessed for such activity. Furthermore, examination of the patent literature suggests that some of the discoveries being made are being investigated. However, the drug discovery and development process is not only expensive and time-consuming (10-12 years), it is also highly speculative (only about 1 in 10,000 compounds synthesized becomes a pharmaceutical product). In addition, when discussing issues relating to drug discovery and development, it is necessary to keep in mind the many disciplines and resources that have to come together to deliver success. Genomics can provide many potential molecular targets, but only those targets that can be readily manipulated and tested are likely to provide an opportunity to discover a lead molecule, for example by high-throughput screening, that is worth taking on to a full medicinal chemistry project. Thus, many potential targets will never be advanced because no chemistry lead is identified.

Once a chemistry effort is started, many factors need to be considered over and above improving enzyme inhibition and efficacy against the parasite in culture. The molecules finally selected for development must also be easy to manufacture (low cost is crucial for antimalarials), stable, readily formulated, bioavailable (that is, extensively adsorbed from the gut and avoiding first pass metabolism in the liver to achieve effective concentrations in the systemic circulation), have an appropriate half-life, and not show any overt toxicity. It is essential that academic scientists better understand these issues, as they increasingly have to take the lead in drug discovery efforts against malaria and other neglected diseases. An analysis of the reasons why candidate drugs fail to achieve registration and reach the market has been undertaken recently. In 39% of cases, drugs failed after entering development because of 'biopharmaceutical' issues such as oral bioavailability and formulation, and in 21% of cases because of toxicity. These issues are equally as important as drug efficacy, for which 29% of cases failed [822].

Current drug discovery and development efforts focus on identifying molecules that will be active in different treatment due to resistance molecular mechanism in uncomplicated malaria. The ideal product profile comprises orally active compounds that can cure the disease with a 3-day regimen using once-a-day dosing. As a strong portfolio develops, however, it is anticipated that other, more specific malarial indications may be targeted, such as adjunct treatments to improve the outcome of severe malaria cases; intermittent treatment of malaria during pregnancy to protect both the mother from the disease and the unborn child from a higher risk of being born underweight; and long half-life drugs to treat malaria with a single dose in complex emergency situations. There is also the prospect of combining compounds from several projects into combination products. This could both enhance efficacy and reduce the likelihood of drug resistance development.

The antimalarial drugs are used in three distinct modes: as causal prophylactics to prevent the development of blood parasitemia; as blood schizonticides to kill blood parasites; and as anti-relapse agents to kill the hypnozoite stages in the liver. Also, there are four species of human malaria parasite, including *P. falciparum* which causes acute disease and is responsible for most malaria deaths; and *P. vivax* which causes a much more chronic disease, typified by frequent relapses. Within the many permutations of these two sets of parameters, most people would agree that there are three key priorities for the pharmaceutical industry.

8.1. Antimalarial Drugs Withdrawn

Change of antimalarial treatment policy in countries requires concerted action among all stakeholders and continuous stewardship by the Ministry of Health. The key evidence of the need for treatment policy change is the failing therapeutic efficacy or severe adverse effects of the antimalarial drugs in use, assessed according to standard WHO protocols. WHO's current recommendation is to change a treatment policy when the:

- Treatment failure of >10% (as assessed through monitoring of therapeutic efficacy at 28 days)
- Similarly, an antimalarial medicine should only be selected as a new policy treatment option only when the medicine has an average cure rate of > 95% as assessed in clinical trials

8.1.1. Drug Withdrawn from Use due to Drug Resistance

Resistance has arisen to all classes of antimalarials excepting to the artemisinin derivatives. This has increased the global malaria burden and is a major threat to malaria control. Widespread and indiscriminate use of antimalarials places a strong selective pressure on malaria parasites to develop high levels of resistance. Current medications (amodiaquine, chloroquine, mefloquine, quinine, and sulfadoxine-pyrimethamine), the degree of resistance varies from drug-to-drug. The geographical distributions and rates of spread have varied considerably. Resistance of *P. falciparum* to chloroquine, the cheapest and the most used drug has been reported in almost all the endemic countries, except in Central America and Hispaniola. Resistance to the combination of sulfadoxine-pyrimethamine, which was already present in South America and in South-East Asia, is now becoming highly prevalent in Africa. Resistance to mefloquine is found mostly in Cambodia, Myanmar, Thailand, and Vietnam. Sporadic cases of prophylactic failure of mefloquine in travelers and therapeutic failure have been reported in Africa, South America, and in other Asian countries. *P. falciparum* has shown the development of resistance towards all antimalarial medicines when used as monotherapy: time periods vary from 12 years (chloroquine), to 5 years (mefloquine), to 1 year (proguanil) up to even less than 1 year (sulfadoxine-pyrimethamine, atovaquone) (Figure 2).

Reasons for a drug to be withdrawn from use, due to resistance, are because the antimalarial resistance result in an enormous public health burden. The prolonged or recurrent illness and progression to severe malaria is associated with increased hospitalization and death. Even lower levels of resistance can cause recrudescence of infection and are associated with return of illness. Prolonged or worsening anemia and increased gametocyte carriage fuels transmission. Particularly for the resistant parasite since this has a higher risk of treatment failure for subsequent infections. Impact of drug resistance is more in areas of low transmission where diagnostic facilities and infrastructure are inadequate and resources not allocated to respond to a sudden increase in treatment needs. The impacts of antimalarial drug resistance on health are listed as follows [4]:

1. Prolonged or recurrent illness;
2. Increased outpatient cases;
3. More progression to severe malaria;
4. Increased cerebral malaria-neurological sequel;
5. Prolonged or worsening anemia and effect of anemia;
6. Increased hospital admissions and death;
7. Increased gametocyte carriage;
8. Increased demand on diagnosis; and
9. Higher cost and fewer cost-effectiveness of combination treatment.

8.1.1.1. Withdrawal of Chloroquine

Chloroquine, an inexpensive, safe and initially highly effective drug, became the antimalarial of choice worldwide near the end of World War II and was the cornerstone of the effort to eradicate malaria in the 1950s and 1960s. Unfortunately, chloroquine-resistant *Plasmodium falciparum* arose and was first detected in Southeast Asia and South America in the late 1950s [823], reaching Africa two decades later [824]. After 1960s, chloroquine resistance began to reach high levels in many parts of Southern and Eastern Africa. In a study of chloroquine efficacy in Kenya and Malawi in the early 1980s, 75% and 82% of infections were found to be resistant to chloroquine *in vivo* [825]. As chloroquine resistance rates increased in Africa, so did morbidity and mortality attributable to malaria [826]. In areas of Southern and Eastern Africa with high levels of antimalarial drug resistance, malaria mortality has nearly doubled in the last decade [827]. In response to unacceptably high rates of chloroquine failure, in 1993 Malawi became the first sub-Saharan African country to discontinue the routine use of chloroquine and to elevate the antifolate combination sulfadoxine-pyrimethamine to the antimalarial of first choice nationwide [825]. Subsequently, Kenya, Botswana, and Tanzania also adopted the policy change from chloroquine to sulfadoxine-pyrimethamine, and other countries have switched from chloroquine to a variety of other drugs and drug combinations.

Resistance of *P. falciparum* to chloroquine appeared almost simultaneously in Colombia and on the frontier between Thailand and Cambodia in 1950s. In Asia, chloroquine resistance was initially confined to the Indochinese peninsula, until the 1970s, when it spread westwards and towards the neighboring islands in the south and east. The advent of chloroquine resistance in Africa occurred much later, and it took a decade to cross the continent. Today, only countries in Central America north of the Panama Canal and on the island of Hispaniola have not documented chloroquine-resistant *P. falciparum* malaria. Despite cross-resistance between chloroquine and amodiaquine, amodiaquine remains more effective than chloroquine in areas of chloroquine resistance. Nevertheless, amodiaquine could rapidly lose its efficacy if it is used intensively in areas in which chloroquine resistance is widespread or very high.

The withdrawal of chloroquine is extended to more countries. Recently, the resistance of *P. falciparum* to chloroquine is determined by the mutation at K76T of the *P. falciparum* chloroquine resistance transporter (*pfcrt*) gene and modified by other mutations in this gene and in the *P. falciparum* multidrug resistance 1 (*pfmdr1*) gene. In China, although CQ has been withdrawn from treating *falciparum* malaria for over two decades, 90.3% of the parasites still carried the *pfcrt K76T* mutation. In contrast, mutations at *pfmdr1* codons 86 and 1246 were rare. Sequencing analysis of the *pfcrt* gene in 34 parasite field isolates revealed CVIET at positions 72-76 as the major type, consistent with the theory of Southeast Asian origin of chloroquine resistance in the parasite. Meanwhile, 28.4% of cases were found to contain mixed clones, which favor genetic recombination. Furthermore, despite a unique history of antimalarial drugs in Yunnan, its geographical connections with three malarious countries facilitate gene flow among parasite populations and evolution of novel drug-resistant genotypes. Therefore, continuous surveillance of drug resistance in this area is necessary for timely adjustment of local drug policies and more effective malaria control [828].

In Cameroon, the availability of epidemiologic data on drug-resistant malaria based on a standardized clinical and parasitological protocol is a prerequisite for a rational therapeutic strategy to control malaria. As part of the surveillance program on the therapeutic efficacy of the first-line (chloroquine and amodiaquine), and second-line (sulfadoxine-pyrimethamine), drugs for the management of uncomplicated *P. falciparum* infections; non-randomized studies were conducted in symptomatic children aged less than 10 years according to the WHO protocol (14-day follow-up period) at 12 sentinel sites in Cameroon between 1999 and 2004. Of 1,407 children enrolled in the studies, 460, 444, and 503 were treated with chloroquine, amodiaquine, or sulfadoxine-pyrimethamine, respectively. Chloroquine treatment resulted in high failure rates (proportion of early and late failures, 48.6%). Therefore, chloroquine is no longer a viable option and has been withdrawn from the official drug outlets in Cameroon in 2005 [829].

In Iran during 2003-2005, the prevalence of *pfcrt* K76T, *pfmdr1* N86Y, *pfdhfr* N51I, C59R, S108N/T and I164L and codons S436F/A, A437G, K540E, A581E, and A613S/T in *pfdhps* genes were genotyped by PCR/RFLP methods in 206 *P. falciparum* isolates. All *P. falciparum* isolates carried the 108N, while 98.5% parasite isolates carried the 59R mutation. 98.5% of patients carried both 108N and 59R. The prevalence of *pfdhps* 437G mutation was 17% (Chabahar) and 33% (Sarbaz) isolates. Present in 20.4% of the samples were *pfdhfr* 108N, 59R with *pfdhps* 437G mutations. The frequency of allele *pfcrt* 76T was 98%, while 41.4% (Chabahar) and 27.7% (Sarbaz) isolates carried *pfmdr1* 86Y allele. Eight distinct haplotypes were identified in all 206 samples, while the most prevalent haplotype was T76/N86/N51R59N108/A437 among both study areas. The finding the fixed level of CQ resistance polymorphisms (*pfcrt* 76T) suggests that chloroquine must be withdrawn from the current treatment strategy in Iran, while sulfadoxine-pyrimethamine may remain the treatment of choice for uncomplicated malaria [830].

8.1.1.2. Withdrawal of Amodiaquine

Amodiaquine (AQ) is a 4-aminoquinoline, similar to chloroquine, has been used widely to treat and prevent malaria. Although Amodiaquine has been widely used to treat malaria, fatal adverse reactions have been reported in adults when used as a prophylaxis. This has led some authorities to suggest it be withdrawn as a first line treatment for malaria. AQ is a cheap alternative to chloroquine, and is available in several countries, some with local production facilities. It is more palatable than chloroquine and therefore easier to administer to children. It has also been suggested that it may be a less toxic alternative to sulphadoxine-pyrimethamine (SP) in people infected with HIV in Sub Saharan Africa. It is also used in combination with the antimalarial drugs artesunate and SP. These combinations are the subject of other Cochrane Reviews [831].

Amodiaquine was first added to the World Health Organization (WHO) Essential Drugs List (EDL) in 1977. In 1979, the committee decided to delete it from the List due to its similarity with chloroquine. However, it was quickly reinstated in the same year [832]. In the mid 1980s, fatal adverse drug reactions were described in travelers using AQ for prophylaxis [833]. As a result, the manufacturer (Parke-Davis) modified the labeling and withdrew prophylaxis as an indication, while, in 1988, the WHO deleted it from the EDL and prevented its use in malaria control programs [834]. The WHO's recommendations confused policy and practice. Several countries banned its use altogether, while others have continued to use the drug as first line treatment for uncomplicated malaria - either giving it alone or in combination with other drugs. In the light of this, the 19th Expert Committee on Malaria, held in 1993, modified their statement to say that "amodiaquine could be used for treatment if the risk of infection outweighs the potential for adverse drug reactions, but still did not recommend AQ as first line treatment [835].

This Cochrane Review, first published in 2003, AQ was found to be significantly more effective than chloroquine in clearing parasites. For SP no difference in parasitological outcomes was observed within 7 days of study. However, SP showed superiority during longer-term follow up. This finding is not unexpected owing to the long half-life of SP. Whether the difference observed is due to recrudescent parasites, or to re-infections, cannot be verified. As reported previously, an improvement in symptomatic amelioration was apparent with AQ. This could be ascribed to the anti-inflammatory/antipyretic effect of the aminoquinolines. Based on the results of this review, AQ (when administered at a dose of up to 35 mg/kg, over 3 days) appears to be no more toxic than chloroquine or SP when used for treating adults and children with uncomplicated *falciparum* malaria. Under these conditions of use, and within the limitations of the sample size, no severe, life-threatening or fatal adverse reaction occurred.

So far, serious and life-threatening adverse drug reactions have been described only during prophylaxis. Based on reported rates, the risk of serious adverse drug reactions associated with the prophylactic use of AQ can be estimated to be approximately 1:2,100 treatments for agranulocytosis; 1:15,500 for hepatotoxicity; and 1:30,000 for aplastic anemia, with a total case fatality rate of 1:15,650 (Phillips-Howard, personal communication). The risk of fatal adverse drug reactions to AQ is in the same order of magnitude to that of SP.

Thus, AQ treatment appears to be safer than AQ prophylaxis. When compared the effectiveness of AQ, chloroquine and SP for treating uncomplicated *falciparum* malaria, AQ was a valuable drug and supported its continued use for the treatment of uncomplicated malaria with the proviso that, due to the partial cross-resistance with CQ, research must continue into both its effectiveness and safety. There is still exists a withdrawal of AQ for chemoprophylaxis. These findings led the WHO to modify its recommendations and reinstate AQ as an option for treating *falciparum* malaria [831, 836].

8.1.1.3. Withdrawal of Sulfadoxine-pyrimethamine

The sulfadoxine-pyrimethamine (SP) combination was used as a replacement for chloroquine in most countries. At the beginning of the 1980s, however, that treatment became almost totally ineffective in Thailand and neighboring countries; and resistance to the treatment spread rapidly in South America. In 1993, Malawi was the first country in East Africa to change from chloroquine to the SP combination as the first-line drug, and other African countries followed this example in the late 1990s. Because of extensive use of the combination, then resistance also spread in East Africa.

The lessons over the past 20 years with the introduction of SP which has been withdrawn as a prophylaxis since 2001, show how sensitive drugs for chemoprophylaxis are to resistance and its side-effects [152, 837]. Based on both laboratory assays and epidemiological studies, mutations occur in a stepwise fashion, with increasing numbers of mutations conferring higher level resistance to SP [838]. In clinical studies in Africa, five mutations have been identified that are most closely associated with clinical sulfadoxine-pyrimethamine failure: DHFR S108N, C59R, N51I, DHPS A437G, and K540E [839]. SP resistance may not have the same natural history as chloroquine resistance when the drug is effectively withdrawn. SP has similar mechanisms of action to trimethoprim-sulfamethoxazole, an antibiotic that is widely used in the developing world both for treatment and increasingly for prophylaxis against opportunistic infections in persons living with HIV. Laboratory experiments with naturally occurring *P. falciparum* isolates as well as genetically engineered DHFR in a yeast expression system demonstrated cross resistance between pyrimethamine and trimethoprim [840], and complete cross resistance has been demonstrated among the sulfas and sulfones in isolates with a variety of DHPS genotypes [841]. In regions of high transmission in Africa, individuals who receive trimethoprim-sulfamethoxazole either for treatment of bacterial infections or prophylaxis for opportunistic infections are also frequently infected with malaria parasites. Thus, antifolate drug pressure will not be eliminated even if sulfadoxine-pyrimethamine is not administered for malaria treatment.

However, recent genetic analyses of SP-resistant parasites has called into question the notion that resistance-conferring DHFR and DHPS mutations arising spontaneously in direct response to drug pressure are primarily responsible for antifolate resistance at the regional level. Studies of microsatellite markers and other genetic sequences flanking the *dhfr* and *dhps* genes among drug-resistant parasites in South America and Africa have demonstrated that allelic haplotypes of the same ancestral origin were driven through large regions in genetic sweeps. Cortese *et al.* analyzed parasites from five regions of the South American Amazon. The haplotypes with mid to high-level mutations (DHFR 50R, DHFR 164L, DHPS 540E and DHPS 581G) appeared to have a common origin. From the lower Amazon, these resistance markers spread in a North–Northwestward direction, perhaps in response to the permissive environment of drug pressure [842].

SP is the only antimalarial drug currently used for IPT during pregnancy. However, SP efficacy for treatment of symptomatic malaria in children has declined in the last 5 years, raising concerns about its longevity for IPT during pregnancy. Although it is not fully understood how IPT works, it is likely that the prophylactic effect of SP, as opposed to the treatment effect alone, plays an important role. Due to increasing drug resistance, the minimum inhibitory concentration at which parasite growth is inhibited increases and the time window for drug concentrations to fall below these levels shortens. This results in a progressive shortening of the duration of the suppressive prophylactic effect post-treatment. Parasites with triple *DHFR* mutations have an approximate 1000-fold reduction in susceptibility to pyrimethamine, and translates into a reduction in the duration of post-treatment prophylaxis of 1 month, compromising the efficacy of the 2-dose regimen, which can have a 3-month interval between doses [304].

Countries with moderate to high levels of SP resistance urgently require guidance on whether to continue using SP for IPT during pregnancy. This requires an improved understanding of the relationship between drug resistance and its implications for IPT during pregnancy. *In vivo* drug resistance in a population is typically monitored by assessing the treatment response in symptomatic children aged 6 months to almost 5 years (4 years and 11 months), who have acquired little or limited immunity [2, 843]. Pregnant women in stable malaria transmission areas have significant levels of acquired immunity to malaria and when infected often have no or few symptoms and low parasite densities. Their response to antimalarial treatment is therefore much better than that in symptomatic young children. A recent meta-analysis illustrated that pregnant women are on average half as likely to fail antimalarial treatment with chloroquine or SP as children [844, 845]. The degree of sulfadoxine- pyrimethamine resistance was determined using the therapeutic response to sulfadoxine-pyrimethamine in children with acute *falciparum* malaria obtained from concomitant treatment studies conducted in the same time frame and country. There is an urgent need to explore alternative drugs that can be used for IPT during pregnancy.

8.1.1.4. Withdrawal of Pyrimethamine/dapsone

In Australia the chloroquine was withdrawn in 1999. In 2005, data was extracted from the online Australian Statistics on Medicines reports published by the Pharmaceutical Benefits Advisory Committee, Drug Utilization Sub-committee, on antimalarials used in Australia from 1998 to 2002. Doxycycline probably remains the malaria chemoprophylaxis of choice prescribed for Australians visiting multiple drug-resistant malarious areas. Over the past 10 years, there has been marked drop in the prescription of less useful antifolate drugs such as pyrimethamine combination antimalarial drugs, especially pyrimethamine plus dapsone, which was withdrawn in 1999. There has also been a reduction in the number of prescriptions for chloroquine, mefloquine, and proguanil. The number of prescriptions for atovaquone and proguanil remain small, although they have increased steadily since its introduction in 2000; in the absence of a recommendation in the prevailing Australian guidelines. The prescription of antimalarials such as proguanil, chloroquine, mefloquine, and the pyrimethamine-containing compounds reduced considerably between 1998 and 2002. This was probably largely influenced by the availability of antimalarials, increasing resistance, the issuing of updated guidelines for malaria chemoprophylaxis, and continuing education. Newer drugs such as atovaquone plus proguanil may displace older antimalarials, particularly in the prevention of *Plasmodium falciparum* infection [846].

8.1.2. Drug Withdrawn from Monotherapy to Prevent Resistance

8.1.2.1. Withdrawal of Mefloquine Monotherapy

Mefloquine resistance was first observed near the Thai-Cambodian border in the late 1980s. The advent of mefloquine resistance in Thailand may have been influenced by the heavy use of the chemically related drug chloroquine and quinine just before the introduction of mefloquine. Mefloquine alone is no longer effective on the Thai-Myanmar and Thai-Cambodian borders, although it was operationally useful in most other endemic areas in and around Thailand during 1992-1997, with field efficacy of more than 75%. Several studies have shown a diminution in sensitivity *in vitro*, and studies *in vitro* in West Africa showed the existence of strains with decreased sensitivity to mefloquine even before its introduction into the region for therapeutic use. Resistance to quinine is often overestimated, as the dosage of 24 mg base per kg for 7 days is rarely respected, and the threshold for resistance *in vitro* has not been clearly defined. So far, no resistance to artemisinin or artemisinin derivatives has been reported, although some decrease in sensitivity *in vitro* has been reported in China and Vietnam. In WHO database [843], covering 1996-2004, of clinical outcomes on day 14 after treatment (which grossly underestimates failures), chloroquine failures >10% were reported in 93% of countries in Africa and 100% in Asia and the Americas. The corresponding values for pyrimethamine-sulfadoxine were 30% in Africa and 80% in Asia and the Americas.

Recent experience in Northwest Thailand with the combination of mefloquine and artesunate indicates that removal of drug pressure might not even be necessary for restoration of susceptibility among malaria parasites. In 1985, mefloquine was introduced in combination with sulfadoxine–pyrimethamine, and then in 1990 mefloquine was used at a higher dose as monotherapy. Within 4 years, clinical cure rates had fallen from 90% to 60%, leading to the deployment of an artesunate-mefloquine combination instead of mefloquine monotherapy [53]. Using this combination, cure rates remained nearly 100% until 1998, when the camp for displaced persons that was under surveillance closed. Of particular interest in this discussion of the reemergence of drug susceptibility, is the pattern of *in vitro* mefloquine resistance. At the time when clinical resistance to mefloquine reached its peak, the geometric mean concentration of mefloquine required to inhibit 50% of schizont formation (IC$_{50}$) was 47–56 ng/ml. By 1999, 5 years after the switch from mefloquine monotherapy to mefloquine-artesunate combination therapy, the IC_{50} decreased to 25 ng/ml. The data from Thailand suggests that if parasites develop resistance to a drug, combining the drug with an appropriate partner drug may in some cases effectively reduce drug pressure and allow for the reemergence of the susceptible forms of the parasite [845].

8.1.2.2. Withdrawal of Artemisinins Monotherapy

Artemisinin-based combination therapies (ACTs) continue to be the mainstay of treatment of uncomplicated *falciparum* malaria. For the next 8–10 years, no alternative medicines are expected to enter the market like the artemisinin derivatives that offer similar high levels of therapeutic efficacy. For this reason, WHO has focused its efforts not only to increase access to quality ACTs, but also to contain the risk of development of *falciparum* resistance, mostly created by the large-scale use of oral monotherapies for treatment of uncomplicated malaria.

In January 2006, WHO appealed to manufacturers to stop marketing oral artemisinin monotherapies and instead to promote quality ACTs in line with WHO policy [1]. The position has been widely disseminated via WHO Offices, WHO briefings to staff and to representatives of national health authorities at regional and inter-country meetings. Major procurement and funding agencies and international suppliers have accepted WHO recommendation and agreed not to fund or procure oral artemisinin monotherapies. In April 2006, the Global Malaria Programme of WHO provided a technical briefing to 25 pharmaceutical companies involved in the production and marketing of artemisinin monotherapies. Out of these, 15 declared their willingness to stop marketing artemisinin monotherapies over a short period of time, but 10 companies did not disclose their marketing plans for the future. In addition, some countries, like China and Pakistan, have been visited by WHO delegations to address multiple domestic manufacturers involved in this sector. The evolving position of manufactures and of National Drug Regulatory Authorities (NDRA) of malaria endemic countries is monitored and displayed on the WHO Global Malaria Programme website front-page: *http://malaria.who.int/*.

In May 2007, the 60th World Health Assembly resolved to take strong action against oral monotherapies and approved the resolution WHA60.18, which:

- urges Member States to cease progressively the provision, in both the public and private sectors, of oral artemisinin monotherapies, to promote the use of artemisinin-combination therapies, and to implement policies that prohibit the production, marketing, distribution, and the use of counterfeit antimalarial medicines;

- requests international organizations and financing bodies to adjust their policies so as progressively to cease to fund the provision and distribution of oral artemisinin monotherapies, and to join in campaigns to prohibit the production, marketing, distribution and use of counterfeit antimalarial medicines;

8.1.3. Drug Withdrawn from Use due to Severe Adverse Effects

In current usage of antimalarial drugs, this theoretical view has been followed in practice whenever toxicity has been measured. Experience over the past 20 years shows that effective malaria drugs such as SP, pyrimethamine/dapsone, and amodiaquine had to be withdrawn due to unexpected side-effects. In 1985, SP was withdrawn as a recommended drug for routine prophylaxis on the basis of an estimated fatal adverse reaction rate of about 1 in 20,000. A year later amodiaquine was withdrawn because of an incidence of fatal neutropenia of about 1 in 2,000. Unfortunately, there are few good techniques available to measure rates of severe adverse effects which are lower than 1 in 10,000. Prospective trials are not large enough to reliably detect side-effects or adverse effects of such frequencies and the much less reliable techniques of post-marketing surveillance must be used. This depends on using isolated case reports, reports to government agencies and to the pharmaceutical industry. These reports have to be assessed in the context of estimates of overall drug usage. Clearly, such estimates are very imprecise and drugs often have to be used for several years before even this imperfect information can be obtained. In contrast, frequent but mild side-effects are much easier to determine. Care is needed in interpreting the nature of mild side-effects because placebo controlled trials have shown that patients often suffer from non-specific side-effects such as nausea, dizziness, and headaches [752].

1) Drug Withdrawn (Amodiaquine, Pyrimethamine/dapsone, and Sulfa doxine-pyrimethamine) or FDA Requires Warning (Mefloquine) for Chemoprophylaxis

Malaria chemoprophylaxis concerns prescribing healthy individuals medication for an infection they have an unknown chance of getting. Sensible use of malaria chemoprophylaxis is a balance between the risk of infection and death, and the risk of side-effects. The risk of malaria varies from one per 35,000 travelers per month in Thailand to one per 150 in West Africa [289]. Taking drugs always implies a risk of adverse events which is the same, no matter what the risk of malaria is and the choice of prophylaxis. Therefore, the risks of infection (disease and death) should be weighed up against the risk of adverse events. Most studies find that between 2 and 5% of travelers experience adverse events, which they consider intolerable, and when the risk is too low, it is not justified to prescribe chemoprophylaxis unless the traveler insists on it. There is no golden rule for choosing where the line should be drawn; however, an operational suggestion could be areas with less than ten malaria cases per 1000 indigenous population per year.

The resistance pattern is known to a certain extent but, for instance, diverging opinion of how much resistance to chloroquine there is in West Africa illustrates the lack of data. There is much debate on rare adverse events, which usually escape Phase III studies prior to registration and are only picked up by passive, post-marketing surveillance. The lessons over the past 20 years with the introduction of amodiaquine, pyrimethamine/dapsone, and sulfadoxine-pyrimethamine, which were all withdrawn for prophylaxis after a few years; show how sensitive drugs for chemoprophylaxis are to side-effects. Three levels of chemoprophylaxis are used: chloroquine in areas with sensitive *P. falciparum*, chloroquine plus proguanil in areas with low level chloroquine resistance, and atovaquone/proguanil, doxycycline or mefloquine in areas with extensive resistance against chloroquine and proguanil. Primaquine and the primaquine analog tafenoquine may be future alternatives but otherwise there are few new drugs for chemoprophylaxis on the horizon.

Amodiaquine was introduced in 1948 and widely used as an analogue of chloroquine for the treatment and prophylaxis of malaria, as well as for the treatment of rheumatoid arthritis.

It was mainly used in Africa and was reintroduced to the UK in 1986 following the development of chloroquine resistance. Observations were made that chloroquine resistant strains of malaria were sensitive to amodiaquine. It is interesting that, at that time, there were a number of case reports of neutropenia associated with amodiaquine at different doses; these included four cases of neutropenia following low dose amodiaquine use. In spite of these anecdotal reports, both WHO and MMWR [757], described the drug as safe, presumably on the basis that anecdotal reports were not in themselves sufficient evidence to class the drug as toxic. After reintroduction in the UK there was a cluster of case reports of neutropenia, and estimates of the frequency of the adverse effects based on the national case reports of neutropenia, together with an assessment of the number of UK prescriptions, suggested that the frequency of adverse effects was 1 in 2,000 [758]. It should be noted that the errors in such estimates are so great that it is impossible to be certain that the frequency of adverse events with amodiaquine is much greater than that with SP or dapsone.

In addition, the Food and Drug Administration (FDA) announced the development of a Medication Guide (FDA-approved patient labeling) to provide better information to consumers about the risks and benefits of Lariam (mefloquine hydrochloride) on July 10, 2003 and to educate patients on the measures to be taken to optimize Lariam's effectiveness. Lariam is a valuable drug in helping to prevent malaria, but in rare instances it has been associated with serious psychiatric adverse events (*http://www.highbeam.com/doc/1G1-105085774.html*). The side-effectswere serious psychiatric adverse events including reported episodes of hallucinations, delusions, suicidal thoughts, sudden, uncontrollable rage, homicidal urges and suicide; these symptoms may persist even after the drug has been stopped.

Mefloquine has now largely replaced earlier malaria prophylaxis drugs that are no longer considered to be effective against all *Plasmodium* species, due to parasite resistance. However, mefloquine may be associated with neuropsychological harmful effects. The objective of this review was to assess the effects of mefloquine in adult travelers. The report searched the Cochrane Infectious Diseases Group trials register, Medline, Embase, Lilacs, Science Citation Index and reference lists of articles. Randomized trials comparing mefloquine with other standard prophylaxis or placebo was conducted in non-immune adult travelers. The two reviewers independently assessed trial quality and extracted data. Ten trials involving 2750 non-immune adult travelers were included. One trial comparing mefloquine with placebo showed mefloquine prevented malaria episodes in an area of drug resistance (odds ratio 0.04). Withdrawals in the mefloquine group were consistently higher in four placebo controlled trials (odds ratio 3. 56, 95% confidence interval 1.67 to 7.60). In five trials comparing mefloquine with other chemoprophylaxis, no difference in tolerability was detected. Mefloquine prevents malaria, but has adverse effects that limit its acceptability. There is evidence from non-randomized studies that mefloquine has potentially harmful effects in tourists and business travelers, and its use needs to be carefully balanced against this. Trials of comparative effects of antimalarial prophylaxis should include episodes of malaria and withdrawn from prophylaxis as outcomes [847].

2) Mepacrine

Mepacrine was first used in 1935, and was widely employed throughout the Second World War. Severe cases of aplastic anemia, transient psychotic reactions and exfoliative dermatitis have been described, together with more minor adverse events including yellow skin pigmentation and gastrointestinal disturbances. The incidence of adverse events is unknown. It is likely that the drug was withdrawn because of the high frequency of minor adverse events, rather than the high frequency of life-threatening events. Also, at the time of withdrawal, the non-toxic drugs chloroquine and proguanil became widely available.

3) Hydroxychloroquine

Acute generalized exanthematous pustulosis (AGEP) is a clinical reaction pattern that is principally drug induced and is characterized by acute, extensive formation of nonfollicular sterile pustules on an erythematous and edematous substrate. Hydroxychloroquine (HCQ), an antimalarial drug widely used to treat rheumatic and dermatologic diseases, has been described as an uncommon cause of AGEP. The US Food and Drug Administration have mandated a change to the labeling for HCQ to include AGEP among potential adverse dermatologic reactions to the drug. HCQ-induced AGEP is a rare but severe, extensive, and acute reaction. No specific therapy is available, and correct diagnosis generally leads to spontaneous resolution once the causative drug has been withdrawn [848].

4) Temafloxacin

Since their introduction in the mid-1980s, the fluoroquinolones have been administered to more than 100 million patients. Generally, adverse effects reported in association with the fluoroquinolones have been those that could have been predicted by previous experience with non-fluorinated derivatives and by animal toxicity studies. Examples of such adverse events are CNS-related toxicity, upper gastrointestinal tract reactions and phototoxicity. Some adverse experiences in animals, notably cartilage toxicity, have been of minimal clinical importance. This should lead to a re-evaluation of the possible paediatric indications for the fluoroquinolones. Temafloxacin, which was withdrawn from clinical use in June 1992, new and serious adverse events were reported at a frequency of about 1 per 3500 patients treated [849].

5) Sulphonamides

The use of sulphonamides was started in the 1930s. The problems of severe skin reactions and neutropenia were well described. Nevertheless, a combination of pyrimethamine and sulfadoxine (SP) was introduced in 1965. Twenty-two cases of Stevens-Johnson syndrome were observed with three deaths. In 1985, 19 reports of severe skin reactions with six fatalities were reported in the United States and a corresponding number of nine cases (four fatal) in UK were also reported [754]. From an estimation of the frequency of the reported reactions and the number of tablets sold within the US, an incidence of fatal reactions of a frequency of 1 in 18 to 1 in 24,000 (with 95% confidence limits about 1 in 10-50,000) has been reported [755]. It is unlikely that this toxicity is due to combination treatment as similar frequencies were observed in Beira when single doses of sulphadoxine were given to 150,000 people. Examples of neutropenia have also been recorded with SP, although the frequency of this has not been properly measured. Many studies suggest that this occurs approximately as frequently as severe skin reactions. One dissenting Swiss study shows a much lower (1 in 150,000) incidence of severe adverse effects. The reason for this difference remains obscure, although it may simply reflect over-estimates of drug usage [756]. It is unclear whether different formulations of sulphonamides have a significantly different incidence of severe adverse effects but, as these effects are so rare, it is unlikely that any high quality data will ever be produced that can be used to disprove this hypothesis.

6) Dapsone

Drug induced agranulocytosis is a rare condition. Yet one hundred and five drugs have been claimed to be associated with agranulocytosis and this list has since been updated. Some drugs are associated with relatively high risk. Dapsone is one of the drugs that were associated with a sufficiently high incidence of fatal agranulocytosis. It was withdrawn from use as prophylaxis against malaria. Dapsone had been used since 1965 as prophylaxis against malaria and, ever since, its use has been associated with neutropenia. Originally, it was used in combination with chloroquine and primaquine at doses of 25 mg a day, and neutropenia occurred in 1 in 10,000 cases, 40% of which died. Since then the combination of 12.5 mg pyrimethamine and 100 mg dapsone (Maloprim) at a dose of two tablets a week has been shown to be associated with agranulocytosisl. A dose of one Maloprim tablet a day has also been associated with four cases of neutropenia, including two deaths. Dapsone is also associated with specific minor side-effects, in particular methemoglobinemia. The toxicity of low dose Maloprim (one a week) is still contentious as only a few reports of neutropenia have been associated with low dose use. Maloprim is not licensed in the US and is only used by a minority of travelers who are advised in Britain and Australia. Thus, in spite of the few cases reported, the frequency of fatal adverse effects is likely to lie in the grey area of 1 in 20 to 1 in 50,000, where the benefits of prophylaxis may not out weight the toxicity [752].

8.1.4. Withdrew Antimalarial Drugs Still in Use

8.1.4.1. Chloroquine Is still Widely Used as First-line Antimalarial Drug

Chloroquine (CQ) was laid to rest some time ago and its use has been discontinued in many countries. CQ is still in quite active use in several African countries such as Togo, Tanzania, Ghana, Sierra Leone, Ivory Coast, Burkina Faso, and in Pakistan [760]. In other countries, in spite of governmental decisions to replace CQ with more effective drugs, CQ (mostly of very low quality) is freely accessible in many African street-markets, and the Africans use it regularly. An inhabitant of a place where malaria is holoendemic immediately recognizes a bout of symptomatic disease from her/his past experience - fever, dizziness, muscular fatigue. He/she then buys a tablet and within a few hours after its absorption, the symptoms are mostly over and the sufferers can get back to their daily chores. The health worker or the scientist working in the same place will produce much evidence (mostly from *ex vivo* drug susceptibility tests) that most of the parasites are resistant to CQ. To back their contention, they will provide evidence that CQ does not protect at all the young children in the same area.

In CQ resistant region of Togo, CQ remains the first-line drug for the treatment of uncomplicated, *P. falciparum* malaria. In the absence of recent data on the level of parasite resistance to antimalarial drugs, Togo's National Malaria Control Programme (NMCP) decided to assess the current efficacy of CQ in the treatment of uncomplicated, *P. falciparum* malaria at three sentinel sites in the north of the country. Between the September and

November of 2001, the World Health Organization's standard 14-day protocol was used to investigate 153 malarious children aged 6–59 months old. Of the subjects from Sokode, Niamtougou, and Dapaong, early treatment failure was observed in 0%, 7%, and 12%; late treatment failure in 0%, 11%, and 17%; and overall parasitological failure in 0%, 45%, and 62%, respectively. Even within northern Togo, there is clearly considerable geographical variation in the level of resistance to CQ. Before an efficient antimalarial drug policy can be developed, there is an urgent need to develop and use the national surveillance system further, to collect relevant data on the efficacies of CQ and other antimalarial drugs, such as amodiaquine and sulfadoxine–pyrimethamine [850].

In the other CQ resistant regions, it has been speculated those years of reliance on antimalarials other than CQ might lead to the reemergence of CQ-sensitive *P. falciparum* and permit the reintroduction of this safe and affordable drug. If this is confirmed in controlled clinical trials, several new opportunities for managing malaria drug resistance become possible. Countries that continue to use CQ despite high levels of resistance will be encouraged to switch now to more effective drugs, or drug combinations, with the knowledge that they may be able to return to CQ use in the future. Countries that have already replaced CQ with other antimalarial drugs could attempt to eliminate CQ use in their population as completely as possible, as Malawi has, with the expectation that if the current drug regimen fails, CQ could be reintroduced [851, 852].

It is important to emphasize that reintroduction of CQ as monotherapy is highly inadvisable and would likely lead to a rapid reemergence of resistance CQ resistant parasites that remain lurking in the neighborhood, as they do in the countries surrounding Malawi. However, where CQ is highly efficacious, it could be an ideal component of combination drug therapies designed to deter the emergence of resistance to the component drugs. And, if withdrawal of both CQ and sulfadoxine-pyrimethamine from Africa were to result in a reemergence of *falciparum* malaria sensitive to these drugs, a combination therapy regimen including an artemisinin derivative combined with CQ and sulfadoxine-pyrimethamine would be very attractive based on cost and safety. In addition, this combination would offer the advantage of including one component that rapidly reduces parasite biomass, reducing the probability of resistant parasites surviving, with two long acting drugs to protect against emergence of resistant parasites during the post-treatment prophylaxis phase. Given the limited number of safe and affordable antimalarial drugs and the compliance and prophylactic benefits of longer acting drugs, CQ and sulfadoxine-pyrimethamine may continue to be important tools in the antimalarial armamentarium [845].

Although CQ is ineffective in most parts of Africa and the onset of resistance has at least doubled childhood malaria death risk, we suggest that it is too premature to lay CQ to rest. Differential use of CQ in holoendemic areas that precludes children and pregnant women could still confer an efficacious protection for semi-immune adults. The recent introduction of artemisinin combination therapy as a chemotherapeutic strategy is certainly justified but at the same time it means a 10-fold increase in the price of treatment, a financial burden that cannot be shouldered by most inhabitants of the African continent. Sequential treatment with CQ and its combination with glutathione depleting drugs is worthy of a serious systematic testing since it could be useful even for treatment of children and adult patients in meso- or hypo-endemic areas where premunition is minimal and is affected seasonally by transmission intensity. No less important is the fact that CQ is still an efficient drug against *Plasmodium vivax* infections in most countries although even their resistance seems to be spreading [760].

8.1.4.2. Re-use of CQ in Malawi

In 1993, Malawi stopped treating patients with CQ for *Plasmodium falciparum* malaria because of a high treatment failure rate (58%). In 1998, the *in vitro* resistance rate to CQ was 3% in the Salima District of Malawi; in 2000, the *in vivo* resistance rate was 9%. We assayed two genetic mutations implicated in CQ resistance (N86Y in the *P. falciparum* multiple drug resistance gene 1 and K76T in the *P. falciparum* CQ resistance transporter gene) in 82 *P. falciparum* isolates collected during studies in 1998 and 2000. The prevalence of N86Y remained similar to that in neighboring African countries that continued to use CQ. In contrast, the prevalence of K76T was substantially lower than in neighboring countries, decreasing significantly from 17% in 1998 to 2% in 2000 ($P < 0.02$). The molecular marker of CQ-resistant *falciparum* malaria subsequently declined in prevalence and was undetectable by 2001, suggesting that CQ might once again be effective in Malawi. Withdrawal of the use of CQ appears to have resulted in the recovery of CQ efficacy and a reduction in the prevalence of K76T. However, other polymorphisms are also expected to contribute to resistance.

A randomized clinical trial was conducted involving 210 children with uncomplicated *P. falciparum* malaria in Blantyre, Malawi. The children were treated with either CQ or sulfadoxine-pyrimethamine and followed for 28 days to assess the antimalarial efficacy of the drug. In analyses conducted according to the study protocol, treatment failure occurred in 1 of 80 participants assigned to CQ, as compared with 71 of 87 participants assigned to sulfadoxine-pyrimethamine. The cumulative efficacy of CQ was 99% (95% confidence interval [CI], 93 to 100), and the efficacy of sulfadoxine-pyrimethamine was 21% (95% CI, 13 to 30). Among children treated with CQ, the mean time to parasite clearance was 2.6 days (95% CI, 2.5 to 2.8) and the mean time to the resolution of fever was 10.3 hours (95% CI, 8.1 to 12.6). No unexpected adverse events related to the study drugs occurred. CQ is again an efficacious treatment for malaria, 12 years after it was withdrawn from use in Malawi.

Different levels of drug pressure are expected to result in different levels of gene stability. Therefore, alleles of genes contributing to CQ resistance in areas of reduced levels of drug pressure ought to be different from those in areas with ongoing drug pressure. The apparently substantial recovery of CQ sensitivity observed in Malawi predicts that some of the mutations contributing to CQ resistance might be difficult to maintain in the absence of drug pressure. The interplay among various mutations in the *pfcrt* gene and the roles of as yet unidentified alleles and genes should be considered in the attempt to understand the molecular mechanisms that contribute to CQ resistance of *P. falciparum* under different levels of drug pressure [851, 852].

8.1.4.3. Withdrawn Drugs Re-used in ACTs

Case management of malaria has relied largely on the use of single antimalarial agents, such as CQ, amodiaquine, SP and, more recently, mefloquine. Since CQ, amodiaquine, and SP are inexpensive and widely available, their misuse has been extensive. Synthesized in 1934, CQ was considered too toxic to be used in humans and did not become the drug of choice for treatment of malaria until 1946 [4]. In 1957, the emergence of CQ-resistant *P. falciparum* was identified in Southeast Asia. Four years later, resistance to the drug was reported in Venezuela and Colombia [360]. In 1973, CQ was no longer recommended as a first-line therapy by Thailand, followed by South America and Asia. In 1993, Malawi was the first African country to replace CQ with SP [360]. Resistance to SP was reported in 1967, the same year in which the drug was introduced, whereas mefloquine resistance was first identified in 1982, 5 years after its introduction [53, 360].

Widepread use of these agents contributed to selection pressure and emergence of *P. falciparum* resistance to almost all currently used antimalarials (i.e., amodiaquine, CQ, mefloquine, quinine, and SP) with the exception of artemisinin and its derivatives. Resistance has also been observed with *P. vivax* and *P. malariae*, with *P. vivax* being resistant to SP and CQ in many areas [4]. As a result, treatment including CQ or SP is often ineffective [360]. To delay or prevent emergence of resistance, WHO now recommends the use of combination therapies for the treatment of uncomplicated *P. falciparum* malaria. Therapy should include two or more schizonticidal drugs with independent modes of action and molecular targets, resulting in synergistic or additive effects. Artemisinin-based combination therapy (ACT) is now the mainstay of malaria treatment. Administration of artemisinins in combination with longer-acting drugs shortens treatment to a 3-day course as opposed to a 7-day course if artemisinins are used as monotherapy [4]. Specifically, ACT combinations (artesunate plus amodiaquine, artemether/lumefantrine, artesunate plus mefloquine and artesunate plus SP) are currently recommended by WHO for the treatment of uncomplicated *P. falciparum* malaria. A combination of amodiaquine and SP may be recommended as an interim option if ACTs are not available, provided the effectiveness of this regimen is deemed high [2]. The combinations are not only to prevent drug resistance but also to reduce the adverse effects of amodiaquine, mefloquine and SP due to daily and total dose to be reduced [245].

1) Amodiaquine

Amodiaquine is a 4-aminoquinoline derivative that is structure ally similar to CQ. Amodiaquine was first introduced as an alternative to CQ since it appeared to have activity against CQ-resistant *P. falciparum* parasites. It has been on the market for more than 60 years and was added to the WHO Essential Drugs List in 1977. The drug was actively used for malaria chemoprophylaxis in travelers between 1948 and 1990 but was removed from the Essential Drugs List due to rare but serious cases of hepatitis and agranulocytosis associated with its long-term use [853]. The use of amodiaquine has since been re-evaluated because of the need for inexpensive drugs to replace CQ in areas of high CQ resistance. The drug is now being actively used in combination with artesunate for the treatment of uncomplicated malaria.

Serious and life-threatening adverse drug reactions of amodiaquine have been described only during prophylaxis. Based on reported rates, the risk of serious adverse drug reactions associated with the prophylactic use of amodiaquine can be estimated to be approximately 1:2,100 treatments for agranulocytosis; 1:15,500 for hepatotoxicity; and 1:30,000 for aplastic anaemia, with a total case fatality rate of 1:15,650 (Phillips-Howard, personal communication). The risk of fatal adverse drug reactions to AQ is in the same order of magnitude to that of SP. Thus, amodiaquine treatment appears to be safer than amodiaquine prophylaxis. When compared the effectiveness of amodiaquine, chloroquine, and SP for treating uncomplicated *falciparum* malaria, amodiaquine was a valuable drug and supported its continued use for the treatment of uncomplicated malaria with the proviso that, due to the partial cross-resistance with CQ, research must continue into both its effectiveness and safety. There was still a withdrawal on amodiaquine as a chemoprophylaxis. These findings led the WHO to modify its recommendations and reinstate amodiaquine as an option for treating *falciparum* malaria [831, 836].

A combination of artesunate and amodiaquine is currently Recommended by WHO for the treatment of uncomplicated *P. falciparum* malaria. It is endorsed as the first-line therapy by 15 African countries and by Indonesia [63]. The recommended treatment dose is 4 mg/kg of artesunate and 10 mg base/kg of amodiaquine administered once daily for 3 days [2]. Separate artesunate 50-mg and amodiaquine 200-mg tablets are currently available with a fixed formulation of artesunate and amodiaquine undergoing evaluation. This advantage of the co-formulation is a further decrease in the daily pill burden and dosage from eight to two tablets [245].

2) Mefloquine

Mefloquine is a 4-quinolinemethanol compound available as a racemic mixture of (+) and (–) enantiomers for oral administration. Mefloquine is structurally related to quinine, the drug effective against all forms of malaria and used for treatment of complicated *P. falciparum* malaria. Mefloquine is soluble in alcohol but only poorly soluble in water. It was developed by the Walter Reed Army Institute of Research (Silver Spring, MD, USA) over 3 decades ago; and initially introduced at a dose of 15 mg/kg in combination with pyrimethamine and sulfadoxine for the treatment of multidrug-resistant *P. falciparum* infection in Thailand in 1984. Worsening resistance led to the replacement of this regimen with mefloquine monotherapy at a higher dose of 25 mg/kg [854]. After an initial improvement in cure rates, further development of resistance resulted in mefloquine monotherapy being replaced by a combination of mefloquine (25 mg/kg) and artesunate (4 mg/kg/day) [855]. Mefloquine has now largely replaced earlier malaria prophylaxis drugs which are no longer considered to be effective against all *Plasmodium* species, due to parasite resistance. However mefloquine may be associated with neuropsychological harmful effects with FDA warning. Mefloquine prevents malaria, but has adverse effects that limit its acceptability. There is evidence from non-randomized studies that mefloquine has potentially harmful effects in tourists and business travelers, and its use needs to be carefully balanced against this. Trials of comparative effects of antimalarial prophylaxis should include episodes of malaria and consider possible withdrawn from prophylaxis as outcomes [847].

Mefloquine is predominantly used in a weekly dose of 250 mg. The most frequently encountered adverse reactions are nausea, abdominal pain, and dizziness. After therapeutic use of mefloquine, often as 1.5 g in three administrations, neuropsychiatric adverse effects have been noted [341]. Similar cases after both therapeutic and prophylactic treatment with mefloquine have been reported in the medical literature on several occasions, but not enough data exists on the frequency of neuropsychiatric effects to mefloquine and other antimalarials. In this prospective cohort study, we investigated the occurrence of adverse events to antimalarials in travelers. The literature and results from our study demonstrate that mefloquine can cause insomnia after prophylactic use of weekly doses of 250 mg. It should be emphasized that despite its neuropsychiatric effects, mefloquine remains a useful antimalarial drug. Even so, the guidelines for prophylaxis in the Netherlands have recently been reconsidered because of an increase in reports of serious neuropsychiatric adverse effects [328, 341]. In these guidelines, it is advised to start mefloquine 3 weeks before departure. As 80% of the serious neuropsychiatric adverse events appear within 3 weeks, this facilitates a timely discontinuation or change of antimalarial prophylaxis. A second reason for starting treatment 3 weeks before departure is that it increases the likelihood of adequate antimalarial blood levels. The neurological adverse effects could be avoided only by prescribing mefloquine within 3 weeks at 250 mg weekly dose [669].

Administration of mefloquine as a single agent for the treatment of uncomplicated malaria led to the development of resistance and a change in recommendations to advocate the use of mefloquine together with artesunate. Mefloquine is administered as a 25-mg/kg dose divided into a 15-mg/kg dose on day 2 followed by a 10-mg/kg dose on day 3 to facilitate absorption and minimize the risk of vomiting and avoiding the neurotoxicity [669]. Artesunate is administered at 4 mg/kg once daily for 3 days. The combination is currently recommended as a first-line therapy by five countries in Asia and Latin America [63]. Recently, a new fixed-dose combination of artesunate with mefloquine, comprising the three daily doses of mefloquine at 8 mg/kg and 4 mg/kg, has been developed [245]. The new regimen will reduce the pill burden and total dosage, and it is hoped that it will increase patient compliance.

3) Sulfadoxine-pyrimethamine

Sulfadoxine-pyrimethamine (SP) combination was used as a replacement for chloroquine in most countries. At the beginning of the 1980s, however, that treatment became almost totally ineffective in Thailand and neighboring countries, and resistance to the treatment spread rapidly in South America. Based on both laboratory assays and epidemiological studies, mutations occur in a stepwise fashion, with increasing numbers of mutations conferring higher level resistance to SP [838]. In clinical studies in Africa, five mutations have been identified that are most closely associated with clinical sulfadoxine–pyrimethamine failure: DHFR S108N, C59R, N51I, and DHPS A437G and K540E [839].

SP resistance may not have the same natural history as chloroquine resistance when the drug is effectively withdrawn. However, recent genetic analyses of SP-resistant parasites has called into question the notion that resistance-conferring DHFR and DHPS mutations arising spontaneously in direct response to drug pressure are primarily responsible for antifolate resistance at the regional level. Studies of microsatellite markers and other genetic sequences flanking the *dhfr* and *dhps* genes among drug-resistant parasites in South America and Africa have demonstrated that allelic haplotypes of the same ancestral origin were driven through large regions in genetic sweeps [842]. Countries with moderate to high levels of SP resistance urgently require guidance on whether to continue using SP for children and IPT during pregnancy. This requires an improved understanding of the relationship between drug resistance and its implications for IPT during pregnancy. *In vivo* drug resistance in a population is typically monitored by assessing the treatment response in symptomatic children aged 6 months to almost 5 years (4 years and 11 months), who have acquired little or limited immunity [2, 843].

This combination is currently available as separate tablets containing artesunate 50 mg and tablets containing both sulfadoxine 500 mg and pyrimethamine 25 mg (20 parts : 1 part). As with other artesunate-based regimens, the recommended dose of artesunate for the treatment of uncomplicated *P. falciparum* malaria is 200 mg for 3 days. The long-acting SP combination is administered at 1500 mg/75 mg (three tablets) on treatment day 1, which regimen could prevent drug resistance and reduce side-effects [281, 296, 320]. SP is available as an intramuscular injection at the same doses as the oral form. Prophylaxis with SP is no longer recommended because of its ability to cause Stevens Johnson syndrome and toxic epidermal necrolysis [263].

8.2. New Targeting in Drug Discovery and Development

Currently, new targeting in antimalarial drugs is focused on the molecular contribution of drug resistant-markers such as *pfcrt* to chloroquine; the mechanism of action of new effective drugs such as artemisinins; and the metabolism of parasite that differ from old drugs and significantly from the human host. For decades, malaria chemotherapy and chemoprophylaxis have relied on a limited number of drugs. However, the acquisition and spread of drug resistance has led to an increase in morbidity and mortality rates in many malaria-endemic regions. This increasing burden caused by drug-resistant parasites has stimulated investigators to seek out novel antimalarial inhibitors and drug targets, and to define the genetic basis of resistance to existing drugs, as a means to facilitate detection and develop novel strategies to overcome resistance.

Additionally, antimalarial drugs rely for their efficacy and specificity on their ability to interfere with aspects of metabolism that differ significantly from the human host. *Plasmodium* infects host erythrocytes during the phase of their life cycle that gives rise to the symptoms of malaria. Parasite survival in this environment requires several metabolic adaptations and innovations that render it susceptible to chemotherapeutic attack. This is manifest both in the mechanism of action of existing drugs and in the plethora of potential drug targets being identified, which is assisted by data from the malaria genome project [856, 857]. Many drug targets can be related to the functions of distinct organellar structures. Of particular interest are the lysosomal food vacuole (the site of extensive hemoglobin degradation), the apicoplast (a plastid organelle thought to originate from a green algal symbiont) and an acrystate mitochondrion with a limited electron transport system. Other aspects of metabolism are also important and are further highlighted.

8.2.1. Targeting on Drug Resistant Markers for Novel Drug Discovery

Multiple laboratory and clinical studies illustrate the central role that PfCRT has in mediating chloroquine resistance (CQR). How this is achieved mechanistically, and to what extent other parasite factors contribute to resistance, remains an active area of exploration. CQ accumulates in the acidic parasite digestive vacuole (DV) by virtue of its weak base properties. There, it is thought may act by binding to toxic heme species, which form during hemoglobin degradation, and inhibiting their detoxification. The CQR phenotype, which is reversible by verapamil, is associated with decreased uptake of CQ in the DV. Previous transfection experiments have shown that mutant *pfcrt* alleles can confer CQR to a CQ-sensitive (CQS) strain (tested with GC03) and that resistance is dependent on the presence of the PfCRT K76T mutation. How do PfCRT mutations lead to CQR? Studies demonstrated that *pfcrt*-mediated CQR is a result of increased drug efflux from the DV, by mechanisms that evoke either a carrier or a channel [858]. Bioinformatic comparisons with members of the drug–metabolite transporter super-family and studies in *Xenopus laevis oocytes* expressing PfCRT indicate that the loss of the positive charge resulting from the K76T mutation enables the protonated drug to traverse the transporter, leading to the movement of drug back into the parasite cytoplasm. This drug translocation is sensitive to verapamil. When the proton pump of the parasite is inhibited by concanamycin A, CQ causes a verapamil sensitive increase in the rate of DV alkalinization of CQR strains, indicating that CQ efflux is associated with protons [859].

Using a transfection-based approach, studies recently have obtained definitive evidence that the genetic background of the host strain determines the extent to which mutant *pfcrt* exerts CQR. This finding confirms earlier indications obtained with genetically diverse field isolates or culture-adapted lines [860]. The most unusual response was observed with the transfected D10 clone engineered to express mutant *pfcrt*, in that its decreased susceptibility to CQ was most pronounced at the IC_{90} level (i.e., the concentration of drug that inhibits 90% of growth relative to untreated control) and not the IC_{50} level. Further experiments revealed that depending on the strain, mutant *pfcrt* confers either resistance or a state of tolerance, implying that other factors must contribute to enable certain strains to acquire high level resistance in the presence of mutant *pfcrt*. Point mutations or changes in the levels of expression of the *pfmdr1* gene product Pgh1 can also affect the *P. falciparum* CQR phenotype in a strain-dependent manner. Recent studies by David Johnson and Steve Ward have provided evidence that Pgh1 expression can be increased upon parasite exposure to the anticonvulsant phenobarbitone, which causes decreased *in vitro* CQ susceptibility in both CQR and CQS parasites. This effect is distinct from other studies in which Pgh1 over expression has been associated with increased CQ susceptibility. Putative nuclear-receptor-binding sites with homology to mammalian sequences activated by phenobarbitone were identified in the *pfmdr1* and *pfcrt* promoter regions, and several putative nuclear receptor genes were identified. These findings raise the novel hypothesis that nuclear-receptor-mediated responses to drug exposure might be a mechanism of gene regulation in *P. falciparum* [859].

Whereas CQ-resistant strains of *P. falciparum* have spread throughout virtually all malaria-endemic areas, CQR in *Plasmodium vivax* remains relatively rare and is restricted mostly to Indonesia, Irian Jaya and other countries in the Pacific region. The genetic basis of *P. vivax* CQR has remained elusive and does not seem to involve pvcg10, the *P. vivax* ortholog of *pfcrt* [861]. A review recently surveyed variation across 36 microsatellite markers in 19 CQ-resistant and 21 CQS *P. vivax* samples from patients in Indonesia. Their preliminary analysis found no evidence of a selective sweep in the resistant isolates and revealed a low level of genetic diversity overall. Study was adopted an alternative approach that employed an optimized *P. vivax in vitro* schizont maturation assay applied to *P. vivax* isolates from Papua and Thailand. Genotyping analysis of *pvmdr1* revealed an association between CQR and a novel Y976F mutation [862]. Their data also supported an association between *pvmdr1* amplification and reduced mefloquine susceptibility, reminiscent of the well-established association between *pfmdr1* amplification and reduced mefloquine susceptibility [22, 863].

These studies found no association with the orthologs of *pfcrt* or *pfmdr1* and resulted in the unexpected finding that mutations in the deubiquitinating enzyme UBP-1 occurred in parasites pressured extensively with chloroquine (or artesunate). Defining the alternative mechanisms of CQR that prevail in other *Plasmodium* species will be important for a more complete understanding of how resistance to this drug is acquired and mediated. Many questions remain about CQR and CQ treatment failures in humans. A study recently reported findings for Guinea-Bissau that CQ, at double doses, was highly effective - curing 78% of uncomplicated malaria patients harboring *pfcrt* K76T genotypes. Questions about the relationship between fitness costs imposed by mutant *pfcrt*, the dynamics of mixed infections and the level of local CQ use have also been raised recently by Qin Cheng, who found that mutant *pfcrt* was decreasing in prevalence far more slowly in Hainan Island, China than in Malawi [852] after the cessation of CQ use to treat *P. falciparum* malaria. Importantly, the returning CQS parasites have wild-type *pfcrt* and are genetically diverse.

The acquisition of drug resistance by *Plasmodium falciparum* has severely curtailed global efforts to control malaria. Researches to define molecular contribution of resistant markers have been greatly enhanced by recent advances in *Plasmodium* genetics and genomics. Sequencing and microarray studies have identified thousands of polymorphisms in the *P. falciparum* genome, and linkage disequilibrium analyses have exploited these to rapidly identify known and novel loci that influence parasite susceptibility to antimalarial drugs such as chloroquine, quinine, and sulfadoxine-pyrimethamine. Genetic approaches have also been designed to predict determinants of *in vivo* resistance to more recent first-line antimalarials such as the artemisinins. Transfection methodologies have defined the role of determinants including *pfcrt*, *pfmdr1*, and *dhfr*. Ever more research in the area of antimalarial resistance mechanisms and new chemotherapies must be supported to sustain the continuous stream of discoveries needed to reduce the devastating impact of malaria across the intertropical regions of the world. This knowledge can be leveraged to develop more efficient methods and drugs of surveillance and treatment [864].

8.2.2. Screen New Drug Targets with Small Molecule Inhibitors

Malaria chemotherapy and chemoprophylaxis have relied on a limited number of drugs. However, the acquisition and spread of drug resistance has led to an increase in morbidity and mortality rates in many malaria-endemic regions. This increasing burden caused by drug-resistant parasites has stimulated investigators to seek out novel antimalarial inhibitors and new drug targets; and to define the genetic basis of resistance to existing drugs, as a means to facilitate detection and develop novel strategies to overcome resistance.

1) Targeting Chromatin-Modifying Enzymes

The post-genomic era has ushered in a renewed focus on mechanisms of gene regulation in *Plasmodium*, driven largely by investigations into the process of allelic exclusion in the var gene family [865]. These have revealed a key role for the *Plasmodium falciparum* histone deacetylase (HDAC) Sir2, the function of which in dynamic chromatin remodeling is intimately associated with var gene silencing. Earlier studies had identified HDAC inhibitors that had moderate potency against cultured *P. falciparum* parasites *in vitro* and rodent malaria parasites *in vivo*. In 2008, a study reported highly potent *in vitro* inhibitors, with IC_{50} values as low as 3 nM, which possess high selectivity for *P. falciparum* compared to mammalian cells. These inhibitors hyperacetylate parasite histones and produce an altered RNA-expression profile for some *P. falciparum* genes. Another study reported similar findings with the HDAC inhibitor apicidin [859].

2) Apicoplast

Additional novel drug targets are thought to reside within the parasite apicoplast, an organelle ancestrally related to chloroplasts [866]. In describing these, a research discussed the apicoplasty type II fatty acid synthesis pathway, which has been postulated to be the target of several microbicides including triclosan and thiolactomycin [866]. A more promising apicoplast pathway for the development of an effective blood-stage antimalarial is that of isoprenoid biosynthesis. Proof of principle has been achieved already with the demonstration that fosmidomycin, an inhibitor of the isoprenoid enzyme 1-deoxy-D-xylulose 5-phosphate (DOXP) reductase, can clinically cure *P. falciparum* infection either with extended monotherapy or as a component of shorter-duration combination therapies. The challenge is to identify fosmidomycin analogs that have a more suitable pharmacokinetic profile. For now, the most clinically useful apicoplast-specific antimalarials are the antibiotics, such as clindamycin, tetracycline, and azithromycin that target protein translation. These antibacterials, which produce a 'delayed death' phenotype with *Plasmodium* parasites, have proven to be fairly effective components of antimalarial combinations [866]. *In vitro* resistance to azithromycin, however, can be acquired readily with *P. falciparum* asexual blood-stage cultures [867].

Recent evidences indicate that the apicoplast possesses a Type II fatty acid biosymthesis and a mevalonate-independent isoprenoid synthesis pathway. These are significantly different from those present in human and other plastid-lacking eukaryotes which possess a single multi-functional protein (Type 1) and a mevalonate-dependent isoprenoid synthesis pathway. These two unique plasmodial pathways can therefore be selectively targeted for antimalarial drug development. Enzymes involved in the apicoplast *de novo* fatty acid biosynthesis which are emerging as drug targets include Acetyl-CoA carboxylase (ACC), ketoacyl-ACP synthase and enoyl-ACP reductase. In the apicomplexan non-mevalonate pathway, 1-Deoxy-D-xyluose-5-phosphate (DOXP) reductoisomerase is inhibited by the antibiotic fosmidomycin and presents a potential target for antimalarial drug development. Close sequential homologues of PFDOXP have been utilized to develop comparative protein models which can be used for structure-based drug design. Another potential target in this pathway is 2C-methyl-D-erythritol 2, 4-cyclodiphosphate synthase (1spF). Crystal structures of this enzyme complexed with various fluorescent inhibitors have been reported, providing opportunities for the development of new inhibitors through structure-based design [856].

3) Parasite Metabolism

Inhibitors that arrest parasite metabolism or transport processes are now attracting considerable interest. One area is vitamin biosynthesis, which occurs in *P. falciparum* but is absent in humans. Carsten Wrenger (Bernhard Nocht Institute for Tropical Medicine, Hamburg, Germany) presented the development of prodrugs that upon activation by pyridoxine kinase can inhibit pyridoxal phosphate. Another pathway currently being targeted is the biosynthesis of coenzyme A from the precursor pantothenate (vitamin B5). Kevin Saliba and his group (Australian National University, Canberra, Australia) have now identified analogs of pantothenate that possess selective antimalarial activity, with IC_{50} values of 15 mM or higher [868].

4) Parasite Transporters

Transporters involved in nutrient acquisition from the host are also under investigation as candidate targets. A study reported that the parasite-encoded folate-biopterin transporters *PfFBT1* and *PfFBT2* can be inhibited by the dihydrofolate reductase inhibitor methotrexate and the organic anion transporter inhibitor probenecid.

Folic acid uptake can also be inhibited by the folate precursor para-amino benzoic acid, indicating a wide spectrum of substrates for these two parasite transporters. An association between energy-dependent uptake of folates and mitochondrial processes has also been observed, based on studies with compounds known to inhibit mitochondrial processes in other systems. Another study also explored inhibitors of the hexose transporter *PfHT* and the *P. falciparum* Vtype H+-ATPase, respectively [859].

5) Choline Biosynthesis

Unquestionably, the most advanced antimalarial drug development program targeting parasite metabolism relates to inhibition of *de novo* phosphatidylcholine biosynthesis. This program has identified choline analogs that inhibit *P. falciparum* asexual blood stages at single-digit nanomolar concentrations [869]. Proof of fully curative antimalarial activity with shortcourse treatments has been obtained in rodents and non-human primates infected at high parasitemias. These compounds are thought to inhibit choline transport and also exert an effect on parasite phospholipid biosynthesis. These multiple modes of action, distinct from current antimalarial agents, are a major strength of these inhibitors [870] because they could help delay the development of resistance.

6) Inhibition of Mitochondrial Enzymes

Compounds that inhibit pyrimidine *de novo* biosynthesis also provide promise for future development. Major efforts in this direction are being applied to dihydroorotate dehydrogenase (DHODH), a key flavin mononucleotide dependent mitochondrial enzyme that utilizes coenzyme Q as the final electron acceptor. A study identified several potent, species selective inhibitors of PfDHODH. These inhibitors display IC_{50} values in the range 20–600 nM and Exhibit 200–20 000-fold selectivity for the malarial over the human enzyme. These completely block the Q-dependent oxidation of flavin mononucleotide, indicating that they act upon the electron transfer pathway between flavin mononucleotide and coenzyme-Q. Giancarlo Biagini and Steve Ward (Liverpool School of Tropical Medicine, Liverpool, UK) have also identified a new class of mitochondrial inhibitors, known as dihydroacridinediones. These inhibitors target the quinol oxidation site (Q_o) of the parasite mitochondrion bc1 complex, causing a collapse of the mitochondrion membrane potential and cell death, and display selectivity against the parasite enzyme that is around 5000-fold higher than for human liver bc1 [871].

7) Inhibition of Parasite Invasion of Red Blood Cells

Drugs could also potentially be designed that inhibit parasite invasion of host red blood cells (RBCs). One mechanism of parasite invasion involves the microneme protein EBA-175, which recognizes host-cell sialic-acid-containing glycophorin A [872]. Using NMR spectroscopic techniques and molecular modeling, the study explored the possibility of using neuraminic acid derivatives to compete for sialic acid recognition by EBA-175. Perturbation of this interaction might decrease or prevent parasite invasion of host RBCs.

8) Inhibition of Parasite Proteases

The *falcipain* and *plasmepsin* families of cysteine and aspartic proteases, respectively, which are involved in parasite-mediated hemoglobin digestion in the digestive vacuole (DV) of intra-erythrocytic parasites, have long been defined as promising drug targets. Scientists at GlaxoSmithKline have identified inhibitors of *falcipains* 2 and 3 that prevent *in vitro P. falciparum* growth at low nanomolar concentrations. These inhibitors can also cure *P. falciparum* infections propagated *in vivo* in human RBCs in a new immuno-compromised mouse model. Drug discovery efforts focused on the *plasmepsin* enzymes have lessened after the discovery of ten *plasmepsin* paralogs in the *P. falciparum* genome and the finding that parasites can survive, albeit with a reduced growth rate, without all four *plasmepsins* of the DV. Recent studies now revealed substantial redundancy within and between these two hemoglobinase families and show that the *plasmepsin* zymogens can be cleaved and activated by *falcipains* or autoactivated in the presence of *falcipain* inhibitors. Interestingly, HIV protease inhibitors, which are believed to target one or more parasite *aspartic* proteases, have been found to be effective against malarial infection. A study presented data with HIV protease inhibitors tested against *P. falciparum in vitro* or *Plasmodium chabaudi in vivo*. These findings are particularly promising in light of the growing use of these protease inhibitors for the treatment of HIV in malaria-endemic regions [859].

9) Emerging Technologies

The ability to identify new drug targets and antimalarial compounds through large-scale screens is rapidly improving with emerging high throughput technologies. A study assayed to screen natural product libraries for whole-cell antimalarial activity and have identified several plant extracts with micromolar activities. Powerful new 'omic' technologies are also being developed to explore drug mode of action and resistance. A quantitative proteomics approach [873] is being used to identify parasite proteins that have expression levels that are altered in response to chloroquine (CQ) or artemisinin treatment. Another study developed high-throughput platforms for selecting drug resistant *P. falciparum* mutants and is screening for transcriptional or sequence changes using microarrays and Solexa sequencing. A study developed an extensive genome-wide map of *P. falciparum* genetic diversity that has identified over 93,000 single nucleotide polymorphisms and can detect selective sweeps of drug resistance determinants (such as *pfcrt)* and define parasite population structures [41]. This work also produced a molecular barcode, based on 24 PCR primer pairs, which ascribes a unique genotype to any strain. This technology has tremendous potential to identify genes associated with emerging drug resistance.

8.2.3. Targeting on Action Artemisinins for New Drug Discovery

Artemisinin derivatives are the most important new class of antimalarials and are in widespread use. Studies into the cytotoxic effects of artemisinin derivatives have indicated that their activity might be propagated by interactions between the endoperoxide bridge of the drug and heme-iron, produced during hemoglobin degradation inside the DV [874]. This iron has been proposed to generate free radicals that alkylate and oxidize proteins and lipids within infected RBCs. This hypothesis is consistent with the finding that artemisinin activity can be potentiated by oxygen and oxidizing agents and attenuated by reducing agents. However, this has been questioned by other studies. One proposed that artemisinins are activated via reductive cleavage of the peroxide bond by intracellular iron-sulfur redox centers, which are common to *Plasmodium* enzymes, and that alkylation of these enzymes could result in parasite death. Interestingly, incubation of parasite lysates with radiolabeled artemisinin identified several interacting proteins, indicating that parasite death might result from the alkylation and inactivation of parasite proteins.

Additional support against the heme hypothesis is that artemisinins are effective against ring-stage parasites that do not seem to have high concentrations of heme. Specific proteins that have been proposed to be the target of artemisinins include cysteine proteases, proteins of the electron transport chain, translationally controlled tumor protein and PfATP6, a SERCA-type Ca^{2+}ATPase. PfATP6 has received the most attention because its activity in transfected *X. laevis* oocytes is abolished by artemisinin but is unaffected by other antimalarials. This includes the inactive compound desoxy-artemisinin, which lacks the endoperoxide bridge.

Additional studies in *X. laevis* have reported that an L263E point mutation can abolish inhibition by artemisinins. Sanjeev Krishna summarized evidence from field studies and heterologous expression systems that support a role for PfATP6 mutations in emerging artemisinin resistance. To test the role of the L263E mutation in parasite response to artemisinins, allelic exchange experiments were carried out in the Fidock laboratory, in collaboration between the two groups. The presentation included Krishna's statistical analyses of IC_{50} values, which showed a skewed distribution of artemisinin and dihydroartemisinin responses in the mutant recombinant parasites expressing the L263E mutation, with some assays showing higher IC_{50} values in the L263E mutants when the data were normalized to wild-type controls. The analysis of these data is ongoing and the results are keenly awaited because the role of PfATP6 in artemisinin susceptibility remains a topic of considerable debate. Additional work investigating the mechanism of action of artemisinins was presented that artemisinin derivatives cause early disruption of the parasite DV and do not seem to affect the structure of the endoplasmic reticulum (where PfATP6 is presumably located). Using a proteomic approach, a study presented evidence that artemisinins might affect parasite endocytosis of host proteins. Whether artemisinin accumulation in the DV relates to its mode of action, and how PfATP6 might contribute to artemisinin action and potential gain of resistance by *Plasmodium* parasites, clearly requires further studies [859].

Today's optimism that artemisinin-based combination therapies provide a sustainable answer as to how to reduce malaria morbidity and mortality finds parallels in earlier campaigns founded on CQ, the efficacy of which was once absolute. The elaboration of a coordinated Worldwide Antimalarial Drug Resistance Network to identify emerging resistance is an important initiative in this regard [37].

8.2.4. Another New Targets with New Action Mechanism

Although conventional targets such as the purine, pyrimidine, and folate pathways are still being investigated in the light of newer knowledge, a new opportunity has emerged from an understanding of certain unique features of the parasite biology. To fight against drug-resistant malaria, the new targets include the food vacuole, hemoglobin catabolism, heme biosynthesis, apicoplasts and their metabolism; as well as macromolecular transactions, import of host proteins, parasite induced alterations in the red cell surface and transport phenomena. This chapter seeks to emphasize the new and emerging targets, while giving a brief account of the targets that have already been exploited.

For Plasmodial targets, *P. falciparum* has developed resistance against almost all antimalarial drugs currently used in clinics. Efficient and potent antimalarial drugs with novel mechanisms of action are therefore urgently needed to counter this menace. Plasmodial metabolic pathways which are not present in human or are fundamentally different from those of the human are excellent targets for the development of selective antimalarial therapies with reduced risks of potential side-effects. Mapping of the *P. falciparum* genome has provided a huge impetus towards this discovery process, and several of these newly discovered metabolic pathways are currently being targeted for the development of novel antimalarial therapies.

1) Food Vacuole

Plasmodium species have a unique digestive organelle, the food vacuole has been widely used as a target for developing antimalarial drugs. This organelle is the site for the degradation of the host's hemoglobin during the trophozoite stage of the parasite's life cycle. Proteases present in the food vacuole are known to play vital roles in the life cycles of malaria parasites since they are involved in hemoglobin hydrolysis, erythrocyte rupture, and erythrocyte invasion. Aspartic (Plasmepsin I, II and IV), metallo and cysteine protease (*Falcipain* 2, 3, *Vivapain* 2, 3) are the major reported hemoglobinases present in the food vacuole of *Plasmodium* species and cause the degradation of heme to produce toxic heme. Hemoglobin hydrolysis supplies the parasite with the amino acids required for protein synthesis. In fact, cysteine protease inhibitors are known to inhibit hemoglobin hydrolysis and parasite development *in vitro* and *in vivo*. Structure-based design efforts against the plasmodial cysteine proteases including *Falcipain* 2 and 3 are being actively pursues as means to the development of selective antimalarial therapy [856].

Free heme is polymerized into a non-toxic crystalline form called hemozoin, a biocrystal of the toxic precursor ferriprotopotphyrin IX (FPIX). There are numerous drugs, including the quinolines, which are known to target the food vacuole. The quinolines such as chloroquine, amodiaquine, piperaquine, pyronaridine, mefloquine, halofantrine, lumefantrine, Primaquine, and tafenoquine are known to prevent the polymerization of toxic heme, leading to its accumulation within the food vacuole and resulting in the death of the parasite. Drug resistance has been reported against these drugs and is attributed to mutations in the food vacuole transporters, resulting in reduced drug concentrations inside the food vacuole [875].

2) Folate Pathway

Several previously published reviews have focused on describing the folate pathway and its inhibitors. Within the parasites, the folate pathway controls DNA synthesis and metabolism of certain amino acids, such as glycine, by producing reduced folate cofactors. The enzymes dihydrofolate reductase (DHFR), dihydropteroate synthase (DHPS), dihydropteridine pyrophosphokinase, dihydropteridine aldolase, and dihydrofolate synthase play important roles at various stages of this pathway. Amongst the antifolate drugs currently in clinical use, DHFR and DHPS are the best defined molecular targets. Within the pathway, DHFR converts dihydrofolate to tetrahydrofolate. The enzymes such as DHPS and dihydroneopterin aldolase are utilized in the biosynthesis of dihydrofolate form a pteridine type precursor. These enzymes are not present in the mammalian systems, thereby making them potential targets for antimalarial therapy. The sulfa drugs are also known to act as inhibitors of the DHPS enzyme. Antifolates are still used as first-line agents for the treatment of chloroquine-resistant malaria in African countries although there have been several reported instances of resistance development [875]. A combination drug of DHFR and DHPS inhibitors is often used for the purpose of enhancing their synergistic effect.

3) Shikimate Pathway

The Shikimate pathway for aromatic biosynthesis produces a key intermediate, *papaaminobenzoate* (pABA) which is used for the folate biosynthesis in *P. falciparum*. The Shikimate pathway is present in plants, algae, bacteria, and fungi for chorismate production and has recently been identified in several apiconplexan parasites. Seven different enzymes are involved in this pathway. The enzymes in this pathway help in the generation of the common aromatic precursor chorismate from erythrose-4-phosphate and its conversion into various aromate acids, pABA, ubiquinone, vitamin K, and enterochelin (in bacteria) [856]. The Shikimate pathway is absent in mammals, making the enzymes involved in this pathway, excellent targets for the development of new antimalarial chemotherapeutic agents.

4) Protein Prenylation: Protein Farnesyltransferase

The prenyltransferases are responsible for the post-translational modification of proteins involved in the signal transduction pathways, as well as those involved in the regulation of the cell cycle and DNA replication. The enzymes protein farnesyltransferase (PFT) and protein geranylgeranyltransferase (PGGT) catalyze the transfer of the farnesyl (C_{15}) or geranylgeranyl (C_{20}) groups from famesyl or geranylgeranyl pyrophosphate to the carboxyl-terminal CaaX motif of the Ras proteins. Ras mutations are assoiciated with 30% of all human cancers, indicating human PFT as a potential target for the development of anticancer therapies. A potent and selective 4-phenylquinolinine is undergoing phase II/III clinical trials for the treatment of hematological and solid tumors. PFT inhibitors have also received tremendous attention for their roles in preventing the malarial epidemic and the reduction of the drug resistance through the inhibition of PfPET. Recently, structural analogues of Zarnestra designed by the modification of the 2-quinolinone moiety and replacement of the 4-chlorophenyl moiety showed *in vitro* potency with IC_{50} values < 5 nM against PfPET. Many studies involving peptidomimetis and benzophenone-based designs have been investigated for the development of PfPET inhibitors. In addition, selective and highly active acyclic PfPET inhibitors with IC_{50} values as low as 0.5 nM and good pharmacokinetic properties have been reported, validating PfPET as a potential antimalarial target [875].

5) Pyrimidine Biosynthesis

The malaria parasites require purines and pyrimidines for DNA and RNA synthesis during their growth and replication in the intraerythrocytic stage of development. Malaria parasites are solely dependent on the *de novo* synthesis of pyrimidine bases from glutamine, bicarbonate, and aspartate for survival since the alternate pyrimidine salvage pathway is absent in the organism. On the contrary, the mammalian host cells are capable of obtaining pyrimidines from both the *de novo* and the salvage pathways. Enzymes involved in the plasmodial *de novo* pyrimidine synthesis pathway are excellent targets for devising novel antimalarials. The key enzyme in this pathway is dihydroorotate dehydrogenase (DHODH) which converts dihydroorotic acid to orotic acid with the help of the cofactor FMN, which is then introduced into the plasmodial DNA and RNA. Inhibitors such as the orotate analogues and atovaquone interfere with the binding of ubiquinone to cytochrome bcl, leading to the reduction of cellular pyrimidine pools and death of the parasites. Another characterized enzyme of this pathway is the bifunctional dihydrofolate reductase-thymidylate synthase (DHFR-TS) which is different from its bacterial and mammalian isoforms. In the case of the malaria parasites, both the enzymes TS and DHFR coexist in the same polypeptide. Inhibitor design as well as structural analysis studies of DHFR-TS crystal structures from the wild type and antifolate resistant mutants have been reported [856].

The zinc-containing carbonic anhydrases (CA) is a family of metalloenzymes which reversibly catalyzes the rapid hydration of carbon dioxide to carbonic acid and bicarbonate and forms an essential component of the *de novo* pyrimidine biosynthesis in *P. falciparum*. At least three CA isoforms exist in *P. falciparum* with PfCA1 being the major isozyme. PfCA1 inhibitors, acetazolamide and sulfanilam derivatives show good antimalarial effect in the *in vitro* growth of *P. falciparum* [875].

6) Purine Salvage Pathway

Like all parasitic protozoa, the malaria parasite *P. falciparum* depends upon the purine nucleotides salvaged from the host because of the absence of a *de novo* pathway for the synthesis of purine nucleotides. Enzymes involved in this salvage pathway are critical to the parasites and are potential antimalarial targets. Hypoxanthine-guanine phosphoribosyltransferase (HGPRT) is the key enzyme involved in the purine salvage pathway of the parasites. HGPRT converts a purine base (hypoxanthine, guanine, or xanthine) into a purine nucleotide (IMP, GMP, or XMP) by removing the pyrophosphate group from α-D-5-phosphoriboxyl 1-pyrophosphate (PRPP) and subsequently adding it to a purine base. Structures of the human HGPRT and plasmodial PfHGPRT have been determined and are very similar, sharing 44% of sequence identity. Human HGPRT efficiently catalyzes the transfer of a phosphoribosyl group from PRPP to the purine base, but unlike PfHGPRT conversion of xanthine to XMP is insignificant. Recently, QM/MM hybrid potential free-energy simulations have shown that the transfer of a phosphoribosyl group from PRPP proceeds via a dissociative mechanism in both the enzymes but the geometries of the critical structures during the reaction paths are significantly different. These differences are indicative of different substrate specificities and subtle variations in the active site of the two enzymes, thereby opening a window for the development of selective antimalarial drugs. It might be added that the lack of HGPRT function in humans results in gouty arthritis and Lesch-Nyhan syndrome which may lead to hyperuricemia and a number of central nervous system disorders.

Adenylosuccinate synthetase (AdSS) is another target of interest in the purine salvage pathway. This enzyme catalyzes the conversion of IMP to succinyl-AMP by the hydrolysis of GTP to GDP, which is absent in the human erythrocyte. The crystal structure of PfAdSS has been determined, and the enzyme shows unique properties and kinetics, thereby making it ideal for structure-based drug design. Along with the purine salvage pathway, purine transporters in *Plasmodium* species are considered as potential targets for development of antimalarial chemotherapy. The *P. falciparum* equilibrative nucleoside transporter (PfENT 1) is localized in the plasma membrane of intraerythrocytic parasite and transports purines across the plasma membrane. PfENT1 is different from the human ENT in terms of protein sequence and functional characteristics such as enzyme kinetics and ligand specificity. The enzyme serves as a major route of purine uptake and is a viable target for drug design [875].

7) Polyamine Biosynthesis

Cellular proliferation and differentiation processes require polyamines for proper functioning. All naturally occurring polyamines (putrescine, spermidine, and spermine) are obtained through a biosynthetic pathway, a catabolic pathway or an active transport system and are essential for cell growth. Polyamine biosynthesis in *Plasmodium* is different from that in the mammalian host cells. In the host cell, two separate enzymes, ornithine decarboxylase (ODC) and S-adenosylmethionine decarboxylase (AdoMetDC), are involved in the polyamine synthesis. ODC catalyzes the conversion of the amino acid ornithine into putrescine by removing a carboxyl group in the first step of the polyamine biosynthesis while AdoMetDC generates decarboxylated S-adenosylmethionine in the second step. Decarboxylated S-adenosylmethionine is subsequently used by the spermidine and spermine synthase as the aminopropyl group donor to form spermidine and spermine respectively. *P. falciparum* however, has a unique bifunctional ODC-AdoMetDC enzyme for polyamine biosynthesis. Inhibition of this enzyme by a known ODC inhibitor, α-Difluoromethyl ornithine (DFMO), is known to deplete the levels of putrescine and spermidine in the parasite infested erythrocytes. Polyamine biosynthesis can therefore be safely classified as a validate target for antimalarial drug discovery. Recently, it has been reported that *P. falciparum* is capable of salvaging putrescine and spermidine using highly specific transporters that differ from the transporters present in uninfected red blood cells, suggesting that inhibition of both the polyamine biosynthesis and the salvage pathway by antimalarial agents would significantly affect parasitic growth [856].

8) Protein Kinases

Protein kinases utilize ATP to modify other proteins (up to 30% of all proteins) by adding phosphate groups to amino acids which then act as regulatory factors for the majority of cellular pathways, especially those involved in calcium signaling and signal transduction. These signaling proteins are fundamental to the processes of cellular proliferation and differentiation. Protein kinases inhibitors have been shown to block the invasion and intraerythrocytic development of *P. falciparum*. Plasmodial protein kinases can be sub classified into several families based on their functional roles [856]:

- Cyclin-dependent protein kinases (CDK) are involved in the cell cycle regulation of humans as well as the malarial parasites. The kinases PfPK5, PfPK6, PfmrK, PfcrK-1, PfcrK-3, and Pfcrk-4 are present in *P. Falciparum* and can be used as antimalarial drug targets. Although CDKs are highly conserved, structural differences between the active sites of the human and parasitic CDKs have been observed and can be exploited to develop specific inhibitors targeting the plasmodial CDKs. Only the crystal structure of PfPK5 and its mutant have been determined so far. Known inhibitors of PfPK5 include NU6102, purvalanol B, and indirubin-5-sulphonate.
- *P. falciparum* mitogen-activated protein kinases (MAPK), such as *Pfmap-1* and *Pfmap-2*, are involved in regulation of cell proliferation.
- The glycogen-synthase kinase-3 (GSK-3) family includes GSK-3, casein kinase 2 (CK-2)-type enzyme and PfPK1.
- The Ca^{2+}-dependent protein kinase (CDPK) family includes PfCDPK1, PfCDPK2, PfKIN, and PfPK2.

- Cyclic nucleotide-dependent kinases (PKA and PKG) play an important role in various signal transduction processes. The PKGs from higher eukaryotes have two cGMP binding sites in their regulatory domain; whereas the Apicomplexan enzymes in *P. falciparum, Eimeria tenella,* and *T. gondii* possess three such sites.

The structural and functional difference between the protein kinases from the parasites and their homologues in the host cell provides an opportunity for development of specific inhibitors of the parasitic protein kinases. Two other alternative strategies utilizing protein kinase inhibition were also discussed. The first alternative approach targets the parasitic kinases involved in the arthropod vector stage of the parasite's life cycle, thereby preventing its transmission. The second approach focuses on targeting the host cell protein kinases that are required for the development of the intracellular stages of the parasites [875].

9) Glutathione (GSH)-dependent Processes

GSH, a cysteine-containing tripeptide, plays an important role in antioxidative defense by protecting the cells from oxidative damage. *P. falciparum* exhibits intense GSH metabolism when present in the trophozoite form within human erythrocytes. GSH is involved in the degradation of toxic ferriprotoporphyrin IX and is implicated in the development of chloroquine resistance. GSH-dependent processes involve the enzymes glutathione reductase (GR), glutaredoxine, glyoxalase I and II, glutathione S-transferases, and thioredoxins. Within the parasite, PfGR functions by reducing oxidized glutathione (GSSH) to GSH. Certain functions of PfGH such as glutathione disulfide reductase activity, oxidoreductase activity, disulfide oxidoreductase activity oxidoreductase activity, disulfide oxidoreductase activity and metal ion binding are very similar to those of the human isoforms which create challenges for attaining selectivity for the PfGR inhibitors. When compared to the human GR, the crystal structure of PfGR shows three unique domains. The major difference between the two enzymes is in the shape and electrostatics of a large cavity at the dimer interface, which is a potential target for non-competitive inhibitor design. Methylene blue is a selective and non-competitive inhibitor of PfGR which causes GSH depletion in the cytosol. GR inhibitors are usually used as drug sensitizers because they do not have intrinsic antimalarial activity when use alone. A combination of methylene blue with chloroquine has a synergistic effect since methylene blue sensitizes the parasites for chloroquine action. Recently, a potent, reversible and uncompetitive inhibitor of both PfGR and human GR, 6-[2'-(3'methyl)-1', 4'-naphthoquinolyl] hexanoic acid (M_5) and mechanism-based inhibitors such as the fluoromethyl M_5 analogues have been discovered [875].

10) P. Falciparum *S-Adenosyl-L-homocysteine Hydrolase (PfSAHH)*

Reversible hydrolysis of S-adenosyl-L-homocysteine (SAH) to adenosine and L-homocysteine is catalyzed by the enzyme S-adenosyl-L-homocysteine hydrolase (SAHH) whose crystal structure has been determined. PfSAHH, a homo-tetramer of identical subunits, consists of a tightly bound NAD cofactor domain, a substrate binding domain, a C-terminal domain as well as a unique 41-amino acid insertion not present in the *Homo sapiens* SAHH (HsSAHH). Since the active site is well conserved between PfSAHH and HsSAHH, the substrate adenosine shows similar binding modes in both the enzymes. Interestingly, a single substitution of PfSAHH (Cys59) or HsSAHH (Thr60) accounts for the differential interactions with respect to nucleoside inhibitors, such as 2-fluoronoraristeromycin. Inhibition of the mutants of PfSAHH (Cys59) or HsSAHH (Thr60) by 2-fluronoraristeromycin suggests that steric hindrance may be a factor for design of selective inhibitors between PfSAHH and HsSAHH. Nucleoside-type PfSAHH inhibitors such as neplanocin A and noraristeromycin cause cellular accumulation of S-adenosyl-L-homocysteine (SAH) in biological methylation, but show cytotoxicity against mammalian cells [856].

11) Glycolysis

Amongst the novel *plasmodial* targets, lactate dehydrogenase from *P. falciparum* (PfLDH) is a potential target for the development of antimalarial drugs. The malaria parasites depend on anerobic glycolysis for energy production and consume glucose at a rate 30-50 fold higher than their host cells since they lack a TCA cycle for ATP production. Targeting enzymes involved in glucolysis would therefore stop ATP production and cause the death of the parasites. Amongst the enzymes involved in glucolysis, PfLDH is the most abundant and essential enzyme and produces L-lactate as a major product in glucose metabolism. The enzyme reduces pyruvate to lactate using NADH as a cofactor. PfLDH possesses certain unique residues and shows some differences in enzyme kinetics when compared to the human LDH isoforms (hLDH) suggesting that PfLDH can be selectively targeted for development of antimalarial drugs have uncovered several inhibitors including gossypol derivatives. Oxamic acid is a competitive inhibitor of pyruvate binding in LDH. Furthermore, cytosolic *P. falciparum* Malate dehydrogenase (PfMDH) may complement the glucose metabolism function of PfLDH when the later is specifically inhibited during the parasite's asexual reproduction and growth phase. It is therefore suggested that development of dual inhibitors of PfMDH and PfLDH may be needed to stop the energy generation. Gossypol inhibits both PfMDH and PfLDH in the low micromolar level supporting the possibility of PfLDH/PfMDH dual inhibitor development [875].

12) Other New Malaria Drug Targets

The presence of an N-terminal extension on each of the *P. falciparum* fatty acid biosynthetic proteins was identified, a characteristic that is consistent with plastid targeting. In addition, inhibitors known to inhibit enzymes involved in fatty acid biosynthesis and prokaryote translation were shown to be active against parasite growth *in vitro*. These data support the suggestion that the malaria plastid could be a potential chemotherapeutic target. Calcium homeostasis mechanisms might provide other novel targets. Lisa Alleva (ANU) presented evidence that the endoplasmic reticulum (ER) Ca21 pump in the *P. falciparum* trophozoite is insensitive to thapsigargin. This is consistent with Ca21 homeostasis in the parasite being more like that in yeast and plants than that in the cells of higher eukaryotes, and might have implications for the design of new inhibitors.

Considerable care will need to be taken with new therapeutics such as atovaquone (ATQ), an analogue of ubiquinone that is currently employed in a combination therapy. Qin Chen (Australian Army Malaria Institute, Enoggera, Australia) showed that mutations in the gene encoding cytochrome b have been detected in laboratory-selected and field isolates of ATQ-resistant *P. falciparum*. Some mutations reduced ATQ susceptibility 9354-fold and were located in a putative drug-binding site. Anton Dluzewski (King's College, London, UK) has already shown that *P. falciparum* invasion requires the red blood cell (RBC) to contain millimolar levels of ATP51, and now considers that the merozoite has a mechanism for sensing and/or utilizing ATP (and/or related nucleotides). He presented evidence for a purine receptor on the merozoite surface, although permeabilization of the merozoite membrane, or the action of an ecto-ATPase, could also contribute to the sensing mechanism. Eliezer Flescher (Tel-Aviv University, Israel) presented evidence suggesting that the antimalarial effect of ART and mefloquine is mediated by sphingomyelinase (SMase)-generated ceramide. These data point toSMase and the ceramide synthetic pathway in *P. falciparum* have been potential novel targets for antimalarial drug development [856].

8.3. Contribution of Old Targets in Drug Discovery and Development

The prevalence of resistance to known anti-malarial drugs has resulted in the expansion of antimalarial drug discovery efforts. Academic and non-profit institutions are partnering with the pharmaceutical industry to develop new anti-malarial drugs. Several new antimalarial agents are undergoing clinical trials, mainly those resurrected from previous antimalarial drug targets and drug discovery programs. Novel antimalarials are being advanced through the drug development process, of course with the anticipated high failure rate typical of drug discovery. Based on the fact that values of IC_{50} (theconcentration of compound that causes 50% growth inhibition of cultured malaria parasites) of well established antimalarial drugs, such as chloroquine and artemisinin, are in the low nanomolar range and emergence of clinically significant resistant parasites occurs when the IC_{50} for the drug increases by ~10-fold, it is probably the case that only compounds that have values of IC_{50} in the low nanomolar range can be considered as significant antimalarial leads. Many of these compounds are summarized in this section. Mechanisms for funding antimalarial drug discovery and genomic information to aid drug target selection have never been better. It remains to be seen whether ongoing efforts will be sufficient for reducing malaria burden in the developing world.

8.3.1. Newer Artemisinin-type Antimalarial Drugs

8.3.1.1. First Generation Artemisinin Analogs

To overcome the solubility problems associated with artemisinin, the first generation semi-synthetic analogs of artemisinin (Figure 20) were prepared. Following sodium borohydride reduction of artemisinin to dihydroartemisinin (DHA), the lactol could be converted to its ether artemether, arteether, artelinic acid and artesunate derivatives, which are widely using and defining as first generation artemisinin analogs [44].

8.3.1.2. Second Generation Artemisinin Analogs

The second generation semi-synthetic analogs were designed and prepared either from artemisinin or its naturally more abundant precursor, artemisinic acid, with some special considerations to overcome the downsides associated with the first generation ether derivatives in metabolic stability, short half-life, poor bioavailability, and chemical stability. As mentioned earlier, these parameters were not necessarily independent as low metabolic or chemical stability would potentially lead to short half-life and poor bioavailability. These second generation analogs may also incorporate other desirable features such as lower toxicity. From the structures, these analogs could be classified into two groups; one still contains acetal functionality at C-10 while the other replaces the exocyclic oxygen of the first generation analogs with carbon, leaving these analogs to contain only an ether linkage at this position.

1) Second Generation 10-acetal Analogs

The 10-phenoxy derivatives (Figure 29A) were designed by O'Neill *et al.* [877] to mainly block the P-450 mediated oxidative metabolic formation of DHA, thereby increasing the metabolic stability of these derivatives. It was reasoned that the 10-phenoxy derivatives, which now possessed an aryl group in place of the alkyl group present in the first generation analogs, would be resistant to the P-450 mediated oxidative dearylation to DHA and its subsequent *O*-glucuronidation to form the inactive polar conjugate. In addition, drug metabolism studies revealed that some of the 10-phenoxy derivatives were expectedly not metabolized to DHA as evidenced by the lack of the corresponding DHA glucoronides excreted in bile [878].

The aza analogs were prepared in order to increase the effective concentration of the drug by the virtue of pH gradient and the weak base "ion-trapping" effect. It has been shown that the pH inside the parasite food vacuole was substantially lower (more acidic) than the external medium. In the case of the nitrogen-containing analogs, it was expected that the more acidic environment inside the parasite food vacuole would protonate the compounds at the nitrogen(s). It was assumed then that the parasite food vacuole membrane would be virtually impermeable (or much less permeable) to these protonated nitrogen-containing species, rendering high accumulated effective concentration of the drug and providing greater quantity of the drug for bioactivation by heme. Several of this class of analogs were antimalarially active, with some even more potent than artemisinin and its first-generation derivatives. However, these derivatives as well as the 10-phenoxy analogs still contain the 10-acetal functionality and may still be acid-sensitive at that position [879].

2) Second Generation 10-carba Analogs

The 10-carba analogs were designed to be more chemically robust toward acidic conditions such as those found in the stomach due to their lack of the C-10 acetal functionality. As a result, these de oxoartemisinin analogs were expected to exhibit much longer half-lives than their C-10 acetal counterparts. Analogs in this group are generally monomers but some of their dimers have recently attracted much attention from researchers not only for their expected high antimalarial activity but also for their unanticipated antiproliferative and antitumor activities.

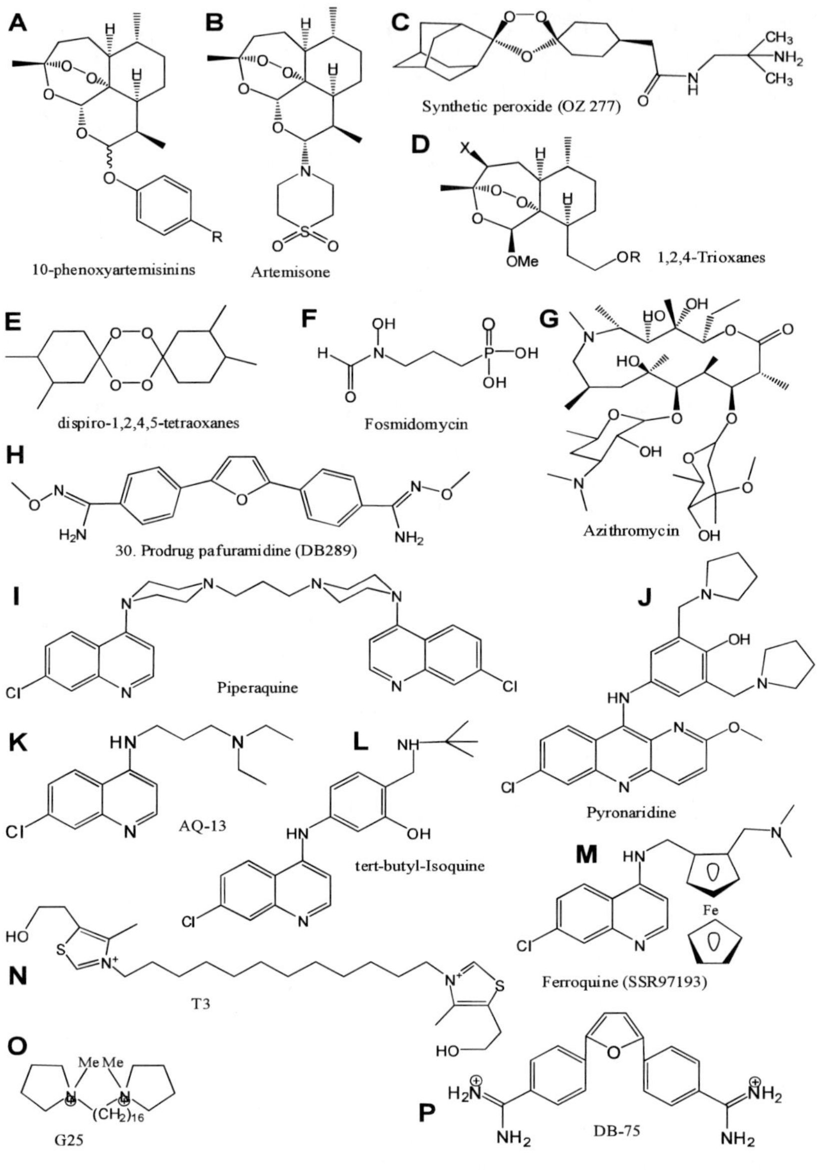

Figure 29. Selected chemical structures of new antimalarial drugs in current discovery and developments by synthesis and semisynthesis. In order to manipulate the C-10 position, it was necessary to convert the lactone functionality of artemisinin into a leaving group (e.g., alkyloxy, acetoxy, and halide) since direct nucleophilic addition of various groups to the lactone carbonyl was unsuccessful with only a few exceptions. The nature of compounds containing these leaving groups is similar to that of the pyranose derivatives in that the leaving groups require complexation/coordination by Lewis acids to generate the corresponding 10-oxonium ion for subsequent nucleophilic addition. However, the reported decomposition of the pharmacologically important 1,2,4-trioxane unit under some acidic/Lewis acid conditions restricted the use of artemisinin and its first generation ether derivatives as starting materials in the preparation of 10-carba analogs. Bearing these restrictions in mind at the time, artemisinic acid was used by Jung and his co-workers [880] to prepare the first series of 10-carba analogs which were later shown to be antimalarially potent as well as exhibiting expected longer half-lives under simulated stomach acid conditions. Jung's synthesis essentially installed the relatively acid-labile 1, 2, 4-trioxane moiety at a later stage after the incorporation of the required carbon tethers at C-10 position. Recently, a hydrolytically stable and water-soluble (+)-deoxoartelinic acid (16) was also prepared by a similar chemistry [881].

Dimers of 10-acetal containing analogs of artemisinin and the structurally related simplified trioxanes were found to be potent antimalarials [882]. However, as with the monomers, the 10-acetal containing analogs were only moderately stable, especially under acidic conditions. The acetal functionality could be replaced with the more stable carbon tethers using chemistry similar to that of the 10-carba monomers to produce the corresponding 10-carba dimers. Biological evaluations of these 10-aryl and 10-heteroaryl dimers revealed that they are not only highly antimalarial but also highly antiproliferative and antitumor [882]. Recently, synthesis and biological evaluations of the new dimeric 10-carba analogs in the deoxoartemisinin family were reported. *In vitro* antimalarial evaluation also showed that the water-soluble deoxoartemisinin dimmers [883] are at least three times more potent than artemisinin. More importantly, when evaluated *in vivo* against *P. berghei* NY in rodents, these water-soluble analogs are more efficacious than artelinic acid administered intravenously and more efficacious than sodium artesunate administered orally. No overt toxicity was observed in these mice during administration. These carboxylic acid dimers are at least as soluble in water buffered at pH 7.4 as is artelinic acid, with the best solubility of dimmer close to 30 times that of artelinic acid [883].

3) Other semi-Synthetic Artemisinin Analogs

A number of other structurally modified semisynthetic analogs in the artemisinin family have also been synthesized by various research groups notably. The groups reported synthesis and antimalarial evaluation of 11-azaartemisinin analogs whose structures contain a lactam, a more chemically robust functionality toward acid, in place of the more acid-sensitive lactone. Such analogs were designed to decrease the extent of drug hydrolysis found in the stomach. *In vitro* and *in vivo* antimalarial evaluations of these analogs showed that some 11-azaartemisinin derivatives are at least as efficacious as artemisinin also reported syntheses and *in vitro* antimalarial evaluations of C-14 and, more recently, C-16 modified artemisinin analogs. Some of these analogs were evaluated for both *in vitro* and *in vivo* antimalarial activity and were found to be reasonably potent antimalarials. However, these analogs, despite their efficacy, still suffered from limiting solubility issues [879].

8.3.1.3. Artemisone

The isolation of qinghaosu (artemisinin) and the discovery of its antimalarial property represent one of the great events in medicine in the latter third of the 20th century. Following on the extraordinary circumstances surrounding the discovery of artemisinin and its conversion into the current clinically-used derivatives dihydroartemisinin (DHA), artemether, and artesunate by the Chinese; these compounds have now become the most important for the treatment of malaria, where they are used in combination therapies with longer half-life drugs. However, the artemisinins do have problems of chemical instability, high cost, and neurotoxicity in experimental animal species which raises difficulties in formulation and compliance with drug regulatory guidelines [884].

This new and highly cost-effective antimalarial drug was developed by scientists at the Hong Kong University of Science and Technology (HKUST). The new drug, called artemisone (Figure 29B) referred to as 10-aminoartemisinins [885], was published on March 20, 2006. Artemisone has compared in terms of effectiveness and neurotoxicity with other artemisinin-type antimalarial drugs commonly used today. Artemisone inhibites *Pf*AT-Pase6 with a K_I of 1.7 nM (artesunate K_I = 167 nM) [184]. The introduction of the polar heterocycle was seemingly guided by the idea to adjust the polarity of artemisone to a desirable value in order to improve pharmacokinetic properties, as well as to eliminate neurotoxicity. Indeed, artemisone did not show any neuro- or cytotoxicity. In animal experiments, artemisone was about two to five times more efficient than artesunate [184]. Artemisone was found to exhibit no neurotoxicity at all. The high effectiveness of artemisone also means that patients require relatively low doses. Artemisone completed clinical Phase I studies in Germany, and Phase II trials in Thailand in non-severe malaria patients; it is curative at one-third the dose of artesunate, and it is active against other parasitic infections [886]. The first phase I and II clinical trials showed that there were no side-effects at all. The preclinical studies were carried out by the Bayer Pharma team within Bayer HealthCare AG in Germany, and other research group Australia, England and the US.

8.3.1.4. 1,2,4-trioxolane (OZ-277)

After it has been known that the activity of artemisinins depends on the endoperoxide substructure, numerous synthetic peroxides have been prepared and evaluated for antimalarial activity. Only OZ-277 (also known as RBx11160) has made it to the stage of clinical development [184]. The demand for plant-derived artemisinin may soon exceed the supply. In a remarkable series of bioengineering experiments, Keasling's lab has transplanted plant biosynthetic genes into yeast to allow production of the artemisinin precursor artemisinic acid in yields that appear suitable for large scale fermentation [887]. Key structural feature is a 1, 2, 4-trioxolane system, which is well known as secondary ozonide, a highly reactive intermediate of the ozonolysis reaction. OZ-277 displayed high activity against field isolates (median IC_{50} = 0.47 nM). In contrast to artemisinins, OZ-277 is only a weak inhibitor of PfATP6 (IC_{50} =7700 nM). Clinical development has been discontinued after it turned out that bioavailability in malaria patients was only one third of the values obtained with healthy volunteers [184].

In a remarkable series of studies and inspired by the presence of the peroxide function of artemisinin, Vennerstrom and their co-workers developed the 1, 2, 4-trioxolane OZ-277 (Figure 29C) as a potent antimalarial that has recently transitioned in to phase II clinical trials [888]. OZ-277 lacks chiral centers and is synthesized in a short and economical fashion by Griesbaum co-ozonolysis involving the joining of O-methyl adamantanone oxime with a substituted cyclohexanone in the presence of ozone followed by post-ozonolysis side chain elaboration [888]. The mode of parasite killing by artemisinin and related peroxides is heavily debated. At one extreme is the idea that carbon-centered radicals derived from metal-induced scission of the peroxide bond lead to alkylation of a large number of intracellular targets. At the other extreme is the proposal that artemisinin acts at a single target, a Ca^{2+}-ATPase known as PfATP6 [193]. Evidence for the former comes from the "loose" structure-activity relationships observed for anti-malarial activity of artemisinin analogs, i.e., activity is maintained even when a large number of diverse structural changes are made to OZ-277. Evidence in support of PfATP6 as the major target comes from recent studies showing that parasites harboring a single point mutant in PfATP6 are resistant to artemisinin [13]. Both sets of data seem compelling, so the controversy continues. It remains possible that artemisinin and OZ-277 kill parasites by different mechanisms [876].

A second generation ozonide, which should provide a single-dose oral cure of uncomplicated malaria, is under development. The lead ozonide (no structure disclosed) has an extended half-life and a higher oral bioavailability in comparison to the parent OZ-277. It displayed a 100% cure rate at a single dose of 30 mg/kg body weight in a murine model and was as effective as mefloquine as a prophylactic agent [184].

8.3.1.5. Simplified Synthetic 1,2,4-trioxane Derivatives and other Endoperoxides

There have been a number of reports on simplified synthetic trioxane and endoperoxide derivatives from various research groups worldwide [879], which was early contribution in investigating reaction conditions towards the synthesis of simplified 1,2,4-trioxanes provided an important basis for other research groups to develop simpler and more economical syntheses of other trioxanes and endoperoxides. Many synthetic analogs have been prepared, aiming to simplify the structures to only those moieties essential for the high antimalarial activity which certainly include the peroxide pharmacophore. The simplified analogs offer a significant advantage in that their scale-up preparations are highly feasible with inexpensive starting materials, a very important consideration given that malaria is most problematic in Africa and other developing countries. In addition, other functionalities crucial for better bioavailability and lower toxicity may be installed as a part of different intermediates involved in the synthesis at various stages leading to the desired analogs, thus not limiting the range of useful reactions and reagents to only those necessarily compatible with the peroxide functionality.

1) Simplified Synthetic Tricyclic 1, 2, 4-trioxane Derivatives

It was shown earlier from various structure-activity relationship studies of artemisinin and its derivatives that the lactone ring may not be essential for the antimalarial activity. A series of simplified synthetic tricyclic 1, 2, 4-trioxane compounds (Figure 29D) were first with the 8a-hydroxyethyl sidechain, the trioxane alcohol. The synthetic approach employed commercially available inexpensive cyclohexanone as the starting material to provide a skeleton for the cyclohexane moiety. The antimalarially important 1, 2, 4-trioxane was then installed via photooxygenation of the electron-richenol ether using singlet oxygen followed by Lewis acid-mediated cyclization to form trioxane. Derivatization of the alcohol on the 8a-hydroxyethyl sidechain resulted in a fairly large number of the first generation ether and ester trioxane derivatives. *In vitro* and *in vivo* antimalarial evaluations of these derivatives showed that some of these compounds are comparably efficacious to clinically used arteether, especially when tested against multidrug-resistant *P. falciparum* in *Aotus* monkeys [883].

More importantly, some of these 3-aryltrioxanes are also orally active when evaluated for antimalarial activity against *P. berghei* in rodents. Recently, 3-*para*-carboxyphenyltrioxane was designed to be thermally stable as well as water soluble. Its synthesis was even further simplified with the practical use of air rather than pure oxygen previously employed in the trioxane synthesis. When compared with clinically used artelinic acid, the analog, despite displaying four-fold lower efficacy when administered both subcutaneously and intravenously in mice, was about six-fold safer, suggesting comparable therapeutic indices between the two compounds. Pyranotrioxane were prepared to address the desirable features of low toxicity, water solubility, thermal stability, and oral efficacy [879].

When compared with artelinic acid, these two compounds showed improvement over their prototype 3-aryltrioxanes as their *in vivo* efficacies are now only two times less than that of artelinic acid. More importantly, they are 14–20 times more soluble in buffered water (pH 7.4) than artelinic acid, making them good antimalarial drug candidates for oral or intravenous administration [883]. A series of sulfur-containing trioxane analogs at different positions on the trioxane skeleton was synthesized. Two factors seem to contribute to the observed antimalarial activity, namely the oxidation state and the spatial proximity of the sulfur-containing substituent. From SAR, it appeared that the sulfones are generally more potent than the corresponding sulfides when the substituents are spatially oriented close to the peroxide unit. This difference diminishes as the sulfur-containing substituents become more structurally remote from the peroxide unit. In fact, when the substituents are sufficiently remote, as in analogs, the antimalarial activities of sulfone- and sulfide containing analogs are comparable. However, the synthesis of these analogs gave products in very low yields, making these analogs impractical for further development.

2) Other Endoperoxides

A series of simplified and easily synthesized cyclic peroxy ketals was prepared and evaluated for their antimalarial activity. Some of these compounds exhibited significant potency, with the activity up to 25% that of artemisinin [889, 890]. Both cycloalkane ring size and the electronic character in the aromatic ring seem to be critical for the high antimalarial activity. However, these compounds suffer from short-plasma half-life when evaluated for their *in vivo* antimalarial activity and also from relatively low chemical stability upon storage even at low temperatures. Various simple symmetric six- and seven-membered endoperoxides were prepared based on (1) the antimalarially active natural endoperoxides isolated from the Thai herb *Amomum krervanh* Pierre ex Gagnep and (2) the proposed mechanism(s) of antimalarial action at the molecular level of artemisinin and related trioxane compounds (Section IV). Some of these compounds exhibited moderate to good antimalarial activity when evaluated *in vitro*. However, their *in vivo* antimalarial potency was much lower than that of artemisinin perhaps also due to short-plasma half-life. Arteflene as well as a series of bicyclic endoperoxides was synthesized based on another Chinese endoperoxide-containing compound, yinghaosu A. Arteflene was a longer lasting antimalarial agent than artemisinin, with its longer half-life and lower rate of recrudescence while the yinghaosu A endoperoxide analogs containing sulfone showed good to excellent *in vitro* and *in vivo* antimalarial activity [879].

Dong and Vennerstrom [891] prepared molecules with two endoperoxide groups, namely the dispiro-1, 2, 4, 5-tetraoxanes (Figure 29E) via acid-mediated per- oxy ketalization. The remarkable *in vitro* and *in vivo* antimalarial efficacy of these tetraoxane analogs, comparable to artemisinin, inspired Nojima and others to prepare other cyclic peroxide systems having two geminal peroxides in the same ring [892]. This class of compounds, 1, 2, 4, 5-tetraoxacycloalkanes was evaluated for antimalarial activity and found to be highly potent antimalarials with activities several fold higher than that of artemisinin. In addition, it appears that the tetraoxacycloalkanes were much less toxic than artemisinin in mouse model 892].

8.3.2. Newer other Antimalarial Drugs with the Clinical Data

1) Dapson/Chlorproguanil/Artesunate

As has been outlined earlier, clinical relevant activity of dapsone/chlorproguanil can only be expected against parasites carrying the triple mutant *dhfr* gene. There are concerns that the quadruple mutation already found in 67% of Southeast Asian isolates may spread further once lapDapX is used extensively. To expand its useful life span, a fixed triple combination with artesunate, named LapDap+, is in clinical development. A dose-ranging study has been completed to establish the optimum ratio of artesunate and the fixed combination of chlorproguanil/dapsone. In a Phase-II clinical study, CDA showed a significant shorter parasite clearance time than dapsone/chlorproguanil alone [184].

2) Fosmidomysin

Fosmidomycin (Figure 29F) is an inhibitor of 1-desoxy-D-xylulose-5-phosphate reductoisomerase (DXR) an enzyme of the mevalonate-independent isoprenoid biosythesis. This pathway is found in many pathogenic microorganisms including *P. spp.* but is absent in humans. Fosmidomycin inhibits *Pf*DXR with an IC_{50} of 35 nM and the growth of four laboratory strains with IC_{50} values of 390-940 nM. There is no cross-resistance with other antimalarials. Although fosmidomycin is generally regarded as an antibiotic, its mechanism of action is different from those antibiotics which inhibit the apicoplast's protein biosynthesis and commonly show a delayed kill effect so that one would not expect such an effect for fosmidomycin. Indeed fosmidomycin is a rapid acting antimalarial. In the *P. vinckeii* infected mouse, the oral ED_{50} is 98 µmol (20 mg/kg). These activities are comparatively low but fosmidomycin has an extraordinary low toxicity, so that sufficiently high doses can be given. In patients, parasites are rapidly cleared (PCT_{50} = 21 h and PCT_{90} = 28 h) and a cure rate of 100% was observed at day seven. However, the recrudescence rate in non-immune patients was unacceptable high. Extensive *in vitro* combination studies revealed synergistic effects with clindamycin . In different clinical studies, the combination fosmidomycin/clindamycin achieved cure rates of 94-100% at day 28 when given twice a day for three days. Only mild gastrointestinal side-effects were reported. A 100% cure rate was obtained in a 3-day regime of a fosmidomycin/artesunate combination, which was also well tolerated. FR900098, a fosmidomycin derivative with higher activity against *P. falciparum*, is in preclinical development [184].

3) Azithromycin

Azithromycin (Figure 29G) is a newer member of the family of macrolide antibiotics. Azithromycin is effective as prophylactic agent against *vivax* malaria, but less effective than doxycycline against *P. falciparum* malaria. In a comparative trial for the treatment of multi-drug resistant *falciparum* malaria the combination of azithromycin (500 mg/d) and dihydroartemisinin was less effective than mefloquine/dihydroartemisinin. Higher doses of 1000 or 1500 mg of azithromycin for 5 and 3 days, respectively, in combination with quinine were equally effective as quinine/doxycycline in the treatment of uncomplicated malaria. Cure rates of 92% were obtained with 1500 mg/d azithromycin in combination with quinine or artesunate. There was no significant difference in the *in vitro* azithromycin susceptibility between the isolates from patients who where cured and therapy failures. In contrast, isolates from patients with therapy failures where less sensitive towards quinine or artesunate [184].

4) Pafuramidine (DB289)

Lead compound of the class of diamidines is pentamidine (Figure 29H), which is, more than 60 years after its introduction in 1945, still one of the most important drugs for the therapy of the early stage of African sleeping sickness (causative agent: *Trypanosoma brucei*) and antimony-resistant visceral leishmaniasis (causative agent: different *leishmania spp.*). Furamidine (DB75), in which the flexible chain of pentamidine has been replaced by a rigid

2.5-furylene residue, displays an IC_{50} value of 15.5 nM against *P. falciparum* and slightly lower activity against *P. vivax*. The antimalarial mechanism of action of diamidines is not exactly known. It has been proposed that diamidines act against malaria parasites through the FPIX binding and the inhibition of mitochondrial functions. Due to its two-fold positive charge, furamidine is insufficiently resorbed from the GI tract. The neutral prodrug pafuramidine (DB289) is taken up from the GI tract and metabolized to furamidine inside liver cells. The first step is a cytochrome P450-catalyzed hydroxylation of the methyl groups of the *O*-methylamidoxime partial structures, followed by spontaneous decomposition of the hemi-acetals, resulting in formaldehyde and the diamidoxim. The latter is reduced to furamidine in a cytochrome b_5-mediated reaction. DB289 has originally been developed for the treatment of the first stage of African sleeping sickness. In a small (23 patients) Phase-II clinical trial against *P. vivax* and uncomplicated *P. falciparum* infections, pafuramidine demonstrated 96% efficiency and good tolerability [184].

8.3.3. Newer other Antimalarial Drugs with the Clinical Trials

1) Piperaquine/Dihydroartemisinin

Piperaquine (Figure 29I), a bisquinoline, has first been synthesized in the 1960s in France and subsequently been further developed and extensively used in China. Widespread resistance has developed in these areas (mean IC_{50} values 240-320 nM). In contract, chloroquine-sensitive and chloroquine-resistant field isolates collected in Africa, where piperaquine has not been used so far, are still susceptible (IC_{50} values 36 and 41 nM, respectively). Piperaquine is reportedly well-tolerated, the most important side-effect being an increase of blood pressure. However, there are indications of a cross-resistance with chloroquine in different clinical trials the combination of piperaquine and dihydroartemisinin has been effective and well tolerated. Against *P. vivax* piperaquine/dihydroartemisinine was more effective than artesunate/amodiaquine [184].

2) Pyronaridine/Artesunate

Pyronaridine (Figure 29J) is another member of the class of Mannich-base schizontocides, although the usual quinoline heterocycle is replaced by an aza-acridine. Like amodiaquine, pyronaridine also displays the aminophenol substructure which can be oxidized to the respective quinoneimine. In contrast to amodiaquine, Pyronaridine contains not one but two Mannich-base side chains. It has been suggested that the second Mannich-base moiery prevents the formation of the hazardous thiol addition products by sterically shielding the quinoneimine from the attack of the sulfur nucleophile. Pyronardine was developed in China in the 1980s, but has not been registered in other countries. In a clinical study performed in Thailand, high recrudescence has been observed. *In vitro* assays revealed the presence of strains resistant to Pyronaridine. In Africa, where the compound has not been used so far, it showed high activity against chloroquine-resistant field isolates (IC_{50} value 0.8-17.9nM). This study also suggests *in vitro* cross-resistance or at least cross-susceptibility between Pyronaridine and chloroquine, and amodiaquine. In clinical studies, high efficiency was observed, but efficiency dropped to 75% when the follow-up time was extended from 14 to 30 days. The combination of Pyronaridine and artesunate is in clinical development, entering

Phase-III stage in 2006 (Medicines for Malaria Venture, MMV). In addition, an intravenous form for treatment of severe malaria is to be developed [184].

3) Tafenoquine

The 8-aminoquinoline tafenoquine (WR 238605) is a more lipophilic derivative of Primaquine. The main structural difference is the trifluoromethylphenyloxy substituent which leads to higher activity against blood and liver stages as well as a higher sporontocidal activity, but reduced gametocidal activity. Against isolates, mean IC_{50} values of 2.68 and 7.28 µM were found for tafenoquine and Primaquine, respectively. Tafenoquine is generally regarded as better tolerated than Primaquine but still has some risk causing hemolysis in glucose-6 phosphate-dehydrogenase-deficient humans. As with Primaquine, nothing firm is known about the mechanism of tafenoquine. In addition to a Primaquine-like effect on the respiratory chain, some heme polymerization inhibitory activity or activity similar to quinine has been postulated to explain the activity of tafenoquine against asexual blood stage. Several clinical studies have shown protective efficacy between 86-100%. Tafenoquine has also been used for the elimination of *P. vivax* hypnozoites and the therapy of acute *vivax* malaria. A clinical trial on long-term prophylaxis with tafenoquine has been suspended due to vortex keratopathy [184].

8.3.4. Newer Antimalarial Drugs in the Pre-clinical Studies

Three of the four compounds mentioned here belong to the class of 4-aminoquinolines, most probably sharing their mechanism of action with older 4-aminoquinolines chloroquine (CQ), amodiaquine (AQ), piperaquine, and also the azaacridine pyronaridine. Due to special molecular features which are a shortened side chain (AQ-13) or lipophilic moieties in the side chain (tertbutyl-isoquine, ferroquine), they retain activity against chloroquine-resistant parasites, possibly through reduced affinity to the substrate binding region of the chloroquine resistance transporter (*Pf*CRT). Only T3 is a drug candidate with a mechanism of action hitherto unexploited in antimalarial therapy. The majority of drugs which are in or about to enter clinical development belong to long-used drug classes especially to the 4-aminoquinolines (piperaquine, Pyronaridine, AQ-13, tert-butyl-isoquine, ferroquine). Since current therapy mainly relies on ACT, the emergence and subsequent spread of resistance against artemisinins would be a disaster. Therefore, it would be necessary, to develop new antimalarials with novel mechanism of actins. Only three compounds (furamidine, T3, and fosmidomycin) fall into this category. Now, great efforts have to be made to identify novel targets and to develop drug candidates against these targets to be prepared for the time when artemisinins begin to lose their effectiveness.

1) AQ-13

AQ-13 (Figure 29K) is a chloroquine derivative with a shortened side chain, which is active against chloroquine-resistant parasites (IC_{50} 59 nM vs. 315 nM for CQ). But a clear correlation between the susceptibility of different isolates toward AQ-13 and CQ points to some degree of cross-resistance. At high doses, AQ-13 is described to be more toxic in rats than chloroquine. Furthermore, the two alkyl residues at the terminal nitrogen are highly susceptible towards oxidative dealkylation. The resulting metabolites are almost inactive. A recently completed dose-ranging trial in healthy volunteers suggests that adverse effects of AQ-13 may not be different from those of chloroquine and that higher doses of AQ-13 compared to chloroquine may be necessary to produce similar blood levels and AUCs [893].

2) tert-butyl-Isoquine

To prevent the undesirable formation of quinonimines from amodiaquine-like molecules the positions of the hydroxyl and the diethylaminpmethyl residue at the phenyl ring have been interchanged. Through this modification, the formation of the toxic quinonimine is no longer possible. Fortunately, this modification has no negative influence on the antimalarial activity. The resulting molecule isoquine (ISQ 1) (Figure 29L) is more active against the chloroquine resistant K1 strain than amodiaquine 2 [894]. Unfortunately, as with AQ-13, the easy access of hydroxylating enzymes to the methylene groups in α-position of the nitrogen atom results in poor bioavailability. Replacement of the diethylamino moiety by a *tert*-butylamine presumably solves the problem of rapid biotransformation. It is expected that *tert*-butyl isoquine advances to clinical trials [184].

3) Ferroquine

Ferroquine (SSR97193) (Figure 29M) bears a ferrocenyl moiety in the side chain, a structural feature rather uncommon in potential drugs. Because of this lipophilic ferrocentyl moiety it has been proposed that ferroquine does not fit into the substrate binding site of the chloroquine resistance transporter (CRT). Ferroquine is active against various chloroquine-sensitive and chloroquine-resistant laboratory strains (IC_{50} values 14-42 nM) as well as field isolates (IC_{50} values 1-62 nM). The mutational status of the *pfcrt* gene did not influence the compounds activity. The selectivity index measured against a lymphoma cell line is about 700. In a mouse model, ferroquine is curative at 19 μmol (8.4 mg)/kg bodyweight reportedly, this drug is about to enter clinical development [184].

4) T3

During the intra-erythrocytic stages of their life cycle, parasites produce large quantities of membrane constituents through the phospholipid metabolism. The class of bis-cationic compounds has been developed to target the phospholipid biosynthesis. The most advanced representative of this class is called T3 (Figure 29N), which is about to enter clinical development (H. Vial, personal communication). The T3 displays IC_{50} values between 2.3 and 6.3 nM against different chloroquine-sensitive and chloroquine-resistant parasite strains. In a rodent model, T3 has an ED_{50} of 0.44 μmol (0.2 mg)/kg when administered i.p. Due to its highly polar nature, bioavailability on oral application is low. The oral ED_{50} is >22 μmol (10mg)/kg. A prodrug has been developed, with a reported bioavailability of about 16% in rats. Its oral ED_{50} is 11 μmol (5mg)/kg in the rodent model. Analogous to the bis-amidines (e.g., furamidine), bis-ammonium compounds enter the infected erythrocyte by the parasite-induced new permeability pathways (NPP) and accumulate inside the erythrocyte up to 270-to 310-fold of the plasma concentration. Apparently, the compound enters the parasites, using the same choline transporter also responsible for the uptake on the bis-amidines. Inside the parasites, the phosphatidylcholine *de nove* synthesis is inhibited (IC_{50} = 0.9 μM), either through the inhibition of the choline uptake, the inhibition of enzymes if this pathway, or the combination of both effects. The precise mechanism of action remains to be unclear. In addition to their effect on the phosphatidylcholine *de novo* synthesis, bis-ammonium compounds have shown to bind to ferriprotoporphyrin IX. This may be important for the observed intraparasitic accumulation as well as for the antimalarial activity [184]. Therefore, the bis-ammonium compounds can be regarded as dual drug by acting on the phosphatidylcholine synthesis and the heme detoxification.

5) GW844520

Gw844520, a 4-(1H)-pyridone, acts on the cytochrome bc_1 complex with an IC_{50} of 2 nM, 10-fold higher than that of atovaquone. However, atovaquone-resistant parasites are inhibited with IC_{50} values of 2.5 – 7.6 nM. In contrast to atovaquone, there is no synergism with proquanil [895]. The development of GW844520 has been discontinued, due to unexpected cardiotoxicity in dogs. A backup pyridone GW308678 (no structure disclosed) is in preclinical development.

6) New Drugs from NIH

It appears that choline uptake is required for this phospholipid biosynthesis since choline uptake in infected erythrocytes is much higher than in non-infected cells. Vial and colleagues have been exploring a number of cationic molecules as choline analogs with the ability to block choline uptake. In early studies, the compound G25 (Figure 29O) emerged as a bis-cationic molecule with exceptional *in vitro* anti-malarial activity (*ED50* = 0.6 nM). G25 is ~1000-fold less toxic to mammalian cells lines. Later studies showed that G25 selectively accumulates in malaria-infected erythrocytes [14]. Treatment of malaria-infected monkeys with low doses of G25 (i.m. 0.15 mg/kg twice daily for 8 days) was sufficient to achieve a cure in all 5 treated animals. Compounds such as G25 display poor oral bioavailability presumably because they contain permanent positive charges. In a remarkable series of studies, Vial and co-workers reported that the bis-cationic compound T3 displays potent *in vitro* anti-malarial activity with ED_{50s} in the 2–9 nM range. In the same report they found that the pro-drug TE3 (Fig. 3) is stable in intestinal fluid, slowly breaks down to T3 in gastric fluid (half-time ~ 8 hr), and rapidly breaks down in human plasma (half-time ~ 5 min). The oral bioavailability of TE3 in rats is 16%, and this pro-drug is able to cure malaria in monkeys when dosed orally in the 3–27 mg/kg range once a day for 4 days. These compounds are exciting leads for anti-malarial drug discovery. Whether a malarial choline transporter is the target of these compounds remains to be conclusively established. Tidwell and co-workers have been studying a family of bis-cationic molecules as anti-microbial agents for several years.

One compound, DB-75 (Figure 29P) is a broad spectrum anti-microbial that displays low nanomolar potency as an inhibitor of the growth of *Trypanosoma brucei*, the parasite that causes African sleeping sickness. DB-75 is structurally related to pentamidine, a compound known to possess anti-protozoal activity since the 1930s. A major breakthrough came when it was discovered that the diamidoxime pro-drug, DB-289 is orally absorbed and then converted to the active DB-75 in the bloodstream. DB-289 is currently undergoing clinical trials for the treatment of African sleeping sickness. In the meantime, DB-289 was found to display good potency against malaria. Given the fact that DB-289 had already gone through phase I clinical trials for the treatment of African sleeping sickness, a clinical trial to test this compound for efficacy in malaria was initiated in Thailand. The study showed that DB-289 exhibited good efficacy in clearing patients of *P. falciparum* and *P. vivax* infections. However, because efficacy required ~7 days, it was decided to halt clinical testing of DB-289 for malaria with the hope that a more active analog can be found that can achieve a cure of malaria in ~3 days (in order to meet the anti-malarial drug target parameters previously listed). The mode of microbial killing by bis-amidines like pentamidine and DB-75 is unknown. It is well established that these compounds accumulate to high concentrations in *Trypanosoma brucei* and likely in *P. falciparum*-infected human red blood cells. Given the high intracellular concentration, multiple targets seem likely [876].

8.4. New Drugs Selected from Herbal Medicines and among Known Drugs

In the case of malaria, quinine and artemisinin are the two major antimalarial drugs widely used today that originally came from indigenous medical systemsused in Peruvian and Chinese ancestral treatments, respectively. There is an urgent need for the discovery of new drugs due to the critical epidemiological situation of this disease. New inexpensive therapies that are simple to use and that will limit the cost of drug research are good justifications for this ethnopharmacological approach. Therefore, the aim of this section is to empirically analyze plants that are used for antimalarial treatment in the whole world, and to determine those with real promising antimalarial activity. The major leads such as those extracted from various species of plants have been highlighted. Indeed, some extracts seem to be promising in future research, but development of new isolation and characterization techniques, for designing new derivatives with improved properties need to be discussed.

8.4.1. Herbal Medicines in Sources of Antimalarial Drugs

8.4.1.1. Antimalarial Herbal Medicines in Asia and Africa

Traditional medicines have been used to treat malaria for thousands of years and are the source of the two main groups (artemisinin and quinine derivatives) of modern antimalarial drugs. Problems with increasing levels of drug resistance is compounded by the difficulties in poor areas being able to afford and gain access to effective antimalarial drugs; traditional medicines could be an important and sustainable source of treatment. The Research Initiative on Traditional Antimalarial Methods (RITAM) was founded in 1999 with the aim of furthering research on traditional medicines for malaria [896]. The initiative now has in excess of 200 members from over 30 countries. It has conducted systematic literature reviews and prepared guidelines aiming to standardize and improve the quality of ethnobotanical, pharmacological, and clinical studies on herbal antimalarials and on plant based methods of insect repellence and vector control. Over 1200 plant species from 160 families are used to treat malaria and fever. On average, one-fifth of patients use traditional herbal remedies for malaria in endemic countries. Larger, more rigorous randomized controlled trials are needed with long term follow up. So far only a few studies have reported on side-effects from preparations. In one trial, some patients stopped treatment due to minor side-effects.

The proportion of patients using traditional herbal remedies for malaria varies widely. A meta-analysis carried out by us of 28 studies on treatment seeking behavior showed that 307 of 315,458 respondents used such remedies. The overall percentage was 20%, but this value is misleading because the range varied widely from 0% to 75%. Many factors influence the use of traditional medicines to treat malaria. To date, 1277 plant species from 160 families used to treat malaria or fever have been listed on a database. More studies are yet to be included in the database. Eleven species were used as antimalarials or antipyretics in all three tropical continents, and 47 species were similarly used in two continents. Most of the species (1213 plants) are not featured in the World Conservation Union's (IUCN) red list of threatened groups: of those that are, 5 were listed as "endangered," 13 were listed as "vulnerable," and 3 were listed as "near threatened."

Eighteen case clinical trials have reported on herbal antimalarials (Table 56). Of cohort studies mentioning herbal treatments, 17 were for *falciparum* malaria, 12 for *vivax* malaria, and 5 for malaria of undefined species. In some cases, this was deliberate, to protect intellectual property rights. Few studies (3 case studies, 13 cohort studies, 4 controlled trials) provided data on side-effects. It seems that in the other studies, patients were not questioned about adverse effects or new symptoms since starting treatment. None of the studies reported serious adverse effects. Only three of the cohort studies and three of the controlled trials reported effects on biochemical variables (most commonly liver function tests), and two studies monitored electrocardiograms. No cases of toxicity were reported. Minor side-effects can, however, be important. Some herbal antimalarials have a bitter taste, making it difficult to give them to children. Doses often need to be taken repeatedly, and the volume may be larger than with conventional drugs. Six of the cohort studies on *falciparum* malaria reported 100% parasite clearance on days 4-7 after treatment, and a further three reported clearance rates above 90%.

Parasite clearance was only one day longer with this remedy than with chloroquine, and the clearance of fever was faster by 12 hours. A larger trial is needed to confirm these results, however, since the number of patients was small. In another trial comparing the treatment of *falciparum* malaria with quinine or with infusions of *Artemisia annua,* the infusions resulted in good parasite clearance at day 7.13 A high proportion of patients experienced a recrudescence, so that by day 28 only 37% of those treated with *Artemisia annua* were still free of parasites compared with 86% of patients treated with quinine. This emphasizes the need for a follow up period of at least 28 days. Nevertheless, adequate clinical response is more important than parasite clearance in endemic areas, which was not reported in this trial. In the trial of the herbal remedy "Malarial" (Suma-Kala), parasites were not cleared completely, but there was a good clinical response, that was sustained for the three weeks of follow up. In the trial of *Cochlospermum tinctorium,* patients were only followed for five days, and clinical outcomes were not used, so it is not possible to comment on long term efficacy. Initial parasite clearance with Ayush-64 was good, but many patients relapsed by day 28. Another trial showed that oil based capsules of *Artemisia annua* cleared parasites and fever more rapidly than did chloroquine, and by day 30 only 8% of those given a course of capsules for six days showed recrudescence [896].

Although traditional medicine is widely used to treat malaria, and is often more available and affordable than Western medicine, it is not without limitations. Firstly, there are few clinical data on safety and efficacy. Secondly, there is no consensus, even among traditional healers, on which plants, preparations, and dosages are the most effective. Thirdly, the concentration of active ingredients in a plant species varies considerably, depending on several factors. None the less, these limitations are all remediable, through research. The Research Initiative on Traditional Antimalarial Methods has written systematic reviews, some of which have been summarized here, and guidelines aiming to standardize and improve the quality of future research. In many areas no ethnobotanical study has been undertaken, and research in these areas is a priority. Traditional medicines are being forgotten with the death of healers who have no successors to their knowledge. The evidence summarized in this section, together with the guidelines proposed, should not only assist researchers already working in this specialty but also inspire other researchers and funding bodies to give serious consideration to the potential of traditional remedies for malaria.

8.4.1.2. Antimalarial Herbal Medicines in West Africa

One of the major threats concerning world public health is malaria. The mortality rate from malaria has been estimated to be approximately between 1.5 and 2.7 million per year, with more than 75%of these deaths occurring among African children. The main reasons that explain this deteriorating situation are:

- Resistance to the current antimalarial drugs by *P. falciparum*.
- Lack of new therapeutic targets [897].
- Unavailability and unaffordability of antimalarial drugs [898].

New drugs against malaria are thus urgently needed, but malaria is one of the diseases, that are commonly treated with natural products, mainly from plants. Natural products and their derivatives have traditionally been the most common source of drugs, and still represent more than 30% of the current pharmaceutical market. Of the 877 new small-molecule chemical entities introduced between 1981 and 2002, roughly half (49%) were natural products, semi-synthetic natural product analogues or synthetic compounds based on natural product pharmacophores [899]. It has long been recognized that natural product structures have the characteristics of high chemical diversity, biochemical specificity, and other molecular properties that make them favorable structures for drug discovery, and serve to differentiate them from libraries of synthetic and combinatorial compounds. In the case of malaria, the new drug discovery approaches need to take into account some specific concerns. In particular, the requirement for new therapies to be inexpensive and simple to use, as well as the need to limit the cost of drug research. Among the currently ongoing efforts is the discovery of new antimalarial targets from natural products. The search for new bioactive plant products can follow two main routes: random or ethnobotanical and ecological research. This present study is an analysis of ethnopharmacological publications describing research into antimalarial treatments.

Forty-eight ethnobotanical studies from 1987 to 2007 in 9 out of 16 West sub-Saharan Africa countries have been collected. The countries were: Benin, Burkina Faso, Ghana, Ivory Coast, Mali, Niger, Nigeria, Sierra Leone, and Togo. The other seven countries (Cape Verde, Gambia, Guinea, Guinea-Bissau, Liberia, Mauritania, Senegal) did not have any ethnobotanical studies published in this data bank. Pharmacological studies have demonstrated some *in vitro* and/or *in vivo* activity and toxicity of extracts from up to 100 species of plants. The Ivory Coast is the country where the most studies have been carried out, particularly *in vitro* and cytotoxicity tests. In the case of *in vivo* tests, the highest number of studies was carried out in Nigeria. Some genus of plants were tested against *Plasmodium falciparum* up to five times by one or several authors in the same country. For example *Alchornea* (four authors in Ivory Coast), or *Croton* (eight times by the same author: Weniger in Benin [900]; whereas *Nauclea*, *Mitragyna*, *Pavetta*, *Acanthospermum*, and

Cochlospermum were studied by different authors from several West Africa countries (data not shown). From the 610 *in vitro* antiplasmodial activity tests carried out in these countries, 94 (15%) showed very good antimalarial activity, with an IC_{50} below 5 g/ml for the crude extracts. Antimalarial plants showing high activity *in vitro* concerned the five countries Ivory Coast, Burkina Faso, Benin, Mali, and Togo. Even though the Ivory Coast showed many reports of *in vitro* antimalarial tests and numerous good antimalarial plants, the best ratio between the number of plants tested and the number of plants with good activity was found in studies from Burkina Faso and Togo.

As stated in the Roll Back Malaria project, a new drug needs to be discovered every 5 years. This is unlikely to be achieved because the development of a new drug is costly and time-consuming. The aim of our review was to report the plants traditionally used in malaria treatment in West Africa and for which antimalarial properties have been already demonstrated by the scientific community. This first step was thus to select plants for which pharmacological studies (*in vitro*, *in vivo* and toxicity) have been completed. The ethnopharmacology approach used in search for new antimalarial compounds appears to be predictive. Indeed, such plants as *Azadirachta, Cryptolepsis, Cochlospermum,* and *Guiera* appeared to open the way to future research on the prevention and/or treatment of malaria. However, it should be emphasized that for any compounds or extracts being recommended for the treatment of malaria it is essential that well-controlled clinical trials are undertaken. Such trials would be invaluable for locating new antimalarial drugs [898].

Another crude acetone/water (50/50) extract of neem leaves (IRAB) was evaluated for activity against the asexual (trophozoites/schizonts) and the sexual (gametocytes) forms of the malaria parasite, *P. falciparum, in vitro*. In separate 72 hour cultures of both asexual parasites and mature gametocytes treated with IRAB (0.5 µg/ml), parasite numbers were less than 50% of the numbers in control cultures, which had 8.0% and 8.5% parasitemia, respectively. In cultures containing 2.5 µg/ml, asexual parasites and mature and immature gametocytes were reduced to 0.1%, 0.2%, and 0% parasitemia, respectively. There were no parasites in the cultures containing 5.0 µg/ml. This extract, if found safe, may provide materials for development of new antimalarial drugs that may be useful both in treatment of malaria as well as the control of its transmission through gametocytes [901].

8.4.1.3. Antimalarial Herbal Medicines in South America

In this section we discuss the ongoing situation of human malaria in the Brazilian Amazon, where it is endemic causing over 610,000 new acute cases yearly, a number which is on the increase. This is partly a result of drug resistant parasites and new antimalarial drugs are urgently needed. The approaches we have used in the search of new drugs during decades are now reviewed and include ethnopharmocology, plants randomly selected, extracts or isolated substances from plants shown to be active against the blood stage parasites in our previous studies. Emphasis is given on the medicinal plant *Bidens pilosa*, proven to be active against the parasite blood stages in tests using freshly prepared plant extracts. The anti-sporozoite activity of one plant used in the Brazilian endemic area to prevent malaria is also described, the so-called "Indian beer" (*Ampelozizyphus amazonicus, Rhamnaceae*). Freshly prepared extracts from the roots of this plant were totally inactive against blood stage parasites, but active against sporozoites of *Plasmodium gallinaceum* or the primary exoerythrocytic stages reducing tissue parasitism.

In contrast with only 1% active plants among 300 randomly selected species which were tested, nearly 20% were found to be active among less than 50 plants based on traditional knowledge, identified up to the present. Antimalarial activities were observed initially with hexane and/or ethanolic extracts of *Esenbeckia febrifuga, Boerhavia hirsuta, Lisianthus* sp., *Acanthospermum* sp., *A. australe, Vernonia* sp., and *Tachia guianensis* [902].*Tachia* is used in the Amazon region as an antimalarial drug. Activities were promising but the plant extracts rarely reduced parasitemia to a degree equivalent to chloroquine tested in parallel. Attempts to isolate and identify purified components from the active plants were disappointing. It could be related to difficulties in purifying glycosides (*Lisianthus*) or to non-retention of the initial activity (*Vernonia* sp. and other plants) in the purified fractions [903].

The ethanolic and methanolic extracts of most *Bidens* species tested were very active [903]. The percentage of reduction of parasitemia in *P. falciparum* cultures *in vitro*, caused by *B. pilosa* extracts, was mostly very high. The number and the aspect of the surviving parasites at day 4 of culturing with the drugs were observed in the blood smears. Lower parasitemia occurred with most extracts of *Bidens sp.* as compared to that found in blood cultures in culture medium only. Ethanol, butanol, and chloroform extracts of *B. pilosa* are active at relatively low concentrations. Ethanolic extracts of *B. pilosa* prepared from leaves, roots, and the whole plant were also tested in mice infected with *P. berghei* showing consistent activity. Regarding the root extracts, a more significant activity was confirmed in further studies [903].

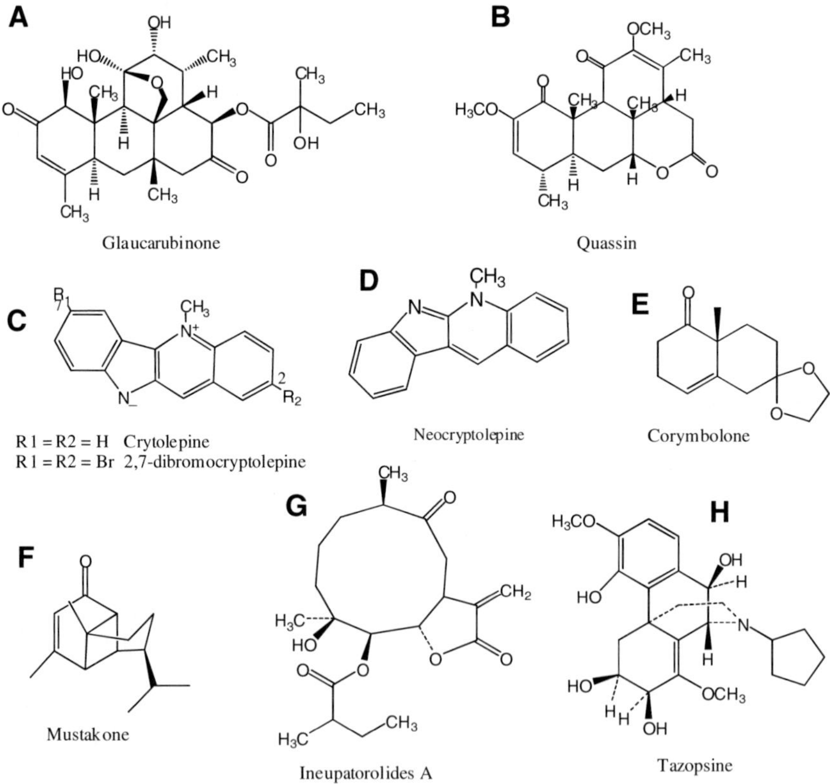

Figure 30. Selected chemical structures of new antimalarial drugs in current discovery and developments extracted and identified from herbal medicines.8.4.2. New Drugs from Herbal Medicine in Malaria Chemotherapy

1) Glaucarubinone and Quassin in Brazil

During the last 25 years, extracts from a large number of plant species including many that are used in traditional medicines have been evaluated for *in vitro* antiplasmodial activities and some have also been tested *in vivo*, usually in mice infected with *P. berghei* or *P. yoelii*. In some cases the constituent(s) responsible for their activities have been isolated but relatively few have been studied further to assess their potential as lead compounds for the development of new antimalarial drugs. One group of compounds that has been explored further is the quassinoids which are degraded triterpenes found in various species of *Simaroubaceae*, some of which are used traditionally for the treatment of malaria and other protozoal diseases. Although quassinoids are generally cytotoxic, a few compounds such as glaucarubinone (Figure 30A), from *Simarrouba amara* are relatively selective against *Plasmodium falciparum in vitro* but the latter was found to be toxic *in vivo* when tested in mice infected with *Plasmodium berghei*. Structure-activity relationship studies of a series of quassinoids as well as structural modifications of quassin from *Quassia amara* (Figure 30B), and brusatol a constituent of *Brucea javanica*, were carried out in an attempt to improve selectivity against *Plasmodium falciparum* but these strategies were not successful. The antiplasmodial and cytotoxic properties of the quassinoids are both due to protein synthesis inhibition and it is likely that parasite and host cell ribosomes are too similar to allow for the development of selective inhibitors [904].

2) Cryptolepine and Neocryptolepine in West Africa

In more recent years, the alkaloid cryptolepine (Figure 30C), a constituent of *Cryptolepis sanguinolenta*, a West African climbing shrub used traditionally for malaria treatment has been investigated as a potential lead to new antimalarials. Although cryptolepine has potent *in vitro* antiplasmodial activity, it failed to cure malaria in mice when given orally and was toxic when given by the intraperitoneal (i.p.) route. In addition, it has cytotoxic properties on account of its DNA-intercalating and topoisomerase II inhibiting activities. These properties would suggest that cryptolepine is not a good candidate for antimalarial drug development but it appears that its antiplasmodial action, is different from its cytotoxic mode of action, and therefore it may be possible to prepare analogues of cryptolepine that retain or have enhanced antiplasmodial activity but with reduced toxicity. The most promising analogue prepared to date is 2,7-dibromocryptolepine (Figure 30C), which is approximately nine-fold more potent than cryptolepine against chloroquine-resistant *Plasmodium falciparum* (strain K1) and when tested in mice infected with *Plasmodium berghei* 25 mg/kg/day given i.p., parasitemia was suppressed by 90% with no apparent toxicity to the mice. Recent studies also suggest that the antimalarial mode of action of this and other cryptolepine analogues involves other mechanisms in addition to the inhibition of hemozoin formation. The cryptolepine isomer, neocryptolepine (Figure 30D), a minor constituent of *Cryptolepis sanguinolenta*, is also currently being pursued as a lead to new antiprotozoal agents [904].

3) New Compounds in Brazilian Amazon

In the Brazilian Amazon, antimalarial evaluation of nine species of *Bidens,* performed in parallel tests, showed intense activity in seven species, partial activity in one and no activity in one species [905]. The ranges of activities observed *in vitro* using 50 µg/ml varied from over 80% (*B. frondosus, B. tripartitus, B. pilosa, and B. ferulaefolia)* to 65-71% (*B.*

bipinnatus, B. maximovicziana, B. campylotheca); 38% with *B. bitternata* or no inhibition of parasite growth with *B. parviflora*. The latter contains no polyacetylene compounds whereas in *B. bitternata* three polyacetylenes were identified; and, all the others have from 6 to 14 polyacetylene compounds identified [902]. Thus, the activity is related to the presence of polyacetylenes, possibly associated with flavonoids also present in *B. pilosa* [903]. In recent tests, two methoxylated flavone glycosides isolated from the roots of *B. pilosa* proved to be highly active *in vitro*. The activity we observed for extracts of *Bidens* spp. contrasts with previous results described in the literature using fenilheptatriyne, a phenylacetylene isolated from the leaves of *Bidens* collected in Africa. The authors found little or no activity possibly because during fractionation or during storage of the fractions the active compounds were degraded. As it is now known, polyacetylenes degrade rather easily and lose activity (our unpublished data) as we observed when preparing them from the crude extracts of *B. pilosa* roots. The purified fraction developed a yellowish color overnight and after two days it became dark brown, even in the refrigerator [905].

Esenbeckia febrifuga (Rutaceae) is a plant traditionally used to treat malaria in the Brazilian Amazon region. Ethanol extract of stems displayed a good antiplasmodial activity against *P. falciparum* strains W-2 ($IC_{50,}$ 15.5 µg/ml) and 3 D7 ($IC_{50,}$ 21.0 µg/ml). Two coumarins (bergaptene and isopimpinellin), five alkaloids (flindersiamine, kokusaginine, skimmiamine, gamma-fagarine and 1-hydroxy-3-methoxy-N-methylacridone), besides a limonoid (rutaevine), have been isolated for the first time from this species. Antiplasmodial activity of compounds 3, 5-8 has been evaluated *in vitro* against *P. falciparum* strains (W-2 and 3D7) and the furoquinolines 5 and 6 were the most potent displaying IC_{50} values <50 µg/ml; flindersiamine showed a weak activity while alkaloid and rutaevine were inactive [906].

4) Sesquiterpenes in Kenya

In Kenya, two sesquiterpenes, corymbolone (Figure 30E) and mustakone (Figure 30F), isolated from the chloroform extract of the rhizomes of Cyperus articulatus, exhibited significant antiplasmodial properties. Mustakone was approximately ten times more active than corymbolone against the sensitive strains of the *P. falciparum* [907].

5) Ineupatorolides A in South Korea

In the previous work, methanol extracts of *Carpesium rosulatum* (Compositae) were found to have high antiplasmodial activity against *P. falciparum in vitro*, this activity being largely attributable to a ineupatorolides A (I-A, Figure 30G). In the present study, encouragingly, I-A was also found to have potential antimalarial activity *in vivo* when tested against *P. berghei* in mice. I-A (2, 5, 10 mg/kg/day) exhibited a significant blood schizonticidal activity in 4-day early infection, repository evaluation, and in established infection with a significant mean survival time comparable to that of the standard drug, chloroquine (5 mg/kg/day). The I-A possesses a promising antiplasmodial activity, which can be exploited in malaria therapy [908].

6) Resveratrol Derivatives in South Korea

Recently, methanol extracts from the dried roots of *Pleuropterus ciliinervis* were found to have high antimalarial activity against *P. falciparum in vitro*, this activity being largely attributable to a (E)-resveratrol-3-O-a-L-rhamnopyranosyl-(1-2)-b-D-xylopyranoside (RRX) [909]. In the present study, encouragingly, RRX was also found to have moderate antimalarial activity *in vivo*, when tested against *P. berghei* in mice.

In both the 4-day test and the test of 'repository' activity, oral RRX produced dose-dependent chemosuppression, with even the lowest dose tested (10 mg/kg/day) giving significant reductions in parasitemia ($P < 0.05$). The highest dose of RRX tested (50 mg/kg/day) did not, however, quite reach the level of chemosuppression seen with the drugs used as positive controls — chloroquine at 5 mg/kg/day or pyrimethamine at 1.2 mg/kg/day. In the mice that were only treated from 72 h post-infection, daily oral doses of RXX or chloroquine led to gradual reductions in parasitemia over time, whereas the parasitemias in the negative-control mice increased with time. All the mice given RRX showed a gradual decrease in body weight from day 7 but this weight loss lasted only for a few days, after which these mice gained in weight each day. Mice in the negative-control group lost weight each day throughout follow-up. One of the five mice given 25 mg RRX/kg/day died before day 29 (on day 23), as did two of the five mice treated with 10 mg/kg/day (on days 20 and 24), and all five of the negative-control mice (on days 11–18). None of the mice given the highest dose of RRX and none of those given chloroquine died during the follow-up period. The mean survival times of the mice given 10, 25, or 50 mg RRX/kg/day, chloroquine and water were 26.0, 28.0, 30.0, 30.0 and 16 days, respectively. The mice still alive on day 29 (all of which had been treated with RRX or chloroquine) then appeared aparasitemic. In the tests of activity against established infection, the highest tested doses of RRX appeared as effective as chloroquine in terms of the day-7 parasitemias and day-29 survival.

In the toxicity tests, all mice given RRX at 5–500 mg/kg showed some signs of toxicity, ranging from writhing and gasping (with the lower doses) to decreased respiratory rate, decreased limb tone and death. The LD_{50} was calculated to be .500 mg/kg, however, and none of the mice given RRX in the tests of antimalarial activity showed any signs of acute toxicity. The present results indicate that RRX possesses useful blood schizontocidal when used at doses that cause no marked toxicity in mice. Although the mechanism of action of this compound has not been elucidated, some plants and/or plant compounds are known to exert antimalarial activity either by causing elevation of erythrocytic oxidation or inhibiting protein synthesis. RRX clearly merits further investigation. [910]

7) 11(13)-dehydroivaxillin in South Korea

The whole plants of *Carpesium* genus are used in traditional medicine as anti-pyretic, analgesic and vermifugic, including a topical application for sores and inflammation. A previous study on *Carpesium* genus suggested that the antiplasmodial activity against *Plasmodium falciparum* was due to the existence of 11(13)-dehydroivaxillin (DDV) from EtOAc extracts of *C. ceruum* (Compositae). Here, the antimalarial activity of DDV was evaluated against *P. berghei* in mice. The LD_{50} of the compound was determined as 51.2 mg/kg, while doses of 124 mg/kg and above were found to be lethal to mice. DDV (2, 5, 10 mg/kg/day) exhibited a significant blood schizontocidal activity in 4-day early infection, repository evaluation and in an established infection with a significant mean survival time comparable to that of the standard drug, chloroquine, 5 mg/kg/day. DDV possesses a promising antiplasmodial activity, which can be exploited in malaria therapy [911].

8) New Stilbene Glycoside in South Korea

A novel stilbene glycoside [piceid-(1-->6)-beta-d-glucopyranoside; PBG] from *Parthenocissus tricuspidata* was tested *in vivo* against *P. berghei*. PBG exhibited significant blood schizontocidal activity in a 4-day early infection, a repository evaluation, and an established infection, with a significant mean survival time comparable to that obtained with the standard drug, chloroquine (5 mg/kg/day) [912].

9) Three Triterpenes in Congo

Cucurbitaceae is traditionally used in Congo Brazzaville for the treatment of malaria. The antiplasmodial activity of the plant and isolated some of the compounds was assessed responsible for this activity. It was the first time that a chemical study of this plant has been undertaken. Three triterpenes were isolated: cucurbitacin B, cucurbitacin D and 20-epibryonolic acid and their structures were assigned from spectroscopic evidence and comparison with published data. The crystallographic structure of 3 was determined. All fractions and compounds obtained in this study were assayed for antiplasmodial activity (on FcM29, a chloroquine-resistant strain of *P. falciparum*) and cytotoxicity (on KB and Vero cell lines). The IC_{50} values of 1, 2, and 3 are 1.6, 4 and 2 µg/ml on FcM29. Both 1 and 2 have a high cytotoxicity whereas 3 show a better selectivity index [913].

10) Eight Novel Compounds in Cameroon

Eight new compounds were isolated from the roots of *Garcinia polyantha*, and identified. Two of them, the xanthone garciniaxanthone I, and the triterpene, named garcinane, are reported as new natural products. The structures of the new compounds were elucidated on the basis of 1D and 2D NMR spectroscopic studies. The structure of compound 1 was confirmed by X-ray crystallography. Among the remaining six known compounds, three were known xanthones (smeathxanthone A, smeathxanthone B, and chefouxanthone), one benzophenone (isoxanthochymol), one triterpene (magnificol), and one sterol (beta-sitosterol). The *in vitro* antimalarial activity of isoxanthochymol against *Plasmodium falciparum* shows strong chemosuppression of parasitic growth. [914]

8.4.3. New Drugs from Herbal Medicine in Malaria Chemoprophylaxis

1) Tazopsine (Figure 30H)

The parasite that causes malaria has quickly developed resistance to many of the drugs that are commonly used to treat this disease. As a result, new drugs and drug combinations are needed. In some parts of the world where antimalarial drugs are failing due to resistance, or are not available to everyone, people often turn to traditional herbal remedies instead. These traditional plant remedies can be a useful starting point for development of new drugs, but the process of developing effective new drugs from plant remedies is long and complicated. An important initial step is to isolate and identify the active compounds from plants and then see how well these compounds perform against malaria parasites in laboratory tests. If the tests are successful, such compounds could then progress to experiments in animals and possibly eventually human trials. One plant used widely in Madagascar for treatment of malaria is *Strychnopsis thouarsii;* the traditional remedy consists of the plant stem bark boiled in water. The researchers suspected that the agents in this plant bark had some activity against the "liver stage" of malaria infection in humans.

After many rounds of separation and testing, the researchers got down to a single, apparently new, molecule that was active against malaria in the laboratory test, and this molecule was named tazopsine (in the Malagasy language the word Tazo refers to malaria). In order to find out how effective the molecule was at killing malaria parasites, the researchers took human or mouse liver cells cultured in the laboratory, infected them with malaria parasites (either the malaria parasite that normally infects humans, or a related species that infects mice), and then added tazopsine at different concentrations. The compound completely killed the malaria parasites even at very low concentrations, and had activity against malaria infecting either liver cells or red blood cells. Tazopsine was then given to mice injected with a species of the malaria parasite. The compound protected most mice against malaria infection when it was used at a dosage level lower than the toxic dose. The researchers then tried making a series of different variants of tazopsine in the hope that some variants would be less toxic, but equally active as, the original compound. They found one variant, named NCP-tazopsine that was much less toxic but just as active as tazopsine, but only against the malaria infecting liver cells.

Using bioassay-guided fractionation based on the parasite's hepatic stage, we have isolated a novel morphinan alkaloid, tazopsine, from a plant traditionally used against malaria in Madagascar. These compound and readily obtained semisynthetic derivatives were tested for inhibitory activity against liver stage development *in vitro* (*P. falciparum* and *P. yoelii*) and *in vivo* (*P. yoelii*). Tazopsine fully inhibited the development of *P. yoelii* (50% inhibitory concentration [IC$_{50}$] 3.1 lM, therapeutic index [TI] 14) and *P. falciparum* (IC$_{50}$ 4.2 lM, TI 7) hepatic parasites in cultured primary hepatocytes, with inhibition being most pronounced during the early developmental stages. One derivative, N-cyclopentyl-tazopsine (NCPtazopsine), with similar inhibitory activity was selected for its lower toxicity (IC$_{50}$ 3.3 lM, TI 46, and IC$_{50}$ 42.4 lM, TI 60, on *P. yoelii* and *P. falciparum* hepatic stages *in vitro*, respectively). Oral administration of NCP-tazopsine completely protected mice from a sporozoite challenge. Unlike the parent molecule, the derivative was uniquely active against *Plasmodium* hepatic stages.

A readily obtained semisynthetic derivative of a plant-derived compound, tazopsine, has been shown to be specifically active against the liver stage, but inactive against the blood forms of the malaria parasite. This unique specificity in an antimalarial drug severely restricts the pressure for the selection of drug resistance to a parasite stage limited both in numbers and duration, thus allowing researchers to envisage the incorporation of a true causal prophylactic in malaria control programs.

In these experiments a new molecule, tazopsine, was discovered from a Malagasy plant, and it was found to be active against liver-stage malaria parasites, in laboratory experiments and in mice. This molecule or variants of it could in future become candidate antimalarial drugs in humans. However, much work would need to be done before testing could get to that stage. Different variants of molecules related to tazopsine would need to be tested to find one that has low toxicity, and these variants would need to be fully evaluated in animals to see how they are handled in the body before any trials could begin in humans [915]. Bioassay-guided fractionation of *S. thouarsii* stem barks extracts, using a rodent *Plasmodium yoelii* liver stage parasites inhibition assay, led to isolate the new morphinan alkaloid tazopsine (1) together with sinococuline (2) and two other new related morphinan analogs, 10-epi-tazopsine (3) and 10-epi-tazoside (4). Structures were characterized by 2D NMR, MS, and CD spectral analysis. Compounds 1-3 were found to fully inhibit the rodent *P. yoelii* liver stage parasites *in vitro*. [916]

2) "Indian Beer" in Brazil

Most medicinal plants used against malaria in endemic areas aim to treat the acute symptoms of the disease such as high temperature fevers with periodicity and chills. In some endemic areas of the Brazilian Amazon region one medicinal plant seems to be an exception: *Ampelozyziphus amazonicus*, locally named "Indian beer" or "Saracura-mira," used to prevent the disease when taken daily as a cold suspension of powdered dried roots. In previous work, we found no activity of the plant extracts against malaria blood parasites in experimentally infected animals (mice and chickens) or in cultures of *Plasmodium falciparum*. However, in infections induced by sporozoites, chickens treated with plant extracts were partially protected against *Plasmodium gallinaceum* and showed reduced numbers of exoerythrocytic forms in the brain. Stronger evidence has been demonstrated that the ethanolic extract of "Indian beer" roots hampers *in vitro* and *in vivo* development of *P. berghei* sporozoites, a rodent malaria parasite. Some mice treated with high doses of the plant extract did not become infected after sporozoites inoculation, whereas others had a delayed prepatent period and lower parasitemia. The data validates the use of "Indian beer" as a remedy for malaria prophylaxis in the Amazon, where the plant exists and the disease represents an important problem which is difficult to control. Studies aiming to identify the active compounds responsible for the herein described causal prophylactic activity are needed and may lead to a new antimalarial prophylactic [917].

3) Multi-herbal Extract 'Agbo-Iba' in Nigeria

To determine the efficacy of a multi-herbal preparation extract of 'Agbo-Iba' on rodent malaria induced in mice. An experimental design in which mice were divided into four groups A, B, C, D representing control, prophylactic, chloroquine, and 'Agbo-Iba' groups respectively. Each mouse was intraperitoneally inoculated with *P. yoelii* nigeriensis and treated with oral herbal extract or chloroquine syrup depending on group. The studies were conducted in College of Medicine of the University of Lagos Medical Microbiology and Parasitology Laboratory. One hundred and twenty male and female albino mice aged 10-12 weeks with an average weight of 25 grams. The herbal extract was effective, preventing the development of parasitemia in the prophylactic group of mice. After intraperitoneal inoculation of *Plasmodium yoelii* nigeriensis, a prepatent period of two days was observed before parasitemia was established in all but the prophylactic group of mice. Induced infection was promptly aborted with oral chloroquine treatment in group C, while in groups A and D, infection terminated fatally. Group B mice appeared normal throughout the duration of investigation with 100% survival rate. 'Agbo-Iba' extract has some prophylactic action against malaria induced in mice with no apparent significant side-effects [918].

Table 56. Trials of herbal medicines for uncomplicated malaria*

Trial, setting	Study design	Species	Age of participants	Treatment (No. of participants)**	Parasite clearance***	Symptems	Slide effects
Boye 1989, Ghana	Open randomized controlled trial	*P. falciparum*	Not available	*Cryptolepis sanguinolenta* aqueous root extract (n=12); chloroquine (n=10)	Clearance time: 3.3 days; 2.2 days	Fever clearance time: 36 hours; 48 hours	Fewer in *C. sanguinolenta* group
Mueller *et al.* 2004, Democratic Republic of Congo	Open randomized controlled trial	*P. falciparum*	≥18	*Artemisia annua* aqueous infusion 5 g/l (n=39); *A. annua* aqueous infusion 9 g/l (n=33); quinine (n=43)	Day 7: 77%; 70%; 91%	Fever clearance on day 3: 91%; 81%; 92%	Fewer in *A. annua* groups; tinnitus with quinine (27%)
Koita 1990, Guindo 198810, Mali	Open randomized controlled trial	*P. falciparum*	5-45	"Malarial" (n=36); chloroquine (n=17)	Day 7: 58% <100; 92% <100	Fever clearance on day 7: 59%; 50%	Constipation (n=1); allergy (n=3)
Benoit-Vical 2003, Burkina Faso	Open randomized controlled trial	*P. falciparum*	12-45	*Cochlospermum planchonii* root decoction (n=46); chloroquine (n=21)	Day 5: 52%; 57%	Not available; 100% fever clearance on day 5	Few, minor in both groups
Tsu 1947, China	Comparative study	*P. falciparum and P. vivax*	Not available	*Dichroa febrifuga* root extract (n=12); placebo (n=8)	Day 7: 100%; 13%	Fever clearance on day 2: 92%; 13%	Abdominal pain, vomiting, D febrifuga

Trial, setting	Study design	Species	Age of participants	Treatment (No. of participants)**	Parasite clearance***	Symptems	Slide effects
CCRAS 1987, India	blind randomized controlled trial	*P. vivax*	Not available >12	Ayush-64 (n=30); chloroquine and primaquine (n=28)	Not available	No significant difference	Not available for either Treatment
			>12	Ayush-64 (n=58); chloroquine and primaquine (n=60)	Day 6: 95%; 100%	Improvements by day 6: 95%; 100%	Not available for either Treatment
				Ayush-64 (n=30); chloroquine and primaquine (n=30)	Day 6: 72%; 100%	Improvements by day 6: 72%; 100%	No effect on full blood, liver function tests,
Valecha *et al.*, 2000, India	Open randomized controlled trial	*P. vivax*	18-60	Ayush-64 (n=54); chloroquine (n=50)	Day 28: 49%; 100%	Slow recovery; fast recovery	No effect on full blood Count, gastrointestinal effects (n=3)
Yeramian *et al.*, 2005, Thailand	Open randomized controlled trial	*P. vivax P. falciparum*	18-32 18-44	DB289 100 mg daily x 5 (n=9 with *P. Vivax*, n=23 *P. falciparum*)	Day 28: 89% in *P. vivax* group; 96% in *P. falciparum*	DB289 was very well tolerated.	All adverse events noted were mild and self-resolved.
Nagelschmitz *et al.*, 2008, Germany	Open randomized controlled trial	none	21-44	Artemisone single (10-80 mg) and multiple (40-80 x 3 mg) n = 24		Artemisone was well tolerated.	No serious adverse events and clinical laboratory tests were normal
Mueller *et al.*, 2004, 2008 Germany, Congo	Randomized controlled trials	*P. falciparum*	≥18	*Artemisia annua* tea 5-9 g/L daily for 7 days	Day 7: 70-74%	Tea preparations were well tolerated.	No significant adverse events were noted

* CCRAS=Central Council for Research in Ayurveda and Sidhha.

** Entries between semicolons refer to different groups of patients receiving different treatments. Order of groups is same in each column.

*** Percentage of patients clear of parasites by specified day. Data were updated from [896].

Table 57. Old and new compounds with greater than 70% inhibition activity against *Plasmodium falciparum* at 1 μM

	Compounds	Current in use	Anti-malri-al activity		Compounds	Current in use	Anti-malri-al activity
1	3,7-dihydroxyflavone			37	emetine	natural product: antiamebic,	
2	amphotericin B	antifungal	(33)	38	gambogic acid	antiproliferative	
3	acivicin	antibiotic, antineoplastic	(34)	39	gentian violet	natural product: antiinflammatory	
4	aclacinomycin A1	antineoplastic		40	heudelottin C	antibacterial, anthelmintic	(42)
5	acriflavinium hydrochloride	antiinfective		41	homidium bromide	natural product	
6	actinomycin D	antineoplastic	(23)	42	hycanthone	antiprotozoal	
7	aklavine hydrochloride	antibiotic, antineoplastic		43	hydroquinidine	anthelmintic, antischistomal	
8	alexidine hydrochloride	antibacterial		44	hydroxychloroquine	antimalarial	
9	amodiaquine	antimalarial	(14)	45	hydroxyprogesterone	antimalarial, lupus suppressant	(13)
10	angolensin	natural product		46	lasolacid sodium	progestogen	
11	anisomycin	antiprotozoal, antifungal	(24)	47	lycorine	antibiotic	(43)
12	atovaquone	antipneumocystic, antimalarial	(18)	48	mefloquine	natural product: mucolytic	(44)
13	avermectin B1 (ivermectin)	antibiotic, antithelmintic	(35)	49	methotrexate	antimalarial	(17)
14	azlocillin	antibiotic		50	methylbenzethonium	antineoplastic, antirheumatic	(25)
15	bebeerine	antimalarial, muscle relaxant		51	chloride	topical antiseptic	
16	benzalkonium chloride	topical anti-infective		52	mitomycin C	antineoplastic	
17	benzethonium chloride	topical anti-infective, antiseptic		53	mitoxantrone	antineoplastic	
18	cadmium acetate			54	monensin sodium	natural product: antibiotic	(43)
19	celastrol	antineoplastic, antiinflamatory	(36)	55	pararosaniline pamoate	anti-Schistosomal	
20	cetrimonium	topical antiseptic, disinfectant		56	pentamidine	anti-trypanasomal, antiprotozoal	(26,27)
21	cetylpyridinium chloride	topical antiinfective		57	perhexiline maleate	coronary vasodilator	
22	chloroquine	antimalarial		58	propafenone	antiarrhythmic	
23	chlorprothixene	antipsychotic, antihistamine		59	puromycin	antineoplastic, antiprotozoal	(24)
24	ciclopiroxolamine	topical antifungal		60	pyrimethamine	antimalarial	(13)
25	cinchonidine	antimalarial	(16)	61	quinacrine	antimalarial	(20)
26	cinchonine	antimalarial	(15)	62	quinidine	antiarrhythmic, antimalarial	(12)
27	coralyne chloride	natural product		63	quinine	antimalarial	(11)
28	cycloheximide	antibiotic	(23, 24)	64	rhodomyrtoxin B		
29	cyclosporin A	natural product: immunosuppressant	(37-39)	65	rutilantinone	antibiotic, anti-neoplastic	
30	cytochalasin R		(40, 41)	66	salinomycin	antibiotic	(45)
31	deoxygedunin	natural product		67	selamectin	veterinary antiparasitic	

32	dequalinium chloride	antibacterial		68	suloctidil	vasodilator	
33	dihydroartemisinin	antimalarial, antiinflammatory	(19)	69	tannic acid	nonspecific enzyme/receptor blocker	
34	dihydrocelastrol derivative	immuosuppressive,		70	tetrandrine	calcium/potassium channel blocker	(21,22)
35	dihydroergotamine	antiinflammatory		71	thimerosal	antiinfective, preservative	(46)
36	dihydrogambogic acid	vasoconstrictor, antimigraine		72	thioridazine	antipsyochtic	
		natural product			tilirone	antiviral	

Author manuscript; available in PMC October 8, 2006. Published in final edited form as: *Chem Biol Drug Des. June 2006; 67(6): 409–416.*

8.4.4. New Antimalarial Therapies from among Known Drugs

The current state of antimalarial chemotherapeutics is particularly bleak for those living in malaria endemic regions of the world because of the low economic incentive for drug development and the rise of resistant strains. Chloroquine-resistant strains of *P. falciparum* are now common in most malaria endemic regions, exacerbating the need for novel, cheap alternatives. Unfortunately, the success rate for new chemotherapeutics to move into clinical use is extremely low. This is true for any new drug; however, the situation is significantly worse for antimalarial therapeutics. The need to discover and develop new antimalarial therapeutics is overwhelming. The annual mortality attributed to malaria, currently approximately 2.5 million, is increasing due primarily to widespread resistance to currently used drugs. One strategy to identify new treatment alternatives for malaria is to examine libraries of diverse compounds for the possible identification of novel scaffolds. Beginning with libraries of drug or drug-like compounds is an ideal starting point because, in the case of approved drugs, substantial pharmacokinetic and toxicologic data should be available for each compound series. A high-throughput screen of the MicroSource Spectrum has been used for screening Killer Collections, a library of known drugs, bioactive compounds, and natural products. Our screening assay identifies compounds that inhibit growth of *P. falciparum* cultured in human erythrocytes.

Because the cost of creating, maintaining, and screening large compound libraries is high, we have chosen the alternative approach of screening a smaller, focused compound library. Weisman and his colleagues used a high-throughput cell-based assay that quantifies parasite growth cultured in human erythrocytes to screen the MicroSource Spectrum and Killer Collections [919]. Together, these make up a library of 2160 known drugs, bioactive compounds, and natural products. This approach aims to leverage the extensive study that has already been performed with the known drugs to ensure their viability as human therapies. Another potentially advantageous aspect to screening this library of diverse compounds is the possibility of identifying novel scaffolds to optimize that are entirely unrelated to known antimalarial drugs. This is particularly important in light of a recent study that highlights the paucity of new antimalarial compound development around novel classes of compounds [920]. Identification of new targets and pathways are a vital aspect of antimalarial discovery in which academia must play a key role [921].

More than 50% of the compounds in this library exhibited some amount of malaria growth inhibition, presumably due to the nature of the library. About 103 compounds inhibited both 3D7 and W2 *P. falciparum* strains, making a total of 251 compounds for further investigation. This subset of compounds was screened again at lower concentrations of 1 μm and 100 nm. Seventy-two compounds exhibited significant activity (defined here as >70% growth inhibition relative to control) at 1 μm and 19 compounds exhibited significant activity at 100 nm. The 72 compounds with >70% growth inhibition relative to control at 1 μm were flagged for further investigation (Table 57). Literature searches were conducted to determine current therapeutic uses, toxicity information, and any prior identification of antimalarial activity. Thirteen of the 72 active inhibitors of *P. falciparum* were previously known antimalarial therapeutic agents. These compounds served as an excellent internal validation of our screening assay, as we were blinded to the identities of the compounds.

The known antimalarial therapeutics we identified was all active at 100 nm and included: quinine, quinidine and hydroquinidine, chloroquine and hydroxychloroquine, amodiaquine, cinchonine, cinchonidine, mefloquine, atovaquone, pyrimethamine, dihydroartemisinin, and quinacrine. Six more members of the MicroSource Collection inhibited *P. falciparum* growth at a concentration of 100 nm. These compounds have been studied previously and confirmed to have antimalarial activity: tetrandrine, actinomycin D, anisomycin, puromycin, methotrexate, and pentamidine. Seventeen additional known inhibitors of *P. falciparum* were identified with significant activity at a concentration of 1 μm. To the knowledge, 36 compounds have not been studied previously for antimalarial activity and inhibited *P. falciparum* growth at 1 μm. Nineteen of these compounds were known therapeutics. This set included known antibiotics, topical anti-infective, natural products, a known antitrypanasomal, antineoplastics, vasodilators, and antipsychotics. Effective 50% inhibitory concentration values ($EC_{50}s$) of these novel antimalarial inhibitors are listed in Table 58.

In conclusion, total 36 novel inhibitors of *P. falciparum*, of which 19 are therapeutics, have been identified and five of these drugs exhibit effective 50% inhibitory concentrations within similar ranges to therapeutic serum concentrations for their recently indicated uses: propafenone, thioridazine, chlorprothixene, perhexiline, and azlocillin. The findings we report here indicate that this is an effective strategy to identify novel scaffolds and therefore aid in antimalarial drug discovery efforts [919]. However, the mode of action of the novel inhibitors described in this study is a question that has not been addressed. While this is beyond the scope of this study, it is extremely important, and we are currently following up on the top leads. Along these lines, questions of protein binding, drug resistance, and combination activity are all avenues we will explore, as they will illuminate the relevance of future study on optimization of these lead compounds.

Table 58. Representative EC_{50} values and distribution of MicroSource antimalarial inhibitors with optical, intravenous and oral delivery routes [919]

Compounds	3D-7 (μM)	W-2 (μM)	Compounds	3D-7 (μM)	W-2 (μM)
Topical			Oral		
acriflavinium	0.03	0.04	chlorprothixene	1.7	1.0
benzalkonium	0.2	0.3	dihydroergotamine	3.0	3.0
benzethonium	0.3	0.2	hycanthone	1.9	0.3
cetrimonium	0.9	0.6	hydroxyprogesterone	1.0	5.4
cetylpyridinium	0.4	0.5	perhexiline	1.1	0.6
ciclopiroxolamine	1.2	0.7	propafenone	1.0	0.2
dequalinium	0.01	0.04	thioridazine	2.6	1.9
methylbenzethonium	0.3	0.2			
tannic acid	1.6	1.3			
Intravenous					
aclarubicin (aclacinomycin A1)				0.4	0.4
azlocillin	5.1	2.5	hydroxyprogesterone	1.0	5.4
dihydroergotamine	3.0	3.0	mitoxantrone	0.1	0.1

Chem Biol Drug Des. Author manuscript; available in PMC October 8, 2006.

8.5. Efforts of WHO/TDR, MMV, and WRAIR on Drug Research

Between 1975 and 1999, only four of almost 1,400 new drugs developed worldwide were antimalarials, and all were at least in part the products of publicly funded research. The pharmaceutical industry largely disengaged from research and development of new drugs for malaria (and other diseases of the poor) in the 1970s because such drugs offered little potential return on investment [922]. This timing was particularly unfortunate because pharmaceutical science was on the brink of major advances and malaria drug development, in large part, missed out on the application of these advances. On the positive side, a sense of opportunity has been heightened by recent decoding of the *Plasmodium* and *Anopheles* genomes. Malaria and other "neglected diseases" (including tuberculosis, filariasis, trypanosomiasis, leishmaniasis, dengue, to mention but a few) - unlike AIDS - are rarities in high-income countries. The world's poor cannot providentially benefit from new drugs developed for wealthier markets. In the case of antimalarials, the only real commercial market is travelers (whose needs differ greatly from those living in malaria-endemic areas). That industrialized market totals only US$200-300 million per year, well below the radar screen for industry.

The market for innovative new drugs for neglected diseases is further depressed by poor regulatory infrastructure in many countries, and by competing counterfeit drugs. The net effect is that about 10% of global drug R&D resources are directed at diseases accounting for 90% of the global disease burden [923]. Malaria is among the most poorly resourced diseases.

As long as malaria persists as a global health problem, new drugs to treat and prevent it will be needed to replace the old ones as they lose effectiveness. Just a few years ago, there were few new drugs in the pipeline and antimalarial resistance was rising. The situation was desperate. It is now healthier than it has been for many decades, with several new combinations and entirely new classes of drug under development. The continued investment in basic science by agencies such as the U.S. National Institutes of Health, the European Commission, and the Wellcome Trust has provided the scientific underpinnings for these developments. But more significantly, genuine progress has resulted from the creation of the Medicines for Malaria Venture (MMV, a public-private partnership devoted to malaria drug development), the continued activities of the Walter Reed Army Institute of Research (WRAIR) in the United States, and the drug development efforts of the World Health Organization (WHO) Special Programme on Research and Training for Tropical Diseases (TDR). These organizations and their partners form what is basically a single international network of collaborators in malaria drug development (in contrast to the competitive character of most profit-driven drug development). This is cause for optimism, but current funding is still very modest and inadequate to complete development as products move downstream [42].

8.5.1. Efforts of WHO/TDR in New Drug Discovery and Development

WHO/TDR is defined as the UNICEF/UNDP/World Bank/WHO Special Programme for Training and Research in Tropical Diseases. WHO/TDR's engagement in antimalarial drug development (as well as drugs for other target diseases) began more than 25 years ago. Its funding for malaria drug development historically ranges around US$2-4 million per year. Although WHO/TDR transferred antimalarial program to the Medicines for Malaria Venture (MMV) in November 1999, WHO/TDR is still supporting some antimalarial drug discovery and development. Early on, WHO/TDR worked with WRAIR and pharmaceutical companies to develop mefloquine and halofantrine. More recently, it has collaborated with pharmaceutical partners to develop injectable artemether and injectable arteether for the treatment of severe malaria. WHO/TDR also sponsored the application for U.S. Food and Drug Administration approval of rectal artesunate in severe malaria (application pending). Other recent accomplishments and ongoing work include:

- TDR worked with the pharmaceutical firm Hoffman-La Roche to develop a more inexpensive way to synthesize mefloquine, and sponsored more than 12 clinical research studies in Latin America, Zambia, and Thailand, leading eventually to registration.
- Introduce new antimalarial drug, artemisinin, to the Western countries. At an early stage by TDR, helped to pave the way for significant breakthroughs in malaria treatment with the artemisinin and its derivatives.
- TDR research confirmed the advantages of combination therapy (CT) and artemisinin-based combination therapy (ACT) and helped define appropriate unit-dosing and packaging of drug combinations
- Mefloquine-artesunate, a proven combination as a non-fixed treatment regimen, is being developed as a fixed-dose combination. A consortium involving the Drugs for Neglected Diseases Initiative (DNDi), TROPIVAL (Bordeaux), and Far Manginhos of Brazil are involved with EU funding, and some WHO/TDR technical support.
- Gaining regulatory approval of chlorproguanil-dapsone in collaboration with GlaxoSmithKline (GSK)
- Developing a fixed-dose combination of chlorproguanil-dapsone plus artesunate with GSK and MMV
- Working with Novartis to extend Coartem use in children down to 5 kg and to develop appropriate, user-friendly, and informative packaging
- A new quinoline antimalarial, similar to amodiaquine, being developed by GSK, and the University of Liverpool with support from MMV, and some technical assistance from WHO/TDR. This compound resembles amodiaquine but potentially has an improved safety profile, and is highly active against chloroquine-resistant strains. This compound may be registered by 2008.
- Fosmidomycin, a new class of antimalarial, is under development by Jomaa Pharmaceuticals in Germany. This also may be used in combination with other agents. The major issue is whether it can be given as a 3-day regimen.
- A new class of synthetic endoperoxide, invented at the University of Nebraska, and initially supported by WHO/TDR, is now being developed by a consortium funded and managed by MMV and is licensed to Ranbaxy of India. A compound entered into full preclinical development late in 2003. The advantage of these compounds is that they have longer half-lives than the artemisinins, and so could serve as better

partners than artemisinins for longer half-life drugs. A compound in this category could be registered by 2008.

- Pediatric Coartem formulation by Novartis in collaboration with MMV, and TDR (MMV funded).
- Recently, new strategy of chemotherapy in children malaria has been issued by WHO/TDR. A rectal application of the inexpensive antimalarial drug artesunate could save the lives of many people who develop severe malaria who live in the world's remotest locations, e.g., rural Africa and Asia.

TDR is an independent global program of scientific collaboration that helps coordinate, support and influence global efforts to combat a portfolio of major diseases of the poor and disadvantaged. Established in 1975, TDR, is sponsored by the United Nations Children's Fund (UNICEF), the United Nations Development Programme (UNDP), the World Bank and the World Health Organization (WHO). The goal of WHO/TDR is to have the priority setting, research and development led and managed by scientific leaders in the countries where the diseases and problems occur. We believe this is a sustainable way of not only creating these tools, but making sure that they are distributed, used, and truly owned by the communities they can help. We have identified a specific set of research goals designed to meet the needs of people in these countries.

Health Research is increasingly seen as critical for poverty alleviation and for achieving the Millennium Development Goals. The UNICEF/UNDP/World Bank/WHO Special Programme for Research and Training in Tropical Diseases (TDR) was created in 1975 to support the development of new tools to fight tropical diseases of poverty and to strengthen the research capacity of affected developing countries. TDR has been effective in delivering its objectives and is proud of its achievements. However, the research environment has changed significantly over the last decades in part due to TDR's efforts:

- the epidemiology of infectious diseases is changing with some diseases moving to elimination and others emerging or re-emerging;
- here are many new initiatives and actors in the field providing new momentum but also leading to a more complex environment;
- disease endemic countries have enhanced research capability but are increasingly left behind in global research planning and priority setting;
- priority research needs are unequally covered and there remain several research areas that are neglected even though they are critical for the ultimate health impact of the global research effort.

TDR's new vision and strategy responds to the new research environment and to the need to make the collective global research effort more effective and responsive to research priorities in disease endemic countries. It also recognizes the need for these countries to play a major role in research and priority setting to ensure relevance, sustainability, and optimal health impact for the poor.

Vision and ten year strategy of WHO/TDR was issued on the achievements over the last 30 years; our work is based now on three major strategic directions:

1) *Stewardship* for research on infectious diseases of poor populations: a major new role for TDR as facilitator and knowledge manager to support needs assessment, priority setting, progress analysis and advocacy, and to provide a neutral platform for partners to discuss and harmonize their activities.

2) *Empowerment* of researchers and public health professionals from disease endemic countries (DECs), to provide support for training and research, and to build leadership at individual, institutional and national levels.

3) *Research on neglected priority needs* that are not adequately addressed by other partners. This focuses on three research functions:
    a.   Innovation for product discovery and development
    b.   Research on how interventions are used in real life settings
    c.   Research to increase access to interventions

TDR's strategy is operating and being implemented through eleven business lines (BLs), each supported by a robust business plan that details deliverables, timelines, milestones, and partnerships. TDR's efforts on drug discovery and development are described here.

1) Mefloquine — A Counterattack on Parasite Resistance

Mefloquine is another significant anti-malarial that TDR was involved in developing. Mefloquine became available at a critical time when other drugs were encountering parasite resistance and artemisinin was not yet widely available. The compound was originally discovered by the US-based Walter Reed Army Institute of Research (WRAIR). But as mefloquine was not covered by a patent and was expensive to produce, there was little initial interest in development. TDR, however, worked with the pharmaceutical firm Hoffman-La Roche to find a more inexpensive way to synthesize the drug, and sponsored more than 12 clinical research studies in Latin America, Zambia, and Thailand, leading eventually to registration. Although mefloquine-based therapies were later superseded by lower-toxicity artemisinin-derived compounds, mefloquine is still used widely as a prophylactic by tourists and travelers to malaria-endemic regions. The development of mefloquine was also an example of the intensive and fruitful collaboration that occurred in this first phase between TDR and WRAIR, which yielded a variety of innovations.

2) Paving the Way for New Malaria Drugs – Artemisinin

Malaria remained the biggest killer worldwide, and it was here that a new initiative, led by Chinese scientists and supported at an early stage by TDR, helped to pave the way for significant breakthroughs in malaria treatment. This initiative was research into the anti-malarial properties of the indigenous Chinese plant known as qinghao or sweet wormwood (*A. annua*). The plant had been used historically in traditional Chinese medicine, and Chinese researchers had identified its active compound, artemisinin, as potentially effective against parasites. Interest in new antimalarials was high as parasite resistance was developing against most other available drugs, some of which had substantial side-effects. TDR, whose far-flung networks already included collaborations in China, would be among the first international institutions to dispatch scientists to China's artemisinin research facilities, appreciate the value of the endeavor, and transmit that to colleagues elsewhere.

Research cooperation into the properties of artemisinin, in an era when the Cold War was still a dominant feature of international politics, involved some delicate diplomacy. But overcoming the obstacles was a shared scientific quest — a better treatment for one of the world's deadliest diseases and a modern use for an ancient Chinese herbal remedy. "The Chinese scientists who were working on artemisinin contacted the malaria section of WHO," recalls Lucas. "They were very anxious to have the drug registered and widely distributed. We said it needed more workup, including pre-clinical laboratory testing of toxicity. Part of that was done through the TDR network."

TDR helped to get researchers outside of China involved in artemisinin and get it on the research agenda.

"In 1979," relates Dr. Wallace Peters, former Chairman of TDR's Steering Committee on Drugs for Malaria (CHEMAL), "...we went to Beijing to meet the Chinese researchers who doing artemisinin research and development at the government's invitation. We visited the labs at the Center for Traditional Medicine and at the Second Military College in Beijing. The Chinese were very open. They were making it and growing it, and wanted to supply and sell it. It was just at a time, after the Cultural Revolution, when the Chinese wanted more dealings with the rest of the world, and wanted international recognition. China recognized that the toxicity standards were not detailed enough and requested assistance. CHEMAL supported inspection of the production plant by the US Food and Drug Administration (FDA) in 1982.

Training fellowships were offered to the Chinese ... from the western regulatory point of view, there were big gaps in the Chinese toxicity and efficacy studies ... but China wanted the drug sold and used, and was uncomfortable about TDR taking over this development work." Although TDR fully recognized the rights of the Chinese to develop artemisinin, validating the safety and efficacy of a drug with potential to improve malaria treatment worldwide required a broader testing effort. TDR sought to obtain a kilogram of purified artemisinin to allow the substance to be tested through its network of partner laboratories. When obtaining even a kilogram of purified artemisinin from Chinese sources proved difficult, Lucas decided to look elsewhere.

TDR commissioned researchers at Mississippi University to grow the plant *A. annua*, and extract the sought-for kilogram of the purified artemisinin. This proved expensive as the yield was quite low and considerable acreage had to be cultivated. However, shortly after TDR had commissioned this work, a kilogram of artemisinin from Chinese collaborators did indeed arrive. By 1992, the first oral artemisinin-based combination therapy (ACT) of artemether and lumefantrine was registered in China. In 1994, Novartis formed a collaborative agreement with the Chinese Academy of Military Medical Sciences (AMMS), Kunming Pharmaceutical Factor (KPF) and the China International Trust and Investment Corporation (CITIC) for further development of the same combination, eventually leading to international registration of Coartem® and Riamet® [924, 925].

3) New Drug Treatment Policy for Malaria – Drug Combination

The chemotherapy equivalent of FIELDMAL was CHEMAL, the steering committee that funded TDR research on malaria chemotherapy. Chaired by Nobel prize winner Professor Gertrude Elion, and subsequently by Professor Dyann Wirth, it was at this committee's 1989 special meeting in Beijing that Chinese researchers first presented their results on artemisinin derivatives, and the dossier for registration of artesunate and artemether in China, therein describing what some regarded as the most significant advance in the treatment of malaria since the entry of quinine into the British Pharmacopoeia in 1677. The next 10 years were dominated by research on these artemisinin derivatives, principally in South East Asia.

The artemisinin derivatives were unprotected by patent and therefore not of wide interest to industry. Consequently, CHEMAL funded the necessary research and development to improve the understanding of the efficacy and safety of various artemisinin derivatives, their mechanism of action and toxicity. In addition, CHEMAL supported drug development and partnerships to register these drugs in developed countries. Together with Kunming Pharmaceutical Factory (KPF) and the French-American pharmaceutical firm Rhone-Poulenc Rorer (now Sanofi-Aventis), CHEMAL would foster the development of injectable artemether for severe malaria, leading to its registration in France on a named-patient basis in 1996, and elsewhere soon thereafter. TDR also collaborated with the Dutch company Artecef and the Walter Reed Army Institute of Research (WRAIR) to develop a second injectable artemisinin derivative for severe malaria, beta-arteether (artemotil®), which was registered in 2000. Particularly in Asia and South East Asia, artemisinin derivatives had rapidly become a popular treatment therapy. Vietnam, still isolated from the global market economy (although this would change rapidly over the course of the decade) had the highest levels of severe malaria in its history, and began producing artemisinin compounds in oral and suppository formulations and distributing them widely within the country in community-based programs.

It was here in South East Asia in the mid-1990s that TDR and its partners would undertake field research on to how to improve the use and distribution of antimalarials in rural areas. The research was led by TDR's Anti-Malaria Task Force (ANTIMALS), in a small but innovative series of trials. Some of the first antimalarial drugs in easy-to-administer unit-dose 'blister packs' came out of this initiative. Such innovations set the stage for broader advocacy of community-based and home-based malaria treatment models, particularly in Africa. A few years later, the approach became a key component of WHO global strategies on malaria. Effectively, unit-dose packaging was the breakthrough that empowered schools, community workers and, most of all, mothers and caregivers to administer life-saving treatment to children without relying on hard-to-access health centers. The ANTIMALS research also led to a subsequent TDR decision to develop for formal registration one popular artemisinin derivative, artesunate (in a rectal formulation), as a pre-referral treatment for severe malaria in children.

While the new artemisinin-based drugs were increasingly popular, concerns were also growing that widespread use of such compounds on their own could foster parasite resistance that would undermine the impact of a remarkable drug discovery. The most common artemisinin derivative, artesunate, was widely sold as a monotherapy in Asia, and seldom regulated. The development of ACTs thus emerged as a global public health priority. In the mid and late 1990s, TDR would lead large-scale, multi-center clinical trials with various international partners to demonstrate the fundamental efficacy of oral ACTs. Trials in South East Asia funded by the Resistance and Policies (RAP) Task Force documented how combination therapies, using two or more drugs at the same time, could improve cure rates and reduce the resistance of malaria parasites to any one drug formulation. TDR-sponsored studies in the mid and late 1990s tested a wide range of ACTs including: amodiaquine plus artesunate; artemether plus lumefantrine (Coartem®); chloroquine plus artesunate; and mefloquine plus artesunate. Ultimately, this research provided an evidence base upon which WHO was able to promote use of ACTs as a core component of malaria treatment policy.

A range of public and private agencies across Asia, Europe, and North America were meanwhile moving various artemisinin derivatives (artemether, artesunate, and arteether) towards international registration, alone and in combinations. In 1998, Novartis, in collaboration with KPF and the Chinese Academy of Military Medical Sciences (AMMS), registered the first formal fixed-dose ACT, lumefantrine plus artemether (Coartem®) in Europe. The drug is available at a preferential price to developing countries under an agreement negotiated by WHO. Oral, fixed-dose ACTs are the fastest-acting anti-malarial available today − destroying parasites in approximately 48 hours on average − with high documented cure rates. Following the initial registration of Coartem®, TDR research helped define appropriate unit-dosing and packaging. And in the program's third decade, TDR helped demonstrate the efficacy and safety of Coartem® for infants over 5 kg in weight, leading to the formal extension of the label in 2004 for such use.

Although cases of malaria in Africa continued to rise markedly throughout the 1990s, the development of a new generation of antimalarials undoubtedly contributed to strategies for halting what would have been even sharper increases in malaria mortality. ACTs can at least blunt the development of parasite drug resistance, effectively buying more time for the development of future disease- and vector-control tools, strategies, and innovations.

4) Drug Discovery Research — Facilitating Global Networks

The huge advances toward understanding the vector of pathogen biology would, in turn, shape and influence TDR discovery research strategies. The challenge is to harness that knowledge to the search for novel lead compounds that can form the basis for innovative disease treatments. While PPPs and other partners assumed greater responsibility for the development of particular drugs and diagnostics for specific diseases, TDR would cast its net into the sea of compounds that had so far not been explored, searching broadly but systematically for novel leads. Alongside traditional whole-parasite screening techniques, this effort would harness the new tools of genomics, combinatorial chemistry and robotics to full advantage. Three decades earlier, the TDR network approach had proved itself with ivermectin. It was a TDR screen of the drug's efficacy in cattle - innovative at the time - that confirmed the potential efficacy of ivermectin against human onchocerciasis. Now, the TDR compound screening network was revitalized and expanded to include a broader range of academic and research institutes, and also industry partners.

Along with this, new research networks for medicinal chemistry, pharmacokinetics, drug target portfolios and helminth drug discovery were created to cover other stages of the drug discovery process in a more integrated manner. TDR's development of formal collaborations for drug discovery with industry has greatly enhanced the capacity of these networks. The high-tech, automated laboratories of industry can screen hundreds or thousands of compounds simultaneously for activity against a target protein or enzyme. Major industry collaborators have opened their vast medicinal chemistry libraries (for example, those at Pfizer, Merck-Serono, and Chemtura) to the research efforts. And certain pharmaceutical firms are training developing country scientists in their laboratories, under TDR auspices. This will help build capacity and foster scientific leaders in the countries where neglected tropical diseases pose the greatest burden. TDR's discovery research program has also supported the creation of a new global research tool, a Drug Target Database, to facilitate research on potential drug targets.

5) Home Management of Malaria - Diagnosis and Treatment

Implementation research on home and community-based management of malaria became another key element of TDR's program in its third decade, continuing into the fourth. Home and community management involves the training of mothers, drug vendors, village volunteers, and teachers, in the first line of care for malaria when health clinics and health care providers are not accessible. The effectiveness of home management has been demonstrated by TDR over the past five years, reducing mortality by 40% or more in some studies [925]. Further research is now underway to determine whether more complex artemisinin-based combination therapies (ACTs) can also be administered at the home and community level. More than 20 studies are examining health outcomes and overall feasibility of ACTs in home management. Preliminary results of a small study in Ghana using Coartem®, the only fixed-dose ACT currently available, show good results for community use. Meanwhile, research has also been initiated to determine whether rapid diagnostic tests for malaria can also be administered at community level by trained caregivers and whether use of diagnostics enhances the effectiveness of ACTs.

6) Rectal Malaria drug Could Save Many Lives in Rural Africa and Asia

A rectal application of the inexpensive antimalarial drug artesunate could save the lives of many people who develop severe malaria that live in the world's remotest locations, e.g., rural Africa and Asia. These are the conclusions of an article first published Online and in an upcoming edition of The Lancet, written by Dr. Melba Gomes, WHO, Geneva, Switzerland, and colleagues from the study 13 Research Group.

Most malaria deaths occur among young children in rural areas. An acute episode of malaria can become so severe that the patient cannot swallow and 'keep down' oral treatment. If patients with severe malaria who cannot be treated orally are several hours away from facilities that can provide injections; rectal artesunate - which acts rapidly on malaria parasites - could be given before setting out on the journey. The authors investigated whether pre-referral treatment reduced mortality and permanent disability.

In Bangladesh, Ghana, and Tanzania, patients with suspected severe malaria who could not be treated orally were allocated randomly to an artesunate (8954 patients) or placebo (8872) suppository (i.e., rectal application).

Most had a drop of blood taken beforehand for later examination, and those with no malaria parasites in it were excluded from the main analysis (as they probably had a viral or bacterial infection, which artesunate cannot affect). For patients who were able to reach the injections clinic within six hours, there was no significant difference in mortality or permanent disability between the two groups. However, in patients still not in clinic after six hours (and half of these were still not there after 15 hours), pre-referral rectal artesunate halved the risk death or permanent disability (29 of 1566 patients for artesunate vs. 57/1519 for placebo).

The authors say: "Death from malaria reflects delay in administration of effective antimalarial treatment. Our results provide strong evidence that if patients with severe malaria cannot be treated orally and referral is likely to take several hours, an immediate rectal dose of artesunate before referral substantially reduces the risk of death or permanent disability." They conclude: "For the foreseeable future, there will be many patients in the community with suspected severe malaria who could, if treated earlier, have been managed orally, but can no longer be treated orally and need immediate help." They say that rectal treatment in the community can offer this.

In an accompanying Comment, Dr. Lorenz von Seidlein, Joint Malaria Project, Tanzania, and Dr. Jacqueline L. Deen, Joint Malaria Project, Tanzania and Vaccine Institute, Kwanakgu, Seoul, Korea, say: "If there are a handful of important papers every decade that will influence the way malaria is treated, this study is one of them...The next important step is to develop widescale deployment strategies, through research, and to assess the effectiveness of artesunate suppositories under various real-life settings." [926].

7) Therapy Strategy to Reduce Children Death from Malaria

Washington D.C.: December 2nd, 2008. Population Services International (PSI) is launching a new comprehensive project that could dramatically reduce the number of African children dying every year from malaria. The program will provide free artemisinin-based combination therapies (ACT) given by local community health workers to poor children suspected of having malaria. This quick-response approach to treatment will be monitored to determine if it reduces the number of African children dying from malaria. It is estimated that this program will avert half a million malaria cases and 11,000 child deaths in each of the four pilot areas.

The project is funded by the Canadian International Development Agency (CIDA) and led by the non-profit organization PSI in collaboration with TDR, the Special Programme for Research and Training in Tropical Diseases. TDR is supported by UNICEF, United Nations Development Program, the World Bank and the World Health Organization. Four African countries will be selected based on a range of criteria, including high malaria burden and the existence of community health networks, through which free anti-malarial medicines and associated health promotion communications will be delivered.

What makes this program groundbreaking is that the effectiveness of the home-based approach for delivering ACT, on child mortality has never before been monitored. ACT is not yet widely available, particularly for the most poor and vulnerable people. By measuring the health impact of home-based management of malaria (HMM) with ACT over the next 2-3 years, PSI and TDR believe that the results of the project will serve as a powerful catalyst for the adoption and expansion of these programs across sub-Saharan Africa. The majority of the 1 million people who die of malaria every year are African children under the age of five. Malaria can progress quickly in children, with as little as 48 hours between the onset of symptoms and death. Preventing malaria-related death in endemic areas involves ensuring that children are rapidly treated with an effective anti-malarial, and that they comply with the full course of treatment.

Now that most endemic countries have switched their drug policies to highly effective artemisinin-based combination therapies, the primary constraint to effective treatment in children is access at community level, especially as almost half of all malaria cases in Africa are treated at home. This problem was recently highlighted in the World Malaria Report 2008 from the WHO, which states, "The procurement of anti-malarial medicines through public health services increased sharply between 2001 and 2006, but access to treatment, especially of artemisinin-based combination therapy, was inadequate in all countries surveyed in 2006."

PSI is one of the world's leading malaria control agencies. Since January 2005, PSI has delivered over 40 million insecticide-treated mosquito nets and over 12.5 million doses of anti-malarial medication. As the international community nears the deadlines for the Millennium Development Goals and Abuja Targets, PSI stands ready to accelerate the scaling up of these interventions to increase and sustain malaria control activities in 32 malaria endemic countries. TDR, the Special Programme for Research and Training in Tropical Diseases, is a global program of scientific collaboration established in 1975, sponsored by the

United Nations Children's Fund, United Nations Development Programme, the World Bank and the World Health Organization. Its focus is research into diseases of the poor -- both improving and developing new approaches, and expanding research capacity in the countries where the diseases are prevalent.

Child survival is a priority for CIDA. Since malaria is the leading cause of death in African children under five, Canada has responded accordingly. Canada has been the largest single country donor in the distribution of free long-lasting bed nets to poor and disadvantaged children under five and pregnant women since 2002 till 2007. CIDA is now investing $20 million in free life-saving artemisinin-based combination therapies (ACT) delivered by community health workers in the home, and evaluating whether this approach is effective in reducing child mortality.

8) TDR Business Plan 2008-2013

The multiplicity of new participants provides new momentum but also leads to a more complex research environment. The views of researchers and policy makers in disease-endemic countries themselves are still under-represented in global priority setting. Research needs are unequally covered and certain critical research areas remain neglected.

More than ever, stewardship is needed to bring diverse organizations and interests together in coherent, operational research programs that are supported by − and inform the decisions of − both disease endemic countries and global health agencies.

More than ever, the global public health sector needs to provide leadership for a coherent approach to priority setting, so that the most serious infectious disease threats are addressed and the needs of the most vulnerable populations are not overlooked. A public sector 'convenor' of disparate groups and interests is essential to ensure independent analysis, evidence-based guidance and meaningful involvement of disease-endemic countries in research that can make a difference to health policy as well as to actual practice in homes and communities. As the main UN agency for research in tropical diseases, TDR has played this kind of role for over three decades. In the coming decade, TDR will scale up its activities significantly in these areas, in respond to the changing health and research environment.

Health research is increasingly seen as critical for poverty alleviation and achieving the Millennium Development Goals. The Special Programme for Research and Training in Tropical Diseases (TDR) created in 1975 to support the development of new tools to fight tropical diseases of poverty and to strengthen the research capacity of affected developing countries, has made a significant contribution to this goal. However, the research environment has changed significantly over the last decades: (i) the epidemiology of infectious diseases is changing with some diseases moving to elimination and others emerging or re-emerging, (ii) there are many new initiatives and actors in the field providing new momentum but also leading to a more complex environment; (iii) disease endemic countries have greater research capability but are increasingly left behind in global research planning and priority setting; (iv) priority research needs are unequally covered with several research areas neglected despite their critical nature. In order to respond to these opportunities and challenges, TDR, through consultations with its stakeholders has developed a renewed vision and strategy.

In order to implement this strategy, TDR will restructure its operations to a limited number of business lines (BLs), each supported by a robust business plan that details deliverables, timelines, milestones, and partnerships. Gender will be mainstreamed into these plans. Their introduction will provide the necessary focus to achieve TDR's objectives and also ensure the desired accountability. Specifically, TDR proposes to introduce eleven business lines in the 2008-2009 biennium based on stakeholder consultations, scientific opportunities in the field and opportunities arising from TDR's current portfolio. Two correspond to the strategic functions of Stewardship (BL1) and Empowerment (BL2) that are core to the TDR strategy. The other nine correspond to the strategic function of Research on Neglected Priority Needs and may change over time. These include Lead discovery for drugs (BL3), Innovation for product development in DECs (BL4), Innovative vector control interventions (BL5), Drug development and evaluation for helminthes and other Neglected Tropical Diseases (BL6), Accessible quality assured diagnostics (BL7), Evidence for treatment policy of HIV and TB co-infection (BL8), Evidence for antimalarial policy and access (BL9), Visceral leishmaniasis elimination (BL10), and Integrated community-based interventions (BL11).

While Stewardship and Empowerment business lines span across all upstream and downstream research areas, the other nine, supporting Research on Neglected Priority Needs, have varying levels of upstream/downstream focus with an increasing overall emphasis on downstream research. Some are functionally specific, while others are focused on specific diseases. From a geographic perspective there will be a strengthened focus on research in DECs with an emphasis on Africa. The scope of the business lines will be reviewed annually by TDR's Scientific and Technical Advisory Committee using clearly defined criteria to ensure optimal use of resources and continued relevance. This review will also allow different business lines to enter and exit the portfolio.

TDR will be evaluated on the impact it creates against each of its three strategic functions. Given the long term nature of its research activities, it is proposed that the impact created by TDR be evaluated over two horizons. In 5 years time (2012), TDR would be evaluated on the overall impact it has created on three specific dimensions, namely: (i) Harmonization of research efforts as an indicator of Stewardship; (ii) Disease endemic country leadership in health research as an indicator of Empowerment; and, (iii) Enhanced access to superior interventions as an indicator of Research for Neglected Priority Needs. This evaluation would entail a qualitative survey of key TDR stakeholders and a comprehensive quantitative review of all research outputs produced by TDR. In the interim, TDR would be evaluated annually on fifteen supporting quantitative impact indicators, which relate to the long term overall impact, and link to business line project milestones. Thus the system will cascade to the eleven business lines and can also be used to monitor their respective progress.

Operationalizing the strategy will require TDR to increase its annual budget of US$ 50 million in 2007 to US$ 80 million in 2013, an increase of ~8% per year which is modest but necessary if TDR is to meet its vision statement and foster effective global research and empowerment efforts in a professional manner. While TDR will attempt to proactively seek greater funding from the private, philanthropy and nongovernmental organization sectors, it is planned that the governmental and international public sector will still provide the majority of resources (70%). Should TDR be unable to mobilize the targeted resources, it plans to prioritize activities both across and within business lines. TDR will operationally restructure its organization to establish clear responsibilities and accountabilities and also foster a sense of entrepreneurship.

Under the revised structure, four functional areas report to the director's office including: (i) Stewardship (BL1), (ii) Empowerment (BL2), (iii) Research on Neglected Priority Needs (BLs 3 to 11) plus, as, an integral part of the organization (iv) a small Portfolio Policy and

Development team will be responsible for metric monitoring and management of an 'innovation fund.' To adequately support these activities, TDR will have to grow its personnel headcount at an annualized rate of 6% from current levels of ~85 to ~120 in 2013. TDR will also strengthen and improve collaboration with all co-sponsoring agencies, notably WHO, including their regional offices. Properly positioned, TDR could become the infectious disease research arm of WHO and its other co-sponsoring agencies. As TDR transitions to this revised operating model, it will ensure adequate emphasis on: (i) coordination and control processes, (ii) mechanisms to ensure accountability, (iii) enhancing capability, and (iv) maintaining motivation of its staff and scientific committees. TDR recognizes the financial, scientific, human resources, external environment and project execution related risks to achieving this business plan and is proactively planning to mitigate them as it transitions to this new model.

The overall emphasis on focused objectives based on analysis and stakeholder consultation, accountability for end-products, optimization of resource requirements, establishment of monitoring mechanisms and an operationalization plan is indicative of TDR's commitment to move towards a more business-like operational model. The anticipated cost-effective research-driven public health impact, with disease endemic countries playing a pivotal role, compares favorably with, and complements, other internationally directed research efforts. TDR believes that this business plan meets the challenges of its exciting new vision and represents a compelling opportunity for investment in research on infectious poverty prevent diseases.

For antimalarial policy and access, WHO estimates that malaria is responsible for over one million deaths per year. These deaths are avoidable using artemisinin-based combination therapies (ACTs), but most malarial fevers are not treated appropriately. Many patients seek treatment too late or do not reach public treatment facilities and obtain health care from the private or informal sector, where they often get inappropriate and poor quality treatment. If Millennium Development Goals 5, 8, and 17 (reducing childhood mortality, halt malaria and provide access to essential drugs) are to be met, there is an urgent need for research to complement existing malaria control policies and strategies through providing evidence for malaria treatment policy and for effective management of malaria at the community level. Several new antimalarial treatments are being developed and increased funds are being made available for DECs to deploy them. Countries need to make policy choices based on objective and comparative evidence on the safety and effectiveness of these antimalarial drugs, and to provide this evidence the different drugs need to be scientifically evaluated under real-life conditions. Research is also needed to determine the usefulness of diagnostics in different epidemiological settings, to improve diagnosis and treatment of malaria to community level and to determine effectiveness of integrated interventions against malaria and other childhood infections at the community level.

To improve access to effective treatment for malaria and case-management of malaria and other childhood infections at all levels of the health system with the aim of reducing childhood mortality with specific objectives:

- To assess the safety and effectiveness of antimalarial drugs in "real-life" conditions of use, at different levels of the health care system, and in high-risk groups including pregnant women and persons infected with HIV.

- To develop a comprehensive package for the diagnosis and treatment of malaria episodes of various degrees of severity at the community level, determining its feasibility, acceptability, safety, and cost-effectiveness.
- To develop an integrated diagnostic and treatment package to be delivered at the community level for the management of malaria and other childhood infections, and measure its impact to reduce mortality in children.

The studies to assess the safety and effectiveness of antimalarial drugs in real life conditions of use will be undertaken by an international consortium, built around the INDEPTH network where the field studies will be undertaken. TDR will be responsible for independent scientific oversight of the studies, and other partner institutions will provide technical support in their area of expertise. The implementation research on home management of malaria and fever will be managed by TDR, and involve multi-country studies to develop and test the effectiveness and health impact of improved home treatment methods and strategies. These studies will be TDR Business Plan 2008-2013.

8.5.2. Efforts of MMV in New Drug Discovery and Development

WHO/TDR's engagement in antimalarial drug development (as well as drugs for other target diseases) began more than 25 years ago. Its funding for malaria drug development historically ranges around US$2-4 million per year. WHO/TDR hosted, matured, and developed an initial portfolio for the Medicines for Malaria Venture (MMV) through 1998-1999 until MMV was established as an independent entity in November 1999.

MMV was established in Geneva in 1999, as a not-for-profit organization with the goal of replenishing the world's portfolio of malaria drugs. As one of the first product development public-private partnerships (PDP) for neglected diseases, MMV also served as a bold experiment in organizational design. It tested whether a "virtual" organization, using philanthropic donations and working in partnership with public and private entities, could succeed in the complex task of developing and registering new drugs. This partnership model has proven enormously successful as MMV has built the largest portfolio of antimalarials in history and is preparing to launch its first set of products. MMV has shown it can solve the problem of market failure in malaria R&D. It has also validated the concept of product development PPPs as vital institutions in the creation of new products for neglected diseases.

In 2000, MMV was a newcomer to the world of antimalarial drug research. More people were dying from malaria than ever before. The malaria parasite had become resistant to widely-used drugs, including two inexpensive medications, chloroquine and sulfadoxine / pyrimethamine. New drugs were desperately needed as malaria continued to afflict countless millions and the death toll in Africa showed an alarming increase. In addition, due to cost, poor health systems, inadequate distribution networks, and policy challenges, existing drugs were often not reaching the poor. The pipeline for new antimalarials was virtually empty. This was to be expected, as a Global Forum for Health Research study had revealed that only 10 % of the world's new drug innovation was targeted at diseases threatening 90 % of the world's population. Motivated by this glaring inequity, and the need to act in the face of a projected public health disaster due to escalating drug resistance, MMV started out modestly with only USD 4 million in its purse and three early-stage projects in its portfolio. However, it was not short on ambition. It aspired to discover and develop at least one new safe, effective, and affordable antimalarial drug before the end of the decade.

MMV started out with the initial goal of developing a single new drug by 2010 and producing an additional antimalarial every five years thereafter. Over the next two years, MMV and its partners could register as many as three *combination* antimalarials, while expanding its pipeline of 40 products in development. MMV has more than 100 R&D partners worldwide, including clinical research sites, academic groups, research institutions, and seven global drug companies. It has raised $323M in commitments from public, private, and philanthropic sources, and has spent around $200M by the end of 2007, building the largest malaria drug portfolio the world has seen. Before 2006, MMV focused exclusively on the development of new antimalarials. However, in recent years the organization has expanded its scope to include issues of access, recognizing the need to remove barriers to the uptake of effective new drugs. MMV understands the importance of partnering with global and country-level health organizations to make drugs accessible to at-risk populations. This shift in emphasis is reflected in MMV's updated mission: DISCOVER, DEVELOP, & DELIVER.

Today, with four new artemisinin-based combination therapies (ACTs) in the last stages of development, MMV is set to exceed that target. The anticipated launch of four new ACTs by 2009 presented a new challenge for MMV. How would we ensure that these lifesaving drugs reach the children and the rural poor, who badly needed them? In response to this question we have added a *deliver* component onto our well-established *discover* and *develop* core functions. We will work to ensure that the drugs expected to soon emerge from our pipeline will swiftly reach patients and have the required health impact.

1)MMV Aspires to Develop in Malaria Chemotherapy

- Antimalarial treatments for USD 1 or less;
- Medicines for high risk groups such as children and pregnant women;
- At least four new ACTs approved by international regulatory authorities before 2010. The first drug could be available for widespread use by 2008;
- A one-dose cure;
- Access strategies to ensure our products reach the vulnerable.

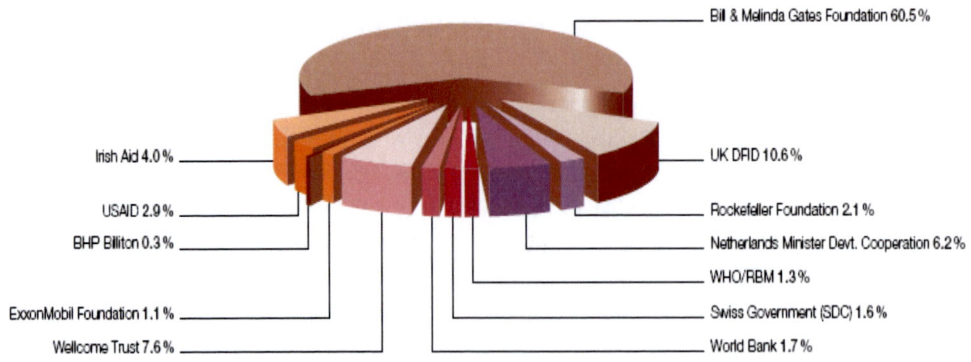

Figure 31. Contributions to MMV, 2000-2010 ($US).

2)MMV: Working in Partnership

Fortunately, the tide is changing. Medicines for Malaria Venture (MMV) was created in 1999, as a public-private partnership to reinvigorate malaria drug research. Acting as a "virtual R&D organization" MMV collaborates with public and private groups to expand and improve the world's arsenal of antimalarial drugs. It currently oversees the largest ever antimalarial drug portfolio. Within the next two years, three new ACTs could gain approval from an international regulatory authority.

MMV's next challenge is to help facilitate the uptake of these medicines to ensure that they make the biggest impact in the population at risk of malaria. This broader remit is reflected in MMV's new aspiration: "Discover, Develop, & Deliver." Our overriding goal is to make the greatest possible public health impact in malaria-endemic countries.

To achieve this goal, MMV follows a set of operating principles:

- *Partnership.* Collaborative working relationships require clarity on roles, responsibilities and goals. MMV's partners are chosen for their technical capabilities, resources and facilities, and their compatibility with the MMV vision.
- *Global access.* R&D partners commit to preferential pricing in disease endemic countries, and often to granting royalty-free rights to associated intellectual property for use in malaria research.
- *Quality.* MMV-funded projects are expected to achieve stringent international regulatory approval and meet WHO pre-qualification standards.
- *Transparency.* MMV employs a range of vehicles to convey accurate and up-to-date information about its work, plans, performance, and finances.

Over 90% of the funding MMV receives is channeled directly into research and associated access activities. By leveraging its resources, pooling its talents and focusing its mission, MMV's performance benchmarks for cost and speed compare favorably not only to other PDPs but also to for-profit industry.

3)Public-Private Partnerships (PDP) and the MMV

Public-private partnerships for drug development are relatively new. While they are still few, and not exclusively focused on neglected diseases of poor countries (some have formed around rare, or orphan, diseases), such collaborations are becoming very important, reviving R&D in areas that have lain fallow too long. A number of administrative arrangements are possible, but the key is tapping into a variety of skills and resources from institutions in the public and private sectors, often in a "virtual" organization. One of the most successful public-private partnerships involved in neglected diseases is the MMV, begun in 1999, and is profiled. A new public-private partnership, the Drugs for Neglected Diseases Initiative (DNDi) has taken on late-stage development of two artemisinin co-formulations as pilot projects (although the long-term focus of the DNDi will be on diseases other than malaria, e.g., African sleeping sickness).

MMV is among the first public-private partnership established to tackle a major global disease. The initiative arose from discussions between WHO (primarily through WHO/TDR), a number of other partners, and the International Federation of Pharmaceutical Manufacturers Associations (IFPMA). Early partners in these exploratory discussions were the Global

Forum for Health Research, the Rockefeller Foundation, the World Bank, and the Swiss Agency for Development and Cooperation, the Association of the British Pharmaceutical Industry, and the Wellcome Trust. The idea was to combine the expertise of the pharmaceutical industry in drug discovery and development, and the public sector, with its depth of expertise in basic biology, clinical medicine, field experience, and above all, its responsibility to the public. MMV was officially launched in November 1999 as a Swiss foundation.

The level of financial contributions to MMV has grown annually, as have expenditures. Donors include the Bill & Melinda Gates Foundation, Rockefeller Foundation, The Wellcome Trust, Swiss Government, U.K. Government, Dutch Government, World Bank via the Global Forum for Health Research, and Exxon Mobil (Figure 31). As well as direct financial support, contributions in kind have been made by all of the corporate partners in specific projects. Total Research & Development annual expenditures (not including in kind support) have been:

- US$ 2,280,748 in 2000
- US$ 6,709,653 in 2001
- US$ 10,353,468 in 2002
- US$ 16,950,454 in 2003
- US$ 23,805,412 in 2004
- US$ 26,844,576 in 2005
- US$ 46,646,940 in 2006
- US$ 47,946,314 in 2007
- US$ 55,773,594 in 2008
- US$ 55,398,309 in 2009

8.5.2.1. Accomplishments and Ongoing Works (2000-2009) in Drug Discovery Include

1) Inhibitors of *P. falciparum* lactate dehydrogenase {LDH} as novel antimalarials
2) Drug discovery on cysteine protease inhibition
3) Synthetic peroxide (OZ-277/RBx 11160)
4) Anti-malarial drug development focused on inhibition of *P. falciparum* fatty acid biosynthesis (FAS)
5) *P. falciparum* protein farnesyltransferase inhibitors as drugs against malaria
6) Development of *falcipain* inhibitors as antimalarial drugs (GSK/MMV Mini-portfolio)
7) Synthesis and selection of inhibitors of malarial dihydrofolate reductase (DHFR)
8) Dihydroorotate dehydrogenase inhibition
9) Protein farnesyltransferase inhibitors (Pf-PFT)
10) *Falcipain* (cysteine protease) inhibitors
11) New dicationic molecules
12) 4(1H)-pyridones (GSK/MMV Mini-portfolio)
13) Novel tetracyclines
14) Manzamine alkaloids
15) Glyceraldhyde-3-phosphate dehydrogenase (GAPDH)

16) Peptide deformylase (PDF) (GSK/MMV Mini-portfolio)
17) Fatty acid biosynthesis (Fab I) (GSK/MMV Mini-portfolio)
18) Enantioselective 8-aminoquinolines
19) Next generation OZ (synthetic peroxide)
20) 4 (1H)- pyridones – Preclinical/ Back-up
21) Novel liver stage antimalarials
22) Next generation pyridones
23) Whole parasite screening platform
24) Novartis mini-portfolio
25) *P. falciparum* enoyl-ACP reductase (FabI)
26) *Plasmodial* surface anion channel (PSAC) antagonists
27) Novel liver stage antimalarials
28) Queensland natural products
29) Dihydrooratate dehydrogenase (DHODH)*
30) Mississippi natural products
31) TDR hit to lead studies
32) New peroxy heterocycles
33) Dihydroorotate dehydrogenase (DHODH)*
34) Novel macrolides
35) Ozonides : Next-generation synthetic peroxides
36) Dihydrofolate reductase (DHFR)
37) TDR / Pharmacopoeia screening hits
38) Queensland Natural Products

8.5.2.2. Accomplishments and Ongoing Works (2000-2009) in Drug Development Include

1) Development of an Improved Semisynthetic Artemisinin Derivative
2) Development of an intravenous artemisinin treatment for severe malaria
3) Pyronaridine Artesunate
4) Development of a novel and superior 4-aminoquinoline – Isoquine
5) Chlorproguanil-dapsone-artesunate
6) Chlorproguanil-dapsone (Lapdap™) + artesunate (CDA)
7) Pyronaridine-artesunate
8) Artemisone (semi-synthetic endoperoxid
9) Isoquine (improved aminoquinoline)
10) Intravenous artesunate
11) Third generation antifolate
12) Artemether – lumefantrine (Pediatric Coartem®)
13) DB289 (accelerated project)
14) Dihydroartemisinin-piperaquine (Artekin®)
15) 8-aminoquinoline
16) Artemifone (semi-synthetic endoperoxide)
17) 4(1H)-pyridones
18) Eurartekin™ (dihydroartemisinin-piperaquine) – Phase I/II/III
19) Chlorproguanil-dapsone (Lapdap™) - artesunate (CDA) – Phase II/III

20) Pyronaridine-artesunate – Phase I/II
21) DB289 (an improved pentamidine) – Phase II
22) RBx11160 – Phase II and partner drug selection
23) Chlorproguanil- dapsone (Lapdap™)-artesunate (CDA) – Phase III
24) Coartem® dispersible tablets — Phase III
25) Dihydroartemisinin-piperaquine – Phase III
26) Pyronaridine-artesunate (PYRAMAX ®) – Phase III
27) Tablet, pediatric and intravenous RBx11160
28) Isoquine (N-tert butyl form)
29) Coartem ® Dispersible — Submission
30) DacartTM (chlorproguanil-dapsone-artesunate) – Phase III
31) Eurartesim ® (dihydroartemisinin-piperaquine) – Phase III
32) Pyramax ® (pyronaridine-artesunate) – Phase III
33) Artesunate (intravenous form) – Phase II
34) 4 (1H)-pyridone
35) Tafenoquine – Phase I
36) Coartem ® Dispersible (artemether-lumefantrine) dispersible tablets — Approved
37) Pyramax ® (pyronaridine-artesunate) – Phase III
38) Eurartesim ® (dihydroartemisinin-piperaquine) – Phase III
39) Dacart™ (chlorproguanil-dapsone-artesunate) – Phase III — Discontinued
40) Intravenous artesunate – Phase II
41) Artemisone – Phase II
42) Tafenoquine – Phase I
43) Isoquine (GSK 369796) – Phase I

8.5.2.3. Accomplishments and Ongoing Works (2003-2009)in Drug Exploratory Include

1) Glyceraldehyde-3-phosphate dehydrogenase inhibitors (GAPDH)
2) Dihydroorotate dehydrogenase inhibitors (DHOD)
3) Heme polymerization inhibitors
4) *P. falciparum* peptide deformylase (PDF)
5) *P. falciparum* enoyl-ACP reductase (Fab i)
6) *Plasmodial* surface anion channel (PSAC) antagonists
7) Antimalarial constituents of Cameroonian medicinal plants
8) GSK 932121A was selected for preclinical development.
9) OZ-439 is ready to start the 'first in human' studies in the spring of 2009
10) MK 4815, first-in-human studies are expected to start in the summer of 2009
11) (+) Mefloquine was planned to start the Phase I safety study in the first half of 2009
12) Mirincamycin is reassembling the IND and preclinical package, with a view to starting studies in normal human volunteers in 2010
13) BCX 4945 is now in preclinical evaluation

8.5.2.4. Project Support Programs (2005-2009)

1) Artemisinin embryotoxicity
2) Consensus meeting on the design of Phase III clinical trials

3)   A new resource for breeding *A. annua* with high yield of artemisinin
4)   Artemisinin in pregnancy
5)   Albert Schweitzer Clinic, Gabon
6)   Antimalarial drug screening at the Swiss Tropical Institute (STI)
7)   Paediatric Antimalarial Combination Therapy Advisory Committee (PACT)
8)   Artemisinin Extraction and Supply Program
9)   Artemisinin extraction benchmarking study

8.5.2.5. MMV Business Plan 2008-2012

If there ever was a golden age for malaria innovation, it is now. By growing confidence and momentum, the malaria community has set its sights on the very ambitious long term goal of malaria eradication. This audacious goal has galvanized the malaria field and encouraged stakeholders to redouble their efforts in research and malaria control. In 2008, this renewed momentum was captured within a strategic document, the Global Malaria Action Plan (GMAP), by the community's overarching body: the RBM Partnership, to which MMV's mission and goals are aligned. Through bold leadership, sustained funding, and effective on-the-ground implementation of existing tools and strategies, dramatic reductions in mortality and morbidity from malaria are within our grasp. The agenda for MMV during the next five years reflects its exciting leadership role in delivering a new arsenal of antimalarial medicines. This plan portrays a significant expansion of the organization's R&D efforts to develop new drugs, target specific patient groups and prepare alternatives to the artemisinin class of drugs. The plan also marks a more explicit movement into new areas of work, such as post-registration studies and global access − to enable the most appropriate use of MMV's products so that they can produce the biggest health impact. Achieving these goals will require a significant expansion in funding, partnerships, staff and management capacity; however, the return on these investments − lives saved, increase in productivity and reduction of burden on the health system − is enormous [927].

1) Aiming for Eradication: The R&D Challenge

MMV's R&D priorities in the next five years include:

- Achieving international stringent regulatory licensure for three new ACTs: Coartem® *Dispersible,* Eurartesim™ and Pyramax® and provide clinical support to expand their use in special patient groups such as pregnant women and infants.
- Accelerating the development of new Non-Artemisinin-based Combination Therapies as alternative treatments and replacements of ACTs if and when this class of compounds falls to resistance.
- Developing a radical cure for *P. vivax* and registering an artemisinin-based therapy for severe malaria.
- Expanding the drug development pipeline, including novel mechanisms of action, molecules that target the challenges of eradication − compounds that can cure malaria and prevent transmission which are synergistic with vaccine strategies.

2) Clinical Development

Today, MMV has some 40 projects in its portfolio. Beginning in 2008, it will launch more clinical studies in special patient groups such as expectant mothers and small children to gain a richer understanding of the optimal treatment for these patients. To optimize the roll-out of the new medicines and ensure their wide usage, MMV will also participate in post-licensure activities, such as resistance monitoring, pharmacovigilance, and head-to-head comparisons. In addition, MMV will expand its medicines' indication beyond treatment, to *prevention and transmission blocking* of malaria infections, and to monitoring the safety and effectiveness of ACTs in different population settings. Two other exciting projects also in clinical trials are: tafenoquine, a radical cure for liver stage *P. vivax*; and intravenous artesunate, an artemisinin-based treatment for severe malaria.

3) Translational Research

Currently, there are two projects in Phase I clinical trials and six others that could enter this phase over the next two years. MMV continues to accelerate the development of radically innovative products with entirely different modes of action than the current ACTs. The transition between laboratory tests and studies conducted in human is a giant milestone where of the many promising compounds only a few succeed. We are working with our partners to streamline the time it takes from the first decision to develop a molecule through to the first evidence that the molecule has good antimalarial activity in human. This area of translational research will be a key part of our progress in the next decade. The work in this particular area will inform the work of other PDPs that face similar challenges in shaping their product portfolio.

4) Drug Discovery

Discovering the next generations of antimalarials will require that MMV move beyond the well trodden set of targets and chemical classes that have been the mainstay of malaria drug research over the past few decades. MMV's partners are deploying new approaches and technologies including bioinformatics, genomics, and high content screening. The ultimate goal is a safe, effective and affordable one-dose transmission-blocking cure. MMV takes two parallel approaches in discovery research. First, the mini-portfolio approach – collaborations with biopharma partners – provides a highly efficient and flexible model for managing drug development candidates. The second approach is individual projects driven by world-class academic research. More projects are needed at the early end of the pipeline to compensate for the inevitable attrition in early-stage drug development. MMV's objective is not only to accelerate the development of individual projects but also to foster cutting-edge innovation in research approaches.

There are currently 9 projects targeting novel mechanisms and 19 new classes of compounds in development in MMV's antimalarial drug portfolio. A snapshot of the portfolio demonstrates the level of genuine innovation. To accelerate these innovations into true medical advances will require significant resources (Table 59).

MMV's main contribution to the antimalarial arsenal will be the development of new antimalarial combination therapies, such as the synthetic peroxides, that do not rely on artemisinin production. In principle, these synthetic drugs will be superior drugs not only in safety and efficacy, but will also cost less and be easier to manufacture, thus mitigating problems of artemisinin shortage (*http://www.mmv.org/article.php3?id_article=185*).

5) A Highly Cost-effective Model: Reducing the Cost of Drug Development

MMV provides a significant proportion of funding to R&D projects. Its partners also co-invest in the research, often bearing much of the true development costs through "in-kind" contributions of expertise and access to technology and facilities. The result is a highly leveraged and efficient model for drug development. Based on MMV's historic costs, timelines and attrition statistics, we have now developed a model to cost a new combination malaria drug based on two new chemical entities (NCEs). As shown, developing a *combination* therapy with two NCEs will cost approximately $420M, including the cost of attrition (see Table 59). This is a fraction of what industry spends on average for the development of a new drug, which is estimated at beyond US$1 billion.

6) Enhancing Capacity: Organization and Resourcing to Meet Renewed Vision

Since 1999, MMV raised over $323M from 14 donors. Key funders include the Bill & Melinda Gates Foundation, Wellcome Trust, and a number of donor governments. A further ~$600M is needed over the next five years to finance planned activities. These funds will be used to develop new drug combinations based on new chemical entities (~$300M), to design medicines for Intermittent Preventive Treatments ($22M), and to develop tafenoquine as a radical cure for *P. vivax* (~$39M). Annual spending on access activities will rise from an estimated $5M in 2007 to $8M in 2012, driven in part by the growing demands to support the launch of new products coming through MMV's pipeline (Table 60).

7) Conclusion

National and international eagerness and ability to tackle malaria is at an unprecedented high level. New funding, tools, and partners from all sectors, combined with new leadership, have emerged. We must take advantage of this momentum. Eradication of malaria will inevitably take many decades. Given the long lead times involved in product development, solid investments in R&D *now* are critical to ensuring that the world has the right healthcare technologies for eradication (including medicines, vaccines, and vector control tools). Operational and market research must continue so that an understanding of current dynamics in the antimalarial market can inform the effective implementation of future tools. Antimalarials play a critical role in ending suffering and saving lives – it is the tip of the spear in the pursuit of eradicating this ancient scourge. Medicines for Malaria Venture is committed to playing a leading role developing the next generation of medicines and ensuring that these innovations make a significant public health impact.

8.5.3. Efforts of WRAIR in New Drug Discovery and Development

The Walter Reed Army Institute of Research (WRAIR), through the United States Army Medical Research and Material Command (USAMRMC), is the leader in the United States for the development of new antimalarial drugs. Drug resistant malaria is considered a major military threat, as well as an American public health issue. Military personnel, tourists, consultants, Peace Corps volunteers, and State Department employees traveling to, or residing in, malaria-endemic areas are at risk of illness and death from malaria.

Table 59. MMV's pipeline

Research Lead Opt	Translational			Development	
	Pre-clinical	Phase I	Phase II	Phase III	Registration
Pyridones GSK	MK4815*	Tafenoquine	Artemifone	Euratesim ™	Coartem® *Dispersible*
Falcipains GSK	Pyridone 932121	Isoquine	IV Artesunate	Pyramax ®	
Macrolide GSK	OZ439			Azithromycin/ Chloroquine IPTp*	
DHODH	Immucillins Bcx4945				
Aminoindole Broad/Genzyme	Mirincamy-cin*				
HSP90 Broad/Genzyme	(+) Erythro-Mefloquine				
DHFR NITD/Biotec					
Nat Product NITD					

* Under negotiation.
** Final draft document for board approval. [927]

Table 60. MMV Financial Plan (2008-2012)

Problem	2008	2009	2010	2011	2012	2008-2012
Research and development						
• NCE combinations (treat + IPT)	43.7	40.8	48.9	71.0	95.1	299.6
• Elimination / eradication	0.5	14.0	19.0	18.5	17.5	69.5
• Phase IV	2.0	4.0	6.0	6.0	6.0	24.0
• Enabling projects	2.5	3.5	3.5	4.0	4.0	17.5
• Science related variable	6.0	7.0	8.0	9.0	10.0	40.0
Total	$54.7M	$69.3M	$85.4M	$108.5M	$132.6M	$450.6M
• % Total budget	82.0%	82.2%	84.6%	87.1%	88.9%	
Access and delivery	$6.0M	$8.0M	$8.0M	$8.0M	$8.0M	$38.0M
• % Total budget	9.0%	9.5%	7.9%	6.4%	5.4%	
Management and admin	$6.0M	$7.0M	$7.5M	$8.0M	$8.5M	$37.0M
• % Total budget	9.0%	8.3%	7.4%	6.4%	5.7%	
MMV Baseline Budget	$66.7M	$84.3M	$100.9M	$124.5M	$149.1M	$525.6M
Security Margin						
• Inflation 3.5% p.a.	$2.3M	$3.0M	$3.5M	$4.4M	$5.2M	$18.4M
• Dollar Exchange Effect (10%)	$6.7M	$8.4M	$10.1M	$12.4M	$14.9M	$52.6M
MMV Final Budget	$75.7M	$95.7M	$114.6M	$141.3M	$169.3M	$596.5M

Staffing will rise from 24 FTEs at the beginning of 2008 to around 50 by the end of 2012. Growth will allow MMV to bolster its capacity in the science and access team and bring in new skills and expertise, including more scientific staff to manage an expanded portfolio and Africa-based representation for global access activities.

* Final draft document for board approval [927].

Table 61. Accomplishments of the WRAIR Malaria Drug Development Program

Drugs	WRAIR Role	Collaborators
Chloroquine-primaquine combination tablets	Clinical trials and FDA approval for prophylaxis	Sterling Winthrop
Sulfadoxine/pyrimethamine (Fansidar)	Clinical trials and FDA approval for prophylaxis	Hoffman-LaRoche
Mefloquine (Lariam)	Invented and developed	WHO, Hoffman-LaRoche
Halofantrine (Halfan)	Invented and developed	GlaxoSmithKline
Arteether (Artemotil)	Co-developed, clinical trials (Injectable, limited use because of potential neurotoxicity, discovered by WRAIR scientists in animal models)	WHO, Artecef Approved in the Netherlands only
Doxycycline	Phase II challenge and clinical trials, FDA approval for prophylaxis	Pfizer, FDA
Azithromycin	Efficacy trials for prophylaxis Ongoing research on combinations	NIAID, Pfizer
Atovaquone-proguanil (Malarone)	Dose-ranging studies and efficacy trials	GlaxoSmithKline
Primaquine	Investigation for prophylaxis (intended to lead to application for added FDA-approved indication to label, which now includes only treatment)	Sterling Winthrop Naval Medical Research Institute
Tafenoquine	Discovered and in Phase III for prophylaxis	GlaxoSmithKline NIAID
Tinidazole	Phase I for radical cure in patients with *P. vivax*	
Intravenous artesunate	Phase I and II completed for severe malaria treatment; Pre-NDA	Sigma-Tau Pharmaceuticals (Rome)
Imidazolinedione derivatives	Discovered for prophylaxis	MMV
Next generation of quinoline Methanols	Discovered for IPT and Chemoprophylaxis	MMV
Malarone	Phase IV for chemoprophylaxis in weekly dosing	

*Based largely on Wilbur Milhous and Peter Weina (2008).

When drug resistant malaria was first encountered by U.S. troops during the Vietnam conflict, the U.S. Army established a malaria research program to develop new prophylactic and therapeutic drugs for military use, coordinated through the Division of Experimental Therapeutics at the Walter Reed Army Institute of Research in Washington, D.C. The program was expected to maintain the expertise and laboratory capability to manage an experimental compound from the chemist's bench, through clinical trials, and on to approval by the Food and Drug Administration. WRAIR's accomplishments are summarized in Table

61. Funding for the program reached a peak of more than US$12 million in 1969 (equivalent to about US$70 million in 2003, adjusted for inflation) (Figure 32), but dropped off precipitously once the Vietnam conflict ended.

The WRAIR malaria drug development program never closed down completely, but government funding began to increase again only in the early 1990s (from about US$2 million in 1990 to about US$14 million in 2009) and stable funding from 1994 to 2009 (Figure 32). Currently, WRAIR has Department of Defense directives to develop a new malaria prophylactic agent, and an intravenous artesunate formulation for severe and complicated malaria. Perhaps more importantly, WRAIR has become an integral component of the global malaria drug development network, having established strategic alliances with MMV, and the National Institute of Allergy and Infectious Diseases Challenge Grant Program, as well as the pharmaceutical industry. Through MMV projects, WRAIR is supported by a wide range of donors. This has allowed the research to benefit both military (drugs mainly for non-immune adults entering malarious areas for short periods or moderate-length stays) and broader global applications that focus on residents of malaria-endemic countries.

An advantage of the WRAIR program is its in-house capabilities to take a compound from early discovery through to early clinical testing, much like a for-profit pharmaceutical company (although on a small scale). Final drug development requires teaming with a commercial partner who contributes to financing in the late stages, before or at the time of clinical trials. WRAIR has its own suite of new antimalarial drugs. Several families of compounds have progressed into pre-clinical and clinical development, in accomplishment and ongoing works including [928, 929]:

1) Quinine and Quinidine

Quinine profoundly influenced history, because it enabled missionaries, explorers, colonists, and militaries to travel and live where malaria was endemic. The drug was commonly used by the US military by 1830. During the Second Seminole War in Florida (1838–1842); Dr. Benjamin Harney (a career army medical officer) demonstrated the efficacy of large doses of quinine to treat remittent fevers. This led to improved results with quinine by military physicians around the globe. During the US Civil War, the Union Army used 125,000 kg of quinine or other cinchona products. Quinine was a key to the completion of the Panama Canal, because it prevented malarial illness in canal workers. Quinine was eventually approved in the United States as an oral drug (after the US Food and Drug Administration (FDA) was created), but an intravenous form of quinine was never approved.

Quinidine is a natural component of cinchona bark that was first described in 1848 by Van Heymingen and was prepared and given its present name by Louis Pasteur in 1853. In approximately 1918, W. Frey noted that patients with auricular fibrillation taking quinidine for malaria developed regular heart rhythms. The drug was originally approved for use in the United States as a drug to treat cardiac dysrhythmias and was manufactured by Lilly. The intravenous formulation was noted to be a life-saving remedy for severe malaria (ironically, quinidine must be given in intensive care unit settings because of cardiotoxicity). In 1991, the FDA granted labeling revision of quinidine for antimalarial purposes. Intravenous quinidine is the only approved drug for treatment of severe malaria available to US troops and civilians, but continued availability is uncertain because newer cardiac drugs are more frequently used.

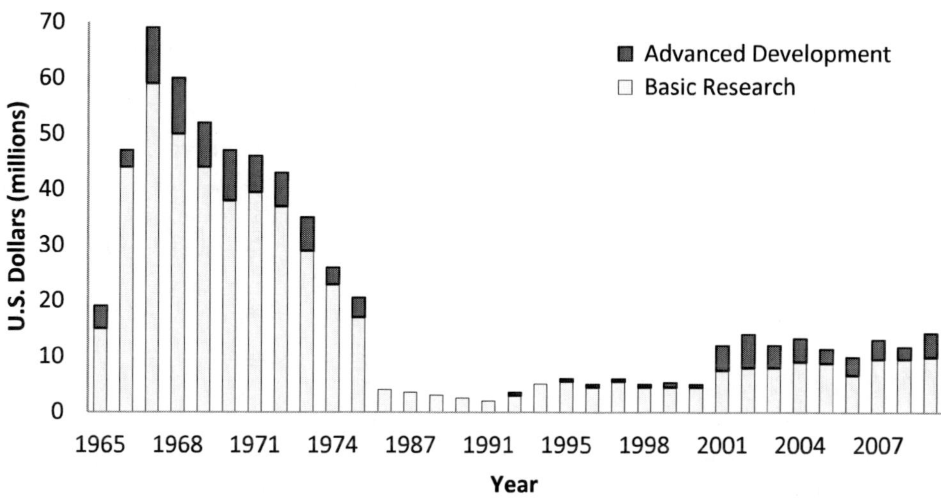

Figure 32. WRAIR malaria drug program budget from 1965 to 2009 in US$ millions adjusted for inflation.

2) Primaquine

The US Army began development of primaquine in 1944 and undertook large-scale safety and efficacy studies in the early 1950s, when relapsing *P. vivax* malaria emerged as a major problem in veterans returning from the Korean War. Compliance with chloroquine prophylaxis was good, so malaria symptoms developed only after troops departed from Korea. Dr. Alf Alving (University of Chicago) led a research team under contract to the US Army that evaluated the safety and efficacy of new antimalarial agents in inmate volunteers at the Illinois State Penitentiary. Primaquine prevents relapse from *P. vivax* and *Plasmodium ovale* hypnozoites and therefore cures relapsing malaria [12].

Primaquine short half-life (4–6 h) requires daily administration for 14 days. During the Korean War, primaquine was given as directly observed therapy during the voyage from Korea to the United States. FDA approval (January 1952 for military use and staff at the Walter Reed Army Institute of Research and the Naval Medical Research Center reviewed extensive data and concluded that primaquine could be useful for malaria prophylaxis. The Navy laboratory in Jakarta subsequently completed a pivotalPhase 3 efficacy studies. The CDC now recommends primaquine as a possible option for prevention of malaria.

3) Tafenoquine

Tafenoquine (WR 238605), an 8-aminoquinoline analogue of primaquine with a half-life of 2–3 weeks, is under development by the US Army in collaboration with GlaxoSmithKline for prevention and treatment of *P. falciparum* and *P. vivax* malaria. Phase 2 and 3 studies have been completed in partnership with GlaxoSmithKline Pharmaceuticals. This product has a very long half-life, and may be effective for prophylaxis when taken monthly.

4) Chloroquine-primaquine

Chloroquine-primaquine was used successfully by United Nations troops during the Korean conflict. Chloroquine and primaquine were not available in a single fixed-dose combination tablet until December 1969, when regulatory approval of combination tablets manufactured by Sanofi-Synthelabo was the first major accomplishment of the Division of Experimental Therapeutics at Walter Reed Army Institute of Research, established in 1961 to develop new antimalarial drugs. Chloroquine-resistant malaria was noted in Colombia in 1959 and later in Thailand and Vietnam. US troops in Vietnam experienced 81,000 cases of malaria, with 1.4 million malarial sick days and 133 deaths. Discontinuation of DDT use because of insecticide-resistant mosquitoes and adverse environmental effects contributed to the problem. Chloroquine-primaquine (commercial name: Aralen phosphate with primaquine phosphate; Winthrop-Stearns) was given to troops during the Vietnam War, but it was ineffective against chloroquine-resistant malaria. The combination drug is no longer commercially available, although each drug component is still available separately.

5) Sulfadoxine-Pyrimethamine

US Army scientists explored other folic acid-blocking drug combinations. Walter Reed Army Institute of Research participated in clinical trials and FDA approval for the sulfadoxine-pyrimethamine combination (Fansidar; Hoffman-LaRoche) for malaria prevention. After approval was granted in other countries, Fansidar was approved by the FDA in 1983. Although Fansidar is still used as malaria treatment in some countries that lack alternative antimalarial drugs, use of Fansidar is limited because of infrequent but serious adverse reactions, including Stevens-Johnson syndrome, and the widespread emergence of Fansidar-resistant malaria strains.

6) Mefloquine

US scientists discovered the prototype drug for mefloquine, SN10275, during World War II. The first-generation synthetic quinoline methanols caused unacceptable phototoxicity. Mefloquine (WR 149240) was developed by the Division of ExperimentalTherapeutics at Walter Reed Army Institute of Research in the late 1960s in collaboration with the World Health Organization Special Programme for Training and Research in Tropical Diseases (WHO/TDR) and Hoffman-LaRoche. Mefloquine, a synthetic 4-quinoline methanol (quinine analogue), was approved by the FDA in May 1989 under the commercial name Lariam (Hoffman-LaRoche) for the prevention and treatment of malaria. The mechanism of antimalarial action is unknown. Mefloquine is effective against *P. vivax* and *P. falciparum,* and its long half-life permits weekly dosing. Mefloquine was not initially recommended for pregnant women. However, some female service members who were unaware that they were pregnant ingested mefloquine during US military operations in Somalia. Follow-up revealed no congenital defects from fetal exposure, although an unexpectedly high rate of spontaneous abortions was observed. Currently, mefloquine is recommended as an antimalarial prophylaxis in pregnant women at risk of multidrug-resistant infection. Mefloquine is an important antimalarial drug for the US military. However, reports of neuropsychiatric effects negatively impact compliance with treatment, and malaria parasites have developed partial resistance to mefloquine, especially in Southeast Asia and East Africa. Walter Reed Army Institute of Research is currently investigating mefloquine analogues, seeking one with similar efficacy but reduced neuropsychiatric toxicity.

7) Halofantrine

Halofantrine (WR 171669) was developed at Walter Reed Army Institute of Research in collaboration with SmithKline Beecham and the WHO/TDR beginning in the late 1960s and early 1970s, as a backup drug to mefloquine to treat chloroquine-resistant *P. falciparum* malaria. Halofantrine was created by replacing the quinoline moiety of the quinoline methanols (quinine-type compounds) with other aromatic groups, to form the aryl (amino) carbinols. Of this class of compounds, halofantrine, a 9-phenanthrenemethanol, is the most potent. Commercial development began in the 1980s, and the drug (Halfan) was approved by the FDA in July 1992. Although halofantrine is still used for treatment of *P. falciparum* malaria outside of the United States in areas where other antimalarial drugs are unavailable, use of halofantrine is limited by its short half-life (1-2 days), slow and variable absorption, and adverse effects (especially potentially fatal cardiotoxicity related to prolonged electrocardiographic QT intervals and embryotoxicity). Its main metabolite appears to be a less toxic and equally efficacious antimalarial, but this was not further developed.

8) Doxycycline

The antibiotic doxycycline (Vibramycin, manufactured by Pfizer) has slow-acting antimalarial properties via binding to malaria ribosomes, thereby inhibiting protein synthesis. Walter Reed Army Institute of Research conducted Phase 2 challenge trials and clinical trials with doxycycline in Thailand, and FDA approval was granted to Pfizer in December 1992 for its use as prophylaxis for *P. falciparum* and *P. vivax* malaria. From June 1992 to November 1993, 12000 Dutch military personnel serving in Cambodia used doxycycline for malaria prophylaxis; only 59 developed malaria.

9) Erythromycin and Azithromycin

The antibacterial drugs erythromycin and azithromycin have antimalarial effects related to inhibition of mitochondrial protein synthesis. Azithromycin's excellent safety profile in children and pregnant women prompted antimalarial prophylaxis evaluations in Kenya, Indonesia, and Thailand by military investigators in the mid-1990s. Prophylaxis efficacy was excellent (99%) for *P. vivax* malaria, but for *P. falciparum* malaria, it ranged from 70% to 83%, even with daily dosing. The US military, in partnership with Pliva Pharmaceuticals, is seeking azithromycin analogues with improved and more-selective antimalarial activity and similar safety and pharmacokinetic characteristics. Combination of azithromycin with other antimalarial agents for malaria is also under investigation. Synergy is noted when azithromycin is combined with chloroquine.

10) Atovaquone-proguanil

Malarone (GlaxoSmithKline) was approved by the FDA for prevention and treatment of *P. falciparum* malaria in 2000. Development of drug combination strategies, dose-ranging preclinical and clinical studies, and pivotal efficacy trials were organized by Walter Reed Army Institute of Research in partnership with GlaxoSmithKline. Atovaquone was consistently effective in clearing parasitemia in uncomplicated *P. falciparum* malaria, but recrudescence rates up to 25% precluded further development as monotherapy. After studies at the Armed Forces Research Institute of Medical Sciences confirmed synergy between atovaquone and proguanil, co-administration trials conducted at various sites, including

Walter Reed Army Institute of Research laboratories in Brazil and Kenya, demonstrated 99% efficacy for treatment of uncomplicated multidrug resistant malaria and 98% efficacy for prophylaxis in placebo controlled trials. Additional prophylaxis studies were completed by the Naval Medical Research Unit-2 in Jakarta, Indonesia. Although Malarone has few adverse effects other than mild gastrointestinal intolerance, experience with this expensive drug is limited. Rapid development of drug resistance related to mutations mostly arising in the cytochrome b gene of *P. falciparum* has been noted when Malarone has been used as malaria monotherapy.

11) Pyrroloquinazolines

New analogs that show none of the toxicity common to the parent drug and yet retain oral potency against highly resistant malaria in animal studies were in development.

12) Third Generation Antifolates

These show no cross-resistance in studies against multidrug resistant *P. falciparum*. They are under development in collaboration with Jacobus Pharmaceuticals.

13) Imidazolinedione Derivatives

These compounds are unique in that they specifically kill liver stages of malaria but have no activity against blood stages. Because they are not in the 8-aminoquinolone drug class, they are not expected to create hemolysis in G6PD-deficient people. They may enter clinical studies in 2010.

14) Tryptanthrins

Tryptanthrins are extremely potent class of chemicals with *in vitro* activity against trypanosomiasis as well as malaria. Difficulties with oral bioavailability have historically impaired progress with this drug class.

15) New macrolides

Azithromycin nearly meets criteria as an effective malaria prophylactic drug. Many analogs have been identified with superior *in vitro* potency against multidrug-resistant malaria. PLIVA Pharmaceuticals is collaborating with WRAIR to develop these new analogs.

16) Chalcones

WRAIR is just beginning assessments of this drug class in collaboration with LICA Pharmaceuticals.

17) Methylene Blue

This compound very rapidly kills malaria in animal studies, and appears to be nontoxic at effective doses. It may be a good partner for a combination antimalarial regimen. Its potential utility in malaria prophylaxis is still being explored. Mefloquine analogs: an *in vitro* assay to predict neurotoxicity has been established, in order to identify several potent mefloquine analogs which do not have *in vitro* neurotoxic effects.

18) Azithromycin/Chloroquine and Azithromycin/Quinine

These combinations are known to be safe and effective in children and in pregnant women. Pfizer Pharmaceuticals is collaborating with WRAIR in clinical trials in Thailand and Kenya.

19) Artesunate i.v. Injection

Artesunate i.v. formulation for severe malaria began as collaboration between WRAIR and MMV in 2001. In 2004, WRAIR continued work alone until finding a partnership with Sigma-Tau Pharmaceuticals in 2007.

The principal goal in the treatment of severe malaria is the provision of prompt, effective therapy and concurrent supportive care to manage life-threatening complications of the disease. In most of the world, standard therapy has been intravenous or intramuscular quinine. In the United States, intravenous quinidine has been the standard and only therapy since 1991. That year, parenteral quinine was withdrawn by the CDC because quinidine had been shown to be more potent *in vitro* and highly effective against *P. falciparum* when used orally for uncomplicated disease [82] or intravenously for severe *falciparum* malaria [930]. The current recommendations for severe malaria are to administer quinidine as a loading dose followed by continuous infusion; the loading dose may be omitted if quinine or mefloquine were recently administered [2]. The cardiac toxic effects of quinidine are a major concern in its use, and intravenous therapy requires continuous cardiac monitoring, with slowing or discontinuation of the infusion for any prolongation of the QT interval. Significant lately, has been the problem of decreasing availability of quinidine as the use of the drug as an antiarrhythmic agent has virtually disappeared with the development of more effective and safer antiarrhythmics.

The emerging evidence of its superiority for artesunate over quinine (and by extension – quinidine), treatment (or compassionate use) investigational-new-drug (IND) application from the CDC went into effect in the United States on June 21, 2007, to allow investigational use of intravenous artesunate for the treatment of severe malaria. While this drug is not currently approved by the Food and Drug Administration, it can be used in the United States through the Treatment IND in cooperation with the CDC, with the drug supplied at no charge by the Walter Reed Army Institute of Research. Patients are eligible if they have uncomplicated malaria but require parenteral therapy because of an inability to take oral medications or if they have a level of parasitemia of more than 5% or other signs of severe malaria. In addition, the CDC requires that artesunate be at least as rapidly available for administration as quinidine or that there be known intolerance of quinidine, previous failure of such treatment, or a contra-indication. Enrollment requires a telephone call to the CDC Malaria Hotline (Monday through Friday from 8 a.m. to 4:30 p.m. Eastern time, at 770-488-7788; at other times, health care providers may call 770-488-7100 and ask for a clinician in the CDC Malaria Branch). If approved, the drug will be released by the CDC Drug Service or by one of the CDC quarantine stations located around the country (8 of the 20 centers maintain a stock). Response time by the CDC is extremely rapid. In the first 37 uses of the drug, the mean time from artesunate request until treatment being initiated has been 7 hours [95].

Under the IND protocol, intravenous artesunate is administered in four equal doses of 2.4 mg per kilogram of body weight over a period of 3 days. The dosing schedule recommended by the World Health Organization (WHO) entails doses every 12 hours on day 1 and then once daily. Therapy for more than 3 days may occasionally be indicated in very ill patients, but specific guidelines on when to extend therapy are not available. Artesunate dosages need not be changed because of hepatic or renal failure or concomitant or previous therapy with other medications, including previous therapy with mefloquine, quinine, or quinidine. There are no known interactions between artesunate and other drugs.

Cardiac monitoring is not mandatory during treatment with artesunate, and no serious toxic effects due to the drug are anticipated. However, patients with severe malaria often require care in an intensive care unit. Indeed, aggressive supportive care, including mechanical ventilation and hemofiltration or hemodialysis, can be instrumental in successful management of severe malaria [931, 932]. In technologically limited settings, high-quality nursing care, management of fluid balance, and control of seizures are helpful, although anticonvulsant agents that are respiratory depressants should be used with caution if mechanical ventilation is unavailable [241]. Aggressive fluid resuscitation, blood transfusion for moderate anemia, exchange transfusion, and specific treatment for acidosis are of uncertain value. Bacterial infections can coexist with severe malaria, so blood cultures should be obtained from patients with shock or other signs of sepsis despite appropriate antimalarial therapy, and these patients should receive broad-spectrum antibiotic therapy. Hypoglycemia will be less common when artesunate is used rather than quinine or quinidine; nonetheless, it is important to monitor the patient's blood glucose level and provide supplementary glucose as needed.

After the acute stage of the illness, artemisinins should be partnered with longer-acting drugs to ensure a high likelihood of complete cure. Appropriate partner drugs that are available in the United States are a full course of treatment with atovaquone-proguanil or mefloquine (although the neuropsychiatric toxic effects of mefloquine may be increased after cerebral malaria). There are also other agents like CoArtem that have been recently approved for use in the U.S., which may be appropriate and useful. There are also those who advocate the use of the less popular antibiotic agents such as a 1-week course of doxycycline or, in children or pregnant women, clindamycin. All of these drugs should be initiated after the patient can tolerate oral medication.

Appendix I

country/Area	Species	Global AMDP database AFRO						P.vivax
		P. falciparum						
		uncomplicated		treatment failure	severe malaria	pregnancy		treatment
		unconfirmed	lab-confirmed			treatment	prevention	
Algeria	*vivax* only							CQ
Angola	*falciparum*	AL	AL	QN(7d)	QN(7d)	QN(7d)	SP(IPT)	
Benin	*falciparum*	AL	AL	QN(7d)	QN(7d)	QN(7d)	SP(IPT)	
Botswana	*falciparum*	AL*	AL*	QN(7d)	QN(7d)	QN(7d)	CQ+PG	
Burkina Faso	*falciparum*	AL*	AL*	QN(7d)	QN(7d)	QN(7d)	SP(IPT)	
Burundi	*falciparum*	AS+AQ	AS+AQ	QN(7d)	QN(7d)	QN(7d)		
Cameroon	*falciparum*	AS+AQ	AS+AQ	QN(7d)	QN(7d)	QN	SP (IPT)	
Cape Verde	*falciparum*	CQ	CQ	SP	QN(7d)	QN	CQ weekly	
Central African Republic	*falciparum*	AL*	AL*	QN(7d)	QN(7d)	QN(7d)	SP(IPT)	
Chad	*falciparum*	AS+AQ* /AL*	AS+AQ* / AL*	QN(7d)	QN(7d)	QN	SP(IPT)	
Comoros	*falciparum*	AL	AL	QN(7d)	QN(7d)	QN(7d)	SP(IPT)	
Congo	*falciparum*	AS+AQ*	AS+AQ*	AL*	QN(7d)	QN	SP(IPT)	
Cote d'Ivoire	*falciparum*	AS+AQ*	AS+AQ*	AL	QN(7d)	QN	SP(IPT)	
Democratic Republic of the Congo	*falciparum*	AS+AQ	AS+AQ	QN(7d)	QN(7d)	QN	SP (IPT)	
Equatorial Guinea	*falciparum*	AS+AQ	AS+AQ	QN(7d)	QN(7d)	QN		
Eritrea	*falciparum/vivax*	CQ+SP	AS+AQ*	QN(7d)	QN(7d)	QN		CQ+PQ
Ethiopia	*falciparum/vivax*	AL	AL	QN(7d)	QN(7d)	QN		CQ
Gabon	*falciparum*	AS+AQ	AS+AQ	AL	QN(7d)	QN	SP(IPT)	
Gambia	*falciparum*	AL	AL	QN(7d)	QN(7d)	QN	SP(IPT)	
Ghana	*falciparum*	AS+AQ	AS+AQ	QN(7d)	QN(7d)	QN; (AS+AQ -2nd +3rd trimester)	SP (IPT)	

Appendix I. (Continued)

Guinea	*falciparum*	AS+AQ*	AS+AQ*	QN(7d)	QN(7d) QN	SP(IPT)
Guinea-Bissau	*falciparum* AL*	AL*	QN(7d)	QN(7d)	QN	SP(IPT)
Kenya	*falciparum* AL	AL	QN(7d)	QN(7d)	QN; (AL-2nd + 3rd trimester)	SP (IPT)
Liberia	*falciparum* AS+AQ	AS+AQ	QN(7d)	QN(7d)	QN	SP(IPT)
Madagascar	*falciparum* AS+AQ	AS+AQ	QN(7d)	QN(7d)	QN	SP (IPT)
Malawi	*falciparum* AL	AL	AS+AQ*	QN(7d)	SP or QN(7)	SP (IPT)
Mali	*falciparum* AL*	AL*	AS+SP	QN(7d)	QN(7d)	SP(IPT)
Mauritania	*falciparum* AS+AQ*	AS+AQ*		QN(7d)		
Mauritius	*vivax* only					CQ
Mozambique	*falciparum* AL*	AL*	AS+AQ	QN(7d)	QN	SP (IPT)
Namibia	*falciparum* AL	AL	QN(7d)	QN(7d)	QN(7)	SP(IPT)
Niger	*falciparum* AL	AL	QN(7d)	QN(7d)		SP(IPT)
Nigeria	*falciparum* AL	AL	QN(7d)	QN(7d)	QN; (ACT - 2nd + 3rd trimester)	SP(IPT)
Rwanda	*falciparum* AL	AL	QN(7d)	QN(7d)		SP (IPT)
Sao Tome and Principe	*falciparum* AS+AQ	AS+AQ	AL	QN(7d)	QN(7d) or AS+AQN	SP (IPT)
Senegal	*falciparum* AS+AQ	AS+AQ		QN(7d)	QN	SP (IPT)
Sierra Leone	*falciparum* AS+AQ	AS+AQ	QN(7d)	QN(7d)	QN	SP(IPT)
South Africa (KwaZulu Natal)	*falciparum*	AL	QN+CL	QN(7d)	CQ + PG	QN(7d)
South Africa (Mpumalana)	*falciparum* AL	AL	QN+CL	QN(7d)	CQ+PG	QN(7d)
Swaziland	*falciparum* CQ	CQ	SP	QN(7d)	QN	CQ+PG
Togo	*falciparum* AL	AL		QN(7d)		SP(IPT)
Uganda	*falciparum* AL	AL	QN(7d)	QN(7d)	QN(7d)	SP (IPT)
United Republic of Tanzania (Mainland)	*falciparum* AL	AL	QN; (ACT - 2nd + 3rd trimester)	QN(7d)	SP (IPT)	QN(7d)
Zambia	*falciparum* AL	AL	QN(7d)	QN(7d)	QN; (ACT - 2nd + 3rd trimester)	SP (IPT)
Zimbabwe	*falciparum* AL*	AL*	QN(7d)	QN(7d)	QN; (ACT - 2nd + 3rd trimester)	SP (IPT)

* Policy adopted, not currently being deployed, implementation process on-going.

AQ =   Amodiaquine

AL =   Artemether-lumefantrine

AS =   Artesunate

CQ =   Chloroquine

IPT =   Intermittent Preventive Treatment

MQ = Mefloquine
PG = Proguanil
PQ = Primaquine
QN = Quinine
SP = Sulphadoxine-pyrimethamine
Last Updated: May 2008.

Global AMDP database - AMRO/PAHO								
Country/Area	Species	*P. falciparum*						*P.vivax*
		uncomplicated		treatment failure	severe malaria	pregnancy		treat-ment
		uncon-firmed	lab-confirmed			treat-ment	preven-tion	
Argentina	*vivax* only							CQ+PQ
Belize	*vivax* only							CQ+PQ
Bolivia		AS+MQ		QN+CL				CQ+PQ
Brazil		AL			AS or AM or QN	QN; (AS+AQ -2nd +3rd trimester); CQ for *vivax*		CQ+PQ(7d)
Colombia		AQ+SP; AS+MQ, AL		QN(3d)+ CL(5d)	QN(7d)	QN; (AS+AQ -2nd +3rd trimester);		CQ+PQ
Costa Pacifica		AL						
Costa Rica	*vivax* only							CQ+PQ
Dominican Republic		CQ+PQ(3d)						
Ecuador		AS+SP; AL		QN+ T/D/CL				CQ+PQ
El Salvador	*vivax* only							CQ+PQ
French Guiana		QN+T		QN+D				CQ+PQ
Guatemala	*vivax* only							CQ+PQ
Guyana		AL		QN+T				CQ+PQ
Haiti		CQ+PQ						
Honduras	*vivax* only							CQ+PQ
Mexico	*vivax* only							CQ+PQ
Nicaragua		CQ+PQ(7d)		AS+SP*/ AS+MQ*	QN+CL	CQ		CQ+PQ(7d)

Appendix I. (Continued)

Panama	*vivax* only		CQ+PQ
Paraguay	CQ+PQ		CQ+PQ
Peru	AS+MQ / AS+SP		CQ+PQ
Suriname	AL	QN(7d)	CQ+PQ
Venezuela	AS+MQ	QN+T/D/CL	CQ+PQ

* Policy adopted, not currently being deployed, implementation process on-going

AQ = Amodiaquine
AL = Artemether-lumefantrine
AS = Artesunate
CQ = Chloroquine
IPT = Intermittent Preventive Treatment
MQ = Mefloquine
PG = Proguanil
Prim= Primaquine
QN = Quinine
SP = Sulphadoxine-pyrimethamine
Last Updated: May 2008.

Global AMDP database - EMRO

Country/Area	Species	P. falciparum						P.vivax
		uncomplicated		treat-ment failure	severe malaria	pregnancy		treatment
		unconfi-rmed	lab-confirmed			treat-ment	preven-tion	
Afghanistan	*falciparum/ vivax*	CQ+SP	AS+SP	QN (7d)	QN(7d)/ AS+SP	QN; (AS+SP - 2nd +3rd trimester)		CQ
Djibouti	*falciparum* mainly	AS+SP*	AS+SP*	AL*	QN	QN; (AS+SP - 2nd +3rd trimester)		CQ+PQ (14d)
Egypt	imported cases		AL	QN (7d)	QN (7d)			CQ+PQ (14d)
Iran (Islamic Republic of)	*falciparum/vivax*		AS+SP	AL	QN or AS			CQ+PQ (14d)
Iraq	*vivax* mainly		AL	QN+D (7d)	QN (7d)	QN(7d)		CQ+PQ (14d)
Morocco	*vivax* only/imported pf		AL	QN (7d)	QN (7d)			CQ+PQ (14d)
Oman	imported cases		QN(3d)+ SP+PQ (1d)	MQ	QN (7d)	QN		CQ+PQ (14d)
Pakistan	*falciparum/vivax*	AS+SP	AS+SP	QN	QN/AM	CQ		CQ+PQ (5d)
Saudi Arabia	*falciparum* mainly		AS+SP	AL	QN (7d)			CQ+PQ (14d)

Somalia	*falciparum* mainly	AS+SP	AS+SP	QN		QN	QN; (AS+SP - 2nd +3rd trimester)	SP (IPT)
Sudan (North)	*falciparum* mainly	AS+SP	AS+SP	QN; (AS+SP - 2nd + 3rd trime ster)	QN/AM	SP(IPT)		AL
Sudan (South)	*falciparum* mainly	AS+AQ	AS+AQ		QN/AM/ AS+SP		CQ+ PQ(14d)	QN
Syrian Arab Republic	*vivax* only/imported pf		SP	QN+D (7d)	QN(7d)			CQ+PQ (14d)
United Arab Emirates	imported cases		MQ or AL	MQ	QN (7d)			CQ+PQ (14d)
Yemen	*falciparum* mainly	AS+SP	AS+SP		QN (7d)			CQ+PQ (14d)

* Policy adopted, not currently being deployed, implementation process on-going.

AQ = Amodiaquine
AL = Artemether-lumefantrine
AS = Artesunate
CQ = Chloroquine
IPT = Intermittent Preventive Treatment
MQ = Mefloquine
PG = Proguanil
Prim= Primaquine
QN = Quinine
SP = Sulphadoxine-pyrimethamine
Last Updated: May 2008.

Global AMDP database - EURO

Country/ Area	Species	P. falciparum					P.vivax
		uncomplicated	treat-ment failure	severe malaria	pregnancy		treatment
		uncon-firmed	lab-confirmed		treatment	preve ntion	
Armenia	*vivax* only						CQ+PQ(14d)
Azerbaija n	*vivax* only						CQ+PQ(14d)
Georgia	*vivax* only						CQ+PQ(14d)
Turkey	*vivax* only						CQ+PQ(14d)
Turkmeni stan	*vivax* only						CQ+PQ(14d)

* Policy adopted, not currently being deployed, implementation process on-going.

AS = Artesunate
CQ = Chloroquine
PQ = Primaquine
Q   = Quinine
SP = Sulphadoxine-pyrimethamine
Last Updated: May 2008.

Appendix I. (Continued)

Global AMDP database - SEARO							
Country/ Area	Species	**P. falciparum**					**P.vivax**
		uncomplicated		treatment failure	severe malaria	pregnancy		treatment
		unconfirmed	lab-confirmed			treatment	prevention	
Bangladesh		CQ+PQ	AL	QN+T or d	QN/AM	QN; (AL-2nd + 3rd trimester)		CQ+PQ(14d)
Bhutan			AL	QN(7d)	QN/AM	QN; (AL-2nd + 3rd trimester)		CQ+PQ(14d)
Democratic People's Republic of Korea	vivax only							CQ+PQ(14d)
Democratic Republic of Timor-Leste		CQ+PQ	AL	QN(7d)	QN/AM	QN; (AL-2nd + 3rd trimester)		CQ+PQ(14d)
India		CQ+PQ	AS+SP		QN/AM	QN; (AS+SP-2nd + 3rd trimester)		CQ+PQ(14d)
Indonesia		CQ+PQ	AS+AQ+PQ	QN+D+PQ	QN/AM	QN; (AS+AQ - 2nd +3rd trimester)		CQ+PQ(14d)
Maldives		CQ	CQ+PQ	M; SP	QN(5d)			CQ+PQ(14d)
Myanmar		CQ	AS+MQ; AL; DHA+PIP	QN+T or D	QN/AS	QN; (AL-2nd + 3rd trimester)		CQ+PQ(14d)
Nepal		CQ+PQ	AL		QN	QN; (AL-2nd + 3rd trimester)		CQ+PQ(14d)
Sri Lanka			AL*	QN(7d)	QN(7d)			CQ+PQ(14d)
Thailand			AS+MQ	QN+T	QN/AS	QN(7d)		CQ+PQ(14d)

* Policy adopted, not currently being deployed, implementation process on-going

AQ =   Amodiaquine
AL =   Artemehter-Lumefantrine
AS =   Artesunate
C   =   Clindamycin
CQ =   Chloroquine
D   =   Doxycycline
MQ =   Mefloquine
PQ =   Primaquine
QN =   Quinine
SP =   Sulphadoxine-pyrimethamine
T   =   Tetracycline
Last Updated: May 2008.

Global AMDP database - WPRO							
		P. falciparum					**P.vivax**
Country/ AREA	Species	uncomplicated		treatment failure	severe malaria	pregnancy		treatment
		unconfirmed	lab-confirmed			treatment	prevention	
Cambodia	*falciparum/ vivax*	AS+MQ	AS+MQ	QN(7d) + T(7d)	AM+MQ	QN or AS+MQ		CQ
China	*falciparum/ vivax*	CQ+AT(5d)	AM/AS(5d); CQ(3d)/PIP(3d)+ PQ(2d)	AM/AS; DHA; PR; AL; PR+SP; AM+PM; DHA/AM/ AS+PYR + PQ(2d)	AM; PR; AS	Q(7d) / CQ		CQ+PQ(8d), CQ+PQ(5d)
China, Other	*vivax* only						CQ+PQ(5d)	
Lao People's Democratic republic	*falciparum/ vivax*	CQ	AL	QN+D	AS+AL	QN	CQ (weekly)/ SP(IPT)	CQ+PQ(14d)
Malaysia	*falciparum/ vivax*		AS+MQ	QN+T	QN+T	CQ		CQ+PQ(14d)
Papua New Guinea	*falciparum/ vivax*	AL*	AL*	QN(7d)	AS/AM+AL*	QN/AL*	CQ weekly	CQ+PQ(14d)
Philippines	*falciparum/ vivax*	AL*	AL+PQ	QN+T	QN+T	QN	CQ (weekly)/ SP(IPT)	CQ+PQ(14d)
Republic of Korea	*vivax* only							CQ+PQ(14d)
Solomon Islands	*falciparum/ vivax*		AL*	QN(7d)	AS+AL	AL	CQ	AL+PQ(14d)
Vanuatu	*falciparum/ vivax*	AL*	AL*	QN(7d)	QN(7d)	CQ+SP	CQ (weekly)	AL+PQ(14d)

Appendix I. (Continued)

| **Vietnam** *falciparum/ vivax* | AS/CQ | DHA+PIP,
AS7 + PQ, DHA+PIP+
TRI+PQ (CV8) | AS+MQ
or QN7 | AS inj/
QN inj | Q(7d) /AS
in 2+3
trimesters | CQ (weekly) | CQ+PQ(14d) |

* Policy adopted, not currently being deployed, implementation process on-going.

AQ = Amodiaquine
AS = Artesunate
AM = Artemether
AL = Artemether-lumefantrine
CQ = Chloroquine
DA = Dihydroartemisinin
MQ = Mefloquine 25mg/kg
PIP = Piperaquine
PM = Pyrimethamine (used only in China)
PQ = Primaquine (normally used at 15 mg/day, but China policy is 22.5 mg/day)
PR = Pyronaridine
QN = Quinine
SP = Sulphadoxine-pyrimethamine
T   = Tetracycline
TRI = Trimethoprim
Last Updated: May 2008.

References

[1] WHO. WHO informal consultation with manufacturers of artemisinin based pharmaceutical products in use for the treatment of malaria. August 24, 2007. 20 Avenue Appia. World Health Organization, Geneva, Switzerland.at: *http://www.who.int/malaria/publications/atoz/manufacturers_artemisinin_products/en/* . (accessed December 2010).

[2] WHO. Guidelines for the treatment of malaria. Geneva, Switzerland: World Health Organization, November 8, 2006. *http://www.who.int/malaria/publications/atoz/9789241547925/en/* (accessed December 2010).

[3] Noedl, H; Se, Y; Schaecher, K; Smith, BL; Socheat, D; Fukuda, MM; Artemisinin Resistance in Cambodia 1 (ARC1) study consortium: Evidence of artemisinin-resistant malaria in western Cambodia. *N Engl J Med.* (2008). *359*, 2619-2620.

[4] White, NJ; Antimalarial drug resistance. *J Clin Invest.* (2004). *113*, 1084-1092.

[5] Afonso, A; Hunt, P; Cheesman, S; Alves, AC; Cunha, CV; do Rosário, V; Cravo, P; Malaria parasites can develop stable resistance to artemisinin but lack mutations in candidate genes *atp6* (encoding the sarcoplasmic and endoplasmic reticulum Ca^{2+} ATPase), *tctp, mdr1,* and *cg10. Antimicrob Agents Chemother.* (2006). *50*, 480-489.

[6] White, NJ; Assessment of the pharmacodynamic properties of antimalarial drugs *in vivo. Antimicrob Agents Chemother.* (1997). *41*, 1413-1422.

[7] WHO. Global Malaria Control and Elimination: Report of a meeting on containment of artemisinin tolerance. Geneva, Switzerland: World Health Organization, January 19, 2008 (accessed August 2008).

[8] Kerr, C; Companies agree to withdraw artemisinin monotherapy. *Lancet Infect Dis.* (2006). *6*, 397.

[9] Artepal Web site. Inventory of ACT producers: Jan. 14, 2007. *http://www.artepal.org/index.php?option=com_content&task=blogcategory&id=39&Itemid=100.* (accessed December 2010).

[10] Newton, PN; McGready, R; Fernandez, F; Green, MD; Sunjio, M; Bruneton, C; Phanouvong, S; Millet, P; Whitty, CJ; Talisuna, AO; Proux, S; Christophel, EM; Malenga, G; Singhasivanon, P; Bojang, K; Kaur, H; Palmer, K; Day, NP; Greenwood, BM; Nosten, F; White, NJ; Manslaughter by fake artesunate in Asia--will Africa be next? *PLoS Med.* (2006). *3*, e197.

[11] Adjuik, M; Babiker, A; Garner, P; Olliaro, P; Taylor, W; White, N; International Artemisinin Study Group. Artesunate combinations for treatment of malaria: Meta-analysis. *Lancet.* (2004). *363*, 9-17.

[12] Uhlemann, AC; Cameron, A; Eckstein-Ludwig, U; Fischbarg, J; Iserovich, P; Zuniga, FA; East, M; Lee, A; Brady, L; Haynes, RK; Krishna, S. A single amino acid residue can determine the sensitivity of SERCAs to artemisinins. *Nat Struct Mol Biol.* (2005). *12*, 628-629.

[13] Epstein, D; Drug resistance could set back malaria control success US$ 22.5 million grant from Gates Foundation to contain malaria parasites resistant to artemisinin. Feb. 25, 2009. http://www.who.int/topics/malaria/en/ (accessed December 2010).

[14] WHO. Susceptibility of *Plasmodium falciparum* to Antimalarial Drugs. Geneva, Switzerland: World Health Organization. Press, (2005).

[15] Jambou, R; Legrand, E; Niang, M; Khim, N; Lim, P; Volney, B; Ekala, MT; Bouchier, C; Esterre, P; Fandeur, T; Mercereau-Puijalon, O; Resistance of *Plasmodium falciparum* field isolates to *in vitro* artemether and point mutations of the SERCA-type PfATPase6. *Lancet.* (2005). *366*, 1960-1963.

[16] Chretien, JP; Fukuda, M; Noedl, H; Improving surveillance for antimalarial drug resistance. *JAMA.* (2007). *297*, 2278-2281.

[17] Vestergaard, LS; Ringwald, P; Responding to the challenge of antimalarial drug resistance by routine monitoring to update national malaria treatment policies. *Am J Trop Med Hyg.* (2007). *77*(Suppl. 6), 153-159.

[18] WHO. Chemotherapy of malaria and resistance to antimalarials. Report of a WHO Scientific Group. Geneva, WHO, 1973 (World Health Organization Technical Report Series, No. 529).

[19] Baird, JK; Leksana, B; Masbar, S; Fryauff, DJ; Sutanihardja, MA; Suradi; Wignall, FS; Hoffman, SL; Diagnosis of resistance to chloroquine by *Plasmodium vivax*: Timing of recurrence and whole blood chloroquine levels. *Am J Trop Med Hyg.* (1997). *56*, 621-626.

[20] Noedl, H; Wongsrichanalai, C; Wernsdorfer, WH; Malaria drug-sensitivity testing: New assays, new perspectives. *Trends Parasitol.* (2003). *19*, 175-181.

[21] Wernsdorfer, WH; Noedl, H; Molecular markers for drug resistance in malaria: Use in treatment, diagnosis and epidemiology. *Curr Opin Infect Dis.* 2003. 16, 553–558.

[22] Price, RN; Uhlemann, AC; Brockman, A; McGready, R; Ashley, E; Phaipun, L; Patel, R; Laing, K; Looareesuwan, S; White, NJ; Nosten, F; Krishna, S; Mefloquine resistance in *Plasmodium falciparum* and increased *pfmdr1* gene copy number. *Lancet.* (2004). *364*, 438-447.

[23] Duraisingh, MT; von Seidlein, LV; Jepson, A; Jones, P; Sambou, I; Pinder, M; Warhurst, DC; The tyrosine-86 allele of the *pfmdr1* gene of *Plasmodium falciparum* is associated with increased sensitivity to the antimalarials mefloquine and artemisinin. *Mol Biochem Parasitol.* (2000). *108*, 13-23.

[24] Gil, JP; Nogueira, F; Stromberg-Norklit, J; Lindberg, J; Carrolo, M; Casimiro, C; Lopes, D; Arez, AP; Cravo, PV; Rosario, VE; Detection of atovaquone and malarone resistance conferring mutations in *Plasmodium falciparum* cytochrome b gene (cytb). *Mol Cell Probes.* (2003). *17*, 85-89.

[25] Wichmann, O; Muehlen, M; Gruss, H; Mockenhaupt, FP; Suttorp, N; Jelinek, T. Malarone treatment failure not associated with previously described mutations in the cytochrome b gene. *Malar J.* (2004). *3*, 14.

[26] Alifrangis, M; Enosse, S; Pearce, R; Drakeley, C; Roper, C; Khalil, IF; Nkya, WM; Ronn, AM; Theander, TG; Bygbjerg, IC; A simple, high-throughput method to detect

Plasmodium falciparum single nucleotide polymorphisms in the dihydrofolate reductase, dihydropteroate synthase, and *P. falciparum* chloroquine resistance transporter genes using polymerase chain reaction- and enzyme-linked immunosorbent assay-based technology. *Am J Trop Med Hyg.* (2005). *72*, 155-162.

[27] Pearce, RJ; Drakeley, C; Chandramohan, D; Mosha, F; Roper, C; Molecular determination of point mutation haplotypes in the dihydrofolate reductase and dihydropteroate synthase of *Plasmodium falciparum* in three districts of northern Tanzania. *Antimicrob Agents Chemother.* (2003). *47*, 1347-1354.

[28] Roper, C; Pearce, R; Bredenkamp, B; Gumede, J; Drakeley, C; Mosha, F; Chandramohan, D; Sharp, B; Antifolate anti-malarial resistance in southeast Africa: A population-based analysis. *Lancet.* (2003). *361*, 1174-1181.

[29] Noedl, H; Artemisinin resistance: How can we find it? *Trends Parasitol.* (2005). *21*, 404-405.

[30] East African Network for Monitoring Antimalarial Treatment (EANMAT). Monitoring antimalarial drug resistance within national malaria control programs: The EANMAT experience. *Trop Med Int Health.* (2001). *6*, 891-898.

[31] Sibley, CH; Ringwald, P; A database of antimalarial drug resistance. *Malar J.* (2006). *5*, 48.

[32] Sibley, CH; Barnes, KI; Plowe, CV; The rationale and plan for creating a World Antimalarial Resistance Network (WARN). *Malar J.* (2007). *6*, 118.

[33] Price, RN; Dorsey, G; Ashley, EA; Barnes, KI; Baird, JK; d'Alessandro, U; Guerin, PJ; Laufer, MK; Naidoo, I; Nosten, F; Olliaro, P; Plowe, CV; Ringwald, P; Sibley, CH; Stepniewska, K; White, NJ; World Antimalarial Resistance Network I: Clinical efficacy of antimalarial drugs. *Malar J.* (2007). *6*,119.

[34] Bacon, DJ; Jambou, R; Fandeur, T; Le Bras, J; Wongsrichanalai, C; Fukuda, MM; Ringwald, P; Sibley, CH; Kyle, DE; World Antimalarial Resistance Network (WARN) II: *In vitro* antimalarial drug susceptibility. *Malar J.* (2007). *6*, 120.

[35] Plowe, CV; Roper, C; Barnwell, JW; Happi, CT; Joshi, HH; Mbacham, W; Meshnick, SR; Mugittu, K; Naidoo, I; Price, RN; Shafer, RW; Sibley, CH; Sutherland, CJ; Zimmerman, PA; Rosenthal, PJ; World Antimalarial Resistance Network (WARN) III: Molecular markers for drug resistant malaria. *Malar J.* (2007). *6*, 121.

[36] Barnes, KI; Lindegardh, N; Ogundahunsi, O; Olliaro, P; Plowe, CV; Randrianarivelojosia, M; Gbotosho, GO; Watkins, WM; Sibley, CH; White, NJ; World Antimalarial Resistance Network (WARN) IV: Clinical pharmacology. *Malar J.* (2007). *6*, 122.

[37] Sibley, CH; Barnes, KI; Watkins, WM; Plowe, CV; A network to monitor antimalarial drug resistance: A plan for moving forward. *Trends Parasitol.* (2008). *24*, 43-48.

[38] Stepniewska, K; White, NJ; Some considerations in the design and interpretation of antimalarial drug trials in uncomplicated *falciparum* malaria. *Malar. J.* (2006). *5*, 127.

[39] Denis, MB; Tsuyuoka, R; Lim, P; Lindegardh, N; Yi, P; Top, SN; Socheat, D; Fandeur, T; Annerberg, A; Christophel, EM; Ringwald, P; Efficacy of artemether-lumefantrine for the treatment of uncomplicated *falciparum* malaria in northwest Cambodia. *Trop Med Int Health.* (2006). *11*, 1800-1807.

[40] Nair, S; Nash, D; Sudimack, D; Jaidee, A; Barends, M; Uhlemann, AC; Krishna, S; Nosten, F; Anderson, TJ;. Recurrent gene amplification and soft selective sweeps during evolution of multidrug resistance in malaria parasites. *Mol Biol Evol. 24*, 562-573.

[41] Volkman, SK; Sabeti, PC; DeCaprio, D; Neafsey, DE; Schaffner, SF; Milner, DA Jr; Daily, JP; Sarr, O; Ndiaye, D; Ndir, O; Mboup, S; Duraisingh, MT; Lukens, A; Derr, A; Stange-Thomann, N; Waggoner, S; Onofrio, R; Ziaugra, L; Mauceli, E; Gnerre, S; Jaffe, DB;

Zainoun, J; Wiegand, RC; Birren, BW; Hartl, DL; Galagan, JE; Lander, ES; Wirth, DF; A genome-wide map of diversity in *Plasmodium falciparum*. *Nat Genet*. (2007). *39*, 113-119

[42] Arrow, KJ; Panosian, C; Gelband, H (Eds.); Saving lives, buying time: Economics of malaria drugs in an age of resistance. *Natl Acad Press*. (2004).

[43] Hsu, E. The history of qinghao in the Chinese materia medica. *Trans R Soc Trop Med Hyg*. (2006). *100*, 505-508.

[44] Li, Q; Milhous, WK; Weina, P (Eds). Antimalarial in Malaria Therapy. *Nova Science Publishers Inc, New York*; 1 edition (2007). pp.1-133.

[45] Woodrow, CJ; Krishna, S; Antimalarial drugs: Recent advances in molecular determinants of resistance and their clinical significance. *Cell Mol Life Sci*. (2006). *63*, 1586-1596.

[46] Price, RN; Nosten, F; Luxemburger, C; ter Kuile, FO; Paiphun, L; Chongsuphajaisiddhi, T; White, NJ; Effects of artemisinin derivatives on malaria transmissibility. *Lancet*. (1996). *347*, 1654-1658.

[47] WHO. Antimalarial Drug Combination Therapy: Report of a Technical Consultation. 2001, World Health Organization, Geneva. (accessed August 2008).

[48] de Vries, PJ; Dien, TK; Clinical pharmacology and therapeutic potential of artemisinin and its derivatives in the treatment of malaria. *Drugs*. (1996). *52*, 818-836.

[49] Price, RN; Artemisinin drugs: Novel antimalarial agents. *Expert Opin. Investig. Drugs*. (2000). *9*, 1815-1827.

[50] White, NJ; Olliaro, P; Artemisinin and derivatives in the treatment of uncomplicated malaria. *Med. Trop. (Mars.)* (1998). *58* (Suppl. 3), 54-56.

[51] White, NJ; Antimalarial drug resistance and combination chemotherapy. *Philos. Trans. R. Soc. Lond. B. Biol. Sci*. (1999). *354*, 739-749.

[52] White, NJ; Delaying antimalarial drug resistance with combination chemotherapy. *Parassitologia*. (1999). *41*, 301-308.

[53] Nosten, F; vanVugt, M; Price, RN; Luxemburger, C; Thway, KL; Brockman, A; McGready, R; ter Kuile, F; Looareesuwan, S; White, NJ; Effects of artesunate-mefloquine combination on incidence of *Plasmodium falciparum* malaria and mefloquine resistance in western Thailand: A prospective study. *Lancet*. (2000). *356*, 297-302.

[54] Roll Back Malaria. The RBM partnership's global response; a programmatic strategy 2004–2008. Available at: *http://rbm.who.int/partnership/board/meetings/docs/strategy_ rev.pdf* (accessed December 2010).

[55] Bosman, A; Mendis, KN; A major transition in malaria treatment: The adoption and deployment of artemisinin-based combination therapies. *Am J Trop Med Hyg*. (2007). *77*(Suppl. 6), 193-197.

[56] Nosten, F; Luxemburger, C; ter Kuile, FO; Woodrow, C; Eh, JP; Chongsuphajaisiddhi, T; White, NJ; Treatment of multidrug-resistant *Plasmodium falciparum* malaria with 3-day artesunate-mefloquine combination. *J Infect Dis*. (1994). *170*, 971-977.

[57] Hutagalung, R; Paiphun, L; Ashley, EA; McGready, R; Brockman, A; Thwai, KL; Singhasivanon, P; Jelinek, T; White, NJ; Nosten, FH; A randomized trial of artemether-lumefantrine versus mefloquine-artesunate for the treatment of uncomplicated multi-drug resistant *Plasmodium falciparum* on the western border of Thailand. *Malar J*. (2005). *4*, 46.

[58] Nosten, F; Hien, TT; White, NJ; Use of artemisinin derivatives for the control of malaria. *Med. Trop. (Mars.)* (1998). *58* (Suppl. 3), 45-49.

[59] Muheki, C; McIntyre, D; Barnes, KI; Atemisinin-based combination therapy reduces expenditure on malaria treatment in KwaZulu Natal, *South Africa. Trop. Med. Int. Health*. (2004). *9*, 959-966.

[60] Attaran, A; Barnes, KI; Curtis, C; d'Alessandro, U; Fanello, CI; Galinski, MR; Kokwaro, G; Looareesuwan, S; Makanga, M; Mutabingwa, TK; Talisuna, A; Trape, JF; Watkins, WM. WHO, the Global Fund, and medical malpractice in malaria treatment. *Lancet.* (2004). *363*, 237-240.

[61] Malenga, G; Palmer, A; Staedke, S; Kazadi, W; Mutabingwa, T; Ansah, E; Barnes, KI; Whitty, CJ; Antimalarial treatment with artemisinin combination therapy in Africa. *BMJ.* (2005). *331*, 706-707.

[62] WHO. Global AMDP Database. World Health Organization. *http://www.who.int/malaria/amdp/amdp afro.htm.* May 2008. (accessed August 28, 2008).

[63] WHO. Facts on ACTs (artemisinin-based combination therapies), World Health Organization. January 2006 update. *http://www.rbm.who.int/cmcupload/0/000/015/364/ RBMInfosheet 9.htm* (accessed December, 2010).

[64] Ogbonna, A; Uneke, CJ; Artemisinin-based combination therapy for uncomplicated malaria in sub-Saharan Africa: The efficacy, safety, resistance and policy implementation since Abuja 2000. *Trans R Soc Trop Med Hyg.* (2008). *102*, 621-627.

[65] Price, R; van Vugt, M; Phaipun, L; Luxemburger, C; Simpson, J; McGready, R; ter Kuile, F; Kham, A; Chongsuphajaisiddhi, T; White, NJ; Nosten, F; Adverse effects in patients with acute *falciparum* malaria treated with artemisinin derivatives. *Am J Trop Med Hyg.* (1999). *60*, 547-555.

[66] Hastings, IM; Watkins, WM; Tolerance is the key to understanding antimalarial drug resistance. *Trends Parasitol.* (2006). *22*, 71-77.

[67] Martinelli, A; Moreira, R; Ravo, PV; Malaria combination therapies: Advantages and shortcomings. *Mini Rev Med Chem.* (2008). *8*, 201-212.

[68] Lefevre, G; Looareesuwan, S; Treeprasertsuk, S; Krudsood, S; Silachamroon, U; Gathmann, I; Mull, R; Bakshi, R; A clinical and pharmacokinetic trial of six doses of artemether-lumefantrine for multidrug-resistant *Plasmodium falciparum* malaria in Thailand. *Am J Trop Med Hyg.* (2001). 64, 247-256.

[69] Vijaykadga, S; Rojanawatsirivej, C; Cholpol, S; Phoungmanee, D; Nakavej, A; Wongsrichanalai, C; *In vivo* sensitivity monitoring of mefloquine monotherapy and artesunate-mefloquine combinations for the treatment of uncomplicated *falciparum* malaria in Thailand in 2003. *Trop Med Int Health.* (2006). *11*, 211-219.

[70] van Vugt, M; Looareesuwan, S; Wilairatana, P; McGready, R; Villegas, L; Gathmann, I; Mull, R; Brockman, A; White, NJ; Nosten, F; Artemether-lumefantrine for the treatment of multidrug-resistant *falciparum* malaria. *Trans R Soc Trop Med Hyg.* (2000). *94*, 545-548.

[71] Krishna, S; White, NJ; Pharmacokinetics of quinine, chloroquine and amodiaquine. Clinical implications. *Clin Pharmacokinet.* (1996). *30*, 263-299.

[72] Mutabingwa, TK; Artemisinin-based combination therapies (ACTs): Best hope for malaria treatment but inaccessible to the needy! *Acta Trop.* (2005). *95*, 305-315.

[73] WHO. Severe *falciparum* malaria. World Health Organization, Communicable Diseases Cluster. *Trans R Soc Trop Med Hyg.* (2000). *94* (Suppl 1), 1-90.

[74] Jones, KL; Donegan, S; Lalloo, DG; Artesunate versus quinine for treating severe malaria. *Cochrane Database Syst Rev.* (2007). 17, CD005967.

[75] White, NJ; Looareesuwan, S; Warrell, DA; Warrell, MJ; Bunnag, D; Harinasuta, T. Quinine pharmacokinetics and toxicity in cerebral and uncomplicated *Falciparum* malaria. *Am J Med.* (1982). *73*, 564-572.

[76] Lesi, A; Meremikwu, M; High first dose quinine regimen for treating severe malaria. *Cochrane Database Syst Rev.* (2004). CD003341.

[77] Artemether-Quinine Meta-analysis Study Group. A meta-analysis using individual patient data of trials comparing artemether with quinine in the treatment of severe *falciparum* malaria. *Trans R Soc Trop Med Hyg* (2001). 95, 637-650.

[78] Barennes, H; Kailou, D; Pussard, E; Munjakazi, JM; Fernan, M; Sherouat, H; Sanda, A; Clavier, F; Verdier, F; Intrarectal administration of quinine: An early treatment for severe malaria in children? *Sante.* (2001). *11*, 145-153.

[79] Woodrow, CJ; Haynes, RK; Krishna, S. Artemisinins. *Postgrad Med J.* (2005). *81*, 71-78.

[80] Pukrittayakamee, S; Supanaranond, W; Looareesuwan, S; Vanijanonta, S; White, NJ; Quinine in severe *falciparum* malaria: Evidence of declining efficacy in Thailand. *Trans R Soc Trop Med Hyg.* (1994). *88*, 324-327.

[81] Miller, KD; Greenberg, AE; Campbell, CC; Treatment of severe malaria in the United States with a continuous infusion of quinidine gluconate and exchange transfusion. *N Engl J Med.* (1989). *321*, 65-70.

[82] White, NJ; Looareesuwan, S; Warrell, DA; Chongsuphajaisiddhi, T; Bunnag, D; Harinasuta, T; Quinidine in *falciparum* malaria. *Lancet.* (1981). 2, 1069-1071.

[83] Pe Than, Myint; Tin, Shwe; A controlled clinical trial of artemether (qinghaosu derivative) versus quinine in complicated and severe *falciparum* malaria. *Trans R Soc Trop Med Hyg.* (1987). *81*, 559-561.

[84] Ambroise-Thomas, P; Intra-muscular artemether in the treatment of severe malaria: synthesis of current results. *Med Trop (Mars).* (1997). *57*, 289-293.

[85] van Hensbroek, MB; Onyiorah, E; Jaffar, S; A trial of artemether or quinine in children with cerebral malaria. *N Engl J Med.* (1996). *335*, 69-75.

[86] Tran, TH; Day, NP; Nguyen, HP; Nguyen, TH; Tran, TH; Pham, PL; Dinh, XS; Ly, VC; Ha, V; Waller, D; Peto, TE; White, NJ; A controlled trial of artemether or quinine in Vietnamese adults with severe *falciparum* malaria. *N Engl J Med.* (1996). *335*, 76-83.

[87] Hien, TT; Davis, TM; Chuong, LV; Ilett, KF; Sinh, DX; Phu, NH; Agus, C; Chiswell, GM; White, NJ; Farrar, J; Comparative pharmacokinetics of intramuscular artesunate and artemether in patients with severe *falciparum* malaria. *Antimicrob Agents Chemother.* (2004). *48*, 4234-4239.

[88] Seaton, RA; Trevett, AJ; Wembri, JP; Nwokolo, N; Naraqi, S; Black, J; Laurenson, IF; Kevau, I; Saweri, A; Lalloo, DG; Warrell, DA; Randomized comparison of intramuscular artemether and intravenous quinine in adult, Melanesian patients with severe or complicated, *Plasmodium falciparum* malaria in Papua New Guinea. *Ann Trop Med Parasitol.* (1998). *92*, 133-139.

[89] Newton, PN; Angus, BJ; Chierakul, W; Dondorp, A; Ruangveerayuth, R; Silamut, K; Teerapong, P; Suputtamongkol, Y; Looareesuwan, S; White, NJ. Randomized comparison of artesunate and quinine in the treatment of severe *falciparum* malaria. *Clin Infect Dis.* 2003. 37, 7-16.

[90] Dondorp, A; Nosten, F; Stepniewska, K; Day, N; White, N; Artesunate versus quinine for treatment of severe *falciparum* malaria: a randomised trial. *Lancet.* (2005). *366*, 717-725.

[91] Checkley, AM; Whitty, CJ; Artesunate, artemether or quinine in severe *Plasmodium falciparum* malaria. *Expert Rev Anti Infect Ther.* (2007). *5*, 199-204.

[92] Barnes, KI; Mwenechanya, J; Tembo, M; McIlleron, H; Folb, PI; Ribeiro, I; Little, F; Gomes, M; Molyneux, ME. Efficacy of rectal artesunate compared with parenteral quinine in initial treatment of moderately severe malaria in African children and adults: A randomised study. *Lancet.* (2004). *363*, 1598-1605.

[93] Pasvol, G. The treatment of complicated and severe malaria. *Br Med Bull.* (2006). 75-76, 29-47.

[94] Li, Q; Xie, LH; Johnson, TO; Si, Y; Haeberle, AS; Weina, PJ. Toxicity evaluation of artesunate and artelinate in *Plasmodium berghei*-infected and uninfected rats. *Trans R Soc Trop Med Hyg.* (2007). *101*, 104-112.

[95] Arguin, PM; Weina, PJ; Dougherty, CP; Artesunate for malaria. *N Engl J Med.* (2008). *359*, 313; author reply 314-315.

[96] Cao, XT; Bethell, DB; Pham, TP; Ta, TT; Tran, TN; Nguyen, TT; Pham, TT; Nguyen, TT; Day, NP; White, NJ; Comparison of artemisinin suppositories, intramuscular artesunate and intravenous quinine for the treatment of severe childhood malaria. *Trans R Soc Trop Med Hyg.* (1997). *91*, 335-342.

[97] Ilett, KF; Batty, KT; Powell, SM; Binh, TQ; Thu, le TA; Phuong, HL; Hung, NC; Davis, TM. The pharmacokinetic properties of intramuscular artesunate and rectal dihydroartemisinin in uncomplicated *falciparum malaria. Br J Clin Pharmacol.* (2002). *53*, 23-30.

[98] Nealon, C; Dzeing, A; Müller-Römer, U; Planche, T; Sinou, V; Kombila, M; Kremsner, PG; Parzy, D; Krishna, S. Intramuscular bioavailability and clinical efficacy of artesunate in gabonese children with severe malaria. *Antimicrob Agents Chemother.* (2002). *46*, 3933-3939.

[99] Aceng, JR; Byarugaba, JS; Tumwine, JK; Rectal artemether versus intravenous quinine for the treatment of cerebral malaria in children in Uganda: Randomized clinical trial. *BMJ.* (2005). *330*, 334.

[100] Gomes, M; Ribeiro, I; Warsame, M; Karunajeewa, H; Petzold, M; Rectal artemisinins for malaria: A review of efficacy and safety from individual patient data in clinical studies. *BMC Infect Dis.* (2008). *8*, 39.

[101] Karunajeewa, HA; Manning, L; Mueller, I; Ilett, KF; Davis, TM; Rectal administration of artemisinin derivatives for the treatment of malaria. *JAMA.* (2007). *297*, 2381-2390.

[102] Simpson, JA; Agbenyega, T; Barnes, KI; Di Perri, G; Folb, P; Gomes, M; Krishna, S; Krudsood, S; Looareesuwan, S; Mansor, S; McIlleron, H; Miller, R; Molyneux, M; Mwenechanya, J; Navaratnam, V; Nosten, F; Olliaro, P; Pang, L; Ribeiro, I; Tembo, M; van Vugt, M; Ward, S; Weerasuriya, K; Win, K; White, NJ; Population pharmacokinetics of artesunate and dihydroartemisinin following intra-rectal dosing of artesunate in malaria patients. *PLoS Med.* (2006). *3*, e444.

[103] Ashley, EA; White, NJ; Artemisinin-based combinations. *Curr Opin Infect Dis.* (2005). *18*, 531-536.

[104] D'Alessandro, U; Treating severe and complicated malaria. *BMJ.* (2004). *328*, 155.

[105] Rosenthal, PJ; Artesunate for the treatment of severe *falciparum* malaria. *N Engl J Med.* (2008). *358*, 1829-1836.

[106] WHO. Economic costs of malaria, Estimate of world malaria burden. Roll Back Malaria, World Health Organization. 2007. *http://www.rbm.who.int/cmc_upload/0/ 000/015/363/RBMInfosheet_10.pdf* (accessed December, 2010)

[107] Amin, AA; Zurovac, D; Kangwana, BB; Greenfield, J; Otieno, DN; Akhwale, WS; Snow, RW. The challenges of changing national malaria drug policy to artemisinin-based combinations in Kenya. *Malar J.* (2007). *6*, 72.

[108] Malik, EM; Mohamed, TA; Elmardi, KA; Mowien, RM; Elhassan, AH; Elamin, SB; Mannan, AA; Ahmed, ES. From chloroquine to artemisinin-based combination therapy: The Sudanese experience. *Malar J.* (2006). *5*, 65.

[109] Yeung, S; Van Damme, W; Socheat, D; White, NJ; Mills, A. Cost of increasing access to artemisinin combination therapy: The Cambodian experience. *Malar J.* (2008). *7*, 84.

[110] Mulligan, JA; Mandike, R; Palmer, N; Williams, H; Abdulla, S; Bloland, P; Mills, A. The costs of changing national policy: Lessons from malaria treatment policy guidelines in Tanzania. *Trop Med Int Health*. (2006). *11*, 452-461.

[111] Njau, JD; Goodman, CA; Kachur, SP; Mulligan, J; Munkondya, JS; McHomvu, N; Abdulla, S; Bloland, P; Mills, A; The costs of introducing artemisinin-based combination therapy: Evidence from district-wide implementation in rural Tanzania. *Malar J*. (2008). *7*, 4.

[112] Gelband, H; Seiter, A; A global subsidy for antimalarial drugs. *Am J Trop Med Hyg*. (2007). *77*(Suppl. 6), 219-221.

[113] WHO. WHO Policy Brief, March 2008. Global Malaria Programme. World Health Organization 2008.

[114] WHO/WPRO. The Use of Malaria Rapid Diagnostic Tests, ISBN 92 9061 088 3. 2004. *http://rbm.who.int/toolbox/tool_UseOfMalariaRDTs.html* (accessed December, 2010)

[115] Greenwood, B; The use of anti-malarial drugs to prevent malaria in the population of malaria-endemic areas. *Am J Trop Med Hyg*. (2004). *70*, 1-7.

[116] WHO. Global Partnership to Roll Back Malaria. World malaria report: 2005. Geneva: World Health Organization, 2005.

[117] WHO. Malaria vector control and personal protection: Report of a WHO study group. World Health Organization. *Technical Report Series* (2006). p. 936.

[118] UNICEF. Malaria & children. Progress in intervention coverage. New York: United Nations Children's Fund, 2007.

[119] WHO. International travel and health. Vaccination requirements and health advice. Geneva, World Health Organization, 2000.

[120] Schlagenhauf, P; Petersen, E; Malaria chemoprophylaxis: Strategies for risk groups. *Clin Microbiol Rev*. (2008). *21*,466-472.

[121] Baird, JK; Effectiveness of antimalarial drugs. *N Engl J Med*. (2005). *352*, 1565-1577.

[122] Whitty, CJ; Allan, R; Wiseman, V; Ochola, S; Nakyanzi-Mugisha, MV; Vonhm, B; Mwita, M; Miaka, C; Oloo, A; Premji, Z; Burgess, C; Mutabingwa, TK. Averting a malaria disaster in Africa--where does the buck stop? *Bull World Health Organ*. (2004). *82*, 381-384.

[123] WHO. Investing in Health Research and Development: Report of the Ad Hoc Committee on Health Research Relating to Future Intervention Options. TDR/Gen/96.1. 1996. Geneva: World Health Organization.

[124] WHO. Commission on Macroeconomics and Health. Macroeconomics and Health: Investing in Health for Economic Development. 2001. Geneva: World Health Organization.

[125] WHO. 2002. World Health Report 2002. Geneva: World Health Organization.

[126] Wilkins, JJ; Folb, PI; Valentine, N; Barnes, KI; An economic comparison of chloroquine and sulfadoxine-pyrimethamine as first-line treatment for malaria in South Africa: Development of a model for estimating recurrent direct costs. *Trans R Soc Trop Med Hyg*. (2002). *96*, 85-90.

[127] Mubyazi, GM; Gonzalez-Block, MA; Research influence on antimalarial drug policy change in Tanzania: Case study of replacing chloroquine with sulfadoxine-pyrimethamine as the first-line drug. *Malar J*. (2005). *4*, 51.

[128] Sudre, P; Breman, JG; McFarland, D; Koplan, JP; Treatment of chloroquine-resistant malaria in African children: A cost-effectiveness analysis. *Int J Epidemiol*. (1992). *21*, 146-154.

[129] Chanda, P; Masiye, F; Chitah, BM; Sipilanyambe, N; Hawela, M; Banda, P; Okorosobo, T; A cost-effectiveness analysis of artemether lumefantrine for treatment of uncomplicated malaria in Zambia. *Malar J.* (2007). *6*, 21.

[130] Bukirwa, H; Yeka, A; Kamya, MR; Talisuna, A; Banek, K; Bakyaita, N; Rwakimari, JB; Rosenthal, PJ; Wabwire-Mangen, F; Dorsey, G; Staedke, SG; Artemisinin combination therapies for treatment of uncomplicated malaria in Uganda. *PLoS Clin Trials.* (2006). *1*, e7.

[131] Coleman, PG; Morel, C; Shillcutt, S; Goodman, C; Mills, AJ; A threshold analysis of the cost-effectiveness of artemisinin-based combination therapies in sub-saharan Africa. *Am J Trop Med Hyg.* (2004). *71*(Suppl. 2), 196-204.

[132] Wiseman, V; Kim, M; Mutabingwa, TK; Whitty, CJ; Cost-effectiveness study of three antimalarial drug combinations in Tanzania. *PLoS Med.* (2006). *3*, e373.

[133] Smithuis, F; Kyaw, MK; Phe, O; Aye, KZ; Htet, L; Barends, M; Lindegardh, N; Singtoroj, T; Ashley, E; Lwin, S; Stepniewska, K; White, NJ; Efficacy and effectiveness of dihydroartemisinin-piperaquine versus artesunate-mefloquine in *falciparum* malaria: An open-label randomized comparison. *Lancet.* (2006). *367*, 2075-2085.

[134] Yeung, S; Pongtavornpinyo, W; Hastings, IM; Mills, AJ; White, NJ; Antimalarial drug resistance, artemisinin-based combination therapy, and the contribution of modeling to elucidating policy choices. *Am J Trop Med Hyg.* (2004). *71*(Suppl. 2), 179-186.

[135] Felger, I; Beck, HP; Fitness costs of resistance to antimalarial drugs. *Trends Parasitol.* (2008). *24*, 331-333.

[136] Hastings, IM; Donnelly, MJ; The impact of antimalarial drug resistance mutations on parasite fitness, and its implications for the evolution of resistance. *Drug Resist Updat.* (2005). *8*, 43-50.

[137] Holmgren, G; Hamrin, J; Svärd, J; Mårtensson, A; Gil, JP; Björkman, A; Selection of pfmdr1 mutations after amodiaquine monotherapy and amodiaquine plus artemisinin combination therapy in East Africa. *Infect Genet Evol.* (2007). *7*, 562-569.

[138] Hayward, R; Saliba, KJ; Kirk, K; *pfmdr1* mutations associated with chloroquine resistance incur a fitness cost in *Plasmodium falciparum. Mol Microbiol.* (2005). *55*, 1285-1295.

[139] Desai, M; ter Kuile, FO; Nosten, F; McGready, R; Asamoa, K; Brabin, B; Newman, RD; Epidemiology and burden of malaria in pregnancy. *Lancet Infect Dis.* (2007). *7*, 93-104.

[140] Worrall, E; Morel, C; Yeung, S; Borghi, J; Webster, J; Hill, J; Wiseman, V; Mills, A; The economics of malaria in pregnancy--a review of the evidence and research priorities. *Lancet Infect Dis.* (2007). *7*, 156-168.

[141] Goodman, CA; Coleman, PG; Mills, AJ; The cost-effectiveness of antenatal malaria prevention in sub-Saharan Africa. *Am J Trop Med Hyg.* (2001). *64*(Suppl. 1-2), 45-56.

[142] WHO. Recommendations on the use of sulfadoxine-pyrimethamine (SP) for intermittent preventive treatment during pregnancy (IPT) in areas of moderate to high resistance to SP in the African Region. October 2005. http://www.who.int/topics/pregnancy/en/ . (accessed December, 2010)

[143] Breman, JG; Mills, A; Snow, RW, *et al.* Conquering malaria. In: Jamison D, Breman JG, Measham A, *et al.*, (Eds.). Disease control priorities in developing countries. Washington: World Bank and Oxford University Press, 2006.

[144] Newman, RD; Parise, ME; Slutsker, L; Nahlen, B; Steketee, RW; Safety, efficacy and determinants of effectiveness of antimalarial drugs during pregnancy: Implications for prevention programs in *Plasmodium falciparum*-endemic sub-Saharan Africa. *Trop Med Int Health.* (2003). *8*, 488-506.

[145] Mbonye, AK; Hansen, KS; Bygbjerg, IC; Magnussen, P; Intermittent preventive treatment of malaria in pregnancy: The incremental cost-effectiveness of a new delivery system in Uganda. *Trans R Soc Trop Med Hyg.* (2008). *102*, 685-693.

[146] Goodman, CA; Mutemi, WM; Baya, EK; Willetts, A; Marsh, V; The cost-effectiveness of improving malaria home management: Shopkeeper training in rural Kenya. *Health Policy Plan.* (2006). *21*, 275-288.

[147] Coleman, PG; Goodman, CA; Mills, A; Rebound mortality and the cost-effectiveness of malaria control: Potential impact of increased mortality in late childhood following the introduction of insecticide treated nets. *Trop Med Int Health.* (1999). *4*, 175-186.

[148] Mutabingwa, TK; Anthony, D; Heller, A; Hallett, R; Ahmed, J; Drakeley, C; Greenwood, BM; Whitty, CJ; Amodiaquine alone, amodiaquine+sulfadoxine-pyrimethamine, amodiaquine+artesunate, and artemether-lumefantrine for outpatient treatment of malaria in Tanzanian children: A four-arm randomised effectiveness trial. *Lancet.* (2005). *365*, 1474-1480.

[149] Rombo, L; Who needs drug prophylaxis against malaria? My personal view. *J Travel Med.* (2005). *12*, 217-221.

[150] Wolfe, EB; Parise, ME; Haddix, AC; Nahlen, BL; Ayisi, JG; Misore, A; Steketee, RW; Cost-effectiveness of sulfadoxine-pyrimethamine for the prevention of malaria-associated low birth weight. *Am J Trop Med Hyg.* (2001). *64*, 178-186.

[151] Tran, TH; Dolecek, C; Pham, PM; Nguyen, TD; Nguyen, TT; Le, HT; Dong, TH; Tran, TT; Stepniewska, K; White, NJ; Farrar, J; Dihydroartemisinin-piperaquine against multidrug-resistant *Plasmodium falciparum* malaria in Vietnam: Randomized clinical trial. *Lancet.* (2004). *363*, 18-22.

[152] Petersen, E; Malaria chemoprophylaxis: when should we use it and what are the options? *Expert Rev Anti Infect Ther.* (2004). *2*, 119-132.

[153] WHO. Commission on Macroeconomics and Health Working Group 3. 2002. Mobilization of Domestic Resources for Health. Geneva: World Health Organization.

[154] Narasimhan, V; Attaran, A. Roll Back Malaria? The scarcity of international aid for malaria control. *Malaria J.* (2003). *2*, 8.

[155] WHO. WHO briefing on Malaria Treatment Guidelines and artemisinin monotherapies. Geneva, April 19, 2006. World Health Organization. At: http://www.who.int/malaria/publications/atoz/meeting_briefing19april/en/ (accessed December, 2010)

[156] White, NJ. Qinghaosu (artemisinin): The price of success. *Science.* (2008). *320*, 330-334.

[157] Hale, V; Keasling, JD; Renninger, N; Diagana, TT. Microbially derived artemisinin: A biotechnology solution to the global problem of access to affordable antimalarial drugs. *Am J Trop Med Hyg.* (2007). *77*(Suppl. 6), 198-202.

[158] Mårtensson, A; Strömberg, J; Sisowath, C; Msellem, MI; Gil, JP; Montgomery, SM; Olliaro, P; Ali, AS; Björkman, A; Efficacy of artesunate plus amodiaquine versus that of artemether-lumefantrine for the treatment of uncomplicated childhood *Plasmodium falciparum* malaria in Zanzibar, Tanzania. *Clin Infect Dis.* (2005). *41*, 1079-1086.

[159] Guthmann, JP; Cohuet, S; Rigutto, C; Fortes, F; Saraiva, N; Kiguli, J; Kyomuhendo, J; Francis, M; Noël, F; Mulemba, M; Balkan, S; High efficacy of two artemisinin-based combinations (artesunate + amodiaquine and artemether + lumefantrine) in Caala, Central Angola. *Am J Trop Med Hyg.* (2006). *75*, 143-145.

[160] Ashley, EA; Pinoges, L; Turyakira, E; Dorsey, G; Checchi, F; Bukirwa, H; van den Broek, I; Zongo, I; Urruta, PP; van Herp, M; Balkan, S; Taylor, WR; Olliaro, P; Guthmann, JP. Different methodological approaches to the assessment of *in vivo* efficacy of three

artemisinin-based combination antimalarial treatments for the treatment of uncomplicated *falciparum* malaria in African children. *Malar J.* (2008). *7*, 154.

[161] Adjei, GO; Kurtzhals, JA; Rodrigues, OP; Alifrangis, M; Hoegberg, LC; Kitcher, ED; Badoe, EV; Lamptey, R; Goka, BQ; Amodiaquine-artesunate vs. artemether-lumefantrine for uncomplicated malaria in Ghanaian children: A randomized efficacy and safety trial with one year follow-up. *Malar J.* (2008). *7*, 127.

[162] Olliaro, PO; Pinoges, L; Checchi, F; Vaillant, M; Guthmann, JP; Risk associated with asymptomatic parasitemia occurring post-antimalarial treatment. *Trop Med Int Health.* (2008). *13*, 83-90.

[163] Attaran, A; Barnes, KI; Curtis, C; d'Alessandro, U; Fanello, CI; Galinski, MR; Kokwaro, G; Looareesuwan, S; Makanga, M; Mutabingwa, TK; Talisuna, A; Trape, JF; Watkins, WM; WHO, the Global Fund, and medical malpractice in malaria treatment. *Lancet.* (2004). *363*, 237-240.

[164] WHO. World Health Organization and United Nations Children Fund, 2005. World Malaria Report III. Global Financing and Service Delivery. *http://rbm.who.int/wmr2005/html/3-1/htm.* (accessed December, 2010)

[165] Pincock, S; Drug company to offer new malaria drug cheaply in Africa. *BMJ.* (2003). *327*, 360.

[166] Karunajeewa, H; Lim, C; Hung, TY; Ilett, KF; Denis, MB; Socheat, D; Davis, TM. Safety evaluation of fixed combination piperaquine plus dihydroartemisinin (Artekin) in Cambodian children and adults with malaria. *Br J Clin Pharmacol.* (2004). *57*, 93-99.

[167] Denis, MB; Davis, TM; Hewitt, S; Incardona, S; Nimol, K; Fandeur, T; Poravuth, Y; Lim, C; Socheat, D; Efficacy and safety of dihydroartemisinin-piperaquine (Artekin) in Cambodian children and adults with uncomplicated *falciparum* malaria. *Clin Infect Dis.* (2002). *35*, 1469-1476.

[168] Wilairatana, P; Krudsood, S; Chalermrut, K; Pengruksa, C; Srivilairit, S; Silachamroon, U; Treeprasertsuk, S; Looareesuwan, S; An open randomized clinical trial of Artecom vs. artesunate-mefloquine in the treatment of acute uncomplicated *falciparum* malaria in Thailand. *Southeast Asian J Trop Med Public Health.* (2002). *33*, 519-524.

[169] Ashley, EA; Krudsood, S; Phaiphun, L; Srivilairit, S; McGready, R; Leowattana, W; Hutagalung, R; Wilairatana, P; Brockman, A; Looareesuwan, S; Nosten, F; White, NJ; Randomized, controlled dose-optimization studies of dihydroartemisinin-piperaquine for the treatment of uncomplicated multidrug-resistant *falciparum* malaria in Thailand. *J Infect Dis.* (2004). *190*, 1773-1782.

[170] Kakkilaya, BS; History, etiology, pathophysiology, clinical features, diagnosis, treatment, complications and control malaria. Malaria Site, 2006. at: *http://www.malariasite.com/drsk.htm.* (accessed December, 2010)

[171] Hsu, E; Reflections on the 'discovery' of the antimalarial qinghao. *Br J Clin Pharmacol.* (2006). *61*, 666-670.

[172] Biography of Alphonse Laveran. The Nobel Foundation. Retrieved on 2007-06-15. Nobel foundation. (Accessed Oct. 25, 2006.)

[173] Ettore Marchiafava. The Nobel Foundation. Retrieved on 2007-06-15.

[174] Biography of Ronald Ross. The Nobel Foundation. Retrieved on 2007-06-15.

[175] Ross and the Discovery that Mosquito Transmit Malaria Parasites. CDC Malaria website. Retrieved on 2007-06-15.

[176] Kaufman, TS; Rúveda, EA: The quest for quinine: Those who won the battles and those who won the war. *Angew Chem Int Ed Engl.* (2005), *44*, 854-885.

[177] Kyle, RA; Shampe, MA;. Discoverers of quinine. *JAMA.* (1974). *229*, 462.

[178] Raju, TN;. Hot brains: Manipulating body heat to save the brain. *Pediatrics*. (2006). *117*, e320-1.

[179] Krotoski, WA; Collins, WE; Bray, RS; Garnham, PC; Cogswell, FB; Gwadz, RW; Killick-Kendrick, R; Wolf, R; Sinden, R; Koontz, LC; Stanfill, PS; Demonstration of hypnozoites in sporozoite-transmitted *Plasmodium vivax* infection. *Am J Trop Med Hyg*. (1982). *31*, 1291-1293.

[180] Meis, JF; Verhave, JP; Jap, PH; Sinden, RE; Meuwissen, JH; Malaria parasites--discovery of the early liver form. *Nature*. (1983). *302*, 424-426.

[181] Bledsoe, GH; Malaria primer for clinicians in the United States. *South Med J*. (2005). *98*, 1197-1204.

[182] Sturm, A; Amino, R; van de Sand, C; Regen, T; Retzlaff, S; Rennenberg, A; Krueger, A; Pollok, JM; Menard, R; Heussler, VT; Manipulation of host hepatocytes by the malaria parasite for delivery into liver sinusoids. *Science*. (2006). *313*, 1287-1290.

[183] Kaiser, M; Wittlin, S; Nehrbass-Stuedli, A; Dong, Y; Wang, X; Hemphill, A; Matile, H; Brun, R; Vennerstrom, JL; Peroxide bond-dependent antiplasmodial specificity of artemisinin and OZ-277 (RBx11160). Antimicrob Agents Chemother. (2007). *51*, 2991-2993.

[184] Schlitzer, M; Antimalarial drugs - what is in use and what is in the pipeline. *Arch Pharm* (Weinheim). (2008). *341*, 149-163.

[185] Hoppe, HC; van Schalkwyk, DA; Wiehart, UI; Meredith, SA; Egan, J; Weber, BW; Antimalarial quinolines and artemisinin inhibit endocytosis in *Plasmodium falciparum*. *Antimicrob Agents Chemother*. (2004). *48*, 2370-2378.

[186] Duraisingh, MT; Refour, P: Multiple drug resistance genes in malaria -- from epistasis to epidemiology. *Mol Microbiol*. (2005). *57*, 874-877.

[187] O'Neill, PM; Ward, SA; Berry, NG; Jeyadevan, JP; Biagini, GA; Asadollaly, E; Park, BK; Bray, PG; A medicinal chemistry perspective on 4-aminoquinoline antimalarial drugs. *Curr Top Med Chem*. (2006). *6*, 479-507.

[188] Nzila, A. The past, present and future of antifolates in the treatment of *Plasmodium falciparum* infection. *J Antimicrob Chemother*. (2006). *57*, 1043-1054.

[189] Krudsood, S; Imwong, M; Wilairatana, P; Pukrittayakamee, S; Nonprasert, A; Snounou, G; White, NJ; Looareesuwan, S. Artesunate-dapsone-proguanil treatment of *falciparum* malaria: Genotypic determinants of therapeutic response. *Trans R Soc Trop Med Hyg*. (2005). *99*, 142-149.

[190] Parenti, MD; Pacchioni, S; Ferrari, AM; Rastelli, G; Three-dimensional quantitative structure-activity relationship analysis of a set of *Plasmodium falciparum* dihydrofolate reductase inhibitors using a pharmacophore generation approach. *J Med Chem*. (2004). *47*, 4258-4267.

[191] Simpson, JA; Hughes, D; Manyando, C; Bojang, K; Aarons, L; Winstanley, P; Edwards, G; Watkins, WA; Ward, S; Population pharmacokinetic and pharmacodynamic modeling of the antimalarial chemotherapy chlorproguanil/dapsone. *Br J Clin Pharmacol*. (2006). *61*, 289-300.

[192] O'Neill, PM; Posner, GH; A medicinal chemistry perspective on artemisinin and related endoperoxides. *J Med Chem*. (2004). *47*, 2945-2964.

[193] Eckstein-Ludwig, U; Webb, RJ; Van Goethem, ID; East, JM; Lee, AG; Kimura, M; O'Neill, PM; Bray, PG; Ward, SA; Krishna, S; Artemisinins target the SERCA of *Plasmodium falciparum*. *Nature*. (2003). *424*, 957-961.

[194] Golenser, J; Waknine, JH; Krugliak, M; Hunt, NH; Grau, GE; Current perspectives on the mechanism of action of artemisinins. *Int J Parasitol*. (2006). *36*, 1427-1441.

[195] Chotivanich, K; Sattabongkot, J; Udomsangpetch, R; Looareesuwan, S; Day, NP; Coleman, RE; White, NJ; Transmission-blocking activities of quinine, primaquine, and artesunate. *Antimicrob Agents Chemother.* (2006). *50*, 1927-1930.

[196] Goodman, CD; Su, V; McFadden, GI; The effects of anti-bacterials on the malaria parasite *Plasmodium falciparum. Mol Biochem Parasitol.* (2007). *152*, 181-191.

[197] Wiesner, J; Seeber, F; The plastid-derived organelle of protozoan human parasites as a target of established and emerging drugs. *Expert Opin Ther Targets.* (2005). *9*, 23-44.

[198] Ramya, TN; Mishra, S; Karmodiya, K; Surolia, N; Surolia, A; Inhibitors of nonhousekeeping functions of the apicoplast defy delayed death in *Plasmodium falciparum. Antimicrob Agents Chemother.* (2007). *51*, 307-316.

[199] Vaidya, AB; Mather, MW; A post-genomic view of the mitochondrion in malaria parasites. *Curr Top Microbiol Immunol.* (2005). *295*, 233-250.

[200] Looareesuwan, S; Chulay, JD; Canfield, CJ; Hutchinson, DB; Malarone (atovaquone and proguanil hydrochloride): A review of its clinical development for treatment of malaria. Malarone Clinical Trials Study Group. *Am J Trop Med Hyg.* (1999). *60*, 533-541.

[201] Srivastava, IK; Morrisey, JM; Darrouzet, E; Daldal, F; Vaidya, AB; Resistance mutations reveal the atovaquone-binding domain of cytochrome b in malaria parasites. *Mol Microbiol.* (1999). *33*, 704-711.

[202] Srivastava, IK; Vaidya, AB; A mechanism for the synergistic antimalarial action of atovaquone and proguanil. *Antimicrob Agents Chemother.* (1999). *43*, 1334-1339.

[203] Nduati, EW; Kamau, EM; Multiple synergistic interactions between atovaquone and antifolates against *Plasmodium falciparum in vitro*: A rational basis for combination therapy. *Acta Trop.* (2006). *97*, 357-363.

[204] Painter, HJ; Morrisey, JM; Mather, MW; Vaidya, AB; Specific role of mitochondrial electron transport in blood-stage *Plasmodium falciparum. Nature.* (2007). *446*, 88-91.

[205] Efferth, T; Willmar Schwabe Award 2006: Antiplasmodial and antitumor activity of artemisinin--from bench to bedside. *Planta Med.* (2007). *73*, 299-309.

[206] Adam, I; Ali, DM; Abdalla, MA; Artesunate plus sulfadoxine-pyrimethamine in the treatment of uncomplicated *Plasmodium falciparum* malaria during pregnancy in eastern Sudan. *Trans R Soc Trop Med Hyg.* (2006). *100*, 632-635.

[207] Dorsey, G; Njama, D; Kamya, MR; Cattamanchi, A; Kyabayinze, D; Staedke, SG; Gasasira, A; Rosenthal, PJ. Sulfadoxine/pyrimethamine alone or with amodiaquine or artesunate for treatment of uncomplicated malaria: a longitudinal randomised trial. *Lancet.* 2002. 360, 2031-2038.

[208] van den Broek, I; Amsalu, R; Balasegaram, M; Hepple, P; Alemu, E; Hussein, el B; Al-Faith, M; Montgomery, J; Checchi, F; Efficacy of two artemisinin combination therapies for uncomplicated *falciparum* malaria in children under 5 years, Malakal, Upper Nile, Sudan. *Malar J.* (2005). *4*, 14.

[209] Yeka, A; Banek, K; Bakyaita, N; Staedke, SG; Kamya, MR; Talisuna, A; Kironde, F; Nsobya, SL; Kilian, A; Slater, M; Reingold, A; Rosenthal, PJ; Wabwire-Mangen, F; Dorsey, G; Artemisinin versus nonartemisinin combination therapy for uncomplicated malaria: Randomized clinical trials from four sites in Uganda. *PLoS Med.* (2005). *2*, e190.

[210] Rwagacondo, CE; Karema,C; Mugisha, V; Erhart, A; Dujardin, JC; Van Overmeir, C; Ringwald, P; D'Alessandro, U; Is amodiaquine failing in Rwanda? Efficacy of amodiaquine alone and combined with artesunate in children with uncomplicated malaria. *Trop Med Int Health.* (2004). *9*, 1091-1098.

[211] Abacassamo, F; Enosse, S; Aponte, JJ; Gómez-Olivé, FX; Quintó, L; Mabunda, S; Barreto, A; Magnussen, P; Rønn, AM; Thompson, R; Alonso, PL. Efficacy of chloroquine,

amodiaquine, sulphadoxine-pyrimethamine and combination therapy with artesunate in Mozambican children with non-complicated malaria. *Trop Med Int Health*. (2004). *9*, 200-208.

[212] Swarthout, TD; van den Broek, IV; Kayembe, G; Montgomery, J; Pota, H; Roper, C; Artesunate + amodiaquine and artesunate + sulphadoxine-pyrimethamine for treatment of uncomplicated malaria in Democratic Republic of Congo: A clinical trial with determination of sulphadoxine and pyrimethamine-resistant haplotypes. *Trop Med Int Health*. (2006). *11*, 1503-1511.

[213] Sutherland, CJ; Ord, R; Dunyo, S; Jawara, M; Drakeley, CJ; Alexander, N; Coleman, R; Pinder, M; Walraven, G; Targett, GA; Reduction of malaria transmission to Anopheles mosquitoes with a six-dose regimen of co-artemether. *PLoS Med*. (2005). *2*, e92.

[214] Faye, B; Ndiaye, JL; Ndiaye, D; Dieng, Y; Faye, O; Gaye, O; Efficacy and tolerability of four antimalarial combinations in the treatment of uncomplicated *Plasmodium falciparum* malaria in Senegal. *Malar J*. (2007). *6*, 80.

[215] Ibrahium, AM; Kheir, MM; Osman, ME; Khalil, IF; Alifrangis, M; Elmardi, KA; Malik, EM; Adam, I; Efficacies of artesunate plus either sulfadoxine-pyrimethamine or amodiaquine, for the treatment of uncomplicated, *Plasmodium falciparum* malaria in eastern Sudan. *Ann Trop Med Parasitol*. (2007). *101*, 15-21.

[216] Mockenhaupt, FP; Ehrhardt, S; Dzisi, SY; Teun Bousema, J; Wassilew, N; Schreiber, J; Anemana, SD; Cramer, JP; Otchwemah, RN; Sauerwein, RW; Eggelte, TA; Bienzle, U; A randomized, placebo-controlled, double-blind trial on sulfadoxine-pyrimethamine alone or combined with artesunate or amodiaquine in uncomplicated malaria. *Trop Med Int Health*. (2005). *10*, 512-520.

[217] Vugt, MV; Wilairatana, P; Gemperli, B; Gathmann, I; Phaipun, L; Brockman, A; Luxemburger, C; White, NJ; Nosten, F; Looareesuwan, S; Efficacy of six doses of artemether-lumefantrine (benflumetol) in multidrug-resistant *Plasmodium falciparum* malaria. *Am J Trop Med Hyg*. (1999). *60*, 936-942.

[218] Omari, AA; Gamble, C; Garner, P; Artemether-lumefantrine for uncomplicated malaria: A systematic review. *Trop Med Int Health*. (2004). *9*, 192-199.

[219] Falade, CO; Ogundele, AO; Yusuf, BO; Ademowo, OG; Ladipo, SM; High efficacy of two artemisinin-based combinations (artemether-lumefantrine and artesunate plus amodiaquine) for acute uncomplicated malaria in Ibadan, Nigeria. *Trop Med Int Health*. (2008). *13*, 635-643.

[220] Ndayiragije, A; Niyungeko, D; Karenzo, J; Niyungeko, E; Barutwanayo, M; Ciza, A; Bosman, A; Moyou-Somo, R; Nahimana, A; Nyarushatsi, JP; Barihuta, T; Mizero, L; Ndaruhutse, J; Delacollette, C; Ringwald, P; Kamana, J; Efficacy of therapeutic combinations with artemisinin derivatives in the treatment of non complicated malaria in Burundi. *Trop Med Int Health*. (2004). *9*, 673-679.

[221] Bell, D; Winstanley, P; Current issues in the treatment of uncomplicated malaria in Africa. *Br Med Bull*. (2004). *71*, 29-43.

[222] Hasugian, AR; Purba, HL; Kenangalem, E; Wuwung, RM; Ebsworth, EP; Maristela, R; Penttinen, PM; Laihad, F; Anstey, NM; Tjitra, E; Price, RN; Dihydroartemisinin-piperaquine versus artesunate-amodiaquine: Superior efficacy and posttreatment prophylaxis against multidrug-resistant *Plasmodium falciparum* and *Plasmodium vivax* malaria. *Clin Infect Dis*. (2007). *44*, 1067-1074.

[223] Plowe, CV; Combination therapy for malaria: Mission accomplished? *Clin Infect Dis*. (2007). *44*, 1075-1077.

[224] Yeka, A; Dorsey, G; Kamya, MR; Talisuna, A; Lugemwa, M; Rwakimari, JB; Staedke, SG; Rosenthal, PJ; Wabwire-Mangen, F; Bukirwa, H; Artemether-lumefantrine versus dihydroartemisinin-piperaquine for treating uncomplicated malaria: A randomized trial to guide policy in Uganda. *PLoS ONE*. (2008). *3*, e2390.

[225] Zongo, I; Dorsey, G; Rouamba, N; Dokomajilar, C; Séré, Y; Rosenthal, PJ; Ouédraogo, JB; Randomized comparison of amodiaquine plus sulfadoxine-pyrimethamine, artemether-lumefantrine, and dihydroartemisinin-piperaquine for the treatment of uncomplicated *Plasmodium falciparum* malaria in Burkina Faso. *Clin Infect Dis*. (2007). *45*, 1453-1461.

[226] Kamya, MR; Yeka, A; Bukirwa, H; Lugemwa, M; Rwakimari, JB; Staedke, SG; Talisuna, AO; Greenhouse, B; Nosten, F; Rosenthal, PJ; Wabwire-Mangen, F; Dorsey, G; Artemether-lumefantrine versus dihydroartemisinin-piperaquine for treatment of malaria: A randomized trial. *PLoS Clin Trials*. (2007). *2*, e20.

[227] Ratcliff, A; Siswantoro, H; Kenangalem, E; Maristela, R; Wuwung, RM; Laihad, F; Ebsworth, EP; Anstey, NM; Tjitra, E; Price, RN; Two fixed-dose artemisinin combinations for drug-resistant *falciparum* and *vivax* malaria in Papua, Indonesia: An open-label randomized comparison. *Lancet*. (2007). *369*, 757-765.

[228] Grande, T; Bernasconi, A; Erhart, A; Gamboa, D; Casapia, M; Delgado, C; Torres, K; Fanello, C; Llanos-Cuentas, A; D'Alessandro, U; A randomized controlled trial to assess the efficacy of dihydroartemisinin-piperaquine for the treatment of uncomplicated *falciparum malaria* in Peru. *PLoS ONE*. (2007). *2*, e1101.

[229] Janssens, B; van Herp, M; Goubert, L; Chan, S; Uong,S; Nong, S; Socheat, D; Brockman, A; Ashley, EA; Van Damme, W; A randomized open study to assess the efficacy and tolerability of dihydroartemisinin-piperaquine for the treatment of uncomplicated *falciparum malaria* in Cambodia. *Trop Med Int Health*. (2007). *12*, 251-259.

[230] Mayxay, M; Thongpraseuth, V; Khanthavong, M; Lindegårdh, N; Barends, M; Keola, S; Pongvongsa, T; Phompida, S; Phetsouvanh, R; Stepniewska, K; White, NJ; Newton, PN; An open, randomized comparison of artesunate plus mefloquine vs. dihydroartemisinin-piperaquine for the treatment of uncomplicated *Plasmodium falciparum* malaria in the Lao People's Democratic Republic (Laos). *Trop Med Int Health*. (2006). *11*, 1157-1165.

[231] Krudsood, S; Tangpukdee, N; Thanchatwet, V; Wilairatana, P; Srivilairit, S; Pothipak, N; Jianping, S; Guoqiao, L; Brittenham, GM; Looareesuwan, S; Dose ranging studies of new artemisinin-piperaquine fixed combinations compared to standard regimens of artemisisnin combination therapies for acute uncomplicated *falciparum* malaria. *Southeast Asian J Trop Med Public Health*. (2007). *38*, 971-978.

[232] Vivas, L; Rattray, L; Stewart, LB; Robinson, BL; Fugmann, B; Haynes, RK; Peters, W; Croft, SL; Antimalarial efficacy and drug interactions of the novel semi-synthetic endoperoxide artemisone *in vitro* and *in vivo*. *J Antimicrob Chemother*. (2007). *59*, 658-665.

[233] Walker, DJ; Pitsch, JL; Peng, MM; Robinson, BL; Peters, W; Bhisutthibhan, J; Meshnick, SR; Mechanisms of artemisinin resistance in the rodent malaria pathogen *Plasmodium yoelii*. *Antimicrob Agents Chemother*. (2000). *44*, 344-347.

[234] Basco, LK; Le Bras, J; *In vitro* activity of artemisinin derivatives against African isolates and clones of *Plasmodium falciparum*. *Am J Trop Med Hyg*. (1993). *49*, 301-307.

[235] Sahr, F; Willoughby, VR; Gbakima, AA; Bockarie, MJ; Apparent drug failure following artesunate treatment of *Plasmodium falciparum* malaria in Freetown, Sierra Leone: Four case reports. *Ann Trop Med Parasitol*. (2001). *95*, 445-449.

[236] van Vugt, M; Brockman, A; Gemperli, B; Luxemburger, C; Gathmann, I; Royce, C; Slight, T; Looareesuwan, S; White, NJ; Nosten, F; Randomized comparison of artemether-

benflumetol and artesunate-mefloquine in treatment of multidrug-resistant *falciparum* malaria. *Antimicrob Agents Chemother*. (1998). *42*, 135-139.

[237] von Seidlein, L; Bojang, K; Jones, P; Jaffar, S; Pinder, M; Obaro, S; Doherty, T; Haywood, M; Snounou, G; Gemperli, B; Gathmann, I; Royce, C; McAdam, K; Greenwood, B; A randomized controlled trial of artemether/benflumetol, a new antimalarial and pyrimethamine/sulfadoxine in the treatment of uncomplicated *falciparum* malaria in African children. *Am J Trop Med Hyg*. (1998). *58*, 638-644.

[238] Jansen, FH; Lesaffre, E; Penali, LK; Zattera, MJ; Die-Kakou, H; Bissagnene, E; Assessment of the relative advantage of various artesunate-based combination therapies by a multi-treatment Bayesian random-effects meta-analysis. *Am J Trop Med Hyg*. (2007). *77*, 1005-1009.

[239] Adhikari, MR; Severe and complicated malaria. *Indian J Med Sci*. (2002). *56*, 445-448.

[240] White, NJ; Malaria. In: Cook Gorden (Eds.),*Manson's Tropical Diseases*.20th ed., London, W.B. Saunders (1996). 1087-1164.

[241] Day, N; Dondorp, AM; The management of patients with severe malaria. *Am J Trop Med Hyg*. (2007). *77*(Suppl. 6), 29-35.

[242] Mehta, SR; Das, S; Management of malaria: Recent trends. *J Commun Dis*. (2006). *38*, 130-138.

[243] Li, Q; Xie, LH; Haeberle, A; Zhang, J; Weina, P; The evaluation of radiolabeled artesunate on tissue distribution in rats and protein binding in humans. *Am J Trop Med Hyg*. (2006). *75*, 817-826.

[244] Batty, KT; Ilett, KF; Davis, TM; Protein binding and alpha: Beta anomer ratio of dihydroartemisinin *in vivo*. *Br J Clin Pharmacol*. (2004). *57*, 529-533.

[245] German, PI; Aweeka, FT; Clinical pharmacology of artemisinin-based combination therapies. *Clin Pharmacokinet*. (2008). *47*, 91-102.

[246] Batty, KT; Ilett, KF; Edwards, G; Powell, SM; Maggs, JL; Park, BK; Davis, TM; Assessment of the effect of malaria infection on hepatic clearance of dihydroartemisinin using rat liver perfusions and microsomes. *Br J Pharmacol*. (1998). *125*, 159-167.

[247] Colussi, D; Parisot, C; Legay, F; Lefèvre, G; Binding of artemether and lumefantrine to plasma proteins and erythrocytes. *Eur J Pharm Sci*. (1999). *9*, 9-16.

[248] White, NJ; van Vugt, M; Ezzet, F; Clinical pharmacokinetics and pharmacodynamics and pharmacodynamics of artemether-lumefantrine. *Clin Pharmacokinet*. (1999). *37*, 105-125.

[249] Giboda, M; Denis, MB; Response of Kampuchean strains of *Plasmodium falciparum* to antimalarials: *In vivo* assessment of quinine and quinine plus tetracycline; multiple drug resistance *in vitro*. *J Trop Med Hyg*. (1988). *91*, 205-211.

[250] Noedl, H; Faiz, MA; Yunus, EB; Rahman, MR; Hossain, MA; Samad, R; Miller, RS; Pang, LW; Wongsrichanalai, C; Drug-resistant malaria in Bangladesh: An *in vitro* assessment. *Am J Trop Med Hyg*. (2003). *68*, 140-142.

[251] Dua, VK; Dev, V; Phookan, S; Gupta, NC; Sharma, VP; Subbarao, SK; Multi-drug resistant *Plasmodium falciparum* malaria in Assam, India: Timing of recurrence and anti-malarial drug concentrations in whole blood. *Am J Trop Med Hyg*. (2003). *69*, 555-557.

[252] Pasvol, G; Newton, CR; Winstanley, PA; Watkins, WM; Peshu, NM; Were, JB; Marsh, K; Warrell, DA; Quinine treatment of severe *falciparum* malaria in African children: A randomized comparison of three regimens. *Am J Trop Med Hyg*. (1991). *45*, 702-1703.

[253] Eisenhut, M; Omari, AA; Intrarectal quinine for treating *Plasmodium falciparum* malaria. *Cochrane Database Syst Rev,* (2005). 25, CD004009.

[254] McIntosh, HM; Olliaro, P; Artemisinin derivatives for treating severe malaria. *Cochrane Database Syst Rev*. (2000). 2) CD000527.

[255] Adam, I; Idris, HM; Mohamed-Ali, AA; Aelbasit, IA; Elbashir, MI; Comparison of intramuscular artemether and intravenous quinine in the treatment of Sudanese children with severe *falciparum* malaria. *East Afr Med J.* (2002). *79*, 621-625.

[256] Satti, GM; Elhassan, SH; Ibrahim, SA; The efficacy of artemether versus quinine in the treatment of cerebral malaria. *J Egypt Soc Parasitol.* (2002). *32*, 611-623.

[257] Afolabi, BB; Okoromah, CN; Intramuscular arteether for treating severe malaria. *Cochrane Database Syst Rev.* (2004). 18, CD004391.

[258] Birku, Y; Makonnen, E; Bjorkman, A; Comparison of rectal artemisinin with intravenous quinine in the treatment of severe malaria in Ethiopia. *East Afr Med J.* (1999). *76*, 154-159.

[259] Griffith, KS; Lewis, LS; Mali, S; Parise, ME; Treatment of malaria in the United States: A systematic review. *JAMA.* (2007). *297*, 2264-2277.

[260] Orton, L; Garner, P; Drugs for treating uncomplicated malaria in pregnant women. *Cochrane Database Syst Rev.* (2005). 3, CD004912.

[261] WHO. Regional Office for Africa. A strategic framework for malaria prevention and control during pregnancy in the African region. Brazzaville: World Health Organization, 2004.

[262] Nosten, F; McGready, R; The treatment of malaria in pregnancy. In: PE Duffyand M Fried(Eds.), *Malaria in Pregnancy: Deadly Parasite, Susceptible Host.* London: Taylor & Francis, (2001). 223-241.

[263] Taylor, WR; White, NJ; Antimalarial drug toxicity: A review. *Drug Saf.* (2004). *27*, 25-61.

[264] Deen, JL; von Seidlein, L; Pinder, M; Walraven, GE; Greenwood, BM; The safety of the combination artesunate and pyrimethamine-sulfadoxine given during pregnancy. *Trans R Soc Trop Med Hyg.* (2001). *95*, 424-428.

[265] Fakeye, TO; Fehintola, FA; Ademowo, OG; Walker, O; Therapeutic monitoring of chloroquine in pregnant women with malaria. *West Afr J Med.* (2002). *21*, 286-287.

[266] Massele, AY; Kilewo, C; Aden Abdi, Y; Tomson, G; Diwan, VK; Ericsson, O; Rimoy, G; Gustafsson, LL; Chloroquine blood concentrations and malaria prophylaxis in Tanzanian women during the second and third trimesters of pregnancy. *Eur J Clin Pharmacol.* (1997). *52*, 299-305.

[267] McGready, R; Stepniewska, K; Seaton, E; Cho, T; Cho, D; Ginsberg, A; Edstein, MD; Ashley, E; Looareesuwan, S; White, NJ; Nosten, F; Pregnancy and use of oral contraceptives reduces the biotransformation of proguanil to cycloguanil. *Eur J Clin Pharmacol.* (2003). *59*, 553-557.

[268] McGready, R; Stepniewska, K; Edstein, MD; Cho, T; Gilveray, G; Looareesuwan, S; White, NJ; Nosten, F; The pharmacokinetics of atovaquone and proguanil in pregnant women with acute *falciparum* malaria. *Eur J Clin Pharmacol.* (2003). *59*, 545-552.

[269] McGready, R; Stepniewska, K; Ward, SA; Cho, T; Gilveray, G; Looareesuwan, S; White, NJ; Nosten, F; Pharmacokinetics of dihydroartemisinin following oral artesunate treatment of pregnant women with acute uncomplicated *falciparum* malaria. *Eur J Clin Pharmacol.* (2006). *62*, 367-371.

[270] Nosten, F; McGready, R; d'Alessandro, U; Bonell, A; Verhoeff, F; Menendez, C; Mutabingwa, T; Brabin, B; Antimalarial drugs in pregnancy: A review. *Curr Drug Saf.* (2006). *1*, 1-15.

[271] Nosten, F; White, NJ; Artemisinin-based combination treatment of *falciparum* malaria. *Am J Trop Med Hyg.* (2007). *77*(Suppl. 6), 181-192.

[272] Mendis, K; Sina, BJ; Marchesini, P; Carter, R; The neglected burden of *Plasmodium vivax* malaria. *Am J Trop Med Hyg.* (2001). *64*(Suppl. 1-2), 97-106.

[273] Guerra, CA; Snow, RW; Hay, SI;. Mapping the global extent of malaria in 2005.*Trends Parasitol.* (2006). *22*, 353-358.

[274] Price, RN; Tjitra, E; Guerra, CA; Yeung, S; White, NJ; Anstey, NM; *Vivax* malaria: Neglected and not benign. *Am J Trop Med Hyg.* (2007). *77*(Suppl. 6), 79-87.

[275] Bell, D; Peeling, RW; WHO-Regional Office for the Western Pacific/TDR.Evaluation of rapid diagnostic tests: Malaria. *Nat Rev Microbiol.* (2006). *4*(Suppl. 9), S34-38.

[276] Baird, JK; Chloroquine resistance in Plasmodium *vivax. Antimicrob Agents Chemother.* (2004). *48*, 4075-4083.

[277] Hastings, MD; Porter, KM; Maguire, JD; Susanti, I; Kania, W; Bangs, MJ; Sibley, CH; Baird, JK; Dihydrofolate reductase mutations in *Plasmodium vivax* from Indonesia and therapeutic response to sulfadoxine plus pyrimethamine. *J Infect Dis.* (2004). *189*, 744-750.

[278] Burns, M; Baker, J; Auliff, AM; Gatton, ML; Edstein, MD; Cheng, Q; Efficacy of sulfadoxine-pyrimethamine in the treatment of uncomplicated *Plasmodium falciparum* malaria in East Timor. *Am J Trop Med Hyg.* (2006). *74*, 361-366.

[279] Mandi, G; Mockenhaupt, FP; Coulibaly, B; Meissner, P; Müller, O; Efficacy of amodiaquine in the treatment of uncomplicated *falciparum* malaria in young children of rural north-western Burkina Faso. *Malar J.* (2008). *7*, 58.

[280] Maguire, JD; Krisin; Marwoto, H; Richie, TL; Fryauff, DJ; Baird, JK; Mefloquine is highly efficacious against chloroquine-resistant *Plasmodium vivax* malaria and *Plasmodium falciparum* malaria in Papua, Indonesia. *Clin Infect Dis.* (2006). *42*, 1067-1072.

[281] Tjitra, E; Baker, J; Suprianto, S; Cheng, Q; Anstey, NM; Therapeutic efficacies of artesunate-sulfadoxine-pyrimethamine and chloroquine-sulfadoxine-pyrimethamine in *vivax* malaria pilot studies: Relationship to *Plasmodium vivax* dhfr mutations. *Antimicrob Agents Chemother.* (2002). *46*, 3947-3953.

[282] Price, RN; Hasugian, AR; Ratcliff, A; Siswantoro, H; Purba, HL; Kenangalem, E; Lindegardh, N; Penttinen, P; Laihad, F; Ebsworth, EP; Anstey, NM; Tjitra, E; Clinical and pharmacological determinants of the therapeutic response to dihydroartemisinin-piperaquine for drug-resistant malaria. *Antimicrob Agents Chemother.* (2007). *51*, 4090-4097.

[283] WHO. Chapter 7 Malaria in: International Travel and Health. Geneva, Switzerland: World Health Onganization, (2008).p. 146. *http://www.who.int/ith/en/*

[284] Shanks, GD; Edstein, MD; Modern malaria chemoprophylaxis. *Drugs.* (2005). *65*(15):2091-2110.

[285] Wyler, DJ; Malaria chemoprophylaxis for the traveler. *N Engl J Med.* (1993). *329*, 31-37.

[286] Stephen, R; Jong, EC; Travelers' and Immigrants' Health. In: RL Guerrant, PH Walker, PF Wyler, (Eds). Tropical Infectious Diseases: Principles, Pathogens and Practice. Philadelphia: Churchill Livingstone (1999). p.108.

[287] Dondorp, AM; Newton, PN; Mayxay, M; Van Damme, W; Smithuis, FM; Yeung, S; Petit, A; Lynam, AJ; Johnson, A; Hien, TT; McGready, R; Farrar, JJ; Looareesuwan, S; Day, NP; Green, MD; White, NJ; Fake antimalarials in Southeast Asia are a major impediment to malaria control: Multinational cross-sectional survey on the prevalence of fake antimalarials. *Trop Med Int Health.* (2004). 9, 1241-1246.

[288] Marra, F; Salzman, JR; Ensom, MH; Atovaquone-proguanil for prophylaxis and treatment of malaria. *Ann Pharmacother.* (2003). *37*, 1266-1275.

[289] Kofoed, K; Petersen, E; The efficacy of chemoprophylaxis against malaria with chloroquine plus proguanil, mefloquine, and atovaquone plus proguanil in travelers from Denmark. *J Travel Med.* (2003). *10*, 150-154.

[290] Strauss, R; Pfeifer, C; Malaria in Austria 1990-2000. *Euro Surveill.* (2003). *8*, 91-96.

[291] Rosen, JB; Breman, JG; Manclark, CR; Meade, BD; Collins, WE; Lobel, HO; Saliou, P; Roberts, JM; Campaoré, P; Miller, MA; Malaria chemoprophylaxis and the serologic response to measles and diphtheria-tetanus-whole-cell pertussis vaccines. *Malar J.* (2005). *4*, 53.

[292] Gkrania-Klotsas, E; Lever, AM; An update on malaria prevention, diagnosis and treatment for the returning traveler. *Blood Rev.* (2007). *21*, 73-87.

[293] Andersen, SL; Oloo, AJ; Gordon, DM; Ragama, OB; Aleman, GM; Berman, JD; Tang, DB; Dunne, MW; Shanks, GD; Successful double-blinded, randomized, placebo-controlled field trial of azithromycin and doxycycline as prophylaxis for malaria in western Kenya. *Clin Infect Dis.* (1998). *26*, 146-150.

[294] Wiltz, SA; Crawford, P; Nichols, W; Hayes, M; Clinical inquiries. What is the most effective and safe malaria prophylaxis during pregnancy? *J Fam Pract.* (2008). *57*, 51-53.

[295] Garfield, RM; Vermund, SH; Changes in malaria incidence after mass drug administration in Nicaragua. *Lancet.* (1983). *2*, 500-503.

[296] von Seidlein, L; Walraven, G; Milligan, PJ; Alexander, N; Manneh, F; Deen, JL; Coleman, R; Jawara, M; Lindsay, SW; Drakeley, C; De Martin, S; Olliaro, P; Bennett, S; Schim, van der; Loeff, M; Okunoye, K; Targett, GA; McAdam, KP; Doherty, JF; Greenwood, BM; Pinder, M; The effect of mass administration of sulfadoxine-pyrimethamine combined with artesunate on malaria incidence: A double-blind, community-randomized, placebo-controlled trial in The Gambia. *Trans R Soc Trop Med Hyg.* (2003). *97*, 217-225.

[297] Kaneko, A; Taleo, G; Kalkoa, M; Yamar, S; Kobayakawa, T; Björkman, A; Malaria eradication on islands. *Lancet.* (2000). *356*, 1560-1564.

[298] Payne, D; Did medicated salt hasten the spread of chloroquine resistance in *Plasmodium falciparum*? *Parasitol Today.* (1988). *4*, 112-115.

[299] Vallely, A; Vallely, L; Changalucha, J; Greenwood, B; Chandramohan, D; Intermittent preventive treatment for malaria in pregnancy in Africa: What's new, what's needed? *Malar J.* (2007). *6*, 16.

[300] Cot, M; Le Hesran, JY; Miailhes, P; Esveld, M; Etya'ale, D; Breart, G; Increase of birth weight following chloroquine chemoprophylaxis during the first pregnancy: Results of a randomized trial in Cameroon. *Am J Trop Med Hyg.* (1995). *53*, 581-585.

[301] Parise, ME; Ayisi, JG; Nahlen, BL; Schultz, LJ; Roberts, JM; Misore, A; Muga, R; Oloo, AJ; Steketee, RW; Efficacy of sulfadoxine-pyrimethamine for prevention of placental malaria in an area of Kenya with a high prevalence of malaria and human immunodeficiency virus infection. *Am J Trop Med Hyg.* (1998). *59*, 813-822.

[302] Shulman, CE; Dorman, EK; Cutts, F; Kawuondo, K; Bulmer, JN; Peshu, N; Marsh, K (1999). Intermittent sulphadoxine-pyrimethamine to prevent severe anemia secondary to malaria in pregnancy: A randomized placebo-controlled trial. *Lancet.* (1999). *353*, 632-636.

[303] Garner, P; Gülmezoglu, AM; Drugs for preventing malaria in pregnant women. *Cochrane Database Syst Rev.* (2006). 4, CD000169.

[304] White, NJ. Intermittent presumptive treatment for malaria. *PLoS Med.* (2005). *2*, e3.

[305] Schultz, LJ; Steketee, RW; Macheso, A; Kazembe, P; Chitsulo, L; Wirima, JJ; The efficacy of antimalarial regimens containing sulfadoxine-pyrimethamine and/or chloroquine in preventing peripheral and placental *Plasmodium falciparum* infection among pregnant women in Malawi. *Am J Trop Med Hyg.* (1994). *51*, 515-522.

[306] Rogerson, SJ; Chaluluka, E; Kanjala, M; Mkundika, P; Mhango, C; Molyneux, ME; Intermittent sulfadoxine-pyrimethamine in pregnancy: Effectiveness against malaria morbidity in Blantyre, Malawi, in 1997-1999. *Trans R Soc Trop Med Hyg*(2000). *94*, 549-553.

[307] Phillips-Howard, PA; Wood, D; The safety of antimalarial drugs in pregnancy. *Drug Saf.* (1996). *14*, 131-145.

[308] Centers for Disease Control and Prevention Web site. Diseases: Malaria: Prevention, Pregnant Women, Public Info, 2008. Available at: http://wwwnc.cdc. gov/travel/content/diseases.aspx#malaria. (accessed December 2010)

[309] Schlagenhauf, P; Mefloquine for malaria chemoprophylaxis 1992-1998: A review. *J Travel Med.* (1999). *6*, 122-133.

[310] McGregor, IA; Gilles, HM; Walters, JH; Davies, AH; Pearson, FA; Effects of heavy and repeated malarial infections on Gambian infants and children; effects of erythrocytic parasitization. *Br Med J.* (1956). *2*, 686-692.

[311] Geerligs, PD; Brabin, BJ; Eggelte, TA; Analysis of the effects of malaria chemoprophylaxis in children on haematological responses, morbidity and mortality. *Bull World Health Organ.* (2003). *81*, 205-216.

[312] Greenwood, BM; Greenwood, AM; Bradley, AK; Snow, RW; Byass, P; Hayes, RJ; N'Jie, AB; Comparison of two strategies for control of malaria within a primary health care programme in the Gambia. *Lancet.* (1988). *1*, 1121-1127.

[313] Meremikwu, MM; Donegan, S; Esu, E; Chemoprophylaxis and intermittent treatment for preventing malaria in children. *Cochrane Database Syst Rev.* (2008). *2*, CD003756.

[314] Greenwood, B; Review: Intermittent preventive treatment--a new approach to the prevention of malaria in children in areas with seasonal malaria transmission. *Trop Med Int Health.* (2006). *11*, 983-991.

[315] Schellenberg, D; Menendez, C; Kahigwa, E; Aponte, J; Vidal, J; Tanner, M; Mshinda, H; Alonso, P; Intermittent treatment for malaria and anaemia control at time of routine vaccinations in Tanzanian infants: A randomized, placebo-controlled trial. *Lancet.* (2001). *357*, 1471-1477.

[316] Massaga, JJ; Kitua, AY; Lemnge, MM; Akida, JA; Malle, LN; Rønn, AM; Theander, TG; Bygbjerg, IC; Effect of intermittent treatment with amodiaquine on anemia and malarial fevers in infants in Tanzania: A randomized placebo-controlled trial. *Lancet.* (2003). *361*, 1853-1860.

[317] Verhoef, H; West, CE; Nzyuko, SM; de Vogel, S; van der Valk, R; Wanga, MA; Kuijsten, A; Veenemans, J; Kok, FJ; Intermittent administration of iron and sulfadoxine-pyrimethamine to control anemia in Kenyan children: A randomized controlled trial. *Lancet.* (2002). *360*, 908-914.

[318] Desai, MR; Mei, JV; Kariuki, SK; Wannemuehler, KA; Phillips-Howard, PA; Nahlen, BL; Kager, PA; Vulule, JM; ter Kuile, FO; Randomized, controlled trial of daily iron supplementation and intermittent sulfadoxine-pyrimethamine for the treatment of mild childhood anemia in western Kenya. *J Infect Dis.* (2003). *187*, 658-666.

[319] O'Meara, WP; Breman, JG; McKenzie, FE. The promise and potential challenges of intermittent preventive treatment for malaria in infants (IPTi). *Malar J.* (2005). *4*, 33.

[320] Cissé, B; Sokhna, C; Boulanger, D; Milet, J; Bâ el, H; Richardson, K; Hallett, R; Sutherland, C; Simondon, K; Simondon, F; Alexander, N; Gaye, O; Targett, G; Lines, J; Greenwood, B; Trape, JF; Seasonal intermittent preventive treatment with artesunate and sulfadoxine-pyrimethamine for prevention of malaria in Senegalese children: A randomized, placebo-controlled, double-blind trial. *Lancet.* (2006). *367*, 659-667.

[321] Camus, D; Djossou, F; Schilthuis, HJ; Høgh, B; Dutoit, E; Malvy, D; Roskell, NS; Hedgley, C; De Boever, EH; Miller, GB; International Malarone Study Team.Atovaquone-proguanil versus chloroquine-proguanil for malaria prophylaxis in nonimmune pediatric

travelers: Results of an international, randomized, open-label study. *Clin Infect Dis*. (2004). *38*, 1716-1723.

[322] Collee, GG; Samra, GS; Hanson, GC; Chloroquine poisoning: Ventricular fibrillation following 'trivial' overdose in a child. *Intensive Care Med*. (1992). *18*, 170-171.

[323] Otoo, LN; Snow, RW; Menon, A; Byass, P; Greenwood, BM; Immunity to malaria in young Gambian children after a two-year period of chemoprophylaxis. *Trans R Soc Trop Med Hyg*. (1988). *82*, 59-65.

[324] Kobbe, R; Kreuzberg, C; Adjei, S; Thompson, B; Langefeld, I; Thompson, PA; Abruquah, HH; Kreuels, B; Ayim, M; Busch, W; Marks, F; Amoah, K; Opoku, E; Meyer, CG; Adjei, O; May, J; A randomized controlled trial of extended intermittent preventive antimalarial treatment in infants. *Clin Infect Dis*. (2007). *45*, 16-25.

[325] Chandramohan, D; Owusu-Agyei, S; Carneiro, I; Awine, T; Amponsa-Achiano, K; Mensah, N; Jaffar, S; Baiden, R; Hodgson, A; Binka, F; Greenwood, B; Cluster randomized trial of intermittent preventive treatment for malaria in infants in area of high, seasonal transmission in Ghana. *BMJ*. (2005). 331, 727-733.

[326] Schellenberg, D, Menendez, C, Aponte, JJ, Kahigwa, E, Tanner M, Mshinda H, Alonso P.Intermittent preventive antimalarial treatment for Tanzanian infants: Follow-up to age 2 years of a randomized, placebo-controlled trial. Lancet. (2005). *365*, 1481-1483.

[327] Macete, E; Aide, P; Aponte, JJ; Sanz, S; Mandomando, I; Espasa, M; Sigauque, B; Dobaño, C; Mabunda, S; DgeDge, M; Alonso, P; Menendez, C. Intermittent preventive treatment for malaria control administered at the time of routine vaccinations in Mozambican infants: A randomized, placebo-controlled trial. *J Infect Dis*. (2006). 194, 276-285.

[328] Houston, S; Keystone, JS; Kain, KC; Mefloquine to prevent malaria. Mefloquine remains the best drug. *Brit Med J* (1998). *316*, 1980-1981.

[329] Schlagenhauf, P; Steffen, R; Stand-by treatment of malaria in travelers: A review. *J Trop Med Hyg*, (1997). *97*, 151-160.

[330] Kain, KC; Shanks, GD; Keystone, JS; Malaria chemoprophylaxis in the age of drug resistance. I. Currently recommended drug regimens. *Clin Infect Dis*. (2001). *33*, 226-234.

[331] Chen, LH; Keystone, JS; New strategies for the prevention of malaria in travelers. *Infect Dis Clin North Am*. (2005). *19*, 185-210.

[332] Cobelens, FG; Leentvaar-Kuijpers, A; Compliance with malaria chemoprophylaxis and preventative measures against mosquito bites among Dutch travelers. *Trop Med Int Health*. (1997). *2*, 705-713.

[333] Dorsey, G; Gandhi, M; Oyugi, JH; Rosenthal, PJ; Difficulties in the prevention, diagnosis, and treatment of imported malaria. *Arch Intern Med*. (2000). *160*, 2505-2510.

[334] Gyorkos, TW; Svenson, JE; Maclean, JD; Mohamed, N; Remondin, MH; Franco, ED. Compliance with antimalarial chemoprophylaxis and the subsequent development of malaria: A matched case-control study. *Am J Trop Med Hyg*. (1995). *53*, 511-517.

[335] Jong, EC; Nothdurft, HD; Current drugs for antimalarial chemoprophylaxis: A review of efficacy and safety. *J Travel Med*. (2001). *8*(Suppl 3), S48-56.

[336] Baird, JK; Hoffman, SL; Prevention of malaria in travelers. *Med Clin North Am*. (1999). *83*, 923-944.

[337] McKeage, K; Scott, L; Atovaquone/proguanil: A review of its use for the prophylaxis of *Plasmodium falciparum* malaria. *Drugs*. (2003). *63*, 597-623.

[338] Giao, PT; De Vries, PJ; Hung, LQ; Binh, TQ; Nam, NV; Kager, PA; Atovaquone-proguanil for recrudescent *Plasmodium falciparum* in Vietnam. *Ann Trop Med Parasitol*. (2003). *97*, 575-580.

[339] Newton, P; Proux, S; Green, M; Smithuis, F; Rozendaal, J; Prakongpan, S; Chotivanich, K; Mayxay, M; Looareesuwan, S; Farrar, J; Nosten, F; White, NJ; Fake artesunate in Southeast Asia. *Lancet*. (2001). 357, 1948-1950.

[340] Wernsdorfer, WH. Coartemether (artemether and lumefantrine): An oral antimalarial drug. *Expert Rev Anti Infect Ther*. (2004). 2, 181-196.

[341] Rendi-Wagner, P; Noedl, H; Wernsdorfer, WH; Wiedermann, G; Mikolasek, A; Kollaritsch, H. Unexpected frequency, duration and spectrum of adverse events after therapeutic dose of mefloquine in healthy adults. *Acta Trop*. (2002). *81*, 167-173.

[342] Menendez, C; Kahigwa, E; Hirt, R; Vounatsou, P; Aponte, JJ; Font, F; Acosta, CJ; Schellenberg, DM; Galindo, CM; Kimario, J; Urassa, H; Brabin, B; Smith, TA; Kitua, AY; Tanner, M; Alonso, PL. Randomized placebo-controlled trial of iron supplementation and malaria chemoprophylaxis for prevention of severe anemia and malaria in Tanzanian infants. *Lancet*. (1997). *350*, 844-850.

[343] Bottieau, E; Clerinx, J; Van Den Enden, E; Van Esbroeck, M; Colebunders, R; Van Gompel, A; Van Den Ende, J. Imported non-*Plasmodium falciparum* malaria: A five-year prospective study in a European referral center. *Am J Trop Med Hyg*. (2006). *75*, 133-138.

[344] Spudick, JM; Garcia, LS; Graham, DM; Haake, DA. Diagnostic and therapeutic pitfalls associated with primaquine-tolerant *Plasmodium vivax*. *J Clin Microbiol*. (2005). *43*, 978-981.

[345] Pukrittayakamee, S; Imwong, M; Looareesuwan, S; White, NJ; Therapeutic responses to antimalarial and antibacterial drugs in *vivax* malaria. *Acta Trop*. (2004). *89*, 351-356.

[346] Mühlberger, N; Jelinek, T; *et al*. Epidemiology and clinical features of *vivax* malaria imported to Europe: Sentinel surveillance data from TropNetEurop. *Malar J*. (2004). *3*, 5.

[347] Schwartz, IK; Lackritz, EM; Patchen, LC; Chloroquine-resistant *Plasmodium vivax* from Indonesia. *N Engl J Med*. (1991). *324*, 927.

[348] Baird, JK; Schwartz, E; Hoffman, SL; Prevention and treatment of *vivax* malaria. *Curr Infect Dis Rep*. (2007). *9*, 39-46.

[349] Chen, LH; Wilson, ME; Schlagenhauf, P; Controversies and misconceptions in malaria chemoprophylaxis for travelers. *JAMA*. (2007). *297*, 2251-2263.

[350] Skarbinski, J; James, EM; Causer, LM; Barber, AM; Mali, S; Nguyen-Dinh, P; Roberts, JM; Parise, ME; Slutsker, L; Newman, RD; Malaria surveillance--United States, 2004. *MMWR Surveill Summ*. (2006). *55*, 23-37.

[351] Schwartz, E; Parise, M; Kozarsky, P; Cetron, M; Delayed onset of malaria--implications for chemoprophylaxis in travelers. *N Engl J Med*. (2003). *349*, 1510-1516.

[352] Soto, J; Toledo, J; Rodriquez, M; Sanchez, J; Herrera, R; Padilla, J; Berman, J; Primaquine prophylaxis against malaria in nonimmune Colombian soldiers: Efficacy and toxicity. A randomized, double-blind, placebo-controlled trial. *Ann Intern Med*. (1998). *129*, 241-244.

[353] Baird, JK; Hoffman, SL; Primaquine therapy for malaria. *Clin Infect Dis*. (2004). *39*, 1336-1345.

[354] Guerra, CA; Snow, RW; Hay, SI; Mapping the global extent of malaria in 2005. *Trends Parasitol*. (2006). *22*, 353-358.

[355] White, NJ; Why is it that antimalarial drug treatments do not always work? *Ann Trop Med Parasitol*. (1998). *92*, 449-458.

[356] Lipsitch, M; Levin, BR; The population dynamics of antimicrobial chemotherapy. *Antimicrob Agents Chemother*. (1997). *41*, 363-373.

[357] Rathod, PK; McErlean, T; Lee, PC; Variations in frequencies of drug resistance in *Plasmodium falciparum*. *Proc Natl Acad Sci U S A*. (1997). *94*, 9389-9393.

[358] Vieira, PP; Ferreira, MU; Alecrim, MG; Alecrim, WD; da Silva, LH; Sihuincha, MM; Joy, DA; Mu, J; Su, XZ; Zalis, MG; pfcrt Polymorphism and the spread of chloroquine resistance in *Plasmodium falciparum* populations across the Amazon Basin. *J Infect Dis.* (2004). *190*, 417-424.

[359] White, NJ; Antimalarial drug resistance: The pace quickens. *J Antimicrob Chemother.* (1992). *30*, 571-585.

[360] Talisuna, AO; Bloland, P; D'Alessandro, U; History, dynamics, and public health importance of malaria parasite resistance. *Clin Microbiol Rev.* (2004). *17*, 235-254.

[361] Ménard, D; Ratsimbasoa, A; Randrianarivelojosia, M; Rabarijaona, LP; Raharimalala, L; Domarle, O; Randrianasolo, L; Randriamanantena, A; Jahevitra, M; Andriantsoanirina, V; Rason, MA; Raherinjafy, R; Rakotomalala, E; Tuseo, L; Raveloson, A; Assessment of the efficacy of antimalarial drugs recommended by the National Malaria Control Programme in Madagascar: Updated baseline data from randomized and multi-site clinical trials. *Malar J.* (2008). *7*, 55.

[362] Sumawinata, IW; Bernadeta, Leksana, B; Sutamihardja, A; Purnomo, Subianto, B; Sekartuti, Fryauff, DJ; Baird, JK; Very high risk of therapeutic failure with chloroquine for uncomplicated *Plasmodium falciparum* and *P. vivax* malaria in Indonesian Papua. *Am J Trop Med Hyg.* (2003). *68*, 416-420.

[363] Checchi, F; Durand, R; Balkan, S; Vonhm, BT; Kollie, JZ; Biberson, P; Baron, E; Le Bras, J; Guthmann, JP; High *Plasmodium falciparum* resistance to chloroquine and sulfadoxine-pyrimethamine in Harper, Liberia: Results *in vivo* and analysis of point mutations. *Trans R Soc Trop Med Hyg.* (2002). *96*, 664-669.

[364] Kofoed, PE; Có, F; Johansson, P; Dias, F; Cabral, C; Hedegaard, K; Aaby, P; Rombo, L; Treatment of uncomplicated malaria in children in Guinea-Bissau with chloroquine, quinine, and sulfadoxine-pyrimethamine. *Trans R Soc Trop Med Hyg.* (2002). *96*, 304-309.

[365] Plowe, CV; Kublin, JG; Dzinjalamala, FK; Kamwendo, DS; Mukadam, RA; Chimpeni, P; Molyneux, ME; Taylor, TE; Sustained clinical efficacy of sulfadoxine-pyrimethamine for uncomplicated *falciparum* malaria in Malawi after 10 years as first line treatment: Five year prospective study. *BMJ.* (2004). 328, 545.

[366] Anderson, TJ; Mapping drug resistance genes in *Plasmodium falciparum* by genome-wide association. *Curr Drug Targets Infect Disord.* (2004). *4*, 65-78.

[367] Nosten, F; ter Kuile, F; Chongsuphajaisiddhi, T; Luxemburger, C; Webster, HK; Edstein, M; Phaipun, L; Thew, KL; White, NJ; Mefloquine-resistant *falciparum* malaria on the Thai-Burmese border. *Lancet.* (1991). *337*, 1140-1143.

[368] Wernsdorfer, WH; Epidemiology of drug resistance in malaria. *Acta Trop.* (1994). *56*, 143-156.

[369] Wongsrichanalai, C; Sirichaisinthop, J; Karwacki, JJ; Congpuong, K; Miller, RS; Pang, L; Thimasarn, K. Drug resistant malaria on the Thai-Myanmar and Thai-Cambodian borders.Southeast *Asian J Trop Med Public Health.* (2001). *32*, 41-49.

[370] Reed, MB; Saliba, KJ; Caruana, SR; Kirk, K; Cowman, AF; Pgh1 modulates sensitivity and resistance to multiple antimalarials in *Plasmodium falciparum. Nature.* (2000). 403, 906-909.

[371] Looareesuwan, S; Wilairatana, P; Chalermarut, K; Rattanapong, Y; Canfield, CJ; Hutchinson, DB: Efficacy and safety of atovaquone/proguanil compared with mefloquine for treatment of acute *Plasmodium falciparum* malaria in Thailand. *Am J Trop Med Hyg.* (1999). *60*, 526-532.

[372] Rahman, MR; Paul, DC; Rashid, M; Ghosh, A; Bangali, AM; Jalil, MA; Faiz, MA; A randomized controlled trial on the efficacy of alternative treatment regimens for

uncomplicated *falciparum* malaria in a multidrug-resistant *falciparum* area of Bangladesh-- narrowing the options for the National Malaria Control Programme? *Trans R Soc Trop Med Hyg*. (2001). *95*, 661-667.

[373] Hopperus, Buma AP; van Thiel, PP; Lobel, HO; Ohrt, C; van Ameijden, EJ; Veltink, RL; Tendeloo, DC; van Gool, T; Green, MD; Todd, GD; Kyle, DE; Kager, PA; Long-term malaria chemoprophylaxis with mefloquine in Dutch marines in Cambodia. *J Infect Dis*. (1996). *173*, 1506-1509.

[374] Ohrt, C; Richie, TL; Widjaja, H; Shanks, GD; Fitriadi, J; Fryauff, DJ; Handschin, J; Tang, D; Sandjaja, B; Tjitra, E; Hadiarso, L; Watt, G; Wignall, FS; Mefloquine compared with doxycycline for the prophylaxis of malaria in Indonesian soldiers. A randomized, double-blind, placebo-controlled trial. *Ann Intern Med*. (1997). *126*, 963-972.

[375] Lobel, HO; Varma, JK; Miani, M; Green, M; Todd, GD; Grady, K; Barber, AM; Monitoring for mefloquine-resistant *Plasmodium falciparum* in Africa: Implications for travelers' health. *Am J Trop Med Hyg*. (1998). *59*, 129-132.

[376] Avila, JC; Villaroel, R; Marquiño, W; Zegarra, J; Mollinedo, R; Ruebush, TK; Efficacy of mefloquine and mefloquine-artesunate for the treatment of uncomplicated *Plasmodium falciparum* malaria in the Amazon region of Bolivia. *Trop Med Int Health*. (2004). *9*, 217-221.

[377] Schlagenhauf, P; Tschopp, A; Johnson, R; Nothdurft, HD; Beck, B; Schwartz, E; Herold, M; Krebs, B; Veit, O; Allwinn, R; Steffen, R; Tolerability of malaria chemoprophylaxis in non-immune travelers to sub-Saharan Africa:Multicenter, randomized, double blind, four arm study. *BMJ*. (2003). *327*, 1078-1084.

[378] van Riemsdijk, MM; Sturkenboom, MC; Ditters, JM; Ligthelm, RJ; Overbosch, D; Stricker, BH; Atovaquone plus chloroguanide versus mefloquine for malaria prophylaxis: A focus on neuropsychiatric adverse events. *Clin Pharmacol Ther*. (2002). *72*, 294-301.

[379] van Riemsdijk, MM; Sturkenboom, MC; Ditters, JM; Tulen, JH; Ligthelm, RJ; Overbosch, D; Stricker, BH; Low body mass index is associated with an increased risk of neuropsychiatric adverse events and concentration impairment in women on mefloquine. *Br J Clin Pharmacol*. (2004). *57*, 506-512.

[380] Zalis, MG; Pang, L; Silveira, MS; Milhous, WK; Wirth, DF; Characterization of *Plasmodium falciparum* isolated from the Amazon region of Brazil: Evidence for quinine resistance. *Am J Trop Med Hyg*. (1998). *58*, 630-637.

[381] McGready, R; Brockman, A; Cho, T; Cho, D; van Vugt, M; Luxemburger, C; Chongsuphajaisiddhi, T; White, NJ; Nosten, F; Randomized comparison of mefloquine-artesunate versus quinine in the treatment of multidrug-resistant *falciparum* malaria in pregnancy. *Trans R Soc Trop Med Hyg*. (2000). *94*, 689-693.

[382] Pukrittayakamee, S; Wanwimolruk, S; Stepniewska, K; Jantra, A; Huyakorn, S; Looareesuwan, S; White, NJ; Quinine pharmacokinetic-pharmacodynamic relationships in uncomplicated *falciparum* malaria. *Antimicrob Agents Chemother*. (2003). *47*, 3458-3463.

[383] Aché, A; Escorihuela, M; Vivas, E; Páez, E; Miranda, L; Matos, A; Pérez, W; Díaz, O; Izarra, E; *In vivo* drug resistance of *falciparum* malaria in mining areas of Venezuela. *Trop Med Int Health*. (2002). *7*, 737-743.

[384] Roche, J; Guerra-Neira, A; Raso, J; Benito, A; Surveillance of *in vivo* resistance of *Plasmodium falciparum* to antimalarial drugs from 1992 to 1999 in Malabo (Equatorial Guinea). *Am J Trop Med Hyg*. (2003). *68*, 598-601.

[385] Demar, M; Carme, B; *Plasmodium falciparum in vivo* resistance to quinine: Description of two RIII responses in French Guiana. *Am J Trop Med Hyg*. (2004). *70*, 125-127.

[386] Taylor, WR; Richie, TL; Fryauff, DJ; Picarima, H; Ohrt, C; Tang, D; Braitman, D; Murphy, GS; Widjaja, H; Tjitra, E; Ganjar, A; Jones, TR; Basri, H; Berman, J; Malaria prophylaxis using azithromycin: A double-blind, placebo-controlled trial in Irian Jaya, Indonesia. *Clin Infect Dis*. (1999). *28*, 74-81.

[387] Fungladda, W; Honrado, ER; Thimasarn, K; Kitayaporn, D; Karbwang, J; Kamolratanakul, P; Masngammueng, R; Compliance with artesunate and quinine + tetracycline treatment of uncomplicated *falciparum* malaria in Thailand. *Bull World Health Organ*. (1998). *76*(Suppl. 1), 59-66.

[388] Arnold, J; Alving, AS; Clayman, CB; Induced primaquine resistance in *vivax* malaria. *Trans R Soc Trop Med Hyg*. (1961). *55*, 345-350.

[389] Targett, G; Drakeley, C; Jawara, M; von Seidlein, L; Coleman, R; Deen, J; Pinder, M; Doherty, T; Sutherland, C; Walraven, G; Milligan, P; Artesunate reduces but does not prevent posttreatment transmission of *Plasmodium falciparum* to Anopheles gambiae. *J Infect Dis*. (2001). *183*, 1254-1259.

[390] Meshnick, SR; Taylor, TE; Kamchonwongpaisan, S; Artemisinin and the antimalarial endoperoxides: From herbal remedy to targeted chemotherapy. *Microbiol Rev*. (1996). *60*, 301-315.

[391] Gay, F; Ciceron, L; Litaudon, M; Bustos, MD; Astagneau, P; Diquet, B; Danis, M; Gentilini, M; *In vitro* resistance of *Plasmodium falciparum* to qinghaosu derivatives in west Africa. *Lancet*. (1994). *343*, 850-851.

[392] Basco, LK; Ringwald, P; Molecular epidemiology of malaria in Cameroon. X. Evaluation of *PFMDR1* mutations as genetic markers for resistance to amino alcohols and artemisinin derivatives. *Am J Trop Med Hyg*. (2002). *66*, 667-671.

[393] Wongsrichanalai, C; Nguyen, TD; Trieu, NT; Wimonwattrawatee, T; Sookto, P; Heppner, DG; Kawamoto, F; *In vitro* susceptibility *of Plasmodium falciparum* isolates in Vietnam to artemisinin derivatives and other antimalarials. *Acta Trop*. (1997). *63*, 151-158.

[394] Treeprasertsuk, S; Viriyavejakul, P; Silachamroon, U; Vannphan, S; Wilairatana, P; Looareesuwan, S; Is there any artemisinin resistance in *falciparum* malaria? Southeast *Asian J Trop Med Public Health*. (2000). *31*, 825-828.

[395] Peters, W; Robinson, BL; The chemotherapy of rodent malaria. LVI. Studies on the development of resistance to natural and synthetic endoperoxides. *Ann Trop Med Parasitol*. (1999). *93*, 325-329.

[396] Tripathi, AK; Sullivan, DJ; Stins, MF; *Plasmodium falciparum*-infected erythrocytes decrease the integrity of human blood-brain barrier endothelial cell monolayers. *J Infect Dis*. (2007). *195*, 942-950.

[397] Wellems, TE; Plowe, CV; Chloroquine-resistant malaria. *J Infect Dis*. (2001). *184*, 770-776.

[398] Peters, JM; Chen, N; Gatton, M; Korsinczky, M; Fowler, EV; Manzetti, S; Saul, A; Cheng, Q; Mutations in cytochrome b resulting in atovaquone resistance are associated with loss of fitness in *Plasmodium falciparum*. *Antimicrob Agents Chemother*. (2002). *46*, 2435-2441.

[399] Bloland, PB; Drug resistance in malaria. Geneva, Switzerland: World Health Onganization,2001.At: http://www.who.int/malaria/publications/drug-resistance/en/index.html (accessed December 2010)

[400] Anderson, TJ; Roper, C; The origins and spread of antimalarial drug resistance: Lessons for policy makers. *Acta Trop*. (2005). *94*, 269-280.

[401] White, NJ; Nosten, F; Looareesuwan, S; Watkins, WM; Marsh, K; Snow, RW; Kokwaro, G; Ouma, J; Hien, TT; Molyneux, ME; Taylor, TE; Newbold, CI; Ruebush, TK 2nd; Danis,

M; Greenwood, BM; Anderson, RM; Olliaro, P; Averting a malaria disaster. *Lancet.* (1999). *353*, 1965-1967.

[402] Handunnetti, SM; Gunewardena, DM; Pathirana, PP; Ekanayake, K; Weerasinghe, S; Mendis, KN; Features of recrudescent chloroquine-resistant *Plasmodium falciparum* infections confer a survival advantage on parasites and have implications for disease control. *Trans R Soc Trop Med Hyg.* (1996). *90*, 563-567.

[403] Wernsdorfer, WH; Landgraf, B; Wiedermann, G; Kollaritsch, H; Chloroquine resistance of *Plasmodium falciparum*: A biological advantage? *Trans R Soc Trop Med Hyg.* (1995). *89*, 90-91.

[404] Basco, LK; Inefficacy of amodiaquine against chloroquine-resistant malaria. *Lancet.* (1991). *338*, 1460.

[405] Watkins, WM; Mberu, EK; Winstanley, PA; Plowe, CV; The efficacy of antifolate antimalarial combinations in Africa: A predictive model based on pharmacodynamic and pharmacokinetic analyses. *Parasitol Today.* (1997). *13*, 459-464.

[406] Feikin, DR; Dowell, SF; Nwanyanwu, OC; Klugman, KP; Kazembe, PN; Barat, LM; Graf, C; Bloland, PB; Ziba, C; Huebner, RE; Schwartz, B. Increased carriage of trimethoprim/sulfamethoxazole-resistant *Streptococcus pneumoniae* in Malawian children after treatment for malaria with sulfadoxine/pyrimethamine. *J Infect Dis.* (2000). *181*, 1501-1505.

[407] Wernsdorfer, WH; The development and spread of drug-resistant malaria. *Parasitol Today.* (1991). *7*, 297-303.

[408] Watkins, WM; Mosobo, M; Treatment of *Plasmodium falciparum* malaria with pyrimethamine-sulfadoxine: Selective pressure for resistance is a function of long elimination half-life. *Trans R Soc Trop Med Hyg.* (1993). *87*, 75-78.

[409] Trape, JF; Rogier, C; Combating malaria morbidity and mortality by reducing transmission. *Parasitol Today.* (1996). *12*, 236-240.

[410] Babiker, HA; Walliker, D; Current views on the population structure of *plasmodium falciparum*: Implications for control. *Parasitol Today.* (1997). *13*, 262-267.

[411] Paul, RE; Day, KP; Mating patterns of *Plasmodium falciparum*. *Parasitol Today.* (1998). *14*, 197-202.

[412] Mackinnon, MJ; Hastings, IM; The evolution of multiple drug resistance in malaria parasites. *Trans R Soc Trop Med Hyg.* (1998). *92*, 188-195.

[413] Hastings, IM; Mackinnon, MJ; The emergence of drug-resistant malaria. *Parasitology.* (1998). *117*, 411-417.

[414] Paul, RE; Hackford, I; Brockman, A; Muller-Graf, C; Price, R; Luxemburger, C; White, NJ; Nosten, F; Day, KP; Transmission intensity and *Plasmodium falciparum* diversity on the northwestern border of Thailand. *Am J Trop Med Hyg.* (1998). *58*, 195-203.

[415] Molyneux, DH; Floyd, K; Barnish, G; Fèvre, EM; Transmission control and drug resistance in malaria: A crucial interaction. *Parasitol Today.* (1999). *15*, 238-240.

[416] Olivar, M; Develoux, M; Chegou, Abari A; Loutan, L; Presumptive diagnosis of malaria results in a significant risk of mistreatment of children in urban Sahel. *Trans R Soc Trop Med Hyg.* (1991). *85*, 729-730.

[417] van Hensbroek, MB; Morris-Jones, S; Meisner, S; Jaffar, S; Bayo, L; Dackour, R; Phillips, C; Greenwood, BM; Iron, but not folic acid, combined with effective antimalarial therapy promotes hematological recovery in African children after acute *falciparum* malaria. *Trans R Soc Trop Med Hyg.* (1995). *89*, 672-676.

[418] Shakoor, O; Taylor, RB; Behrens, RH; Assessment of the incidence of substandard drugs in developing countries. *Trop Med Int Health.* (1997). *2*, 839-845.

[419] Ballereau, F; Prazuck, T; Schrive, I; Lafleuriel, MT; Rozec, D; Fisch, A; Lafaix, C; Stability of essential drugs in the field: Results of a study conducted over a two-year period in Burkina Faso. *Am J Trop Med Hyg.* (1997). *57*, 31-36.

[420] Hastings, IM; Ward, SA; Coartem (artemether-lumefantrine) in Africa: The beginning of the end? *J Infect Dis.* (2005). *192*, 1303-1304.

[421] Sisowath, C; Strömberg, J; Mårtensson, A; Msellem, M; Obondo, C; Björkman, A; Gil, JP; *In vivo* selection of *Plasmodium falciparum pfmdr1* 86N coding alleles by artemether-lumefantrine (Coartem). *J Infect Dis.* (2005). *191*, 1014-1017.

[422] Palmer, KJ; Holliday, SM; Brogden, RN; Mefloquine. A review of its antimalarial activity, pharmacokinetic properties and therapeutic efficacy. *Drugs.* (1993). *45*, 430-475.

[423] Dye, C; Williams, BG; Multigenic drug resistance among inbred malaria parasites. *Proc Biol Sci.* (1997). *264*, 61-67.

[424] Hastings, IM; Watkins, WM; White, NJ; The evolution of drug-resistant malaria: The role of drug elimination half-life. *Philos Trans R Soc Lond B Biol Sci.* (2002). *357*, 505-519.

[425] Bauchner, H; Pelton, SI; Klein, JO; Parents, physicians, and antibiotic use. *Pediatrics.* (1999). *103*, 395-401.

[426] Wernsdorfer, WH; Chongsuphajaisiddhi, T; Salazar, NP; A symposium on containment of mefloquine-resistant *falciparum* malaria in Southeast Asia with special reference to border malaria. *Southeast Asian J Trop Med Public Health.* (1994). *25*, 11-18.

[427] White, NJ; Preventing antimalarial drug resistance through combinations. *Drug Resist Updat.* (1998). *1*, 3-9.

[428] Curtis, CF; Otoo, LN; A simple model of the build-up of resistance to mixtures of anti-malarial drugs. *Trans R Soc Trop Med Hyg.* (1986). *80*, 889-892.

[429] Hastings, IM; A model for the origins and spread of drug-resistant malaria. *Parasitology.* (1997). *115*, 133-141.

[430] Hastings, IM; D'Alessandro, U; Modeling a predictable disaster: The rise and spread of drug-resistant malaria. *Parasitol Today.* (2000). *16*, 340-347.

[431] Cross, AP; Singer, B; Modeling the development of resistance of *Plasmodium falciparum* to anti-malarial drugs. *Trans R Soc Trop Med Hyg.* (1991). *85*, 349-355.

[432] Myint, HY; Ashley, EA; Day, NP; Nosten, F; White, NJ; Efficacy and safety of dihydroartemisinin-piperaquine. *Trans R Soc Trop Med Hyg.* (2007). *101*, 858-866.

[433] Austin, DJ; White, NJ; Anderson, RM;. The dynamics of drug action on the within-host population growth of infectious agents: Melding pharmacokinetics with pathogen population dynamics. *J Theor Biol.* (1998). *194*, 313-339.

[434] Peters, W; Robinson, BL; The chemotherapy of rodent malaria XXXV. Further studies on the retardation of drug resistance by the use of a triple combination of mefloquine, pyrimethamine and sulfadoxine in mice infected with *P. berghei* and *'P. berghei NS'*. *Ann Trop Med Parasitol.* (1984). *78*, 459-466.

[435] Barnes, KI; Watkins, WM; White, NJ; Antimalarial dosing regimens and drug resistance. *Trends Parasitol.* (2008). *24*, 127-134.

[436] ter Kuile, F; White, NJ; Holloway, P; Pasvol, G; Krishna, S; *Plasmodium falciparum*: *In vitro* studies of the pharmacodynamic properties of drugs used for the treatment of severe malaria. *Exp Parasitol.* (1993). *76*, 85-95.

[437] Landau, I; Chabaud, A; Cambie, G; Ginsburg, H; Chronotherapy of malaria: An approach to malaria chemotherapy. *Parasitol Today.* (1991). *7*, 350-352.

[438] Jiang, JB; Li, GQ; Guo, XB; Kong, YC; Arnold, K; Antimalarial activity of mefloquine and qinghaosu. *Lancet.* (1982). *7*, 2, 285-288.

[439] White, NJ; The assessment of antimalarial drug efficacy. *Trends Parasitol.* (2002). *18*, 458-464.

[440] A-Elbasit, IE; Alifrangis, M; Khalil, IF; Bygbjerg, IC; Masuadi, EM; Elbashir, MI; Giha, HA; The implication of dihydrofolate reductase and dihydropteroate synthetase gene mutations in modification of *Plasmodium falciparum* characteristics. *Malar J.* (2007). *6*, 108.

[441] Gordi, T; Xie, R; Jusko, WJ; Semi-mechanistic pharmacokinetic/pharmacodynamic modeling of the antimalarial effect of artemisinin. *Br J Clin Pharmacol.* (2005). *60*, 594-604.

[442] Hietala, SF; Bhattarai, A; Msellem, M; Röshammar, D; Ali, AS; Strömberg, J; Hombhanje, FW; Kaneko, A; Björkman, A; Ashton, M. Population pharmacokinetics of amodiaquine and desethylamodiaquine in pediatric patients with uncomplicated *falciparum* malaria. *J Pharmacokinet Pharmacodyn.* 2007. 34, 669-686.

[443] Newton, PN; Ward, S; Angus, BJ; Chierakul, W; Dondorp, A; Ruangveerayuth, R; Silamut, K; Teerapong, P; Suputtamongkol, Y; Looareesuwan, S; White, NJ; Early treatment failure in severe malaria resulting from abnormally low plasma quinine concentrations. *Trans R Soc Trop Med Hyg.* (2006). *100*, 184-186.

[444] Barnes, KI; Little, F; Smith, PJ; Evans, A; Watkins, WM; White, NJ; Sulfadoxine-pyrimethamine pharmacokinetics in malaria: Pediatric dosing implications. *Clin Pharmacol Ther.* (2006). *80*, 582-596.

[445] Ringwald, P; Same Ekobo, A; Keundjian, A; Kedy Mangamba, D; Basco, LK. Chemoresistance of *P. falciparum* in urban areas of Yaounde, Cameroon. Part 1: Surveillance of *in vitro* and *in vivo* resistance of *Plasmodium falciparum* to chloroquine from 1994 to 1999 in Yaounde, Cameroon. *Trop Med Int Health.* (2000). *5*, 612-619.

[446] Checchi, F; Piola, P; Fogg, C; Bajunirwe, F; Biraro, S; Grandesso, F; Ruzagira, E; Babigumira, J; Kigozi, I; Kiguli, J; Kyomuhendo, J; Ferradini, L; Taylor, WR; Guthmann, JP; Supervised versus unsupervised antimalarial treatment with six-dose artemether-lumefantrine: Pharmacokinetic and dosage-related findings from a clinical trial in Uganda. *Malar J.* (2006). *5*, 59.

[447] Desai, M; ter Kuile, FO; Nosten, F; McGready, R; Asamoa, K; Brabin, B; Newman, RD; Epidemiology and burden of malaria in pregnancy. *Lancet Infect Dis.* (2007). *7*, 93-104.

[448] Lee, SJ; McGready, R; Fernandez, C; Stepniewska, K; Paw, MK; Viladpai-nguen, SJ; Thwai, KL; Villegas, L; Singhasivanon, P; Greenwood, BM; White, NJ; Nosten, F; Chloroquine pharmacokinetics in pregnant and nonpregnant women with *vivax* malaria. *Eur J Clin Pharmacol.* (2008). *64*, 987-992.

[449] Orton, LC; Omari, AA; Drugs for treating uncomplicated malaria in pregnant women. *Cochrane Database Syst Rev.* (2008). 4, CD004912.

[450] Clark, RL; Arima, A; Makori, N; Nakata, Y; Bernard, F; Gristwood, W; Harrell, A; White, TE; Wier, PJ; Artesunate: Developmental toxicity and toxicokinetics in monkeys. *Birth Defects Res B Dev Reprod Toxicol.* (2008). *83*, 418-434.

[451] McGready, R; Stepniewska, K; Lindegardh, N; Ashley, EA; La, Y; Singhasivanon, P; White, NJ; Nosten, F; The pharmacokinetics of artemether and lumefantrine in pregnant women with uncomplicated *falciparum* malaria. *Eur J Clin Pharmacol.* (2006). *62*, 1021-1031.

[452] Na Bangchang, K; Davis, TM; Looareesuwan, S; White, NJ; Bunnag, D; Karbwang, J; Mefloquine pharmacokinetics in pregnant women with acute *falciparum* malaria. *Trans R Soc Trop Med Hyg.* (1994). *88*, 321-323.

[453] Green, MD; van Eijk, AM; van Ter Kuile, FO; Ayisi, JG; Parise, ME; Kager, PA; Nahlen, BL; Steketee, R; Nettey, H; Pharmacokinetics of sulfadoxine-pyrimethamine in HIV-infected and uninfected pregnant women in Western Kenya. *J Infect Dis.* (2007). *196*, 1403-1408.

[454] Newton, P; Suputtamongkol, Y; Teja-Isavadharm, P; Pukrittayakamee, S; Navaratnam, V; Bates, I; White, N; Antimalarial bioavailability and disposition of artesunate in acute *falciparum* malaria. *Antimicrob Agents Chemother.* (2000). *44*, 972-977.

[455] Menendez, C; Malaria during pregnancy. *Curr Mol Med.* (2006). *6*, 269-273.

[456] Akintonwa, A; Meyer, MC; Yau, MK; Placental transfer of chloroquine in pregnant rabbits. *Res Commun Chem Pathol Pharmacol.* (1983). *40*, 443-455.

[457] Abdelrahim, II; Adam, I; Elghazali, G; Gustafsson, LL; Elbashir, MI; Mirghani, RA; Pharmacokinetics of quinine and its metabolites in pregnant Sudanese women with uncomplicated *Plasmodium falciparum* malaria. *J Clin Pharm Ther.* (2007). *32*, 15-19.

[458] Ward, SA; Sevene, EJ; Hastings, IM; Nosten, F; McGready, R; Antimalarial drugs and pregnancy: Safety, pharmacokinetics, and pharmacovigilance. *Lancet Infect Dis.* (2007). *7*, 136-144.

[459] O'Meara, WP; Smith, DL; McKenzie, FE; Potential impact of intermittent preventive treatment (IPT) on spread of drug-resistant malaria. *PLoS Med.* (2006). *3*, e141.

[460] Alexander, N; Sutherland, C; Roper, C; Cissé, B; Schellenberg, D; Modeling the impact of intermittent preventive treatment for malaria on selection pressure for drug resistance. *Malar J.* (2007). *6*, 9.

[461] van Eijk, AM; De Cock, KM; Ayisi, JG; Rosen, DH; Otieno, JA; Nahlen, BL; Steketee, RW; Pregnancy interval and delivery outcome among HIV-seropositive and HIV-seronegative women in Kisumu, Kenya. *Trop Med Int Health.* (2004). *9*, 15-24.

[462] Cohen, C; Karstaedt, A; Frean, J; Thomas, J; Govender, N; Prentice, E; Dini, L; Galpin, J; Crewe-Brown, H; Increased prevalence of severe malaria in HIV-infected adults in South Africa. *Clin Infect Dis.* (2005). *41*, 1631-1637.

[463] Shah, SN; Smith, EE; Obonyo, CO; Kain, KC; Bloland, PB; Slutsker, L; Hamel, MJ; HIV immunosuppression and antimalarial efficacy: sulfadoxine-pyrimethamine for the treatment of uncomplicated malaria in HIV-infected adults in Siaya, Kenya. *J Infect Dis.* (2006). *194*, 1519-1528.

[464] Khoo, S; Back, D; Winstanley, P; The potential for interactions between antimalarial and antiretroviral drugs. *AIDS.* (2005). *19*, 995-1005.

[465] German, P; Greenhouse, B; Coates, C; Dorsey, G; Rosenthal, PJ; Charlebois, E; Lindegardh, N; Havlir, D; Aweeka, FT; Hepatotoxicity due to a drug interaction between amodiaquine plus artesunate and efavirenz. *Clin Infect Dis.* (2007). *44*, 889-891.

[466] Barradell, LB; Fitton, A; Artesunate, A review of its pharmacology and therapeutic efficacy in the treatment of malaria. *Drugs* (1995). *50*, 714-741.

[467] Anstey, NM; Price, RN; White, NJ; Improving the availability of artesunate for treatment of severe malaria. Artesunate reduces mortality and should now be the treatment of choice in severe malaria in adults: Good news for countries in our region, but registration in Australia must wait. *Med J Aust.* (2006). *184*, 3-4.

[468] Xie, LH; Johnson, TO; Weina, PJ; Si, Y; Haeberle, A; Upadhyay, R; Wong, E; Li, Q; Risk assessment and therapeutic indices of artesunate and artelinate in *Plasmodium berghei*-infected and uninfected rats. *Int J Toxicol.* (2005). *24*, 251-264.

[469] Li, Q; Gerena, L; Xie, L; Zhang, J; Kyle, D; Milhous, W; Development and validation of flow cytometric measurement for parasitemia in cultures of *P. falciparum* vitally stained with YOYO-1. *Cytometry A.* (2007). *71*, 297-307.

[470] Park, BK; O'Neill, PN; Maggs, JL; Pirmohamed, M; Safety assessment of peroxide antimalarials: Clinical and chemical perspectives. *Br J Clin Pharmacol* (1998). *46*, 521-529.

[471] Ashton, M; Hai, TN; Sy, ND; Huong, DX; Van Huong, N; Nieu, NT; Cong, LD; Artemisinin pharmacokinetics is time-dependent during repeated oral administration in healthy male adults. *Drug Metab Dispos*. (1998). *26*, 25-27.

[472] Ashton, M; Sy, ND; Gordi, T; Hai, TN. Thach, DC; Huong, NV; Johansson, M; Coeng, LD; Evidence for time-dependence artemisinin kinetics in adults with uncomplicated malaria. *Pharm Pharmacol Lett*. (1996). *6*, 127-130.

[473] van Agtmael, MA; Cheng-Qi, S; Qing, JX; Mull, R; van Boxtel, CJ; Multiple dose pharmacokinetics of artemether in Chinese patients with uncomplicated *falciparum* malaria. *Int J Antimicrob Agents*. (1999). *12*, 151-158.

[474] Khanh, NX; de Vries, PJ; Ha, LD; van Boxtel, CJ; Koopmans, R; Kager, PA. Declining concentrations of dihydroartemisinin in plasma during 5-day oral treatment with artesunate for *falciparum* malaria. *Antimicrob Agents Chemother*. (1999). *43*, 690-692.

[475] Li, Q; Xie, LH; Si, Y; Wong, E; Upadhyay, R; Yanez, D; Weina, PJ; Toxicokinetics and hydrolysis of artelinate and artesunate in malaria-infected rats. *Int J Toxicol*. (2005). *24*, 241-250.

[476] Salako, LA; Pharmacokinetics of antimalarial drugs: Their therapeutic and toxicological implications. *Ann Ist Super Sanita*. (1985). *21*, 315-325.

[477] Li, QG; Peggins, JO; Brewer, TG; Comparative anoretic toxicity of im arteether in rats using sesame oil or cremophor as vehicle. *AM J Trop Med Hyg*. (1996). *55*, 182.

[478] Alin, MH; Bjorkman, A; Concentration and time dependence of artemisinin efficacy against *Plasmodium falciparum in vitro*. *Am J Trop Med Hyg*. (1994). *50*, 771-776.

[479] White, NJ; Krishna, S; Treatment of malaria: Some considerations and limitations of the current methods of assessment. *Trans Roy Soc Trop Med Hyg*. (1989). *83*, 767-777.

[480] Skinner, TS; Manning, LS; Johnston, WA; Davis, TM; *In vitro* stage-specific sensitivity of *Plasmodium falciparum* to quinine and artemisinin drugs. *Int J Parasit*. (1996). *26*, 519-525.

[481] Karbwang, J; Na-Bangchang, K; Thanavibul, A; Molunto, P; Plasma concentrations of artemether and its major plasma metabolite, dihydroartemisinin, following a 5-day regimen of oral artemether, in patients with uncomplicated *falciparum* malaria. *Ann Trop Med Parasit*. (1998). *92*, 31-36.

[482] Sharma, P; Swarup, D; Saxena, GN; Bhandari, S; Sharma, UB; Tuteja, R; An open study to evaluate the efficacy of artemether in severe *falciparum* malaria. *J Assoc Physicians India*. (1999). *47*, 883-885.

[483] Gachot, B; Eliaszewicz, M; Dupont, B; Artesunate and cerebellar dysfunction in *falciparum* malaria. *N Engl J Med*. (1997). *337*, 792-793.

[484] Li, QG; Brueckner, RP; Peggins, JO; Trotman, KM; Brewer, TG; Arteether toxicokinetics and pharmacokinetics in rats after 25 mg/kg/day single and multiple doses. *Eur J Drug Metab Pharmacokinet*. (1999). *24*, 213-223.

[485] Li, Q; Lugt, CB; Looareesuwan, S; Krudsood, S; Wilairatana, P; Vannaphan, S; Chalearmrult, K; Milhous, WK; Pharmacokinetic investigation on the therapeutic potential of artemotil (beta-arteether) in Thai patients with severe *Plasmodium falciparum* malaria. *Am J Trop Med Hyg*. (2004). *71*, 723-731.

[486] Suputtamongkol, Y; Newton, PN; Angus, B; Teja-Isavadharm, P; Keeratithakul, D; Rasameesoraj, M; Pukrittayakamee, S; White, NJ; A comparison of oral artesunate and

artemether antimalarial bioactivities in acute *falciparum* malaria. *Br J Clin Pharmacol.* (2001). *52*, 655-661.

[487] Marsh, K; Forster, D; Waruiru, C; Mwangi, I; Winstanley, M; Marsh, V; Newton, C; Winstanley, P; Warn, P; Peshu, N.; *et al.*, Indicators of life-threatening malaria in African children. *N Engl J Med* (1995). *332*, 1399-1404.

[488] Horton, RJ; Parr, SN; Bokor, LC; Clinical experience with halofantrine in the treatment of malaria. *Drugs Exp Clin Res.* (1990). *16*, 497-503.

[489] Hussein, Z; Eaves, J; Hutchinson, DB; Canfield, CJ; Population pharmacokinetics of atovaquone in patients with acute malaria caused by *Plasmodium falciparum. Clin Pharmacol Ther.* (1997). *61*, 518-530.

[490] Milton, KA; Edwards, G; Ward, SA; Orme, ML; Breckenridge, AM; Pharmacokinetics of halofantrine in man: Effects of food and dose size. *Br J Clin Pharmacol.* (1989). *28*, 71-77.

[491] ter Kuile, FO; Dolan, G; Nosten, F; Edstein, MD; Luxemburger, C; Phaipun, L; Chongsuphajaisiddhi, T; Webster, HK; White, NJ;. Halofantrine versus mefloquine in treatment of multidrug-resistant *falciparum* malaria. *Lancet.* (1993). *341*, 1044-1049.

[492] Bunnag, D; Viravan, C; Looareesuwan, S; Karbwang, J; Harinasuta, T; Double blind randomized clinical trial of oral artesunate at once or twice daily dose in *falciparum* malaria. *Southeast Asian J Trop Med Public Health.* (1991). *22*, 539-543.

[493] White, NJ; Antimalarial pharmacokinetics and treatment regimens. *Br J Clin Pharmacol.* (1992). *34*, 1-10.

[494] Rehwagen, C; WHO ultimatum on artemisinin monotherapy is showing results. *BMJ.* (2006). *332*, 1176.

[495] Yeates RA; Artemotil Artecef. *Curr Opin Investig Drugs.* (2002). *3*, 545-549.

[496] Kyle, DE; Teja-Isavadharm, P; Li, Q; Leo, K. Pharmacokinetics and pharmacodynamics of qinghaosu derivatives: How do they impact on the choice of drug and the dosage regimens? *Med Trop (Mars).* (1998). *58*(Suppl. 3), 38-44.

[497] Li, QG; Peggins, JO; Fleckenstein, LL; Masonic, K; Heiffer, MH; Brewer, TG; The pharmacokinetics and bioavailability of dihydroartemisinin, arteether, artemether, artesunic acid and artelinic acid in rats. *J Pharm Pharmacol.* (1998). *50*, 173-182.

[498] Dayan, AD; Neurotoxicity and artemisinin compounds do the observations in animals justify limitation of clinical use? *Med Trop (Mars).* (1998). *58*(Suppl. 3), 32-37.

[499] Genovese, RF; Newman, DB; Understanding artemisinin-induced brainstem neurotoxicity. *Arch Toxicol.* (2008). *82*, 379-385.

[500] Li, QG; Mog, SR; Si, YZ; Kyle, DE; Gettayacamin, M; Milhous, WK; Neurotoxicity and efficacy of arteether related to its exposure times and exposure levels in rodents. *Am J Trop Med Hyg.* (2002). *66*, 516-525.

[501] Navaratnam, V; Mansor, SM; Sit, NW; Grace, J; Li, QG; Olliaro, P. Pharmacokinetics of artemisinin-type compounds. *Clin Pharmacokinet.* (2000). *39*, 255-270.

[502] Angus, BJ; Thaiaporn, I; Chanthapadith, K; Suputtamongkol, Y; White, NJ; Oral artesunate dose-response relationship in acute *falciparum* malaria. *Antimicrob Agents Chemother.* (2002). *46*, 778-782.

[503] Newton, PN; Barnes, KI; Smith, PJ; Evans, AC; Chierakul, W; Ruangveerayuth, R; White, NJ; The pharmacokinetics of intravenous artesunate in adults with severe *falciparum* malaria. *Eur J Clin Pharmacol.* (2006). *62*, 1003-1009.

[504] Luxemburger, C; ter Kuile, FO; Nosten, F; Dolan, G; Bradol, JH; Phaipun, L; Chongsuphajaisiddhi, T; White, NJ; Single day mefloquine- artesunate combination in the treatment of multi-drug resistant *falciparum* malaria. *Trans R Soc Trop Med Hyg.* (1994). *88*, 213-217.

[505] Brockman, A; Price, RN; van Vugt, M; Heppner, DG; Walsh, D; Sookto, P; Wimonwattrawatee, T; Looareesuwan, S; White, NJ; Nosten, F; *Plasmodium falciparum* antimalarial drug susceptibility on the north-western border of Thailand during five years of extensive use of artesunate-mefloquine. *Trans R Soc Trop Med Hyg.* (2000). *94*, 537-544.

[506] Nyunt, MM; Plowe, CV; Pharmacologic advances in the global control and treatment of malaria: Combination therapy and resistance. *Clin Pharmacol Ther.* (2007). *82*, 601-605.

[507] Meshnick, SR; Alker, AP; Amodiaquine and combination chemotherapy for malaria. *Am J Trop Med Hyg.* (2005). *73*, 821-823.

[508] Grandesso, F; Hagerman, A; Kamara, S; Lam, E; Checchi, F; Balkan, S; Scollo, G; Durand, R; Guthmann, JP; Low efficacy of the combination artesunate plus amodiaquine for uncomplicated *falciparum* malaria among children under 5 years in Kailahun, Sierra Leone. *Trop Med Int Health.* (2006). *11*, 1017-1021.

[509] Ramharter, M; Oyakhirome, S; Klein Klouwenberg, P; Adégnika, AA; Agnandji, ST; Missinou, MA; Matsiégui, PB; Mordmüller, B; Borrmann, S; Kun, JF; Lell, B; Krishna, S; Graninger, W; Issifou, S; Kremsner, PG; Artesunate-clindamycin versus quinine-clindamycin in the treatment of *Plasmodium falciparum* malaria: A randomized controlled trial. *Clin Infect Dis.* (2005). *40*, 1777-1784.

[510] Dieckmann, A; Jung, A; Stage-specific sensitivity of *Plasmodium falciparum* to antifolates. *Z Parasitenkd.* (1986). *72*, 591-594.

[511] Geary, TG; Divo, AA; Jensen, JB; Stage specific actions of antimalarial drugs on *Plasmodium falciparum* in culture. *Am J Trop Med Hyg.* (1989). *40*, 240-244.

[512] Rieckmann, K; Suebsaeng, L; Rooney, W; Response of *Plasmodium falciparum* infections to pyrimethamine-sulfadoxine in Thailand. *Am J Trop Med Hyg.* (1987). *37*, 211-216.

[513] White, NJ; Chapman, D; Watt, G; The effects of multiplication and synchronicity on the vascular distribution of parasites in *falciparum* malaria. *Trans R Soc Trop Med Hyg.* (1992). *86*, 590-597.

[514] Yayon, A; Vande Waa, JA; Yayon, M; Geary, TG; Jensen, JB; Stage-dependent effects of chloroquine on *Plasmodium falciparum in vitro. J Protozool.* (1983). *30*, 642-647.

[515] Zhang, Y; Asante, KS; Jung, A; Stage-dependent inhibition of chloroquine on *Plasmodium falciparum in vitro. J Parasitol.* (1986). *72*, 830-836.

[516] Hien, TT; White, NJ; Qinghaosu. *Lancet.* (1993). *341*, 603-608.

[517] White, NJ; Clinical pharmacokinetics and pharmacodynamics of artemisinin and derivatives. *Trans R Soc Trop Med Hyg.* (1994). *88*(Suppl. 1), S41-43.

[518] Drusano, GL;. Role of pharmacokinetics in the outcome of infections. *Antimicrob Agents Chemother.* (1988). *32*, 289-297.

[519] Looareesuwan, S; Overview of clinical studies on artemisinin derivatives in Thailand. *Trans Roy Soc Trop Med Hyg.* (1994). *88*(Suppl. 1), S9-11.

[520] McLean, WG; Ward, SA; *In vitro* neurotoxicity of artemisinin derivatives. *Med Trop (Mars).* (1998). *58*(Suppl. 3), 28-31.

[521] Bhatt, KM; Samia, BM; Bhatt, SM; Wasunna, KM; Efficacy and safety of an artesunate/mefloquine combination, (artequin) in the treatment of uncomplicated *P. falciparum* malaria in Kenya. *East Afr Med J.* (2006). *83*, 236-242.

[522] Kaiwa, S; Kano, S; Suzuki, M; Morphologic effects of artemether on *Plasmodium falciparum* in Aotus trivirgatus. *Am J Trop Med Hyg.* (1993). 49, 812-818.

[523] Jung, M; Lee, K; Kim, H; Park, M; Recent advances in artemisinin and its derivatives as antimalarial and antitumor agents. *Curr Med Chem.* (2004). *11*, 1265-1284.

[524] Li, Q; Milhous, W.K; Weina, P.J; Fatal neurotoxicity of the artemisinin derivatives is related to drug pharmacokinetic profiles in animal species. *Curr. Topics Toxicol*, (2006). *3*, 1-16.

[525] Bustos, MD; Gay, F; Diquet, B; *In vitro* tests on Philippine isolates of *Plasmodium falciparum* against four standard antimalarials and four qinghaosu derivatives. *Bull World Health Organ*, (1994). *72*, 729-735.

[526] China Cooperative Research Group on Qinghaosu and Its Derivatives as Antimalarials. Antimalarial efficacy and mode of action of qinghaosu and its derivatives in experimental models. *J Trad Chinese Med*, (1982). *2*, 17-24.

[527] Ittarat, W; Pickard, AL; Rattanasinganchan, P; Wilairatana, P; Looareesuwan, S; Emery, K; Low, J; Udomsangpetch, R; Meshnick, SR. Recrudescence in artesunate-treated patients with *falciparum* malaria is dependent on parasite burden not on parasite factors. *Am J Trop Med Hyg*. (2003). *68*, 147-152.

[528] Li, QG; Peggins, JO; Lin, AJ; Masonic, K; Trotman KM; Brewer, TG; Pharmacology and toxicology of artelinic acid: Preclinical investigation on pharmacokinetics, metabolism, protein and RBC binding, acute and anorectic toxicities. *Trans R Soc Trop Med Hyg*. (1998). *92*, 332-340.

[529] Gu, H.M; Warhurst, DC; Peters, W; Uptake of [^{3}H]-dihydroartemisinin by erythrocytes infected with *Plasmodium falciparum in vitro. Trans. R. Soc. Trop. Med. Hyg*. (1984). *78*, 265-270.

[530] Batty, KT; Thu, LT; Davis, TM; Ilett, KF; Mai, TX; Hung, NC; Tien, NP; Powell, SM; Thien, HV; Binh, TQ; Kim, NV; A pharmacokinetic and pharmacodynamic study of intravenous vs. oral artesunate in uncomplicated *falciparum* malaria. *Br J Clin Pharmacol*. (1998). *45*, 123-129.

[531] Binh, TQ; Ilett, KF; Batty, KT; Davis, TM; Hung, NC; Powell, SM; Thu, LT; Thien, HV; Phuong, HL; Phuong, VD. Oral bioavailability of dihydroartemisinin in Vietnamese volunteers and in patients with *falciparum* malaria. *Br J Clin Pharmacol*. (2001). *51*, 541-546.

[532] Jelinek, T; Intravenous artesunate recommended for patients with severe malaria: Position statement from TropNetEurop. *Euro Surveill*. (2005). 10, E051124.5.

[533] Luxemburger, C; Nosten, F; Raimond, SD; Chongsuphajaisiddhi, T; White, NJ; Oral artesunate in the treatment of uncomplicated hyperparasitemic *falciparum* malaria. *Am J Trop Med Hyg*. (1995). *53*, 522-525.

[534] McGready, R; Cho, T; Keo, NK; Thwai, KL; Villegas, L; Looareesuwan, S; White, NJ; Nosten, F; Artemisinin antimalarials in pregnancy: A prospective treatment study of 539 episodes of multidrug-resistant *Plasmodium falciparum. Clin Infect Dis*. (2001). *33*, 2009-2016.

[535] Dellicour, S; Hall, S; Chandramohan, D; Greenwood, B; The safety of artemisinins during pregnancy: A pressing question. *Malar J*. (2007). *6*, 15.

[536] Nosten, F; Ashley, E; McGready, R; Price, R; We still need artesunate monotherapy. *BMJ*. (2006). *333*, 45.

[537] Weina, PJ; Haeberle, AS; Lowe, MC; Cantilena, L; Milhous, WK; Intravenous artesunate: New product for the treatment of severe and complicated malaria. *Am J Trop Med Hyg*. (2005). 73, 32.

[538] Chongsuphajaisiddhi, T; Sabcharoen, A; Attanath, P; *In vivo* and *in vitro* sensitivity of *Falciparum* malaria to quinine in Thai children. *Ann Trop Paediatr*. (1981). *1*, 21-26.

[539] Hellgren, U; Kihamia, CM; Mahikwano, LF; Björkman, A; Eriksson, O; Rombo, L; Response of *Plasmodium falciparum* to chloroquine treatment: Relation to whole blood

concentrations of chloroquine and desethylchloroquine.*Bull World Health Organ.* (1989). *67*, 197-202.

[540] Slutsker, LM; Khoromana, CO; Payne, D; Allen, CR; Wirima, JJ; Heymann, DL; Patchen, L; Steketee, RW; Mefloquine therapy for *Plasmodium falciparum* malaria in children under 5 years of age in Malawi: *In vivo/in vitro* efficacy and correlation of drug concentration with parasitological outcome. *Bull World Health Organ.* (1990). *68*, 53-59.

[541] Batty, KT; Le, AT; Ilett, KF; Nguyen, PT; Powell, SM; Nguyen, CH; Truong, XM; Vuong, VC; Huynh, VT; Tran,QB; Nguyen, VM; Davis, TM; A pharmacokinetic and pharmacodynamic study of artesunate for *vivax* malaria. *Am J Trop Med Hyg.* (1998). *59*, 823-827.

[542] Wilairatana, P; Chanthavanich, P; Singhasivanon, P; Treeprasertsuk, S; Krudsood, S; Chalermrut, K; Phisalaphong, C; Kraisintu, K; Looareesuwan, S; A comparison of three different dihydroartemisinin formulations for the treatment of acute uncomplicated *falciparum* malaria in Thailand. *Int J Parasitol.* (1998). *28*, 1213-1218.

[543] Wise, R; Maximizing efficacy and reducing the emergence of resistance. *J Antimicrob Chemother.* (2003). *51*(Suppl.), 37-42.

[544] Hassan, Alin M; Ashton, M; Kihamia, CM; Mtey, GJ; Bjorkman, A; Multiple dose pharmacokinetics of oral artemisinin and comparison of its efficacy with that of oral artesunate in *falciparum* malaria patients. *Trans R Soc Trop Med Hyg.* (1996). *90*, 61-65.

[545] Titulaer, HA; Zuidema, J; Lugt, CB;. Formulation and pharmacokinetics of artemisinin and its derivatives. *Int J Pharmaceui.* (1991). *69*, 83-92.

[546] Murphy, S; Watkins, WM; Bray, PG; Lowe, B; Winstanley, PA; Peshu, N; Marsh, K. Parasite viability during treatment of severe *falciparum* malaria: Differential effects of artemether and quinine. *Am J Trop Med Hyg.* (1995). *53*, 303-305.

[547] Thu, LTA; Davis, TME; Binh, TQ; van Phuong, N; Anh, TK; Delayed parasite clearance in a Splenectomized patients with *falciparum* malaria who was treated with artemisinin derivatives. *Clin Infect Dis.* (1997). *25*, 923-925.

[548] Zhao, KC; Song, ZY; Pharmacokinetics of dihydroqinghaosu in human volunteers and comparison with qinghaosu. *Yao Xue Xue Bao.* (1993). *28*, 342-346.

[549] Koopmans, R; Ha, LD; Duc, DD; Dien, TK; Kager, PA; Khanh, NX; van Boxtel, CJ; De Vries, PJ; The pharmacokinetics of artemisinin after administration of two different suppositories to healthy Vietnamese subjects. *Am J Trop Med Hyg.* (1999). *60*, 244-247.

[550] Na-Bangchang, K; Krudsood, S; Silachamroon, U; Molunto, P; Tasanor, O; Chalermrut, K; Tangpukdee, N; Matangkasombut, O; Kano, S; Looareesuwan, S; The pharmacokinetics of oral dihydroartemisinin and artesunate in healthy Thai volunteers. *Southeast Asian J Trop Med Public Health.* (2004). *35*, 575-582.

[551] Li Q, Weina P, Milhous W; Pharmacokinetic and pharmacodynamic profiles of rapid-acting artemisinins in the antimalarial therapy. *Current Drug Therapy.* (2007). *2*, 210-223.

[552] Li, GQ; Guo, XB; Fu, LC; Jian, HX; Wang, XH. Clinical trials of artemisinin and its derivatives in the treatment of malaria in China. *Trans R Soc Trop Med Hyg.* (1994). *88*(Suppl. 1), S5-6.

[553] White, NJ; Artemisinin: Current status. *Trans R Soc Trop Med Hyg.* (1994). *88*(Suppl. 1), S3-4.

[554] Looareesuwan, S; Wilairatana, P; Vanijanonta, S; Pitisuttithum, P; Ratanapong, Y; Andrial, M; Monotherapy with sodium artesunate for uncomplicated *falciparum* malaria in Thailand: A comparison of 5- and 7-day regimens. *Acta Trop.* (1997). *67*, 197-205.

[555] Meshnick, SR; Thomas, A; Ranz, A; Xu, CM; Pan, HZ; Artemisinin (qinghaosu): The role of intracellular hemin in its mechanism of antimalarial action. *Mol Biochem Parasitol.* (1991). *49*, 181-189.

[556] Li, ZL; Gu, HM; Warhurst, DC; Peters, W; Effects of qinghaosu and related compounds on incorporation of [G-^{3}H] hypoxanthine by *Plasmodium falciparum in vitro. Trans R Soc Trop Med Hyg.* (1983). *77*, 522-523.

[557] Brand, VB; Sandu, CD; Duranton, C; Tanneur, V; Lang, KS; Huber, SM; Lang, F; Dependence of *Plasmodium falciparum in vitro* growth on the cation permeability of the human host erythrocyte. *Cell Physiol Biochem.* (2003). *13*, 347-356.

[558] Schrier, SL; Red cell membrane biology--introduction. *Clin Haematol.* (1985). *14*, 1-12.

[559] Asawamahasakda, W; Benakis, A; Meshnick, SR; The interaction of artemisinin with red cell membranes. *J Lab Clin Med.* (1994). *123*, 757-762.

[560] Maguire, PA; Sherman, IW; Phospholipid composition, cholesterol content and cholesterol exchange in *Plasmodium falciparum*-infected red cells. *Mol Biochem Parasitol.* (1990). *38*, 105-112.

[561] Fujioka, H; Aikawa, M; Morphological changes of clefts in *Plasmodium*-infected erythrocytes under adverse conditions. *Exp Parasitol.* (1993). *76*, 302-307.

[562] Pouvelle, B; Spiegel, R; Hsiao, L; Howard, RJ; Morris, RL; Thomas, AP; Taraschi, TF; Direct access to serum macromolecules by intraerythrocytic malaria parasites. *Nature.* (1991). 353, 73-75.

[563] Ginsburg, H; Kutner, S; Zangwil, M; Cabantchik, ZI; Selectivity properties of pores induced in host erythrocyte membrane by *Plasmodium falciparum.* Effect of parasite maturation. *Biochim Biophys Acta.* (1986). *861*, 194-196.

[564] Barnwell, JW; Vesicle-mediated transport of membrane and proteins in malaria-infected erythrocytes. *Blood Cells.* (1990). *16*, 379-395.

[565] Yuthavong, Y; Wilairat, P; Panijpan, B; Potiwan, C; Beale, GH; Alterations in membrane proteins of mouse erythrocytes infected with different species and strains of malaria parasites. *Comp Biochem Physiol B.* (1979). *63*, 83-85.

[566] Asawamahasakda, W; Ittarat, I; Pu, YM; Ziffer, H; Meshnick, SR; Reaction of antimalarial endoperoxides with specific parasite proteins. *Antimicrob Agents Chemother.* (1994). *38*, 1854-1858.

[567] Cofresi, A; Milhous, WK; Komisar, J; Increasing parasitemia modulates *in vitro* concentration response to artemisinin drugs: Can parasite burden alter treatment outcomes in severe and complicated malaria? (Submitted to *Parasitology* 2008)

[568] Navaratnam, V; Mordi, MN; Mansor, SM; Simultaneous determination of artesunic acid and dihydroartemisinin in blood plasma by high-performance liquid chromatography for application in clinical pharmacological studies. *J Chromatogr B Biomed Sci Appl.* (1997). *692*, 157-162.

[569] Na-Bangchang, K; Tippawangkosol, P; Thanavibul, A; Ubalee, R; Karbwang, J; Pharmacokinetic and pharmacodynamic interactions of mefloquine and dihydroartemisinin. *Int J Clin Pharmacol Res.* (1999). *19*, 9-17.

[570] Gordi, T; Xie, R; Jusko, WJ; Semi-mechanistic pharmacokinetic/pharmacodynamic modeling of the antimalarial effect of artemisinin. *Br J Clin Pharmacol.* (2005). *60*, 594-604.

[571] Brocks, DR; Mehvar, R; Stereoselectivity in the pharmacodynamics and pharmacokinetics of the chiral antimalarial drugs. *Clin Pharmacokinet.* (2003). *42*, 1359-1382.

[572] Tett, SE; McLachlan, AJ; Cutler, DJ; Day, RO; Pharmacokinetics and pharmacodynamics of hydroxychloroquine enantiomers in patients with rheumatoid arthritis receiving multiple doses of racemate. *Chirality*. (1994). *6*, 355-359.

[573] Miller, DR; Fiechtner, JJ; Carpenter, JR; Brown, RR; Stroshane, RM; Stecher, VJ; Plasma hydroxychloroquine concentrations and efficacy in rheumatoid arthritis.*Arthritis Rheum.* (1987). *30*, 567-571.

[574] Brocks, DR; Skeith, KJ; Johnston, C; Emamibafrani, J; Davis, P; Russell, AS; Jamali, F; Hematologic disposition of hydroxychloroquine enantiomers. *J Clin Pharmacol*. (1994). *34*, 1088-1097.

[575] Wesche, DL; Schuster, BG; Wang, WX; Woosley, RL; Mechanism of cardiotoxicity of halofantrine. *Clin Pharmacol Ther*. (2000). *67*, 521-529.

[576] Abernethy, DR; Wesche, DL; Barbey, JT; Ohrt, C; Mohanty, S; Pezzullo, JC; Schuster, BG; Stereoselective halofantrine disposition and effect: Concentration-related QTc prolongation. *Br J Clin Pharmacol*. (2001). *51*, 231-237.

[577] Salako, LA, Brieger, WR, Afolabi, BM, Umeh, RE, Agomo, PU, Asa, S, Adeneye, AK, Nwankwo, BO, Akinlade, CO;Treatment of childhood fevers and other illnesses in three rural Nigerian communities. J *Trop Pediatr*. (2001). *47*, 230-238.

[578] WHO. Assessment and Monitoring of Antimalarial Drug Efficacy for the Treatment of Uncomplicated *Falciparum* Malaria. 2003. Geneva: World Health Organization. WHO/HTM/ RBM/2003.50.

[579] Krause, G; Sauerborn, R; Comprehensive community effectiveness of health care. A study of malaria treatment in children and adults in rural Burkina Faso. *Ann Trop Paediatr*. (2000). *20*, 273-282.

[580] Yeboah-Antwi, K; Gyapong, JO; Asare, IK; Barnish, G; Evans, DB; Adjei, S; Impact of prepackaging antimalarial drugs on cost to patients and compliance with treatment. *Bull World Health Organ*. (2001). *79*, 394-399.

[581] Pagnoni, F; Convelbo, N; Tiendrebeogo, J; Cousens, S; Esposito, F; A community-based programme to provide prompt and adequate treatment of presumptive malaria in children. *Trans R Soc Trop Med Hyg*. (1997). *91*, 512-517.

[582] Ansah, EK; Gyapong, JO; Agyepong, IA; Evans, DB; Improving adherence to malaria treatment for children: The use of pre-packed chloroquine tablets vs. chloroquine syrup. *Trop Med Int Health*. (2001). *6*, 496-504.

[583] Gomes, M; Wayling, S; Pang, L; Interventions to improve the use of antimalarials in southeast Asia: An overview. *Bull World Health Organ*. (1998). *76*(Suppl. 1), 9-19.

[584] Shwe, T; Lwin, M; Aung, S; Influence of blister packaging on the efficacy of artesunate + mefloquine over artesunate alone in community-based treatment of non-severe *falciparum* malaria in Myanmar. *Bull World Health Organ*. (1998). *76*(Suppl. 1), 35-41.

[585] Qingjun, L; Jihui, D; Laiyi, T; Xiangjun, Z; Jun, L; Hay, A; Shires, S; Navaratnam, V; The effect of drug packaging on patients' compliance with treatment for *Plasmodium vivax* malaria in China. *Bull World Health Organ*. (1998). *76*(Suppl. 1), 21-27.

[586] Bloland, PB; Kachur, SP; Williams, HA; Trends in antimalarial drug deployment in sub-Saharan Africa. *J Exp Biol*. (2003). *206*, 3761-3769.

[587] Baume, C; Helitzer, D; Kachur, SP; Patterns of care for childhood malaria in Zambia. *Soc Sci Med*. (2000). *51*,1491-1503.

[588] Kofoed, PE; Lopez, F; Aaby, P; Hedegaard, K; Rombo, L; Can mothers be trusted to give malaria treatment to their children at home? *Acta Trop*. (2003). *86*, 67-70.

[589] Minzi, OM; Moshi, MJ; Hipolite, D; Massele, AY; Tomson, G; Ericsson, O; Gustafsson, LL; Evaluation of the quality of amodiaquine and sulphadoxine/pyrimethamine tablets sold

by private wholesale pharmacies in Dar Es Salaam Tanzania. *J Clin Pharm Ther.* (2003). *28*, 117-122.

[590] Marsh, VM; Mutemi, WM; Willetts, A; Bayah, K; Were, S; Ross, A; Marsh, K; Improving malaria home treatment by training drug retailers in rural Kenya. *Trop Med Int Health.* (2004). *9*, 451-460.

[591] Nuwaha, F; People's perception of malaria in Mbarara, Uganda. *Trop Med Int Health.* 2002. 7, 462-470.

[592] Bundy, DA; Lwin, S; Osika, JS; McLaughlin, J; Pannenborg, CO; What should schools do about malaria? *Parasitol Today.* (2000). *16*, 181-182.

[593] Luzzi, GA; Peto, TE; Adverse effects of antimalarials. An update. *Drug Saf.* (1993). *8*, 295-311.

[594] Winstanley, P; Ward, S; Snow, R; Breckenridge, A; Therapy of *falciparum* malaria in sub-saharan Africa: From molecule to policy. *Clin Microbiol Rev.* (2004). *17*, 612-637.

[595] WHO, Severe and complicated malaria. World Health Organization, Division of Control of Tropical Diseases. *Trans R Soc Trop Med Hyg.* (1990). *84*(Suppl. 2), 1-65.

[596] Phillips, RE; Looareesuwan, S; White, NJ; Chanthavanich, P; Karbwang, J; Supanaranond, W; Turner, RC; Warrell, DA; Hypoglycemia and antimalarial drugs: Quinidine and release of insulin. *Br Med J (Clin Res Ed).* (1986). *292*, 1319-1321.

[597] Powrie, JK; Smith, GD; Shojaee-Moradie, F; Sönksen, PH; Jones, RH; Mode of action of chloroquine in patients with non-insulin-dependent diabetes mellitus. *Am J Physiol.* (1991). 260, E897-904.

[598] Davis, TM; Dembo, LG; Kaye-Eddie, SA; Hewitt; BJ; Hislop, RG; Batty, KT; Neurological, cardiovascular and metabolic effects of mefloquine in healthy volunteers: A double-blind, placebo-controlled trial. *Br J Clin Pharmacol.* (1996). *42*, 415-421.

[599] Davis, TM; Antimalarial drugs and glucose metabolism. *Br J Clin Pharmacol.* (1997). *44*, 1-7.

[600] Fatherazi, S; Cook, DL. Specificity of tetraethylammonium and quinine for three K channels in insulin-secreting cells. *J Membr Biol.* (1991). *120*, 105-114.

[601] Taylor, TE; Molyneux, ME; Wirima, JJ; Fletcher, KA; Morris, K; Blood glucose levels in Malawian children before and during the administration of intravenous quinine for severe *falciparum* malaria. *N Engl J Med.* (1988). *319*, 1040-1047.

[602] Seltzer, HS; Drug-induced hypoglycemia. A review of 1418 cases. *Endocrinol Metab Clin North Am.* (1989). *18*, 163-183.

[603] Assan, R; Perronne, C; Chotard, L; Larger, E; Vilde, JL; Mefloquine-associated hypoglycaemia in a cachectic AIDS patient. *Diabete Metab.* (1995). *21*, 54-58.

[604] Karbwang, J; White, NJ; Clinical pharmacokinetics of mefloquine. *Clin Pharmacokinet.* (1990). *19*, 264-279.

[605] Nosten, F; ter Kuile, FO; Luxemburger, C; Woodrow, C; Kyle, DE; Chongsuphajaisiddhi, T; White, NJ; Cardiac effects of antimalarial treatment with halofantrine. *Lancet.* (1993). *341*, 1054-1056.

[606] Brewer, TG; Grate, SJ; Peggins, JO; Weina, PJ; Petras, JM; Levine, BS; Heiffer, MH; Schuster, BG; Fatal neurotoxicity of arteether and artemether. *Am J Trop Med Hyg.* (1994). *51*, 251-259.

[607] Batty, KT; Davis, TM; Thu, LT; Binh, TQ; Anh, TK; Ilett, KF; Selective high-performance liquid chromatographic determination of artesunate and alpha- and beta-dihydroartemisinin in patients with *falciparum* malaria. *J Chromatogr B Biomed Appl.* (1996). *677*, 345-350.

[608] Lai, H; Singh, NP; Selective cancer cell cytotoxicity from exposure to dihydroartemisinin and holotransferrin. *Cancer Lett.* (1995). *91*, 41-46.

[609] Woerdenbag, HJ; Moskal, TA; Pras, N; Malingré, TM; el-Feraly, FS; Kampinga, HH; Konings, AW; Cytotoxicity of artemisinin-related endoperoxides to Ehrlich ascites tumor cells. *J Nat Prod.* (1993). *56*, 849-856.

[610] Kamchonwongpaisan, S; Chandra-ngam, G; Avery, MA; Yuthavong, Y; Resistance to artemisinin of malaria parasites *(Plasmodium falciparum)* infecting alpha-thalassemic erythrocytes *in vitro*. Competition in drug accumulation with uninfected erythrocytes. *J Clin Invest.* (1994). *93*, 467-473.

[611] Meshnick, SR;The mode of action of antimalarial endoperoxides. *Trans R Soc Trop Med Hyg.* (1994). *88*(Suppl. 1), S31-32.

[612] Meshnick, SR; Free radicals and antioxidants. *Lancet.* (1994). 344, 1441-1442.

[613] Hartwig, CL; Rosenthal, AS; D'Angelo, J; Griffin, CE; Posner, GH; Cooper, RA; Accumulation of artemisinin trioxane derivatives within neutral lipids of *Plasmodium falciparum* malaria parasites is endoperoxide-dependent. *Biochem Pharmacol.* (2009). *77*, 322-336.

[614] Meshnick, SR; Yang, YZ; Lima, V; Kuypers, F; Kamchonwongpaisan, S; Yuthavong, Y; Iron-dependent free radical generation from the antimalarial agent artemisinin (qinghaosu). *Antimicrob Agents Chemother.* (1993). *37*, 1108-1114.

[615] Brewer, TG; Peggins, JO; Grate, SJ; Petras, JM; Levine, BS; Weina, PJ; Swearengen, J; Heiffer, MH; Schuster, BG; Neurotoxicity in animals due to arteether and artemether. *Trans R Soc Trop Med Hyg.* (1994). *88* (Suppl. 1), S33-36.

[616] Kamchonwongpaisan, S; McKeever, P; Hossler, P; Ziffer, H; Meshnick, SR; Artemisinin neurotoxicity: Neuropathology in rats and mechanistic studies *in vitro*. *Am J Trop Med Hyg.* (1997). *56*, 7-12.

[617] Wesche, DL; DeCoster, MA; Tortella, FC; Brewer, TG; Neurotoxicity of artemisinin analogs *in vitro*. *Antimicrob Agents Chemother.* (1994). *38*, 1813-1819.

[618] Fishwick, J; McLean, WG; Edwards, G; Ward, SA; The toxicity of artemisinin and related compounds on neuronal and glial cells in culture. *Chem Biol Interact.* (1995). *96*, 263-271.

[619] Smith, SL; Maggs, JL; Edwards, G; Ward, SA; Park, BK; McLean, WG; The role of iron in neurotoxicity: A study of novel antimalarial drugs. *Neurotoxicology.* (1998). *19*, 557-559.

[620] Miller, LG; Panosian, CB; Ataxia and slurred speech after artesunate treatment for *falciparum* malaria. *N Engl J Med.* (1997). *336*, 1328.

[621] Davis, TM; Edwards, GO; McCarthy, JS; Artesunate and cerebellar dysfunction in *falciparum* malaria. *N Engl J Med.* (1997). *337*, 792.

[622] Exon, JH; A review of the toxicology of acrylamide. *J Toxicol Environ Health B Crit Rev.* (2006). *9*, 397-412.

[623] Karbwang, J; Na-Bangchang, K; Thanavibul, A; Bunnag, D; Chongsuphajaisiddhi, T; Harinasuta, T; Comparison of oral artesunate and quinine plus tetracycline in acute uncomplicated *falciparum* malaria. *Bull World Health Organ.* (1994). *72*, 233-238.

[624] Karbwang, J; Sukontason, K; Rimchala, W; Namsiripongpun, W; Tin, T; Auprayoon, P; Tumsupapong, S; Bunnag, D; Harinasuta, T; Preliminary report: A comparative clinical trial of artemether and quinine in severe *falciparum* malaria. *Southeast Asian J Trop Med Public Health.* (1992). *23*, 768-772.

[625] Win, K; Than, M; Thwe, Y; Comparison of combinations of parenteral artemisinin derivatives plus oral mefloquine with intravenous quinine plus oral tetracycline for treating cerebral malaria. *Bull World Health Organ.* (1992). *70*, 777-782.

[626] von Seidlein, L; Jaffar, S; Greenwood, B. Prolongation of the QTc interval in African children treated for *falciparum* malaria. *Am J Trop Med Hyg.* (1997). *56*, 494-497.

[627] White, NJ; Cardiotoxicity of antimalarial drugs. *Lancet Infect Dis.* (2007). *7*, 549-558.

[628] Yap, YG; Camm, AJ; Drug induced QT prolongation and torsades de pointes. *Heart.* (2003). 89, 1363-1372.

[629] Bakshi, R; Hermeling-Fritz, I; Gathmann, I; Alteri, E; An integrated assessment of the clinical safety of artemether-lumefantrine: A new oral fixed-dose combination antimalarial drug. *Trans R Soc Trop Med Hyg.* (2000). *94*, 419-424.

[630] Supanaranond, W; Davis, TM; Pukrittayakamee, S; Nagachinta, B; White, NJ; Abnormal circulatory control in *falciparum* malaria: the effects of antimalarial drugs. *Eur J Clin Pharmacol.* (1993). *44*, 325-329.

[631] Harada, T; Abe, J; Shiotani, M; Hamada, Y; Horii, I; Effect of autonomic nervous function on QT interval in dogs. *J Toxicol Sci.* (2005). *30*, 229-237.

[632] Diedrich, A; Jordan, J; Shannon, JR; Robertson, D; Biaggioni, I; Modulation of QT interval during autonomic nervous system blockade in humans. *Circulation.* (2002). *106*, 2238-2243.

[633] Uyarel, H; Okmen, E; Cobanoğlu, N; Karabulut, A; Cam, N; Effects of anxiety on QT dispersion in healthy young men. *Acta Cardiol.* (2006). *61*, 83-87.

[634] Rautaharju, PM; Why did QT dispersion die? *Card Electrophysiol Rev.* (2002). *6*, 295-301.

[635] Costedoat-Chalumeau, N; Hulot, JS; Amoura, Z; Delcourt, A; Maisonobe, T; Dorent, R; Bonnet, N; Sablé, R; Lechat, P; Wechsler, B; Piette, JC; Cardiomyopathy related to antimalarial therapy with illustrative case report. *Cardiology.* (2007). *107*, 73-80.

[636] Azie, NE; Adams, G; Darpo, B; Francom, SF; Polasek, EC; Wisser, JM; Fleishaker, JC; Comparing methods of measurement for detecting drug-induced changes in the QT interval: Implications for thoroughly conducted ECG studies. *Ann Noninvasive Electrocardiol.* (2004). *9*, 166-174.

[637] Makanga, M; Premji, Z; Falade, C; Karbwang, J; Mueller, EA; Andriano, K; Hunt, P; De Palacios, PI; Efficacy and safety of the six-dose regimen of artemether-lumefantrine in pediatrics with uncomplicated *Plasmodium falciparum* malaria: A pooled analysis of individual patient data. *Am J Trop Med Hyg.* (2006). *74*, 991-998.

[638] White, NJ; The pharmacokinetics of quinine and quinidine in malaria. *Acta Leiden.* (1987). *55*, 65-76.

[639] Shibata, K; Hirasawa, A; Foglar, R; Ogawa, S; Tsujimoto, G; Effects of quinidine and verapamil on human cardiovascular alpha1-adrenoceptors. *Circulation.* (1998). *97*, 1227-1230.

[640] Clark, RB; Sanchez-Chapula, J; Salinas-Stefanon, E; Duff, HJ; Giles, WR; Quinidine-induced open channel block of K+ current in rat ventricle. *Br J Pharmacol.* (1995). *115*, 335-343.

[641] White, NJ; Looareesuwan, S; Warrell, DA; Quinine and quinidine: A comparison of EKG effects during the treatment of malaria. *J Cardiovasc Pharmacol.* (1983). *5*, 173-175.

[642] Suzuki, S; Murakami, S; Tsujimae, K; Findlay, I; Kurachi, Y; In silico risk assessment for drug-induction of cardiac arrhythmia. *Prog Biophys Mol Biol.* (2008). 98, 52-60.

[643] Touze, JE; Heno, P; Fourcade, L; Paule, P; Effects of antimalarial drugs and cardiomyocytes. Pathogenic approach and new therapeutic recommendations. *Bull Acad Natl Med.* (2006). *190*, 439-449.

[644] Malvy, D; Receveur, MC; Ozon, P; Djossou, F; Le Metayer, P; Touze, JE; Longy-Boursier, M; Le Bras, M; Fatal cardiac incident after use of halofantrine. *J Travel Med.* (2000). *7*, 215-216.

[645] Traebert, M; Dumotier, B; Antimalarial drugs: QT prolongation and cardiac arrhythmias. *Expert Opin Drug Saf.* (2005). *4*, 421-431.

[646] Fonteyne, W; Bauwens, A; Jordaens, L; Atrial flutter with 1:1 conduction after administration of the antimalarial drug mefloquine. *Clin Cardiol.* (1996). *19*, 967-968.

[647] Bukauskas, FF; Kreuzberg, MM; Rackauskas, M; Bukauskiene, A; Bennett, MV; Verselis, VK; Willecke, K; Properties of mouse connexin 30.2 and human connexin 31.9 hemichannels: Implications for atrioventricular conduction in the heart. *Proc Natl Acad Sci USA.* (2006). *103*, 9726-9731.

[648] Supanaranond, W; Suputtamongkol, Y; Davis, TM; Pukrittayakamee, S; Teja-Isavadharm, P; Webster, HK; White, NJ; Lack of a significant adverse cardiovascular effect of combined quinine and mefloquine therapy for uncomplicated malaria. *Trans R Soc Trop Med Hyg.* (1997). *91*, 694-696.

[649] Na-Bangchang, K; Tan-Ariya, P; Thanavibul, A; Riengchainam, S; Shrestha, SB; Karbwang, J; Pharmacokinetic and pharmacodynamic interactions of mefloquine and quinine. *Int J Clin Pharmacol Res.* (1999). *19*, 73-82.

[650] Messant, I; Jérémie, N; Lenfant, F; Freysz, M; Massive chloroquine intoxication: Importance of early treatment and pre-hospital treatment. *Resuscitation.* (2004). *60*, 343-346.

[651] van Vugt, M; Ezzet, F; Nosten, F; Gathmann, I; Wilairatana, P; Looareesuwan, S; White, NJ; No evidence of cardiotoxicity during antimalarial treatment with artemether-lumefantrine. *Am J Trop Med Hyg.* (1999). *61*, 964-967.

[652] Orta-Salazar, G; Bouchard, RA; Morales-Salgado, F; Salinas-Stefanon, EM; Inhibition of cardiac Na+ current by primaquine. *Br J Pharmacol.* (2002). *135*, 751-763.

[653] Matsuo, S; Ruiz, R; Smith, J; Jr, Aviado, DM; Cardiopulmonary effects of antimalarial drugs. 3. Diaminopyrimidines: Trimethoprim (WR 5949) and 5-piperonyl-2,4-diaminopyrimidine (WR 40,070). *Toxicol Appl Pharmacol.* (1970). *17*, 130-150.

[654] Gupta, RK; Van Vugt, M; Paiphun, L; Slight, T; Looareesuwan, S; White, NJ; Nosten, F. Short report: No evidence of cardiotoxicity of atovaquone-proguanil alone or in combination with artesunate. *Am J Trop Med Hyg.* (2005). *73*, 267-268.

[655] Karbwang, J; Laothavorn, P; Sukontason, K; Thiha, T; Rimchala, W; Na-Bangchang, K; Bunnag, D; Effect of artemether on electrocardiogram in severe *falciparum* malaria. *Southeast Asian J Trop Med Public Health.* (1997). *28*, 472-475.

[656] Phillips-Howard, PA; ter Kuile, FO; CNS adverse events associated with antimalarial agents. Fact or fiction? *Drug Saf.* (1995). *12*, 370-383.

[657] Idro, R; Ndiritu, M; Ogutu, B; Mithwani, S; Maitland, K; Berkley, J; Crawley, J; Fegan, G; Bauni, E; Peshu, N; Marsh, K; Neville, B; Newton, C; Burden, features, and outcome of neurological involvement in acute *falciparum* malaria in Kenyan children. *JAMA.* (2007). *297*, 2232-2240.

[658] Mohan, D; Mohandas, E; Rajat, R; Chloroquine psychosis: A chemical psychosis? *J Natl Med Assoc.* (1981). *73*, 1073-1076.

[659] Ragan, E; Wilson, R; Li, F; Spasoff, R; Bigelow, G; Spinner, N; Psychotic symptoms in volunteers serving overseas. *Lancet.* (1985). *2*, 37.

[660] Williams, ML; Wainer, IW; Role of chiral chromatography in therapeutic drug monitoring and in clinical and forensic toxicology. *Ther Drug Monit.* (2002). *24*, 290-296.

[661] Patchen, LC; Campbell, CC; Williams, SB; Neurologic reactions after a therapeutic dose of mefloquine. *N Engl J Med.* (1989). *321*, 1415-1416.

[662] Weinke, T; Trautmann, M; Held, T; Weber, G; Eichenlaub, D; Fleischer, K; Kern, W; Pohle, HD; Neuropsychiatric side-effects after the use of mefloquine. *Am J Trop Med Hyg.* (1991). *45*, 86-91.

[663] Sowunmi, A; Salako, LA; Oduola, AM; Walker, O; Akindele, JA; Ogundahunsi, OA; Neuropsychiatric side-effects of mefloquine in Africans. *Trans R Soc Trop Med Hyg.* (1993). *87*, 462-463.

[664] Magnussen, P; Bygbjerg, IC; Treatment of *Plasmodium falciparum* malaria with mefloquine alone or in combination with i.v. quinine at the Department of Communicable and Tropical Diseases, Rigshospitalet, Copenhagen 1982-1988. *Dan Med Bull.* (1990). *37*, 563-564.

[665] Petras, JM; Kyle, DE; Ngampochjana, M; Young, GD; Bauman, RA; Webster, HK; Corcoran, JD; Peggins, JO; Vane, MA; Brewer, TG; Arteether: Risks of two-week administration in macaca mulatta. *Am J Trop Med Hyg.* (1997). *56*, 390-396.

[666] Nosten, F; Artemisinin: Large community studies. *Trans R Soc Trop Med Hyg.* (1994). *88*(Suppl. 1), S45-46.

[667] Nosten, F; Price, RN; New antimalarials. A risk-benefit analysis. *Drug Saf.* (1995). *12*, 264-273.

[668] Spencer, CM; Goa, KL; Atovaquone. A review of its pharmacological properties and therapeutic efficacy in opportunistic infections. *Drugs.* (1995). *50*, 176-196.

[669] WHO. Review of central nervous system adverse events related to the antimalarial drug, mefloquine (1985-1990). World Health Organization/MAL/91.1063. (1991).

[670] Li, QG; Peggins, JO; Anorectic toxicity of dihydroartemisinin, arteether and artemether in rats following multiple intramuscular doses. *Int J Toxicol.* (1998). *17*, 663-676.

[671] Genovese, RF; Newman, DB; Brewer, TG; Behavioral and neural toxicity of the artemisinin antimalarial, arteether, but not artesunate and artelinate, in rats. *Pharmacol Biochem Behav.* (2000). *67*, 37-44.

[672] Li, QG; Carpenter, C; Trotman, C; Kathcart, AK; Donahue, R; Milhous, WK; Pharmacokinetic comparison of artesunate and dihydroartemisinin in beagle dogs. *Am. J. Trol. Med. Hyg.* (1999). *61*, 274.

[673] Li, QG; Bossone, CA; Ohrt, C; Chung, H; Harris, D; Harre, J; Lee, P; Kathcart, AK; Brueckner, RP; Early signs and biomarkers of the neurotoxicity: Toxicokinetic and toxicodynamic evaluation following 15-mg/kg arteether daily intramuscular injections for two weeks in beagle dogs. *Am. J. Trol. Med. Hyg.* (2000). 62, 142.

[674] Classen, W; Altmann, B; Gretener, P; Souppart, C; Skelton-Stroud, P; Krinke, G. Differential effects of orally versus parenterally administered qinghaosu derivatives artemether in dogs. *Exp Toxicol Pathol* (1999). *51*, 507-516.

[675] China Cooperative Research Group on Qinghaosu and Its Derivatives as Antimalarials. Metabolism and pharmacokinetics of qinghaosu and its derivatives. *J Trad Chin Med,* (1982). 2, 25-30.

[676] Niu, XY; Ho, LY; Ren, ZH; Song, ZY. Metabolic fate of Qinghaosu in rats; A new TLC densitometric method for its determination in biological material. *Eur J Drug Metab Pharmacokinet.* (1985). *10*, 55-59.

[677] Zhao, KC; Song, ZY; Distribution and excretion of artesunate in rats. *Proc Chin Acad Med Sci Peking Union Med Coll.* (1989). *4*, 186-188.

[678] Jiang, JR; Zou, CD; Shu, HL; Zeng, YL; Assessment of absorption and distribution of artemether in rats using a thin layer chromatography scanning technique. *Zhongguo Yao Li Xue Bao.* (1989). *10*, 431-434.

[679] Kearney, BP; Aweeka, FT; The penetration of anti-infectives into the central nervous system. *Neurol Clin.* (1999). *17*, 883-900.

[680] Davis, TM; Phuong, HL; Ilett, KF; *et al.* Pharmacokinetics and pharmacodynamics of intravenous artesunate in severe *falciparum* malaria. *Antimicrob Agents Chemother* (2001). *45*, 181-186.

[681] Kager, PA; Schultz, MJ; Zijlstra, EE; van den Berg, B; van Boxtel, CJ; Arteether administration in humans: Preliminary studies of pharmacokinetics, safety and tolerance. *Trans R Soc Trop Med Hyg.* (1994). *88*(Suppl. 1), S53-54.

[682] Brewer, TG; Genovese, RF; Newman, DB; Li, Q; Factors relating to neurotoxicity of artemisinin antimalarial drugs "listening to arteether." *Med Trop (Mars).* (1998). *58* (Suppl. 3), 22-27.

[683] Nontprasert, A; Pukrittayakamee, S; Nosten-Bertrand, M; Vanijanonta, S; White, NJ; Studies of the neurotoxicity of oral artemisinin derivatives in mice. *Am J Trop Med Hyg.* (2000). *62*, 409-412.

[684] Rowland, M; Tozer, TN; In: M Rowland and TN Tozer (Eds.)., *Clinical Pharmacokinetics: Concepts and applications*, Williams & Wilkins, Baltimore, (1995).p. 83.

[685] Mayer, PR; Absorption, metabolism, and other factors that influence drug exposure in toxicology studies. *Toxicol Pathol.* (1995). *23*, 165-169.

[686] Shargel, L; Yu, ABC; In: L Shargel and ABC Yu , (Eds.), *Applied Biopharmaceutics and Pharmacokinetics.* Appleton & Lange, Norwalk, CN,(1993). p. 353.

[687] Li, QG; Peggens, JO; Brown, LD; Brewer, TG; Pharmacokinetics and hydrolysis of sodium artelinate in dogs. *AM J Trop Med Hyg.* (1994). *51*, 260.

[688] Li, QG; Xie, LH; Peggins, JO; Pharmacokinetics and tolerant dose range studies of AL and AS following daily intramuscular dosing for 7 days in rats. *Am J Trop Med Hyg.* (2001). *65*, 214.

[689] Giao, PT; de Vries, PJ; Pharmacokinetic interactions of antimalarial agents. *Clin Pharmacokinet.* (2001). *40*, 343-373.

[690] Li, QG; Milhous, WK; Therapeutic potency of arteether via pharmacokinetic and pharmacodynamic evaluation in humans and animals. *Am. J. Trol. Med. Hyg.* (2002). *67*, 255.

[691] Klaassen, CD; Rozman, K; Absorption, distribution, and excretion of toxicants. In: MO Amdur, J Doull, and D Cortis, (Eds.), *Casarett and Doull's Toxicology, The Basic Science of Poisons*, Chap. 3, Klaassen. McGraw-Hill, Inc. Press, (1993). p. 53.

[692] Genovese, RF; Newman, DB; Li, QG; Peggins, JO; Brewer, TG; Dose-dependent brainstem neuropathology following repeated arteether administration in rats. *Brain Res. Bull.* (1998). *45*, 199-202.

[693] Perouansky, M; General anesthetics and long-term neurotoxicity. *Handb Exp Pharmacol.* (2008). *182*, 143-157.

[694] Si, Y; Li, Q; Xie, L; Bennett, K; Weina, PJ; Mog, S; Johnson, TO; Neurotoxicity and toxicokinetics of artelinic acid following repeated oral administration in rats. *Int J Toxicol.* (2007). *26*, 401-410.

[695] Davidson, DE; Role of arteether in the treatment of malaria and plans for further development. *Trans. R Soc. Trop. Med. Hyg.* (1994). *88*(Suppl. 1), S51-52.

[696] Looareesuwan, S; Wilairatana, P; The rational use of qinghaosu and its derivatives: What is the future of new compounds? *Med Trop (Mars).* (1998). *58* (Suppl. 3), 89-92.

[697] Geyer, HJ; Scheuntert, I; Rapp, K; Kettrup, A; Korte, F; Greim, H; Rozman, K; Correlation between acute toxicity of 2,3,7,8-tetrachlorodibenzo-p-dioxin (TCDD) and total body fat content in mammals. *Toxicology.* (1990). *65*, 97-107.

[698] Kimbrough, RD; How toxic is 2,3,7,8-tetrachlorodibenzodioxin to humans? *J Toxicol Environ Health.* (1990). *30*, 261-271.

[699] Culotta, E; Koshland, DE Jr; DNA repair works it works its way to the top. *Science.* (1994). *266*, 1926-1929.

[700] Myint, HY; Tipmanee, P; Nosten, F; Day, NP; Pukrittayakamee, S; Looareesuwan, S; White, NJ; A systematic overview of published antimalarial drug trials. *Trans R Soc Trop Med Hyg.* (2004). *98*, 73-81.

[701] Tagbor, HK; Chandramohan, D; Greenwood, B; The safety of amodiaquine use in pregnant women. *Expert Opin Drug Saf.* (2007). *6*, 631-635.

[702] McGready, R; Ashley, EA; Moo, E; Cho, T; Barends, M; Hutagalung, R; Looareesuwan, S; White, NJ; Nosten, F; A randomized comparison of artesunate-atovaquone-proguanil versus quinine in treatment for uncomplicated *falciparum* malaria during pregnancy. *J Infect Dis.* (2005). *192*, 846-853.

[703] Wolfe, MS; Cordero; JF; Safety of chloroquine in chemosuppression of malaria during pregnancy. *Br Med J (Clin Res Ed).* (1985). *290*, 1466-1467.

[704] Clerk, CA; Bruce, J; Affipunguh, PK; Mensah, N; Hodgson, A; Greenwood, B; Chandramohan, D; A randomized, controlled trial of intermittent preventive treatment with sulfadoxine-pyrimethamine, amodiaquine, or the combination in pregnant women in Ghana. *J Infect Dis.* (2008). *198*, 1202-1211.

[705] Keuter, M; van Eijk, A; Hoogstrate, M; Raasveld, M; van de Ree, M; Ngwawe, WA; Watkins, WM; Were, JB; Brandling-Bennett, AD; Comparison of chloroquine, pyrimethamine and sulfadoxine, and chlorproguanil and dapsone as treatment for *falciparum* malaria in pregnant and non-pregnant women, Kakamega District, Kenya. *BMJ.* (1990). *301*, 466-470.

[706] Sulo, J;Chimpeni, P; Hatcher, J; Kublin, JG; Plowe, CV; Molyneux, ME; Marsh, K; Taylor, TE; Watkins, WM; Winstanley, PA; Chlorproguanil-dapsone versus sulfadoxine-pyrimethamine for sequential episodes of uncomplicated *falciparum* malaria in Kenya and Malawi: A randomized clinical trial. *Lancet.* (2002). *360*, 1136-1143.

[707] American Academy of Pediatrics Committee on Drugs: The transfer of drugs and other chemicals into human milk. *Pediatrics.* (1994). *93*, 137-150.

[708] China Cooperative Research Group on qinghaosu and its derivatives as antimalarials. Studies on the toxicity of qinghaosu and its derivatives. *J Trad Chin Med.* (1982). *2*, 31-38.

[709] Longo, M; Zanoncelli, S; Manera, D; Brughera, M; Colombo, P; Lansen, J; Mazué, G; Gomes, M; Taylor, WR; Olliaro, P; Effects of the antimalarial drug dihydroartemisinin (DHA) on rat embryos *in vitro. Reprod Toxicol.* (2006). *21*, 83-93.

[710] White, NJ; McGready, RM; Nosten, FH; New medicines for tropical diseases in pregnancy: Catch-22. *PLoS Med.* (2008). *5*, e133.

[711] Wang, TY; Follow-up observation on the therapeutic effects and remote reactions of artemisinin (Qinghaosu) and artemether in treating malaria in pregnant woman. *J Trad Chin Med.* (1989). *9*, 28-30.

[712] McGready, R; Cho, T; Cho, JJ; Simpson, JA; Luxemburger, C; Dubowitz, L; Looareesuwan, S; White, NJ; Nosten, F; Artemisinin derivatives in the treatment of *falciparum* malaria in pregnancy. *Trans R Soc Trop Med Hyg.* (1998). *92*, 430-433.

[713] Chen, LJ; Wang, MY; Sun, WK; Liu, MZ; Embryotoxicity and teratogenicity studies on artemether in mice, rats and rabbits. *Zhongguo Yao Li Xue Bao.* (1984). *5*, 118-122.

[714] Li, ZL; Teratogenicity of sodium artesunate. *Zhong Yao Tong Bao,* (1988). *13*, 42-44; 63-64.

[715] Si, YZ; Zeng, Q; Zhang, J; Johnson, TO; Xie, LH; Weina, PJ; Milhous, WK; Li, QG; Evaluation of the embryonic toxicity of artesunate in rats. *Am J Trop Med Hyg.* (2004). *71*, 260.

[716] Xu, JH; Zhang, YP; Contra-gestational effects of dihydroartemisinin and artesunate. *Yao Xue Xue Bao*. (1996). *31*, 657-661.

[717] Clark, RL; White, TE; Clode, SA; Gaunt, I; Winstanley, P; Ward, SA; Developmental toxicity of artesunate and an artesunate combination in the rat and rabbit. *Birth Defects Res B Dev Reprod Toxicol*. (2004). *71*, 380-394.

[718] Longo, M; Zanoncelli, S; Torre, PD; Riflettuto, M; Cocco, F; Pesenti, M; Giusti, A; Colombo, P; Brughera, M; Mazue, G; Navaratman, V; Gomes, M; Olliaro, P. *In vivo* and *in vitro* investigations of the effects of the antimalarial drug dihydroartemisinin (DHA) on rat embryos. *Reprod Toxicol*. (2006). *22*, 797-810.

[719] Ejiofor, JI; Kwanashie, HO; Anuka, JA; Some pregnancy-related effects of artemether in laboratory animals. *Pharmacol* (2006). *77*, 166-170.

[720] Lou, XE; Zhou, HJ; Effects of artesunate on progesterone estrogen content and decidua in rats. *Yao Xue Xue Bao*. (2001). *36*, 254-257.

[721] Lou, XE; Zhou, HJ; Huang, HB; Lipid-peroxidantion damage of embryo and placenta induced by artesunate in rats. *Zhejiang Da Xue Xue Bao Yi Xue Ban*. (2003). *32*, 41-45.

[722] Chen, HH; Zhou, HJ; Wu, GD; Lou, XE; Inhibitory effects of artesunate on angiogenesis and on expressions of vascular endothelial growth factor and VEGF receptor KDR/flk-1. *Pharmacol* (2004). *71*, 1-9.

[723] Chen, HH; Zhou, HJ; Wang, WQ; Wu, GD; Antimalarial dihydroartemisinin also inhibits angiogenesis. *Cancer Chemother Pharmacol*. (2004). *53*, 423-432.

[724] Anfosso, L; Efferth, T; Albini, A; Pfeffer, U; Microarray expression profiles of angiogenesis-related genes predict tumor cell response to artemisinins. *Pharmacogenomics J*. (2006). *6*, 269-278.

[725] Efferth, T; Mechanistic perspectives for 1,2,4-trioxanes in anti-cancer therapy. *Drug Resist Updat*. (2005). *8*, 85-97.

[726] Dell'Eva, R; Pfeffer, U; Vene, R; Anfosso, L; Forlani, A; Albini, A; Efferth, T; Inhibition of angiogenesis *in vivo* and growth of Kaposi's sarcoma xenograft tumors by the anti-malarial artesunate. *Biochem Pharmacol*. (2004). *68*, 2359-2366.

[727] Huan-huan, C; Li-Li, Y; Shang-Bin, L. Artesunate reduces chicken chorioallantoic membrane neovascularisation and exhibits antiangiogenic and apoptotic activity on human microvascular dermal endothelial cell. *Cancer Lett*. (2004). *211*, 163-173.

[728] Chen, HH; Zhou, HJ; Inhibitory effects of artesunate on angiogenesis. *Yao Xue Xue Bao*. (2004). *39*, 29-33.

[729] Chen, HH; Zhou, HJ; Fang, X; Inhibition of human cancer cell line growth and human umbilical vein endothelial cell angiogenesis by artemisinin derivatives *in vitro*. *Pharmacol Res*. (2003). *48*, 231-236.

[730] Zon, LI; Developmental biology of hematopoiesis. *Blood*. (1995). *8*, 2876-2891.

[731] Choe, H; Hansen, JM; Harris, C; Spatial and temporal ontogenesis of glutathione peroxidase, and glutathione disulfide reductase during development of the prenatal rat. *J Biochem Mol Toxicol* (2001). *15*, 197-206.

[732] Li, Q; Si, Y; Smith, KS; Zeng, Q; Weina, PJ; Embryotoxicity of artesunate in animal species related to drug tissue distribution and toxicokinetic profiles. *Birth Defects Res B Dev Reprod Toxicol*. (2008). *83*, 435-445.

[733] McGready, R; Cho, T; Samuel; Villegas, L; Brockman, A; van Vugt, M; Looareesuwan, S; White, NJ; Nosten, F; Randomized comparison of quinine-clindamycin versus artesunate in the treatment of *falciparum* malaria in pregnancy. *Trans R Soc Trop Med Hyg*. (2001). *95*, 651-656.

[734] McGready, R; Keo, NK; Villegas, L; White, NJ; Looareesuwan, S; Nosten, F; Artesunate-atovaquone-proguanil rescue treatment of multidrug-resistant *Plasmodium falciparum* malaria in pregnancy: A preliminary report. *Trans R Soc Trop Med Hyg.* (2003). *97*, 592-594.

[735] Phillips-Howard, PA; Wood, D; The safety of antimalarial drugs in pregnancy. *Drug Saf.* (1996). *14*, 131-145.

[736] Si, YZ; Li, QG; Haeberle, A; Milhouse, WK; Weina, P; [^{14}C] artesunate tissue distribution in pregnant rats following a single intravenous dose with whole-body autoradiography. *Am J Trop Med Hyg.* (2006). *75*, 18.

[737] WHO. Assessment of the safety of artemisinin compounds in pregnancy. Report of two informal consultations convened by WHO in 2002 (RBM and TDR). World Health Organization /CDS/MAL/2003.1094.

[738] Steketee, RW; Wirima, JJ; Campbell, CC; Developing effective strategies for malaria prevention programs for pregnant African women. *Am J Trop Med Hyg.* (1996). *55*(Suppl. 1), 95-100.

[739] Steketee, RW; Nahlen, BL; Parise, ME; Menendez, C; The burden of malaria in pregnancy in malaria-endemic areas. *Am J Trop Med Hyg.* (2001). *64*(Suppl. 1-2), 28-35.

[740] WHO, The importance of pharmacovigilance: Safety monitoring of medicinal products. Geneva: World Health Organization. 2002.

[741] WHO, Assessment of the safety of artemisinin compounds in pregnancy. Report of two joint informal consultations convened in 2006. Geneva: World Health Organization. (2006).

[742] Simooya, O; The WHO 'Roll Back Malaria Project': Planning for adverse event monitoring in Africa. *Drug Saf.* (2005). *28*, 277-286.

[743] Beyens, MN; Guy, C; Ratrema, M; Ollagnier, M; Prescription of drugs to pregnant women in France: The HIMAGE study. *Therapie.* (2003). *58*, 505-511.

[744] AlKadi, HO; Antimalarial drug toxicity: A review. *Chemotherapy.* (2007). *53*, 385-391.

[745] Zuckner, J; Drug-related myopathies. *Rheum Dis Clin North Am.* (1994). *20*, 1017-1032.

[746] Wasay, M; Wolfe, GI; Herrold, JM; Burns, DK; Barohn, RJ; Chloroquine myopathy and neuropathy with elevated CSF protein. *Neurology.* (1998). *51*, 1226-1227.

[747] Ribeiro, IR; Olliaro, P; Safety of artemisinin and its derivatives. A review of published and unpublished clinical trials. *Med Trop (Mars).* (1998). *58*(Suppl. 3), 50-53.

[748] Sun, HY; Fang, CT; Wang, JT; Kuo, PH; Chen, YC; Chang, SC. Successful treatment of imported cerebral malaria with artesunate-mefloquine combination therapy. *J Formos Med Assoc.* (2006). *105*, 86-89.

[749] Itoda, I; Yasunami, T; Kikuchi, K; Yamaura, H; Totsuka, K; Yoshinaga, K; Teramura, M; Mizoguchi, H; Hatabu, T; Kano, S; Severe *falciparum mal*aria with prolonged hemolytic anemia after successful treatment with intravenous artesunate. *Kansenshogaku Zasshi.* (2002). *76*, 600-603.

[750] Same-Ekobo, A; Lohoue, J; Essono, E; Ravinet, L; Ducret, JP; Rapid resolution of *Plasmodium ovale* malarial attacks using artesunate (Arsumax). *Med Trop (Mars).* (1999). *59*, 43-45.

[751] Yoshizawa, S; Hike, K; Kimura, K; Matsumoto, T; Furuya, N; Tateda, K; Kano, S; Yamaguchi, K; A case of *falciparum* malaria successfully treated with intravenous artesunate. *Kansenshogaku Zasshi.* (2002). *76*, 888-892.

[752] Peto, TE; Toxicity of antimalarial drugs. *J R Soc Med.* (1989). *82* (Suppl. 17), 30-33; discussion 33-34.

[753] Cooper, RG; Magwere, T; Chloroquine: Novel uses & manifestations. *Indian J Med Res.* (2008). *127*, 305-316.

[754] Phillips-Howard, PA; West, LJ; Serious adverse drug reactions to pyrimethamine-sulphadoxine, pyrimethamine-dapsone and to amodiaquine in Britain. *J R Soc Med.* (1990). *83*, 82-85.

[755] Hernborg, A; Stevens-Johnson syndrome after mass prophylaxis with sulfadoxine for cholera in Mozambique. *Lancet.* (1985). *2*, 1072-1073.

[756] Steffen, R; Somaini, B; Severe cutaneous adverse reactions to sulfadoxine-pyrimethamine in Switzerland. *Lancet.* (1986). *1*, 610.

[757] Centers for Disease Control (CDC). Revised recommendations for preventing malaria in travelers to areas with chloroquine-resistant *Plasmodium falciparum. MMWR Morb Mortal Wkly Rep.* (1985). *34*, 185-190, 195.

[758] Hatton, CS; Peto, TE; Bunch, C; Pasvol, G; Russell, SJ; Singer, CR; Edwards, G; Winstanley, P; Frequency of severe neutropenia associated with amodiaquine prophylaxis against malaria. *Lancet.* (1986). *1*, 411-414.

[759] Chattopadhyay, R; Mahajan, B; Kumar, S; Assessment of safety of the major antimalarial drugs. *Expert Opin Drug Saf.* (2007). *6*, 505-521.

[760] Ginsburg, H; Should chloroquine be laid to rest? *Acta Trop.* (2005). *96*, 16-23.

[761] Fish, DR; Espir, ML; Convulsions associated with prophylactic antimalarial drugs: Implications for people with epilepsy. *BMJ.* (1988). *297*, 526-527.

[762] Jaeger, A; Sauder, P; Kopferschmitt, J; Flesch, F. Clinical features and management of poisoning due to antimalarial drugs. *Med Toxicol Adverse Drug Exp.* (1987). *2*, 242-273.

[763] Riou, B; Barriot, P; Rimailho, A; Baud, FJ; Treatment of severe chloroquine poisoning. *N Engl J Med.* (1988). *318*,1-6.

[764] White, NJ; Miller, KD; Churchill, FC; Berry, C; Brown, J; Williams, SB; Greenwood, BM; Chloroquine treatment of severe malaria in children. Pharmacokinetics, toxicity, and new dosage recommendations. *N Engl J Med.* (1988). *319*, 1493-1500.

[765] Croft, A; Garner, P; Mefloquine to prevent malaria: A systematic review of trials. *BMJ.* (1997). *315*, 1412-1416.

[766] Millard, TP; Smith, HR; Black, MM; Barker, JN; Bullous pemphigoid developing during systemic therapy with chloroquine. *Clin Exp Dermatol.* (1999). *24*, 263-265.

[767] Nicolas, X; Granier, H; Laborde, JP; Martin, J; Talarmin, F; Danger of malaria self-treatment. Acute neurologic toxicity of mefloquine and its combination with pyrimethamine-sulfadoxine. *Presse Med.* (2001). *30*, 1349-1350.

[768] Edstein, MD; Veenendaal, JR; Hyslop, R; Excretion of mefloquine in human breast milk. *Chemotherapy.* (1988). *34*, 165-169.

[769] Genovese, RF; Petras, JM; Brewer, TG; Arteether neurotoxicity in the absence of deficits in behavioral performance in rats. *Ann Trop Med Parasitol.* (1995). *89*, 447-449.

[770] Meshnick, SR; Artemisinin: Mechanisms of action, resistance and toxicity. *Int J Parasitol.* (2002). *32*, 1655-1660.

[771] Nontprasert, A; Pukrittayakamee, S; Prakongpan, S; Supanaranond, W; Looareesuwan, S; White, NJ; Assessment of the neurotoxicity of oral dihydroartemisinin in mice. *Trans R Soc Trop Med Hyg.* (2002). *96*, 99-101.

[772] Nontprasert, A; Nosten-Bertrand, M; Pukrittayakamee, S; Vanijanonta, S; Angus, BJ; White, NJ; Assessment of the neurotoxicity of parenteral artemisinin derivatives in mice. *Am J Trop Med Hyg.* (1998). *59*, 519-522.

[773] Wilairatana, P; Krudsood, S; Treeprasertsuk, S; Chalermrut, K; Looareesuwan, S; The future outlook of antimalarial drugs and recent work on the treatment of malaria. *Arch Med Res.* (2002). *33*, 416-421.

[774] Hien, TT; An overview of the clinical use of artemisinin and its derivatives in the treatment of *falciparum* malaria in Vietnam. *Trans R Soc Trop Med Hyg.* (1994). *88* (Suppl. 1), S7-8.

[775] Kar, K; Shankar, G; Bajpai, R; Dutta, GP; Vishwakarma, RA; Artemisinin: A potent antimalarial agent, general pharmacological properties. *Ind J Parasitol.* (1988). *12*, 209-212.

[776] Myint, PT; Shwe, T; The efficacy of artemether (qinghaosu) in *Plasmodium falciparum* and *P. vivax* in Burma. *Southeast Asian J Trop Med Public Health* (1986). *17*, 19-22.

[777] Myint, PT; Shwe, T; A controlled clinical trial of artemether (qinghaosu derivative) versus quinine in complicated and severe *falciparum* malaria. *Trans R Soc Trop Med Hyg.* (1987). *81*, 559-561.

[778] Causer, LM; Filler, S; Wilson, M; Papagiotas, S; Newman, RD; Evaluation of reported malaria chemoprophylactic failure among travelers in a US University Exchange Program, 2002. *Clin Infect Dis.* (2004). *39*, 1583-1588.

[779] Keystone, JS; The sound of hoof beats does not always mean that it is a zebra. *Clin Infect Dis.* (2004). *39*, 1589-1590.

[780] Franco-Paredes, C; Dismukes, R; Nicholls, D; Kozarsky, PE; Neurotoxicity due to antimalarial therapy associated with misdiagnosis of malaria. *Clin Infect Dis.* (2005). *40*, 1710-1711.

[781] Newton, PN; Day, NPJ; White, NJ; Misattribution of central nervous system dysfunction to artesunate. *Clin Infect Dis.* (2005). *41*, 1687-1688.

[782] Miller, LG; Panosian, CB; Artesunate and cerebellar dysfunction in *falciparum* malaria-reply. *N Engl J Med.* (1997). *337*, 793.

[783] Toovey, S; Jamieson, A; Audiometric changes associated with the treatment of uncomplicated *falciparum* malaria with co-artemether. *Trans R Soc Trop Med Hyg.* (2004). *98*, 261-267.

[784] Toovey, S; Effects of weight, age, and time on artemether-lumefantrine associated ototoxicity and evidence of irreversibility. *Travel Med Infect Dis.* (2006). *4*, 71-76.

[785] Hutagalung, R; Htoo, H; Nwee, P; Arunkamomkiri, J; Zwang, J; Carrara, VI; Ashley, E; Singhasivanon, P; White, NJ; Nosten, F; A case-control auditory evaluation of patients treated with artemether-lumefantrine. *Am J Trop Med Hyg.* (2006). *74*, 211-214.

[786] Hoffman, SL; Artemether in severe malaria – still to many deaths. *N Engl J Med.* (1996). *335*, 124-126.

[787] Clark, RL; Lerman, SA; Cox, EM; Gristwood, WE; White, TE; Developmental toxicity of artesunate in the rat: Comparison to other artemisinins, comparison of embryotoxicity and kinetics by oral and intravenous routes, and relationship to maternal reticulocyte count. *Birth Defects Res B Dev Reprod Toxicol.* (2008). *83*, 397-406.

[788] Li, GQ; Guo, XB; Jin, R; Wang, ZC; Jian, HX; Li, ZY; Clinical trials on qinghaosu and its derivatives: *Guangzhou College Trad Chin Med* (1990). *13*, 74-79.

[789] Bounyasong, S; Randomized trial of artesunate and mefloquine in comparison with quinine sulfate to treat *P. falciparum* malaria pregnant women. *J Med Assoc Thai.* (2001). *84*, 1289-1299.

[790] Brabin, BJ; Ramagosa, C; Abdelgalil, S; Menendez, C; Verhoeff, FH; McGready, R; Fletcher, KA; Owens, S; d'Alessandro, U; Nosten, F; Fischer, PR; Ordi, J; The sick placenta: The role of malaria. *Placenta* (2004). *25*, 359-378.

[791] Kitchener, SJ; Nasveld, PE; Gregory, RM; Edstein, MD; Mefloquine and doxycycline malaria prophylaxis in Australian soldiers in East Timor. *Med J Aust.* (2005). *182*, 168-171.

[792] Chiou, CS; Lin, SM; Lin, SP; Chang, WG; Chan, KH; Ting, CK; Clindamycin-induced anaphylactic shock during general anesthesia. *J Chin Med Assoc.* (2006). *69*, 549-551.

[793] Na-Bangchang, K; Ruengweerayut, R; Karbwang, J; Chauemung, A; Hutchinson, D; Pharmacokinetics and pharmacodynamics of fosmidomycin monotherapy and combination therapy with clindamycin in the treatment of multidrug resistant *falciparum* malaria. *Malar J.* (2007). *6*, 70.

[794] Peters, PJ; Thigpen, MC; Parise, ME; Newman, RD. Safety and toxicity of sulfadoxine/pyrimethamine: implications for malaria prevention in pregnancy using intermittent preventive treatment. *Drug Saf.* (2007). *30*, 481-501.

[795] Lell, B; Kremsner, PG; Clindamycin as an antimalarial drug: Review of clinical trials. *Antimicrob Agents Chemother.* (2002). *46*, 2315-2320.

[796] Kremsner, PG; Krishna, S; Antimalarial combinations.*Lancet.* (2004). *364*, 285-294.

[797] Kokwaro, G; Mwai, L; Nzila, A; Artemether/lumefantrine in the treatment of uncomplicated *falciparum* malaria. *Expert Opin Pharmacother.* (2007). *8*, 75-94.

[798] Adaramoye, OA; Osaimoje, DO; Akinsanya, AM; Nneji, CM; Fafunso, MA; Ademowo, OG; Changes in antioxidant status and biochemical indices after acute administration of artemether, artemether-lumefantrine and halofantrine in rats. *Basic Clin Pharmacol Toxicol.* (2008). *102*, 412-418.

[799] Sagara, I; Diallo, A; Kone, M; Coulibaly, M; Diawara, SI; Guindo, O; Maiga, H; Niambele, MB; Sissoko, M; Dicko, A; Djimde, A; Doumbo, OK; A randomized trial of artesunate-mefloquine versus artemether-lumefantrine for treatment of uncomplicated *Plasmodium falciparum* malaria in Mali. *Am J Trop Med Hyg.* (2008). *79*, 655-661.

[800] Kissinger, E; Hien, TT; Hung, NT; Nam, ND; Tuyen, NL; Dinh, BV; Mann, C; Phu, NH; Loc, PP; Simpson, JA; White, NJ; Farrar, JJ; Clinical and neurophysiological study of the effects of multiple doses of artemisinin on brain-stem function in Vietnamese patients. *Am J Trop Med Hyg.* (2000). *63*, 48-55.

[801] van Vugt, M; Angus, BJ; Price, RN; Mann, C; Simpson, JA; Poletto, C; Htoo, SE; Looareesuwan, S; White, NJ; Nosten, F; A case-control auditory evaluation of patients treated with artemisinin derivatives for multidrug-resistant *Plasmodium falciparum* malaria. *Am J Trop Med Hyg.* (2000). *62*, 65-69.

[802] McCall, MB; Beynon, AJ; Mylanus, EA; Ven, AJ van der; Sauerwein, RW; No hearing loss associated with the use of artemether-lumefantrine to treat experimental human malaria. *Trans R Soc Trop Med Hyg.* (2006). *100*, 1098-1104.

[803] Gürkov, R; Eshetu, T; Miranda, IB; Berens-Riha, N; Mamo, Y; Girma, T; Krause, E; Schmidt, M; Hempel, JM; Löscher, T; Ototoxicity of artemether/lumefantrine in the treatment of *falciparum* malaria: A randomized trial. *Malar J.* (2008). *7*, 179.

[804] Sirima, SB; Gansané, A. Artesunate-amodiaquine for the treatment of uncomplicated malaria. *Expert Opin Investig Drugs.* (2007). *16*, 1079-1085.

[805] Orrell, C; Little, F; Smith, P; Folb, P; Taylor, W; Olliaro, P; Barnes, KI; Pharmacokinetics and tolerability of artesunate and amodiaquine alone and in combination in healthy volunteers. *Eur J Clin Pharmacol.* (2008). *64*, 683-690.

[806] Oduro, AR; Anyorigiya, T; Anto, F; Amenga-Etego, L; Ansah, NA; Atobrah, P; Ansah, P; Koram, K; Hodgson, A; A randomized, comparative study of supervised and unsupervised artesunate-amodiaquine, for the treatment of uncomplicated malaria in Ghana. *Ann Trop Med Parasitol.* (2008). *102*, 565-576.

[807] Kobbe, R; Klein, P; Adjei, S; Amemasor, S; Thompson, WN; Heidemann, H; Nielsen, MV; Vohwinkel, J; Hogan, B; Kreuels, B; Bührlen, M; Loag, W; Ansong, D; May, J; A randomized trial on effectiveness of artemether-lumefantrine versus artesunate plus amodiaquine for unsupervised treatment of uncomplicated *Plasmodium falciparum* malaria in Ghanaian children. *Malar J.* (2008). *7*, 261.

[808] Sarrassat, S; Senghor, P; Le Hesran, JY; Trends in malaria morbidity following the introduction of artesunate plus amodiaquine combination in M'lomp village dispensary, south-western Senegal. *Malar J.* (2008). *7*, 215.

[809] Orrell, C; Taylor, WR; Olliaro, P; Acute asymptomatic hepatitis in a healthy normal volunteer exposed to 2 oral doses of amodiaquine and artesunate. *Trans R Soc Trop Med Hyg.* (2001). *95*, 517-518.

[810] Adjuik, M; Agnamey, P; Babiker, A; Borrmann, S; Brasseur, P; Cisse, M; Cobelens, F; Diallo, S; Faucher, JF; Garner, P; Gikunda, S; Kremsner, PG; Krishna, S; Lell, B; Loolpapit, M; Matsiegui, PB; Missinou, MA; Mwanza, J; Ntoumi, F; Olliaro, P; Osimbo, P; Rezbach, P; Some, E; Taylor, WR; Amodiaquine-artesunate versus amodiaquine for uncomplicated *Plasmodium falciparum* malaria in African children: A randomized, multicentre trial. *Lancet.* (2002). *359*, 1365-1372.

[811] Agomo, PU; Meremikwu, MM; Watila, IM; Omalu, IJ; Odey, FA; Oguche, S; Ezeiru, VI; Aina, OO; Efficacy, safety and tolerability of artesunate-mefloquine in the treatment of uncomplicated *Plasmodium falciparum* malaria in four geographic zones of Nigeria. *Malar J.* (2008). *7*, 172.

[812] El-Sayed, B; El-Zaki, SE; Babiker, H; Gadalla, N; Ageep, T; Mansour, F; Baraka, O; Milligan, P; Babiker, A; A randomized open-label trial of artesunate- sulfadoxine-pyrimethamine with or without primaquine for elimination of sub-microscopic *P. falciparum* parasitemia and gametocyte carriage in eastern Sudan. *PLoS ONE.* (2007). *2*, e1311.

[813] Kalilani, L; Mofolo, I; Chaponda, M; Rogerson, SJ; Alker, AP; Kwiek, JJ; Meshnick, SR; A randomized controlled pilot trial of azithromycin or artesunate added to sulfadoxine-pyrimethamine as treatment for malaria in pregnant women. *PLoS ONE.* (2007). *2*, e1166.

[814] Rulisa, S; Gatarayiha, JP; Kabarisa, T; Ndayisaba, G; Comparison of different artemisinin-based combinations for the treatment of *Plasmodium falciparum* malaria in children in Kigali, Rwanda, an area of resistance to sulfadoxine-pyrimethamine: Artesunate plus sulfadoxine/pyrimethamine versus artesunate plus sulfamethoxypyrazine/pyrimethamine. *Am J Trop Med Hyg.* (2007). *77*, 612-616.

[815] Elamin, SB; Malik, EM; Abdelgadir, T; Khamiss, AH; Mohammed, MM; Ahmed, ES; Adam, I; Artesunate plus sulfadoxine-pyrimethamine for treatment of uncomplicated *Plasmodium falciparum* malaria in Sudan. *Malar J.* (2005). *4*, 41.

[816] Doherty, JF; Sadiq, AD; Bayo, L; Alloueche, A; Olliaro, P; Milligan, P; von Seidlein, L; Pinder, M; A randomized safety and tolerability trial of artesunate plus sulfadoxine--pyrimethamine versus sulfadoxine-pyrimethamine alone for the treatment of uncomplicated malaria in Gambian children. *Trans R Soc Trop Med Hyg.* (1999). *93*, 543-546.

[817] Sarkar, M; Woodland, CC; Koren, G; Einarson, AR; Pregnancy outcome following gestational exposure to azithromycin. *BMC Pregnancy and Childbirth.* (2006). *6*, 81918.

[818] Bojang, KA; Novel therapies for the prevention of malaria. *Expert Opin Investig Drugs.* (2008). *17*, 1839-1847.

[819] Davis, TM; Karunajeewa, HA; Ilett, KF; Artemisinin-based combination therapies for uncomplicated malaria. *Med J Aust.* (2005). *182*, 181-185.

[820] van Vugt, M; Leonardi, E; Phaipun,□ ; Slight, T; Thway, KL; McGready, R; Brockman, A; Villegas, L; Looareesuwan, S; White, NJ; Nosten, F; Treatment of uncomplicated multidrug-resistant *falciparum* malaria with artesunate-atovaquone-proguanil. *Clin Infect Dis*. (2002). *35*, 1498-1504

[821] Tran, TH; Day, NP; Ly, VC; Nguyen, TH; Pham, PL; Nguyen, HP; Bethell, DB; Dihn, XS; Tran, TH; White, NJ; Blackwater fever in southern Vietnam: A prospective descriptive study of 50 cases. *Clin Infect Dis*. (1996). *23*, 1274-1281.

[822] Venkatesh, S; Lipper, RA; Role of the development scientist in compound lead selection and optimization. *J Pharm Sci*. (2000). *89*, 145-154.

[823] Peters, W; Drug resistance in malaria--a perspective. *Trans R Soc Trop Med Hyg*. (1969). *63*, 25-45.

[824] Fogh, S; Jepsen, S; Effersøe, P; Chloroquine-resistant *Plasmodium falciparum* malaria in Kenya. *Trans R Soc Trop Med Hyg*. (1979). *73*, 228-229.

[825] Bloland, PB; Lackritz, EM; Kazembe, PN; Were, JB; Steketee, R; Campbell, CC; Beyond chloroquine: Implications of drug resistance for evaluating malaria therapy efficacy and treatment policy in Africa. *J Infect Dis*. (1993). *167*, 932-937.

[826] Zucker, JR; Ruebush, TK 2nd; Obonyo, C; Otieno, J; Campbell, CC; The mortality consequences of the continued use of chloroquine in Africa: Experience in Siaya, western Kenya. *Am J Trop Med Hyg*. (2003). *68*, 386-390.

[827] Korenromp, EL; Williams, BG; Gouws, E; Dye, C; Snow, RW; Measurement of trends in childhood malaria mortality in Africa: An assessment of progress toward targets based on verbal autopsy. *Lancet Infect Dis*. (2003). *3*, 349-358.

[828] Yang, Z; Zhang, Z; Sun, X; Wan, W; Cui, L; Zhang, X; Zhong, D; Yan, G; Cui, L; Molecular analysis of chloroquine resistance in *Plasmodium falciparum* in Yunnan Province, China. *Trop Med Int Health*. (2007). *12*, 1051-1060.

[829] Basco, LK; Ngane, VF; Ndounga, M; Same-Ekobo, A; Youmba, JC; Abodo, RT; Soula, G; Molecular epidemiology of malaria in Cameroon. XXI. Baseline therapeutic efficacy of chloroquine, amodiaquine, and sulfadoxine-pyrimethamine monotherapies in children before national drug policy change. *Am J Trop Med Hyg*. (2006). *75*, 388-395.

[830] Zakeri, S; Afsharpad, M; Raeisi, A; Djadid, ND; Prevalence of mutations associated with antimalarial drugs in *Plasmodium falciparum* isolates prior to the introduction of sulphadoxine-pyrimethamine as first-line treatment in Iran. *Malar J*. (2007). *6*, 148.

[831] Olliaro, P; Mussano, P; Amodiaquine for treating malaria. *Cochrane Database Syst Rev*. (2003). *2*, CD000016.

[832] WHO. The selection and use of essential medicines: Report of the WHO Expert Committee (including the 12th Model List of Essential Medicines). Geneva: World Health Organization,2002.   (accessed   December   2010) http://www.who.int/medicines/publications/essentialmeds_committeereports/en/

[833] Neftel, KA; Woodtly, W; Schmid, M; Frick, PG; Fehr, J; Amodiaquine induced agranulocytosis and liver damage. *Br Med J (Clin Res Ed)*. (1986). *292*, 721-723.

[834] WHO. Practical chemotherapy of malaria. World Health Organization Technical Report Series (1990). p.805.

[835] WHO. WHO expert committee on malaria. Nineteenth report. Geneva: World Health Organization (1993).

[836] WHO. WHO/MAL/96.1075. World Health Organization (1996).

[837] Hyde, JE; Antifolate resistance in Africa and the 164-dollar question. *Trans R Soc Trop Med Hyg*. (2008). *102*, 301-303.

[838] Plowe, CV; Cortese, JF; Djimde, A; Nwanyanwu, OC; Watkins, WM; Winstanley, PA; Estrada-Franco, JG; Mollinedo, RE; Avila, JC; Cespedes, JL; Carter, D; Doumbo, OK; Mutations in *Plasmodium falciparum* dihydrofolate reductase and dihydropteroate synthase and epidemiologic patterns of pyrimethamine-sulfadoxine use and resistance. *J Infect Dis.* (1997). *176*, 1590-1596.

[839] Kublin, JG; Dzinjalamala, FK; Kamwendo, DD; Malkin, EM; Cortese, JF; Martino, LM; Mukadam, RA; Rogerson, SJ; Lescano, AG; Molyneux, ME; Winstanley, PA; Chimpeni, P; Taylor, TE; Plowe, CV; Molecular markers for failure of sulfadoxine-pyrimethamine and chlorproguanil-dapsone treatment of *Plasmodium falciparum* malaria. *J Infect Dis.* (2002). *185*, 380-388.

[840] Iyer, JK; Milhous, WK; Cortese, JF; Kublin, JG; Plowe, CV; *Plasmodium falciparum* cross-resistance between trimethoprim and pyrimethamine. *Lancet.*(2001). *358*, 1066-1067.

[841] Triglia, T; Menting, JG; Wilson, C; Cowman, AF; Mutations in dihydropteroate synthase are responsible for sulfone and sulfonamide resistance in *Plasmodium falciparum. Proc Natl Acad Sci USA.* (1997). *94*, 13944-13949.

[842] Cortese, JF; Caraballo, A; Contreras, CE; Plowe, CV; Origin and dissemination of *Plasmodium falciparum* drug-resistance mutations in South America. *J Infect Dis.* (2002). *186*, 999-1006.

[843] WHO. Susceptibility of *Plasmodium falciparum* to Antimalarial Drugs: Report on Global Monitoring 1996-2004. Geneva, Switzerland: World Health Organization; 2005.

[844] Kalanda, GC; Hill, J; Verhoeff, FH; Brabin, BJ; Comparative efficacy of chloroquine and sulfadoxinepyrimethamine in pregnant women and children: A meta-analysis. *Trop Med Int Health.* (2006). *11*, 569-577.

[845] Laufer, MK; Plowe, CV; Withdrawing antimalarial drugs: Impact on parasite resistance and implications for malaria treatment policies. *Drug Resist Updat.* (2004). *7*, 279-288.

[846] Leggat, PA; Trends in antimalarial prescriptions in Australia from 1998 to 2002. *J Travel Med.* (2005). *12*, 338-342.

[847] Croft, AM; Garner, P; Mefloquine for preventing malaria in non-immune adult travelers. *Cochrane Database Syst Rev.* (2000). *2*, CD000138.

[848] Paradisi, A; Bugatti, L; Sisto, T; Filosa, G; Amerio, PL; Capizzi, R; Acute generalized exanthematous pustulosis induced by hydroxychloroquine: Three cases and a review of the literature. *Clin Ther.* (2008). *30*, 930-940.

[849] Norrby, SR; Lietman, PS; Safety and tolerability of fluoroquinolones. *Drugs.* (1993). *45*(Suppl. 3), 59-64.

[850] Sodahlon, YK; Agbo, K; Morgah, K; Adjogble, K; Avodagbe, A; Djadou, KE; Dekou, K; Pignandi, A; Kassankogno, Y; Sukwa, T; Penali, KL; Millet, P; Malvy, JM; Chloroquine efficacy in the treatment of uncomplicated malaria at three sentinel sites in northern Togo. *Ann Trop Med Parasitol.* (2003). *97*, 775-782.

[851] Mita, T; Kaneko, A; Lum, JK; Bwijo, B; Takechi, M; Zungu, IL; Tsukahara, T; Tanabe, K; Kobayakawa, T; Björkman, A; Recovery of chloroquine sensitivity and low prevalence of the *Plasmodium falciparum* chloroquine resistance transporter gene mutation K76T following the discontinuance of chloroquine use in Malawi. *Am J Trop Med Hyg.* (2003). *68*, 413-415.

[852] Laufer, MK; Thesing, PC; Eddington, ND; Masonga, R; Dzinjalamala, FK; Takala, SL; Taylor, TE; Plowe, CV; Return of chloroquine antimalarial efficacy in Malawi. *N Engl J Med.* (2006). *355*, 1959-1966.

[853] D'Alessandro, U; ter Kuile, FO; Amodiaquine, malaria, pregnancy: The old new drug. *Lancet.* (2006). *368*, 1306-1307.

[854] Simpson, JA; Price, R; ter Kuile, F; Teja-Isavatharm, P; Nosten, F; Chongsuphajaisiddhi, T; Looareesuwan, S; Aarons, L; White, NJ; Population pharmacokinetics of mefloquine in patients with acute *falciparum* malaria. *Clin Pharmacol Ther.* (1999). *66*, 472-484.

[855] Ashley, EA; Stepniewska, K; Lindegårdh, N; McGready, R; Hutagalung, R; Hae, R; Singhasivanon, P; White, NJ; Nosten, F; Population pharmacokinetic assessment of a new regimen of mefloquine used in combination treatment of uncomplicated *falciparum* malaria. *Antimicrob Agents Chemother.* (2006). *50*, 2281-2285.

[856] Macreadie, I; Ginsburg, H; Sirawaraporn, W; Tilley, L; Antimalarial drug development and new targets. *Parasitol Today.* (2000). *16*, 438-444.

[857] Padmanaban, G; Rangarajan, PN; Emerging targets for antimalarial drugs. *Expert Opin Ther Targets* (2001). *5*, 423-441.

[858] Klonis, N; Tan, O; Jackson, K; Goldberg, D; Klemba, M; Tilley, L; Evaluation of pH during cytostomal endocytosis and vacuolar catabolism of hemoglobin in *Plasmodium falciparum. Biochem J.* (2007). *407*, 343-354.

[859] Fidock, DA; Eastman, RT; Ward, SA; Meshnick, SR; Recent highlights in antimalarial drug resistance and chemotherapy research. *Trends Parasitol.* (2008). *24*, 537-544.

[860] Cooper, RA; Hartwig, CL; Ferdig, MT; pfcrt is more than the *Plasmodium falciparum* chloroquine resistance gene: A functional and evolutionary perspective. *Acta Trop.* (2005). *94*, 170-180.

[861] Nomura, T; Carlton, JM; Baird, JK; del Portillo, HA; Fryauff, DJ; Rathore, D; Fidock, DA; Su, X; Collins, WE; McCutchan, TF; Wootton, JC; Wellems, TE; Evidence for different mechanisms of chloroquine resistance in 2 *Plasmodium* species that cause human malaria. *J Infect Dis.* (2001). *183*, 1653-1661.

[862] Russell, B; Chalfein, F; Prasetyorini, B; Kenangalem, E; Piera, K; Suwanarusk, R; Brockman, A; Prayoga, P; Sugiarto, P; Cheng, Q; Tjitra, E; Anstey, NM; Price, RN; Determinants of *in vitro* drug susceptibility testing of *Plasmodium vivax. Antimicrob Agents Chemother.* (2008). *52*, 1040-1045.

[863] Lim, P; Alker, AP; Khim, N; Shah, NK; Incardona, S; Doung, S; Yi, P; Bouth, DM; Bouchier, C; Puijalon, OM; Meshnick, SR; Wongsrichanalai, C; Fandeur, T; Le Bras, J; Ringwald, P; Ariey, F; *Pfmdr1* copy number and arteminisin derivatives combination therapy failure in *falciparum* malaria in Cambodia. *Malar J.* (2009). *8*, 11.

[864] Ekland, EH; Fidock, DA; Advances in understanding the genetic basis of antimalarial drug resistance. *Curr Opin Microbiol.* (2007). *10*, 363-370.

[865] Scherf, A; Rivière, L; Lopez-Rubio, JJ; SnapShot: Var gene expression in the malaria parasite. *Cell.* (2008). *134*, 190.

[866] Ralph, SA; van Dooren, GG; Waller, RF; Crawford, MJ; Fraunholz, MJ; Foth, BJ; Tonkin, CJ; Roos, DS; McFadden, GI; *Tropical* infectious diseases: Metabolic maps and functions of the *Plasmodium falciparum* apicoplast. *Nat Rev Microbiol.* (2004). *2*, 203-216.

[867] Sidhu, AB; Sun, Q; Nkrumah, LJ; Dunne, MW; Sacchettini, JC; Fidock, DA. *In vitro* efficacy, resistance selection, and structural modeling studies implicate the malarial parasite apicoplast as the target of azithromycin. *J Biol Chem.* (2007). *282*, 2494-2504.

[868] Spry, C; Chai, CL; Kirk, K; Saliba, KJ. A class of pantothenic acid analogs inhibits *Plasmodium falciparum* pantothenate kinase and represses the proliferation of malaria parasites. *Antimicrob Agents Chemother.* (2005). *49*, 4649-4657.

[869] Hamzé, A; Rubi, E; Arnal, P; Boisbrun, M; Carcel, C; Salom-Roig, X; Maynadier, M; Wein, S; Vial, H; Calas, M. Mono- and bis-thiazolium salts have potent antimalarial activity. *J Med Chem.* (2005). *48*, 3639-3643.

[870] Richier, E; Biagini, GA; Wein, S; Boudou, F; Bray, PG; Ward, SA; Precigout, E; Calas, M; Dubremetz, JF; Vial, HJ; Potent antihematozoan activity of novel bisthiazolium drug T16: Evidence for inhibition of phosphatidylcholine metabolism in erythrocytes infected with Babesia and *Plasmodium spp. Antimicrob Agents Chemother.* (2006). *50*, 3381-3388.

[871] Biagini, GA; Fisher, N; Berry, N; Stocks, PA; Meunier, B; Williams, DP; Bonar-Law, R; Bray, PG; Owen, A; O'Neill, PM; Ward, SA. Acridinediones: Selective and potent inhibitors of the malaria parasite mitochondrial bc1 complex. *Mol Pharmacol.* (2008). *73*, 1347-1355.

[872] von Itzstein, M; Plebanski, M; Cooke, BM; Coppel, RL. Hot, sweet and sticky: The glycobiology of *Plasmodium falciparum. Trends Parasitol.* (2008). *24*, 210-218.

[873] Washburn, MP; Ulaszek R; Deciu, C; Schieltz, DM; Yates, JR 3rd; Analysis of quantitative proteomic data generated via multidimensional protein identification technology. *Anal Chem.* (2002). *74*, 1650-1657.

[874] Hartwig, CL; Rosenthal, AS; D'Angelo, J; Griffin, CE; Posner, GH; Cooper, RA; Accumulation of artemisinin trioxane derivatives within neutral lipids of *Plasmodium falciparum* malaria parasites is endoperoxide-dependent. *Biochem Pharmacol.* (2009). *77*, 322-336.

[875] Choi, SR; Mukherjee, P; Avery, MA; The fight against drug-resistant malaria: Novel plasmodial targets and antimalarial drugs. *Curr Med Chem.* (2008). *15*, 161-171.

[876] Gelb, MH; Drug discovery for malaria: A very challenging and timely endeavor. *Curr Opin Chem Biol.* (2007). *11*, 440-445.

[877] O'Neill, PM; Searle, NL; Kan, KW; Storr, RC; Maggs, JL; Ward, SA; Raynes, K; Park, BK; Novel, potent, semisynthetic antimalarial carba analogues of the first-generation 1,2,4-trioxane artemether. *J Med Chem.* (1999). *42*, 5487-5493.

[878] O'Neill, PM; Miller, A; Bishop, LP; Hindley, S; Maggs, JL; Ward, SA; Roberts, SM; Scheinmann, F; Stachulski, AV; Posner, GH; Park, BK; Synthesis, antimalarial activity, biomimetic iron(II) chemistry, and *in vivo* metabolism of novel, potent C-10-phenoxy derivatives of dihydroartemisinin. *J Med Chem.* (2001). *44*, 58-68.

[879] Ploypradith, P; Development of artemisinin and its structurally simplified trioxane derivatives as antimalarial drugs. *Acta Trop.* (2004). *89*, 329-342.

[880] Jung, M; Li, X; Bustos, DA; ElSohly, HN; McChesney, JD; Milhous, WK; Synthesis and antimalarial activity of (+)-deoxoartemisinin. *J Med Chem.* (1990). *33*, 1516-1518.

[881] Jung, M; Lee, K; Kendrick, H; Robinson, BL; Croft, SL; Synthesis, stability, and antimalarial activity of new hydrolytically stable and water-soluble (+)-deoxoartelinic acid. *J Med Chem.* (2002). *45*, 4940-4944.

[882] Posner, GH; Northrop, J; Paik, IH; Borstnik, K; Dolan, P; Kensler, TW; Xie, S; Shapiro, TA; New chemical and biological aspects of artemisinin-derived trioxane dimers. *Bioorg Med Chem.* (2002). *10*, 227-232.

[883] Posner, GH; Paik, IH; Sur, S; McRiner, AJ; Borstnik, K; Xie, S; Shapiro, TA; Orally active, antimalarial, anticancer, artemisinin-derived trioxane dimers with high stability and efficacy. *J Med Chem.* (2003). *46*, 1060-1065.

[884] Haynes, RK; From artemisinin to new artemisinin antimalarials: Biosynthesis, extraction, old and new derivatives, stereochemistry and medicinal chemistry requirements. *Curr Top Med Chem.* (2006). *6*, 509-537.

[885] Haynes, RK; Chan, WC; Lung, CM; Uhlemann, AC; Eckstein, U; Taramelli, D; Parapini, S; Monti, D; Krishna, S; The Fe2+-mediated decomposition, PfATP6 binding, and antimalarial activities of artemisone and other artemisinins: The unlikelihood of C-centered radicals as bioactive intermediates. *ChemMedChem.* (2007). *2*, 1480-1497.

[886] Haynes, RK; Fugmann, B; Stetter, J; Rieckmann, K; Heilmann, HD; Chan, HW; Cheung, MK; Lam, WL; Wong, HN; Croft, SL; Vivas, L; Rattray, L; Stewart, L; Peters, W; Robinson, BL; Edstein, MD; Kotecka, B; Kyle, DE; Beckermann, B; Gerisch, M; Radtke, M; Schmuck, G; Steinke, W; Wollborn, U; Schmeer, K; Römer, A; Artemisone--a highly active antimalarial drug of the artemisinin class. *Angew Chem Int Ed Engl.* (2006). *45*, 2082-2088.

[887] Ro, DK; Paradise, EM; Ouellet, M; Fisher, KJ; Newman, KL; Ndungu, JM; Ho, KA; Eachus, RA; Ham, TS; Kirby, J; Chang, MC; Withers, ST; Shiba, Y; Sarpong, R; Keasling, JD; Production of the antimalarial drug precursor artemisinic acid in engineered yeast. *Nature.* (2006). *440*, 940-943.

[888] Vennerstrom, JL; Arbe-Barnes, S; Brun, R; Charman, SA; Chiu, FC; Chollet, J; Dong, Y; Dorn, A; Hunziker, D; Matile, H; McIntosh, K; Padmanilayam, M; Santo Tomas, J; Scheurer, C; Scorneaux, B; Tang, Y; Urwyler, H; Wittlin, S; Charman, WN; Identification of an antimalarial synthetic trioxolane drug development candidate. *Nature.* (2004). *430*, 900-904.

[889] del Pilar Crespo, M; Avery, TD; Hanssen, E; Fox, E; Robinson, TV; Valente, P; Taylor, DK; Tilley, L; Artemisinin and a series of novel endoperoxide antimalarials exert early effects on digestive vacuole morphology. *Antimicrob Agents Chemother.* (2008). *52*, 98-109.

[890] Laurent, D; Pietra, F; Antiplasmodial marine natural products in the perspective of current chemotherapy and prevention of malaria: A review. *Mar Biotechnol (NY).* (2006). *8*, 433-447.

[891] Dong, Y; Vennerstrom, JL; Mechanisms of *in situ* activation for peroxidic antimalarials. *Redox Rep.* (2003). *8*, 284-288.

[892] Kim, HS; Nagai, Y; Ono, K; Begum, K; Wataya, Y; Hamada, Y; Tsuchiya, K; Masuyama, A; Nojima, M; McCullough, KJ; Synthesis and antimalarial activity of novel medium-sized 1,2,4,5-tetraoxacycloalkanes. *J Med Chem.* (2001). *44*, 2357-2361.

[893] Mzayek, F; Deng, H; Mather, FJ; Wasilevich, EC; Liu, H; Hadi, CM; Chansolme, DH; Murphy, HA; Melek, BH; Tenaglia, AN; Mushatt, DM; Dreisbach, AW; Lertora, JJ; Krogstad, DJ; Randomized dose-ranging controlled trial of AQ-13, a candidate antimalarial, and chloroquine in healthy volunteers. *PLoS Clin Trials.* (2007). *2*, e6.

[894] O'Neill, PM; Mukhtar, A; Stocks, PA; Randle, LE; Hindley, S; Ward, SA; Storr, RC; Bickley, JF; O'Neil, IA; Maggs, JL; Hughes, RH; Winstanley, PA; Bray, PG; Park, BK; Isoquine and related amodiaquine analogues: A new generation of improved 4-aminoquinoline antimalarials. *J Med Chem.* (2003). *46*, 4933-4945.

[895] Xiang, H; McSurdy-Freed, J; Moorthy, GS; Hugger, E; Bambal, R; Han, C; Ferrer, S; Gargallo, D; Davis, CB; Preclinical drug metabolism and pharmacokinetic evaluation of GW844520, a novel anti-malarial mitochondrial electron transport inhibitor. *J Pharm Sci.* (2006). *95*, 2657-2672.

[896] Willcox, ML; Bodeker, G; Traditional herbal medicines for malaria. *BMJ.* (2004). *329*, 1156-1159.

[897] Bathurst, I; Hentschel, C; Medicines for Malaria Venture: Sustaining antimalarial drug development. *Trends Parasitol.* (2006). *22*, 301-307.

[898] Soh, PN; Benoit-Vical, F; Are West African plants a source of future antimalarial drugs? *J Ethnopharmacol.* (2007). *114*, 130-140.

[899] Newman, DJ; Cragg, GM; Snader, KM; Natural products as sources of new drugs over the period 1981-2002. *J Nat Prod.* (2003). *66*, 1022-1037.

[900] Weniger, B; Lagnika, L; Vonthron-Sénécheau, C; Adjobimey, T; Gbenou, J; Moudachirou, M; Brun, R; Anton, R; Sanni, A; Evaluation of ethnobotanically selected Benin medicinal plants for their *in vitro* antiplasmodial activity. *J Ethnopharmacol.* (2004). *90*, 279-284.

[901] Udeinya, JI; Shu, EN; Quakyi, I; Ajayi, FO; An antimalarial neem leaf extract has both schizonticidal and gametocytocidal activities. *Am J Ther.* (2008). *15*, 108-110.

[902] Carvalho, LH; Brandão, MG; Santos-Filho, D; Lopes, JL; Krettli, AU; Antimalarial activity of crude extracts from Brazilian plants studied *in vivo* in *Plasmodium berghei*-infected mice and *in vitro* against *Plasmodium falciparum* in culture. *Braz J Med Biol Res.* (1991). *24*, 1113-1123.

[903] Krettli, AU; Andrade-Neto, VF; Brandão, MG; Ferrari, WM; The search for new antimalarial drugs from plants used to treat fever and malaria or plants ramdomly selected: A review. *Mem Inst Oswaldo Cruz.* (2001). *96*, 1033-1042.

[904] Wright, CW; Traditional antimalarials and the development of novel antimalarial drugs. *J Ethnopharmacol.* (2005). *100*, 67-71.

[905] Brandão, MG; Krettli, AU; Soares, LS; Nery, CG; Marinuzzi, HC. Antimalarial activity of extracts and fractions from Bidens pilosa and other Bidens species (*Asteraceae*) correlated with the presence of acetylene and flavonoid compounds. *J Ethnopharmacol.* (1997). *57*, 131-138.

[906] Dolabela, MF; Oliveira, SG; Nascimento, JM; Peres, JM; Wagner, H; Póvoa, MM; de Oliveira, AB; *In vitro* antiplasmodial activity of extract and constituents from Esenbeckia febrifuga, a plant traditionally used to treat malaria in the Brazilian Amazon. *Phytomedicine.* (2008). *15*, 367-372.

[907] Rukunga, GM; Muregi, FW; Omar, SA; Gathirwa, JW; Muthaura, CN; Peter, MG; Heydenreich, M; Mungai, GM; Anti-plasmodial activity of the extracts and two sesquiterpenes from Cyperus articulatus. *Fitoterapia.* (2008). *79*, 188-190.

[908] Chung, IM; Kim, MY; Moon, HI; Antiplasmodial activity of sesquiterpene lactone from Carpesium rosulatum in mice. *Parasitol Res.* (2008). *103*, 341-344.

[909] Lee, SI; Yang, HD; Son, IH; Moon, HI; Antimalarial activity of a stilbene glycoside from *Pleuropterus ciliinervis. Ann Trop Med Parasitol.* (2008). *102*, 181-184.

[910] Moon, HI; Sim, J; Antimalarial activity in mice of resveratrol derivative from *Pleuropterus ciliinervis. Ann Trop Med Parasitol.* (2008). *102*, 447-450.

[911] Kim, JJ; Chung, IM; Jung, JC; Kim, MY; Moon, HI. *In vivo* antiplasmodial activity of 11(13)-dehydroivaxillin from Carpesium ceruum. *J Enzyme Inhib Med Chem.* (2008). *7:1*.

[912] Park, WH; Lee, SJ; Moon, HI; Antimalarial activity of a new stilbene glycoside from *Parthenocissus tricuspidata* in mice. *Antimicrob Agents Chemother.* (2008). *52*, 3451-3453.

[913] Banzouzi, JT; Soh, PN; Mbatchi, B; Cavé, A; Ramos, S; Retailleau, P; Rakotonandrasana, O; Berry, A;Benoit-Vical, F; Cogniauxia podolaena: Bioassay-guided fractionation of defoliated stems, isolation of active compounds, antiplasmodial activity and cytotoxicity. *Planta Med.* (2008). *74*, 1453-1456.

[914] Lannang, AM; Louh, GN; Lontsi, D; Specht, S; Sarite, SR; Flörke, U; Hussain, H; Hoerauf, A; Krohn, K; Antimalarial compounds from the root bark of Garcinia polyantha Olv. *J Antibiot (Tokyo).* (2008). *61*, 518-523.

[915] Carraz, M; Jossang, A; Franetich, JF; Siau, A; Ciceron, L; Hannoun, L; Sauerwein, R; Frappier, F; Rasoanaivo, P; Snounou, G; Mazier, D. A plant-derived morphinan as a novel lead compound active against malaria liver stages. *PLoS Med.* (2006). *3*, e513.

[916] Carraz, M; Jossang, A; Rasoanaivo, P; Mazier, D; Frappier, F; Isolation and antimalarial activity of new morphinan alkaloids on *Plasmodium yoelii* liver stage. *Bioorg Med Chem.* (2008). *16*, 6186-6192.

[917] Andrade-Neto, VF; Brandão, MG; Nogueira, F; Rosário, VE; Krettli, AU; Ampelozyziphus amazonicus Ducke (*Rhamnaceae*), a medicinal plant used to prevent malaria in the Amazon Region, hampers the development of *Plasmodium berghei* sporozoites. *Int J Parasitol.* (2008). May 27.

[918] Nwabuisi,C; Prophylactic effect of multi-herbal extract 'Agbo-Iba' on malaria induced in mice. *East Afr Med J.* (2002). *79*, 343-346.

[919] Weisman, JL; Liou, AP; Shelat, AA; Cohen, FE; Guy, RK; DeRisi, JL; Searching for new antimalarial therapeutics amongst known drugs. *Chem Biol Drug Des.* (2006). *67*, 409-416.

[920] Mackey, ZB; Baca, AM; Mallari, JP; Apsel, B; Shelat, A; Hansell, EJ; Chiang, PK; Wolff, B; Guy, KR; Williams, J; McKerrow, JH; Discovery of trypanocidal compounds by whole cell HTS of Trypanosoma brucei. *Chem Biol Drug Des.* (2006). *67*, 355-363.

[921] Barkan, D; Ginsburg, H; Golenser, J; Optimization of flow cytometric measurement of parasitemia in *plasmodium*-infected mice. *Int J Parasitol.* (2000). 30, 649-653.

[922] Veeken, H; Pécoul, B; Drugs for 'neglected diseases': A bitter pill. *Ned Tijdschr Geneeskd.* (2000). *144*, 1253-1256.

[923] Walgate, R; Reflectors could reduce road deaths ninefold, says Global Forum for Health Research. *Bull World Health Organ.* (2002). *80*, 989-990.

[924] Olliaro, PL; Taylor, WR; Antimalarial compounds: From bench to bedside. *J Exp Biol.* (2003). *206*, 3753-3759.

[925] TDR, 30 Years of Research and Capacity Building in Tropical Diseases, http://www.who.int/tdrold/about/history_book/anniversary_book.htm (accessed Dec. 2010)

[926] TDR, TDR Business Plan 2008-2013. http://apps.who.int/tdr/svc/publications/about-tdr/business-plans/tdr-businessplan08-13 (accessed December 2010)

[927] MMV, MMV Business Plan 2008-2012, http://www.mmv.org/newsroom/press-releases/mmv-business-plan-2008-2012-now-available . (accessed December 2010)

[928] Ridley, RG; Product R&D for neglected diseases. Twenty-seven years of WHO/TDR experiences with public-private partnerships. *EMBO Rep.* (2003). *4* Spec No, S43-6.

[929] Kitchen, LW; Vaughn, DW; Skillman, DR; Role of US military research programs in the development of US Food and Drug Administration--approved antimalarial drugs. *Clin Infect Dis.* (2006). *43*, 67-71.

[930] Phillips, RE; Warrell, DA; White, NJ; Looareesuwan, S; Karbwang, J; Intravenous quinidine for the treatment of severe *falciparum* malaria: Clinical and pharmacokinetic studies. *N Engl J Med* (1985). *312*, 1273-1278.

[931] Bruneel, F; Hocqueloux, L; Alberti, C; The clinical spectrum of severe imported *falciparum* malaria in the intensive care unit: Report of 188 cases in adults. *Am J Respir Crit Care Med.* (2003). *167*, 684-689.

[932] White, NJ; The management of severe *falciparum* malaria. *Am J Respir Crit Care Med.* (2003). *167*, 673-674.

Index

#

20th century, 205, 343, 499

A

abstraction, 396
academic performance, 438
access, xi, xiii, 3, 7, 18, 28, 31, 39, 47, 50, 51, 54,
    62, 64, 66, 76, 77, 82, 88, 102, 103, 105, 106,
    111, 144, 155, 169, 179, 184, 186, 190, 196,
    198, 224, 227, 235, 256, 260, 320, 321, 323,
    403, 473, 506, 508, 528, 530, 533, 535, 536,
    538, 539, 543, 545, 546, 569, 572, 597
accessibility, 89, 203
accommodation, 399
accountability, 535, 536
accounting, 80, 112, 158, 328, 525
acetaminophen, 143
acetone, 511
acetylcholinesterase, 358
acid, 113, 114, 119, 130, 143, 164, 168, 283, 299,
    314, 315, 320, 334, 359, 360, 366, 367, 375,
    388, 426, 486, 487, 491, 492, 494, 495, 496,
    497, 498, 499, 500, 501, 502, 516, 521, 524,
    540, 541, 593, 595, 597, 604, 615, 616
acidic, 42, 167, 410, 483, 497, 498, 499
acidity, 158
acidosis, 161, 165, 351, 422, 451, 454, 554
acne, 357
acquired immunity, 15, 53, 56, 85, 205, 209, 256,
    269, 381, 471
action potential, 344, 347, 349
active compound, xiv, 466, 514, 516, 518, 528, 617
active site, 492, 493, 494
active transport, 493
acute brain syndrome, 354, 398
acute infection, 131, 209
acute kidney failure, 158

acute renal failure, 163, 165, 166, 167, 173, 189,
    192, 344, 415
acute respiratory distress syndrome, 158, 189, 192
adaptation(s), 12, 75, 483
adenine, 332
adenosine, 494
adjunctive therapy, 160
adjustment, 144, 165, 301, 445, 468
adolescents, 106, 432
ADP, 120
adrenaline, 164, 165, 346, 406, 448
adrenoceptors, 601
adverse conditions, 597
adverse event, 22, 77, 92, 93, 107, 150, 204, 205,
    207, 216, 220, 222, 231, 237, 269, 317, 319,
    344, 350, 353, 354, 356, 357, 358, 381, 383,
    384, 385, 395, 396, 398, 399, 401, 402, 417,
    418, 422, 425, 428, 430, 431, 432, 434, 435,
    436, 437, 438, 441, 443, 444, 449, 474, 475,
    476, 479, 481, 520, 584, 586, 602, 603, 607
advocacy, 528, 530
Afghanistan, 26, 238, 558
African Americans, 414
agencies, 18, 28, 67, 95, 96, 102, 103, 108, 144, 324,
    341, 401, 473, 474, 525, 530, 533, 534, 536
agranulocytosis, 92, 204, 330, 383, 397, 402, 413,
    415, 416, 423, 424, 425, 426, 430, 436, 439,
    451, 470, 477, 480, 612
agricultural sector, 103
AIDS, 21, 63, 67, 71, 96, 112, 357, 525, 591, 599
airways, 165
alanine, 397, 451, 458
alanine aminotransferase, 397, 451
albumin, 168, 397, 451
albuminuria, 135
alcohols, 16, 197, 587
aldolase, 190, 397, 490
algae, 119, 491
Algeria, 113, 555
alkaloids, 114, 129, 161, 336, 343, 347, 514, 540,
    618

alkylation, 338, 340, 488, 500

allantois, 389, 390

allele, 83, 106, 250, 257, 469, 564

allelic exclusion, 485

allergic reaction, 92, 249, 333, 405, 416, 424, 459, 460

allergy, 410, 519

alopecia, 400

ALT, 433, 443, 458

alternative medicine, x, 1, 28, 473

alternative treatments, 543

alveolitis, 424

amino, 10, 16, 120, 122, 126, 289, 487, 490, 493, 494, 551, 564, 587

amino acid(s), 10, 120, 289, 490, 493, 494, 564

ammonium, 506

amplitude, 356

amputation, 120

analgesic, 515

anaphylactic shock, 610

anaphylaxis, 416, 422, 423, 428

anesthetics, 604

anger, 250

angina, 411

angioedema, 422

angiogenesis, 388, 391, 420, 606

Angola, 26, 555, 572

ankles, 458

anorexia, 143, 317, 407, 416, 419, 424, 426, 433, 456, 457, 459

antacids, 414, 462

antagonism, 356

anthropologists, 236

antibiotic, 23, 125, 134, 162, 163, 251, 265, 356, 383, 429, 470, 486, 503, 521, 551, 554, 589

antibiotic resistance, 265

antibody, 164, 230, 410, 424

anti-cancer, 606

anticonvulsant(s), 452, 484, 554

antiemetics, 143, 345

antiepileptic drugs, 352

antigen, 40, 236, 261, 397, 451

antihistamines, 345

antimony, 503

antioxidant, 388, 391, 610

antipsychotic, 521

antipyretic, 23, 165, 321, 326, 400, 470

antiretrovirals, 437

antitumor, 497, 499, 575, 594

antitumor agent, 594

anxiety, 154, 317, 330, 345, 352, 354, 358, 408, 418, 601

anxiety disorder, 317, 354

aorta, 389

apathy, 352

aplastic anemia, 401, 406, 424, 470, 475

appetite, 177

ARDS, 158

Argentina, 557

Armenia, 559

arousal, 345

arrest, 317, 346, 352, 407, 449, 486

arterioles, 397

arthralgia, 408, 433

arthritis, 319, 399, 406

ascites, 600

aseptic, 425

aseptic meningitis, 425

Asia, 6, 7, 9, 23, 26, 28, 31, 37, 39, 40, 41, 42, 43, 44, 50, 57, 62, 63, 65, 68, 75, 76, 85, 112, 123, 124, 126, 138, 146, 155, 156, 160, 167, 169, 170, 171, 180, 189, 190, 194, 198, 206, 213, 214, 223, 225, 233, 237, 247, 249, 250, 258, 325, 326, 384, 407, 440, 446, 450, 468, 472, 479, 482, 508, 527, 530, 532, 563, 598

Asian countries, x, 4, 81, 196, 467

aspartate, 383, 397, 416, 491

asphyxia, 438

aspiration, 449, 539

assessment, 10, 16, 55, 57, 99, 103, 108, 154, 156, 161, 162, 163, 177, 180, 191, 193, 223, 238, 245, 287, 291, 292, 295, 305, 323, 336, 344, 348, 394, 395, 402, 437, 456, 461, 475, 528, 572, 578, 590, 591, 592, 601, 612, 614

assets, 86

asthma, 112, 422

asymmetry, 319

asymptomatic, 13, 56, 176, 179, 206, 209, 217, 243, 254, 259, 260, 271, 346, 380, 381, 406, 426, 437, 440, 444, 449, 459, 573, 611

ataxia, 24, 350, 408, 416, 418, 425, 457

ATP, 120, 131, 335, 336, 493, 495, 496

atrial flutter, 409

atrioventricular block, 346, 348, 409, 416, 449

atrioventricular node, 348

at-risk populations, 88, 538

atrophy, 406

attachment, 116

audit, 327

auditory nerve, 382

auditory stimuli, 356

Austria, 201, 580

authority(ies), 5, 28, 78, 96, 97, 99, 102, 104, 106, 145, 182, 215, 224, 226, 232, 330, 343, 354, 356, 403, 469, 473, 538, 539

autonomic nervous system, 601

autopsy, 612
avermectin, 521
avoidance, 37, 52, 159, 350
awareness, 226, 235, 251, 319, 327, 351
axial skeleton, 420
Azerbaijan, 559

B

back pain, 441
bacteria, 134, 246, 265, 422, 491
bacterial infection, 163, 182, 470, 532
bacteriostatic, 134
Bangladesh, 26, 40, 171, 250, 251, 272, 430, 532, 560, 578, 586
barbiturates, 452
barriers, 37, 79, 88, 320, 321, 323, 361, 538
base, 7, 45, 57, 97, 127, 129, 137, 145, 148, 149, 167, 170, 175, 183, 191, 192, 201, 202, 203, 217, 228, 230, 240, 246, 292, 343, 344, 380, 405, 407, 414, 415, 440, 442, 472, 481, 483, 492, 497, 504, 530, 565
base pair, 292
BBB, 362, 363
beer, 120, 511, 518
Beijing, 124, 529
Belgium, 98, 216, 240
benchmarking, 543
benchmarks, 72, 539
beneficial effect, 445, 457
benefits, xi, 25, 28, 29, 40, 46, 50, 53, 61, 67, 74, 75, 76, 86, 87, 88, 94, 144, 162, 165, 179, 180, 192, 205, 208, 213, 226, 257, 261, 271, 287, 322, 323, 326, 395, 402, 475, 477, 478
benign, 192, 291, 356, 414, 421, 580
benzoyl peroxide, 125
Bhutan, 26, 560
bias, 396
bicarbonate, 159, 161, 491, 492
bilateral aid, 95
bile, 497
bilirubin, 135, 159
bioassay, 282, 359, 517
bioavailability, xiv, 23, 44, 137, 237, 269, 278, 279, 280, 281, 308, 310, 319, 338, 341, 342, 368, 417, 440, 446, 466, 496, 500, 501, 506, 507, 552, 569, 591, 593, 595
biochemistry, 161
bioinformatics, 544
biomarkers, 603
biomass, 22, 29, 34, 62, 82, 119, 157, 243, 245, 258, 262, 281, 285, 286, 290, 293, 294, 299, 300, 303, 317, 318, 478

biosynthesis, 119, 127, 131, 486, 487, 489, 490, 491, 492, 493, 495, 503, 506, 507, 540, 541
biotechnology, 572
birds, 3, 114, 397, 451
Birmingham, Alabama, 366
birth weight, 56, 84, 85, 86, 88, 94, 116, 176, 179, 181, 189, 209, 211, 212, 214, 261, 380, 381, 391, 394, 420, 426, 442, 444, 447, 572, 581
births, 382, 444
blackwater fever, 317
blame, 9, 327
bleeding, 120, 159, 165, 281, 302, 333, 416
blindness, 38, 167, 177, 355, 397, 411, 450
blood cultures, 512, 554
blood dyscrasias, 426
blood flow, 199, 451
blood plasma, 597
blood pressure, 39, 158, 172, 411, 504
blood smear, 135, 188, 512
blood supply, 47, 49
blood transfusion, 47, 161, 164, 177, 188, 415, 450, 554
blood transfusions, 47, 188
blood urea nitrogen, 458
blood vessels, 116, 390
blood-brain barrier, 351, 362, 587
bloodstream, 63, 116, 507
body fat, 604
body mass index, 586
body weight, 160, 166, 175, 202, 216, 227, 230, 270, 285, 303, 435, 501, 515, 554
Bolivia, 26, 557, 586
bone, 56, 123, 210, 362, 391, 398, 413, 422, 423, 439, 450
bone growth, 56, 210, 422, 423
bone marrow, 123, 362, 391, 398, 413, 439, 450
bones, 181, 221
Botswana, 31, 468, 555
bounds, 64, 73
bowel, 24, 400
bradycardia, 39, 173, 346, 348, 349, 398, 400, 408, 409, 415, 416, 449
brain, 116, 176, 178, 331, 339, 340, 350, 355, 361, 362, 363, 367, 368, 374, 377, 379, 453, 455, 460, 461, 518, 574, 610
brain damage, 176, 379, 461
brain stem, 178, 339, 340, 355, 374, 455, 460
brain structure, 362
brainstem, vii, 24, 374, 377, 379, 418, 434, 593, 604
Brazil, 26, 112, 223, 234, 251, 430, 513, 518, 526, 552, 557, 586
breakdown, 120, 332, 339

breast milk, 181, 211, 217, 331, 381, 386, 407, 410,
      412, 415, 424, 426, 427, 608
breastfeeding, 155, 211, 215, 331, 386, 407, 412,
      413, 426
breathing, 158, 160, 161, 189, 327, 441
breeding, 259, 543
Britain, 114, 402, 477, 608
bronchopneumonia, 449
bronchospasm, 410
bundle branch block, 349
Burkina Faso, 26, 322, 323, 477, 510, 519, 555, 577,
      580, 589, 598
Burma, 54, 80, 252, 323, 325, 353, 384, 449, 458,
      609
Burundi, 26, 152, 395, 555, 576
business travelers, 475, 481
buyers, 72

C

Ca^{2+}, 489, 493, 500, 563
cadmium, 521
calcium, 24, 119, 128, 346, 347, 385, 449, 493, 521
calcium channel blocker, 346, 449
Cambodia, 4, 6, 10, 26, 30, 54, 81, 102, 109, 123,
      125, 196, 223, 229, 234, 250, 252, 282, 325,
      326, 421, 467, 468, 551, 561, 563, 565, 577,
      586, 614
Cameroon, 26, 107, 125, 249, 468, 516, 555, 581,
      587, 590, 612
campaigns, 28, 95, 96, 114, 144, 323, 473, 489
cancer, 33, 293, 388, 599, 606
cancer cells, 388
candida, 398
candidates, 2, 134, 240, 258, 388, 501, 505, 544
capillary, 161, 199, 390, 454
capillary refill, 454
capsule, 56, 201, 368, 369
carbamazepine, 452
carbohydrate, 335
carbon, 119, 253, 339, 492, 497, 498, 499, 500
carbon dioxide, 492
carbon-centered radicals, 119, 500
carboxyl, 491, 493
carboxylic acid, 499
carboxymethyl cellulose, 367
cardiac arrest, 346, 410, 411, 448
cardiac arrhythmia, 39, 173, 406, 601
cardiac glycoside, 346, 449
cardiomyopathy, 348, 397
cardiovascular disease, 344, 448
cardiovascular system, 344, 420
caregivers, 86, 325, 327, 530, 532

Caribbean, 247
cartilage, 476
case studies, 509
casein, 493
cash, 67, 86, 88, 108
cash flow, 67
catabolism, 489, 614
catalyst, 533
category a, 382
cation, 597
cattle, 531
Caucasians, 350, 351
causal relationship, 356, 397, 434, 451
causality, 350, 354
CDC, 93, 203, 209, 213, 214, 226, 549, 553, 573,
      608
cell culture, 388, 420
cell cycle, 491, 493
cell death, 389, 487
cell line(s), 340, 506, 516, 606
cell membranes, 313, 363, 597
cell surface, 116, 489
Central African Republic, 26, 555
central nervous system (CNS), 24, 204, 284, 302,
      330, 331, 349, 350, 351, 352, 353, 355, 356,
      357, 358, 361, 380, 404, 406, 408, 416, 417,
      425, 449, 450, 470, 492, 602, 603, 609
cerebrospinal fluid, 24, 161, 362
certification, 41, 97, 170
Chad, 26, 555
challenges, vii, 18, 30, 47, 78, 151, 200, 269, 290,
      337, 394, 494, 534, 536, 537, 543, 544, 569,
      582
channel blocker, 334, 521
chemical(s), xiii, 102, 113, 114, 115, 118, 252, 329,
      396, 465, 496, 498, 499, 502, 510, 512, 516,
      544, 545, 552, 592, 602, 605, 615
chemical stability, 496, 502
chemical structures, xiii, 115, 465, 498, 512
chemoprevention, 32
chemotherapeutic agent, 21, 491
chemotherapy, vii, 58, 117, 127, 135, 150, 195, 237,
      243, 293, 313, 388, 483, 485, 492, 527, 529,
      566, 574, 584, 587, 589, 594, 612, 614, 616
Chicago, 549
chicken, 49, 64, 71, 606
child bearing, 210
child development, 86
child mortality, 46, 86, 533, 534
childhood, 44, 81, 88, 89, 143, 176, 207, 215, 219,
      221, 235, 238, 351, 385, 478, 536, 537, 569,
      572, 582, 598, 612

China, 7, 23, 28, 30, 31, 41, 42, 46, 57, 63, 69, 96, 97, 99, 100, 106, 109, 112, 121, 124, 125, 126, 134, 170, 187, 223, 249, 323, 337, 386, 387, 400, 420, 421, 442, 446, 447, 468, 472, 473, 485, 504, 519, 528, 529, 561, 562, 595, 596, 598, 603, 605, 612

Chinese government, 101

Chinese medicine, 528

chiral center, 500

chloroform, 512, 514

cholera, 201, 608

cholesterol, 597

choline, 487, 506, 507

chorion, 389

chromatography, 361, 602, 603

chromosome, 344

chronotherapy, 268, 289

circulation, xiv, 49, 116, 117, 160, 163, 200, 276, 290, 316, 389, 390, 391, 466

cirrhosis, 451

cities, 141, 187

Civil War, 548

clarity, 539

classes, ix, x, 1, 4, 114, 118, 120, 161, 252, 260, 265, 335, 343, 344, 345, 467, 505, 523, 525, 544

classification, 252

cleavage, 488

clinical application, 66

clinical assessment, 44, 45, 160, 162, 166, 403

clinical diagnosis, 47, 190, 260, 261, 453

clinical examination, 161

clinical presentation, 238, 245

clinical symptoms, 34, 85, 117, 135, 179, 200, 413

clone, 63, 312, 314, 484

closure, 119

clusters, 390

CMC, vii, 367

coagulopathy, 165

coding, 589

codon, 16, 197

coenzyme, 486, 487

cognitive development, 47, 176

cognitive impairment, 438

Cold War, 528

colitis, 422, 429

collaboration, 29, 46, 108, 144, 395, 489, 526, 527, 528, 531, 533, 536, 550, 551, 552, 553

Colombia, 6, 31, 112, 123, 246, 247, 468, 479, 550, 557

color, 333, 514

coma, 38, 116, 158, 161, 163, 165, 171, 172, 177, 178, 281, 287, 301, 317, 340, 351, 355, 418, 450, 462

combination therapy, ix, x, xi, xii, 1, 3, 18, 21, 22, 25, 29, 31, 32, 33, 35, 37, 47, 48, 51, 59, 61, 63, 65, 78, 128, 136, 137, 146, 151, 153, 156, 157, 164, 173, 194, 196, 205, 208, 232, 237, 257, 259, 262, 286, 420, 423, 432, 435, 440, 455, 473, 478, 480, 495, 526, 529, 533, 545, 566, 567, 569, 570, 571, 575, 576, 607, 610, 614

commerce, 47

commercial, 43, 104, 525, 548, 550

common sense, 271

communication, 144

community(ies), x, xiii, 7, 18, 19, 29, 33, 47, 48, 49, 54, 65, 77, 78, 86, 87, 103, 144, 155, 175, 186, 194, 206, 207, 208, 219, 256, 260, 261, 301, 323, 324, 327, 351, 354, 380, 465, 511, 527, 530, 531, 532, 533, 534, 535, 536, 537, 543, 581, 598, 603

compatibility, 539

competition, 49, 70, 71, 72, 84, 99, 259

complement, 495, 536

complexity, 38, 47

compliance, xii, 8, 22, 34, 77, 78, 87, 94, 97, 103, 136, 144, 151, 195, 196, 200, 207, 209, 217, 219, 227, 235, 243, 251, 255, 256, 258, 260, 261, 264, 269, 323, 324, 325, 358, 404, 429, 436, 478, 482, 499, 551, 598

complications, 84, 85, 116, 139, 164, 170, 172, 173, 177, 178, 215, 236, 239, 269, 270, 280, 281, 301, 326, 327, 447, 448, 553, 573

composition, 108, 597

comprehension, 305

computer, 274, 275, 306, 307, 308, 309, 364, 367, 369, 370, 372, 375, 378

conception, 427

conditioning, 93

conductance, 313, 344, 348

conduction, 345, 346, 348, 409, 411, 434, 449, 602

conductive hearing loss, 434

conflict, 101, 265, 548, 550

confounding variables, 319

congenital malformations, 384, 410

Congo, 26, 150, 516, 520, 555

conjunctivitis, 445

consciousness, 158, 161, 163, 164, 170, 172, 173, 188, 327, 350, 351, 356, 408, 454

consensus, 18, 31, 59, 146, 152, 166, 271, 421, 509

consent, 443

constituents, 506, 542, 617

construction, 114, 206, 433

consulting, 67, 214

consumers, 50, 62, 325, 326, 475

consumption, 62, 206, 399, 435

containers, 216, 217

contraceptives, 579

control group, 216, 218, 219, 221, 376, 418, 434, 515

control measures, 8, 206

controlled studies, 201, 213

controlled trials, 170, 250, 251, 401, 409, 445, 474, 475, 508, 509, 520, 552

controversial, 23, 63, 418

conversion rate, 276, 294, 359, 360

convulsion, 163

cooling, 165

cooperation, 19, 64, 528, 553

coordination, 358, 409, 410, 498, 536

coping strategies, 68

corneal opacities, 406

correlation, 15, 348, 376, 411, 433, 505, 596

corticosteroids, 164, 165

cost effectiveness, 80, 220

cost saving, 79

Costa Rica, 557

Côte d'Ivoire, 26

cough, 412, 454

coumarins, 514

counseling, 236

counterfeiting, 9, 256, 324

covalent bond, 24

covering, 7, 70, 219, 472

creatine, 397

creatinine, 159, 230, 397, 443, 451, 454, 458

critical period, 390, 393

criticism, 79

CRM, 277, 366

crops, 68

crystal structure, 492, 493, 494

crystalline, 490

crystallization, 102, 197

crystalluria, 424

crystals, 424

CSF, 362, 363, 607

cultivation, 65, 69, 100

culture, xiv, 11, 19, 84, 161, 162, 191, 193, 245, 295, 311, 312, 313, 339, 388, 389, 391, 465, 484, 512, 594, 600, 617

culture conditions, 19, 391

culture medium, 388, 389, 512

cures, 131, 142, 304, 342, 549

current limit, 33

current prices, 65

curricula, 328

cyanosis, 415

cycles, 22, 116, 136, 199, 243, 264, 266, 281, 286, 291, 294, 300, 304, 318, 490

cyclohexanone, 500, 501

cyclosporine, 451

cysteine, 487, 489, 490, 494, 540

cytochrome, 16, 120, 131, 197, 272, 437, 447, 492, 495, 504, 507, 552, 564, 575, 587

cytokines, 346

cytometry, 311

cytoplasm, 484

cytotoxicity, 374, 495, 500, 510, 516, 599, 617

D

danger, 229, 296

data analysis, 377

data collection, 17

database, 7, 15, 18, 19, 20, 21, 156, 354, 357, 472, 509, 555, 557, 558, 559, 560, 561, 565

death rate, 37, 172, 221

deaths, xi, xiv, 3, 33, 40, 44, 49, 61, 62, 77, 78, 105, 112, 158, 170, 171, 172, 174, 176, 177, 179, 184, 187, 211, 213, 216, 219, 220, 225, 227, 230, 238, 239, 280, 381, 382, 387, 394, 401, 402, 442, 444, 445, 448, 466, 476, 477, 510, 532, 533, 536, 550, 609, 618

decision makers, 59

decoding, 525

decomposition, 324, 339, 363, 498, 504, 616

deduction, 362

defects, 179, 331, 353, 355, 382, 383, 384, 396, 407, 414, 426, 551

defense mechanisms, 304

deficiency(ies), 67, 125, 126, 133, 192, 203, 211, 240, 317, 385, 386, 411, 414, 415, 424, 431, 461, 463

deficit, 52, 159, 457, 460, 461

degradation, 197, 329, 339, 422, 452, 483, 488, 490, 494

dehydration, 449, 451, 462

delayed gastric emptying, 376, 377, 380

delirium, 317, 462

delusions, 475

dementia, 449

Democratic Republic of Congo, 26, 395, 519, 576

dengue, 525

Denmark, 580

Department of Defense, 18, 548

deployments, vii

depolarization, 336, 343, 344, 345, 347

deposition, 398

depressants, 554

depression, 231, 250, 330, 352, 353, 356, 398, 406, 418, 423, 425, 449, 450

depth, 540

dermatitis, 141, 401, 409, 424, 453, 475

designers, 328
destruction, 127, 158
detectable, 74, 82, 288, 339, 374, 376, 377, 379, 461
detection, 9, 10, 16, 50, 59, 190, 222, 228, 288, 290,
     340, 374, 394, 396, 483, 485
detection system, 222
detoxification, 127, 129, 197, 429, 483, 507
developed countries, 200, 324, 530
developing countries, 65, 67, 71, 72, 95, 99, 104,
     108, 174, 215, 322, 380, 394, 450, 501, 527,
     531, 534, 571, 588
developing nations, 66
development assistance, 94, 95
deviation, 346
diabetes, 397, 452, 599
dialysis, 164, 165, 173, 299
diamonds, 370
diarrhea, 177, 234, 317, 333, 341, 403, 405, 406,
     407, 408, 409, 410, 412, 414, 421, 422, 423,
     424, 425, 427, 428, 429, 433, 441, 443, 446,
     454, 456, 457, 460
differential diagnosis, 162, 451
diffusion, 348, 390
digestion, 118, 487
diploid, 230
diplomacy, 528
direct action, 29
direct cost(s), 47, 82, 85, 86, 570
directives, 121, 548
disability, xi, 62, 72, 74, 82, 87, 88, 111, 112, 350,
     532
disaster, 2, 9, 22, 67, 157, 253, 394, 505, 537, 570,
     588, 589
disbursement, 99
discomfort, 345, 424, 427, 441
discrimination, 410
diseases, xii, xiv, 43, 47, 71, 72, 88, 95, 105, 112,
     140, 176, 177, 195, 287, 346, 396, 399, 403,
     449, 461, 466, 476, 510, 513, 525, 526, 527,
     528, 529, 531, 534, 535, 536, 537, 539, 582,
     605, 614, 618
disequilibrium, 485
disorder, 331, 345, 408, 410, 438
dispersion, 345, 601
displaced persons, 3, 472
disposition, 270, 271, 278, 316, 319, 591, 598
dissatisfaction, 51, 58
disseminated intravascular coagulation, 411
distillation, 167
distress, 158, 161, 163, 188, 414, 452
distribution, 20, 21, 28, 58, 66, 67, 96, 168, 181, 207,
     225, 267, 269, 270, 276, 277, 279, 281, 296,
     298, 299, 300, 313, 314, 316, 319, 355, 361,

362, 363, 371, 373, 377, 391, 393, 411, 439,
     446, 473, 489, 524, 530, 534, 537, 578, 594,
     603, 604, 606, 607
diuretic, 165
divergence, 418
diversity, 18, 21, 105, 116, 510, 566, 588
dizziness, 24, 38, 149, 169, 317, 330, 333, 350, 351,
     352, 353, 356, 357, 358, 382, 383, 398, 400,
     401, 405, 406, 407, 408, 409, 410, 415, 416,
     419, 421, 425, 431, 432, 433, 436, 438, 440,
     441, 443, 449, 456, 457, 458, 459, 460, 474,
     477, 481
DNA, 12, 16, 127, 131, 253, 339, 490, 491, 513, 605
DNA repair, 605
doctors, 227
dogs, 308, 339, 349, 355, 360, 362, 363, 366, 367,
     368, 369, 370, 371, 377, 379, 416, 417, 460,
     507, 601, 603, 604
domestic chores, 177
Dominican Republic, 223, 557
donations, 537
donors, 30, 67, 94, 95, 108, 545, 548
dopamine, 165
dorsal aorta, 389
dosage, 7, 44, 56, 97, 101, 102, 107, 165, 182, 202,
     203, 204, 215, 239, 263, 264, 265, 267, 269,
     270, 271, 273, 274, 283, 285, 295, 303, 305,
     307, 308, 309, 313, 318, 323, 341, 346, 350,
     352, 354, 355, 356, 357, 371, 372, 385, 387,
     392, 397, 414, 449, 452, 456, 457, 472, 481,
     482, 517, 590, 593, 608
dose-response relationship, 284, 303, 593
DOT, 261
double-blind trial, 151, 576, 582
draft, 546
drug action, 265, 289, 304, 589
drug delivery, 9
drug design, 486, 492
drug discovery, vii, xiii, xiv, 465, 466, 493, 496, 507,
     510, 524, 526, 528, 530, 531, 540
drug efflux, 483
drug half-life, 263, 265, 268, 290
drug interaction, 234, 256, 272, 345, 354, 577, 591
drug metabolism, 181, 497, 616
drug reactions, 350, 394, 416, 422, 424, 426, 438,
     469, 470, 480, 608
drug safety, 184, 269, 381
drug targets, 11, 137, 483, 485, 486, 488, 493, 496,
     531
drug testing, 325
drug therapy, 33
drug toxicity, xii, xiii, 329, 330, 338, 340, 397, 403,
     404, 420, 451, 460, 461, 579, 607

drug treatment, xiii, 18, 20, 36, 39, 169, 243, 259, 289, 290, 295, 300, 316, 345, 404, 405, 457, 584
dyes, 331
dysarthria, 413
dysphagia, 422
dysphoria, 149, 407, 410, 430
dysplasia, 56, 210
dyspnea, 458

E

early warning, 9, 15, 16, 59
East Asia, 4, 37, 38, 42, 51, 57, 81, 142, 147, 160, 162, 190, 191, 192, 196, 232, 234, 247, 249, 250, 251, 252, 254, 255, 259, 381, 425, 467, 529, 530
East Timor, 4, 422, 580, 610
Eastern Europe, 57
economic consequences, 76, 230, 239
economic development, 46, 48, 65, 189
economic growth, 46, 47, 86
economic losses, 46
economic performance, 46
economic policy, 46
economic well-being, 68
economics, 65, 254, 324, 571
Ecuador, 26, 557
edema, 38, 158, 169, 178, 397
education, 144, 324, 326, 327, 328, 394, 472
educational attainment, 86
educational programs, 327
egg, 49, 64, 71
Egypt, 114, 558, 579
EKG, 601
El Salvador, 557
elaboration, 489, 500
electricity, 321
electrocardiogram, 386, 408, 432, 440, 602
electroencephalogram, 406
electroencephalography, 438
electrolyte, 164, 281, 301, 344, 345, 437
electron, 120, 131, 133, 339, 483, 487, 489, 501, 575, 616
elucidation, 339
emergency, xiv, 45, 57, 124, 158, 162, 166, 167, 178, 224, 227, 228, 232, 234, 238, 280, 447, 466
emigration, 225
emission, 435
employees, 433, 548
employment, 68, 433
employment levels, 68

empowerment, 535
enabling environments, 108
enamel, 422, 423
enantiomers, 319, 320, 481, 598
encephalopathy, 199, 354, 407, 409
encoding, 134, 252, 447, 495, 563
encouragement, 42
endothelial cells, 388
energy, 338, 487, 492, 495
England, 120, 121, 500
enlargement, 135
enrollment, 218, 443
enteritis, 398, 443
entrepreneurs, 253
environment, 49, 72, 83, 84, 190, 322, 471, 483, 497, 527, 534
environmental effects, 550
environmental factors, 396
enzyme(s), xiv, 11, 15, 119, 127, 130, 131, 197, 252, 272, 275, 332, 358, 371, 385, 397, 400, 414, 415, 425, 437, 446, 451, 452, 459, 465, 484, 486, 487, 488, 490, 491, 492, 493, 494, 495, 503, 506, 521, 531, 565
enzyme induction, 452
enzyme-linked immunosorbent assay, 15, 565
eosinophilia, 416, 422, 423, 424
epidemic, 33, 59, 84, 95, 152, 178, 184, 206, 261, 264, 491
epidemiology, 190, 225, 527, 534, 564, 574, 587, 612
epilepsy, 176, 177, 214, 230, 333, 404, 418, 449, 452, 458, 608
epinephrine, 406
epistasis, 574
Equatorial Guinea, 251, 555, 586
equilibrium, 312, 313, 314, 315, 417
equipment, 9, 11, 12, 48, 69, 74, 177, 260
Eritrea, 31, 555
erythema multiforme, 426
erythema nodosum, 424
erythrocyte membranes, 338
erythrocytes, 119, 131, 159, 199, 267, 275, 290, 298, 300, 305, 313, 316, 338, 339, 483, 493, 494, 507, 523, 578, 587, 595, 597, 600, 615
erythroid cells, 390
esophagitis, 421
ester, 128, 501
estrogen, 606
ethers, 337
ethics, 171
ethnic groups, 351, 354
ethnographic study, 327
etiology, 410, 573

EU, 526

Europe, 9, 115, 121, 122, 187, 226, 227, 239, 240, 261, 530, 584

European Commission, 525

evolution, 7, 73, 83, 257, 265, 266, 468, 565, 571, 588, 589

examinations, 368, 377, 439, 457

exchange transfusion, 164, 554, 568

excitability, 425, 452

excitation, 426

excretion, 20, 270, 363, 415, 603, 604

execution, 536

expenditures, 101, 540

experimental condition, 15

experimental design, 518

expertise, 177, 537, 540, 545, 546, 548

exports, 102

external environment, 536

extraction, 12, 49, 69, 70, 71, 99, 100, 102, 112, 113, 121, 543, 615

extracts, 253, 343, 488, 508, 510, 511, 512, 513, 514, 515, 517, 518, 617

extrapyramidal effects, 353, 449

F

faith, 78

false positive, 439

families, 57, 64, 68, 80, 126, 207, 237, 322, 416, 487, 493, 508, 509, 548

family history, 408

family income, 67, 322

family planning, 85

farmers, 65, 69, 99

FAS, 540

fasting, 335, 345

fat, 148, 280, 317, 361, 362, 428, 446

fat intake, 148

fat soluble, 428

fatal arrhythmia, 386

FDA, 39, 41, 43, 170, 286, 301, 331, 474, 475, 481, 529, 547, 548, 549, 550, 551, 552

FDA approval, 41, 170, 301, 547, 549, 550, 551

fear(s), 61, 126, 179, 207, 209, 225

febrile seizure, 178, 327, 351

female rat, 385, 389

fermentation, 500

fetal abnormalities, 43, 410

fetal demise, 386

fetal development, 24, 56, 419

fetal growth, 179, 210, 391

fetal growth retardation, 391

fetal nutrition, 85, 179

fetus, 43, 55, 56, 140, 179, 180, 182, 184, 209, 210, 212, 213, 214, 229, 380, 381, 382, 383, 385, 386, 387, 392, 395, 404, 415, 426, 445, 447

fibrillation, 346, 449, 549, 583

field trials, 22, 35, 69

films, 15, 40, 172, 288

financial, 65, 70, 84, 94, 102, 108, 144, 207, 208, 322, 478, 536, 540

financial resources, 208

financial support, 207, 540

first generation, 337, 496, 497, 498, 501

fitness, 83, 84, 254, 485, 571, 587

flavonoids, 113, 514

flavor, 430

flexibility, 219

flooding, 206

flora, 422

fluctuations, 102, 103

fluid, 161, 164, 165, 177, 281, 301, 507, 554

fluid balance, 161, 165, 554

fluorine, 134

fluoroquinolones, 114, 134, 345, 476, 613

focal seizure, 453

folate, 182, 249, 356, 399, 442, 486, 487, 489, 490, 491

folic acid, 130, 425, 426, 427, 550, 588

food, 55, 127, 129, 148, 153, 192, 197, 203, 216, 221, 223, 237, 267, 338, 412, 414, 428, 433, 435, 483, 489, 490, 497, 593

food intake, 267

force, 85, 95, 207, 246, 257, 264, 266

forecasting, 103

foreign investment, 47

formaldehyde, 504

formation, 24, 27, 127, 131, 340, 388, 389, 390, 391, 397, 410, 422, 424, 451, 472, 476, 497, 504, 506, 513

formula, 108, 122, 126, 156

foundations, 95

France, 239, 446, 504, 530, 607

fraud, 325

free radicals, 23, 253, 338, 339, 488

fresh frozen plasma, 165

fulminant hepatitis, 414

funding, vii, 17, 28, 49, 53, 67, 68, 69, 70, 95, 96, 101, 103, 108, 160, 473, 496, 510, 525, 526, 535, 537, 539, 543, 545, 548

funds, 65, 67, 69, 71, 95, 99, 100, 102, 103, 105, 108, 207, 394, 536, 545

fungi, 491

fusion, 389

G

Gabon, 26, 218, 282, 423, 431, 543, 555
gait, 355, 418
gamma globulin, 216
gastric lavage, 426
gastrointestinal tract, 476
GDP, 72, 73, 492
GDP per capita, 72, 73
gel, 125
gene amplification, 11, 253, 565
gene expression, 614
gene regulation, 484, 485
gene silencing, 485
general anesthesia, 610
general practitioner, 78, 225
generalized seizures, 453
genes, 8, 17, 82, 187, 197, 254, 258, 262, 266, 469,
    471, 479, 482, 484, 485, 488, 500, 563, 565,
    574, 585, 606
genetic background, 484
genetic diversity, 17, 484, 488
genetic factors, 351, 354
genetic marker, 16, 59, 106, 587
genetic mutations, 479
genetics, 485
genome, 21, 483, 485, 488, 490, 566, 585
genomics, 21, 485, 531, 544
genotype, 488
genotyping, 58, 90, 107, 108, 155, 156, 190, 193,
    245, 247, 248
genre, 113
genus, 112, 158, 187, 510, 515
Georgia, 559
Germany, 121, 122, 239, 246, 352, 486, 500, 520,
    526
gestation, 24, 182, 385, 387, 388, 389, 390, 391, 392,
    396, 410, 419, 420, 421, 435, 444, 445, 447
gestational age, 395, 442, 444
glial cells, 600
glioma, 340
global demand, 100
glossitis, 422, 425
glucagon, 334
gluconeogenesis, 335
glucose, 57, 126, 158, 159, 160, 163, 165, 176, 177,
    182, 188, 203, 214, 238, 240, 289, 329, 332,
    333, 334, 335, 385, 404, 411, 414, 424, 431,
    452, 461, 495, 505, 554, 599
glucose tolerance, 452
glutamate, 17
glutamine, 491
glutathione, 127, 388, 420, 478, 494, 606

glycine, 490
glycogen, 493
glycolysis, 391, 495
glycoside, 516, 617
GNP, 46
goods and services, 85
gouty arthritis, 492
governments, 33, 62, 64, 67, 78, 81, 84, 94, 95, 99,
    109, 545
grades, 290, 438
grading, 367
grants, 30, 67, 96, 108
graph, 257
Greece, 114
green alga, 483
gross domestic product, 86
gross national product, 46
grouping, 73
growth, 15, 19, 23, 46, 85, 112, 117, 200, 208, 217,
    255, 266, 288, 295, 385, 389, 423, 427, 447,
    471, 484, 488, 491, 492, 493, 495, 496, 503,
    507, 514, 516, 523, 524, 589, 597, 606
growth factor, 389, 606
growth rate, 266, 488
Guangzhou, 109, 609
guanine, 492
Guatemala, 557
guidance, 15, 29, 57, 108, 237, 471, 482, 534
guidelines, xi, 4, 13, 31, 42, 57, 58, 59, 72, 77, 90,
    94, 144, 166, 215, 234, 238, 245, 327, 358, 395,
    403, 421, 433, 472, 481, 499, 508, 509, 554,
    570
guiding principles, 54
Guinea, 4, 26, 57, 192, 226, 238, 240, 246, 249, 250,
    282, 321, 324, 392, 484, 510, 555, 556, 561,
    568, 585
Guyana, 26, 57, 223, 557

H

hair, 333, 405, 406, 408, 423, 427, 435
hair loss, 333, 405, 408, 423, 427
Haiti, 223, 226, 557
hallucinations, 330, 475
halos, 406
Han dynasty, 123
haplotypes, 469, 471, 482, 565, 576
harbors, 197
harmful effects, 56, 181, 210, 475, 481
hazards, 318, 329, 338
HDAC, 485
headache, 24, 143, 317, 333, 351, 352, 353, 354,
    356, 357, 358, 385, 400, 405, 406, 407, 408,

410, 416, 419, 421, 422, 424, 425, 426, 427, 432, 433, 440, 441, 444, 449, 450, 457, 460
healing, 121, 330
health care, xi, 30, 61, 62, 64, 71, 73, 85, 105, 111, 135, 168, 174, 228, 235, 251, 305, 321, 325, 327, 332, 532, 536, 553, 582, 598
health care professionals, 228
health care programs, 251
health care system, xi, 30, 61, 536
health education, 326
health expenditure, 87
health information, 79, 82
health practitioners, 231, 326
health problems, 85, 94
health promotion, 533
health services, 3, 26, 30, 64, 321, 324
hearing impairment, 355, 433, 434, 450
hearing loss, 24, 350, 355, 410, 418, 432, 433, 434, 435, 450, 610
heart block, 24, 341, 347, 400, 413, 459
heart disease, 348
heart rate, 345, 346, 390, 459
height, 140, 462
hemangioma, 445
hematocrit, 19, 159, 160, 177, 441, 461
hematology, 426
hematopoietic stem cells, 390
hematopoietic system, 391
hematuria, 424, 427
heme, 23, 118, 119, 127, 128, 129, 197, 253, 339, 340, 429, 483, 488, 489, 490, 497, 505, 507
hemodialysis, 164, 165, 166, 173, 189, 554
hemoglobin, 118, 127, 153, 158, 159, 160, 165, 177, 187, 197, 211, 216, 329, 414, 438, 441, 443, 483, 487, 488, 489, 490, 614
hemoglobinopathies, 253, 414
hemoglobinopathy, 414
hemolytic anemia, 56, 125, 182, 210, 332, 414, 415, 422, 424, 446, 607
hemorrhage, 445
hemorrhoids, 123
hemozoin, 127, 338, 339, 490, 513
hepatic failure, 163, 397, 451
hepatitis, 47, 204, 333, 383, 397, 405, 409, 413, 424, 426, 435, 437, 451, 463, 480, 611
hepatitis a, 384, 397, 451, 480
hepatitis c, 437
hepatocytes, 116, 199, 517, 574
hepatotoxicity, 272, 385, 421, 422, 425, 470, 480
herbal medicine, 512, 519, 616
heterogeneity, 41
hexane, 512
high-risk populations, 20

histamine, 344
histidine, 15, 40
histone(s), 485
histone deacetylase, 485
historical reason, 167
history, x, 10, 18, 38, 112, 120, 122, 145, 177, 198, 199, 205, 208, 214, 230, 231, 293, 333, 350, 351, 352, 354, 394, 404, 406, 408, 410, 418, 436, 438, 451, 453, 454, 456, 458, 468, 470, 482, 530, 537, 548, 566, 618
HIV/AIDS, 21, 33, 52, 63, 95, 112, 139, 270, 271, 395
homeostasis, 495
homes, 534
homocysteine, 494
Honduras, 557
Hong Kong, 499
hookworm, 177
hormones, 334
hospitalization, 8, 85, 171, 196, 230, 239, 350, 408, 429, 461, 467
host, xii, 8, 11, 15, 71, 82, 116, 119, 140, 142, 143, 195, 197, 199, 200, 239, 243, 252, 254, 259, 269, 299, 322, 335, 341, 483, 484, 486, 487, 489, 490, 491, 492, 493, 494, 495, 513, 574, 589, 597
host population, 589
human behavior, 254
human body, 115, 116
human capital, 67, 68, 86
human exposure, 343
human health, ix, 1
human immunodeficiency virus (HIV), xiii, 20, 21, 33, 47, 52, 63, 87, 94, 95, 102, 112, 130, 135, 139, 177, 209, 212, 243, 261, 262, 270, 271, 329, 395, 404, 426, 438, 444, 469, 470, 488, 535, 536, 581, 591
human milk, 407, 428, 605
human resources, 536
human subjects, 373, 379
humidity, 232
hybrid, 300, 316, 492
hydrogen, 339
hydrogen peroxide, 339
hydrolysis, 120, 275, 359, 360, 361, 490, 492, 494, 499, 592, 604
hydroxyl, 357, 506
hyperinsulinemia, 188, 335, 382
hyperkalemia, 454
hyperplasia, 412
hypersensitivity, 23, 42, 137, 383, 397, 400, 411, 415, 422, 423, 424, 425, 426, 439, 451
hypertension, 356, 408, 421

630 Index

hyperuricemia, 422, 492
hypoglycemia, 41, 43, 85, 140, 158, 163, 165, 169, 170, 176, 178, 180, 182, 183, 184, 188, 213, 214, 281, 301, 333, 334, 335, 336, 351, 382, 387, 404, 410, 412, 425, 441, 452, 453, 454, 462, 599
hypokalemia, 39, 173, 345, 406, 422
hypomagnesemia, 39, 173, 345
hyponatremia, 351, 427, 449, 453, 454
hypophosphatemia, 422
hypoplasia, 422, 423
hypoprothrombinemia, 424
hypotension, 162, 163, 167, 173, 344, 345, 346, 347, 348, 406, 411, 449
hypotensive, 343, 347
hypothermia, 85
hypothesis, 24, 91, 152, 221, 351, 358, 391, 401, 445, 477, 484, 488, 489
hypothyroidism, 397
hypovolemia, 161, 165
hypoxemia, 165, 345, 448
hypoxia, 163

I

iatrogenic, 345, 403, 456
ideal, xiv, 22, 39, 136, 157, 258, 262, 285, 286, 305, 341, 466, 478, 492, 523
identification, 194, 287, 394, 523, 615
identity, 219, 492
idiosyncratic, 249, 355, 382
illiteracy, 320
images, 319
immune defense, 267, 291
immune response, 219, 254, 293, 304, 351
immune system, 116, 254
immunity, 8, 12, 13, 38, 56, 76, 82, 84, 125, 141, 158, 176, 177, 178, 179, 187, 189, 192, 207, 208, 209, 218, 219, 221, 226, 227, 232, 246, 259, 267, 270, 288, 292, 293, 318, 326, 381, 471, 482
immunization, 219, 221
immunosuppression, 591
impairments, 158
imports, 102
improvements, 32, 49, 72, 80, 101, 322
in utero, 56, 210, 383, 387, 445
inadmissible, 295, 310
incidence, 3, 8, 22, 27, 29, 32, 41, 53, 57, 85, 86, 87, 89, 149, 150, 164, 173, 205, 206, 211, 214, 216, 218, 219, 220, 221, 222, 223, 224, 233, 262, 286, 320, 329, 330, 350, 351, 353, 354, 357, 384, 396, 397, 398, 399, 400, 401, 411, 416, 418, 426, 432, 439, 440, 449, 451, 452, 453, 457, 459, 474, 476, 477, 566, 581, 588
income, 46, 64, 67, 71, 72, 87, 94, 106, 322, 394, 396, 525
increased access, 87
incubation period, 192, 230, 232
independence, 67, 177
independent variable, 318
India, 7, 10, 26, 40, 57, 97, 98, 112, 114, 171, 233, 235, 250, 272, 520, 526, 560, 578, 592
indirect effect, 336, 351
individuals, 3, 9, 47, 64, 73, 76, 163, 169, 204, 214, 217, 224, 240, 241, 254, 258, 271, 285, 303, 333, 345, 347, 349, 358, 403, 404, 409, 414, 431, 432, 439, 470, 474
Indonesia, 4, 26, 40, 41, 57, 171, 191, 192, 193, 238, 249, 272, 432, 440, 481, 484, 551, 552, 560, 577, 580, 584, 587
induction, 275, 284, 302, 305, 361, 371, 377, 379, 380, 390, 393, 426, 601
industry, xiii, 9, 99, 100, 323, 401, 465, 466, 474, 496, 525, 530, 531, 539, 540, 545, 548
ineffectiveness, 81, 109, 115
inequity, 537
infancy, 185, 218, 235, 238, 271, 317, 386
infant mortality, 116
infants, 20, 52, 87, 91, 95, 101, 105, 138, 141, 176, 177, 179, 184, 185, 186, 208, 211, 212, 215, 216, 217, 218, 219, 221, 222, 235, 238, 270, 271, 332, 356, 381, 384, 385, 386, 391, 410, 420, 422, 426, 427, 430, 442, 444, 445, 447, 531, 543, 582, 583, 584
infectious disorders, 327
inferiority, 107
inflammation, 515
inflation, 548, 549
informal sector, 536
infrastructure, 8, 87, 108, 168, 176, 225, 261, 321, 467, 525
ingest, 113, 221
ingestion, 353, 424, 426, 434, 444
ingredients, 103, 256, 324, 509
inhibition, xiv, 15, 56, 129, 210, 255, 268, 281, 289, 295, 347, 376, 388, 398, 425, 437, 439, 450, 465, 487, 489, 491, 493, 494, 496, 504, 506, 513, 514, 517, 521, 523, 540, 551, 594, 615
inhibitor, 272, 388, 485, 486, 493, 494, 495, 500, 503, 507, 616
initiation, 44, 107, 169, 174, 187, 344
injections, ix, 1, 164, 184, 276, 284, 302, 327, 373, 392, 410, 420, 532
injure, 330
injury, 167, 177, 350, 355, 361, 368, 376, 397

inoculation, 518

insecticide, 54, 67, 87, 88, 89, 95, 105, 155, 206, 215, 436, 533, 550, 572

insertion, 175, 494

insomnia, 154, 250, 317, 353, 357, 358, 383, 406, 407, 408, 427, 432, 440, 441, 443, 449, 481

inspectors, 103

institutions, 18, 81, 95, 103, 496, 528, 537, 539

insulin, 334, 335, 336, 411, 452, 599

insurgency, 155

integrity, 587

intellectual property, 509, 539

intellectual property rights, 509

intensive care unit, 39, 163, 172, 189, 227, 239, 549, 554, 618

interface, 494

interference, 409, 425

international financial institutions, 67

interstitial nephritis, 424

intervention, 33, 54, 73, 79, 87, 89, 93, 94, 211, 220, 221, 222, 236, 238, 251, 261, 324, 441, 570

intestinal tract, 361, 362

intestine, 361

intoxication, 354, 426, 450, 602

intracranial pressure, 422

intramuscular injection, 38, 42, 161, 162, 167, 168, 271, 278, 279, 280, 284, 295, 302, 305, 360, 361, 363, 364, 365, 367, 368, 369, 371, 372, 373, 374, 377, 379, 380, 392, 393, 411, 417, 419, 442, 460, 482, 603

intravenous fluids, 165

intravenously, 39, 40, 43, 169, 170, 274, 275, 347, 371, 392, 462, 499, 501, 553

investment(s), 47, 65, 69, 95, 206, 271, 322, 325, 525, 536, 543, 545

ions, 344, 348

Iran, 26, 469, 558, 612

Iraq, 238, 558

iron, 23, 119, 238, 253, 329, 338, 339, 340, 488, 582, 584, 600, 615

islands, 6, 390, 391, 468, 581

isolation, 499, 508, 617

isozyme, 492

Israel, 496

issues, vii, xiii, xiv, 88, 160, 219, 225, 238, 251, 254, 326, 332, 342, 395, 465, 466, 499, 538, 576

Italy, 109, 205, 206

Ivory Coast, 477, 510

J

jaundice, 135, 188, 192, 214, 385, 423, 424, 425, 426, 442, 443, 445, 463

Java, 122

joint pain, 441

Jordan, 601

justification, 37, 84

K

K^+, 336, 601

Kenya, 26, 77, 85, 171, 212, 218, 226, 255, 265, 324, 327, 405, 426, 427, 431, 432, 440, 454, 468, 514, 551, 552, 553, 556, 569, 572, 581, 582, 591, 594, 599, 605, 612

kidney(s), 230, 231, 361, 362

kill, xii, xiv, 23, 25, 111, 118, 119, 140, 260, 266, 298, 304, 309, 311, 312, 331, 436, 466, 500, 503, 552

kinetics, 175, 181, 182, 301, 492, 495, 592, 609

Korea, 533, 549, 560, 561

Krebs cycle, 391

L

labeling, 97, 217, 325, 469, 475, 476, 549

labor force, 3, 55, 222

laboratory studies, 23, 349

laboratory tests, 516, 520, 544

lactate dehydrogenase, 15, 397, 495, 540

lactate level, 161

lactation, 181, 382, 386, 391

Laos, 26, 193, 325, 432, 440, 577

larvae, 327

Latin America, 226, 326, 425, 482, 526, 528

law enforcement, 145

laws, 102

lead, xi, xiv, 13, 16, 42, 56, 67, 70, 82, 84, 89, 105, 111, 126, 151, 176, 181, 208, 209, 217, 227, 228, 230, 234, 239, 252, 255, 260, 265, 305, 322, 334, 339, 379, 380, 381, 382, 385, 386, 394, 396, 397, 399, 406, 429, 431, 440, 448, 449, 462, 465, 466, 476, 478, 483, 492, 497, 500, 513, 518, 524, 530, 531, 541, 545, 547, 612, 618

leadership, 528, 534, 535, 543, 545

leakage, 321

learning, 86, 155, 177, 384

legs, 333

leishmaniasis, 132, 503, 525, 535

leisure, 68, 230

leisure time, 68

lens, 399

leprosy, 33, 132, 262, 383, 424

lesions, vii, 24, 204, 340, 350, 355, 360, 367, 409, 450
leukocytosis, 414
leukopenia, 408, 425, 426
Lewis acids, 498
Liberia, 26, 249, 510, 556, 585
lichen, 411
lichen planus, 411
life cycle, 25, 115, 116, 199, 252, 268, 269, 285, 288, 289, 290, 291, 303, 304, 483, 490, 494, 506
ligand, 492
light, 38, 137, 229, 232, 335, 397, 399, 469, 488, 489, 523
lipid metabolism, 329
lipid peroxidation, 338
lipids, 488
lipoproteins, 168
liquid chromatography, 597
litigation, 215
liver, xiv, 17, 66, 115, 116, 117, 118, 122, 123, 124, 125, 131, 135, 141, 145, 189, 192, 197, 199, 200, 204, 228, 230, 231, 239, 240, 251, 264, 266, 267, 286, 361, 362, 385, 391, 397, 400, 408, 412, 413, 415, 422, 426, 427, 430, 436, 437, 438, 440, 441, 451, 452, 459, 465, 466, 487, 504, 505, 509, 517, 520, 541, 544, 552, 574, 578, 612, 618
liver cells, 504, 517
liver damage, 397, 430, 451, 612
liver disease, 141
liver enzymes, 414, 426, 427, 459
liver failure, 204, 397, 451
liver function tests, 413, 438, 440, 509, 520
liver transplant, 451
liver transplantation, 451
local community, 533
local conditions, 11
loci, 21, 123, 285, 293, 438, 485
locus, 266
longevity, 325, 471
loss of appetite, 333
low risk, 225, 228, 232, 250
low temperatures, 502
low-density lipoprotein, 168
lower prices, 49, 71
loyalty, vii
LTA, 97, 596
lumbar puncture, 161, 162, 163
lupus, 521
lying, 201, 389
lymphoma, 506
lysine, 275

M

machinery, 119, 399
macrolide antibiotics, 39, 173, 503
macromolecules, 397, 451, 597
magnetic resonance imaging, 351
magnitude, 82, 288, 292, 347, 349, 396, 470, 480
major issues, 56, 209, 393
majority, 20, 33, 41, 48, 50, 57, 62, 63, 79, 82, 93, 141, 145, 148, 170, 174, 176, 185, 187, 190, 237, 259, 300, 303, 316, 354, 355, 390, 435, 450, 454, 493, 505, 533, 535
malaise, 135, 403, 405, 408
Malaysia, 26, 282, 561
malnutrition, 47, 176, 177, 327, 395, 438, 450
mammal, 455
mammalian cells, 338, 485, 495, 507
mammals, 3, 491, 604
man, 187, 340, 397, 418, 458, 593
management, xii, 13, 30, 33, 50, 52, 57, 86, 88, 90, 101, 129, 139, 141, 143, 145, 162, 165, 166, 175, 186, 187, 193, 194, 195, 217, 228, 237, 247, 281, 290, 301, 327, 393, 408, 468, 479, 531, 533, 536, 537, 543, 554, 572, 578, 608, 618
manufacturing, 9, 31, 45, 100, 101, 138, 144, 160, 256, 301, 324
manufacturing companies, 102
market economy, 530
market failure, 537
marketing, 9, 28, 29, 96, 101, 104, 145, 341, 380, 384, 394, 398, 401, 439, 473, 474
marketplace, 323
marrow, 362, 398, 450
Mars, 566, 568, 593, 594, 604, 607
marsh, 112
Maryland, vii
mass, 53, 108, 181, 205, 206, 208, 249, 250, 256, 261, 326, 387, 427, 444, 446, 581, 608
mass media, 326
materials, 103, 144, 498, 501, 511
mathematics, 265
matter, 42, 63, 119, 474
Mauritania, 26, 510, 556
Mauritius, 114, 556
measles, 218, 221, 581
measurement(s), 14, 19, 157, 160, 193, 267, 282, 292, 303, 306, 311, 344, 346, 356, 364, 439, 591, 601, 618
mechanical ventilation, 188, 346, 448, 554
Médecins Sans Frontières, 72
media, 78, 250

median, 152, 156, 279, 280, 281, 304, 318, 390, 392, 419, 434, 438, 500

medical, vii, 13, 33, 37, 45, 49, 57, 66, 67, 86, 90, 93, 95, 112, 113, 114, 121, 144, 158, 162, 166, 178, 203, 215, 224, 225, 227, 231, 232, 234, 236, 237, 253, 280, 322, 326, 327, 328, 331, 332, 333, 351, 357, 396, 408, 441, 447, 454, 481, 508, 544, 548, 567, 573

medical care, 49, 95, 203, 227, 232, 234, 236, 237

medical history, 113, 332, 454

medication, xii, 14, 40, 48, 53, 56, 66, 104, 143, 175, 183, 186, 187, 188, 195, 200, 205, 217, 221, 225, 235, 236, 237, 287, 331, 412, 443, 444, 452, 458, 474, 533, 554

medicine, vii, 52, 58, 99, 102, 105, 137, 138, 139, 142, 147, 164, 173, 185, 187, 225, 321, 327, 331, 333, 334, 343, 394, 466, 499, 509, 515, 540

Mediterranean, 190, 414

melanin, 399

mellitus, 452

membrane permeability, 298

membranes, 129, 313, 338

memory, 236

meningitis, 161, 162, 163, 165

mental disorder, 332

mental illness, 352

mentor, vii

mesenchyme, 389, 390

messages, 326

meta-analysis, 9, 19, 39, 94, 154, 156, 162, 169, 220, 221, 409, 471, 508, 568, 578, 613

Metabolic, 165, 603, 614

metabolic acidosis, 158, 160, 189, 281, 301, 345, 448, 454

metabolic pathways, 490

metabolism, xiv, 12, 20, 158, 168, 270, 305, 329, 334, 335, 342, 363, 414, 425, 426, 429, 452, 465, 483, 486, 487, 489, 490, 494, 495, 506, 595, 599, 604, 615

metabolites, 20, 168, 266, 284, 302, 336, 348, 358, 362, 380, 438, 505, 591

metabolized, 120, 128, 131, 272, 318, 347, 349, 446, 452, 497, 504

metabolizing, 275, 371

metal ion, 494

metalloenzymes, 492

methanol, 124, 130, 334, 336, 357, 384, 440, 514, 550

methemoglobinemia, 402, 415, 424, 431, 477

methodology, 68, 83, 86, 94, 156, 265, 319

methyl groups, 504

methylation, 495

methylene blue, 494

Mexico, 233, 557

mice, 123, 277, 278, 339, 340, 388, 390, 392, 416, 417, 419, 420, 460, 499, 501, 512, 513, 514, 515, 517, 518, 589, 604, 605, 608, 617, 618

microcirculation, 451

microdialysis, 314, 315

microorganisms, 503

microscope, 135, 288

microscopy, 11, 14, 30, 145, 190, 261, 288

microsomes, 578

Middle East, 4, 58, 149, 187, 189, 223, 442

migrants, 55

migration, 259, 388

military, vii, 101, 547, 548, 549, 550, 551, 618

miscarriage, 224, 331, 432

misconceptions, 326, 327, 584

mission(s), 101, 155, 172, 211, 538, 539, 543

misuse, 34, 235, 247, 322, 479

mitochondria, 338, 397

mitogen, 493

models, 73, 76, 77, 78, 83, 265, 266, 271, 289, 318, 339, 341, 342, 347, 388, 393, 417, 486, 530, 547, 595

modifications, 12, 94

mole, 311, 312

molecular mass, 362

molecular structure, 113

molecular weight, 164, 299

molecules, xiv, 311, 312, 348, 465, 466, 502, 506, 507, 517, 540, 543

momentum, 67, 527, 534, 543, 545

monoclonal antibody, 391

monomers, 497, 499

mood change, 330

Moon, 617

morbidity, xi, 10, 31, 37, 41, 46, 47, 48, 51, 53, 54, 58, 61, 62, 72, 74, 76, 79, 81, 82, 84, 86, 88, 89, 138, 150, 176, 187, 196, 200, 205, 207, 208, 213, 216, 218, 219, 220, 221, 235, 260, 264, 271, 322, 328, 404, 412, 415, 448, 468, 483, 485, 489, 543, 581, 582, 588, 611

morning stiffness, 319

Morocco, 105, 436, 558

morphological abnormalities, 390, 400, 415

morphology, 616

mortality rate, ix, 1, 38, 41, 65, 86, 89, 90, 158, 160, 176, 322, 335, 352, 407, 449, 483, 485, 510

mosquito bites, 3, 215, 222, 224, 226, 229, 232, 583

mosquitoes, xii, 10, 63, 105, 114, 116, 117, 150, 195, 200, 209, 225, 226, 235, 251, 327, 550, 576

motif, 491

motivation, 226, 536

mouth ulcers, 214, 332, 404, 406, 431

Mozambique, 26, 32, 218, 222, 395, 418, 432, 556, 608

MSF, 95

mucous membranes, 426

multidimensional, 615

multiple factors, 181

multiplication, 7, 131, 254, 259, 264, 266, 267, 268, 281, 289, 290, 299, 300, 316, 594

muscle relaxant, 521

muscles, 332, 373, 406

mutant, 21, 22, 83, 84, 108, 119, 136, 249, 252, 254, 259, 260, 263, 273, 281, 285, 286, 293, 439, 483, 484, 485, 489, 493, 500, 502

mutation(s), 7, 8, 11, 16, 23, 27, 34, 83, 84, 118, 119, 136, 151, 187, 193, 196, 197, 250, 251, 252, 253, 255, 256, 257, 258, 260, 263, 264, 266, 273, 285, 292, 446, 468, 469, 470, 471, 479, 482, 483, 484, 489, 490, 491, 495, 502, 552, 563, 564, 571, 575, 580, 587, 590, 612, 613

myalgia, 408, 433

Myanmar, 4, 6, 26, 36, 40, 42, 43, 57, 80, 81, 171, 196, 223, 229, 234, 250, 252, 272, 337, 353, 409, 432, 442, 449, 467, 472, 560, 585, 598

myasthenia gravis, 452

myocardial ischemia, 344, 345

myocarditis, 424

myoglobin, 397

myopathy, 397, 406, 607

N

Na$^+$, 602

NAD, 494

NADH, 332, 495

Namibia, 26, 556

National Institutes of Health, 525

national policy, 15, 91, 141, 187, 232, 570

natural selection, 83

nausea, 24, 135, 140, 143, 149, 251, 317, 333, 335, 341, 352, 356, 357, 385, 400, 401, 403, 405, 406, 407, 410, 412, 414, 416, 419, 421, 422, 423, 424, 425, 426, 427, 428, 430, 431, 432, 433, 436, 437, 440, 441, 443, 446, 456, 457, 459, 460, 474, 481

necrosis, 284, 302, 360, 410, 417

needy, 567

negative externality, 50

neglect, 46

neonates, 383, 386, 423, 425, 427

Nepal, 31, 560

nephritis, 424

nerve, 38, 42, 167, 410, 452

nervous system, 179, 425, 441

Netherlands, vii, 239, 358, 481, 547

neural function, 417

neuroblastoma, 340

neurological disability, 37

neuronal cells, 339, 340

neurons, 340

neuropathy, 607

neurotoxicity, vii, 24, 145, 284, 296, 302, 310, 339, 340, 341, 349, 351, 359, 360, 361, 362, 363, 365, 368, 374, 375, 376, 377, 379, 380, 416, 417, 418, 419, 439, 460, 461, 482, 499, 547, 553, 593, 594, 595, 599, 600, 603, 604, 608

neutral, 504, 528, 600, 615

neutral lipids, 600, 615

neutropenia, 24, 333, 384, 398, 400, 401, 402, 405, 414, 415, 416, 422, 427, 437, 438, 439, 461, 474, 475, 476, 477, 608

next generation, 86, 544, 545

NGOs, 324

Nicaragua, 206, 557, 581

niche market, 43

nicotinamide, 332

Nigeria, 26, 33, 182, 211, 256, 324, 327, 352, 383, 431, 510, 518, 556, 576, 611

nightmares, 408, 432, 440

Nile, 575

nitric oxide, 344

nitrogen, 497, 505, 506

NMR, 487, 516, 518

Nobel Prize, 114

non-nucleoside reverse transcriptase inhibitors, 272

normal distribution, 253

North Africa, 57, 122

North America, 187, 198, 227, 261, 342, 530

nuclei, 374, 455, 460

nucleic acid, 127

nucleoside inhibitor, 495

nucleotides, 492, 496

nucleus, 338, 339

nursing, 163, 217, 331, 428, 554

nursing care, 554

nutrient, 486

nutritional deficiencies, 177

nystagmus, 355, 439, 457, 458

O

obstacles, 79, 80, 155, 157, 286, 529

obstruction, 345, 448, 451

occupational groups, 232

Oceania, ix, 1, 4, 57, 189, 191, 192, 223, 225, 247, 249, 251

odynophagia, 421
oedema, 164, 165
oesophageal, 231, 421, 422, 423
oil, 24, 25, 26, 41, 43, 128, 168, 269, 277, 278, 280,
    294, 308, 309, 310, 355, 359, 362, 363, 364,
    365, 366, 367, 368, 369, 370, 371, 372, 373,
    374, 376, 378, 380, 392, 393, 416, 417, 450,
    460, 509, 592
omission, 266
operations, 9, 95, 120, 534, 550
operon, 134
ophthalmologist, 406
opportunities, 30, 102, 103, 177, 478, 486, 534, 535
optic nerve, 412
optic neuritis, 317
optimism, 489, 525
optimization, 318, 446, 524, 536, 573, 612
oral hypoglycemic agents, 452
organ(s), 38, 135, 158, 165, 169, 173, 199, 361, 397,
    424, 448, 451
organism, 491
ornithine, 493
orthostatic hypotension, 343
otitis media, 443
ototoxicity, 433, 434, 609
outpatients, 88, 429
ovarian cancer, 388
overhead costs, 87
overlap, 143, 162, 240
oversight, 537
oxidation, 338, 487, 502, 515
oxidative damage, 339, 494
oxygen, 163, 165, 338, 390, 391, 488, 497, 501
ozone, 500
ozonolysis, 500

P

Pacific, 484, 580
pain, 24, 46, 333, 341, 355, 357, 385, 397, 400, 403,
    407, 408, 410, 412, 414, 416, 421, 422, 423,
    424, 425, 428, 433, 441, 446, 460, 462, 481,
    519
pairing, 35
Pakistan, 28, 31, 96, 473, 477, 558
palpitations, 317, 408, 418, 432, 433, 440, 458
Panama, 6, 114, 206, 223, 468, 548, 558
pancreatitis, 421, 422
panic attack, 330
pantothenic acid, 614
Paraguay, 558
parallel, 155, 244, 336, 398, 450, 512, 513, 544
paranoia, 330

parasitic infection, 208, 500
parents, 56, 177, 204, 221, 327, 445
parity, 87, 211, 212, 213
participants, 93, 152, 154, 170, 211, 213, 216, 220,
    433, 441, 442, 443, 479, 519, 520, 534
patents, 49, 70
pathogenesis, 351, 358, 414
pathogens, 255
pathology, 143, 405
pathophysiological, 363
pathophysiology, 573
pathways, 434, 486, 489, 490, 491, 493, 506, 523
PCR, 12, 16, 17, 25, 58, 80, 84, 106, 107, 108, 151,
    152, 153, 155, 156, 157, 318, 462, 469, 488
PCT, 25, 26, 273, 275, 279, 280, 285, 288, 290, 292,
    294, 295, 303, 318
peptide, 128, 542
perfusion, 267, 351, 454
pericarditis, 422
perinatal, 211, 213, 214, 444
periodicity, 120, 518
peripheral blood, 159, 288, 290, 338, 373, 398, 450
peripheral neuropathy, 424
permeability, 313, 506, 597
permission, 166
permit, xii, 9, 111, 180, 478
peroxidation, 338
peroxide, 23, 113, 118, 253, 341, 488, 500, 501, 502,
    540, 541, 592
personal communication, 357, 426, 470, 480, 506
personal history, 332
pertussis, 581
Peru, 4, 26, 33, 115, 121, 122, 153, 191, 250, 558,
    577
pH, 127, 159, 161, 167, 168, 299, 410, 497, 499,
    501, 614
phagocytosis, 439
pharmaceutical(s), xi, xiii, 9, 28, 43, 61, 96, 97, 99,
    100, 101, 102, 103, 104, 109, 118, 144, 215,
    325, 342, 401, 465, 466, 473, 474, 496, 510,
    525, 526, 528, 530, 531, 540, 548, 563
pharmacokinetics, 20, 141, 155, 168, 181, 184, 185,
    204, 243, 254, 257, 258, 265, 266, 269, 270,
    271, 275, 276, 283, 284, 285, 302, 303, 305,
    311, 318, 372, 415, 531, 567, 568, 569, 578,
    579, 589, 590, 591, 592, 593, 594, 595, 596,
    597, 599, 601, 603, 604, 614
pharmacology, 19, 20, 205, 565, 566, 578, 591
phenobarbitone, 163, 462, 484
phenothiazines, 345
phenotype(s), 12, 119, 253, 453, 483, 484, 486
phenytoin, 452
Philadelphia, 580

Philippines, 26, 561

phosphate, 55, 56, 126, 168, 183, 201, 202, 203, 240, 241, 313, 329, 332, 385, 414, 424, 431, 461, 486, 491, 493, 503, 505, 540, 542, 550

phosphatidylcholine, 487, 506, 615

phospholipids, 313

phosphorus, 397

phosphorylation, 391

photophobia, 162, 406

photosensitivity, 398, 406, 411, 421, 424

physicians, 112, 121, 198, 227, 236, 284, 300, 302, 329, 331, 548, 589

physicochemical properties, 319

pigmentation, 401, 413, 476

pigs, 419

piloerection, 355

pilot study, 219

placebo, 92, 93, 206, 222, 240, 384, 385, 401, 409, 414, 428, 474, 475, 519, 532, 552, 576, 581, 582, 583, 584, 586, 587, 599

placenta, 85, 179, 212, 361, 383, 385, 388, 412, 606, 609

plants, 70, 491, 495, 508, 509, 510, 511, 512, 515, 516, 518, 542, 617

plasma levels, 181, 346, 410, 449

plasma membrane, 492

plasma proteins, 168, 397, 451, 578

plasmodium, 199, 588, 618

plasticity, 255

plastid, 483, 486, 495, 575

platelet count, 161, 441, 458

platelets, 135, 165

platform, 528, 541

playing, 536, 545

PM, 561, 562, 569, 572, 574, 576, 601, 615, 616

pneumonia, 132, 162, 163, 357, 435

point mutation, 16, 197, 265, 489, 564, 565, 585

polar, 497, 500, 506

polarity, 500

police, 55, 222

policy, xi, xiii, 4, 8, 11, 12, 13, 14, 17, 19, 27, 28, 29, 30, 31, 34, 38, 40, 47, 48, 50, 51, 58, 59, 61, 64, 67, 73, 75, 77, 78, 81, 83, 84, 96, 155, 156, 169, 170, 179, 180, 183, 184, 192, 212, 238, 243, 258, 321, 382, 395, 404, 436, 466, 468, 469, 473, 478,530, 534, 535, 536, 537, 562, 567, 569, 570, 571, 577, 587, 599, 612

policy choice, 536, 571

policy makers, 13, 78, 258, 534, 587

policy options, 81

policymakers, 77

politics, 67, 528

polyamine, 493

polyamines, 493

polydipsia, 422

polymer, 127

polymerase, 106, 127, 129, 565

polymerase chain reaction, 106, 565

polymerization, 118, 490, 505, 542

polymorphism(s), 24, 83, 252, 469, 479, 485, 488, 565

polypeptide, 492

polyuria, 422

pools, 492

population, 3, 12, 13, 15, 16, 21, 30, 47, 51, 54, 73, 74, 76, 77, 82, 84, 89, 92, 93, 95, 112, 145, 151, 152, 155, 189, 190, 205, 206, 208, 218, 219, 221, 231, 233, 243, 244, 254, 256, 257, 264, 265, 266, 267, 285, 287, 288, 289, 299, 300, 303, 316, 318, 324, 353, 356, 384, 399, 418, 420, 456, 459, 471, 474, 478, 482, 488, 537, 539, 544, 565, 570, 584, 588, 589

population density, 190, 324

population group, 30

population structure, 21, 488, 588

portfolio, xiv, 466, 527, 535, 537, 538, 539, 540, 541, 543, 544, 546

positive correlation, 320, 411

positive externalities, 50

postural hypotension, 344, 408, 462

potassium, 335, 344, 347, 348, 397, 521

potential benefits, 86, 213, 395

poverty, 62, 88, 95, 176, 325, 527, 534, 536

poverty alleviation, 527, 534

precursor cells, 439

predicate, 265

premature death, 72

prematurity, 210

preparation, 57, 113, 134, 336, 359, 498, 518

preschool, 54, 216, 220, 221

preschool children, 54, 216, 220, 221

preservative, 521

president, 105

prevention, x, xi, xii, 3, 18, 22, 41, 46, 52, 54, 55, 56, 58, 67, 86, 89, 93, 94, 107, 112, 130, 133, 157, 159, 170, 179, 187, 195, 197, 198, 200, 203, 204, 205, 209, 212, 218, 220, 222, 225, 228, 234, 235, 237, 239, 268, 285, 293, 299, 304, 329, 385, 393, 394, 403, 407, 426, 472, 511, 544, 549, 550, 552, 559, 560, 571, 572, 579, 581, 582, 583, 584, 607, 610, 611, 616

primary prophylaxis, 201, 202, 203, 240

primary school, 176

primate, 3

priming, 354

principles, xii, 53, 58, 187, 195, 263, 266, 293, 539

prior knowledge, 230

probability, 8, 21, 77, 78, 87, 136, 157, 240, 255, 259, 261, 262, 263, 266, 268, 273, 281, 285, 391, 404, 478

prodrome, 143

prodrugs, 11, 486

producers, 49, 64, 69, 70, 71, 99, 100, 325, 563

production costs, 49, 69, 70, 71, 72, 99

professionals, 324, 327, 394, 528

profit, 106, 496, 525, 533, 537, 539, 548

progesterone, 388, 420, 606

prognosis, 165

project, xiv, 104, 118, 220, 465, 483, 511, 533, 535, 536, 541

proliferation, 18, 340, 388, 493, 614

promoter, 484

prophylactic, x, 3, 4, 55, 56, 81, 87, 92, 93, 126, 131, 134, 144, 153, 155, 196, 197, 198, 199, 202, 208, 223, 226, 227, 228, 234, 237, 250, 255, 261, 271, 329, 332, 333, 334, 335, 336, 343, 346, 352, 357, 358, 397, 399, 403, 404, 405, 406, 409, 413, 426, 430, 436, 446, 448, 449, 451, 452, 467, 470, 471, 478, 480, 481, 501, 503, 517, 518, 528, 548, 552, 608

prophylactic agents, 197, 403

propranolol, 334, 346, 449

prostration, 140, 188, 454

protease inhibitors, 272, 488, 490

protection, xi, xii, 9, 22, 27, 53, 54, 55, 61, 71, 93, 123, 136, 137, 157, 195, 196, 205, 208, 214, 215, 217, 218, 222, 224, 227, 229, 232, 234, 235, 236, 253, 255, 261, 262, 264, 285, 381, 452, 478, 570

protective role, 391

protein kinases, 493, 494

protein sequence, 492

protein synthesis, 134, 340, 490, 513, 515, 551

proteins, 17, 24, 116, 119, 313, 329, 339, 488, 489, 491, 493, 495, 597

proteinuria, 422

proteolysis, 127

proteomics, 488

protons, 484

prototype, 19, 117, 127, 200, 337, 343, 501, 550

pruritus, 214, 399, 406, 408, 412, 413, 424, 430, 432, 438, 440

pseudomembranous colitis, 398, 422, 423, 425

psoriasis, 230, 399

psychiatric disorders, 350, 355

psychological stress, 82

psychosis, 93, 317, 352, 354, 355, 398, 407, 408, 418, 424, 425, 602

psychotropic drugs, 408

public awareness, 29, 144

public concern, 78, 211

public health, 3, 8, 18, 27, 29, 30, 31, 44, 46, 51, 77, 83, 95, 137, 138, 144, 151, 178, 196, 207, 215, 243, 260, 320, 321, 328, 404, 467, 510, 528, 530, 533, 534, 536, 537, 539, 545, 547, 585

public sector, 28, 30, 31, 33, 47, 69, 82, 99, 100, 106, 321, 322, 534, 535, 540

public service, 72

public-private partnerships, 537, 539, 618

pulmonary edema, 158, 182, 192

purchasing power, 100

purification, 69

purines, 131, 491, 492

purpura, 317, 426

pyridoxine, 486

pyrimidine, 131, 487, 489, 491, 492

pyrophosphate, 491, 492

Q

QHS, 128, 275, 283, 297, 298, 305, 306, 308, 309, 310, 361, 371, 379, 392, 393, 419, 420

QRS complex, 173, 343, 347

QT interval, 336, 343, 344, 345, 346, 347, 348, 349, 386, 398, 400, 410, 415, 449, 459, 551, 553, 601

quality assurance, 13, 20, 324

quality control, 20

quality of service, 88

quality standards, 99, 104

quantification, 20, 394

quassinoids, 513

Queensland, 541

quinacrine, 122, 319, 521, 524

R

race, 134, 319, 481

radar, 525

radicals, 119, 338, 339, 600, 616

radio, 326

rash, 333, 341, 412, 422, 424, 433, 458

raw materials, 99, 100

reaction rate, 400, 474

reactions, xii, 55, 57, 122, 137, 145, 200, 204, 223, 224, 238, 329, 330, 331, 339, 350, 352, 357, 382, 383, 394, 395, 396, 397, 398, 400, 401, 409, 410, 411, 412, 413, 415, 416, 422, 423, 424, 425, 426, 436, 443, 451, 455, 459, 461, 469, 470, 475, 476, 480, 481, 501, 550, 602, 605, 608 reactive oxygen, 133, 391

reading, 240, 325, 346

reagents, 501

real time, vii, 97

reality, 50, 75, 93, 178, 179

recall, 102, 104, 396

receptor sites, 358

receptors, 391

recognition, 3, 21, 58, 224, 232, 487, 529

recombination, 116, 468

recommendations, xi, 4, 50, 53, 54, 55, 57, 59, 78, 97, 102, 114, 137, 138, 146, 151, 160, 166, 179, 180, 181, 182, 183, 185, 186, 187, 188, 192, 198, 214, 222, 226, 237, 252, 298, 305, 326, 382, 386, 393, 403, 469, 470, 480, 482, 553, 601, 608

recovery, 27, 153, 171, 172, 177, 264, 272, 287, 290, 294, 317, 340, 343, 345, 346, 355, 404, 411, 414, 418, 441, 443, 450, 458, 460, 479, 520, 588

rectum, 175, 186

recurrence, 146, 153, 156, 188, 193, 194, 245, 453, 454, 564, 578

red blood cells, 114, 116, 158, 199, 295, 296, 298, 299, 314, 332, 388, 420, 487, 493, 508, 517

redundancy, 488

reflexes, 178, 355, 397

refugee camps, 249

refugees, 3

regions of the world, x, 126, 137, 238, 485, 523

registries, 396

regression, 86, 320, 454

regression analysis, 86

regression model, 454

regulations, 50

regulatory bodies, 325

regulatory requirements, 342

regulatory systems, 324

rehydration, 161, 163

reintroduction, 402, 475, 478

relapses, xii, xiv, 116, 133, 142, 189, 192, 194, 195, 199, 201, 239, 240, 448, 466

relatives, 112, 226, 227, 235

relevance, 119, 187, 194, 455, 524, 527, 535

reliability, 185, 190

REM, 216, 220

renal dysfunction, 459

renal failure, 141, 159, 164, 165, 173, 188, 199, 204, 281, 300, 301, 315, 317, 410, 411, 415, 421, 422, 451, 554

repair, 379, 394

replication, 491

reproduction, 213, 495

Republic of the Congo, 555

reputation, 382

requirements, xii, 17, 55, 56, 99, 100, 103, 109, 153, 195, 209, 222, 341, 536, 570, 615

resale, 321

research facilities, 528

research institutions, 538

researchers, ix, x, xi, 1, 3, 43, 46, 77, 78, 108, 123, 254, 255, 359, 497, 510, 517, 528, 529, 534

residues, 495, 505

resolution, 28, 34, 62, 96, 108, 136, 317, 354, 409, 413, 425, 430, 443, 473, 476, 479, 607

resources, xiii, 8, 17, 43, 71, 73, 79, 85, 94, 95, 108, 465, 467, 525, 535, 539, 544

respiration, 447

respiratory arrest, 317, 407

respiratory rate, 135, 163, 515

responsiveness, 251

restoration, 472

restrictions, 229, 498

resveratrol, 515, 617

retail, xi, 61, 72, 79

retardation, 419, 427, 589

reticulum, 24, 128, 338, 489, 495, 563

retinopathy, 396, 397, 399, 406, 413

rheumatic diseases, 348

rheumatoid arthritis, 319, 353, 399, 402, 406, 414, 449, 474, 598

rheumatoid factor, 319

rhythm, 38, 344

ribosome, 134, 429

rights, 529, 539

rings, 252, 268, 289, 291, 295, 300, 316

risk assessment, 55, 223, 272, 601

risk factors, 225, 354, 356, 357, 394, 396, 453, 454, 456

risks, 43, 47, 56, 70, 102, 103, 135, 143, 160, 172, 179, 180, 182, 184, 192, 208, 213, 221, 226, 245, 247, 270, 332, 381, 386, 396, 398, 413, 429, 430, 436, 438, 443, 456, 474, 475, 490, 536

RNA, 485, 491

robotics, 531

rodents, vii, 3, 413, 420, 435, 487, 499, 501, 593

root(s), 511, 512, 514, 516, 518, 519, 617

routes, 44, 167, 171, 371, 419, 510, 524, 609

royalty, 539

rural areas, 3, 41, 49, 185, 186, 226, 228, 232, 301, 305, 530, 532

rural women, 323

Russia, 187

Rwanda, 26, 556, 575, 611

S

saliva, 116
salivary gland(s), 114, 116, 199, 362
salmonella, 398
salts, 141, 161, 167, 395, 410, 430, 453, 615
sampling error, 346
Saudi Arabia, 26, 558
saving lives, 32, 545
savings, 79, 80
scabies, 445
scaling, 108, 533
scarce resources, xi, 61, 85
scarcity, 222, 572
schizophrenia, 317
school, vii, 7, 47, 85, 86, 176, 177, 326, 328, 438, 530, 599
school performance, 86, 177
schooling, 189
science, 73, 394, 525, 546
scope, 92, 344, 524, 535, 538
second generation, 496, 500
Second World, 401, 475
secretion, 334, 335, 336, 386, 411, 427, 452
security, 320
seed, 168
seizure, 143, 185, 189, 406, 418, 438, 453, 458
selectivity, 338, 485, 487, 494, 506, 513, 516
sellers, 321, 325, 327
semi-structured interviews, 327
sensation, 135, 333
sensing, 496
sensitivity, 7, 10, 11, 12, 15, 16, 20, 36, 57, 59, 62, 80, 82, 87, 119, 120, 126, 139, 140, 190, 193, 197, 229, 234, 238, 249, 250, 251, 255, 257, 265, 287, 288, 292, 318, 319, 390, 391, 392, 446, 472, 479, 564, 567, 585, 592, 594, 595, 613
sepsis, 163, 554
sequencing, 488
serum, 159, 164, 257, 304, 334, 336, 341, 352, 361, 383, 388, 389, 397, 412, 420, 424, 443, 452, 459, 524, 597
service provider, 58
services, 44, 50, 67, 79, 80, 85, 95, 174, 190, 207, 320, 321, 322, 533
sex, 344, 345, 352, 445
sexually transmitted diseases, 385
shape, 264, 276, 494, 531
shelf life, 101, 102
shigella, 398
shock, 158, 161, 188, 407, 454, 554
shortage, xiii, 69, 70, 99, 100, 109, 312, 544

shortfall, 1
showing, 7, 36, 42, 43, 114, 120, 253, 310, 340, 369, 428, 489, 500, 511, 512, 593
sialic acid, 487
side chain, 500, 504, 505, 506
side effects, 107, 203
Sierra Leone, 26, 312, 418, 477, 510, 556, 577, 594
signal transduction, 329, 491, 493, 494
signals, 395, 396
signs, 10, 11, 13, 14, 135, 143, 162, 163, 284, 290, 302, 346, 350, 355, 379, 413, 416, 417, 433, 437, 439, 449, 450, 452, 456, 457, 460, 462, 515, 553, 554, 603
simulation(s), 77, 366, 492
single-nucleotide polymorphism, 10, 21
sinus arrhythmia, 409
skeletal muscle, 361
skeleton, 501, 502
skin, 38, 116, 122, 145, 158, 169, 204, 231, 333, 398, 399, 401, 405, 406, 408, 409, 410, 411, 412, 413, 422, 426, 444, 476
sleep disturbance, 149, 418, 422, 433
sleeping sickness, 503, 507, 539
smuggling, 325
social costs, 81, 82
society, 85
socioeconomic status, 88
sodium, 165, 168, 284, 302, 347, 349, 359, 417, 452, 496, 499, 521, 596, 604, 605
solid tumors, 491
Solomon I, 26, 57, 112, 226, 561
solubility, 23, 42, 336, 359, 362, 496, 499, 501
solution, 67, 168, 258, 260, 389, 572
Somalia, 26, 384, 421, 550, 559
somnolence, 407
sorption, 435
South Africa, 26, 29, 32, 33, 59, 77, 78, 79, 105, 205, 206, 233, 395, 430, 431, 556, 566, 570, 591
South America, x, 4, 6, 37, 51, 57, 62, 63, 81, 122, 123, 124, 138, 149, 156, 189, 190, 194, 196, 197, 223, 225, 233, 234, 235, 238, 247, 249, 250, 251, 254, 422, 429, 440, 442, 446, 467, 470, 471, 479, 482, 511, 613
South Asia, 149, 442
South Korea, 514, 515, 516
South Pacific, 198
Southeast Asia, ix, xi, 1, 23, 25, 29, 30, 33, 37, 39, 40, 61, 63, 80, 123, 124, 169, 170, 190, 193, 205, 206, 226, 235, 237, 238, 240, 246, 247, 249, 250, 254, 325, 342, 386, 395, 429, 432, 440, 467, 468, 479, 502, 551, 573, 577, 580, 584, 585, 587, 589, 593, 596, 600, 602, 609
spasticity, 177, 413

specifications, 93, 97, 144

spectroscopic techniques, 487

speculation, 287

speech, 24, 177, 353, 418, 600

spending, 73, 88, 95, 545

spleen, 116, 135, 192, 362, 391

splenomegaly, 327

spontaneous abortion, 56, 84, 179, 209, 381, 382, 383, 384, 387, 407, 410, 444, 551

Spring, vii, 481

Sri Lanka, 31, 112, 254, 560

SSA, 89

stability, 97, 104, 479, 496, 497, 615

stabilization, 344, 441

stakeholders, 58, 466, 534, 535, 543

state(s), 23, 54, 62, 106, 163, 167, 259, 323, 334, 343, 358, 363, 371, 373, 484, 502, 523, 533

statistics, 19, 190, 545

status epilepticus, 453

stereomicroscope, 389

sterile, 389, 476

Stevens-Johnson syndrome, 204, 214, 401, 424, 426, 442, 476, 550, 608

stillbirth, 43, 180, 210, 224, 331, 384, 387, 391, 394, 409, 420, 421, 432

stomach, 57, 204, 333, 376, 380, 414, 497, 498, 499

stomatitis, 317, 422

storage, 69, 256, 502, 514

strategic planning, 258

stress, 334, 345, 414

structural changes, 500

structural modifications, 513

structure, 7, 83, 113, 114, 118, 196, 256, 265, 319, 337, 342, 353, 358, 433, 449, 480, 486, 489, 492, 500, 501, 507, 516, 535, 574

style, 225

subcutaneous injection, 343, 392, 393, 407, 420

sub-Saharan Africa, 10, 31, 32, 48, 50, 54, 55, 62, 71, 75, 87, 89, 90, 92, 94, 105, 108, 112, 123, 136, 150, 181, 189, 196, 219, 225, 226, 236, 245, 247, 249, 320, 323, 381, 391, 403, 422, 432, 468, 510, 533, 567, 571, 586, 598

subsidy, ix, 1, 47, 49, 50, 63, 71, 91, 101, 323, 570

subsistence, 80

substitutes, 122

substitution, 10, 86, 494

substrate(s), 119, 129, 351, 476, 487, 492, 494, 505, 506

success rate, 523

Sudan, 26, 33, 85, 150, 172, 432, 559, 575, 576, 611

suicide, 93, 475

sulfa drugs, 124, 249, 334, 490

sulfate, 55, 56, 167, 201, 202, 204, 609

sulfonamide(s), 92, 131, 149, 197, 291, 424, 442, 446, 613

sulfur, 120, 488, 502, 504

sulphur, 356

Sun, 605, 607, 612, 614

supernatural, 327

supervision, 66, 81, 144, 231

supplementation, 182, 238, 582, 584

supplier, 97, 100

suppliers, 28, 72, 96, 97, 103, 473

suppository, 44, 143, 175, 188, 305, 342, 530, 532

suppression, 23, 51, 138, 191, 209, 272, 346, 361, 399, 448

surface area, 185

surveillance, ix, 1, 8, 9, 15, 17, 18, 19, 20, 29, 59, 82, 89, 90, 107, 109, 116, 166, 184, 206, 226, 266, 341, 380, 384, 394, 396, 401, 410, 439, 468, 472, 474, 478, 485, 564, 584

survival, 19, 39, 40, 80, 84, 152, 154, 156, 170, 179, 187, 212, 253, 254, 259, 260, 264, 268, 271, 296, 342, 380, 483, 491, 514, 515, 516, 518, 534, 588

survival rate, 518

survivors, 38, 342

susceptibility, 10, 14, 20, 21, 22, 29, 59, 149, 176, 190, 193, 196, 238, 245, 251, 252, 253, 262, 265, 268, 281, 286, 289, 290, 300, 303, 351, 381, 442, 460, 471, 472, 477, 484, 485, 489, 495, 503, 504, 505, 565, 587, 594, 614

sustainability, 92, 527

Sweden, 93, 411

swelling, 129, 333

Switzerland, 98, 124, 532, 563, 564, 580, 587, 608, 613

syndrome, 143, 251, 317, 332, 340, 345, 352, 354, 355, 356, 383, 397, 409, 410, 411, 418, 422, 423, 424, 439, 450, 451, 482, 492

synergistic effect, 32, 316, 491, 494, 503

synovitis, 319

synthesis, 109, 114, 118, 122, 130, 131, 249, 329, 339, 423, 425, 486, 490, 491, 492, 493, 498, 499, 501, 502, 506, 568

syphilis, 115

systemic lupus erythematosus, 382, 422

T

tachycardia, 345, 346, 408, 441, 448

Tajikistan, 26

Tanzania, 26, 33, 67, 77, 90, 106, 109, 218, 238, 322, 325, 326, 327, 383, 430, 431, 468, 477, 532, 556, 565, 570, 571, 572, 582, 599 target, 7, 10, 11, 23, 56, 69, 88, 100, 107, 119, 131, 154, 196, 197, 199, 209, 219, 252, 270, 271, 327, 339, 345, 382, 389, 442, 486, 487, 488, 489, 490, 491, 492, 493, 494, 495, 496, 500, 506, 507, 508, 526, 531, 537, 538, 543, 574, 575, 614

target population(s), 56, 88, 107, 209, 270, 271, 382

teachers, 328, 532

technical assistance, 526

technical support, 103, 526, 537

techniques, 185, 401, 474, 508, 531

technology(ies), 58, 260, 323, 488, 544, 545, 565, 615

teeth, 56, 181, 210, 221, 330, 385, 398, 422, 423

telephone, 431, 553

temperature, 14, 135, 143, 144, 150, 177, 346, 454, 518

tendon, 397

tension, 338

territory, 5

test data, 20

testing, xi, 16, 17, 18, 19, 61, 97, 125, 203, 240, 245, 303, 325, 396, 418, 434, 457, 478, 507, 517, 529, 548, 564, 614

testis, 444

tetanus, 581

tetrachlorodibenzo-p-dioxin, 604

tetracyclines, 117, 125, 136, 141, 164, 173, 180, 181, 210, 231, 343, 380, 398, 422, 423, 429, 448, 540

therapeutic agents, 523

therapeutic effects, 267, 605

therapeutic targets, 510

therapeutic use, 1, 7, 136, 184, 349, 350, 352, 353, 357, 410, 426, 449, 472, 481, 523

therapeutics, 495, 523, 524, 618

thermal stability, 501

thimerosal, 521

thin films, 177

thoughts, 475

threats, 510, 534

thrombocytopenia, 135, 192, 382, 408, 410, 411, 422, 423, 424, 425, 426, 441

thyroid, 334

time frame, 80, 471

time periods, x, 4, 81, 219, 467

tincture, 121

tinnitus, 22, 38, 169, 251, 317, 354, 356, 400, 410, 415, 419, 430, 434, 441, 450, 519

tissue, 33, 117, 133, 142, 146, 158, 199, 200, 251, 258, 267, 276, 296, 314, 361, 362, 363, 389, 391, 420, 511, 578, 606, 607

Togo, 26, 477, 510, 556, 613

Tonga, 77

tonic, 406, 418, 458

tonic-clonic seizures, 418

tooth, 385

total costs, 76, 88

toxic effect, 340, 380, 392, 399, 414, 553, 554

toxic waste, 118

toxicology, 34, 341, 358, 396, 427, 595, 600, 602, 604

toxicology studies, 341, 396, 427, 604

toxoplasmosis, 132, 427

trade, 47, 123, 125, 126, 198, 204, 260

trade-off, 260

training, vii, 11, 77, 88, 103, 225, 260, 323, 327, 332, 395, 528, 531, 532, 572, 599

trajectory, 155

tranquilizers, 395, 462

transactions, 489

transaminases, 341, 409, 412

transfection, 23, 483, 484

transferrin, 391

transformation, 21

transfusion, 57, 182, 238, 451, 463

transition period, 99

translation, 486, 495

translocation, 484

transplant recipients, 451

transport, 11, 77, 85, 118, 120, 131, 133, 168, 329, 483, 486, 487, 489, 575, 597, 616

transport processes, 486

transportation, 67, 321

trauma, 445

treatment methods, 537

tremor, 355, 358, 413

tricyclic antidepressants, 345

trypanosomiasis, 525, 552

tuberculosis, xiii, 21, 33, 63, 71, 95, 96, 102, 112, 135, 243, 262, 324, 395, 404, 525

tumor(s), 120, 164, 489, 600, 606

tumor cells, 600

tumor necrosis factor, 164

Turkey, 559

Turkmenistan, 559

turnover, 327

typhoid, 201, 231, 334

tyrosine, 564

U

umbilical cord, 412

umbilical hernia, 444

underlying mechanisms, 193

unique features, 489

unit cost, 80
United Kingdom (UK), 108, 122, 125, 202, 203, 213,
     356, 401, 402, 413, 475, 476, 487, 495
United Nations (UN), 95, 96, 102, 144, 238, 324,
     328, 421, 527, 533, 534, 550, 570, 573
United Nations Development Programme, 527, 534
United States, vii, 38, 41, 93, 97, 101, 106, 122, 123,
     126, 170, 172, 216, 226, 229, 239, 240, 401,
     476, 481, 525, 547, 548, 549, 551, 553, 554,
     568, 574, 579, 584, 602, 613
updating, 64, 88, 104
urban, 67, 224, 225, 256, 321, 322, 323, 588, 590
urban areas, 590
urea, 164
urinary tract, 426
urine, 135, 158, 159, 163, 177, 333, 416, 424
urticaria, 137, 408, 409, 410, 411, 416, 422, 423
usual dose, 141, 424, 427
uterus, 396, 412

Vietnam, 4, 7, 26, 29, 30, 42, 63, 69, 81, 99, 100,
     101, 109, 112, 123, 125, 126, 129, 196, 205,
     223, 233, 234, 272, 282, 325, 337, 434, 467,
     472, 530, 548, 550, 562, 572, 583, 587, 609,
     612
vision, 330, 333, 352, 354, 356, 358, 397, 405, 406,
     410, 450, 527, 534, 535, 536, 539
visual field, 397
vitamin K, 165, 491
vitamins, 395
Volunteers, 359
vomiting, 24, 38, 48, 57, 105, 135, 140, 143, 149,
     152, 169, 177, 185, 188, 234, 243, 267, 301,
     333, 335, 352, 354, 356, 385, 400, 403, 406,
     407, 408, 410, 412, 413, 414, 416, 419, 421,
     422, 423, 424, 425, 426, 427, 428, 432, 433,
     434, 436, 437, 440, 441, 443, 446, 449, 450,
     455, 456, 457, 459, 460, 482, 519
vulnerability, 47, 265, 423
vulnerable people, 533

V

vaccinations, 222, 582, 583
vaccine, x, 3, 150, 219, 230, 231, 334, 543
vacuole, 118, 119, 127, 129, 197, 338, 483, 487, 489,
     490, 497, 616
vaginitis, 229
validation, 15, 523, 591
Vanuatu, 31, 57, 206, 561
variables, 75, 77, 80, 300, 316, 322, 509
variations, 116, 298, 441, 492
vascular endothelial growth factor (VEGF), 388
vascularization, 388
vasculature, 304, 389
vasculitis, 409, 411, 424
vasoconstriction, 143
vasodilation, 344, 346, 354, 448, 450
vasodilator, 521
vector, 8, 32, 54, 80, 131, 190, 194, 199, 254, 494,
     508, 531, 535, 545, 570
vehicles, 277, 278, 539
vein, 388, 606
Venezuela, 26, 247, 251, 479, 558, 586
ventilation, 189, 554
ventricle, 601
ventricular arrhythmias, 344, 346, 449
ventricular fibrillation, 411
ventricular tachycardia, 173, 345, 346, 449
venules, 116, 199
vertigo, 251, 317, 350, 408, 410, 441
vesicle, 197
vessels, 116, 389, 390
videos, 144

W

wage rate, 80
wages, 68
walking, 171
war, 155, 573
Washington, 533, 548, 571
water, vii, 23, 24, 41, 42, 44, 55, 113, 128, 143, 162,
     169, 201, 204, 223, 231, 269, 272, 280, 284,
     302, 321, 327, 359, 362, 367, 380, 410, 411,
     417, 421, 423, 460, 481, 498, 499, 501, 511,
     515, 517, 615
weakness, 135, 143, 158, 177, 383, 397, 406, 408,
     432, 437, 440, 441, 444, 458
wealth, 71
weapons, 105, 123
weight loss, 515
well-being, ix, 1
West Africa, 7, 91, 155, 189, 200, 226, 235, 472,
     474, 510, 511, 513, 617
Western countries, 198, 410, 526
white blood cell count, 458
wholesale, 47, 64, 65, 74, 325, 599
wild type, 250, 265, 492
windows, 286
witchcraft, 327
withdrawal, ix, 1, 69, 99, 101, 326, 401, 426, 430,
     468, 470, 476, 478, 480
WMD, 170, 213
workers, 57, 87, 105, 114, 206, 207, 209, 216, 217,
     238, 265, 301, 325, 327, 434, 498, 500, 507,
     530, 533, 534, 548

World Bank, 95, 97, 102, 328, 525, 527, 533, 534, 540, 571
World Health Organization (WHO), vii, ix, xii, 5, 58, 63, 79, 93, 113, 121, 124, 166, 195, 209, 211, 212, 214, 218, 226, 245, 301, 328, 354, 469, 478, 525, 527, 533, 534, 550, 554, 563, 564, 566, 567, 569, 570, 572, 573, 579, 598, 599, 603, 607, 612, 613
World War I, x, 122, 123, 124, 198, 467, 550
worldwide, xi, 3, 18, 21, 55, 61, 63, 105, 112, 158, 169, 189, 208, 222, 226, 467, 501, 525, 528, 529, 538

X

xanthones, 516
X-axis, 257

Y

Yaounde, 590
yeast, 470, 495, 500, 616
yellow fever, 114
Yemen, 26, 559
yield, 19, 69, 70, 116, 119, 322, 325, 327, 529, 543
yolk, 388, 389, 390, 391, 392, 420
young adults, 334, 352, 386

Z

Zimbabwe, 26, 556
zinc, 492
Zulu, 26
zygote, 199